T0140471

High-Performance Ferrous Alloys

Radhakanta Rana
Editor

High-Performance Ferrous Alloys

 Springer

Editor
Radhakanta Rana
Tata Steel
IJmuiden, Noord-Holland, The Netherlands

ISBN 978-3-030-53824-8   ISBN 978-3-030-53825-5   (eBook)
https://doi.org/10.1007/978-3-030-53825-5

This Springer imprint is published by the registered company Springer Nature Switzerland AG.
The registered company address is: Gewerbestrasse 11, 6330 Cham, Switzerland

Dedicated to my family, particularly to my wife Basabdutta and son Apratim. Without their accommodative behaviour and support, it would have been difficult to spend time in realizing this interesting piece of work.

Foreword

It is a great pleasure to contribute some remarks in advance of the publication of *High-Performance Ferrous Alloys*, a project led by Dr. Radhakanta Rana from Tata Steel Europe R&D in the Netherlands. I have known Radhakanta for several years now, and he has been associated with globally recognized leading institutions in ferrous metallurgy through his education at IIT Kharagpur in India and RWTH Aachen University in Germany, and professionally as a researcher with us at Colorado School of Mines and at Tata Steel Europe. Radhakanta meets every challenge with energy and enthusiasm, and it is an honor to participate in this newest project which is a follow-on to his recent collection of work in *Automotive Steels: Design, Metallurgy, Processing and Applications*.

Steel is a ubiquitous material in the modern economy and enables many aspects of our daily lives. Steel developments continue unabated, to meet society's need for enhanced performance in countless applications, at low cost. These developments are often evolutionary and sometimes revolutionary, but progress is always impressive as we look back at any given time over a few decades of contributions from the steel research community. In the project at hand, however, the emphasis is clearly more *forward-looking*, including families of steels that are either newcomers in important applications, or where there is future potential for implementation of new approaches involving iron alloys that are substantially different than the high-volume products of the present, such as steels with reduced density or increased elastic modulus, iron-based high-entropy alloys (or MPEAs, multi-principal-element alloys) and iron-based intermetallics. These topics, along with the latest developments in other high-alloy ferrous systems such as stainless and electrical steels and cast iron alloys, are covered by a high-quality team of experts from Europe, Asia, and North America. I hope and expect that the book will be a significant resource for graduate students, researchers, and engineers with interests in these topics.

The book also provides an update of developments in microalloyed steels, automotive advanced high strength steels, and nanostructured steels, where much attention of the international steel research community has been focused. Additional background is provided in relation to thermodynamics and phase equilibria in relation to steels and ferrous alloys, and steel processing. Processing technology is a critical enabler of cost-effective developments, and process understanding and development often control the implementation of new physical metallurgy concepts.

I look forward to seeing the book in print, and a positive response of the steel community to its newest resource.

ABS Professor, Colorado School of Mines, and   John G. Speer
Director, Advanced Steel Processing and
Products Research Center
Golden, CO, USA
July 2020

Preface

Steels are engineered materials produced and used in large quantity across various industrial and commodity sectors. While most conventional steels are produced at industrial scale routinely today, many advanced high-strength steels and iron-based alloys are still in the laboratory stage. Iron-based emerging alloys can yield high levels of mechanical and physical properties due to novel alloy concepts and microstructure developments leading to multiple potential benefits for their use. Today most of the innovative steel developments are taking place in the area of automotive steels where demand for lightweighting is ever-growing due to stringent emission laws. Considering these facts, a book entitled *Automotive Steels: Design, Metallurgy, Processing and Applications* was published a few years ago with the editor of the current book as its lead editor. That book on automotive steels focused on summarizing the state of the art as well as new advances in various existing automotive sheet steels, such as extra deep drawable (EDD), bake hardenable (BH), high-strength low alloy (HSLA), dual phase (DP), transformation-induced plasticity (TRIP)-aided, complex phase (CP), bainitic, quenching and partitioning (Q&P) and elevated manganese (medium manganese and twinning induced plasticity (TWIP)) steels, and design of autobody and formability of automotive sheets. However, the developments of new steel concepts and high-alloyed ferrous materials were not covered well and there is no book currently available assembling the knowledge in these areas in one place. Therefore, a need was felt to fill this void so that researchers and professionals can benefit from it. Hence, the work on the current book entitled *High-Performance Ferrous Alloys* was undertaken to bridge the gap. This book aims to compile the current state of understanding of emerging iron-rich alloys and some high-alloy ferrous systems, in comparison with traditional alloys, in one volume to further their development. It covers the new alloy systems such as low-density steels, high-modulus steels, high-entropy iron alloys as well as iron-rich alloy systems of cast irons, stainless steels, electrical steels and iron intermetallics. It also presents processing-driven high-potential microstructures in steels (i.e. nanostructured steels) as well as compact overviews of advanced high-strength steels (AHSS) and microalloyed steels to give a comparative foundation. The basics of the long chain of processing of steel products and the thermodynamics and phase equilibria of iron-based systems are also covered, offering the readers some of the necessary background knowledge.

The motto of this book is not only to review and present the knowledge available in literature on the above novel iron alloys, their design, processing, microstructural evolution, and existing and potential applications but also to identify the gaps in their understanding and to suggest future research and developments necessary to show some of the emerging alloy systems the light of applications and improve the performance of the existing alloy systems. To serve this purpose, leading experts in the concerned areas from reputed companies, academic institutions and research laboratories from around the globe were invited as contributors for the individual chapters. I sincerely hope this book will be highly useful to researchers, industry professionals, academicians and students as a great knowledge source and will foster new research directions in the broad field of iron and steels.

I am grateful to my employer, Tata Steel, for giving me kind permission to undertake this book project and I acknowledge the moral support of my nearest colleagues. Finally, I would like to thank all the participating experts whose contributions have made publication of this book possible.

IJmuiden, Noord-Holland, The Netherlands   Radhakanta Rana

Contents

About the Editor and Contributors

Editor

Radhakanta Rana, PhD, FIMMM, has undergraduate and postgraduate academic background in Metallurgical Engineering. After a research stint at the Central Glass and Ceramic Research Institute in Kolkata, following his Master of Engineering at the then Bengal Engineering College (D.U.), Shibpur, West Bengal, India, he earned his PhD in 2009 in the same discipline from the Indian Institute of Technology Kharagpur (IIT Kharagpur), India, with a 2 years DAAD Fellowship for doctoral research at RWTH Aachen University, Germany. Dr Rana started his professional career in early 2008 in the Aluminium Metallurgy Group of the then Corus R&D in the Netherlands following the conclusion of his doctoral research. Currently, he is a senior member of the metallurgy group, High Strength Strip Steels, at Tata Steel R&D division based in IJmuiden, the Netherlands. He is experienced in experimental research on the development of innovative hot- and cold-formable steels and aluminium alloys particularly for automotive applications, with a particular focus on the understanding of their chemistry-processing-structure-property paradigm. Dr Rana has also pursued his postdoctoral research for 2 years (2013–2015) at the Advanced Steel Processing & Products Research Center (ASPPRC), Colorado School of Mines, Golden, CO. He has published some of his work on precipitation hardening steels, low density steels, high modulus steels, 3rd generation advanced high-strength steels, press-hardening steels and research techniques such as thermoelectric power and resistivity measurements in various reputed international journals and conference proceedings, and he also filed a number of patents in the areas of new steel products, their processing and properties. Dr Rana has also edited special themed issues on low-density steels, high-modulus steels and medium manganese steels in *JOM*, *Canadian Metallurgical Quarterly* and *Materials Science and Technology*, respectively. Previously, he has published another book *Automotive Steels* as the Lead Editor for Elsevier with Woodhead imprint. Dr Rana serves as an Editor of *Materials Science and Technology* (an IOM3 journal), an Editorial Board Member of *Scientific Reports* (a Nature journal) and a

Key Reader of *Metallurgical and Materials Transactions A* (a TMS journal). He is member of two professional organizations – the Minerals, Metals & Materials Society (TMS) and the Institute of Materials, Minerals and Mining (IOM3). On the personal front, Dr. Rana is married and the couple has a son.

Contributors

Hamid Azizi Department of Materials Science and Engineering, McMaster University, Hamilton, ON, Canada

Christian Baron Department Microstructure Physics and Alloy Design, Max-Planck-Institut für Eisenforschung GmbH, Düsseldorf, Nordrhein-Westfalen, Germany

Francisca G. Caballero Physical Metallurgy Department, National Centre for Metallurgical Research (CENIM-CSIC), Madrid, Spain

Carlos Capdevila Physical Metallurgy Department, National Centre for Metallurgical Research (CENIM-CSIC), Madrid, Spain

Carola Celada-Casero Tata Steel, IJmuiden, Noord-Holland, The Netherlands

Shangping Chen Tata Steel, IJmuiden, Noord-Holland, The Netherlands

Sourav Das Department of Metallurgical and Materials Engineering, Indian Institute of Technology Roorkee, Roorkee, Uttarakhand, India

Emmanuel De Moor Advanced Steel Processing and Products Research Center (ASPPRC), George S. Ansell Department of Metallurgical and Materials Engineering, Colorado School of Mines, Golden, CO, USA

Olga A. Girina ArcelorMittal, East Chicago, IN, USA

K. C. Hari Kumar Department of Metallurgical and Materials Engineering, Indian Institute of Technology Madras, Chennai, Tamil Nadu, India

Anish Karmakar Department of Metallurgical and Materials Engineering, Indian Institute of Technology Roorkee, Roorkee, Uttarakhand, India

Fritz Körmann Department Computational Materials Design, Max-Planck-Institut für Eisenforschung GmbH, Düsseldorf, Nordrhein-Westfalen, Germany

Zhiming Li School of Materials Science and Engineering, Central South University, Changsha, Hunan, China

Dawid Myszka Department of Metal Forming and Foundry Engineering, Warsaw University of Technology, Warsaw, Mazowsze, Poland

Martin Palm Department Structure and Nano-/Micromechanics of Materials, Max-Planck-Institut für Eisenforschung GmbH, Düsseldorf, Nordrhein-Westfalen, Germany

Damon Panahi ArcelorMittal, East Chicago, IN, USA

Dierk Raabe Department Microstructure Physics and Alloy Design, Max-Planck-Institut für Eisenforschung GmbH, Düsseldorf, Nordrhein-Westfalen, Germany

Radhakanta Rana Tata Steel, IJmuiden, Noord-Holland, The Netherlands

Rosalía Rementeria Additive Manufacturing – New Frontier, ArcelorMittal Global R&D, Avilés (Asturies), Spain

David San-Martin Physical Metallurgy Department, National Centre for Metallurgical Research (CENIM-CSIC), Madrid, Spain

Shiv Brat Singh Department of Metallurgical and Materials Engineering, Indian Institute of Technology Kharagpur, Kharagpur, West Bengal, India

Marcel H. F. Sluiter Department of Materials Science and Engineering, Delft University of Technology, Delft, Zuid-Holland, The Netherlands

Hauke Springer Department Microstructure Physics and Alloy Design, Max-Planck-Institut für Eisenforschung GmbH, Düsseldorf, Nordrhein-Westfalen, Germany

Frank Stein Department Structure and Nano-/Micromechanics of Materials, Max-Planck-Institut für Eisenforschung GmbH, Düsseldorf, Nordrhein-Westfalen, Germany

Javier Vivas Joining Processes, IK4 Lortek, Gipuzkoa, Ordizia, Spain

Tihe Zhou Stelco Inc., Hamilton, ON, Canada

Hatem S. Zurob Department of Materials Science and Engineering, McMaster University, Hamilton, ON, Canada

List of Figures

List of Tables

Chapter 1
Thermodynamics and Phase Equilibria of Iron-Base Systems

K. C. Hari Kumar

Symbols

A_s	Austenite start temperature
a_s	Relative number of sites in sublattice s
b_{ij}	Stoichiometric coefficient of component j in species i
c	Number of components
C_p	Heat capacity at constant pressure
G	Gibbs energy
$g(\tau)$	Magnetic ordering function
H	Enthalpy
H_i^{SER}	Enthalpy of element i according to the Stable Element Reference (SER)
$I(z)$	Constituent array I of order z
$^{\nu}L_{i,j}$	Binary interaction (Redlich-Kister) parameter of the order ν
$L_{i,j,k}$	Ternary interaction parameter
M_s	Martensite start temperature
N_s	Total number of sites in sublattice s
$N_{i\#s}$	Number of sites occupied by constituent species i in sublattice s
p	Pressure
P	Number of phases
R	Gas constant
S	Entropy
T	Temperature
x_i	Mole fraction of component i
$y_{i\#s}$	Site fraction of constituent i in sublattice s

K. C. Hari Kumar (✉)
Department of Metallurgical and Materials Engineering, Indian Institute of Technology Madras, Chennai, Tamil Nadu, India
e-mail: kchkumar@iitm.ac.in

© Springer Nature Switzerland AG 2021
R. Rana (ed.), *High-Performance Ferrous Alloys*,
https://doi.org/10.1007/978-3-030-53825-5_1

1

β	Magnetic moment in expressed in Bohr magneton
$\mu_i^{\phi_j}$	Chemical potential of the component i in phase ϕ_j
τ	The dimensionless quantity given by the ratio T/T_C

1.1 Introduction

Engineering properties of materials primarily depend upon their phase constitution and microstructure. Phase diagrams serve as roadmaps for arriving at a specific phase constitution and microstructure. They aid in the selection of right composition and appropriate heat treatment process for an engineering material to arrive at the right phase constitution and microstructure. When combined with chemical thermodynamics, it is a powerful approach to understand the behaviour of materials in various service conditions.

Traditionally, phase diagrams are determined by meticulous and time-consuming experiments. Phase diagrams of most binary systems and certain ternary systems of technological interest are already established this way. These are available in the form of handbooks. Engineering materials, however, are often multicomponent and multiphase in nature, making it quite impractical to establish their phase diagrams by experiments alone.

Computational thermodynamics provides an alternate approach to obtain phase diagrams, namely, the CALPHAD (the acronym for calculation of phase diagrams) method, which works by combining experimental information and thermodynamic models for phases. Such an approach describes the Gibbs energy of each phase present in a system by means of mathematical functions of temperature and composition. These functions, used in conjunctions with specialised computer programs, are able to generate phase diagrams and thermochemical properties of multicomponent-multiphase materials.

1.2 Thermodynamic Equilibrium

When a system is in thermodynamic equilibrium, it simultaneously meets conditions of mechanical, thermal and chemical equilibria. Such a system is incapable of any further changes by itself. When a system is in thermodynamic equilibrium, there will not be any perceptible changes in the value of any state variable with time. Hence, state variables have definite values when a system is in equilibrium. The thermodynamic analysis is possible only for systems that are in equilibrium.

A system in equilibrium may be in a stable or metastable state (Fig. 1.1). The stable state corresponds to the lowest energy state, whereas a metastable state is one that appears to be stable but truly not at its lowest energy state. When a system is in a metastable state, it is trapped in a local energy minimum. With external

Fig. 1.1 Stable, metastable and unstable states

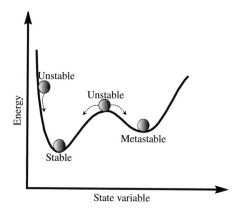

influence, it can overcome the energy barrier to reach the stable state. For example, graphite is a stable state of carbon, whereas diamond is metastable. On the other hand, an unstable state refers to a state that is either changing or just on the verge of doing so. A system in the unstable state, when left alone for sufficient time, tends towards equilibrium state without any external influence. An undercooled liquid is an example of an unstable state.

1.3   Phase Diagrams and Gibbs Energy-Composition Diagrams

A phase diagram is a graphical representation of changes that take place in a system under equilibrium conditions when at least two non-conjugate intensive state variables are independently varied. Typically the state variables such as pressure, temperature, composition, etc. are chosen, as these can be easily handled in an experiment. Equilibrium conditions can be arrived at by the criterion that Gibbs energy (G) of the system has to be minimum at equilibrium. At constant pressure (p) and temperature (T), it turns out that minimum Gibbs energy criterion is fulfilled when

$$\mu_1^{\phi_1} = \mu_1^{\phi_2} = \mu_1^{\phi_3} = \cdots\cdots = \mu_1^{\phi_P}$$

$$\mu_2^{\phi_1} = \mu_2^{\phi_2} = \mu_2^{\phi_3} = \cdots\cdots = \mu_2^{\phi_P}$$

$$\mu_c^{\phi_1} = \mu_c^{\phi_2} = \mu_c^{\phi_3} = \cdots\cdots = \mu_c^{\phi_P} \tag{1.1}$$

where $\mu_i^{\phi_j}$ is the chemical potential of the component i in phase ϕ_j, P is the number of phases in equilibrium and c is the number of components. The maximum number of phases that can coexist at equilibrium is governed by the Gibbs phase rule.

Fig. 1.2 Common-tangent
construction

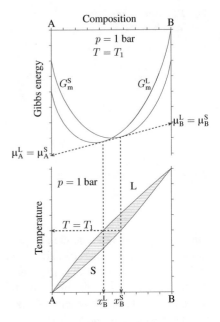

The requirement that Gibbs energy must be at its minimum when the system is in equilibrium can be better understood by the common-tangent construction. This is shown in Fig. 1.2 for a two-phase equilibrium in the case of a hypothetical binary system A-B. The figure also illustrates the link between common-tangent construction and the phase diagram. The construction is based on the fact that intercepts made by the tangents to the Gibbs energy-composition curve $(G_m - x)$ of a phase at any chosen composition give chemical potentials of the respective components. Thus common-tangent represents the equilibrium conditions given in Equation 1.1. Compositions corresponding to the points of tangency define the tie line at the selected temperature of the phase diagram.

1.4 CALPHAD: CALculation of PHAse Diagrams

CALPHAD [1, 2] is a classical thermodynamics method, devised in the 1960s primarily to calculate phase diagrams in a reliable and efficient manner. The approach is based on the principle that state variables corresponding to thermodynamic equilibrium can be computed by the Gibbs energy minimization.

Meijering [3] was the first to try out this idea to calculate phase diagrams in his pioneering work on Ni-Cu-Cr. In 1970, Kaufman and Bernstein [4] authored the first book on the subject. They have extensively discussed the concept of lattice stability (Gibbs energy difference between two structural states of elements) and proposed a method to establish Gibbs energy functions for stable and metastable modifications of several elements. They demonstrated that phase diagrams of many binary and ternary systems can be computed using the CALPHAD approach. Another milestone in the development of the CALPHAD was the advent of sublattice formalism [5], which allowed thermodynamic modelling of many type of phases using a general approach [6, 7]. This paved way to the development of several computer programs for Gibbs energy minimization [8–12]. Since then CALPHAD has grown significantly. Its application spans not only in the field of calculation of phase diagrams of multicomponent systems but also in areas such as thermochemistry, phase transformation, microstructure simulation, etc.

It is apparent that for the CALPHAD method to work Gibbs energy functions for each phase of the system should be available. Fortunately, one can build the Gibbs functions for a multicomponent system by combining the Gibbs energy functions of the constituent unary and binary systems using a geometrical extrapolation scheme [13]. Correction terms are required to account for the higher-order effects, usually not exceeding ternaries. For elements, most of the functions are already made available by SGTE (Scientific Group Thermodata Europe) [14]. Thus, problem of modelling a multicomponent system is reduced to finding Gibbs energy functions of various phases of the constituent binary and ternary systems in a mutually consistent manner. Model parameters present in the Gibbs energy functions are established through a computer-assisted optimization procedure using carefully selected thermochemical and constitutional data as input. Procedure for doing the thermodynamic optimization (also known as thermodynamic assessment, thermodynamic modelling) is shown in Fig. 1.3 in the form of a flowchart [15].

Software packages such as Thermo-Calc [9, 16], PANDAT [11, 17], FactSage [12, 18], JMatPro [19], etc. implement various aspects of the CALPHAD method. For many class of engineering materials, including ferrous alloys, optimised functions are available in the form of commercial Gibbs energy databases that can used with these software packages. Some details of the databases for ferrous materials are given in Table 1.1.

CALPHAD method has many advantages that contributed to its success. It is a combinatorial approach (limited experiments + modelling). It has the ability to interpolate experimental data in a reliable manner using Gibbs energy models. This drastically reduces the dependence on experimental data to generate phase diagrams and thermochemical information. One of the most powerful features of the method is its ability to extrapolate to higher-order systems in a reliable manner using the Gibbs energy functions of constituent unaries, binaries and ternaries. The Gibbs energy functions can also be used to calculate metastable phase diagrams. Lastly, it is computationally very efficient, even in the case of multicomponent systems.

Most notable disadvantage of the CALPHAD method is that the underlying Gibbs energy modelling requires high-quality experimental thermochemical and constitutional data as input. To some extent, the dependency on thermochemical data can be addressed by making use of first principles approach such as the one based on Density Functional Theory (DFT), whose results can be used as virtual data [20] in the thermodynamic modelling. Another disadvantage of the CALPHAD method is that it cannot predict the existence of a completely new phase for which

K. C. Hari Kumar

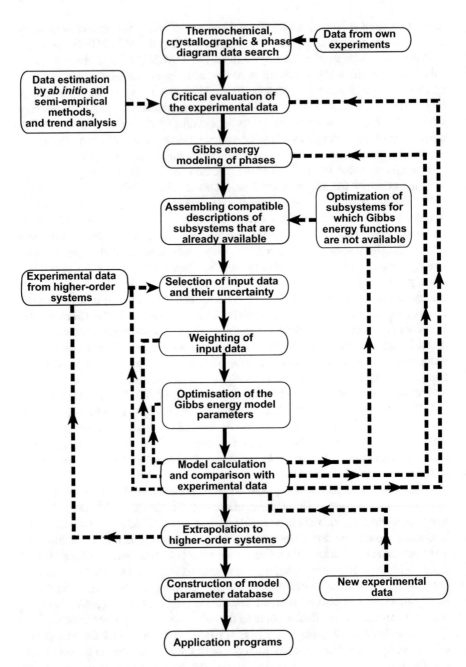

Fig. 1.3 Flowchart for thermodynamic optimization [15]

Table 1.1 Some details of the commercial Gibbs energy databases for ferrous materials

Software:	Thermo-Calc
Database:	TCFE9
Components:	28 (Ar, Al, B, C, Ca, Ce, Co, Cr, Cu, Fe, H, Mg, Mn, Mo, N, Nb, Ni, O, P, S, Si, Ta, Ti, V, W, Y, Zn, Zr)
Systems:	Binaries-255, Ternaries-256, Quaternaries-77
Phases:	402
Software:	PANDAT
Database:	PanIron
Components:	27 (Al, As, B, C, Ca, Co, Cr, Cu, Fe, Mg, Mn, Mo, N, Nb, Ni, P, Pb, S, Si, Sn, Ta, Ti, V, W, Y, Zn, Zr)
Systems:	Binaries-278, Ternaries-278
Phases:	534
Software:	FactSage
Database:	FSstel
Components:	32 (Al, B, Bi, C, Ca, Ce, Co, Cr, Cu, Fe, Hf, La, Mg, Mn, Mo, N, O, Nb, Ni, P, Pb, S, Sb, Si, Sn, Ta, Te, Ti, V, W, Zn, Zr)
Systems:	Binaries-205, Ternaries-100, Quaternaries-20
Phases:	572
Software:	JMatPro
Database:	Fe alloys
Components:	21 (Al, B, C, Co, Cr, Cu, Fe, Mg, Mn, Mo, N, Nb, Ni, O, P, S, Si, Ta, Ti, V, W)
Phases:	53

the Gibbs function has not been modelled. The approach also lacks suitable models to handle short-range order, which may be important in some systems.

1.5 Thermodynamic Modelling of Iron-Base Systems

1.5.1 Elements and Stoichiometric Compounds

Elements and stoichiometric compounds have fixed composition. Hence, their Gibbs energies depend only on temperature and pressure, which may be split into three parts:

$$G(T, p) = G^{ch} + G^{mo} + G^{pr} \tag{1.2}$$

where G^{ch} is the chemical (lattice) contribution, G^{mo} is the magnetic ordering contribution and G^{pr} is the pressure contribution. For most cases, we only need to consider G^{ch}. In the case of ferromagnetic substances, contribution from magnetic ordering can be quite significant. Figure 1.4 shows the effect of ferromagnetic

Fig. 1.4 Heat capacity of α-Fe showing the effect of ferromagnetic ordering

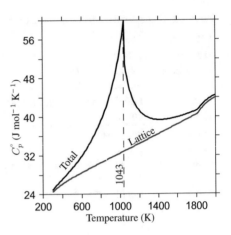

ordering, for example, on the heat capacity of α-Fe. We shall ignore G^{pr} term as mostly we are concerned only with equilibrium at atmospheric pressure.

In the case of elements, the functions recommended by Scientific Group Thermodata Europe (SGTE) [14] are generally used for representing G^{ch}. It has the following format:

$$G_i^{\circ}(T) = H_i^{SER} + a + bT + cT\ln T + dT^2 + eT^{-1} + fT^3 + \cdots \qquad (1.3)$$

where a, b, c. . . are the model parameters. ° indicates that standard pressure (1 bar) is used. The abbreviation SER in Equation 1.3 refers to the Stable Element Reference, i.e. with respect to the enthalpy of the element at 298.15 K and 1 bar ($H_i^{SER} = H_i^{\circ}(298) - H_i^{\circ}(0)$) and its entropy at 0 K, which is zero according to the third law of thermodynamics. Note that these functions are not valid at low temperatures, since they do not satisfy the requirement that heat capacity should approach zero as the temperature approaches absolute zero. The $G_i^{\circ}(T)$ for an element may be split into more than one temperature range for preserving accuracy while data fitting. It should be emphasised that we not only need expressions for $G_i^{\circ}(T)$ for all stable structural states of an element but also for several metastable states (e.g.: Al in bcc state).

The same approach can be extended to model the Gibbs energy function of a stoichiometric compound (θ). Thus,

$$G_{\theta}^{\circ}(T) = \sum_i v_i H_i^{SER} + A + BT + CT\ln T + DT^2 + ET^{-1} + FT^3 + \cdots \qquad (1.4)$$

where, A, B, C. . . are the model parameters and v_i are the stoichiometric coefficients for the elements that make up the compound. For compounds with zero C_p of formation (Neumann and Kopp rule, $\Delta_f C_p^{\circ} = 0$), one can use a simpler expression.

$$G_\theta^\circ(T) = \sum_i v_i G_i^{\circ,\text{ref}} + \Delta_f G_\theta^\circ \tag{1.5}$$

$$\Delta_f G_\theta^\circ = h + sT \tag{1.6}$$

where $h = \Delta_f H_\theta^\circ$ and $-s = \Delta_f S_\theta^\circ$.

The Gibbs energy due to magnetic ordering $(G^{\text{mo}}(T))$ is modelled using the expression due to [21], which is given below.

$$G^{\text{mo}}(T) = RT\ln(1+\beta)g(\tau) \tag{1.7}$$

where $\tau = T/T_C$, T_C is the critical temperature for magnetic transition, β is the magnetic moment in Bohr magneton and $g(\tau)$ is the magnetic ordering function. The function $g(\tau)$ is given by

$$g(\tau) = \begin{cases} 1 - \frac{1}{A}\left(\frac{79\tau^{-1}}{140f} + \frac{474}{497}\left(\frac{1}{f}-1\right)\left(\frac{\tau^3}{6} + \frac{\tau^9}{135} + \frac{\tau^{15}}{600}\right)\right) & \tau \leq 1 \\ -\frac{1}{A}\left(\frac{\tau^{-5}}{10} + \frac{\tau^{-15}}{315} + \frac{\tau^{-25}}{1500}\right) & \tau \geq 1 \end{cases} \tag{1.8}$$

where $A = \frac{518}{1125} + \frac{11692}{15975}\left(\frac{1}{f}-1\right)$ and f is a constant related to the coordination number ($f = 0.4$ for bcc and it is 0.28 for other structures).

1.5.2 Solutions

In the case of solutions, Gibbs energy is not only a function of pressure and temperature but also of the composition. In its most general form, Gibbs energy of a solution phase consists of four parts.

$$G = G^{\text{ref}} + G^{\text{conf}} + G^{\text{mo}} + G^{\text{E}} \tag{1.9}$$

where G^{ref} is the reference term which can be regarded as Gibbs energy of a mechanical mixture of the constituents, G^{conf} accounts for the contribution to the Gibbs energy from configurational entropy $(-TS^{\text{conf}})$ and G^{E} is the excess Gibbs energy term, which accounts for deviation from ideal solution behaviour. The exact form of these terms depend on the Gibbs energy model chosen for a phase.

A substitutional solution is one of the simplest mixing models, where the solute atoms substitute the solvent atoms randomly. For a multicomponent random substitutional solution at standard pressure, G^{ref} can be written as

$$G^{\text{ref}} = \sum_i x_i(G_i^\circ - H_i^{\text{SER}}) \tag{1.10}$$

where x_i is the mole fraction of component i. The G^{conf} has the following form in the case of a multicomponent random substitutional solution.

$$G^{\text{conf}} = RT \sum_i x_i \ln x_i \tag{1.11}$$

It is common to use Redlich-Kister polynomial [22] for representing G^{E} of a binary solution. For a binary substitutional solution with the random mixing of constituents, one can write

$$G^{\text{E}} = x_{\text{A}} x_{\text{B}} L_{\text{A, B}}(x_{\text{A}}, x_{\text{B}}) \tag{1.12}$$

$$= x_{\text{A}} x_{\text{B}} \sum_{\nu=0}^{n} {}^{\nu}L_{\text{A, B}}(x_{\text{A}} - x_{\text{B}})^{\nu} \tag{1.13}$$

where ${}^{\nu}L_{\text{A, B}}$ are the Redlich-Kister parameters. Their T dependence can be described using

$$^{\nu}L_{\text{A, B}} = {}^{\nu}a_{\text{A, B}} + {}^{\nu}b_{\text{A, B}} T \tag{1.14}$$

The model parameters ${}^{\nu}a_{\text{A, B}}$, ${}^{\nu}b_{\text{A, B}}$, etc. have to be established by the thermodynamic optimization of relevant binary systems, as described in Sect. 1.4.

Excess Gibbs energy of a multicomponent solution ($G^{\text{E}}_{\text{m}-\text{c}}$) can be written in terms of binary excess Gibbs energies using an extrapolation model. Many schemes are available to do this, but Muggianu extrapolation scheme [23] is preferred for its simplicity when used along with Redlich-Kister polynomial. Accordingly, for a multicomponent solution with random mixing one can write

$$G^{\text{E}}_{\text{m}-\text{c}} = \sum_{i=1}^{c-1} \sum_{j=i+1}^{c} G^{\text{E}}(x_i, x_j) + \text{Correction terms} \tag{1.15}$$

where c is the number of components. Correction terms account for contributions from higher-order systems, usually not exceeding ternaries. In the case of a ternary random substitutional solution, the Redlich-Kister-Muggianu scheme gives the following expression for the Gibbs energy.

$$G = \sum_{i=1,2,3} x_i (G_i^{\circ} - H_i^{\text{SER}}) + RT \sum_{i=1,2,3} x_i \ln x_i + \sum_{i=1}^{2} \sum_{j=i+1}^{3} G^{\text{E}}(x_i, x_j) + x_1 x_2 x_3 L_{1,2,3} \tag{1.16}$$

The ternary correction term $L_{1,2,3}$ may be made composition dependent as follows.

$$L_{1,2,3} = x_1 {}^{0}L_{1,2,3} + x_2 {}^{1}L_{1,2,3} + x_3 {}^{2}L_{1,2,3} \tag{1.17}$$

1.5.2.1 Sublattice Formalism

A simple mixing model such as random substitutional solution is usually not adequate, since many phases have more than one type of sites in which mixing takes place. This led to a general modelling concept known as sublattice formalism. The approach is very suitable for modelling variety of phases such as random substitutional solutions, interstitial solutions, stoichiometric compounds, intermediate phases with homogeneity range, ionic melts, etc.

The concept of subdividing configuration space into sublattices was originally introduced by Temkin [24] to describe the thermodynamics of reciprocal salt systems. It was proposed that the anions and cations occupy separate sublattices and the mixing is random in each sublattice. The two-sublattice model of Temkin was adapted for an arbitrary number of constituents by Hillert and co-workers [5, 25, 26]. This was later generalized for multiple sublattices with an arbitrary number of constituent species occupying the configuration space [6]. It is represented as

$$(A, B, C \ldots)_{a_1} : (A, C, E \ldots)_{a_2} : \ldots (B, D, E \ldots)_{a_n} \tag{1.18}$$

where the constituent species A, B, C... may be atoms, molecules, charged species or vacancies (Va) and a_s is the relative number of sites in sublattice s. Each pair of parenthesis in the above representation denotes a sublattice, a colon (:) separates each sublattice and a comma (,) separates each constituent within a sublattice. It should be noted that a constituent species is allowed only once in each sublattice. It is also assumed that there is random mixing among the constituents in each sublattice.

A composition variable known as site fraction ($y_{i\#s}$) is introduced in order to formulate the Gibbs energy using the sublattice formalism. It is defined as the mole fraction of a constituent species i in the sublattice s, i.e.

$$y_{i\#s} = \frac{N_{i\#s}}{N_s} \tag{1.19}$$

where $N_{i\#s}$ is the number of constituent species i and N_s is the total number of sites in sublattice s. N_s is given by

$$N_s = N_{Va\#s} + \sum_i N_{i\#s} \tag{1.20}$$

where $N_{Va\#s}$ is the number of vacant sites in sublattice s. The total number of sites N among all sublattices is given by

$$N = \sum_{s=1}^{n} N_s \tag{1.21}$$

The relative number of sites in a sublattice can be expressed as

$$a_s = \frac{N_s}{N} \tag{1.22}$$

It should be noted that according to this definition, $\sum_{s=1}^{n} a_s = 1$. Usually a_s are multiplied by a common factor, so that they are all integers. The sum of the site fractions in a given sublattice is unity, i.e. $y_{Va\#s} + \sum_i y_{i\#s} = 1$. The mole fraction of a component j can be calculated from the site fractions of the constituent species using the following relation.

$$x_j = \sum_i \left(\frac{b_{ij}}{\sum_k b_{ik}} \left(\frac{\sum_s a_s y_{j\#s}}{\sum_s a_s (1 - y_{Va\#s})} \right) \right) \tag{1.23}$$

where b_{ij} is the stoichiometric coefficient of component j in species i. When all the specie are monoatomic, the above equation can be simplified to yield

$$x_j = \frac{\sum_s a_s y_{j\#s}}{\sum_s a_s (1 - y_{Va\#s})} \tag{1.24}$$

Using the sublattice formalism, the Gibbs energy per mole of a formula unit (mfu) of a phase is given as

$$G_{mfu} = \sum_{I(0)} \left(\prod_{I(0)} y_{i\#s} \right) G^{\circ}_{I(0)} + RT \sum_s a_s \sum_i y_{i\#s} \ln y_{i\#s} + \sum_{I(z)} \left(\prod_{I(z)} y_{i\#s} \right) L_{I(z)} \tag{1.25}$$

where $I(z)$ is the constituent array of zth order. A 0th order constituent array $(I(0))$ has only one species occupying each sublattice. It is commonly referred to as an end-member, and it resembles a stoichiometric compound. The first term in Equation 1.25 represents the reference energy term that consists of the Gibbs energies of all the end-members $I(0)$. Hence, when applied to solid phases, the sublattice formalism is known as Compound Energy Formalism (CEF) [7]. The second term denotes the contribution of configurational entropy to Gibbs energy due to random mixing of constituents within each sublattice. The excess contribution to Gibbs energy is given by the third term. It consists of several interaction energy terms $L_{I(z)}$ which arise due to the interaction of constituents within a sublattice. The interaction terms beyond the second-order constituent array are not considered since their contribution to Gibbs energy is negligible. Note that when $s = 1$ and constituents are all elements, Equation 1.25 reduces to that of a random substitutional solution (Equation 1.16). More details of the sublattice formulation can be found in [1].

Sublattice formalism provides an appropriate method to formulate realistic mixing models for a variety of phases. Since this formalism is straightforward and general, it is highly suitable for computer implementation. Care must be taken to choose the right number of sublattices. In the case of solids, it must agree with the crystallography and homogeneity range of the phase that is considered. Use of too many sublattices is not practical in the case of solids with complex crystal structures since the Gibbs energy functions needs to be determined for a large number of end-members. Hence, for phases with complex crystal structures, one may need to simplify the model by combining sublattices based on point group symmetry, coordination number, etc.

Some examples of application of sublattice formalism to model Gibbs energies of phases relevant to steels are given below.

1.5.2.2 Interstitial Solid Solutions

Gibbs energy models for ferrite (bcc, α) and austenite (fcc, γ) must distinguish between substitutional and interstitial positions, without which configurational entropy would be unrealistic. Smaller atoms such as C and N occupy the octahedral voids of the lattice, while bigger atoms like Ni and Mn substitute the Fe atoms (Fig. 1.5). A sublattice formulation with two sublattices is sufficient to take care of this requirement.

The ratio of substitutional sites to octahedral voids is 1:3 for bcc, and it is 1:1 for fcc. Note that only a fraction of octahedral voids will be occupied by the interstitial atoms and remaining will be vacant. Hence vacancy (Va) must be included as a constituent species in the interstitial sublattice. The corresponding sublattice formulation that reflects these crystallographic requirements can be represented by

$$(\text{Fe, Mn, Ni, } \ldots)_{a_1} (\text{C, N, Va, } \ldots)_{a_2}$$

Fig. 1.5 Substitutional and interstitial positions

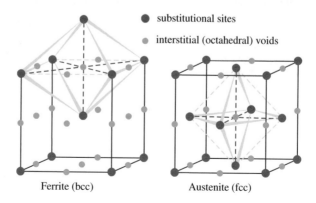

● substitutional sites

● interstitial (octahedral) voids

Ferrite (bcc)   Austenite (fcc)

where $a_1 = 1$, $a_2 = 3$ for bcc and $a_1 = 1$, $a_2 = 1$ for fcc.

As an example, sublattice representation of ferrite in Fe-Mn-C system can be denoted by $(Fe, Mn)_1 : (C, Va)_3$. The corresponding expression for the Gibbs energy can be readily obtained from Equation 1.25 as

$$
\begin{aligned}
G_{\mathrm{mfu}} = {} & y_{\mathrm{Fe\#1}}\left(y_{\mathrm{C\#2}}G^{\circ}_{\mathrm{Fe}:\mathrm{C}} + y_{\mathrm{Va\#2}}G^{\circ}_{\mathrm{Fe}:\mathrm{Va}}\right) + y_{\mathrm{Mn\#1}}\left(y_{\mathrm{C\#2}}G^{\circ}_{\mathrm{Mn}:\mathrm{C}} + y_{\mathrm{Va\#2}}G^{\circ}_{\mathrm{Mn}:\mathrm{Va}}\right) \\
& + RT[1(y_{\mathrm{Fe\#1}}\ln(y_{\mathrm{Fe\#1}}) + y_{\mathrm{Mn\#1}}\ln(y_{\mathrm{Mn\#1}})) + 3(y_{\mathrm{C\#2}}\ln(y_{\mathrm{C\#2}}) + y_{\mathrm{Va\#2}}\ln(y_{\mathrm{Va\#2}}))] \\
& + y_{\mathrm{Fe\#1}}y_{\mathrm{Mn\#1}}\left(y_{\mathrm{C\#2}}L_{\mathrm{Fe,\,Mn}:\mathrm{C}} + y_{\mathrm{Va\#2}}L_{\mathrm{Fe,\,Mn}:\mathrm{Va}}\right) \\
& + y_{\mathrm{C\#2}}y_{\mathrm{Va\#2}}\left(y_{\mathrm{Fe\#1}}L_{\mathrm{Fe}:\mathrm{C,\,Va}} + y_{\mathrm{Mn\#1}}L_{\mathrm{Mn}:\mathrm{C,\,Va}}\right) \\
& + y_{\mathrm{Fe\#1}}y_{\mathrm{Mn\#1}}y_{\mathrm{C\#2}}y_{\mathrm{Va\#2}}L_{\mathrm{Fe,\,Mn}:\mathrm{C,\,Va}} + G^{\mathrm{mo}}
\end{aligned}
$$

$$(1.26)$$

One can identify various constituent arrays of the model: $I(0) = $ Fe:C, Fe:Va, Mn:C and Mn:Va; $I(1) = $ Fe:(C,Va), Mn:(C,Va), (Fe,Mn):C and (Fe,Mn):Va; and $I(2) = $ (Fe,Mn):(C,Va). Composition domain of the model is fixed by the compositions of end-members as $x_{\mathrm{Fe}} = 0 \rightarrow 1$, $x_{\mathrm{Mn}} = 0 \rightarrow 1$ and $x_{\mathrm{C}} = 0 \rightarrow 0.75$. This shown in Fig. 1.6. Reference term in the expression for Gibbs energy can be visualised as a surface defined by the Gibbs energies of end-members (Fig. 1.7). Note that both β and T_C are composition dependent in the case of non-stoichiometric phases.

1.5.2.3 Carbides, Nitrides and Carbonitrides

Certain alloying elements in steels have a strong tendency to form carbides and nitrides. In many instances when C and N are present together along with such elements, carbonitrides are formed. Some carbides are quite similar to corresponding nitrides and may exhibit considerable mutual solubilities.

Carbides, nitrides and carbonitrides relevant to steels can be classified into one of the following types: MX (M = Ti, V, Nb), M_2X (M = W, Mo), M_3X (M = Fe), M_7X_3 (M = Cr, Mn) and $M_{23}X_6$ (M = Cr, Mn), where X = C, N. MX is modelled identical to austenite, $(M)_1 : (X, Va)_1$, but the interstitial site is almost fully occupied by X. M_2X is modelled similar to hcp, i.e. $(M)_1 : (X, Va)_{0.5}$. Sublattice formulation for

Fig. 1.6 Composition domain for the ferrite phase in the Fe-Mn-C system

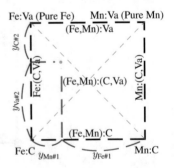

Fig. 1.7 Reference energy
surface

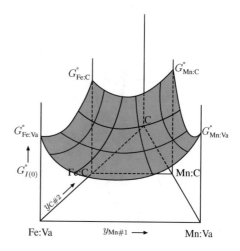

M_3X is $(M)_3 : (X)_1$ and that for M_7X_3 is $(M)_7 : (X)_3$. $M_{23}X_6$ is modelled using
three sublattices as $(M)_{21} : (M)_2 : (X)_6$.

1.5.2.4 Non-stoichiometric Intermetallic Phases

Here we consider σ phase as an example since it is found in several iron binaries
(Fe-Cr, Fe-V, Fe-Mo, etc.). It belongs to the Frank-Kasper type (topologically
close-packed) of phases. The unit cell consists of 32 atoms, distributed over 5
sublattices (Fig. 1.8). Details of its crystal structure are given in Table 1.2. The ideal
crystal structure indicates that the preferred stoichiometry is AB_2. However, there
is considerable non-stoichiometry in most cases due to the mixed occupancy of the
lattice sites. Larger B atoms prefer lattice sites with higher coordination number.

In order to model the Gibbs energy of this phase, one would require 5 sublattices.
However, such a model would be very complex since it results in 32 end-members.
Therefore, several simplified sublattice formulations for modelling σ phase exist
[27–31]. The model proposed by Joubert [31] is most appropriate, which is
a simplified two-sublattice model given by $(\underline{A},B)_{10}:(A,\underline{B})_{20}$. It is obtained by
combining sites with higher coordination number into one sublattice and rest into
another. It follows from Equation 1.25 that the corresponding Gibbs energy function
can be written as

$$
\begin{aligned}
G_{mfu} = {} & y_{A\#1}\left(y_{A\#2}G^\circ_{A\,:\,A} + y_{B\#2}G^\circ_{A\,:\,B}\right) + y_{B\#1}\left(y_{A\#2}G^\circ_{B\,:\,A} + y_{B\#2}G^\circ_{B\,:\,B}\right) \\
& + RT\left[10\left(y_{A\#1}\ln(y_{A\#1}) + y_{B\#1}\ln(y_{B\#1})\right) + 20\left(y_{A\#2}\ln(y_{A\#2}) + y_{B\#2}\ln(y_{B\#2})\right)\right] \\
& + y_{A\#1}y_{B\#1}\left(y_{A\#2}L_{A,\,B\,:\,A} + y_{B\#2}L_{A,\,B\,:\,B}\right) \\
& + y_{A\#2}y_{B\#2}\left(y_{A\#1}L_{A\,:\,A,\,B} + y_{B\#1}L_{B\,:\,A,\,B}\right) \\
& + y_{A\#1}y_{B\#1}y_{A\#2}y_{B\#2}L_{A,\,B\,:\,A,\,B}
\end{aligned}
$$

$$(1.27)$$

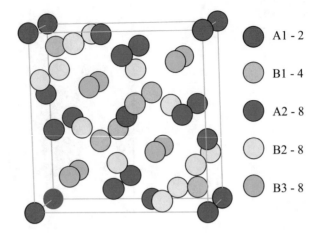

Fig. 1.8 Crystal structure of σ phase

Table 1.2 Crystallographic
details of σ phase

Pearson symbol: tP30		
Strukturbericht: D8$_b$		
Prototype: Cr$_{0.49}$Fe$_{0.51}$		
Site	WP^a	CN^b
A1	2a	12
B1	4f	15
A2	8i	12
B2	8i	14
B3	8j	14

^{a} Wyckoff position
^{b} Coordination number

Other important non-stoichiometric phases of interest in the context of steels are
the Laves phase (Strukturbericht: C14, Pearson symbol: hP12, prototype: MgZn$_2$)
and the μ phase (Strukturbericht: D8$_5$, Pearson symbol: hR13, prototype: Fe$_7$W$_6$).
For the details on Gibbs energy modelling of these phases, the reader may refer to
[15, 32].

1.5.2.5 Chemical Ordering

Some solid phases that are chemically less ordered may tend to become more
ordered with lowering of temperature and/or variation in composition. Crystal
structures of the less-ordered and the more-ordered phases are related, the more-
ordered phase being a superstructure of the less-ordered phase.

There are many ordered phases whose unit cells can be thought of superstructures
of unit cells of disordered phases. For example, the ordered phases B2 and D0$_3$
are superstructures derived from the unit cell of the disordered A2 (bcc) phase,

and ordered phases L1$_2$ and L1$_0$ are superstructures derived from the unit cell of disordered A1 (fcc) phase. In certain systems, an ordered phase and its disordered counterpart may be joined together through an order-disorder transformation. As examples from iron binaries, Fe-Si and Fe-Al systems show A2-B2 transformation, and Fe-Ni system has A1-L1$_2$ transformation.

When an ordered phase becomes disordered, the distinction between sublattices disappears. In other words, the sublattices of the ordered phase merge on disordering as indicated below for a simple case.

ordered : $(A, B)_{a_1} : (A, B)_{a_2}$ $\searrow \nearrow$ disordered : $(A, B)_{(a_1+a_2)}$

In the Gibbs modelling of such phases, it is desirable to treat an ordered phase and its disordered counterpart using a unified Gibbs energy model using the sublattice framework [33, 34]. Therefore, a combined Gibbs energy function consisting of two parts is used to model order-disorder transformations.

$$G_m = G_m^{\text{dis}}(x_i) + \Delta G_m^{\text{ord}}$$ (1.28)

where $G_m^{\text{dis}}(x_i)$ is the molar Gibbs energy of the disordered phase and ΔG_m^{ord} is the ordering contribution.

$$\Delta G_m^{\text{ord}} = G_m^{\text{ord}}(y_{i\#s}) - G_m^{\text{ord}}(y_{i\#s} \text{ replaced by } x_i)$$ (1.29)

In order to evaluate G_m^{ord}, the molar Gibbs energy of ordered phase needs to be calculated twice, i.e. using site fractions and mole fractions. When the phase is disordered $y_{i\#1} = y_{i\#2} = x_i$, then $\Delta G_m^{\text{ord}} = 0$. Due to crystal symmetry, it may be necessary to relate some of the model parameters of $G_m^{\text{ord}}(y_{i\#s})$. It is also possible to establish constraining relations between the model parameters of $G_m^{\text{dis}}(x_i)$ and $G_m^{\text{ord}}(y_{i\#s})$, thus reducing the number of independent model parameters. More details on modelling of chemical ordering can be found in [1, 2, 20].

1.6 Phase Equilibria in Iron-Base Systems

In this section computed phase diagrams of selected Fe-containing systems are presented. All the phase diagrams presented here are generated using the Thermo-Calc [16] software. Most of the diagrams are calculated using the PrecHiMn [35] thermodynamic database.

1.6.1 Pure Iron

Iron exhibits allotropy, which is the key to its versatility as an engineering material. Some important characteristics of allotropes of iron are given in Table 1.3. The temperature-dependent phase changes in pure iron at ambient pressure are shown in Fig. 1.9 along with the metallurgical designations of the transformation temperatures. The *p-T* phase diagram of iron is given in Fig. 1.10. It shows that both α-Fe (bcc) and γ-Fe (fcc) can transform to ε-Fe (hcp) under high pressure.

Table 1.3 Some characteristics of allotropes of iron

Properties					
Property	α-Fe	β-Fe	γ-Fe	δ-Fe	ε-Fe
Crystal structure	bcc	bcc	fcc	bcc	hcp
Lattice parameters (pm) [36]	286.65^a	–	364.67^b	293.15^c	$a = 284.5$ $c = 399.0$
Space group	$Im\bar{3}m$	$Im\bar{3}m$	$Fm\bar{3}m$	$Im\bar{3}m$	$P6_3/mmc$
Pearson symbol	cI2	cI2	cI4	cI2	hP2
Strukturbericht	A2	A2	A1	A2	A3
Stability range (°C) [36]	< 770	770–912	912–1394	1394–1538	> 10.4 GPa

[a] at 25°C
[b] at 915°C
[c] at 1394°C

Fig. 1.9 Phase transformations of iron at 1 bar

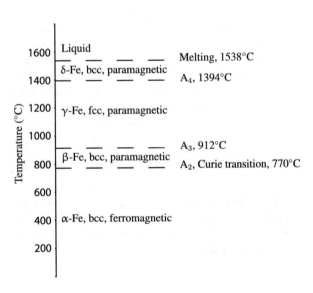

Fig. 1.10 *p-T* phase diagram of iron

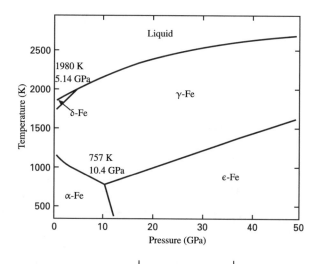

Fig. 1.11 Types of iron binary phase diagrams

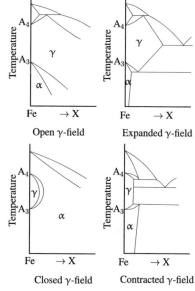

1.6.2 Fe-Base Binary Systems

Based on the appearance of γ phase field, Wever [37] classified Fe-X phase diagrams into fours types: open γ-field, expanded γ-field, closed γ-field and contracted γ-field. These are shown schematically in Fig. 1.11.

Alloying elements such as Ni, Mn, Co, etc. lower A_3 and raise A_4 temperatures of iron and exhibit extended solubility in the γ phase. Thus, the corresponding Fe-X phase diagrams belong to open γ-field type. Though alloying elements such as C, N, Cu, etc. are like Ni and Mn, they have only limited solubility in the γ

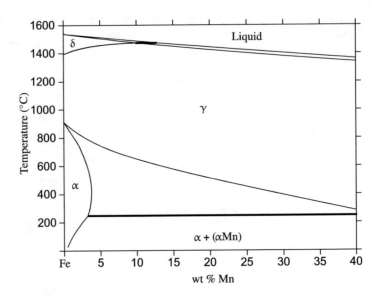

Fig. 1.12 Fe-rich part of Fe-Mn phase diagram (open γ-field type)

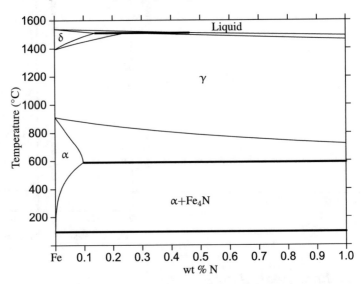

Fig. 1.13 Fe-rich part of Fe-N phase diagram (expanded γ-field type)

phase. Their phase diagram with iron belong to the expanded γ-field type. Both these categories of alloying elements are austenite (γ) stabilisers (gammagenous). As representative examples, the iron-rich part of Fe-Mn and Fe-N phase diagrams are shown in Figs. 1.12 and 1.13, respectively.

When alloyed with iron, elements such as Al, Si, Mo, etc. raise A$_3$ and lower A$_4$ temperatures. In the phase diagrams of such systems, the γ-field is constrained

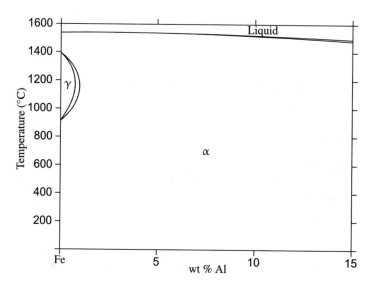

Fig. 1.14 Fe-rich part of Fe-Al phase diagram (closed γ-field type)

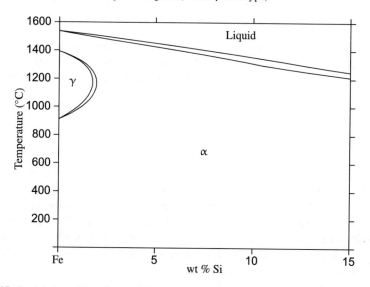

Fig. 1.15 Fe-rich part of Fe-Si phase diagram (closed γ-field type)

to a small region that resembles a loop (γ-loop) and thus belongs to the closed γ-field type. As representative examples, the iron-rich part of Fe-Al and Fe-Si phase diagrams are shown in Figs. 1.14 and 1.15, respectively.

Alloying elements such B, Nb, Ta, etc. also raise A₃ and lower A₄ but have much less solubility in γ. Their behaviour somewhat similar to the γ-loop formers, but the γ-loop is interrupted by compound formation. Fe-X phase diagrams with

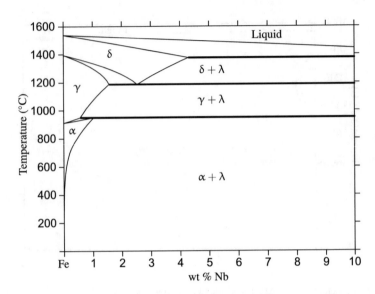

Fig. 1.16 Fe-rich part of Fe-Nb phase diagram (contracted γ-field type)

these alloying elements belong to the contracted γ-field type. Alloying elements of both these groups belong to the class of ferrite (α) stabilisers (alphageneous). As representative example, the iron-rich part of the Fe-Nb phase diagram is shown in Fig. 1.16.

1.6.2.1 Fe-C System

Most important among the iron binaries is the Fe-C system. Carbon being an austenite stabiliser lowers the A_3 temperature and raises A_4 temperature of iron. Carbon is an effective solid solution strengthener for martensite, austenite and ferrite phases. Microalloying elements such as Ti, V and Nb readily combine with carbon forming very fine MC carbides. Carbon distribution between the main microstructural constituents controls the properties of steel. When weldability is critical, it is desirable to keep carbon content below 0.2 wt %.

Two versions of the phase diagrams are useful in the context of physical metallurgy of steels. The stable one corresponds to the iron-graphite equilibrium, whereas the metastable one corresponds to the iron-cementite (Fe_3C) equilibrium. Figure 1.17 is a combined representation, in which the dashed lines are applicable to the metastable case. Note that eutectoid temperature (A_1) is about 13°C lower for the metastable case than for the stable case. Similarly, the solubility of carbon in austenite (represented by the A_{cm} phase boundary) is slightly higher in the case of the metastable one.

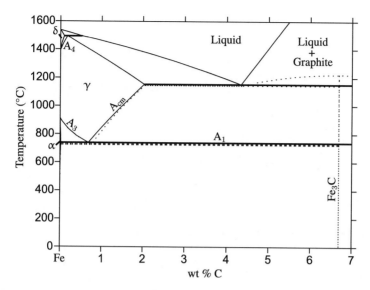

Fig. 1.17 Fe-rich part of the Fe-C phase diagram (stable and metastable versions)

1.6.3 Fe-Base Ternary Systems

Full representation of phase diagram of a c-component system requires $(c + 1)$ dimensions. It is common practice to reduce the dimension of a ternary phase diagram to two by fixing pressure and temperature. These sections are known as isothermal sections. Another way to obtain a two-dimensional ternary phase diagram is to fix the pressure and impose a composition constraint. Such sections are known as vertical sections (isopleths). Here we present some selected examples of Fe-X-C phase diagrams (X = Mn, Al or Si) in the form of isothermal and vertical sections.

Manganese is an important alloying in all generations of advanced high strength steels. Being an austenite stabiliser, manganese promotes retained austenite. It also increases hardenability of steels. The phase equilibria of the Fe-Mn-C ternary system is, therefore, very important for understanding the physical metallurgy and heat treatment of these steels. Figures 1.18 and 1.19 are some representative isothermal and vertical sections of Fe-Mn-C system, respectively.

Aluminium and silicon are ferrite stabilisers. They suppress cementite formation and enhances diffusion of carbon from martensite to retained austenite during the heat treatment of advanced high strength steels. Silicon when present in excess of 1 wt % causes problems during hot-dip galvanising. Aluminium when present along with nitrogen readily forms AlN, which is also not desirable. Figures 1.20, 1.21, 1.22, and 1.23 are some representative isothermal and vertical sections of Fe-Al-C and Fe-Si-C systems.

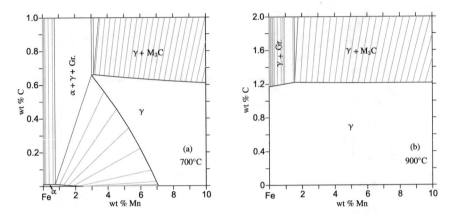

Fig. 1.18 Fe-Mn-C: Fe-rich part of representative isothermal sections

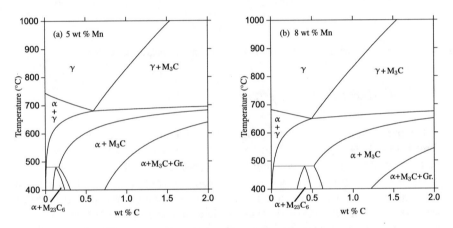

Fig. 1.19 Fe-Mn-C: Fe-rich part of representative vertical sections

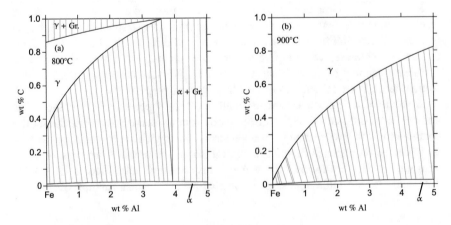

Fig. 1.20 Fe-Al-C: Fe-rich part of representative isothermal sections

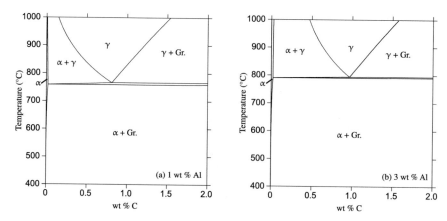

Fig. 1.21 Fe-Al-C: Fe-rich part of representative vertical sections

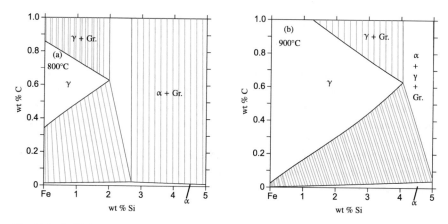

Fig. 1.22 Fe-Si-C: Fe-rich part of representative isothermal sections

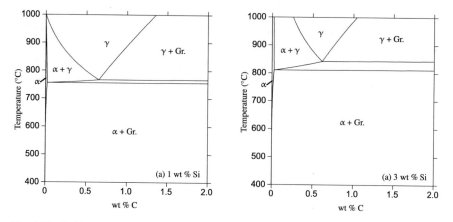

Fig. 1.23 Fe-Si-C: Fe-rich part of representative vertical sections

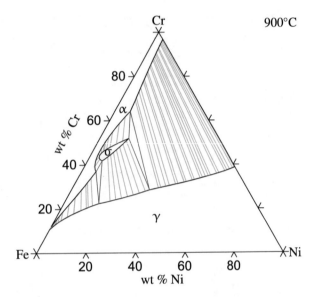

Fig. 1.24 Fe-Cr-Ni: isothermal section at 900°C

The phase equilibria in Fe-Cr-Ni system is of importance in understanding the constitution of stainless steels. Basis for stainless steel is the Fe-Cr system, properties of which are modified by the addition of alloying elements such as Ni, Mo, etc. Figure 1.24 is the isothermal section of the system at 900°C. A vertical section of the system taken at 70 wt % of Fe is shown in Fig. 1.25. The vertical section is particularly useful to understand the solidification behaviour of stainless steel melts. Both the diagrams were calculated using the TCFE9 database from Thermo-Calc [16].

1.6.4 Fe-Base Multicomponent Systems

Phase diagrams of multicomponent systems are best represented by isopleths in two dimensions by imposing $(c - 2)$ composition constraints. Consequently, several isopleths are possible. As examples, Fig. 1.26a, b are representative isopleths of the quaternary systems Fe-Al-Mn-C and Fe-Mn-Si-C, respectively. Similarly, Fig. 1.27a, b are representative isopleths of the quinary system Fe-Al-Mn-Si-C.

1.6.5 Iron Containing High Entropy Alloys

High entropy alloys (HEAs) [38, 39], also known as multi-principal element alloys (MPEAs), are a new class of material with five or more components in equimolar or near equimolar proportions. The design philosophy behind HEAs is

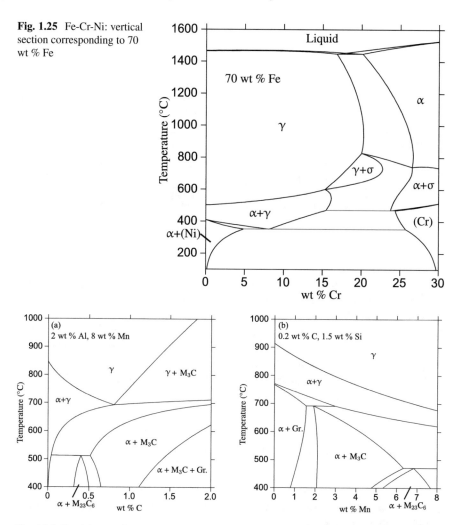

Fig. 1.25 Fe-Cr-Ni: vertical section corresponding to 70 wt % Fe

Fig. 1.26 Fe-rich part of representative vertical sections of Fe-Al-Mn-C and Fe-Mn-Si-C systems

that the high configurational entropy present in the system is expected to promote a predominantly single-phase microstructure consisting of a simple solid solution (fcc, bcc, or hcp).

Iron is a component in some of the earliest of HEAs, namely, CoCrFeMnNi [38] and AlCoCrCuFeNi [39]. About 85 % of the reported HEAs so far contain Fe [40] as one of the components. The most studied among these is the CoCrFeMnNi alloy, which is a single-phase fcc microstructure. Though fcc phase in this alloy is stable over a wide temperature range, after prolonged heat treatments at intermediate temperatures (700°C, 12000 h), small amount of σ phase is seen. AlCoCrFeNi is another well-studied HEA. This alloy has a duplex microstructure consisting

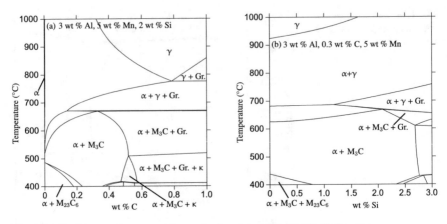

Fig. 1.27 Fe-rich part of representative vertical sections of Fe-Al-Mn-Si-C system

of disordered and ordered bcc phases. The σ phase was reported to appear after heat treatment in this alloy also [41]. Though the design idea of HEA is to use equimolar ratios, "massive solid solutions/massively alloyed solid solutions" can be made through non-equimolar compositions as well. For example, 5Co-2Cr-40Fe-27Mn-26Ni alloy has single-phase fcc microstructure with attractive mechanical properties [42, 43]. Massive fcc solid solutions with the major element being Fe and Mn, Al, Si and C as potential alloying additions is considered as the key for exploring alloys termed as high-entropy steels [44].

1.7 Applications of the CALPHAD Method

In this section, some examples of the application of the CALPHAD method to compute other types of diagrams relevant to phase transformations in ferrous materials are presented.

1.7.1 Phase Fraction Plots

Phase fraction plots, although do not qualify as phase diagrams, are quite handy when one is interested in temperature-dependent phase changes in a specific alloy. Figure 1.28 is the phase fraction plot for the alloy Fe-3Al-5Mn-2Si-0.3C.

Figure 1.29 is the phase fraction plot for Fe-25Mn-2.7Al-2.7Si-0.05C high-entropy steel. It shows that the alloy has a wide temperature range in which γ phase is the only stable phase.

Fig. 1.28 Phase fraction plot for the alloy Fe-3Al-5Mn-2Si-0.3C

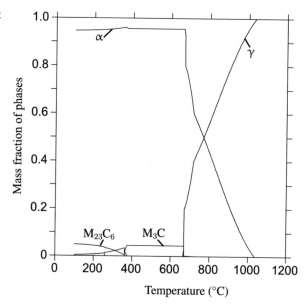

Fig. 1.29 Phase fraction plot for the alloy Fe-25Mn-2.7Al-2.7Si-0.05C

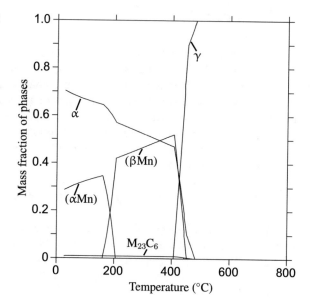

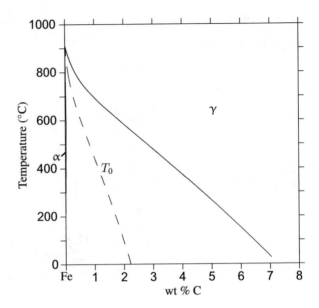

Fig. 1.30 $(\alpha+\gamma)$ phase-field of Fe-C system showing T_0 boundary

1.7.2 T_0 Boundary

The loci of points within the two-phase field of a phase diagram, where the Gibbs energy of both the solution phases of the same composition are equal, can be given as a relation between temperature and composition at constant pressure; this is known as the T_0 boundary. It is of great significance in understanding the nature of phase transformations. T_0 can be used to separate martensitic regions from diffusion-controlled phase transformation regions. Figure 1.30 shows the T_0 boundary within the $(\alpha+\gamma)$ phase field of Fe-C system. Along the T_0 boundary, the following condition is satisfied:

$$G_m^\alpha(T, p, x_C) = G_m^\gamma(T, p, x_C) \tag{1.30}$$

The above equation can be solved for T_0 temperature for a given carbon content. In a Gibbs energy-composition plot, composition coordinate of the T_0 is given by the point of intersection of the Gibbs energy boundarys of the participating solution phases. For two-phase regions that are narrow, T_0 is located approximately along the midpoints of the tie lines.

Since the driving force for γ to transform to α of the same composition (i.e. without any diffusion) is zero along the T_0 boundary, it can be regarded as the upper limit of the compositions capable of undergoing a diffusion-less transformation. Such a transformation is very similar to allotropic transformation in elements, and therefore T_0 boundary is sometimes known as the allotropic phase boundary. Growth without diffusion can occur only if the carbon content of austenite lies left of the T_0 boundary in Fe-C diagram. Above T_0 boundary, transformation without composition

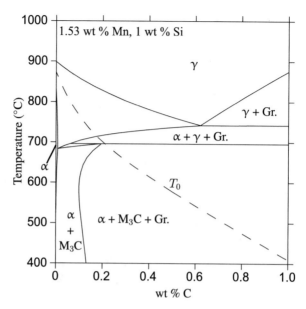

Fig. 1.31 Isopleth of Fe-(0-1)C-1.53Mn-1Si overlaid with T_0 boundary in the $(\alpha+\gamma)$ phase field

change is not possible. The martensitic start temperature (M_s) is located somewhat below the T_0, since higher driving forces are required for the transformation, primarily due to strain energy associated with the martensite. Likewise, the austenite start temperature (A_s) is located above T_0. As an approximation, one may use

$$T_0 \approx \frac{1}{2}(M_s + A_s) \tag{1.31}$$

to estimate the T_0 temperature of a given steel composition.

For a multicomponent steel, the thermodynamic condition along the T_0 boundary in the $(\alpha + \gamma)$ field is given by

$$G_m^\alpha (T, p, x_{i=2..c}) = G_m^\gamma (T, p, x_{i=2..c}) \tag{1.32}$$

The above equation can be solved for T_0 temperature for different values of composition of a selected component. As an example, Fig. 1.31 shows T_0 boundary overlaid on the isopleth of Fe-(0-1)C-1.53Mn-1Si.

1.7.3   Paraequilibrium

Paraequilibrium refers to a kinetically constrained local equilibrium at the parent/product interface involving a phase transformation between two solid solutions. It was first proposed by Hultgren [45] to explain peculiarities in austenite/ferrite

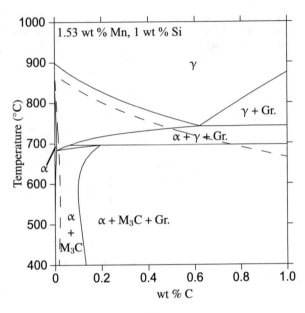

Fig. 1.32 Isopleth of Fe-(0-1)C-1.53Mn-1Si overlaid with paraequilibrium phase boundaries of the ($\alpha + \gamma$) phase field

phase transformation in ternary and multicomponent steels. The basis of paraequilibrium hypothesis is that diffusion of substitutional elements is very slow compared to interstitial elements. Hence it is possible that when ferrite forms rapidly from austenite under paraequilibrium conditions, where compositions of substitutional elements in austenite and ferrite remain the same, while there is a redistribution of interstitial elements.

The thermodynamic conditions for paraequilbrium are given by [46]

$$\mu_i^\alpha = \mu_i^\gamma$$
$$\sum_j x_j^\alpha \mu_j^\alpha = \sum_j x_j^\gamma \mu_j^\gamma \qquad (1.33)$$
$$x_j^\alpha = x_j^\gamma$$

where i refers to interstitial elements such as carbon, nitrogen, etc. and j refers to all substitutional elements including iron. Figure 1.32 shows isopleth of Fe-(0-1)C-1.53Mn-1Si overlaid with paraequilibrium phase boundaries of the ($\alpha + \gamma$) phase field.

Paraequilibrium should be regarded only as a limiting case since in reality some redistribution of substitutional elements always takes place.

1.8 Summary and Outlook

Determination of phase equilibria and thermodynamic properties of materials are efficiently done using a combinatorial approach that involves the experimental methods and the CALPHAD approach. The CALPHAD approach relies on finding the equilibrium constitution of material by Gibbs energy minimisation, for which Gibbs energy functions of all participating phases must be known. Gibbs energy functions are established through thermodynamic modelling. Several Gibbs energy models for phases often found in ferrous materials were discussed. Examples of computed phase equilibria in selected iron-base systems relevant to advanced high-strength steels for automotive applications were presented. Finally, some applications of the CALPHAD method relevant to the phase transformations in ferrous materials were given.

The reliability of the results obtained for a multicomponent system obtained by the CALPHAD approach is limited by the accuracy of the Gibbs energy functions of the binary, ternary and quaternary sub-systems. Current generation Gibbs energy databases for ferrous materials still require improvements with respect to increased composition limits for some of the alloying elements. Most notably, it is required that future generation databases should be improved to handle calculations involving more amount of zinc, aluminium and nickel than presently possible. Such databases will offer possibilities to do calculations involving zinc coating on steels, low-density steels and joining of steels with superalloys. Not all commercially available databases has the capability of handling dissolved gases such as oxygen, nitrogen and hydrogen. Besides, inclusion of elements such as tin, arsenic, antimony and tellurium will enhance the capability of databases to handle problems associated with recycling and free-cutting steels. Present generations databases are recommended for doing calculations above 300 K. Efforts are underway to extend the validity of Gibbs energy functions to cryogenic temperatures. Lastly, more phases need to be added to the database. An example is the FeTiP phase, which is important for the interstitial-free steels, is still missing in the thermodynamic databases for ferrous materials.

Acknowledgments The author wishes to acknowledge Professor G.D. Janaki Ram for his meticulous reading of the manuscript and suggesting several modifications.

References

1. N. Saunders, A.P. Miodownik, *CALPHAD (Calculation of Phase Diagrams): A Comprehensive Guide*, vol. 1 (Elsevier Science Ltd, Oxford, UK, 1998)
2. H.L. Lukas, S.G. Fries, B. Sundman, *Computational Thermodynamics: The Calphad Method* (Cambridge University Press, Cambridge, 2007)
3. J.L. Meijering, Calculation of the Nickel-Chromium-Copper phase diagram from binary data. Acta Metall. **5**(5), 257–264 (1957)

4. L. Kaufman, H. Bernstein, *Computer Calculation of Phase Diagrams* (Academic Press, New York, 1970)
5. M. Hillert, L.-I. Steffansson, Regular-solution model for stoichiometric phases and ionic melts. Acta Chem. Scand. **24**(10), 3618–3626 (1970)
6. B. Sundman, J. Ågren, A regular solution model for phases with several components and sublattices suitable for computer application. J. Phys. Chem. Solids **42**(4), 297–301 (1981)
7. M. Hillert, The compound energy formalism. J. Alloys Compd. **320**(2), 161–176 (2001)
8. H.L. Lukas, E.T. Henig, B. Zimmermann, Optimization of phase diagrams by a least squares method using simultaneously different types of data. Calphad **1**(3), 225–236 (1977)
9. B. Sundman, B. Jansson, J.-O. Andersson, The Thermo-Calc databank system. Calphad **9**(2), 153–190 (1985)
10. G. Eriksson, K. Hack, ChemSage-A computer program for the calculation of complex chemical equilibria. Metall. Trans. **21B**(6), 1013–1023 (1990)
11. S.L. Chen, S. Daniel, F. Zhang, Y.A. Chang, X.Y. Yan, F.Y. Xie, R. Schmid-Fetzer, W.A. Oates, The PANDAT software package and its applications. Calphad **26**(2), 175–188 (2002)
12. C.W. Bale, P. Chartrand, S.A. Degterov, G. Eriksson, K. Hack, R. Ben Mahfoud, J. Melançon, A.D. Pelton, S. Petersen, FactSage thermochemical software and databases. Calphad **26**(2), 189–228 (2002)
13. Z.-C. Wang, R. Lück, B. Predel, A general regular-type geometric model for quaternary and higher-order systems. Calphad **17**(3), 303–333 (1993)
14. A. Dinsdale, SGTE data for pure elements. Calphad **15**(4), 317–425 (1991)
15. K.C. Hari Kumar, P. Wollants, Some guidelines for the thermodynamic optimization of phase diagram. J. Alloys Compd. **320**(2), 189–198 (2001)
16. Thermo-Calc Software, Thermodynamic software & databases. [Online; Accessed 15 Sept 2018] (2018). https://www.thermocalc.com/
17. CompuTherm, Thermodynamic databases. [Online; Accessed 15 Sept 2018] (2018). http://www.computherm.com/
18. FactSage Software, Thermodynamic software & databases. [Online; Accessed 15 Sept 2018] (2018). https://www.factsage.com/
19. JMatPro Software, Thermodynamic software & databases. [Online; Accessed 15 Sept 2018] (2018). https://www.sentesoftware.co.uk/jmatpro
20. Z.-K. Liu, Y. Wang, *Computational Thermodynamics of Materials* (Cambridge University Press, Cambridge, 2016)
21. M. Hillert, M. Jarl, A model for alloying in ferromagnetic metals. Calphad **2**(3), 227–238 (1978)
22. O. Redlich, A.T. Kister, Algebraic representation of thermodynamic properties and the classification of solutions. Ind. Eng. Chem. **40**(2), 345–348 (1948)
23. Y.-M. Muggianu, M. Gambino, J.-P. Bros, Enthalpy of formation of liquid Bi-Sn-Ga alloys at 723 K: choice of an analytical expression of integral and partial excess quantities of mixing. J. Chim. Phys. **72**(1), 83–88 (1975)
24. M. Temkin, Mixtures of fused salts as ionic solutions. Acta Phys. Chem. USSR **20**, 411 (1945)
25. H. Harvig, Extended version of the regular solution model for stoichiometric phases and ionic melts. Acta Chem. Scand. **25**, 3199–3204 (1971)
26. M. Hillert, M. Waldenstorm, Isothermal sections of the Fe-Mn-C system in the temperature range 873-1373 K, Calphad **1**(2), 97–132 (1977)
27. A. Guillermet, The Fe-Mo (iron-molybdenum) system. Bull. Alloy Phase Diagrams **3**(3), 359–367 (1982)
28. J.-O. Andersson, A thermodynamic evaluation of the iron-vanadium system. Calphad **7**(4), 305–315 (1983)
29. J.-O. Andersson, B. Sundman, Thermodynamic properties of the Cr-Fe system. Calphad **11**(1), 83–92 (1987)
30. A. Watson, F. Hayes, Some experiences modelling the sigma phase in the Ni-V system. J. Alloys Compd. **320**, 199–206 (2001)

31. J.-M. Joubert, Crystal chemistry and Calphad modeling of the σ phase. Prog. Mater. Sci. **53**(3), 528–583 (2008)
32. K.C. Hari Kumar, I. Ansara, P. Wollants, Sublattice modeling of the μ-phase. Calphad **22**(3), 323–334 (1998)
33. N. Dupin, Contribution evaluation thermodynamique des alliages polyconstitués à base de nickel, Ph.D. thesis, Institut National Polytechnique de Gernoble (1995)
34. I. Ansara, N. Dupin, H.L. Lukas, B. Sundman, Thermodynamic assessment of the Al-Ni system, J. Alloys Compd. **247**(1–2), 20–30 (1997)
35. B. Hallstedt, A. Khvan, B. Lindahl, M. Selleby, S. Liu, PrecHiMn-4-A thermodynamic database for high-Mn steels. Calphad **56**(1), 49–57 (2017)
36. T. Massalski (Ed.), *Binary Alloy Phase Diagrams* (ASM International, Metals Park, 1990)
37. F. Wever, On the influence of elements on the polymorphism of iron. Arch. Eisenhüttenwes. **2**, 739–746 (1929)
38. B. Cantor, I. Chang, P. Knight, A. Vincent, Microstructural development in equiatomic multicomponent alloys. Mater. Sci. Eng. A **375–377**, 448–511 (2004)
39. J. Yeh, S. Chen, S. Lin, J. Gan, T. Chin, T. Shun, C. Tsau, S. Chang, Nanostructured high-entropy alloys with multiple principal elements: Novel alloy design concepts and outcomes. Adv. Eng. Mater. **6**, 299–303 (2004)
40. D. Miracle, O. Senkov, A critical review of high entropy alloys and related concepts. Acta Mater. **122**, 448–511 (2017)
41. Z. Tang, O. Senkov, C. Parish, C. Zhang, F. Zhang, L. Santodonato, G. Wang, G. Zhao, F. Yang, P. Liaw, Tensile ductility of an AlCoCrFeNi multi-phase high-entropy alloy through hot isostatic pressing (HIP) and homogenization. Mater. Sci. Eng. A. **647**, 229–240 (2015)
42. M. Yao, K. Pradeep, C. Tasan, D. Raabe, A novel, single phase, non-equiatomic FeMnNiCoCr high-entropy alloy with exceptional phase stability and tensile ductility. Scr. Mater. **72–73**, 5–8 (2014)
43. K. Pradeep, C. Tasan, M. Yao, Y. Deng, H. Springer, D. Raabe, Non-equiatomic high entropy alloys: approach towards rapid alloy screening and property-oriented design. Mater. Sci. Eng. A. **648**, 183–192 (2015)
44. D. Raabe, C. Tasan, H. Springer, M. Bausch, From high-entropy alloys to high-entropy steels. Steel Res. Int. **86**(10), 1127–1138 (2015)
45. A. Hultgren, Isothermal transformation of Austenite. Trans. ASM **39**, 915–1005 (1947)
46. G. Ghosh, G. Olson, Simulation of para equilibrium growth in multicomponent systems. Metall. Mater. Trans. **32A**(3), 455–467 (2001)

Chapter 2
Processing of Ferrous Alloys

Hamid Azizi, Olga A. Girina, Damon Panahi, Tihe Zhou, and Hatem S. Zurob

Abbreviations

AF	Acicular ferrite
AHSS	Advanced high strength steel
ART	Austenite reversion transformation
BOF	Basic oxygen furnace
Bs	Bainite start temperature
CAL	Continuous annealing line
CAPL	Continuous annealing with presence of isotheral holding line
CFB	Carbide-free bainite
CP	Complex phase
CQ	Commercial quality
CRMLQ	Cold-roll magnetic lamination quality
DDQ	Deep drawing quality
DP	Dual phase
DQ	Drawing quality
EAF	Electric arc furnace
EDDQ	Extra-deep-drawing quality
FRT	Finish rolling temperature
GOES	Grain-oriented electrical steel

H. Azizi · H. S. Zurob (✉)
Department of Materials Science and Engineering, McMaster University, Hamilton, ON, Canada
e-mail: zurobh@mcmaster.ca

O. A. Girina · D. Panahi
ArcelorMittal, East Chicago, IN, USA

T. Zhou
Stelco Inc, Hamilton, ON, Canada

© Springer Nature Switzerland AG 2021
R. Rana (ed.), *High-Performance Ferrous Alloys*,
https://doi.org/10.1007/978-3-030-53825-5_2

37

HDCGL	Hot-dip continuous galvanizing line
HER	Hole expansion ratio
HSLA	High strength low alloy
IF	Interstitial free steel
M	Martensite
M_s	Martensite start temperature
MS	Martensitic steel
NOES	Non-oriented electrical steel
P	Pearlite
PF	Polygonal ferrite
Q&P	Quench and partition
RA	Retained austenite
RCR	Recrystallization controlled rolling
ROT	Run-out table
SEM	Scanning electron microscope
SK	Semi-killed
SRP	Solute retardation parameter
TBF	TRIP steels with bainitic ferrite
TEM	Transmission electron microscope
TMCP	Thermomechanical controlled process
TMP	Thermomechanical processing
Tnr	No recrystallization temperature
TRIP	Transformation-induced plasticity
ULSAB	Ultra-light steel body consortium
UTS	Ultimate tensile strength
YS	Yield strength

2.1 Steelmaking Process: A Long Way to the Final Products

Figure 2.1 depicts the key steps in the processing of steels from ores to final products [1]. As shown in this figure, a final product requires a long list of processing steps before reaching customers. To start the raw materials are prepared for the key step of ironmaking using the blast furnace. Once the oxide has been reduced to molten metal, additional refining is carried out using a basic oxygen furnace (BOF) or an electric arc furnace (EAF). Ladle refining is followed by a casting operation to make semi-finished products in the shape of slabs, billets, or blooms. A hot/cold rolling process brings these semi-finished products close to the final shape that is ready for shaping/cutting operations. These close-to-final products are in the shape of long products (rods, bars, and sections) or flat products (strip, plate, and sheet). These products are used in many industries including automotive, construction, transportation, and energy. These illustrations highlight the key steps of a lengthy process, which is further complicated by the many interacting microstructure evolution phenomena that take place during processing. This chapter aims to shed

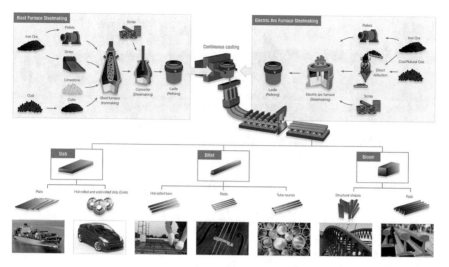

Fig. 2.1 Overview of steelmaking process (www.worldsteel.org)

light on some of the recent developments in (solid-state) steel processing, with an emphasis on thermomechanical processing (TMP) of steels in conjunction with an up-to-date understanding of the process-microstructure-property relationships in modern steels.

2.2 Hot Working of Steels: An Introduction

In this section, the main emphasis is on the hot rolling of flat products, with few examples of other processing techniques such as forging and wire drawing. Rolling products comprise a large fraction of the steel market, and the underlying metallurgical aspects of the process discussed in this chapter can be largely extended to the other hot working processes even though they offer different temperature/strain/strain rate paths. Figure 2.2 is a schematic illustration of a typical hot rolling mill [2]. It includes a reheating furnace, roughing and finishing mills, a runout table with laminar cooling, and down coiler(s). Design changes can be made to accommodate specific product requirements.

A modern thermomechanical controlled process (TMCP) requires a clear understanding of the underlying thermodynamics and kinetics of the phenomena that contribute to microstructure evolution including recovery, recrystallization, precipitation, and phase transformations, specifically austenite decomposition. Figure 2.3 highlights the key microstructure evolution events taking place during thermomechanical processing. Important temperatures to pay attention to are the *no recrystallization temperature* (T_{nr}), the *finish rolling temperature* (FRT), and the phase transformation start temperatures (the austenite to ferrite transformation starts

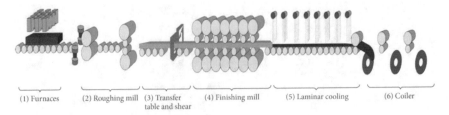

(1) Furnaces   (2) Roughing mill (3) Transfer table and shear   (4) Finishing mill   (5) Laminar cooling   (6) Coiler

Fig. 2.2 A schematic illustration of a steel hot rolling mill [2]

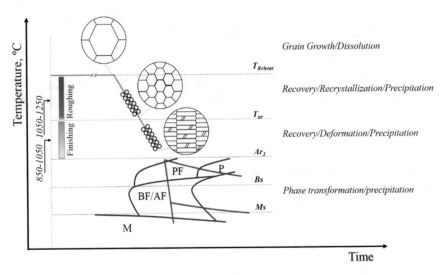

Fig. 2.3 A schematic diagram of a TMCP with corresponding microstructures and metallurgical events taking place in each segment (T_{Reheat}, reheat temperature; T_{nr}, non-recrystallization temperature; Ar_3, austenite decomposition temperature upon cooling; M_s: martensite start temperature; B_s, bainite start temperature; PF, polygonal ferrite; P, pearlite; BF, bainitic ferrite; AF, acicular ferrite; M, martensite)

upon cooling below Ar_3, bainite formation starts at B_s, and martensite formation starts at M_s). These temperatures depend upon the deformation history, as well as the chemical composition of the alloy. In the following sections, a brief description of the metallurgical phenomena taking place in each segment of the TMCP is reviewed.

2.3 Reheating Stage

The main purpose of the reheating stage is to make the initial material ready for the deformation process. The reheating temperature for the roughing stage generally lies between 1100 and 1250 °C, and the holding time may vary from one to several hours, depending on the dimensions and chemistry of the slab.

In microalloyed steels, the reheating stage is accompanied by the dissolution of carbides and carbonitride precipitates. It is worth noting that, depending on the chemistry of the steel, some of the nitrides (e.g., TiN and AlN) don't dissolve during reheating. The presence of these nitrides can effectively retard austenite grain growth during the reheating stage. A large body of work in the literature is devoted to the effect of different alloying elements and their precipitates on austenite grain evolution in the reheating stage [3, 4]. Depending on the chemistry of the steel and the reheating temperature and time, normal and abnormal grain growth may take place. During normal grain growth, the average grain size increases, but the size distribution remains self-similar. In contrast, grains can grow "abnormally," where a larger grain grows at the expense of many smaller ones. This phenomenon can occur when the reheating temperature of microalloyed steels is not carefully selected, because the dissolution of precipitates may result in the removal of local particle pinning and selective grain growth. Therefore, a proper understanding of the dissolution of precipitates and their solubility limits is of great importance when trying to avoid abnormal grain growth, as abnormal grain growth may result in an inhomogeneous grain distribution in the final product and lead to inferior toughness.

2.4 Roughing Stage

The main purpose of the roughing step is to reduce the cross section of the castings and break down their dendritic structure. Roughing stands are usually in the form of 2-high or 4-high roughers. During the roughing stage, austenite grains recrystallize statically or dynamically depending upon the composition, roughing strain, strain rate, and temperature. Repeated recrystallization has the desired effect of reducing the average grain size. The high temperatures during roughing (1050–1250 °C) can lead to grain growth, which is a concern for developing high strength steels with a fine final microstructure. The rolling of plates often requires large forces and is carried in many, relatively, light passes because of mill load limitations. A reversing mill is usually used for plate roughing with interpass times of up to 10–30 s at relatively high temperatures [5]. This time scale is long enough for the interaction between precipitation, recrystallization, and grain growth. The presence of TiN particles that exert a pinning force on austenite grain boundaries results in the suppression of austenite grain growth while allowing further recrystallization to take place. This is the basis for recrystallization controlled rolling (RCR), in which the recrystallization process results in grain refinement before the onset of phase transformations [6]. Solutes might also contribute to the retardation of grain growth as a result of the solute drag effect. Table 2.1 allows one to compare the relative effectiveness of different solutes (Mo, Ti, V, and Nb) in the retardation of austenite recrystallization [7]. The solute retardation parameter (SRP) is a parameter comparing the influence of different alloying elements on the recrystallization rate. It is evident that Nb is comparatively more effective than other alloying elements in retarding austenite recrystallization, even at low concentrations.

Table 2.1 Solute retardation parameter for static recrystallization of austenite for different alloying elements [7]

Element	SRP	
	01·at.-%	0·01 wt-%
V	12	13
Mo	33	20
Ti	70	83
Nb	325	222

Table 2.2 Summary of finish rolling mill for different product geometries

Mill	Plate	Strip	Bar
Reduction per pass, %	15	50	25
Total strain	< 10	10–200	1000
Interpass, sec	20	0.1–1	<0.1
Finish temperature, °C	750–950	800–900	1000
Number of passes	15–27	6–7	8–15

Adapted from [8]

In contrast to the plate rolling, in a strip mill, the interpass times are usually 1–2 s. Hot strip mills are usually designed to be faster than plate rolling with more reduction per pass to avoid temperature loss and intercritical rolling. Thus, grain growth is of less concern in hot strip mills compared to plate mills.

2.5 Finish Rolling

While roughing deformation is carried out at temperatures well above 1000 °C, leading to repeated steps of austenite grain refinement via recrystallization, the finishing stage can be extended to just above the start of the phase transformation temperature, Ar_3, as shown in Fig. 2.3. Table 2.2 summarizes the finishing mill operational parameters for plate, strip, and bar products [8]. The finishing schedule of plates and strips differs significantly with respect to interpass time, strain rate, and number of passes. The resulting microstructure is very sensitive to the interactions between the processes of deformation, recovery, recrystallization, and precipitation. As the rolling temperature decreases, the kinetics of recovery and recrystallization slow down. Simultaneously, the solubility of the microalloying elements decreases leading to the precipitation of fine-scale carbonitrides. The combination of these factors can result in the accumulation of deformation in austenite prior to phase transformation resulting in the so-called "pancaked" austenite microstructure. Depending on the desired microstructure, the temperature, strain rate, and rolling reductions need to be carefully controlled in order to produce either a fully recrystallized or a fully pancaked austenite microstructure at the end of finish rolling. Partial recrystallization during finish rolling is not desirable as it results in an inhomogeneous final microstructure that is detrimental to toughness [9].

From the above discussion, microalloying elements play an important role in controlling the microstructure during finish rolling. As an example, small additions

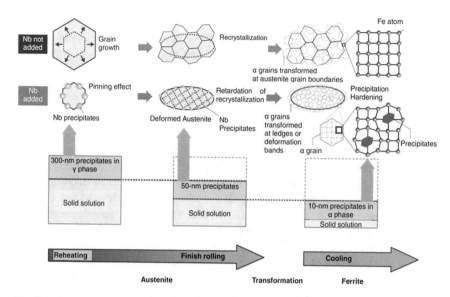

Fig. 2.4 Schematic representation of niobium precipitates at each stage of TMCP and their effects on microstructure refinement in comparison with no niobium addition [11]

of niobium and titanium (wt.% ≤ 0.1) can help to refine the microstructure during slab reheating, controlled rolling, and accelerated cooling, thereby enhancing the strength of the final products [10]. The effects of niobium are schematically shown in Fig. 2.4 [11]. Here, austenite recrystallization does not take place due to the presence of fine Nb(C, N) precipitates which retard dislocation rearrangement and the nucleation of recrystallization. For practical purposes, it is convenient to introduce the no-recrystallization temperature, T_{nr}, which is the temperature below which static recrystallization can no longer go to completion in the interpass interval [12, 13]. Deformation below T_{nr} leads to the accumulation of strain and results in the appearance of elongated grains and the formation of ledges on austenite grain boundaries. These extra surfaces act as nucleation sites for ferrite, resulting in significant grain refinement upon transformation by accelerated cooling, and ultimately leading to improved properties. The no-recrystallization temperature, T_{nr}, strongly depends on the chemistry of the alloy and increases as the (micro)alloying content increases [13]. In fact, T_{nr} is a complex concept since it is influenced by the interaction between different mechanisms, namely, deformation, recovery, recrystallization, solute drag, and precipitation. Among the different microalloying elements, Nb has the largest impact on T_{nr}, approximately extending it into the roughing temperature range. In addition to being used to control T_{nr}, microalloying additions are also used for precipitation hardening. Very fine dispersions of microalloyed carbides can form through interphase precipitation at the austenite/ferrite interface during the phase transformation or within ferrite after the phase transformation. These fine carbonitrides make an important contribution to the final strength of the steel. Therefore, it is necessary to design hot rolling schedules that leave a sufficient amount of the microalloying elements in solid solution to allow subsequent precipitation of fine carbonitrides during or after the phase transformation.

2.6   Cooling and Coiling

Steels are commonly cooled using either air or water. The air cooling rate is approximately 10 °C/s, while water cooling rates vary from 18 to 50 °C/s depending on the dimensions of the steels [14]. The cooling pattern after rolling, together with the chemical composition and the austenite condition, determines the final microstructure and mechanical properties. A wider range of cooling rates and patterns is available on strip mills compared to plate mills. Depending on the cooling rate, a variety of microstructures can be obtained as can be seen in Fig. 2.3. Higher cooling rates on the runout table (ROT) result in finer ferritic structures and, depending on the steel, a higher volume fraction of martensite/retained austenite. Similarly, coiling temperature plays an important role in determining the final microstructure and mechanical properties of the steel. The selection of coiling temperature depends on the chemical composition of the steel being processed and typically varies between 400 and 650 °C. The cooling rate and coiling temperatures are important process variables with regard to microstructure and properties. Phase transformation of austenite during cooling/coiling fundamentally changes the microstructure and properties of steel. This decomposition can result in formation of new constituents such as bainite, polygonal ferrite, pearlite, bainitic ferrite, acicular ferrite, and martensite. The final microstructure is usually a mixture of abovementioned phases depending on steel chemistry, austenite deformation state, and cooling rate.

Coiling temperature can also affect mechanical properties by controlling precipitate formation and coarsening, as well as impurity segregation. The precipitation of $V(C,N)$ in ferrite during coiling can lead to significant strength increases depending not only on the rolling schedule used but also on the chemical composition. In fact, vanadium forms almost no precipitates in austenite and is, therefore, available for precipitation hardening during or after austenite decomposition [9]. Similarly, retained solute Nb after finish rolling can also form very fine precipitates in ferrite, 1–2 nm, resulting in considerable precipitate strengthening, ~ 90 MPa [15].

2.7   Cold Rolling and Annealing

To further reduce steel strip gauge and better control dimensional tolerance, hot rolled steel strips are further processed by cold rolling. This has the additional advantages of improving the surface finishes, surface flatness, and mechanical

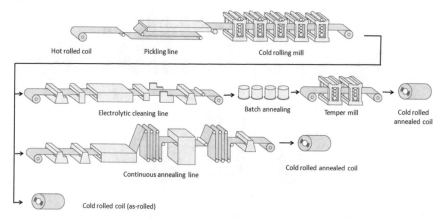

Fig. 2.5 Sketch of high-level process map of cold mill [16]

properties. The cold processing unit includes cold reduction stands, as well as an annealing line and temper rolling. A layer of oxides or scales form on the surface of hot rolled steel strips during the hot rolling process, and these oxide and scales must be removed prior to further processing at the cold mill. This is usually accomplished by a pickling line. The operation of a pickling line is usually continuous with entering coils welded head to tail. The strips are pickled in a hydrochloric or sulfuric acid solution and then subjected to a rinsing and passivation step, followed by drying and side trimming. Pickled oil is usually applied on the top surface of the strip just before recoiling to act as a lubricant in the first cold reduction pass. Figure 2.5 displays a typical, high-level process map for cold rolling and annealing that would be used at most industrial facilities [16]. The microstructure of the final product depends on the steel itself (chemistry) as well as the configuration of the processing line. Some lines are designed to achieve a very specific target of mechanical or magnetic properties.

Some steels are sold in the cold rolled steel. More commonly, however, the cold rolling step is followed by either batch or continuous annealing. Batch annealing products include commercial quality and drawing quality carbon steels, interstitial free steels, high strength low alloy steels, and electrical steels. Continuous annealing products include advanced high strength steels, interstitial steels, conventional carbon steels, and electrical steels. It is worth noting that some steels could be produced using both batch and continuous annealing. The next section will focus on cold rolling, followed by a discussion of continuous and batch annealing.

2.7.1 Cold Rolling

Cold rolling operations are carried out at either a tandem cold rolling mill or a reversing cold reduction mill. Most operations are at room temperature. Figure 2.6 shows examples of both a tandem cold rolling mill (a) and a reversing cold reduction

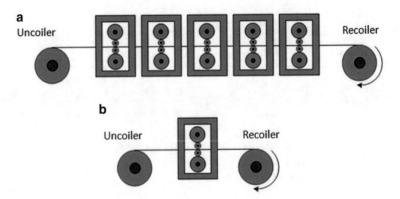

Fig. 2.6 (a) Schematic of tandem cold rolling mill, (b) reversing cold rolling mill [17]

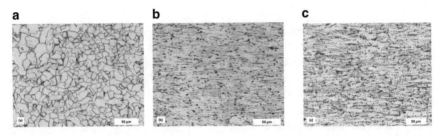

Fig. 2.7 Microstructure evolution of low carbon steel: (a) original hot band, (b) 70% and (c) 80% reduction ratios

mill (b). Skin rolled products are only subjected to 0.5–1% reduction to correct the hot rolled steel strip shape and to eliminate the yield point elongation, thus avoiding stretcher strain (Lüder lines) in the final products. Most other products are subjected to reductions in the range of 35–90%. In general, the lowest percentage reduction takes place in the last pass to obtain better control of flatness, gauge, and surface finishes.

Most energy expended in cold work during the cold reduction operation is converted into heat, but only a small fraction (<10%) is stored in the reduced steel strip as strain energy [18]. The microstructure of a hot rolled steel strip experiences dramatic change during the cold rolling process. An example of the microstructural evolution of a low carbon steel (0.044 C, 0.25 Mn wt.%) is summarized in Fig. 2.7 [19], with Fig. 2.7a displaying the original microstructure of the hot rolled strip and Fig. 2.7b showing the microstructure after 70% reduction and Fig. 2.8c after 80% reduction ratio. The cold-reduced optical microstructures consist of ferrite (bright phase) with a small amount of cementite (dark phase). Figure 2.7c displays highly deformed ferrite grains and a preferential grain boundary distribution of the carbides

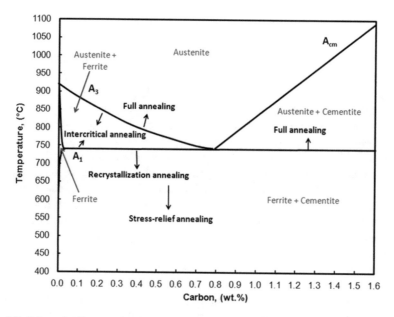

Fig. 2.8 Schematic diagram showing approximate temperature ranges for different annealing processes superimposed on the partial Fe-Fe$_3$C phase diagram [22]

along the rolling direction. Not only have the elongated grains increased the total grain boundary area; there is also an accumulation of dislocations within the grains. The sum of the energy of all the new interfaces and all the dislocations represents the stored energy of deformation. The release of this stored energy provides the driving force for recovery and recrystallization during the subsequent annealing process, and the nature of the deformed microstructure controls the development, growth, and orientation of the nuclei which will become recrystallized grains.

Cold rolling products are divided based on formability. Full-hard products can be bent up to 45°, while half-hard and quarter-hard can be bent up to 90° and 180°, respectively. In the case of half- and quarter-hard products, the purpose of cold rolling is to further reduce strip thickness, improve dimensional tolerance, and increase both the yield and tensile strength of high strength low alloy steel for automobile applications. As an example, HSLA Gr80 is cold reduced 45% to meet 700B specification (100 ksi or 700 MPa yield strength) [20]. Due to the lack of formability, cold rolled full-hard strips find very limited commercial applications. A notable exception is full-hard carbon steel, which is utilized in strapping applications. Carbon steel (0.17–0.33 C wt.%), for example, is reduced by 50–60% to meet SAE J403 1017 full-hard strapping [21]. To restore the formability, most full-hard strips are submitted to annealing or otherwise heat treated in order to meet different mechanical and magnetic specifications.

2.7.2   Basic Metallurgical Aspects of Annealing

The cold rolled material is annealed to recover the ductility and achieve the final properties and microstructures required by the customer. The annealing process consists of heating to a designed temperature, soaking at the designed temperature, and cooling. Figure 2.8 shows the different annealing processes used at a cold mill complex, superimposed on the Fe-C phase diagram [22]. Based on the full-hard strip chemistry and soaking temperature, the annealing process is divided into the following subgroups: (1) full annealing, (2) intercritical annealing, (3) recrystallization (subcritical) annealing, and (4) stress-relief annealing. Full annealing, i.e., annealing above A_{c3} with full austenite formation, is used to produce martensitic steel (MS) and complex-phase (CP) steel and involves the austenite transforming to martensite or bainite during fast cooling or quenching. Intercritical annealing, i.e., heating between the A_{c1} and A_{c3} temperatures to produce a small percentage of austenite within the ferrite matrix, is employed to produce dual-phase (DP) and transformation-induced plasticity (TRIP) steels. Recrystallization annealing, i.e., annealing below A_{c1} temperatures without austenite transformation, is usually utilized to produce formable recrystallized or partially recrystallized steels with ferritic microstructure. Stress-relief annealing is used to restore a certain degree of formability for cold rolled products, and generally there is no recrystallization during reheating and cooling. These annealing processes, which are associated with different products, will be discussed in the next sections on batch annealing and continuous annealing.

Annealing is accomplished by either a batch annealing process or continuous annealing process. In both cases, one must consider the recovery and recrystallization of the cold-deformed microstructure. The main difference is the additional time available for microstructure evolution during batch annealing compared to continuous annealing.

2.7.3   Batch Annealing

Batch annealing is usually performed at a subcritical temperature, below A_{c1}, which is below ~725 °C for low carbon steel. Figure 2.9 shows a typical single-stack gas-fired annealing furnace, on the left, and a multi-stack annealing furnace, on the right [23]. In the batch process, the cold-reduced strip coils are stacked on top of each other on fixed bases in a bell-type furnace. The space between the inner cover and the furnace is heated by natural gas/air combustion. Mixture of 5% H_2 and 95% N_2 or 100% H_2 (by volume) is fan circulated inside the cover to transfer the heat to the coils. After soaking, the above gas is cooled by a heat exchanger to speed up the cooling step. An example of batch annealing products is the electrical steels used for motors, transformers, and generators. These are annealed in a multi-stack annealing furnace. Electrical heating is used for the furnace, and pure hydrogen is used as the

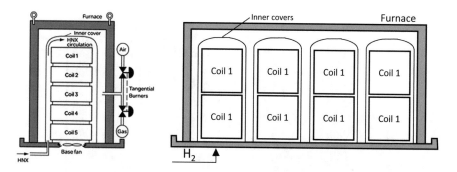

Fig. 2.9 Single-stack annealing furnace (left) and multi-stack annealing furnace (right) [23]

gas in contact with the coils in order to maintain cleanliness and eliminate carbon and nitrogen pickup [24].

Batch annealing processing has been widely applied in producing interstitial free (IF) steels, ultra-low carbon steels, low carbon steels, and HSLA steels for automotive, structural, and appliance industries. In these examples, the aim of the batch annealing process is to lower the hardness and increase ductility and formability.

It is important to keep in mind that batch annealing is characterized by slow heating and cooling rates and very long soak time. A full batch annealing cycle could take days to complete. In addition, the outer surfaces of the coil, referred to as hot spots, are heated faster and achieve the designed annealing temperature in a shorter time as compared to the inner core of the coil, commonly referred to as a cold spot. Such thermal lag during heating results in a variation in microstructure and mechanical properties within a coil. In addition, due to the axial temperature variation along the furnace from the bottom to top, there is the potential for coil-to-coil variations in their microstructure and mechanical properties [25].

To improve productivity and eliminate mechanical property variations, as well as to produce different advanced high strength steels for automobile applications, great efforts have been put into the development of continuous annealing, and a variety of continuous annealing lines have appeared, worldwide, since the first application of continuous annealing was used to produce hot-dip galvanized sheet in 1936 [26].

2.7.4 Continuous Annealing

A typical continuous annealing line (Fig. 2.10) contains entry-side equipment, the furnace section, and the exit-side equipment [27]. The main entry-side equipment comprises an un-coiler, a welder, an electrolytic cleaning tank, and an entry accumulator. The furnace section comprises a heating zone, a soaking zone, and a cooling zone. The cooling zone is divided into different sub-zones based on the

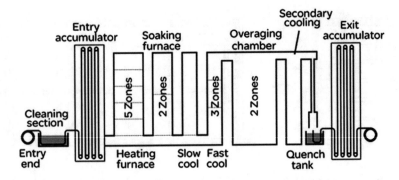

Fig. 2.10 Schematic drawing of continuous annealing line process section [27]

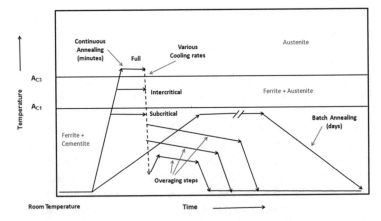

Fig. 2.11 Temperature-time processing schedules for cold rolled and annealed sheet steels. Continuous and batch annealing are schematically compared, and full and intercritical annealing temperatures used to produce advanced high strength steels are indicated (Adapted from [28])

line design. The exit equipment comprises an exit accumulator, shears, and coilers. To impart the required levels of hardness, evenness, and surface finish to the steel, it may be linked to a temper rolling mill or other equipment, such as a galvanizing line, as part of a larger combined continuous line. A controlled mixture of H_2 and N_2 (5 vol.% H_2 and 95 vol.% N_2), and sometimes pure hydrogen (100 vol.% H_2), is used inside the furnace to prevent the oxidation of the strip. The furnaces themselves are sealed gas tight, and each of them is maintained under slight positive pressure. The operation of the continuous annealing lines can be different based on desired products and the line design.

The annealing cycles applied to cold-reduced strips by continuous annealing differ from product to product. Figure 2.11 shows different annealing cycles that can be applied to produce cold rolled and annealed steel strip. Annealing to produce formable recrystallized ferritic structures is accomplished by subcritical annealing with different overaging steps. On the other hand, DP and TRIP steels are subjected to intercritical annealing. Martensitic (MS), multiphase (MP), and some third-generation advanced high strength steels, such as quenched and partitioned (Q&P) steels, are subjected to full or intercritical annealing with different cooling rates (water, nitrogen, or hydrogen quenching) and different overaging steps.

To avoid coarse carbides, which form at high batch annealing temperatures and damage the formability, the batch annealing soaking temperature is usually limited to about 700–730 °C, which is close to the A_1 equilibrium temperature. The very slow cooling in batch annealing results in near complete precipitation of carbon once the furnace cools down to room temperature. In contrast, it takes only a few minutes to complete the total annealing cycle during the continuous process; the single strip strand is heated to the designed annealing temperature very quickly, the soak time is very short, and the cooling rate is rapid. This rapid and non-equilibrium cooling during continuous annealing inhibits the complete precipitation of carbon from the ferrite solid solution, and therefore, overaging at an intermediate temperature is required to promote such precipitation. Without overaging after rapid cooling, strain aging would impair the ductility and formability of the steel strip. Therefore, to minimize the deleterious effects of potential strain aging on ductility and formability, it is important to control the cooling rates and overaging temperature for different products. In addition, it is necessary to couple these process parameters with the design of the steel chemistry in order to keep carbon in solid solution. For example, titanium and niobium are added in aluminum-killed continuously annealed steels to combine with the residual carbon (IF Steels).

2.7.5 Products Processed by Cold Rolling and Annealing

Low Carbon Steel for Commercial and Drawing Quality Applications Generally low carbon steels may be considered to contain less than 0.1 wt.% carbon, and manganese in the range of 0.20–0.50 wt.%. In some cases, the steels may be aluminum-killed. Low carbon steels are widely made into hot rolled or cold rolled commercial quality (CQ) and drawing quality (DQ), semi-killed (SK) sheet and tin mill products. The hot rolled gauge can be rolled to 1.0 mm on some mini-mill operations. In comparison, cold rolled and annealed products routinely achieve a gauge of less than 0.4 mm with tighter gauge tolerance and improved shape control. Normally the hot band is reduced 50–70%, and the annealing temperature is between 560 and 600 °C for CQ applications and 600 and 700 °C for DQ applications. Commercially, both batch and continuous annealing are used to anneal cold rolled low carbon steels. However, during continuous annealing, it is essential to overage the ferrite microstructure to promote coarse carbide precipitation and remove free carbon from solid solution to improve formability for DQ products. Typical steel strip for CQ applications would have a yield strength between 140 MPa and 275 MPa and elongation of at least 30%. DQ applications target a yield strength between 150 MPa and 240 MPa, a total elongation of at least 36%, as well as a

Table 2.3 Typical chemical composition ranges of low carbon aluminum-killed steel (wt.%) [29]

	C	Mn	Si	P	S	O	Al	N
Min	0.03	0.15	0.01	0.01	0.008	0.0026	0.018	0.001
Max	0.05	0.28	0.02	0.02	0.020	0.0051	0.100	0.005

strain hardening exponent (n values) value in the range of 0.17–0.22 and a normal anisotropy parameter (r value) value between 1.3 and 1.7.

Low Carbon Aluminum-Killed Steel for Deep Drawing Quality Applications Low carbon aluminum-killed steels are known for their excellent formability which is achieved through a cold rolling and annealing process and are extensively used for deep drawing quality (DDQ) applications such as automotive bodies. A typical chemical composition of a low carbon aluminum-killed steel is shown in Table 2.3 [29]. To keep aluminum and nitrogen in solid solution, the finishing hot rolling temperature should be around 890 °C; in addition, to prevent aluminum nitride precipitation in ferrite, low coiling temperatures around 580 °C are recommended. These conditions allow aluminum and nitrogen to stay in supersaturated solid solution during cold rolling. In the subsequent annealing process, the precipitates of aluminum nitride (AlN) retard recrystallization and slow down the nucleation of ferrite grains with randomly orientated crystal orientations. Thus, the nucleation and growth of ferrite grains with {111} planes parallel to the rolling plane are favored leading to (111) [110] texture. This texture is an important property in steel sheets as it induces plastic anisotropy that can be beneficial to the drawability of the steel.

The anisotropy is normally measured in terms of the r value, which is the ratio of true width strain to true thickness strain that is determined by standard tensile testing. Since the r value varies with respect to rolling direction of the sample, an average of r values is calculated, which is expressed as:

$$r_m = \frac{1}{4} (r_o + 2r_{45} + r_{90}) \tag{2.1}$$

where r_o is the r value determined in specimens aligned in the rolling direction, r_{45} is the r value at 45° to the rolling direction, and r_{90} is the r value in the sheet transverse direction. Isotropic steels have an r_m value around 1. Typically, low carbon aluminum-killed steels have r_m values between 1.5 and 1.8 after cold rolling and annealing. According to ASTM A1008 standard, the r_m range for cold rolled steel sheet designation DDQ requires values between 1.4 and 1.8. Steels having high r_m values indicate excellent drawability. Good drawability also diminishes the edge splitting tendency during hole expansion tests.

Interstitial Free (IF) Steels Interstitial free (IF) steels with very low carbon and nitrogen in solution exhibit excellent ductility and high n values and, when cold rolled and annealed, give high r_m values ranging from 1.8 to 2.2 [30]. Therefore, they are widely used for extra deep drawing (EDDQ) and combined drawing and stretching applications. The chemical composition of a typical IF steel is

Table 2.4 Chemical composition ranges of IF steel (wt.%) [30]

	C	Mn	Si	P	S	Ti	Nb	Al	N
Min	0.002	0.10	0.01	0.01	0.004	0.01	0.005	0.03	0.001
Max	0.008	0.34	0.03	0.02	0.010	0.11	0.040	0.07	0.005

Table 2.5 Typical composition used for cold rolled and annealed HSLA steels (wt.%) [32]

		C	Mn	Si	P	S	Nb	Al
Grade 50	Min	0.04	0.60	0.15	–	–	0.020	0.030
	Max	0.07	0.80	0.25	0.01	0.008	0.040	0.070
Grade 60	Min	0.07	0.70	0.20	–	–	0.040	0.030
	Max	0.10	0.90	0.30	0.01	0.008	0.060	0.070

listed in Table 2.4 [30]. IF steels should be aluminum-killed to ensure the oxygen level is less than 70 ppm, and titanium and niobium are added to further remove carbon and nitrogen from solid solution through the precipitation of carbides and nitrides. During the hot rolling process, the finishing temperature should be above the transformation temperature (890 °C), and the coiling temperature should be above 650 °C to ensure that all precipitations are on a coarse scale in order to minimize the carbon and nitrogen in the ferrite matrix [31]. Due to their very low carbon content, IF steels can be annealed at high temperatures from 800 to 850 °C in the single ferrite phase field, and this high annealing temperature can promote recrystallization. Thus, cold rolled and annealed IF steels have very strong (111) [111] recrystallization texture and high r values. However, IF steels have low strength and poor dent resistance when applied to outer panels. Generally, two strengthening approaches are used to increase the strength: solid solution strengthening and bake hardening.

High-Strength Low-Alloy (HSLA) Steel The minimum yield strength of hot rolled HSLA steel is in the range of 45–100 ksi (310–690 MPa). The yield strength of HSLA steels is derived from several strengthening mechanisms, which include solid solution strengthening, precipitation strengthening, transformation strengthening, dislocation hardening, and grain size refinement. Solid solution strengthening is introduced by substitutional and/or interstitial alloy elements such as Mn, Mo, Cr, Ni, Si, N, and C; however, in order to improve formability and weldability, C concentration is usually restricted to 0.10 wt.% maximum. Precipitation strengthening is realized by the addition of V, Nb, and/or Ti in combination with C and N to form complex precipitates, namely, V(C,N), Ti(N,C), and Nb(N,C). Transformation strengthening is achieved by controlling the cooling rate during the austenite to ferrite transformation to obtain low-temperature transformation products such as acicular ferrite, pearlite, bainite, and even very small volume fraction of martensite (<3%). Dislocation hardening is accomplished by an increase in the dislocation density, through hot work or cold work. Additional strengthening can be achieved through grain refinement by careful control of recrystallization and grain growth [9]. Typical composition ranges used for cold rolled and annealed HSLA steels are present in Table 2.5 [32].

In the processing of cold rolled HSLA steels, subsequent annealing would reduce the precipitation strengthening effects, and it is difficult to exceed 50 ksi (~ 345 MPa) yield strength in batch annealing and 60 ksi (~ 415 MPa) in continuous annealing [33]. For batch annealed HSLA grades, the hot band strength should be about 10–15 ksi (~ 70–105 MPa) higher than that required in the final cold rolled and annealed condition, while for the continuous annealing process, the strength of the hot band should be 5–10 ksi (~ 35–70 MPa) higher than that required in the final product. To retain the strength, recovery or non-recrystallization annealing, where a high dislocation density survives the annealing process, might be used. For an HSLA 0.12 C-0.04 Nb-0.04 V (wt.%) steel, the yield strength can be 85–135 ksi (~ 595–945 MPa) when it is annealed at 600 °C and 500 °C, respectively. The trade-off is that this low-temperature annealing leads to lower formability and ductility.

Another important factor that should be considered for processing cold rolled and annealed HSLA steels is the desire to minimize the mechanical property variation along the coil's length. The continuous annealing process is more suitable than that of batch annealing for producing uniform HSLA steel sheet. Furthermore, the alloy design also plays a role in the property variation. For example, it has been known since the early 1980s that titanium microalloyed steels were more difficult to recrystallize and had larger property variations than niobium steels. It is for this reason that HSLA steels with Ti additions are rarely processed by batch annealing [34].

Electrical Steels Electrical steels are Fe-Si alloys which are used for generators and motors. These steels are discussed in detail in Chap. 12. There are two types of electrical steels: grain-oriented electrical steels (GOES) and non-oriented electrical steels (NOES). GOES have much better power and permeability properties in the direction of rolling than they do at 90° to this direction. GOES are based on Fe-Si alloys with 3.0–3.5 wt.% of Si. They are characterized by excellent magnetization behavior and low values of specific magnetic losses along one direction (i.e., the rolling direction). The thickness of the grain-oriented electrical steels is typically in the range of 0.23–0.35 mm. The non-oriented electrical steels (NOES) have magnetic properties that are practically the same in any direction in the plane of the material. They are the most used material among all soft magnetic materials. Non-oriented electrical steels consist of Fe-Si steels with 3.0–3.2 wt.% of Si. The thickness of the fully processed material is in the range of 0.35–0.65 mm, although recently reductions lower than 0.35 mm have been achieved. Due to the intricacies of developing adequate magnetic properties, GOES are produced fully processed. NOES are either fully processed or semi-processed.

The essential requirements for electrical steels are easy magnetization (high magnetic permeability), low hysteresis loss, and low eddy current loss. Easy magnetization and low hysteresis are controlled by chemical composition, grain orientation, and purity. Si increases the volume resistivity of the steel and thereby reduces the core loss and increases the magnetic flux density; however, high Si alloys are brittle and cannot be cold rolled. High magnetic permeability is found in the <100> direction in which the steel strip has been rolled. This is called

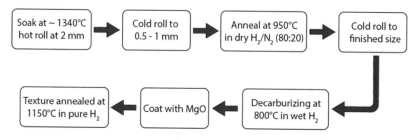

Fig. 2.12 Outline of two-stage cold rolling method for producing GOES

Goss orientation {110}<001>. Eddy current loss can be reduced by decreasing the electrical conductivity of the material and by laminating the material. It would also be desirable to have the smallest possible domain size and maximum domain wall mobility to reduce anomalous loss [35].

To develop a strong {110}<001> texture, provisions must be made for the nucleation of {110}<001> grains, these grains must be able to grow, and grains of other orientations must not grow. Figure 2.12 shows a two-stage cold rolling method to process GOES.

During the hot rolling process, the desired Goss {110}<001> orientation first appears as a friction-induced shear texture near and at the strip surface. However, in the second step of the cold rolling process, most of Goss {110}<001> orientation would be lost and would be replaced by {112}<110> and {111}<110> orientation textures. The employment of two stages of light cold rolling assures that even though {112}<110> and {111}<110> orientation textures are produced, their presence is still under control. In addition, the Goss {110}<001> orientations are still surviving at the centers of the transition bands which are separating from the above two orientations. In this kind of texture arrangement, {110}<001> orientated grains tend to nucleate and grow in the annealing texture after the subsequent decarburizing anneal. These {110}<001> orientated grains are larger than those of the other orientations, and they can grow through the abnormal grain growth to dominate the final texture during the final texture annealing [36]. The imperative conditions for Goss {110}<001> orientation grain growth are provided by both process and microstructural control. Before hot rolling, a fine manganese sulfide (MnS) dispersion is precipitated during the rapid slab cooling. These fine particles can retard normal grain growth, keeping the matrix grain size uniform and small during the early stages of high-temperature annealing. During the final texture annealing process, the dissolution of these fine MnS particles will lead to the unpinning of some grains, which will then consume the original fine grains by an abnormal grain growth process; thus, the desired Goss-oriented grains will grow and dominate the final microstructure. Abnormal grain growth is also promoted by the presence of a sharp texture. The possibility of generating undesirable orientations through surface nucleation processes is eliminated by the addition of sulfur to the MgO

Table 2.6 Primary cooling rates associated with different cooling media used in continuous annealing lines [28]

Cooling media	Gas jet cooling	Boiling water	Roll cooling	Gas water spray	H2 quenching	Cold water quenching
Cooling rate, °C/s	5–50	25–150	80–150	80–300	150–250	>1000

surface coating. This prevents the growth of surface grains, and the sulfide formed is eventually lost by reaction with the hydrogen atmosphere during annealing.

Among NOES, cold roll magnetic lamination quality (CRMLQ) steel is used to replace the more expensive silicon steels. CRMLQ steel is sold as batch annealed and heavily temper rolled, with around 4–8% extension, so that the customer can develop the required texture during annealing, after stamping out the motor laminations. These steels have very low C and high P and may have up to 0.5 wt.% Si. The low C, S, and N levels are important to minimize the precipitation of any fine particles in the microstructure that can retard grain growth during final critical annealing at the customer's facilities. The control of Al and N is especially critical because solubility products of AlN above 2×10^{-5} are detrimental to the final magnetic properties.

Advanced High Strength Steels (AHSS) First-generation advanced high strength steels, such as DP, TRIP, MS, and MP, and third-generation AHSS, such as quenched and partitioned steels, can be produced using modern continuous annealing lines by employing different cooling rates during the primary cooling stage, different overaging treatments, and in most cases temper rolling processes. The continuous annealing lines differ essentially in the cooling media used when cooling from the annealed soaking temperature. Table 2.6 [28] summarizes the different cooling media used and the associated cooling rates. Different grades of AHSS are reheated at or above the intercritical annealing temperature, quenched at different cooling rates, and tempered at different temperatures to achieve different austenite, ferrite, bainite, and martensite microstructure combinations so that they can meet different specifications. The metallurgical and process design of different AHSS grades is reviewed in Sect. 4.2. AHSS will be discussed in Chap. 4 of this book.

2.7.6 Galvanizing

Thin and lighter weight sheet steel grades that are used for automotive panel applications are usually aluminum or zinc coated and offer outstanding corrosion protection. If a zinc coating is used, the coating is often applied as a hot-dip galvanized or hot-dip galvannealed coating. Galvannealed steel goes through additional annealing processing after hot dipping which greatly improves the formability and paint adhesion. Modern continuous hot-dip galvanizing lines are usually

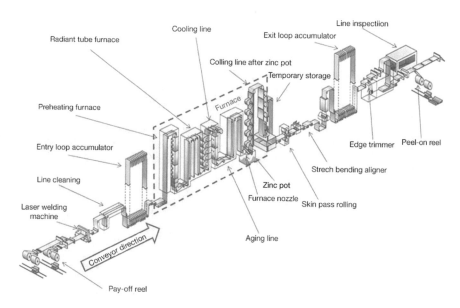

Fig. 2.13 Outline of hot-dip galvanizing line [37]

incorporated in the continuous annealing line prior to the overaging treatment. Figure 2.13 [37] shows the outline of a modern continuous hot-dip galvanizing process. The cold-reduced steel sheet is uncoiled and laser welded to the tail end of the coil ahead of it in the processing line. Then it is cleaned in a cleaning process unit. After cleaning and drying, the strip passes into the high-temperature annealing furnace where it is annealed as a full-hard strip based on the steel grade. In the annealing furnace, the strip is maintained under a reducing gas atmosphere to remove any oxide on the steel surface. Leaving the annealing furnace, the strip passes through the coating bath, and a set of gas knives are used to wipe off excess molten metal and obtain the required coating weight. The coating is then cooled to allow the metal to freeze on the steel surface. Galvannealed strip is produced from galvanized strip by reheating the coated sheet above the wiping knives in order to alloy the zinc-coated surface with the iron in the steel. Once the strip cools down to room temperature, it feeds into the exit-end equipment, which includes the temper mill, the tension leveler, and the chemical treatment section, and then is recoiled on an exit-end mandrel.

Most of the cold rolled annealed steels, including low carbon CQ, DQ, low carbon aluminum-killed DDQ, IF, EDDQ, HSLA steels, and AHSS, can be coated by using a continuous hot-dip galvanizing process. An example of the temperature profile for CQ, HSLA (50 ksi), and EDDQ steels during the hot-dip galvanizing process is shown in Fig. 2.14. Usually hot bridles are used following the heating/soaking zone to provide tension control to minimize shape distortion and maintain the low-tension operation necessary for EDDQ steels. To produce AHSS,

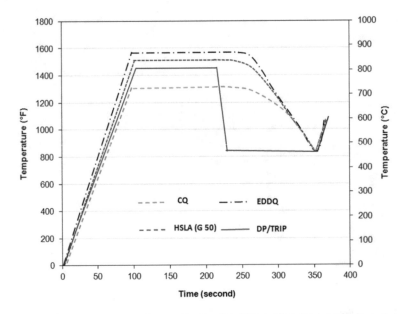

Fig. 2.14 Typical heating and cooling profiles for CQ, HSLA (50 ksi), and EDDQ steels in the hot-dip galvanizing process

a rapid jet cooler is required to cool the strip. The required cooling rate depends on chemical composition and for some grades could reach up to 60 °C/s.

2.8 Process-Structure-Properties Relationships

AHSS are considered the dominant future light weighting material options for automobiles to meet the stringent fuel efficiency and carbon emissions targets in the coming years. Development concepts for the different product classifications of automotive steels are based on the strengthening mechanism employed to get the desired strength level. The increased strength typically leads to an inverse relationship with ductility, which in turn determines the ability of the metal to be formed. The new generation of AHSS are characterized by the distinct microstructural approaches used to obtain the unique strength-ductility balance to satisfy the required design characteristics of modern automakers.

Mechanical properties are defined by the microstructure that is formed during processing of the steel and are affected by the amount and distribution of microstructural constituents and the presence of imperfections such as solutes, precipitates, inclusions, dislocations, and grain boundaries. Phase transformations contribute to strengthening through the coexistence of different microstructural constituents whose individual mechanical behaviors, along with their mutual interaction, could

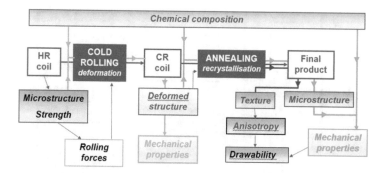

Fig. 2.15 Interaction between chemical composition, processing parameters, and final properties

give superior strength-ductility balance. It is well-known that microstructures are often, if not always, complex and heterogeneous in nature. Proper control of those multiphase microstructures through alloying systems and processing can enhance the final mechanical properties of steels and satisfy the complicated requirements of automotive customers. For example, the requirement for adequate coating needs to be considered during alloy design, in addition to meeting the specific mechanical properties requirements, such as hole expansion, bendability, resistance to delayed fracture, etc.

Figure 2.15 shows how chemical composition and subsequent processing strongly affect the final microstructure and, through that, the final mechanical properties of the steel. Each step of the manufacturing process is affecting the following step and, through this complex interaction, affecting the final mechanical properties. For example, the microstructure and properties of steel after hot rolling influence the capability of the steel to undergo cold rolling and influence the kinetics of austenitization during the following heat treatment.

2.8.1 Processing of AHSS for Automotive

Several definitions of steel classes have been developed. The American Iron and Steel Institute (AISI) and the International Ultra-Light Steel Auto Body Consortium (ULSAB) defined the strength level for steel classification [38]. The accepted practice involves specification of both yield strength (YS) and ultimate tensile strength (UTS). For example, AHSS classification designates that the steel has a minimum yield strength of 80 ksi /550 MPa and a minimum ultimate tensile strength of 100 ksi/700 MPa. AHSS derive their properties from a multiphase complex microstructure of different phases: ferrite, bainite, martensite, and retained austenite. As opposed to the conventional steels, where ductility decreases with increasing strength, AHSS can combine both high strength and ductility. In Fig. 2.16 [39], the families of AHSS are shown with respect to the strength-ductility

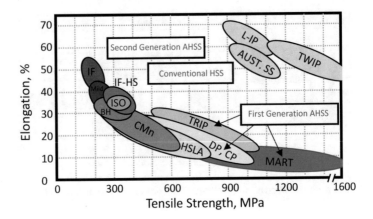

Fig. 2.16 Strength-elongation relationship for various families of steel [39]

combination. AHSS can be divided into three generations. The first generation of AHSS includes four high strength grades that were developed decades ago and are fully commercialized: dual-phase (DP) steels, complex-phase or multiphase (CP) steels, transformation-induced plasticity (TRIP) steels, and martensitic steels.

2.8.2 Dual-Phase Steels

The microstructure of DP steels is composed of a soft ferrite matrix and 10–70% volume fraction of hard martensite islands. Figure 2.17a and b shows examples of nonuniformly distributed and uniformly distributed martensite islands, respectively, within the ferrite matrix of a DP steel. The ultimate strength ranges from 500 to 1200 MPa depending on the volume fraction of martensite [40]. The ductility of DP steels is determined by the volume fraction of the ferrite phase, its size, and its distribution. For some applications, for example, the presence of a small volume fraction of bainitic constituent in addition to the martensitic one is desirable, if the higher level of YS or hole expansion ratio (HER) is required.

The distinctive features of DP steels are:

- Continuous yielding behavior
- Low YS/UTS ratio (<0.6) and high strain hardening (high n value >0.2)
- Wide range of strength from 500 to 1200 MPa
- Improved balance of strength and ductility
- No room temperature aging
- High bake hardenability (BH$_0$ and BH$_2$) values

Typical properties of commercially produced DP grades are presented in Fig. 2.18 [41]. To achieve the unique combination of mechanical properties for DP steel, the following alloying elements are required [42]: 0.05–0.15 wt.% C; 1.0–2.5 wt.% Mn; up to 0.4 wt.% Cr and/or Mo to improve the hardenability of

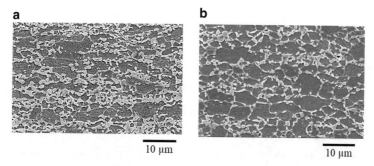

Fig. 2.17 Typical microstructure of DP780 steel with different martensite morphologies: (**a**) nonuniform morphology, (**b**) uniform necklace morphology, typically resulting in improved ductility. Scanning electron microscope (SEM) images. Etch: 2% Nital

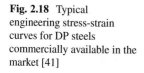

Fig. 2.18 Typical engineering stress-strain curves for DP steels commercially available in the market [41]

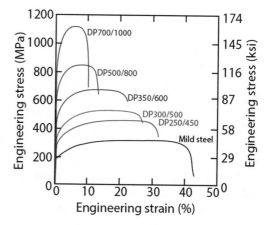

the steel and strengthen the ferrite; 0.3–0.7 wt.% Si to strengthen the ferrite; small (<0.10 wt.%) additions of Nb, Ti, and/or V to refine the microstructure; and up to 20 ppm B, to suppress ferrite formation, if needed.

Dual-phase steels with ferrite-martensite microstructure can be obtained by thermomechanical processing, either during hot rolling or after annealing of cold rolled sheets. The microstructure of the most common commercial DP hot rolled grade with 600 MPa strength contains up to 80% of ferrite, which is formed during cooling of the strip on the runout table. The remaining austenite transforms to martensite during coil cooling due to the high hardenability of austenite which has a high carbon content as a result of ferrite formation and appropriate alloying additions [43]. Several works have demonstrated that deformation accelerates the nucleation of ferrite. When the deformed microstructure is combined with austenite grain refinement and low carbon content, it is sometimes difficult to avoid the ferritic constituent in the microstructure, even when it is not desirable [44].

The modern mill construction is equipped with a powerful cooling system on the runout table that is capable of interrupted controllable cooling which allows

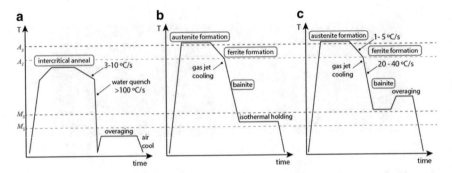

Fig. 2.19 Continuous annealing cycles to achieve dual-phase microstructure (dash lines, avoidable transformation; solid lines, preferable transformation): (**a**) direct with water quench; (**b**) with isothermal holding after cooling; (**c**) coating lines

one to manipulate the phase transformation of austenite to achieve the required microstructure in the final hot rolled product.

The dual-phase microstructure could be achieved in cold rolled product after batch annealing for steels with a high amount of Mn (>2.5 wt.%) [45]; however, the most cost-efficient method is annealing in continuous annealing lines (CAL). This can be accomplished with direct quenching to room temperature (CAL) or with isothermal holding (CAPL) above martensite start temperature (M_s) [46] and at hot-dip coating lines (HDCGL). The types of continuous annealing with the corresponding phase transformation of austenite can be seen in Fig. 2.19.

During continuous annealing, the sheet is heated to the dual-phase region to achieve the required volume fraction of ferrite and austenite. Alternatively, it could be heated to the fully austenitic region, and then the ferrite can be formed instead during an appropriate slow cooling step. It should be mentioned that the morphology of the microstructural constituents after cooling differs in both cases. To obtain a dual-phase ferrite-martensite structure, the alloying system of the steel should be chosen while considering the following processing parameters: soaking temperature, holding time, and cooling rate of the applicable annealing line.

2.8.3 TRIP Steels

The transformation-induced plasticity (TRIP) effect was discovered in stainless steels in the 1960s, but real commercialization began only recently [47]. The TRIP effect refers to a deformation mechanism where metastable austenite of high carbon content so-called retained austenite (RA) transforms to martensite during plastic deformation at room temperature and results in a steel with excellent formability combined with high strength. TRIP steels are used for applications requiring high crash energy absorption. The typical properties of commercially produced grades are plotted in Fig. 2.20 [41].

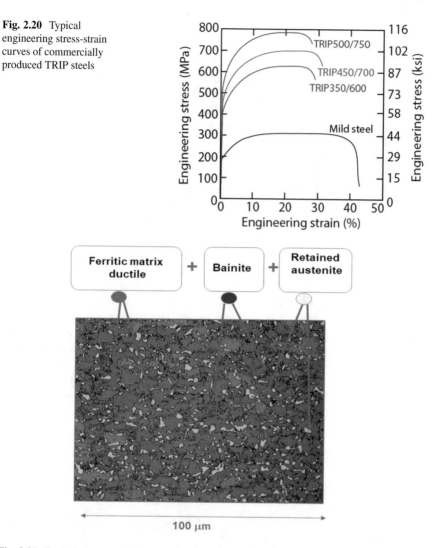

Fig. 2.20 Typical engineering stress-strain curves of commercially produced TRIP steels

Fig. 2.21 Typical structure of TRIP steel: dark brown, bainite; light brown, ferrite; yellow, RA. Etch: color etching

The excellent combination of strength and ductility in TRIP steels is a result of the unique microstructural components present: metastable retained austenite (RA), carbide-free bainite, martensite, and ferrite. The typical microstructure of a TRIP steel is shown in Fig. 2.21. Depending on the strength required, the proportion of different microstructural components can be different, but the key point of the TRIP concept is the presence of metastable RA at room temperature. Therefore, the chemical composition and processing parameters are aligned toward achievement of high carbon content in retained austenite.

The carbon content in TRIP steels can vary from 0.1 to 0.4 wt.% with typical commercial content around 0.2 wt.%. The presence of high levels of Si and/or Al (>1 wt.%) is necessary to prevent carbide formation during the bainitic reaction, which is desirable as this promotes the enrichment of carbon to retained austenite. A high content of Si is typically used for cold rolled, non-coated versions of TRIP products, and high Al or Al-Si combinations are used for coated TRIP steels due to the damaging effect of Si on coating quality in high Si steels.

TRIP steels can be produced as a hot rolled product or as a cold rolled product after continuous annealing. Hot rolled TRIP steels typically undergo heavy reductions at low finishing temperatures close to Ac_3 in order to promote ferrite formation, which is followed by rapid cooling to the coiling temperature located in bainitic transformation temperature region. Optimization of the coiling temperature depends on the chemical composition of the steel and the capability of the mill to control phase transformation of austenite to ferrite and bainite, targeting a high volume fraction of RA with a high carbon content [48].

Cold rolled TRIP steels are produced in continuous annealing lines (CAPL) with the presence of an isothermal holding section or in hot-dip galvanizing lines equipped with an equalizing section of adequate length needed for bainite formation. The phase transformation happening during continuous annealing in TRIP steel is described in Fig. 2.22. To achieve the desired microstructure of ferrite, carbide-free bainite, and metastable retained austenite, the M_s should be below room temperature or at least below the isothermal holding temperature. Carbon is the main factor in decreasing M_s. The formation of pearlite during cooling should be prevented by using a faster cooling rate when cooling into the isothermal holding section; this

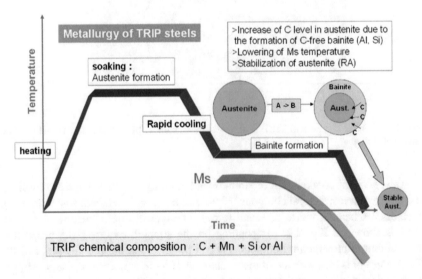

Fig. 2.22 Schematic presentation of phase transformation during continuous annealing of TRIP steels

allows carbide-free bainite to form and consequently results in the formation of high carbon RA. The temperature and time for isothermal holding, which defines the volume fraction of formed bainite, are key parameters for the carbon enrichment in RA.

2.8.4 Complex-Phase Steels

The microstructure of DP and TRIP steels, which are characterized by a combination of high strength and ductility, reveals some shortcomings, such as high sensitivity to edge fracture and low HER. The presence of a soft phase and a hard phase in the dual-phase structure, which is responsible for the typical features of DP steels, is a major factor contributing to local stresses and the damage occurring during punching holes. To overcome the low flangeability of DP steels, the microstructure can be changed to reduce the differences between the soft and hard structural constituents [42]. This is done by softening the martensite through reduced carbon content and/or tempering, reducing the volume fraction of martensite, replacing ferrite with bainite, and strengthening of ferrite by Si, small precipitates, and/or grain refinement. Reducing the fraction of soft constituents in CP steels results in a high YS, a high YS to UTS ratio, a low level of ductility, but higher value of HER. The typical properties of CP grades are shown in Fig. 2.23 [41].

Numerous reports have demonstrated the critical role of microstructure homogeneity in the improvement of HER [49, 50]. The replacement of ferrite and martensite by bainite can be achieved by the proper combination of alloying elements and processing parameters. The most effective element at suppressing ferrite formation from austenite is boron [51]. Mo and, to a lesser extent, Cr are also well-known for suppressing ferrite formation. In general, an increase in C content

Fig. 2.23 Typical engineering stress-strain curves of commercially produced CP steels [41]

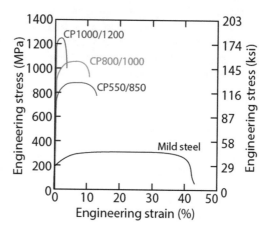

Fig. 2.24 Typical bainitic microstructure of CP1000 steel. SEM micrograph; Etch condition: 2% Nital

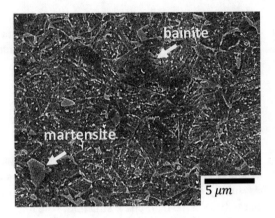

increases the hardenability of austenite, promoting martensite formation, but, due to the non-homogeneous distribution of carbon in austenite, the effect of alloying elements could be more complicated. For example, Nb has been shown, on the one hand, to accelerate ferrite and delay bainite formation and on the other hand to promote a more homogeneous structure due to strong grain refinement. A typical microstructure of CP1000 steel is shown in Fig. 2.24.

Processing parameters play another important role for the achievement of a homogeneous structure with low local stresses. Hot rolled CP steels are produced using thermomechanical processing that focuses on the replacement of martensite by bainite or high strength refined ferrite. The most important processing parameters for CP steels are the reheating temperature, runout table cooling rate, and coiling temperature. The reheating temperature is responsible for the type and size of precipitates strengthening the ferrite. The runout table cooling rate determines the volume fraction of ferrite formed, and the coiling temperature impacts the size of precipitates. For cold rolled CP steels, that are processed in a continuous annealing line, a high soaking temperature close to Ac_3 and a high cooling rate, to prevent ferritic transformation from austenite, are key annealing parameters. The type of bainite formed during cooling or isothermal holding in continuous annealing lines determines the final mechanical properties. For example, it is well-known that TRIP steels featuring the isothermal formation of carbide-containing bainite exhibit low levels of hole expansion ratio [52].

2.8.5 Martensitic Sheet Steels

Martensitic steels are the strongest group of AHSS, featuring both a high UTS and a high YS to UTS ratio. The engineering stress-strain curves of commercially available martensitic steels are shown in Fig. 2.25 [41]. The microstructure of martensitic steels consists of more than 85% of martensite. The strength is controlled mostly by carbon content, but some alloying elements such as Mn, Cr, Si, and Mo can

Fig. 2.25 Typical
engineering stress-strain
curves of commercially
produced martensitic steels
[41]

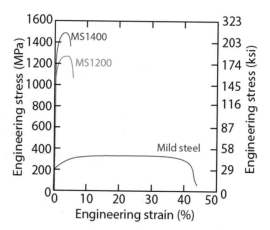

also be added to achieve the required strength during processing and to affect other properties such as ductility, bendability, and delayed fracture. The strength of modern martensitic steels ranges from 750 MPa up to 2000 MPa, with total elongation ranging from 3% to 15%.

Martensitic steels can be produced as hot rolled and cold rolled steels in a continuous annealing line, and their microstructures are created as a result of austenite transformation to martensite during cooling. The alloying system should be adjusted to accommodate the cooling capability of the processing mill or line. The processing of martensitic grades in hot rolling mills requires fast cooling after the last finishing stand to temperatures below the M_s temperature to prevent or minimize pearlite, ferrite, and/or bainite formation. Martensite formed on the runout table undergoes tempering during the coiling process, and therefore, the coiling temperature should be optimized with respect to M_s and tempering temperatures.

The cycle of a typical annealing line with water quench capability used for martensitic grades is shown in Fig. 2.19a. One of the key processing requirements for martensitic grades is that annealing is performed in the fully austenitic region with a soaking temperature above the Ac_3. Another important parameter is the cooling capability of the line, which in turn determines the alloying elements in martensitic grades which can be produced in the line. The third key component is the tempering step that is required mostly for any martensitic product to regain ductility, as martensite is such a brittle microstructural component. Improved ductility through the tempering process is a result of stress relief and carbide formation. The final strength and ductility depend on the applied tempering temperature.

The high UTS and typically low ductility of martensitic steels restrict their utilization in automotive structural parts. To overcome the low ductility of martensitic steels and to stamp parts of complex shape without cracks and excessive press forces, the hot stamping process was developed. The scheme of the hot stamping process is shown in Fig. 2.26 [53].

Cold rolled steel sheet, typically consisting of a pearlite-ferrite structure with a low UTS, is austenitized in the oven at a temperature above Ac_3. Then the sheet is

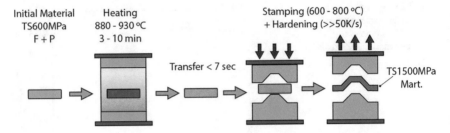

Fig. 2.26 Schematic of hot forming process to produce martensitic steels. (Adapted from [53])

transferred over a short time to a die for hot stamping followed by press quenching to form martensite in a water-cooled die.

The most commonly used steel grade for hot stamping is 22MnB5. Initially, before stamping, the material has a ferrite-pearlite structure with the UTS around 600 MPa. After austenitization and martensite formation in the water-cooled die, the UTS reaches 1500 MPa. High carbon and Mn determine the strength level of martensite, and additions of boron are required to prevent ferrite formation while the sheet is being transferred from the oven to the die [54]. Finally, one of the challenges when using martensitic steels of high strength (>1000 MPa) is the susceptibility to delayed fracture due to hydrogen embrittlement; this problem can be solved by appropriate alloying.

2.8.6 Third-Generation AHSS

As one of the main factors defining fuel efficiency, the reduction of vehicle weight through the introduction of high and ultra-high strength steels with adequate formability has been one of the major objectives for steel manufacturing companies. These efforts were significantly accelerated by the agreement of 13 large automakers in July 2011 to increase the fuel efficiency of cars and light-duty trucks to 54.4 mpg, which resulted in the new CAFE regulations for every vehicle that will be on US roads in 2025 [55]. Steels with tensile strength (UTS) of 980 MPa and total elongation (TE) > 21%, or UTS of 1180 MPa and TE > 14%, or even UTS of >1470 MPa and TE > 20% are among the popular targets which are currently either under development or in the commercialization stages. In most cases, the classical first-generation AHSS such as DP, TRIP, or martensitic steels cannot deliver such combinations of properties. As a result, new metallurgical concepts and processing routes should be employed.

Steels with carbide-free bainite (CFB) or TRIP steels with bainitic ferrite (TBF), quenched and partitioned (Q&P) steels, and medium manganese steels are among the third-generation AHSS that can simultaneously satisfy the high strength and high ductility requirements combined with reasonably high stretch-flangeability (i.e., high hole expansion ratio). In all these cases, the microstructural design strategy is mainly focused on the introduction of hard phases like martensite or bainite in combination with retained austenite. Here, it should be emphasized that in addition to target tensile properties, in most cases, other requirements like bendability, flangeability, resistance to hydrogen embrittlement, good weldability, etc. should be satisfied too. In this regard, the complex chemistries that are usually employed in the new generation of steels leave a rather narrow window for successful production through all different upstream and downstream processing steps.

Obtaining the desired morphology and fraction of phases requires a very careful evaluation of the effect of critical variables, not only in the final annealing step but also in all other preceding stages like hot rolling, batch annealing, pickling, etc. For example, temperature and time in reheating furnaces, finishing temperature, cooling patterns on runout tables, and coiling temperatures are important variables during hot rolling processing which can significantly affect the processability and final properties of steels. Intermediate processes like hot band annealing and pickling should also be evaluated carefully as each one can have non-reversible impacts on the microstructure and surface quality of the products. Time, temperature, atmosphere, and uniformity of the process throughout the coils should be adjusted in each step with respect to the microstructure and property requirements in the next processing steps without sacrificing the desired final properties.

2.8.7 Carbide-Free Bainite (CFB)

In general, the design of these steels is based on the introduction of fine carbide-free bainite with a substantial fraction of retained austenite in the form of interlath films in the microstructure, which is achieved through the isothermal holding of steels in the bainite transformation temperature range. Steels with CFB structures typically feature a good balance of high hole expansion, high ductility, and high tensile strength (> 980 MPa), which cannot be achieved in the first generation of AHSS like DP, TRIP, and CP steels.

As discussed before, DP and classical bainite-containing TRIP steels are characterized by their excellent strength-ductility balance, but large differences in the hardness of microstructural constituents result in low flangeability (i.e., low hole expansion values). CP steels are used in the production of parts requiring both high strength and high flangeability, but their low level of total elongation limits their wide application. Elimination/minimization of ferrite, suppression of cementite precipitation, and replacement of martensite with fine bainite laths in CFB steels lead to achieve a more uniform microstructure and, as a result, a higher hole expansion. At the same time, existence of substantial amount of very fine retained austenite islands in CFB structure helps to achieve TRIP effect and high elongation values. This combination of properties (i.e., strong matrix and good ductility) makes these steels good candidates for parts where energy absorption is important.

The general strategy for the chemistry design of CFB steels is similar to that of TRIP steels, wherein elements that retard cementite formation, such as Si, and Al, are an essential component of the composition. Alloying with austenite hardening elements, such as Mn, Cr, Mo, Ni, and B, and microalloying with Nb and Ti, should help to achieve the target mechanical properties. The choice of certain alloying elements should be made very carefully due to their complex effect on phase transformation. For example, austenite grain refinement by Nb addition can potentially accelerate the transformation kinetics of both ferrite (during initial cooling) and bainite (during isothermal holding) by providing more nucleation sites. This means that if an industrial line is able to limit formation of undesired ferrite by very fast initial cooling from the austenitizing temperature, then the positive effect of Nb addition on bainite transformation kinetics can be seen. Otherwise, ferrite formation and rejection of carbon into the austenite retard bainite growth kinetics during the isothermal holding stage [56–60]. Similarly, Al addition on the one hand might be beneficial for reducing the sensitivity of some properties to the annealing parameters and improving the coatability of the steel, but, on the other hand, due to a significant increase of the Ac_3 temperature in Al-added steel, annealing in the fully austenitic region is limited in most industrial lines (e.g., above 870 °C). The addition of Mn, Cr, and Mo can also have several advantages like increasing strength, limiting internal oxidation at high temperatures, or improving hardenability of the steel by preventing ferrite formation during initial cooling. However, they can also damage the final properties by retarding bainite transformation kinetics through restraining the nucleation stage or causing solute drag effect during growth [59, 61]. The alloying element strategy for achieving the necessary target mechanical properties should be coordinated with the existing processing capabilities of the industrial lines. In addition to target mechanical properties and existing processing limitations, the effect of alloying elements on in-use properties like coatability and weldability should also be examined.

With regard to processing considerations, the available time and temperature for austenitization, the minimum and maximum cooling rates from the soak section to the bainite formation region, and the available isothermal holding temperature and time are among the major processing factors which should be considered in the design of CFB steels for industrial annealing lines. For the development of CFB steels, it would be ideal to have high soaking times and temperatures in the fully austenitic region, fast initial cooling rates to limit undesired ferrite precipitation, and flexibility in controlling the isothermal holding temperature and holding times to allow for sufficient bainite formation. It should be noted that most of the existing industrial continuous annealing lines are originally designed to produce the soft, highly formable steels with ferritic structures. Therefore, the maximum soaking temperature is usually around 850 °C, and the average initial cooling rates might be limited to ~10–20 °C/s. In addition, overaging (equalizing) zones in continuous galvanizing lines (CGL) are not very long (i.e., short isothermal holding section) with temperature usually limited to Zn pot temperatures (~460 °C).

Recently, new lines with substantially different capabilities like higher soaking temperatures (e.g., > 900 °C), faster cooling rates (e.g., ~40–50 °C/s), and variable isothermal holding temperatures (e.g., 250 °C–460 °C) and holding times (e.g., ~5–10 min) have been built for the production of new AHSS. Usually, low isothermal holding temperatures (<400 °C) result in the formation of favorable bainite and retained austenite morphologies. However, slow bainite growth kinetics at low temperature increases the required holding times for formation of desired bainite fraction. In some cases, this may take more than 30 min, which might be beyond the capability of the industrial lines. It is worth noting that some industrial lines (e.g., quench and partitioning lines) might be able to go slightly below the M_s temperature for a short time and take advantage of martensite transformation which can drastically accelerate bainite formation kinetics at low temperatures. Similarly, application of stresses can be considered for enhancing bainite transformation kinetics [62, 63]. Depending on composition, and processing times and temperatures, fresh martensite can be formed during the final cooling stage. In this case, tempering through post-annealing processing should be considered to minimize the harmful effect of fresh martensite on the final mechanical properties (e.g., on stretch-flangeability).

2.8.8   Quenched and Partitioned Steels

Quenching and partitioning (Q&P) is one of the latest concepts in the design of third-generation AHSS of high strength and high ductility. This concept, which was first introduced by Speer et al. in 2003 [64], is based on the carbon enrichment of austenite from martensite through a partitioning treatment [64, 65]. By using this concept, unique mechanical properties such as UTS > 980 MPa with TE > 21% and UTS >1500 MPa with TE > 20% can be achieved. A schematic of the Q&P cycle with major phase transformations and corresponding microstructure changes is presented in Fig. 2.27.

As shown in Fig. 2.28, the final microstructure of Q&P steels is a complex mixture of tempered martensite, fresh martensite, bainite, ferrite, and retained austenite Examples of thin films of retained austenite, blocky austenite, and fine carbides are shown in Fig. 2.28a, b and c, respectively.

The martensite formed during quenching from the soaking temperature to a desired temperature between M_s and M_f controls the fraction of untransformed austenite before the isothermal holding step. At this stage, depending upon the soaking temperature, initial cooling rate, and of course the chemical composition of the steel, the microstructure may also contain some fraction of ferrite. The rejection of carbon into remaining austenite during isothermal holding step (i.e., partitioning) is the key for austenite stabilization. The partitioning stage practically starts below the M_s temperature (e.g., in one-step Q&P) and continues during heating and holding at higher temperatures above M_s. Initial publications employed the Koistinen-Marburger equation [66] to determine an optimum quenching temperature to obtain the maximum retained austenite fraction in the final microstructure. However, in many cases, it is seen that this approach cannot accurately explain the experimental

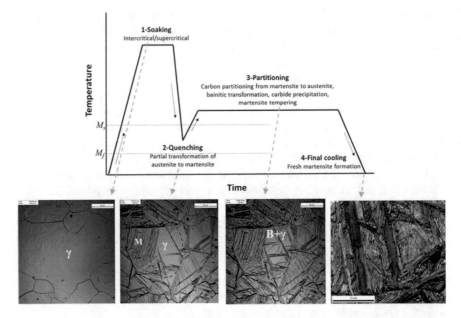

Fig. 2.27 Schematic of a two-step Q&P cycle and an example of the evolution of microstructure during different steps of a Q&P cycle: 1, 2, and 3 are confocal laser microscope images, 4 is electron backscatter diffraction (EBSD) image

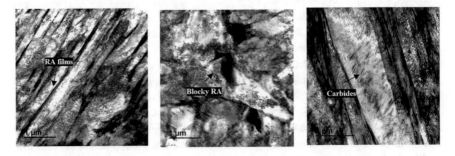

Fig. 2.28 Transmission electron microscope (TEM) images of Q&P microstructure: (**a**) film-like RA, (**b**) blocky RA, and (**c**) carbides

observations. It is confirmed that other coexisting mechanisms, like carbide precipitation, bainite transformation, and martensite tempering, play important roles in determining the fraction of retained austenite and final mechanical properties of Q&P steels. Depending on the chemical composition and the annealing parameters, stabilization of austenite is controlled by both bainite formation, as in TRIP and CFB steels, and direct partitioning of carbon to the remaining austenite from martensite laths. Some later publications suggest that the role of martensite/austenite interface migration on carbon partitioning should also be considered [67, 68].

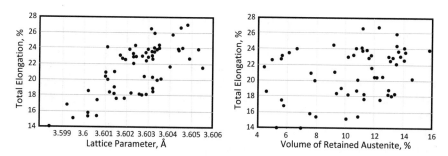

Fig. 2.29 An example of the relationship of ductility with retained austenite characteristics from two different industrially produced TRIP grades: (**a**) total elongation vs. volume fraction, (**b**) total elongation vs. lattice parameter of retained austenite

The precipitation of stable and metastable carbides during both the quenching and partitioning steps is another important process that can significantly reduce the volume fraction of retained austenite through the consumption of a high fraction of the available carbon for austenite stabilization. For example, it is often seen that the rather high partitioning temperatures (> 450 °C) that are required for galvanizing processes of coated sheet steels promote carbide precipitation and reduce the retained austenite fraction in the final microstructure. As seen, the coexistence of competing mechanisms combined with unavoidable inhomogeneities in the distribution of alloying elements during industrial production makes the prediction of the final fraction of retained austenite a very challenging task.

It is worth mentioning that even though retained austenite is a key phase for the strength-ductility balance in Q&P structures, the relationship between the volume fraction of retained austenite and the final properties is not fully understood. In fact, in addition to volume fraction, other parameters, like lattice parameter, morphology, size, and location of austenite, can play a role in determining the final properties. An example provided in Fig. 2.29 shows a strong correlation between the lattice parameter and total elongation, Fig. 2.29(a), while the volume fraction of retained austenite exhibits a weak influence on total elongation, Fig. 2.29(b).

Despite the significant number of publications on Q&P steels during the past 10–15 years, there are a limited number of systematic studies on the effect of different alloying elements on the properties of Q&P steels. Carbon and silicon are the most critical elements in the metallurgical design of Q&P steels as they have the biggest effect on volume fraction, morphology, and stability of retained austenite. In addition, both elements affect transformation temperatures (e.g., A_{c3}, M_s, and M_f) and the strength and morphology of martensite. Si plays a similar role as in TRIP steel, acting as a cementite formation-retarding element and controlling undesired carbide precipitation in Q&P steels. Usually, Si-bearing Q&P steels contain around 1–2.5 wt.% Si. Such a high content of Si strongly influences the strengthening level in Q&P steels. In some cases, replacement of Si with Al might be considered for improving hydrogen embrittlement resistance and coatability. However, one should note that Al raises A_{c3} temperature significantly and is less effective than

Si in strengthening and in retarding cementite precipitation. The addition of other alloying elements, such as Mn, Cr, Mo, Ni, Nb, B, and Ti, should be considered with respect to the target properties and limitations in processing lines. For example, on the one hand, Cr and Mo increase the final strength of the steel by increasing the hardenability, i.e., preventing ferrite and pearlite formation during cooling from the soak to the quenching section, and by retarding the tempering of martensite during the partitioning step. On the other hand, incomplete dissolution of stable Cr and Mo carbides, which form during upstream processes like hot rolling, can damage properties by consuming carbon and reducing the fraction of retained austenite. As in CFB and TRIP steels, the addition of Cr or Mo can reduce the fraction of RA by retarding carbide-free bainite formation. As seen, each one of these elements has a compound effect on the phase transformation of austenite at different stages of the Q&P process. A full understanding of the role of each element individually and their synergetic effect in the presence of other elements is a very complicated but necessary task for successful production of this family of steels.

2.8.9 Medium Manganese Steels

Medium-Mn steels are another major category of third-generation AHSS that are designed based on the TRIP effect and the stabilization of austenite at room temperature through the addition of relatively moderate levels of Mn. Potentially, a combination of high strength and high ductility (e.g., UTS = 1200 MPa and TE = 25%) can be achieved in medium-Mn steels. There is very limited data available on stretch-flangeability of these steels due to the low volume of commercial production. Reported YS/UTS ratios suggest that reasonable hole expansion ratios (HER >20%) can be expected under the right annealing conditions.

In comparison to expensive TWIP steels with very high levels of Mn (> 22 wt.%) and a fully austenitic structure at room temperature, medium-Mn steels have a much lower Mn content (e.g., 4–10 wt.%) with a partially austenitic structure at room temperature. Therefore, the stabilization of austenite in medium-Mn steels requires partitioning and the enrichment of austenite with stabilizing elements (i.e., Mn and C) at the expense of formation of other phases (e.g., ferrite) in the microstructure. In this regard, different approaches, like full austenitization at high temperatures followed by cooling and holding in the intercritical region, or austenite formation by heating the initial martensitic structure into the intercritical region and nucleation of fine austenite grains (aka austenite reverted transformation – ART) during annealing, might be employed to create the desired microstructure. Full austenitization is an extremely slow process. As a result, industry focuses on characteristics of the initial microstructures (i.e., upstream processes) in order to control the fraction, stability, and morphology of the austenite in the final microstructure. For example, the nucleation of fine austenite grains along laths in the initial martensitic structure results in a different morphology of RA compared to the nucleation of austenite at high angle grain boundaries of prior austenite. Depending upon the target final properties, other paths, like intercritical annealing followed by isothermal holding at bainite transformation temperatures, or Q&P cycles, may be also considered. In general, the optimum austenite fraction and stability should be obtained through both chemistry and control of process parameters. Different studies show that the stabilization of high fractions of austenite in medium-Mn structures cannot be explained solely by chemical composition of the austenite phase (e.g., amount of Mn and C) in the intercritical region. The size and morphology of austenite should also be considered. For example, reducing the size of austenite grains increases the austenite stability by reducing the M_s temperature and lowering the probability of martensite nucleation. The intercritical annealing time and temperature should be optimized toward the best balance between partitioning of austenite-stabilizing elements (e.g., Mn and C), austenite grain growth, and the dissolution of Mn-rich carbides [69–74].

The final mechanical properties of medium-Mn steels usually demonstrate a strong dependency on the processing and annealing conditions. One of the biggest concerns with respect to the manufacturability of medium-Mn steels is their narrow processing window for obtaining the desired properties. Due to a high content of Mn, for example, the annealing window is very narrow and should be well controlled. Chemistry modifications to expand the processing windows for the robust production of these steels are needed. For example, Al (up to 3 wt.%) is added to medium-Mn steels to shift the annealing temperatures upward and to expand the intercritical annealing region.

2.8.10 *Importance of Steel Cleanliness*

In general, the development of a new generation of steels brings a new level of sophistication to all development stages, from upstream processes, like steelmaking, to downstream processes like spot welding. As a result, new tools, knowledge, and expertise should be developed to address these complexities and manage different aspects of the industrialization processes. For example, some of the existing criteria and technologies which are acceptable for the first-generation AHSS may not be able to address the required quality in the third-generation AHSS, which have much higher strengths and elongations. Steel cleanliness is one of these industrially challenging areas where new tools, techniques, and expertise, with a high level of attention, are required.

It is well-known that the formation of inclusions during steelmaking and casting processes not only has damaging impacts on almost all important final mechanical properties of the steel but also could cause several manufacturability issues, such as poor surface quality or edge cracking, during downstream processes [75–78]. Steel strength, elongation, bendability, fatigue life, delayed fracture, and stretch-flangeability are among the properties that can suffer due to the presence of inclusions. Usually, these inclusions, which can form by chemical reactions in the molten steel or by mechanical incorporation of slags, are chemical compounds

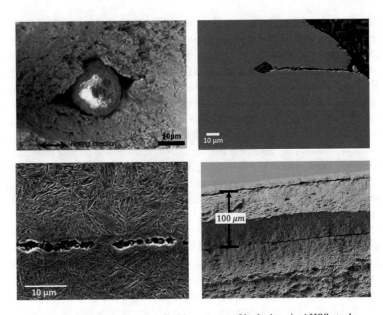

Fig. 2.30 Example of damages associated with presence of inclusions in AHSS steels

of metals (e.g., Fe, Mn, Al, Si) with undesirable non-metals (e.g., O, S, P, N). These non-metallic inclusions are classified as oxides, sulfides, silicates, nitrides, and phosphides [79]. In general, the main recognized mechanisms for the damaging effects of inclusions are associated with the stress amplification around the inclusions due to their morphology, the entrapped gas around the inclusions, and the residual stresses due to differences in thermal expansions of the metal matrix and the inclusions. Recent literature [76] demonstrates that size, shape, volume fraction, distribution, location, thermal and elastic properties, adhesion to the matrix, and composition are among the key parameters that can be manipulated to control the damaging effects of inclusions. For example, the formation of elongated chains of inclusions and the coalescence of nucleated voids around them create a potentially dangerous combination for crack formation and propagation. In these cases, the manipulation of size, type, and distribution of inclusions during upstream processes may be considered as an effective way to reduce the damaging impacts. Also, the location of the inclusions can become a critical part of the solution for avoiding failure under certain conditions like bending. It is well-known that damaging effects become stronger as inclusions get closer to the surface. Slight changes in the subsurface microstructure can significantly help to decrease the damaging effects. For example, Kaijalainen et al. recently showed that the introduction of a soft layer of polygonal ferrite and granular bainite can significantly improve the bendability of the steel [80]. Some examples of damages associated with presence of inclusions in AHSS steels are presented in Fig. 2.30.

Decreasing the undesirable elements (e.g., O, S, P, N) in steel is the most straightforward solution to minimize the negative effect of inclusions on final properties. However, the complexity of the new steel chemistries combined with the restrictions associated with upgrading the existing steelmaking and casting facilities makes this approach very costly for most steel manufacturers. Consequently, the systematic study of inclusion impact on final mechanical properties of AHSS of different microstructures, with an aim to control the damaging effects of inclusions through optimization of the chemical composition and processing parameters, becomes one of the key objectives for the successful production of new classes of AHSS steels.

2.9 Process Modelling

Physics-based models are introduced in order to better understand the process-structure-property relationships and the interactions between different metallurgical phenomena and to avoid costly and lengthy experiments, physics-based models are introduced to capture the complexity of these metallurgical processes with process parameters. A large portion of the literature on modelling of hot working of steels is devoted to empirical approaches, where mathematical models are fitted to the experimental data without a proper linkage to the microstructure. However, there have been attempts to model steel processing via physics-based models, where processing parameters and materials-related parameters are linked and validated using experimental data. These models deal with different stages of processing and with the relevant metallurgical events. Figure 2.31 shows an example of such an approach where process parameters (strain, strain rate, and temperature) are linked to materials-related variables, such as dislocation density, precipitate number density, radius and volume fraction, recrystallization nuclei, solute content, etc., to describe the phenomena of recovery, recrystallization, and precipitation as well as their interactions. New models have been developed in recent years to better explain microstructure evolution of complex alloy systems, such as microalloyed and highly alloyed steels. With regard to the precipitation of microalloying elements, Dutta et al. [81] proposed a precipitation model in which the precipitation kinetics of the Nb-bearing steels during low-temperature finish rolling practices was modelled. This model is based on dislocation-induced accelerated precipitation. Zurob et al. [82] extended the model to capture the interaction between precipitation, recovery, and recrystallization. Modifications were introduced to both recovery and recrystallization models to fit a large body of kinetic data. Similar treatments have been introduced for phase transformation and austenite decomposition. In these models, the nucleation and growth of ferrite/bainite from austenite were related to the chemistry of the steel, process parameters, and the undercooling in isothermal or continuous cooling setups.

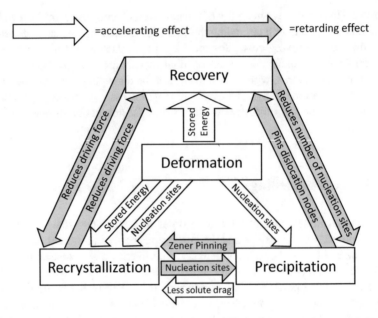

Fig. 2.31 Schematic illustration of interactions between deformation, recovery, recrystallization, and precipitation

2.10 Summary and Outlook

Successful processing of steels requires simultaneous adjustment of the steel composition and the various thermomechanical and annealing steps. The optimum combination of chemistry and processing route will depend on the target properties and the specific capabilities of the infrastructure (e.g., hot mill or annealing line) being used to produce the steel. These concepts will be explored in more detail in subsequent chapters. The aim of this introductory chapter was to shed some light on the complex considerations and interactions that need to be considered during steel processing of different families of steels.

In terms of the future evolution of the steel processing, several important trends are worth highlighting as these would potentially have a very important impact in the coming years. The first of these is the development of new infrastructure that would expand the range of processing conditions. Many steel producers are investing in more powerful rolling mills and improved cooling and coiling capabilities. In addition, annealing lines that can access a wider range of annealing temperatures and times as well as faster cooling rates will become available in the coming years. This would provide new opportunities for developing new steel grades such as third-generation steels, as well as reducing the cost of existing first-generation steels. As an example, the new accelerated cooling capabilities may allow steel producers to reduce alloying additions such as those used for increasing the hardenability of some first-generation steels. Another emerging trend is the increased use of batch annealing lines. The batch annealing process can be used to produce a wide range of steels including electric steels. Given the anticipated increase in demand for electric steels in the automotive industry, steel producers have made significant new investment in batch annealing. This new capacity could also be used for the processing of third-generation steels such as medium-Mn steels which require long intercritical treatments and/or softening treatments. New processes for producing multilayer steels using batch annealing furnaces are also being explored. The third trend that could shape the industry in the coming years is the development of a new generation of physics-based process models which take into account the specific mill layout and capabilities. Unlike earlier mill models, this new generation would have the ability to explore "what if" scenarios due to the ability of physics-based models to predict microstructure evolution. This along with recent advances in computational power and machine learning would make it possible to tune the models using data from a specific mill and then use the models to simultaneously design the chemistry and processing schedules given the specific capabilities of a given mill.

All of the above factors point to an acceleration in the development of new steels and an important improvement in the range and combinations of steel properties. The steel industry will become increasingly more dynamic and offer materials solutions for an increasing number of applications in the automotive and energy sectors.

References

1. Reproduced with permission: www.worldsteel.org
2. L. Ma, J. Dong, K. Peng, K. Zhang, A novel data-based quality-related fault diagnosis scheme for fault detection and root cause diagnosis with application to hot strip mill process. Control. Eng. Pract. **67**, 43–51 (2017)
3. Q.Y. Sha, Z.Q. Sun, Microstructure and precipitation in as cast low carbon Nb–V–Ti microalloyed medium thin slab. Ironmak. Steelmak. **37**, 320–325 (2010)
4. T.H. Zhou, H.S. Zurob, Abnormal and post-abnormal austenite grain growth kinetics in Nb–Ti microalloyed steels. Can. Metall. Q. **50**, 389–395 (2011)
5. D. Porter, Thermomechanical processing on hot strip and plate mills. Ironmak. Steelmak. **28**, 164–169 (2001)
6. A.J. de Ardo, Metallurgical basis for thermomechanical processing of microalloyed steels. Ironmak. Steelmak. **28**, 138–144 (2001)
7. Andrade H L, Akben M G and Jonas J J; "Effect of molybdenum, niobium, and vanadium on static recovery and recrystallization and on solute strengthening in microalloyed steels"; Metall. Trans. A, v 14A, n 10, (1983), pp. 1967-1977
8. A.J. de Ardo, Metallurgical basis for thermomechanical processing of microalloyed steels. Ironmak. Steelmak. **28**(2), 138–144 (2001)
9. T.N. Baker, Microalloyed steels. Ironmak. Steelmak. **43**(4), 264–307 (2016)
10. B. Dutta, C.M. Sellars, Mater. Sci. Technol. **2**(2), 146–153 (1985)
11. K. Nishioka, K. Ichikawa, Sci. Technol. Adv. Mater. **13**, 023001 (2012)
12. J.J. Jonas, Mater. Sci. Forum **3**, 284–286 (1998)

13. S. Vervynckt, K. Verbeken, B. Lopez, J.J. Jonas, Modern HSLA steels and role of non-recrystallisation temperature. Int. Mater. Rev. **57**, 187–207 (2012)
14. J. Zhao, Z. Jiang, Thermomechanical processing of advanced high strength steels. Prog. Mater. Sci. **94**, 174–242 (2018)
15. ref. on strengthening by Nb and V
16. Reproduced with permission from JFE Steel Corporation
17. Reproduced with permission from Dr. Dmitri Kopeliovich, substech.com (http://www.substech.com/dokuwiki/doku.php?id=steel_strip_processing)
18. J.W. Martin, R.D. Doherty, B. Cantor, *Stability of Microstructure in Metallic Systems*, 2nd edn. (Cambridge University Press, Cambridge, 1997)
19. T. Zhou, H. Zurob, P. Zhang, S. Cho, Heterogeneous grain microstructure reducing/eliminating edge breaks in low carbon steel, TMS 2020. Mat. Proc. Fundament.
20. Algoma Inc., Internal Research Report, *Development of a Light Gauge Superior High Strength Low Alloy Steel with Minim Yield Strength 700MPa* (2012)
21. Stelco Inc., Internal Research Report, *Thermal Cleaning of Full Hard Strapping* (2000)
22. *Calculated Using the TCFE6 Database of ThermoCacl.* www.thermocalc.com
23. Reproduced with permission from Eurotherm. https://www.eurotherm.com/metals-processing-applications-us/single-and-multi-stack-batch-annealing/
24. M.W. McDonald, D.L. Weaver, *P.E. Hydrogen Batch Annealing, Performance Comparison of Convection Systems, AISTech 2015 Proceedings* © 2015 by AIST 2800–2808
25. S.S. Sahay, A.M. Kumar, Applications of integrated batch annealing furnace simulator. Mater. Manuf. Process **17**, 439–453 (2002)
26. R. Prandhan (ed.), *Technology of Continuous-Annealed Sheet Steel* (TMS-AIME, Warrendale, 1985)
27. Reproduced with permission from Eurotherm. http://www.eurotherm.nl/continuous-annealing-and-galvanizing-lines
28. P.R. Mould, An Overview of Continuous-Annealing Technology for Steel-Sheet Products, Metallurgy of Continuous-Annealed Sheet Steel, B.L. Bramfitt and P.L. Mangonon, Jr., Ed., TMS-AIME, Warrendale, 1982, p 3–33
29. S.K. Paul, U. Ahmed, G.M. Megahed, Effect of hot rolling process on microstructure and properties of low carbon Al-killed steels produced through TSCR technology. J. Mater. Eng. Perform. **20**(7), 1163–1170 (2010)
30. *Physical Metallurgy of IF Steels* (The Iron and Steel Institute of Japan, Tokyo, 1994)
31. L. William, *Roberts: Flat Processing of Steel* (Marcel Dekker Inc, New York, 1988)
32. Stelco Inc. Internal Report, *Development of Cold Roll High-Strength Low-Alloy Steel* (2019)
33. L. William, *Roberts: Flat Processing of Steel* (Marcel Dekker Inc, New York, 1988)
34. S.R. Goodman, H.S.L.A. Steels, *Technology & Applications* (ASM, Materials Park, 1984), pp. 239–252
35. J.J. Kramer, Metall. Trans. **23A** (1992)
36. F.J. Humphreys, M. Hatherly, *Recrystallization and Related Annealing Phenomena*, 2nd edn. (Elsevier Ltd, Oxford, 2004)
37. Reproduced with permission from www.salzgitter-flachstahl.de
38. S. Bhat, in *Advances in High Strength Steels for Automotive Application*, Great Design in Steels" (Livonia-Michigan, AUTOSTEEL, 2008)
39. M.Y. Demeri, *Advanced High Strength Steels* (ASM International, 2013), p. 301
40. G.R. Speich Physical metallurgy of dual-phase steels. Fundament. *Dual Phase Steel* (1981), pp. 3–46
41. Reproduced with permission from "*Advanced High Strength Steels Applications Guidelines*", WorldAutosteel (2009)
42. N. Fonstein, "*Advanced High Strength Steels. Physical Metallurgy, Design, Processing and Properties* (Springer, 2015), p. 397
43. T. Kato, K. Hashiguchi, I. Takahashi, T. Irie, N. Ohashi, Development as hot rolled dual-phase steel sheet. Fundament. Dual Phase Steels. (1981), pp. 199–220

44. A.P. Coldren, G.T. Eldis, in *Using CCT Diagrams to Optimize the Composition of an As-Rolled Dual-Phase Steel*, vol. 32, issue 3 (JOM March, 1980), pp. 41–48
45. P.R. Mould, C.C. Skena, in *Structure and Properties of Cold-Rolled Ferrite-Martensite (Dual-Phase) Steel Sheets, Formable HSLA and Dual-Phase Steels* (The Metallurgical Society of AIME, 1977), pp. 183–205
46. O. Girina, N.M. Fonstein, D. Bhattacharya, in *Effect of Annealing Parameters on Austenite Decomposition in a Continuously Annealed Dual-Phase Steel*, 45th MWSP Conference on Proceedings of ISS, vol. XLI (2003), pp. 403–414
47. V. Zackey, E. Parker, D. Fahr, R. Bush, Enhancement of ductility in high strength steels. Trans. ASM **60**, 252–259
48. I. Tsukatani, S. Hashimoto, T. Inoue, Effects of silicon and manganese addition on mechanical properties of high-strength hot-rolled sheet steel containing retained austenite. ISIJ Int., 992–1000 (1991)
49. A. Karelova, C. Krempaszky, M. Duenckelmeyer, E. Werner, T. Hebesberger, A. Pichler, *Formability of Advanced High Strength Steels determined by Instrumented Hole Expansion testing* (MST2009, Pittsburgh), pp. 1358–1368
50. *540–780 MPa Grade Hot-rolled Steel with Excellent Formability*, JFE TECHNICAL REPORT No. 18 (March, 2013), pp. 132–134
51. O. Girina, N. Fonstein, O. Yakubovsky, D. Panahi, D. Bhattacharya, S. Jansto, *The Influence of Mo, Cr and B Alloying on Phase Transformation and Mechanical Properties in Nb Added High Strength Dual Phase Steels*, HSLA Steels 2015, Microalloying 2015& Offshore Engineering Steels 2015, Hangzhou, Zhejiang Province, China, November 11–13th, 2015 (The Chinese Society for Metals (CSM) and Chinese Academy of Engineering (CAE) TMS, 2016), pp. 237–245
52. F. Hairer, C. Krempaszky, P. Sipouridis, E. Werner, T. Hebesberger, A. Pichler, *Effects of Heat Treatment on Microstructure and Mechanical Properties of Bainitic Single- and Complex-Phase Steel* (MST2009, Pittsburgh, 2009), pp. 1391–1401
53. H. Karbasian, A.E. Tekkaya, A review on hot stamping. J. Mater. Process. Technol. **210**, 2103–2118 (2010)
54. M. Naderi, M. Ketabchi, M. Abbasi, W. Bleck, Analysis of microstructure and mechanical properties of different boron and non-boron alloyed steels after being hot stamped. Proc. Eng. **10**, 460–465 (2011)
55. FR 48758—*2017-2025 Model Year Light-Duty Vehicle GHG Emissions and CAFE Standards: Supplemental Notice of Intent—Content Details—2011-19905*. (n.d.). Retrieved November 25, 2019, from: https://www.govinfo.gov/app/details/FR-2011-08-09/2011-19905/summary
56. A. Ali, H.K.D.H. Bhadeshia, Growth rate data on bainite in alloy steels. Mater. Sci. Technol. **5**(4), 398–402 (1989)
57. C.-A. Däcker, O. Karlsson, C. Luo, K. Zhu, J.-L. Collet, M. Green, … Z.I. Olano, Bainitic hardenability – Effective use of expensive and strategically sensitive alloying elements in high strength steels (BainHard) (EUROPEAN COMMISSION No. EUR 25072) (2012)
58. K. Hausmann, D. Krizan, K. Spiradek-Hahn, A. Pichler, E. Werner, The influence of Nb on transformation behavior and mechanical properties of TRIP-assisted bainitic–ferritic sheet steels. Mater. Sci. Eng. A **588**, 142–150 (2013)
59. D. Quidort, Y. Bréchet, The role of carbon on the kinetics of bainite transformation in steels. Scr. Mater. **47**(3), 151–156 (2002)
60. F. Xu, Y. Wang, B. Bai, H. Fang, CCT curves of low-carbon Mn-Si steels and development of water-cooled bainitic steels. J. Iron Steel Res. Int. **17**(3), 46–50 (2010)
61. S.K. Liu, J. Zhang, The influence of the Si and Mn concentrations on the kinetics of the bainite transformation in Fe-C-Si-Mn alloys. Metall. Trans. A. **21**(6), 1517–1525 (1990)
62. H. Guo, X. Feng, A. Zhao, Q. Li, J. Ma, Influence of prior martensite on bainite transformation, microstructures, and mechanical properties in ultra-fine bainitic steel. Materials **12**(3) (2019)
63. P.H. Shipway, H.K.D.H. Bhadeshia, The effect of small stresses on the kinetics of the bainite transformation. Mater. Sci. Eng. A **201**(1), 143–149 (1995)

64. J. Speer, D.K. Matlock, B.C. De Cooman, J.G. Schroth, Carbon partitioning into austenite after martensite transformation. Acta Mater. **51**(9), 2611–2622 (2003)
65. A.M. Streicher-Clarke, J.G. Speer, D.K. Matlock, B.C. De Cooman, in *Quenching and Partitioning Response of a Si-Added TRIP Sheet Steel*, ed. by R. E. Ashburn, (Association for Iron & Steel Technology, Warrendale, 2004), pp. 51–62
66. D.P. Koistinen, R.E. Marburger, A general equation prescribing the extent of the austenite-martensite transformation in pure iron-carbon alloys and plain carbon steels. Acta Metall. **7**(1), 59–60 (1959)
67. M.J. Santofimia, J.G. Speer, A.J. Clarke, L. Zhao, J. Sietsma, Influence of interface mobility on the evolution of austenite–martensite grain assemblies during annealing. Acta Mater. **57**(15), 4548–4557 (2009)
68. J.G. Speer, R.E. Hackenberg, B.C. Decooman, D.K. Matlock, Influence of interface migration during annealing of martensite/austenite mixtures. Philos. Mag. Lett. **87**(6), 379–382 (2007)
69. A. Arlazarov, M. Gouné, O. Bouaziz, A. Hazotte, G. Petitgand, P. Barges, Evolution of microstructure and mechanical properties of medium Mn steels during double annealing. Mater. Sci. Eng. A **542**, 31–39 (2012)
70. S. Lee, B.C. De Cooman, Tensile behavior of intercritically annealed 10 pct Mn multi-phase steel. Metall. Mater. Trans. A **45**(2), 709–716 (2014)
71. S. Lee, S.-J. Lee, B.C. De Cooman, Work hardening behavior of ultrafine-grained Mn transformation-induced plasticity steel. Acta Mater. **59**(20), 7546–7553 (2011)
72. D.-W. Suh, S.-J. Park, H.N. Han, S.-J. Kim, Influence of Al on microstructure and mechanical behavior of Cr-containing transformation-induced plasticity steel. Metall. Mater. Trans. A **41**(13), 3276–3281 (2010)
73. D.-W. Suh, S.-J. Park, T.-H. Lee, C.-S. Oh, S.-J. Kim, Influence of Al on the microstructural evolution and mechanical behavior of low-carbon, manganese transformation-induced-plasticity steel. Metall. Mater. Trans. A **41**(2), 397 (2009)
74. R. Sun, W. Xu, C. Wang, J. Shi, H. Dong, W. Cao, Work hardening behavior of ultrafine grained duplex medium-Mn steels processed by ART-annealing. Steel Res. Int. **83**(4), 316–321 (2012)
75. H. Colpaert, *Metallography of Steels: Interpretation of Structure and the Effects of Processing* (ASM International, 2018)
76. A.L.V. da Costa e Silva, The effects of non-metallic inclusions on properties relevant to the performance of steel in structural and mechanical applications. J. Mater. Res. Technol. **8**(2), 2408–2422 (2019)
77. T. Gladman, G. Fourlaris, M. Talafi-Noghani, Grain refinement of steel by oxidic second phase particles. Mater. Sci. Technol. **15**(12), 1414–1424 (1999)
78. H. Suito, R. Inoue, Thermodynamics on control of inclusions composition in ultraclean steels. ISIJ Int. **36**(5), 528–536 (1996)
79. A.L.V. da Costa e Silva, Non-metallic inclusions in steels – Origin and control. J. Mater. Res. Technol. **7**(3), 283–299 (2018)
80. A.J. Kaijalainen, P.P. Suikkanen, L.P. Karjalainen, D.A. Porter, Influence of subsurface microstructure on the bendability of ultrahigh-strength strip steel. Mater. Sci. Eng. A **654**, 151–160 (2016)
81. B. Dutta, E.J. Palmiere, C.M. Sellars, Modelling the kinetics of strain induced precipitation in Nb microalloyed steels. Acta Mater. **49**, 785–794 (2001)
82. H.S. Zurob, C.R. Hutchinson, Y. Brechet, G. Purdy, Modeling recrystallization of microalloyed austenite: Effect of coupling recovery, precipitation and recrystallization. Acta Mater. **50**, 3077–3094 (2002)

Chapter 3
Microalloyed Steels

Sourav Das, Anish Karmakar, and Shiv Brat Singh

Abbreviations

BH	Bake hardening
CE	Carbon equivalent
DBTT	Ductile to brittle transition temperature
DP	Dual phase
FLD	Forming limit diagram
HAZ	Heat affected zone
HSLA	High strength low alloy
IF	Interstitial free
IIW	International Institute of Welding
IP	Interphase precipitation
MA	Microalloyed steel
MMAW	Manual metal arc welding
MPa	Megapascal
SAW	Submerged arc welding
TMCP	Thermo-mechanical controlled processing
TRIP	Transformation-induced plasticity
TTT	Time-temperature-transformation
VHN	Vickers hardness number

S. Das (✉) · A. Karmakar
Department of Metallurgical and Materials Engineering, Indian Institute of Technology Roorkee, Roorkee, Uttarakhand, India
e-mail: sourav.das@mt.iitr.ac.in

S. B. Singh
Department of Metallurgical and Materials Engineering, Indian Institute of Technology Kharagpur, Kharagpur, West Bengal, India

© Springer Nature Switzerland AG 2021
R. Rana (ed.), *High-Performance Ferrous Alloys*,
https://doi.org/10.1007/978-3-030-53825-5_3

Symbols

a_M	Activity of dissolved microalloying element M
A_M	Atomic mass of M
a_{MX}	Activity of dissolved compound
a_X	Activity of dissolved interstitial element X
A_X	Atomic mass of X
A, B	Constants with the product of the activity coefficients
c_i	Concentration of solute i
$[C]$	Concentration of C in solution in ferrite
D	Grain size
D_γ	Prior austenite grain size
f	Volume fraction of precipitate
h	Planck constant
I	Nucleation rate
k	Boltzmann constant
k_y	Locking parameter which measures the relative hardening contribution of the grain boundaries
k'	Fitting constant
k_i	Strengthening coefficient of solute i
k_s	Solubility product
$[M]$	Microalloying solute element M dissolved in primary phase
N	Number density of sites supporting nucleation
Q	Activation energy per atom
R	Universal gas constant
S_V	Grain boundary area per unit volume
T	Temperature in absolute scale
t_i and t_f	Initial and final thickness of the sheet, respectively
T_{NR}	No recrystallization temperature
T_{RX}	Recrystallization temperature
T_{sol}	Dissolution temperature of MX
w_i and w_f	Initial and final width of the sheet, respectively
$\overline{X}$	Mean linear intercept diameter of a precipitate particle
$[X]$	Interstitial solute element X dissolved in primary phase
α	Ferrite
γ	Austenite
γ_i	Activity coefficient of solute i
ΔG_{crit}	Critical activation energy for nucleation
ΔG^0	Standard free energy change for the forward reaction
ΔH^0	Change in enthalpy
ΔS^0	Change in entropy
$\Delta\sigma$	Strength increment
$\Delta\sigma_{Disl}$	Dislocations strengthening
$\Delta\sigma_{GS}$	Strengthening due to grain refinement

$\Delta\sigma_{Pptn}$	Precipitation strengthening
$\Delta\sigma_{SS}$	Solid solution strengthening
$\Delta\sigma_{Tex}$	Strengthening due to textural effect during rolling
λ	Perpendicular sheet spacing of vanadium precipitates
σ_y	Yield stress
σ_0	Friction stress representing the overall resistance of the crystal to dislocation movement

3.1 Introduction

Microalloyed steels are one of the most important genres of steels that are available for various applications in challenging environmental and operational conditions all over the world currently. After the crude oil crisis of the 1970s, almost all the industries started using thinner and lighter steels. Such demands led to the development of hot rolled high-strength low-alloy (HSLA) steels which have the required properties (e.g., yield strength > 400 MPa) after cooling and/or coiling [1]. Apart from the common alloying elements like Mn and Si, this type of steels typically contains a small amount of carbide and nitride-forming alloying elements like Nb, Ti, or V either singly or in combination. As the total concentration of these alloying elements is very minute, up to 0.1 wt% for a single addition or up to 0.15 wt% for a combination of elements [2], this grade of steels came to be widely known as microalloyed (MA) steels or HSLA steels in recognition of the fact that even in small concentrations, these elements engender dramatic effect on the microstructure and properties of the steel. The primary role of such alloying addition is to produce finer grains and/or precipitation strengthening by the formation of relatively stable carbides or nitrides. The effects of addition of such alloying elements are greatly influenced by the thermal and thermo-mechanical cycles followed during the processing of steels [3].

The commercial application of microalloying started in the 1960s [4, 5]. Along with the concept of microalloying, advancements in hot rolling processing aided the development of microalloyed steels. This enabled the steel researchers to obtain fine ferrite grains, needed for MA grade steels, from fine austenite grains by closely controlling the rolling operation. Carrying this momentum forward, further development in the MA concept, e.g., roles and effectiveness of different microalloying elements (like Nb, V, Ti, etc.), fracture behavior, effects of different nonmetallic inclusions, possibilities of reducing carbon, etc. led to the development of steels with even better properties. By the end of the 1970s, the steels could deliver an excellent combination of mechanical properties: yield strength 350–420 MPa, ultimate tensile strength 450–700 MPa, total elongation 14–27%, and Charpy impact energy value of 150 J and 100 J at room temperature and − 40 °C, respectively [6–8].

The concept of microalloying has subsequently been extended to different genres of steels like dual-phase (DP) steels where additions of microalloying elements led to improved work hardening behavior due to the refinement of martensite islands [9]. With the development of steel making technology for producing steels with very low carbon and nitrogen content (typically <0.005 wt%), another application of microalloying was utilized in scavenging the solute carbon and nitrogen present in the matrix, and thus the development of *interstitial free* (IF) steel was realized [10]. Further developments and innovations in controlling the carbon led to the invention of *bake hardening* (BH) steel where some amount of carbon is kept in solution to cause strain aging at a later stage of production cycle and thereby improving the strength of the final component [11]. Some good work on this family of steels can be found elsewhere [12, 13]. It is worth mentioning here that all these three types of steels (DP, IF, and BH) are extensively used in automotive industry. As the focus in the present chapter is on MA or HSLA family of steels in general, no further discussion on other automotive steels will be made here.

It is thus apparent from the above discussion that the MA/HSLA family of steels could find application in many different segments which include pipeline, automotive, ship building, and other structural materials. A large number of conferences have been fully devoted to discuss the progress of this class of steels [14–16]. The interest among the researchers around the world about this particular type of steel is so high that it led to the publication of a number of review articles [1, 4, 6, 17]. The role of different alloying elements has been researched [18, 19], and also an excellent reference book has been published [3]. The interest in the past few decades has pushed the microalloying concept to develop steels for sectors like pipeline, automotive, shipbuilding, construction, etc. with even better set of mechanical properties. Application in newer areas certainly has increased the tonnage-wise production of this type of steels worldwide. It is estimated that currently this type of steels comprises almost 12% (~ 200 million tons) of the world steel production (1631 million tons) [20, 21]. The mechanical properties have also improved significantly: yield strength well in excess of 690 MPa with tensile strength above 1000 MPa [6, 7, 22, 23] and Charpy impact energy of 220 J at room temperature and 150 J at −40 °C [7, 20, 24] are now attainable for microalloyed HSLA steels.

Alloying with carbon is perhaps the cheapest and the easiest method to attain high strength and was in fact practiced in the past when hot rolled steels with carbon content up to 0.4 wt.% and Mn up to 1.5 wt.% were used to give yield strength of 300–400 MPa. However, high pearlite content associated with high carbon results in poor impact toughness (as measured by impact energy and ductile-brittle transition temperature (DBTT)), as can be seen from Fig. 3.1 [25, 26]. Similarly, high carbon steels have high hardenability (which increases with carbon content) that results in poor weldability. The solution obviously is low-carbon steel that would improve toughness and weldability substantially and attain the required strength and toughness by microstructure engineering. HSLA steels are perhaps one of the earliest examples of engineered material where the microstructure was designed using well-understood scientific principles [27]. Grain refinement of ferrite is the obvious choice for improving the strength and toughness simultaneously [27] where the well-known Hall-Petch equation is exploited. There are however theoretical and practical limitations to the extent of grain refinement that can be

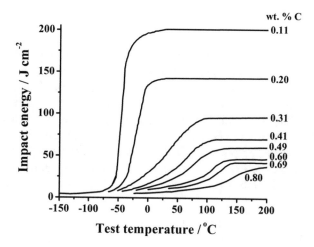

Fig. 3.1 Effect of carbon content (wt. %) on the impact transition temperature curves of ferrite/pearlite steels [25, 26]

attained [28]. Therefore if higher strengths are required, then the next most suitable alternative is precipitation strengthening which of course comes with a small penalty to toughness [7, 18]. The microstructure of this type of steels thus consists of fine grains of ferrite with fine dispersion of carbides /nitrides/carbonitrides, and as will be discussed in subsequent sections, addition of microalloying elements and special processing make it possible to attain this microstructure during industrial rolling practice.

Grain refinement can be achieved by increasing the nucleation rate of ferrite that forms from austenite. Nucleation rate (I) in general can be given by [29, 30]:

$$I = N \frac{k\,T}{h} \exp \left\{ \frac{-(\Delta G_{crit} + Q)}{k\,T} \right\} \tag{3.1}$$

where N is the number density of sites supporting nucleation, k is the Boltzmann constant, h is the Planck constant, T is the absolute temperature, ΔG_{crit} is critical activation energy for nucleation and Q is the activation energy per atom for the atoms to cross the austenite/ferrite nucleus interface. It is clear from the above equation that grain refinement can be achieved in two ways: (i) by increasing the number density of nucleation sites, that is, by increasing the austenite grain boundary area per unit volume (S_V) and introducing other heterogeneous nucleation sites in austenite grain, known as "*austenite conditioning*" and (ii) by increasing the driving force for ferrite transformation that would lower the activation barrier for nucleation. In practice, (i) is achieved by controlled rolling which is briefly described in Sect. 3.4 and (ii) by accelerated cooling that increases the undercooling at which austenite to ferrite transformation takes place. Together they constitute what is known as *thermo-mechanical controlled processing (TMCP)* which together with the addition of microalloying elements represents a major technological breakthrough.

3.2 Role of Microalloying Elements

Microalloying elements play an important role in governing the evolution of microstructure during processing of HSLA steels, in spite of their small concentration. Depending on temperature and other processing conditions, the elements could be in solid solution in the primary phase (austenite or ferrite) or precipitate out. The solubility of carbon in ferrite or austenite in equilibrium with cementite (Fe₃C) and the fraction of cementite in binary Fe-C alloy can simply be obtained from the $\alpha/\alpha + Fe_3C$ or $\gamma/\gamma + Fe_3C$ phase boundaries of Fe-Fe₃C phase diagram and mass balance. For example, the solubility of carbon in ferrite in equilibrium with cementite is given by [31]:

$$\log [C] = -\frac{2120}{T} + 0.41 \tag{3.2}$$

where $[C]$ represents wt. % C in solution in ferrite and T is the temperature in Kelvin. The situation is somewhat complex in ternary or multicomponent alloys with two or more alloying elements where the phase boundaries cannot be visualized so easily. It is convenient in such cases to work out and quantify the partitioning of the alloying elements between the primary solution phase (α or γ in this case) and the precipitate using the concept of *solubility product*.

Consider the equilibrium between the dissolution of precipitate MX and its formation given by the following equation:

$$(MX) \leftrightarrow [M] + [X] \tag{3.3}$$

where $[M]$ and $[X]$ represent the microalloying solute element M and interstitial solute X, respectively, dissolved in the primary phase which in the present case could be austenite or ferrite, and (MX) is the compound that precipitates out. If the activities of dissolved microalloying element, interstitial, and the compound are denoted by a_M, a_X, and a_{MX} respectively, then the equilibrium constant k_s of the above reaction (Eq. 3.3) can be expressed as:

$$k_s = \frac{a_M . a_X}{a_{MX}} \tag{3.4}$$

The activities of metal and interstitial are essentially the function of elemental concentrations in solution ($[i]$) and the corresponding activity coefficient (γ_i). The activity a_{MX} of the precipitate MX can be considered as unity so that

$$k_s = a_M . a_X \tag{3.5}$$

Now if ΔG^0 is the standard free energy change for the forward reaction (Eq. 3.3), then:

Table 3.1 Solubility product constants (A and B) of some of the most common carbides and nitrides summarized and collected from Gladman [3]

Compound	A	B	Temperature range (°C)	Phase
VN	8330	3.40	900–1350	Austenite
NbN	8500	2.80	900–1350	Austenite
	12,230	4.96	600–800	Ferrite
TiN	15,020	3.82	900–1350	Austenite
VC	9407	5.65	900–1100	Austenite
	12,265	8.05	900	Ferrite
NbC	7900	2.96	900–1200	Austenite
	10,960	5.43	–	Ferrite
TiC	10,300	5.12	1000–1350	Austenite
	9575	4.40	–	Ferrite

$$\Delta G^0 = -R\,T \ln k_s \qquad (3.6)$$

where T is the temperature in Kelvin, and R is the universal gas constant. Since the concentrations of the solute elements M and X are very small, dilute solution approximation can be used so that their activities in the solution are directly proportional to their mole fractions $[M]$ and $[X]$, respectively. Since $\Delta G^0 = \Delta H^0 - T\Delta S^0$, Eq. 3.6 above can in general be rewritten to give the solubility product $[M][X]$ as:

$$\log\,[[M]\,[X]] = -\frac{A}{T} + B \qquad (3.7)$$

where A and B are constants with the product of the activity coefficients subsumed in the constant B; $[M]$ and $[X]$ can be presented in terms of atom % or mass %. The solubility product in austenite and ferrite and the constants A and B for each phase have been painstakingly evaluated for the microalloy precipitates that form in microalloyed steel using a variety of techniques (precipitate extraction, gaseous equilibrium, thermodynamic calculation, etc.). Although these values can be found out from elsewhere [3], the constants and solubility product values of some of the most common carbides and nitrides are listed in Table 3.1 for ready reference. It is to be kept in mind that the values of A and B reported by different investigators vary to an extent presumably due to the variations in the method used, subtle variations in the alloy composition and the size of the precipitate. Solubility product is usually plotted as a function of inverse of absolute temperature as suggested by Eq. 3.7 (Fig. 3.2a). Alternatively, solubility relationships are presented as isotherms; some examples for vanadium, niobium, and titanium are shown in Fig. 3.2b–d.

The solubility product in ferrite or austenite in a ternary or multicomponent alloy can be treated in the same way as solubility of carbon in the matrix phase in a binary Fe-C alloy. The solubility product $[M][X]$ in a given matrix phase increases with temperature so that more of M and X go into solution, i.e., the stability of

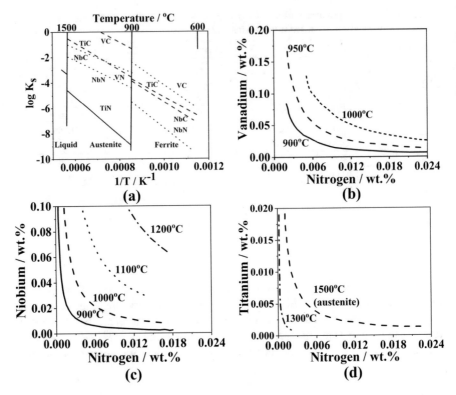

Fig. 3.2 (a) Evaluation of the temperature dependence of the solubility product of different microalloyed carbide and nitride precipitates and (**b, c & d**) limits of mutual solubility of vanadium, niobium, and titanium with nitrogen in austenite at various temperatures [3]

the precipitate MX decreases, and at sufficiently high temperature, T_{sol}, the entire amount of M and X in the alloy might dissolve. Nitrides of titanium are very stable and may dissolve completely only in liquid, whereas vanadium carbide has the highest solubility and might precipitate only during or after austenite to ferrite transformation. The stability of some common precipitates in microalloyed steels decreases in the following order: TiN >AlN>NbN>VN>NbC>TiC>VC [32]. They all have a face-centered cubic (FCC) structure and are completely soluble in each other. As a result, complex carbonitride precipitates can also be formed.

At a given temperature, precipitation becomes possible when the product of the total concentrations of M and X in the alloy is larger than the solubility product of MX at the temperature concerned. Thus, if $\{M\}$ and $\{X\}$ are the total concentration of M and X in the alloy and $[M]$ and $[X]$ are the respective (solute) concentration in solution in the matrix phase at any given temperature T, then the amount of M and X in the precipitate can be worked out from stoichiometry as:

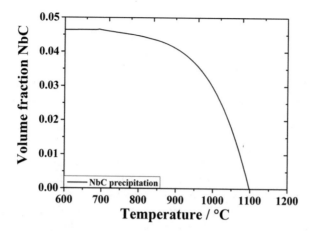

Fig. 3.3 Calculated evolution of phase fraction of NbC for a steel with nominal composition (in wt%): 0.08 C-1.30 Mn-0.30 Si-0.04 Nb using TCFE8 database

$$\frac{\{M\} - [M]}{\{X\} - [X]} = \frac{A_M}{A_X} \tag{3.8}$$

where A_M and A_X are the atomic mass of M and X, respectively. If the alloy is at equilibrium at T, then the amounts of M and X in solution and in precipitate can be found out by simultaneously solving Eqs. 3.7 and 3.8, and the fraction of precipitate can be calculated from mass balance or stoichiometry. These quantities can also be calculated graphically from solubility product isotherms like the ones shown in Fig. 3.2b–d [3, 27].

The solubility product relationships discussed above are empirical where the interaction between solutes in a multicomponent alloy is ignored. More accurate calculations can be done with CALPHAD method where the precipitate dissolution temperature, the equilibrium amount of solute at any temperature and the fraction of precipitate can be calculated directly using a commercially available program like Thermo-Calc or MTDATA along with a suitable database. An example calculation using TCFE8 database is provided in Fig. 3.3 for NbC for a steel with nominal composition (in wt%): 0.08 C-1.30 Mn-0.30 Si-0.04 Nb. It has to be kept in mind that solubility depends on the size of the precipitate as well (Gibbs-Thomson capillarity effect), especially when the precipitates are extremely fine, on a nanometric scale [30]. The solubility increases if the precipitates are finer and the dissolution temperature would be lower. Kinetics of precipitate dissolution and formation are equally important aspects that have to be taken into consideration since actual processing conditions deviate from equilibrium. It has been shown, for example, that when the heating rate increases from 1.3×10^{-3} °C/s to 1.3 °C/s, the dissolution temperature of NbC in austenite increases by about 100 °C [33]. The equilibrium solubility product relationships or equilibrium calculations can therefore provide only broad guidelines.

At any temperature below T_{sol}, the solutes are partitioned between the matrix and the precipitate phase. The kinetics of precipitation is accelerated significantly when austenite is deformed below the recrystallization temperature. The solute atoms and the strain-induced precipitates of carbide and/or carbonitrides of these elements retard the recrystallization of austenite (the latter being much more effective than solute) and enable pancaking of austenite grains even at higher temperatures which augments the ferrite nucleation rate. The complex interaction between precipitation and recrystallization is an interesting aspect that governs the evolution of microstructure in this case.

The solubility product decreases with decreasing temperature, and there is a sharp drop in solubility when austenite transforms to ferrite. This results in extensive precipitation at austenite/ferrite transformation interface when the transformation temperature and the precipitation temperature are synchronized. This is known as *interphase precipitation* [34]. The precipitates, especially the fine ones, formed during or after austenite to ferrite transformation, provide additional strengthening (Sect. 3.3.2). In addition, the precipitates of these elements inhibit grain coarsening during reheating and restrict the growth of recrystallized austenite grains in stage 1 of controlled rolling described later. Coarse precipitates that remain undissolved during soaking are not effective for retardation of recrystallization and strengthening. They must therefore be taken into solution to exploit their full potential. Dissolved solute atoms, particularly of Nb, retard austenite to ferrite transformation, that is, they lower the transformation temperature during cooling and therefore facilitate finer ferrite grains and interphase precipitate; precipitates themselves, on the other hand, accelerate the transformation because they remove C and N from solution in austenite, and coarse ones might provide heterogeneous nucleation sites for ferrite.

In general, Nb is the most effective microalloying element (among the three: Ti, Nb, and V) for grain refinement because of its effectiveness in retarding the recrystallization of austenite by strain-induced precipitation. Vanadium carbide on the other hand is highly soluble in austenite, but it precipitates out during or after austenite to ferrite transformation and contributes mostly by way of precipitation strengthening. Thus, the largest contribution toward strengthening from grain size refinement comes from Nb, whereas that for precipitation strengthening comes from V (VC); Ti has an intermediate contribution in both these aspects [35].

3.3 Strengthening Mechanisms

It is well-known that the strength can only be improved if the dislocation motion is restricted. The observed strength of MA steel is the result of contributions from grain refinement ($\Delta\sigma_{gs}$), precipitation of carbides/carbonitrides ($\Delta\sigma_{pptn}$), solid solution strengthening due to alloying with different elements (Si, Mn, and others while in solid solution) ($\Delta\sigma_{ss}$), textural effect during rolling ($\Delta\sigma_{tex}$), or generation of dislocations ($\Delta\sigma_{disl}$). Although there are some questions regarding the appropriate way of expressing the effects of these variables on the observed yield strength [36, 37], DeArdo concluded that [18] the best way to represent their effect is through linear addition as in:

$$\sigma_y = \Delta\sigma_{gs} + \Delta\sigma_{pptn} + \Delta\sigma_{ss} + \Delta\sigma_{tex} + \Delta\sigma_{disl} \qquad (3.9)$$

In the following sections, the primary source of high strength in these steels, namely, the *grain refinement* and the *precipitation hardening*, will be discussed in details.

3.3.1   Grain Refinement in MA Steels

The contribution toward strengthening from grain size was first postulated independently by Hall [38] and Petch [39]. According to this model, the yield stress (σ_y) can be expressed with the help of grain size (D) in the following manner:

$$\sigma_y = \sigma_0 + k_y\, D^{-1/2} \qquad (3.10)$$

where, σ_0 and k_y are the friction stress representing the overall resistance of the crystal to dislocation movement and the locking parameter which measures the relative hardening contribution of the grain boundaries, respectively. The mathematical expression is the resultant of suitable treatment of stress accumulation in front of a pile up of dislocations. It is thus important to measure the grain size properly. The most common and reliable method of determining the grain size [3] is the mean linear intercept method where a known length of line is drawn on the planar metallographic section and the mean intercept length is estimated from the number of grains it intersects. It is considered that this mean intercept length is the diameter (D) of a spherical grain. The factors k_y and σ_0 are the slope of the straight line and intercept along the ordinate, respectively, if σ_y is plotted against $D^{-1/2}$. The theoretical value of k_y of ferrite, 21.9 MPa mm$^{0.5}$, was proposed by Dingley and McLean [40], and this matches quite well with other experimental observations [3]. It is worth mentioning here that the value of k_y (of ferrite) was calculated considering no free dislocation in the system. Thus, the presence of unlocked dislocation will lead to a lower value of k_y. The k_y value depends on the temperature as well [3]. However, different researchers have reported different values for σ_0, as summarized by Gladman [3]. The differences primarily originate from the effects of the presence of substitutional solute atoms. Consequently, the intercept of the Hall-Petch plot (i.e., σ_0) can be considered as the summation of the friction stress of pure iron and the solid solution hardening effects due to the presence of solute atoms [41]. It is thus appropriate to express the yield stress in the following form:

$$\sigma_y = \sigma_0 + \sum k_i\, c_i + k_y\, D^{-1/2} \qquad (3.11)$$

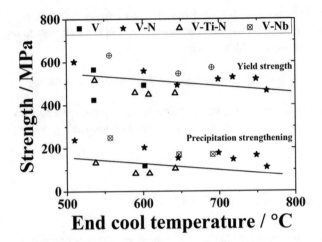

Fig. 3.4 Variation of yield strength (the upper line) and contribution from precipitation strengthening (the lower line) as a function of end water cool temperature (or coiling temperature) for microalloyed steel containing different precipitating elements [43]. The difference between the upper and lower lines represents the contribution from other sources of strengthening

where σ_0 is the friction stress of pure iron, and k_i and c_i are, strengthening coefficient and the concentration of solute i respectively. One of the earliest studies in this direction was carried out by Pickering and Gladman [42]. They have proposed the variation of yield strength (in MPa) as a function of both composition and grain size (Eq. 3.8).

$$\sigma_y = 105 + 84 \times (wt\%Si) + 33 \times (wt\%Mn) + 17.5 \, D^{-0.5} \qquad (3.12)$$

However, Li et al [43] have proposed a different expression to calculate the yield strength, and this is given in Eq. 3.9.

$$\sigma_y = 45 + 84 \times (wt\%Si) + 32 \times (wt\%Mn) + 680 \times (wt\%P)$$
$$+ 38 \times (wt\%Cu) + 43 \times (wt\%Ni) + 18.1 \, D^{-0.5} \qquad (3.13)$$

It has been estimated that the major contribution to the strengthening (~ 250 MPa) comes from grain refinement (considering 5 μm ferrite grain), another 100–220 MPa might be due to precipitate particles, and the remainder (i.e., the difference between the yield strength and the precipitation contribution) is from solid solution hardening [44]. This can be seen from Fig. 3.4 where variations in yield strength and contribution from precipitation strengthening for different types of microalloyed steel are plotted against end cool temperature (or simply the coiling temperature) which is equivalent to coiling temperature in industrial condition. Apart from directly contributing to solid solution strengthening, the solute elements affect the austenite to ferrite transformation temperature, and therefore the ferrite grain size as well and this can further affect the strength. Mn, for example, lowers the transformation temperature and therefore refines the ferrite grain size [45].

3.3.2 Precipitation Hardening in MA Steels

It has been discussed earlier that depending on the philosophy of alloy design, microalloying additions are chosen in such a way that carbides, nitrides, or carbo-nitrides particles are dissolved in the solid solution at high temperature but re-precipitate during the cooling up to room temperature. If the precipitate particles are uniformly distributed, they provide barriers to dislocation motion. The dislocations need to then overcome such barriers to continue its motion, and strength is increased in the process. The increase of strength can be represented as [35]:

$$\sigma\ (MPa) = \frac{5.9\sqrt{f}}{\overline{X}} \ln\left(\frac{\overline{X}}{2.5 \times 10^{-4}}\right) \qquad (3.14)$$

where, f = Volume fraction of precipitate

$\overline{X}$ = mean linear intercept diameter of a precipitate particle

It is thus obvious that the finer the precipitates, the more effective strengthener they are. It is noteworthy to mention here that the calculation for the increase of strength due to precipitation (Eq. 3.14) was carried out for random distribution only, as can be seen in Fig. 3.5a. Apart from this general type, a special kind of precipitation is also encountered in MA steel, and that is called interphase precipitation which has also been studied extensively [46]. In this case, carbide particles of the same orientation are precipitated on a plane which is generally parallel to the γ/α interface. It was concluded that the precipitation of sheets of microalloyed carbides (of vanadium or niobium or titanium) platelets occurs on the γ/α interface inside ferrite [18]. If the γ/α interface is microscopically planar and highly coherent, the precipitation is thought to be associated with incoherent, mobile interface ledges (Fig. 3.5b) [47, 48]. On the other hand, if the incoherent γ/α interface is curved (Fig. 3.5c), Rickes and Howell [49] have suggested a quasi-ledge mechanism for the precipitation of particles. However, the separation of precipitate particles in both the instances is controlled by the rate of diffusion of the substitutional element within the interface. It is worth mentioning here that although some reports have claimed the interphase precipitation in Nb-alloyed steel [50, 51], DeArdo [18] has argued that only very specific thermal paths would intersect the precipitation region as can be seen from Fig. 3.6. It is primarily because of this reason that such kind of precipitation is likely to occur when high supersaturation of NbC is available along with slow growth rate. Sakuma and Honeycombe have also suggested earlier that the balance between the ferrite growth and the supersaturation of NbC in ferrite is essential for such precipitations to occur [50, 51].

(a)   (b)

(c)

Fig. 3.5 TEM micrographs showing precipitates in microalloyed steel. (**a**) ordinary precipitate of NbC (C-0.09, Nb-0.07, Fe-balance, all in wt. %, reheated to 1250 °C, hot rolled to 1000 °C, and air cooled to room temperature) [18], (**b** and **c**) interphase precipitates of different kinds [48]: (**b**) parallel sheet and (**c**) curved sheet (C-0.073, Ti-0.227, Fe-balance, all in wt. %; isothermal holding at 800 °C for 1000 s)

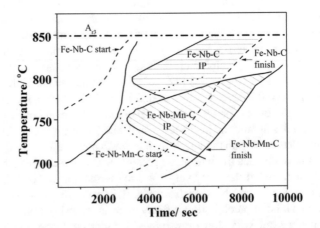

Fig. 3.6 Schematic TTT curves for Fe–0·036Nb–0·09C and Fe–0·036Nb–1·07Mn-0·09C alloys (solid lines): interphase precipitation (IP) occurs in the shaded areas only. The dotted line represents the IP region for a different steel composition: 0·025Nb–0·07C–1·1Mn. All the compositions are in wt. % only [18]

In MA steels, combined contributions from precipitation and grain refinement are utilized by suitable addition of microalloying elements like Nb, V, Ti, etc. As already discussed, the effectiveness of these different elements depends on the stability of the corresponding carbides and/carbonitrides or in other words their solubility in austenite. It is to be kept in mind that precipitate particles formed at high temperature are generally coarse, and therefore, they do not contribute much to the strengthening directly. However, on cooling, the particles which had been dissolved earlier now start precipitating during the $\gamma \rightarrow \alpha$ transformation at the corresponding interface leading to a row of interphase precipitate particles. Being lower-temperature particles, they are very fine and thus are very much effective for strengthening. Consequently, the particles that precipitate at low temperatures in austenite or at the γ/α interface during transformation or in ferrite during cooling are effective for strengthening because they are dispersed on a much finer scale (e.g., VN). Batte and Honeycombe [52] have deduced the following empirical relationship between the strength increment ($\Delta\sigma$) and the perpendicular sheet spacing (λ) for vanadium added steel:

$$\Delta\sigma = k'.\lambda^{-0.7} \tag{3.15}$$

where the value of $k' = 0.0468$ N/mm$^{3/2}$ (0.468 kgf/mm$^{3/2}$). Similar observation was also made for other microalloying elements as well. However, the above expression is true for a fixed volume fraction of precipitates only. If the total volume fraction increases, the strength will obviously increase further. It thus can be concluded that high volume fraction of finely dispersed precipitates can maximize the strength increment. For steels containing different concentrations of Nb, it was shown that the increase in yield stress can take place for both the increase in the total amount of Nb content and the size of the NbC particles (Fig. 3.7) [25].

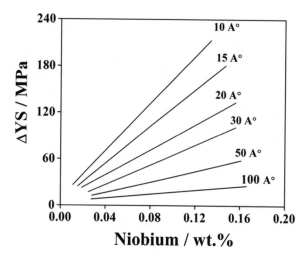

Fig. 3.7 Effect of Nb content and the size of NbC particles on the increase in yield strength of microalloyed steel [27]

On the other hand, the precipitates that form at high temperatures (e.g., TiN) are not very suitable for contributing to precipitation hardening because of their comparatively larger size though they might be effective grain refiner. These particles become very much useful in operations like welding where one of the prime importance is to restrict the grain coarsening to minimize the deterioration of properties in the heat-affected area.

3.4 Processing of Microalloyed Steels

The biggest contribution to the development of MA steel probably has come from the significant improvement and innovations in the hot rolling technology. In this process, the feedstock (slab/ingot) is heated in a furnace to a temperature which depends on the chemical composition of the steel but remains usually in the range of 1100–1250 °C. This is called the reheating/soaking temperature, and the corresponding microstructure becomes fully austenitic. After removing the oxide scales, which are formed on the surface due to the exposure to such high temperature, the feedstock is subjected to deformation in the roll gap. The deformation is carried out in a number of steps, called passes, in a reversing or continuous mill with a predetermined amount of thickness reduction in each pass. Finally, after the required thickness has been achieved, the product is allowed to cool freely in air or is subjected to accelerated cooling using air or water sprays. For a given chemical composition, the final microstructure and hence the mechanical properties depend on the microstructure of austenite before it begins to transform and on the rate of cooling. It is possible to control the time-temperature-deformation sequence (the "rolling schedule") to leave the austenite in a state which will induce it to transform into the required microstructure upon cooling. This is known as "*austenite conditioning*."

It has been reported earlier that the ferrite grain structure is determined by both the austenite grain size and the amount of deformation [53–55]. Figure 3.8 shows such a behavior for two different microalloyed steels (Nb and Nb + V-added compositions) with varying prior austenite grain size and deformed to different extent but cooled to a specific rate only. In general, the rolling operation is started at a temperature region where the austenite can recrystallize readily. During this stage, the coarse grains of austenite, produced during the high-temperature soaking, are refined by repeated deformation and recrystallization. Precipitates of microalloying elements, if present, restrict the coarsening of recrystallized austenite grains. However, there is a limit to the degree of austenite grain refinement that can be achieved in this way. Another possibility is to leave the austenite in a deformed state before cooling it to induce ferrite transformation. This is done by introducing a "delay" period after a few initial passes during rolling when the rolling operation is suspended to allow the temperature of the ingot to fall below a certain temperature known as the recrystallization temperature, T_{RX}. Rolling is again resumed when the temperature falls to a predetermined temperature below the recrystallization

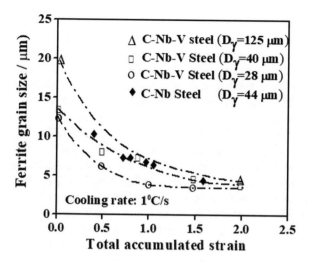

Fig. 3.8 Influence of accumulated strain on the final ferrite grain size for different prior austenite grain sizes (D_γ) under a constant cooling rate of 1 °C s^{-1} [55]

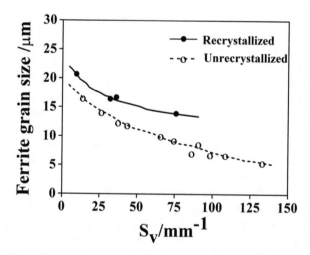

Fig. 3.9 Relationship between ferrite grain size and austenite grain boundary area (S_v) for recrystallized and unrecrystallized grains. For the same S_v, a finer ferrite grain structure is obtained if the austenite is in an unrecrystallized state [56]

temperature usually known as the no recrystallization temperature or T_{NR}. Rolling at such low temperatures leads to elongated or *pancaked* austenite grains with larger austenite grain boundary area per unit volume (S_V) that provides larger number of nucleation sites and finer ferrite grains as can be seen from Fig. 3.9 [56, 57].

It was observed that when Nb-bearing steels were hot rolled under conventional rolling conditions, the toughness deteriorated remarkably [53, 54]. However, when the same steel was finish rolled at lower temperatures (typically below 800 °C) or was given a total reduction of more than 30% at temperatures lower than 900 °C, the corresponding toughness increased to a great extent [53, 54]. This possibility of carrying out the hot rolling operation at lower temperature gives the opportunity of obtaining "pancaked" austenite grains (after hot working) which in turn resulted in finer ferrite grains in transformed product. This type of rolling is called *controlled rolling* which differs from the conventional hot rolling in terms of the formation of ferrite. In ordinary hot rolling, ferrite nucleation takes place on the austenite grain boundaries only, whereas in controlled rolling, they can nucleate within the deformed austenite grains as well, resulting in a finer ferrite grain size. A significant amount of research work has been carried out in the past to gain fundamental understanding regarding the recrystallization [58, 59], retardation in recrystallization due to the presence of microalloying elements [60], changes in the microstructure due to deformation in the non-recrystallized region [61], and so on.

The effect of deformation of austenite in the non-recrystallization region is more than just an increase in S_V. It has been found that for the same S_V, the ferrite grains formed from deformed and unrecrystallized austenite are finer than that formed from recrystallized austenite (Fig. 3.9) [57]. The austenite grain boundary area per unit volume increases by a factor of only about two even when the total amount of deformation reaches 80–90%. However, detailed studies by a number of researchers suggest that apart from an increase in S_V, the deformation enhances the nucleation potency of austenite grain surfaces [62–65]. Deformation bands and other such defects that act as ferrite nucleation sites are also introduced within the austenite grains during such rolling. However, this effect does not occur at relatively smaller strains. The density of such deformation bands increases rapidly when at least 30% of deformation is given, and it reduces sharply if the deformation is carried out at or above 1000 °C (Fig. 3.10). The increased density of deformation bands leads to the nucleation of very fine ferrite grains. Umemoto and Tamura [66] have estimated that a rolling deformation of 30% and 50% increases the ferrite nucleation rate per unit area of austenite grain surface by a factor of 740 and 4200, respectively. This finding can explain the behavior observed by early researchers [53, 54]. However, the kinetics of recrystallization is very fast for plain C-Mn steel, and the T_{NR} is also quite low where rolling operation is difficult due to the requirement of high rolling load. Addition of very small concentration of elements such as Nb, V, and Ti plays an important role here. The solute atoms and the strain-induced precipitates of carbide and/or carbonitrides of these elements retard the recrystallization of austenite. As a result T_{NR} goes up so that rolling can be performed at even high temperatures in order to get non-recrystallized deformed austenite grains at a lower rolling load [3] which in turn augments ferrite nucleation rate to give very fine ferrite grains in the final product [53, 54]. Nb is the most effective microalloying element that raises the T_{NR} sharply in concentrations up to 0.04 wt%, and therefore, it has the greatest influence on ferrite grain refinement [3, 27, 35].

However, the deformation need not be limited to single phase austenite region only. Research revealed that the deformation region can be extended to intercritical range $(\gamma + \alpha)$ as well [67]. Besides a continuation of the earlier processes, the deformation in this case introduces plastic strain in the ferrite. Subsequent recovery processes in the ferrite produce a sub-grain structure which is another source of

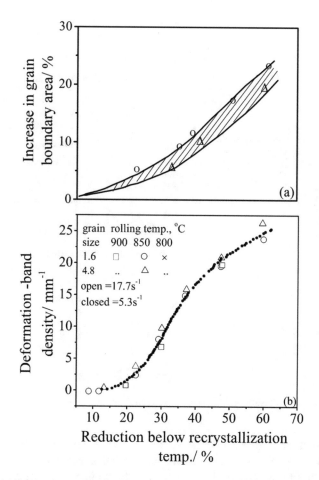

Fig. 3.10 (**a**) Variation in austenite grain boundary area and (**b**) introduction of deformation bands resulting from reduction in non-recrystallization region; in this 0.03 wt% Nb steel, initial grain size was varied by rolling in two passes at higher temperatures after reheating to 1250 °C [54, 56]

strengthening in these steels. In order to further augment the properties, the cooling is accelerated at a suitable temperature. This increases the extent of supercooling at which transformation can take place, and a highly refined structure can be obtained. Following Tanaka and co-workers, deformation during controlled rolling described above can be classified into:

(i) Deformation in recrystallized γ region
(ii) Deformation in non-recrystallized γ region
(iii) Deformation in $\gamma - \alpha$ two phase region, as illustrated schematically in Fig. 3.11 [53, 54]

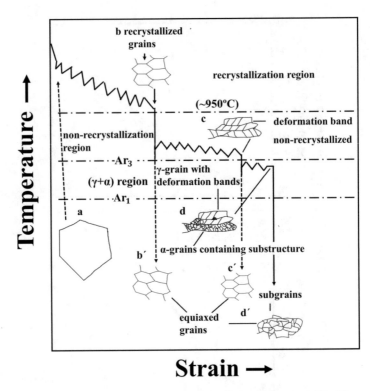

Fig. 3.11 Schematic illustration of three stages of controlled rolling process [53, 54]

The advent of different cooling strategies like water spray cooling, laminar flow cooling, etc. [3] after hot rolling has further enabled the metallurgists to extend the rolling regime by lowering down the coiling temperature (and hence the corresponding $\gamma \rightarrow \alpha$ transformation temperature) in order to obtain the desired microstructure. The thermomechanical processing can thus be divided into three distinct zones: rough rolling, finish rolling, and accelerated cooling with a possible "delay" period between rough and finish rolling. The rough rolling can be considered as the deformation given in the temperature region where recrystallization can take place. On the other hand, the finish rolling can be considered as the deformation given in the non-recrystallization or in the two-phase region. The combined extent of deformation in the two distinct stages of hot rolling is highly suitable for achieving fine grain and pancaked austenite through repeated recrystallization-deformation cycle and eventually results in fine ferrite grains [53]. Adoption of newer cooling strategies helps accelerated cooling of the rolled material on the run out table so quickly that the MA carbides/carbonitrides do not get sufficient time to be precipitated and remain in the solid solution. However, during holding of rolled material at the coiling temperature, precipitation of corresponding carbides/carbonitrides of Nb and/V takes place which leads to a great enhancement of strength. Considering the effects of such deformation on the final microstructure after rolling, Tamura et al. [53] have suggested the following:

- To use a low enough reheating temperature just to solutionize the MA carbides/carbonitrides but to restrict the significant coarsening of austenite grains;
- To allow a time delay between the roughing and finishing passes so that the temperature can be lowered below the recrystallization temperature;
- Lowering down the finish rolling temperature to close to 800 °C and increasing the total deformation up to 30% below 900 °C;

3.5 Welding of MA Steels

Welding is one of the most important fabrication techniques which provides the final structure by joining different smaller pieces of the materials. High-strength steels that are easily weldable can expand the targeted market for any industry sector: be it for line pipe sector where thousands of kilometers of pipeline are to be joined for installation or automotive sector where several smaller parts are to be welded to build up a full component. Ease of welding or good weldability is thus always a preferred parameter in the final selection of steel grade for a particular application. However, the improvements in weldability of carbon steels were not the initial incentives for the development of microalloyed steels, but the opportunities that microalloying give for improving the weldability by reducing the carbon content were fairly quickly recognized and became another major driving force for research [68].

For low-carbon microalloyed steels used for structural applications, among all the available welding processes, the submerged arc welding (SAW) process and the manual metal arc welding (MMAW) are the best suited processes [3]. The details of these welding processes can be found elsewhere [69].

From a metallurgical point of view, carbon equivalent (CE) is one of the most crucial factors that govern the weldability of the microalloyed steels. It represents the effects of carbon and other alloys on the hardenability of the steel. The severe effect of different alloying elements on the critical cooling rate after welding has been taken into account by the International Institute of Welding (IIW), and the equation covering a wide range of steels is shown in Eq. 3.16 below [3, 70]:

$$CE = C + \frac{Mn}{6} + \frac{Cr + Mo + V}{5} + \frac{(Ni + Cu)}{15} \qquad (3.16)$$

However, for modern low carbon steels, the modified equation (Eq. 3.17) has been suggested to be more appropriate [71]:

$$CE = C + \frac{Si}{30} + \frac{Mn + Cu + Cr}{20} + \frac{Ni}{60} + \frac{Mo}{15} + \frac{V}{10} + 5B \qquad (3.17)$$

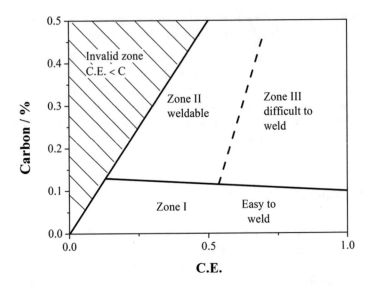

Fig. 3.12 Graville diagram differentiating the zones in terms of feasibility for welding [71]

The effect of composition on the weldability can be represented on Graville diagram where total carbon is plotted against the carbon equivalent, and the plot-area is divided into three regions marked as Zone I, Zone II, and Zone III according to the ease of weldability with Zone I representing the most easily weldable region and Zone III the most difficult one (Fig. 3.12) [72]. The above diagram further elaborates the beneficial effect of low carbon equivalents which permits significant improvement of the properties of the weld fusion zone and heat- affected zone (HAZ). Considering the chemical composition of microalloyed steels, usually, the carbon level does not exceed 0.1 wt. % [17] which indicates that the MA family of steels typically falls within Zone I, the easy-to-weld region.

Cracking in the HAZ is one of the oldest problems in welding. Different microalloying elements have a significant role in that respect as well. Vanadium has a decreasing effect on steel hardenability which leads to a reduction in risk of cracking [73, [74]]. However, in order to resist different forms of environmental cracking, HAZ hardening is important [69, 75, 76]. Some interesting trends have been reported by Hart and Harrison on both the positive and negative effects of vanadium on the HAZ hardness [69]. It was observed for plain C-Mn steel (without V addition) that higher HAZ hardness levels (> 325 VHN) can be achieved due to faster cooling rate in the weld zone, while longer cooling time is responsible for softer HAZ (hardness values ~250–300 VHN). Addition of vanadium was found to reduce the HAZ hardness values for faster cooling cycles but improve the hardness for slower cooling cycle compared with plain C-Mn steels. The reason can be explained from a viewpoint on the effect of vanadium on austenite to ferrite transformations. Incomplete and fine precipitation of vanadium carbide /nitride/carbonitride, as the steel passes through the high-temperature region quickly

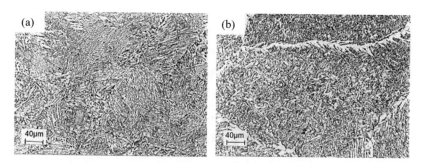

Fig. 3.13 HAZ microstructure with 5 kJ/mm heat input in steels having compositions (wt. %) of (**a**) 0.07C, 1.7Mn, 0.43Si, 0.026Al and (**b**) 0.07C, 1.7Mn, 0.43Si, 0.029Al, 0.16 V [72]

because of faster cooling rate, can pin down the austenite grain boundaries and prevent austenite grain growth. Thus, it can decrease the hardenability and thereby the formation of strong and hard phases like martensite and/bainite becomes less likely which is reflected in lower hardness of HAZ. Precipitate pinning is possible in case of fast cooling cycle in weld joints where there is a less chance for precipitate growth as well as dissolution. In case of slower cooling, the vanadium-added material spends comparatively more time in the high-temperature domain. This facilitates complete precipitation of carbides/nitrides/carbonitrides of vanadium which can enhance the HAZ hardness compared with the previous cycle [73].

The addition of V can further change the morphology of microstructure constituents in plain C-Mn steels from predominantly grain boundary nucleated products to a significant degree of intragranular nucleated products [77, 80]. Figure 3.13 depicts how the addition of V changes the morphology of microstructure in HAZ for identical heat input (5 kJ/mm). This type of microstructure in the HAZ of V-added steel is responsible for its improved crack propagation resistance and ductile to brittle transition temperature [74]. In this fashion, both HAZ and weld metal microstructures can be modified by the addition of vanadium [73].

On the contrary, Nb doesn't have any influence on CE, i.e., decreasing the weldability, as well as cold cracking due to hydrogen. Indeed it improves the performance of steel in this context [78]. This improvement comes with faster thermal cycles during welding, where fine distribution of niobium rich precipitates restricts grain growth in the HAZ resulting in a better microstructure (without hard martensite) by decreasing the hardenability of austenite. This benefit applies to all niobium-treated steels with an optimized composition of carbon <0.06 wt. % and niobium >0.07 wt. % irrespective of the processing schedules [79, 80]. Finer grain sizes in the HAZ result in improved microstructure and toughness. For a given chemical composition, finer austenite grain size tends to reduce hardenability which reduces the formation of brittle martensite after cooling in the weld zone and results in improvement of toughness. Addition of Nb can help maintain a good Charpy impact toughness over a large range of heat input as compared with steel without Nb addition as can be seen in Fig. 3.14 [81].

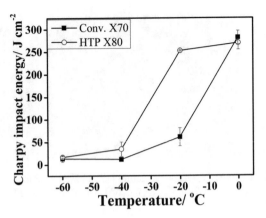

Fig. 3.14 HAZ Charpy impact toughness data from a 0.11 wt% Nb steel directly compared with that from a conventional X70 steel (weld thermal simulation data) [80]

3.6 Forming of MA Steels

One of the major aspects of steel that needs to be considered in order to give them the desired shape suitable for applications in different industry segments is the formability. Formability can be defined as the ability of a material to undergo plastic deformation without being damaged. The ability of plastic deformation in any polycrystalline materials like microalloyed steels is limited to a certain extent. Beyond that, the material could experience fracture.

Generally the term formability is applicable to sheet metal forming operations like deep drawing, cup drawing, bending, etc. where the complicated path of loading and subsequent complex deformation take place. Sometimes, one part can experience tensile, and the adjacent part can be under compressive loading or one surface can be under compression, while the other is in tension, etc. Uniaxial tensile tests are not a true representation of the formability of sheet metals due to the involvements of such complex path of deformation acting on the material during forming operations. Therefore, specific formability tests have been developed for more appropriate representation of actual loading conditions. The extent of formability depends on material parameters (like microstructure, anisotropy, etc.) and the loading conditions.

One of the key tests to understand and optimize the formability, especially for sheet metal, is the construction of forming limit diagram (FLD) where a grid of tiny circles or squares is first etched on the surface of the work piece followed by stretching over a dome-shaped die. In order to induce different state of stress (e.g., uniaxial, biaxial, or triaxial), different widths can be taken. Finally, the results are plotted as a function of major strain versus minor strain [82–88]. A representative FLD of a typical microalloyed steel (with nominal composition in wt%: C-0.07, Mn 0.3, Si 0.02, Cr 0.03, Mo 0.02, Ti 0.1, Al 0.05) is shown in Fig. 3.15 where the hatched region is the zone of safe working where no failure is expected. However, different modes of failure (wrinkling or shearing) can appear at different

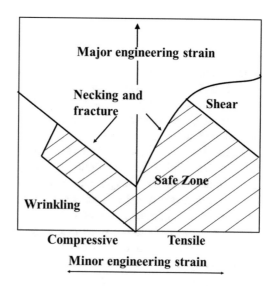

Fig. 3.15 A schematic forming limit diagram (FLD) for MA steel for a nominal composition (wt. %): C ~ 0.07, Mn ~ 0.3, Si ~ 0.02, Cr ~ 0.03, Mo ~ 0.02, Ti ~ 0.1, Al ~ 0.05 showing different modes of failure at different regions

combinations of strain outside this zone. The upper part of the safe zone represents necking and fracture.

A material with higher anisotropy is preferred in forming operation like deep drawing because of the difference in the straining conditions between the flange and the cup wall sections. Furthermore, the average flow stress is decreased with the increment of normal anisotropy [89]. Microstructural features (like volume fraction and morphology of different phases, their orientation, grain size, etc.), another set of parameters which can control the anisotropy and thus can influence the final formability to a great extent, are heavily dependent on the chemical composition of the steel as well as on the thermomechanical processing. For example, coiling temperature is an important parameter which can influence the formability of steel. Thus, the formability would deteriorate for Nb-microalloyed steel if it is processed at a low coiling temperature as that will increase the strength and alter the mechanism of strain aging [90]. On the contrary, V-microalloyed steels show good formability at lower coiling temperatures. This situation arises due to the different rate of removal of interstitial elements like C and N during the coiling treatment. In the vanadium grades, the precipitation temperatures of the corresponding carbides or nitrides are in the temperature range of the coiling operation. This leads to significant removal of carbon and nitrogen atoms from the matrix during coiling. On the other hand, such precipitation temperatures for Nb-microalloyed steel are comparatively higher. This phenomenon gives a good formability in V-microallyed steels compared with the Nb grades [90].

However, fine niobium carbide (NbC) refines the prior austenite grains leading to the promotion of uniform fine mixed ferrite-bainite matrix by decreasing the hardenability of austenite. This microstructure shows excellent stretch flangeability of the steel which is another measure of formability. Transformation-induced plasticity (TRIP) steels are also well recognized for their high formability without sacrificing strength due to the effect of deformation-induced transformation of retained austenite to martensite [91, 92]. It was observed that the addition of Nb in the range of only 0.08–0.11 wt% followed by austempering at low temperatures (300–350 °C) can improve the stretch flangeability in TRIP steels significantly. It is considered that the improvement in flangeability is associated with the refinement of austenite grain structure and an increase in volume fraction of proeutectoid ferrite in the final microstructure [93].

Grain size and grain orientation of the final microstructure also play important roles in determining the formability. It was observed that in Nb-Ti-added steels, good combination of strength and formability is a cumulative contribution of fine grain size and the significant intensity of the desired {332} < 112> texture component that nullifies the undesirable {110} < 011> texture components [94]. During some manufacturing situations, edge formability (defined as the ability of steel to be stamped into a part without failure by fracture or excessive thinning) is influenced by a number of factors which can include inclusions, microstructural banding, grain size distribution, and constituents of the microstructure. In such situations as well {332} < 112> texture component was proven as the most beneficial among [95] all the transformation textures in order to achieve good deep drawability without hampering strength and toughness.

3.7 Summary

In this chapter, the evolution and development of microalloyed steels have been presented. The roles of different microalloying elements have been discussed in terms of tailoring the microstructures. The concept of solubility product has been described in detail for calculation of precipitation of some common carbides and nitrides of typical microalloying elements (Nb, V, and Ti) at any temperature. Formation of different types of precipitates, ordinary or interphase, has been discussed briefly. Different types of strengthening mechanisms prevailing in this type of steel, their individual contribution toward final strength of the material, and how this can be quantified were shown in detail. Different microalloying elements actively participate in precipitation strengthening depending upon their formation temperatures which is governed by the solubility product of the elements along with the interstitials. On the other hand, these precipitates enhance grain boundary strengthening by pinning down the grain boundaries and restricting the grain growth. The industrial hot rolling for processing of the microalloyed steels, the evolution of microstructure, and the effect of different microalloying elements have been elaborated at length. Two major application-oriented characteristics, namely, the formability and weldability were discussed briefly to highlight the advantages of this genre of steel.

References

1. A.J, De Ardo, in *International Conference on "HSLA Steels 2015, Microalloying 2015 and Offshore Engineering Steels 2015", 11-13th November*, (Hangzhou, China), p. 2015
2. W.B. Morrison, Microalloy steels - the beginning. Mater. Sci. Technol. **25**, 1066–1073 (2009)
3. T. Gladman, *The Physical Metallurgy of Microalloyed Steels* (Maney Publishing, London, 2002)
4. C.L. Altenburger, F.A. Bourke, U.S. Patent number, US 3010822, Priority date **23**, 01 (1961)
5. R. Phillips, US Patent number: 3472707, Patented on 14.10. 1969, Priority date: 09.04.1964
6. M.S. Rashid, High-strength, low-alloy steels. Science **206**, 862–869 (1980)
7. A.J. De Ardo, M.J. Hua, K.G. Cho, C.I. Garcia, On strength of microalloyed steels: An interpretive review. Mater. Sci. Technol. **25**, 1074–1082 (2009)
8. G. Tither, *The Proceedings of the 2nd International Conference on "HSLA Steels: Processing, Properties and Applications"* (Beijing, China, 1990)
9. M.S. Rashid, GM 980x-a unique high strength sheet steel with superior formability. SAE Tech. Rep. **760206** (1976)
10. N. Ohashi, T. Irie, S. Satoh, O. Hashimoto, I. Takahashi, Development of cold rolled high strength steel sheet with excellent deep drawability. SAE Tech. Rep. **810027** (1981)
11. N. Hanai, N. Takemoto, Y. Takunaga, Y. Misuyama, Effect of grain size and solid solution strengthening elements on the bake hardenability of low carbon aluminium-killed steel. Trans. Iron Steel Inst. Jpn. **24**, 17–23 (1984)
12. S. Das, S.B. Singh, O.N. Mohanty, A phenomenological model for bake hardening in minimal carbon steels. Philos. Mag. **94**(18), 2046–2061 (2014)
13. S. Das, S.B. Singh, *O.N* (Mohanty, "Bake hardening" Published in Encyclopedia of Iron, Steel and Their Alloys (Taylor and Francis, 2016), pp. 306–319
14. Proceedings of Microalloying '75 International Symposium on High Strength Low Alloy Steels, 1-3rd October, 1975, Washington, DC
15. Second International Symposium on Microalloyed Bar and Forging Steel: 8-10 July (Golden, Colorado, 1996)
16. Proceedings of the International Symposium Niobium 2001, 2-5th December 2001, Orlando, USA
17. T.N. Baker, M.s. Ironmak, Steelmak. **43**, 264–307 (2016)
18. A.J. DeArdo, Niobium in modern steels. Int. Mater. Rev. **48**, 371–402 (2003)
19. R. Lagneborg, T. Siwecki, S. Zajac, B. Hutchinson, The role of vanadium in microalloyed steels. Scand. J. Metall. **28**, 186–241 (1999)
20. World Steel in Figures 2018. World Steel Association
21. D. Price, Materials World, August, 2016
22. G. Jha, S. Das, A. Lodh, A. Haldar, Mater. Sci. Eng. A **522**, 457–463 (2012)
23. G. Jha, S. Das, S. Sinha, A. Lodh, A. Haldar, Design and development of precipitate strengthened advanced high strength steel for automotive application. Mater. Sci. Eng. A **561**, 394–402 (2013)
24. S.Y. Han, S.Y. Shin, S. Lee, N.J. Kim, J.H. Bae, K. Kim, Effects of cooling conditions on tensile and charpy impact properties of api x80 linepipe steels. Metall. Mater. Trans. A 41A, 329–340 (2009)
25. F.B. Pickering, Microalloying 75, 1-3rd October (Union Carbide Corporation, New York, 1975)
26. K.W. Burns, F.B. Pickering, Deformation and fracture of ferrite-pearlite structures. J. Iron Steel Inst. 202, 899–906 (1964)
27. F.B. Pickering, *Physical Metallurgy and the Design of Steels* (Applied Science Publishers Ltd, London, 1978)
28. T. Yokota, C.G. Mateo, H.K.D.H. Bhadeshia, Formation of nanostructured steels by phase transformation. Scr. Mater. **51**, 767–770 (2004)

29. R.C. Reed, H.K.D.H. Bhadeshia, Kinetics of reconstructive austenite to ferrite transformation in low alloy steels. Mater. Sci. Technol. **8**, 421–435 (1992)
30. J.W. Christian, *The Theory of Transformation in Metals and Alloys* (Pergamon Press, Oxford, 1975)
31. L.S. Darken, R.W. Gurry, *Physical Chemistry of Metals* (McGraw- Hill Book Company, New York, 1953)
32. K. Narita, Physical chemistry of the groups iv a (ti, zr), va (v, nb, ta) and the rare earth elements in steel. Trans. Iron Steel Inst. Jpn. **15**, 145–152 (1975)
33. H. Lee, K.S. Park, J.H. Lee, Y.U. Heo, D.W. Suh, H.K.D.H. Bhadeshia, Dissolution behaviour of nbc during slab reheating. ISIJ Int. **54**, 1677–1681 (2014)
34. H.K.D.H. Bhadeshia, R.W.K. Honeycombe, *Steels: Microstructures and Properties* (Butterworth-Heinemann, Amsterdam, 2006)
35. W.C. Leslie, *The Physical Metallurgy of Steels* (McGraw-Hill Book Company, New York, 1981)
36. J. Irvine, T.N. Baker, The influence of rolling variables on the strengthening mechanisms operating in niobium steels. Mater. Sci. Eng. **64**, 123–134 (1984)
37. H. Kejian, T.N. Baker, The effects of small titanium additions on the mechanical properties and the microstructures of controlled rolled niobium-bearing hsla plate steels. Mater. Sci. Eng. **A169**, 53–65 (1993)
38. E.O. Hall, The deformation and ageing of mild steel: Iii discussion of results. Proc. Phys. Soc. B **64**, 747–753 (1951)
39. N.J. Petch, The cleavage strength of polycrystals. J. Iron Steel Inst. **174**, 25–28 (1953)
40. D.J. Dingley, D. McLean, Components of the flow stress of iron. Acta Metall. **15**, 885–901 (1967)
41. A. Cracknell, N.J. Petch, Frictional forces on dislocation arrays at the lower yield point in iron. Acta Metall. **3**, 186–189 (1955)
42. F.B. Pickering, T. Gladman, Metallurgical Developments in Carbon Steels (Iron and Steel Institute, London, 1963), pp. 10–20
43. Y. li, J.A. Wilson, D.N. Crowther, P.S. Mitchell, A.J. Craven, T.N. Baker, The effects of vanadium, niobium, titanium and zirconium on the microstructure and mechanical properties of thin slab cast steels. ISIJ Int. **44**, 1093–1102 (2004)
44. W.B. Morrison, R.C. Cochrane, P.S. Mitchell, The influence of precipitation mode and dislocation substructure on the properties of vanadium-treated steels. ISIJ Int. **33**, 1095–1103 (1993)
45. S.B. Singh, H.K.D.H. Bhadeshia, D.J.C. Mackay, H. Carey, I. Martin, Neural network analysis of steel plate processing. Ironmak. Steelmak. 25, 355–365 (1998)
46. A.T. Davenport, R.W.K. Honeycombe, Precipitation of carbides at γ–α boundaries in alloy steels. Proc. R. Soc. A **322**, 191–205 (1971)
47. K. Campbell, R.W.K. Honeycombe, Transformation from austenite in alloy steels. Met. Sci. **8**, 197–203 (1974)
48. G.M. Smith, *Ph.D Thesis* (Newnham College, Cambridge University, Cambridge, 1984)
49. R.A. Ricks, P.R. Howell, The formation of discrete precipitate dispersions on mobile interphase boundaries in iron-base alloys. Acta Metall. **31**, 853–861 (1983)
50. T. Sakuma, R.W.K. Honeycombe, Microstructures of isothermally transformed fe-nb-c alloys. Met. Sci. **18**, 449–454 (1984)
51. T. Sakuma, R.W.K. Honeycombe, Effect of manganese on microstructure of an isothermally transformed fe-nb-c alloy. Mater. Sci. Technol. **1**, 351–356 (1985)
52. A.D. Batte, R.W.K. Honeycombe, Strengthening of ferrite by vanadium carbide precipitation. Met. Sci. **7**, 160–168 (1973)
53. I. Tamura, H. Sekine, T. Tanaka, C. Ouchi, *Thermomechanical Processing of HSLA Steel* (Butterworth, London, 1988)
54. T. Tanaka, Controlled rolling of steel plate and strip. Int. Met. Rev. **26**, 185–212 (1981)
55. R. Bengochea, B. Lopez, I. Gutierrez, Influence of the prior austenite microstructure on the transformation products obtained for c-mn-nb steels after continuous cooling. ISIJ Int. **39**, 583–591 (1999)

56. S.B. Singh, H.K.D.H. Bhadeshia, Mater. Sci. Technol. **14**, 832–834 (1998)
57. I. Kozasu, C. Ouchi, T. Sanpei, T. Okita, *'Microalloying '75', 1-3rd October* (Union Carbide Corporation, New York, 1975)
58. R. Priestner, C.C. Earley, J.H. Rendall, Observations on behaviour of austenite during hot working of some low-carbon steels. J. Iron Steel Inst. **206**, 1252–1262 (1968)
59. K.J. Irvine, T. Gladman, J. Orr, F.B. Pickering, Controlled rolling of structural steels. J. Iron Steel Inst. 208, 717–726 (1970)
60. J.J. Jonas, I. Weiss, Effect of precipitation on recrystallization in micro alloyed steels. Met. Sci. **13**, 238–245 (1979)
61. H. Sekine, T. Maruyama, Retardation of recrystallization of austenite during hot-rolling in nb-containing low-carbon steels. Trans. Iron Steel Inst. Jpn. **16**, 427–436 (1976)
62. D.J. Walker, R.W.K. Honeycombe, Effects of deformation on the decomposition of austenite: Part 1–the ferrite reaction0 met. Sci. **12**, 445–452 (1978)
63. A. Sandberg, W. Roberts, *Thermomechanical Processing of Microalloyed Austenite* (TMS-AIME, Warrendale, 1982)
64. R.K. Amin, F.B. Pickering, *Thermomechanical Processing of Microalloyed Austenite* (TMS-AIME, Warrendale, 1982)
65. S.B. Singh, P.D. Thesis, *Darwin College* (Cambridge University, UK, 1998)
66. M. Umemoto, I. Tamura, *HSLA Steels: Metallurgy and Applications* (ASM International, Metals Park, 1986)
67. T. Gladman, D. Dulieu, Grain-size control in steels. Met. Sci. **8**, 167–176 (1974)
68. P.H.M. Hart, International Symposium on Vanadium Application Technology (Gui Lin, China, November, 2000)
69. L. Juffus, *Welding Principles and Applications*, 4th edn. (Delmar Publishers, New York, 1999)
70. H. Cotton, *The Contribution of Physical Metallurgy in Engineering Practice* (The Royal Society, London, 1976)
71. Y. Ito, *Weldability Formula of High Strength Steels: Related to Heat-Affected Zone Cracking* (International Institute of Welding, Paris, 1968)
72. R. Kurji, N. Coniglio, Towards the establishment of weldability test standards for hydrogen-assisted cold cracking. Int. J. Adv. Manuf. Technol. **77**, 1581–1597 (2015)
73. P.H.M. Hart, P.L. Harrison, The effect of vanadium on welds in structural and the toughness of pipeline steels. Weld. Int. **5**, 521–536 (1991)
74. P.L. Harrison, P.H.M. Hart, 2nd International Conference on Trends in Welding Research (ASM International, Gatlinburg, 1989)
75. N. Yurioka, T. Kasuya, A chart method to determine necessary preheat temperature in steel welding. Q. J. Jpn. Weld. Soc. **13**, 347–357 (1995)
76. N. Yurioka, Physical metallurgy of steel weldability. ISIJ Int. **41**, 566–570 (2001)
77. P.J. Boothby, P.H.M. Hart, Welding 0.45% vanadium linepipe steels', metal construction. Met. Constr. **13**, 560–569 (1981)
78. P. Kirkwood, *Proceedings of HSLA Steels 2015, Microalloying 2015 and Offshore Engineering Steels 2015* (Hangzhou, 2015)
79. A.D. Batte, P.J. Boothby, B. Rothwell, *Proceedings of the International Conference on Niobium: Science and Technology* (Orlando, 2001)
80. *Modern Niobium Microalloyed Steels and their Weldability, Technical Report (CBMM Report, 2018)*, pp. 1–8
81. F.J. Barbaro, Z. Zhu, L. Kuzmikova, H. Li, *Proceedings of the 10th International "Pipeline Conference IPC2014", Vol. 3, Paper no: V003T07A029* (Calgary, Canada, 2014)
82. N.J. Den Uijl, L.J. Carless, *Advanced Materials in Automotive Engineering* (Elsevier, 2012), pp. 28–56
83. A.A. Zadpoor, J. Sinke, *Failure Mechanisms of Advanced Welding Processes* (Elsevier, 2010), pp. 258–288
84. C.D. Horvath, *Materials, Design and Manufacturing for Lightweight Vehicles* (Elsevier, 2010), pp. 35–78

85. M. Ramezani, Z.M. Ripin, Analysis of deep drawing of sheet metal using the marform process. Int. J. Adv. Manuf. Technol. **59**, 491–505 (2012)
86. G. Centeno, A.J. Martínez-Donaire, D. Morales-Palma, C. Vallellano, M.B. Silva, P.A.F. Martins, *Materials Forming and Machining* (Elsevier, 2015), pp. 1–24
87. K.H. Chang, *Product Manufacturing and Cost Estimating Using Cad/Cae* (Elsevier, Amsterdam, 2013), pp. 133–190
88. S. Hashmi, *Comprehensive Materials Processing* (Elsevier Science, Amsterdam, 2014)
89. D. Banabic, H.-J. Bunge, K. Pöhlandt, A.E. Tekkaya, *Formability of Metallic Materials* (Springer, Berlin/Heidelberg, 2000)
90. R. Riva, C. Mapelli, R. Venturini, Effect of coiling temperature on formability and mechanical properties of mild low carbon and hsla steels processed by thin slab casting and direct rolling. ISIJ Int. **47**, 1204–1213 (2007)
91. K. Sugimoto, M. Mukherejee, in *Automotive Steels: Design, Metallurgy, Processing and Applications*, ed. by R. Rana, (Woodhead Publishing, Oxford, 2016)
92. T. Bhattacharyya, S.B. Singh, S. Das, A. Haldar, D. Bhattacharjee, Development and characterisation of c-mn-al-si-nb trip aided steel. Mater. Sci. Eng. A **528**, 2394–2400 (2011)
93. K.I. Sugimoto, T. Muramatsu, S.I. Hashimoto, Y. Mukai, Formability of nb bearing ultra-high-strength trip-aided sheet steels. J. Mater. Process. Technol. **177**, 390–395 (2006)
94. R.D.K. Misra, H. Nathani, F. Siciliano, T. Carneiro, Effect of texture and microstructure on resistance to cracking of high-strength hot-rolled nb-ti microalloyed steels. Metall. Mater. Trans. A **35**, 3024–3029 (2004)
95. R.K. Ray, J.J. Jonas, Transformation textures in steels. Int. Mater. Rev. **35**, 1–36 (1990)

Chapter 4
Advanced High-Strength Sheet Steels for Automotive Applications

Emmanuel De Moor

Abbreviations

AHSS	Advanced high-strength steels
AUST SS	Austenitic stainless steels
BH	Bake hardening
CCT	Continuous cooling transformation
CP	Complex phase
CS	Cold spot
DP	Dual phase
DQ	Drawing quality
EDS	Energy-dispersive spectroscopy
HER	Hole expansion ratio
HS	Hot spot
HSLA	High-strength low alloy
IA	Intercritical annealing
IF	Interstitial free
L-IP	Lightweight steels with induced plasticity
Low C	Low carbon
M	Martensite
PHS	Press-hardenable steels
PT	Partitioning temperature
Q&P	Quenching and partitioning
Q&T	Quench and tempered

E. De Moor (✉)
Advanced Steel Processing and Products Research Center (ASPPRC), George S. Ansell
Department of Metallurgical and Materials Engineering, Colorado School of Mines, Golden, CO,
USA
e-mail: edemoor@mines.edu

© Springer Nature Switzerland AG 2021
R. Rana (ed.), *High-Performance Ferrous Alloys*,
https://doi.org/10.1007/978-3-030-53825-5_4

QT	Quenching temperature
RA	Reduction in area
SAZ	Shear-affected zone
SFE	Stacking fault energy
TBF	TRIP bainitic ferrite
TE	Total elongation
TEl	Total elongation
TRIP	Transformation-induced plasticity
TS	Tensile strength
TWIP	Twinning-induced plasticity
UEl	Uniform elongation
YPE	Yield point elongation
YS	Yield strength

Symbols

M_s	Martensite start temperature
M_f	Martensite finish temperature
$M_{initial\ quench}$	Initial martensite fraction based on Koistinen-Marburger equation
M_d	Temperature at which no transformation of austenite occurs
M_d^{30}	Temperature at which 50 pct of the originally present austenite volume fraction transforms to martensite at an applied strain of 30 pct
M_s^{σ}	Deformation temperature where the yield strength of the austenite equals the required stress for transformation to occur
$f_{\alpha'}$	Fraction of martensite
ε_{true}	True strain
C_i	Initial alloy carbon content
C_γ	Austenite carbon content
C_m	Martensite carbon content
T_A	Isothermal annealing temperature
γ_{final}	Austenite fraction following Q&P heat treating and assuming "idealized" partitioning

4.1 Introduction

Substantial research and implementation efforts are underway toward the development and application of advanced high-strength steels (AHSS) for automotive applications. These steels enable low-cost strategies to vehicle lightweighting by use of thinner sections that guarantee vehicle crash-worthiness and occupant protection while contributing to increased fuel efficiency. Consumer expectations

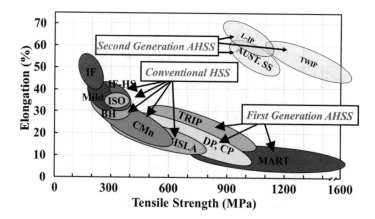

Fig. 4.1 Situation of tensile properties envelopes for conventional, first-generation AHSS transformation-induced plasticity (TRIP), dual-phase (DP), complex phase (CP), and martensitic (MART) steels, and second-generation twinning-induced plasticity (TWIP), lightweight steels with induced plasticity (L-IP), and austenitic stainless (AUST SS) steel grades on a total elongation versus tensile strength diagram [6, 7]

and regulatory pressure are driving increased emphasis on fuel efficiency and reduced tailpipe emissions [1–5].

The steel industry has responded to these needs and challenges by developing a variety of AHSS exhibiting varying strength and formability, alloy compositions, and microstructures. An overview of various AHSS grades is depicted in Fig. 4.1 on a total elongation versus tensile strength diagram [6, 7]. Dual-phase (DP), transformation-induced plasticity (TRIP), complex (CP), or multiphase steels are examples of AHSS commercially available at a variety of strength levels. The nomenclature has developed to further classify the grades into subcategories – first and second generation AHSS with DP, TRIP, CP and martensitic steels making up the first generation. These grades are characterized by lean alloying and heat-treating strategies to develop high-strength and high-ductility microstructures. First-generation AHSS have made substantial inroads into vehicle architectures as reflected in the pie charts shown in Fig. 4.2 for 2004 and 2017 steel usage breakdown for the same vehicle model, namely the Chevrolet Malibu [8, 9]. A reduction in the overall use of mild steel and replacement with AHSS is clearly observed. Lightweight metals also see an increase albeit at a more modest level in this selected example. Other vehicle architectures have chosen more intensive use of lightweight metals, in particular, aluminum for the body-in-white (BIW) [10–12].

The so-called second-generation AHSS employ high alloying levels to develop fully austenitic microstructures exhibiting high strength and exceptional ductility. Examples of these types of steels are twinning-induced plasticity (TWIP) and austenitic stainless [13]. Overall these grades have seen much less widespread implementation, predominantly due to increased cost and difficulties in processing. This chapter will review AHSS typically used for structural components in vehicle

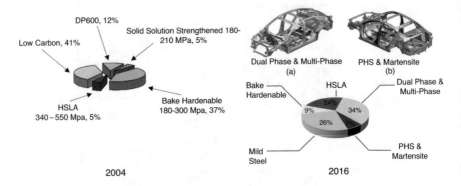

Fig. 4.2 Comparison of breakdown of steel grades used in a Chevrolet Malibu over time from 2004 to 2016 [8, 9]

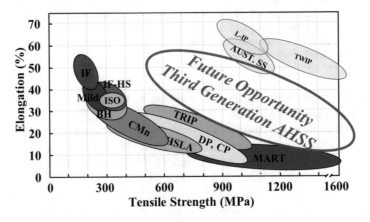

Fig. 4.3 Situation of the future opportunity third-generation AHSS tensile properties' envelope with respect to other steel grades on an elongation versus tensile strength diagram [6, 7]

architectures. A review of automotive steels for closures such as interstitial free (IF), bake hardening (BH), and drawing quality (DQ) grades can be found elsewhere [14].

More recently research and development efforts have been dedicated toward the development of so-called third-generation AHSS [6, 15]. This generation of steels is being developed to address the properties gap situated between the first-generation AHSS (i.e., DP, TRIP, CP, etc.) and martensitic steels, and the fully austenitic second-generation AHSS as shown in Fig. 4.3 by use of leaner chemistries than typically employed for the later-generation AHSS.

The above-discussed steel grades enable or are being developed for stamping operations where no additional thermal processing is required and the stamping takes place at room temperature. Another route which has seen widespread implementation is the so-called hot stamping or press hardening where the material is deformed at high temperature and accelerated cooled to set the final high-strength martensitic microstructure. This chapter will review various AHSS grades in addition to hot stamping. Alloying, microstructures, and processing will be discussed.

4.2   Dual-Phase Steels

Dual-phase (DP) steels [16–18] have seen the most widespread use as AHSS in current vehicle architectures as illustrated in Fig. 4.2 [9]. The microstructure consists of ferrite and martensite as shown in Fig. 4.4 [19], and varying volume fractions of martensite are developed by using different reheating temperatures, which result in varying achievable strength levels and tensile properties. A range of strength levels can be achieved and are commercially available, e.g., DP590, DP780, DP980, and DP1180. The heat treatment involves reheating in the intercritical phase field and cooling to form martensite. Figure 4.5 shows processing thermal paths for cold-rolled (Fig. 4.6a) and hot-rolled (Fig. 4.6b) DP steels [20].

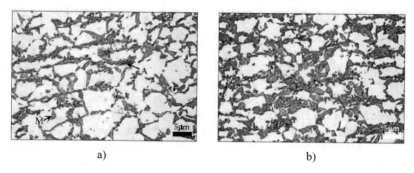

Fig. 4.4   Microstructures of DP steels quenched from (a) 775 °C and (b) 800 °C. Martensite (M) is straw color and ferrite is white, fine-retained austenite (A) also indicated [19]

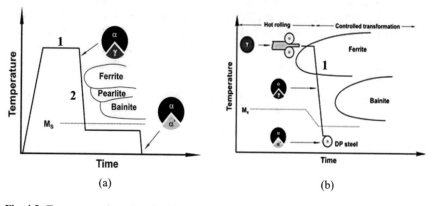

Fig. 4.5   Temperature-time plots for (a) cold-rolled and (b) hot-rolled DP steels [20]

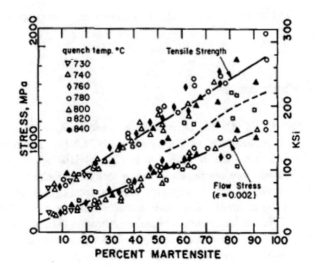

Fig. 4.6 Tensile strength and 0.2 pct offset flow stress as a function of percent (by volume) martensite for various Fe-C-Mn alloys and quenched from different intercritical annealing temperatures [22]

Judicious alloying is required to set hardenability for the desired microstructural requirement toward mechanical properties, and the levels of required alloying are processing dependent. Carbon, manganese, chromium, and molybdenum are prevalent additions [21]. Some fractions of bainite may be present depending on the effectiveness of the cooling step and the corresponding alloy design. Dual-phase steels are usually characterized by lean alloying, in particular, compared to TRIP steels and good weldability results, which has contributed to the widespread use of DP steels. Furthermore, the change in the ferrite fraction present at the intercritical annealing temperature allows for flexibility in setting strength levels. Data compiled by Davis [22] shown in Fig. 4.6 for a variety of Fe-C-Mn chemistries constituting in particular carbon alloying modifications show that strength levels increase with martensite level present or, conversely reduced intercritical ferrite levels. As shown in Fig. 4.6, the 0.2 pct offset flow stress and tensile strength for a variety of intercritical annealing temperatures and DP chemistries increase with increasing martensite fraction.

An example engineering stress-strain curve of a dual-phase steel is shown in Fig. 4.7 [23] and is compared to the engineering stress-strain curves of a plain carbon and high-strength low-alloy (HSLA) steel. Clearly greater tensile strength levels develop in a DP steel compared to the plain carbon predominantly ferritic steel and in this example similar to the included HSLA steel. The yielding behavior and strain hardening of the DP steel are however clearly different. Yielding at strength levels intermediate between the yield strength of the HSLA and plain carbon steel is followed by substantial strain hardening to equivalent ultimate tensile strength levels as the HSLA steel in this example. Round-house yielding behavior

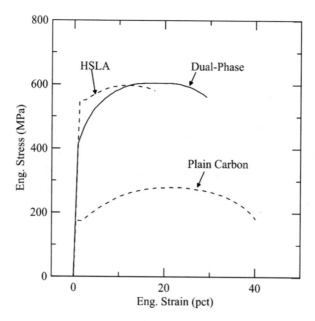

Fig. 4.7 Engineering stress-strain curves for plain carbon, HSLA, and DP steels. (Redrawn from [23])

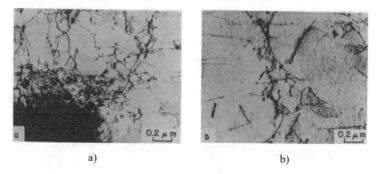

Fig. 4.8 TEM micrographs of DP microstructures showing dislocations (**a**) near a martensite-ferrite interface and (**b**) in a single ferrite grain away from the interface [24]

is observed characteristic of DP steels whereas the HSLA and plain carbon steels show yield point elongations. The round-house yielding results from the presence of mobile dislocations in DP microstructures introduced in the microstructure from the martensitic shear transformation and volume expansion–generating dislocations in the ferrite which remain unpinned during quenching and are mobile upon tensile testing. Figure 4.8 shows transmission electron micrographs of a DP steel where dislocations are abundantly present near the martensite-ferrite interface whereas a much lower density is observed away from the interface in the ferrite [24].

4.3 Transformation-Induced Plasticity Steels

Another family of first-generation AHSS is the transformation-induced plasticity or TRIP steels albeit these steels have seen less widespread applications in vehicle designs compared to DP steels. The TRIP effect refers to the transformation upon straining of meta-stable austenite [25], which positively impacts strain hardening and postpones necking to higher levels of strain. Metastable austenitic stainless steels such as AISI 304 exhibit the TRIP effect, and martensite progressively develops in the initial fully austenitic microstructure with increased strain resulting in increased strain hardening and superior formability [25]. Low-carbon TRIP steels for automotive BIW applications were developed to employ the TRIP effect in leaner and more cost-effective alloy chemistry. For instance, nickel is usually not added and manganese is added in limited quantities for hardenability rather than for austenite retention purposes. An austenite fraction of 10–20 vol pct is retained at room temperature by enrichment with carbon, which takes place during bainitic transformation. A heat-treating schematic to produce a low-carbon TRIP steel is provided in Fig. 4.9 [26].

Following reheating in the intercritical temperature range, the steel is accelerated cooled and held at a temperature in the bainitic transformation regime to develop a microstructure containing intercritical ferrite, bainitic ferrite, and retained austenite. As carbon is supersaturated in the bainitic ferrite, it needs to be rejected from the ferrite to alleviate the supersaturation. In the absence of alloying with Si and/or Al, this is done by the precipitation of cementite in the bainitic laths or in between the laths. The supersaturation can also be alleviated by stimulating the enrichment of the austenite remaining in the microstructure that requires alloying with Si and/or Al to suppress the formation of cementite due to its insolubility in cementite which reduces its growth kinetics [27]. Cementite suppression results in the remaining carbon available to enrich in austenite and as carbon is rejected from the growing

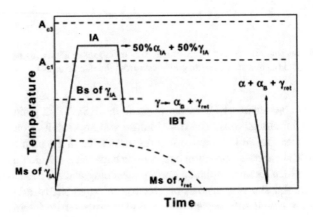

Fig. 4.9 Schematic of the thermal cycle for low-carbon TRIP steel processing, from [26]

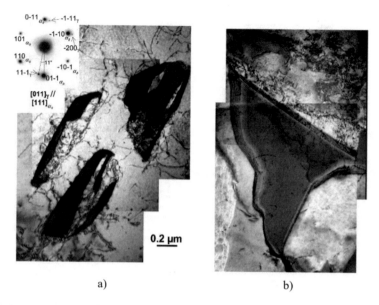

Fig. 4.10 Low-carbon TRIP steel microstructures showing (**a**) laths of bainitic ferrite and retained austenite and (**b**) a blocky retained austenite grain. Reproduced from [28]

bainitic laths, it enriches the remaining austenite and stabilizes it by lowering its martensite start temperature (M_s) as illustrated schematically in Fig. 4.9.

Representative TRIP steel microstructures are shown in Fig. 4.10 [28] where bainitic laths can be observed containing dislocations and where the darker laths were identified as austenite in Fig. 4.10a. A triangular grain of austenite is also included in Fig. 4.10b. The austenite is metastable, i.e., the austenite is retained at room temperature due to carbon enrichment but will transform to martensite when strain (or stress) is applied and is referred to as strain-induced (or stress-assisted) transformation. The degree of strain-induced transformation is strain dependent, and the evolution of the martensite fraction is shown in Fig. 4.11 as a function of true strain for three TRIP steel chemistries containing Si, Al, and a combination thereof [28]. The strain-induced martensite evolution with respect to strain exhibits a sigmoidal behavior that has been plotted using the Olson-Cohen model which assumes martensite embryos to form following the generation of nucleation sites such as shear-band intersections in the form of mechanical twins, dense stacking-fault bundles, and ε martensite [29]. A martensitic embryo is then able to nucleate on these intersections as more mechanical work is imparted. The sigmoidal behavior has been fitted with the Olson-Cohen model as expressed by according to [29]:

$$f_{\alpha'} = 1 - \exp\left[-\beta\left[1 - \exp\left(-\alpha\varepsilon\right)\right]^n\right] \qquad (4.1)$$

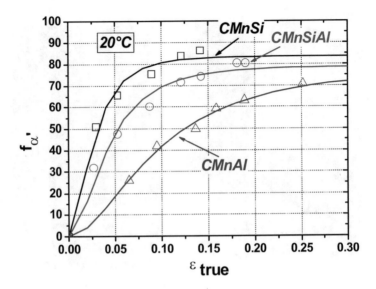

Fig. 4.11 Strain-induced martensite fraction ($f_{\alpha'}$) as a function of true strain (ε_{true}) for 0.24C-1.61Mn-1.45Si (CMnSi), 0.25C-1.70Mn-0.55Si-0.69Al (CMnSiAl), and 0.22C-1.68Mn-1.49Al (CMnAl) steels, plotted from data in [28]. Steel chemistries are in wt pct

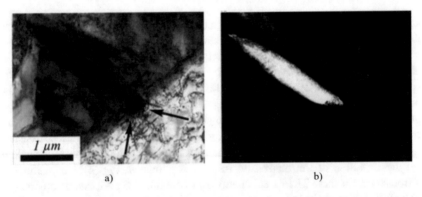

Fig. 4.12 (**a**) Bright-field and (**b**) dark-field TEM micrographs of a mechanical partially transformed austenite grain in a TRIP steel. Dark-field reflection taken from a martensite spot to illuminate the strain-induced martensite. From [30]

where $f_{\alpha'}$ is the fraction of martensite; α, β, and n are fitting parameters with α related to the generation of nucleation sites; β relates to the driving force for the $\gamma \rightarrow \alpha_M$ transformation; and n is a stereological parameter. Figure 4.12 shows micrographs of a TRIP steel following deformation with Fig. 4.12a showing a bright field TEM micrograph and Fig. 4.12b a corresponding dark field micrograph where reflection from a martensite/ferrite spot was used to illuminate the strain-induced martensite formed within the austenite [30]. It can further be noted that dislocations

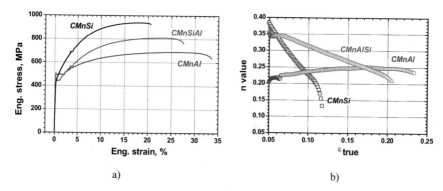

Fig. 4.13 (**a**) Engineering stress-strain curves and (**b**) corresponding instantaneous strain hardening (n-value) as a function of true strain for 0.24C-1.61Mn-1.45Si (CMnSi), 0.25C-1.70Mn-0.55Si-0.69Al (CMnSiAl), and 0.22C-1.68Mn-1.49Al (CMnAl) steels [31]. Steel chemistries are in wt pct

were generated at the strain-induced martensite/ferrite boundary in the ferrite. These dislocations help to harden the steel and improve strain hardening. Engineering stress-strain curves and corresponding instantaneous strain hardening (n-value) are shown from three TRIP steels in Fig. 4.13 [31]. Note that the corresponding retained austenite evolution with strain was presented in Fig. 4.11 for these steels. The degree of improved strain hardening is dependent on the austenite stability as reflected by comparing Figs. 4.13 and 4.11. It can be seen that the austenite in the CMnSi steel shows the least stability, i.e., the greatest rate of martensite formation with strain (Fig. 4.11) resulting in a continuously decreasing rate of strain hardening (Fig. 4.13). The CMnAl steel exhibits the slowest transformation with strain, i.e., greatest mechanical stability, and results in n-values increasing with strain up to high levels that postpone necking to greater levels of strain, and the highest uniform elongation is observed for this steel compared to the other steels. The CMnSiAl alloy exhibits intermediate mechanical stability of the austenite and strain hardening.

In addition to the M_s temperature as a measure for the thermal stability of the austenite against transformation to martensite, the M_s^σ, M_d^{30}, and M_d temperatures have been introduced [32–36] to assess the mechanical stability of the austenite against transformation upon mechanical loading. A schematic of stress required for austenite to martensite transformation to occur as a function of temperature situating these various temperature parameters is shown in Fig. 4.14 [37]. Undercooling below the M_s temperature is sufficient to induce martensite transformation without the application of load or requiring deformation to transform the austenite. At temperatures above the M_s temperature, the austenite may be transformed by loading the material. If transformation occurs prior to the yielding of austenite, the transformation is stress assisted, and increased stress levels are required with increasing testing temperatures. At greater temperatures, yielding of the austenite will occur prior to martensite formation, and the transformation is referred to as strain induced. The M_s^σ temperature delineates the stress-assisted from the

Fig. 4.14 Schematic
representation of (a) stress
required for transformation to
occur as a function of
temperature where the M_S,
M_S^σ, and M_d temperatures are
identified [37], (b)
engineering stress-strain
curve when stress-assisted
transformation occurs, and (c)
yielding associated with
strain-induced transformation
[31]

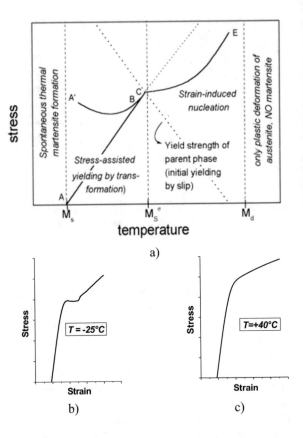

strain-induced transformation temperature range. Stress-assisted transformation of austenite into martensite occurs upon applying stress, and martensite nucleates on preexisting nucleation sites; transformation occurs prior to plastic deformation and yielding of the austenite. Increased stress levels are required for transformation with temperature at which the material is loaded. Strain-induced transformation is preceded by yielding of the austenite and the generation of nucleation sites for martensite. The M_S^σ temperature corresponds to the deformation temperature where the yield strength of the austenite equals the required stress for transformation to occur. If the steel is loaded at temperatures greater than the M_S^σ temperature, the stress required to induce austenite transformation exceeds the yield strength of the austenite, and, therefore, plastic deformation of the austenite occurs prior to transformation. Increased stress levels for transformation are again required with increasing testing temperature albeit a nonlinear relationship between the required stress and temperature developing. The transition from stress-assisted to strain-induced transformation is characterized by a change in yielding behavior where a plateau is observed in the stress-assisted transformation temperature regime below M_S^σ and round-house yielding at temperatures exceeding M_S^σ as illustrated in Fig. 4.14. The plateau in the stress-strain curves directly relates to the transformation from austenite into martensite, which is associated with a sudden volume increase. The difference in yielding behavior enables determining the M_S^σ temperature experimentally using the single-specimen temperature-variable tension test (SS-TV-TT) [33, 34]. At a certain temperature, referred to as M_d, no transformation of austenite occurs, and only plastic deformation exists. An intermediate temperature, the so-called M_d^{30} temperature, has been defined as the temperature at which 50 pct of the originally present, i.e., in the unstrained condition, austenite volume fraction will transform to martensite at an applied strain of 30 pct. The M_d^{30} temperature is a measure of the overall austenite stability where a high M_d^{30} temperature is related to low mechanical stability of the austenite and has been shown to be related to the chemical composition of the austenite, in particular, carbon content and overall TRIP steel chemistry [28].

As vehicles in service may experience deformation at high strain rates (~102/s) during, e.g., vehicle collisions, it is important to assess tensile properties at strain rates far exceeding typical quasi-static strain rates employed in tensile tests (on the order of 10^{-4}/s). These strain rates are so elevated that insufficient time is available for heat dissipation and adiabatic heating may occur. The austenite transformation is susceptible to relatively small changes in temperature, and adiabatic heating of the steel results in reduced transformation behavior, i.e., greater austenite stability. This in turn affects the dynamic stress-strain behavior and amount of energy absorbed during a crash. A comparison of energy absorption during dynamic testing as a function of dynamic tensile strength for a variety of steel grades is shown in Fig. 4.15 [38]. The energy values were obtained by assessing the surface area under dynamic stress-strain curves generated on a split-Hopkinson bar test in tension [38, 39]. TRIP steels show superior behavior compared to structural, low-carbon (low C),

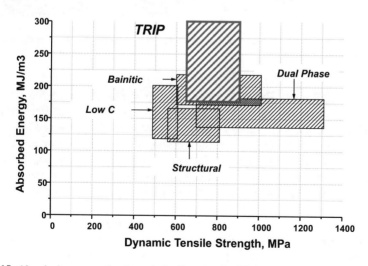

Fig. 4.15 Absorbed energy under dynamic loading situating TRIP steel performance with respect to structural, low-carbon (low C), and dual-phase steels [38]

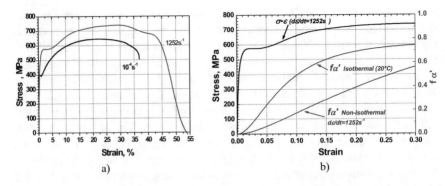

Fig. 4.16 (**a**) Engineering stress-strain curves obtained by testing at a quasi-static strain rate of 10^{-4}/s and dynamic at 1252/s by split-Hopkinson bar testing in tension, (**b**) strain-induced transformation as a function of strain for isothermal testing at room temperature and calculated effect of adiabatic heating during dynamic testing on martensite evolution superimposed on the dynamic stress-strain curve shown in (**a**) [40]

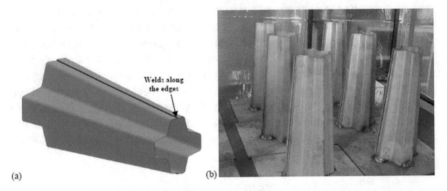

Fig. 4.17 (**a**) Schematic of a 12-sided component for drop-weight testing, and (**b**) welded components prior to testing [41]

bainitic, and DP steels. Examples of engineering stress-strain curves under quasi-static and dynamic testing are shown in Fig. 4.16a [40]. Superior strength levels, elongation, and absorbed energy develop under dynamic loading. Figure 4.16b shows the calculated effect of adiabatic heating on the strain-induced martensite evolution with strain during dynamic loading. Reduced kinetics are clearly observed compared to the martensite evolution at room temperature, and these reduced kinetics are believed to be involved in the superior dynamic energy absorption by TRIP steels as shown in Fig. 4.15.

In addition to dynamic tensile properties, components may be tested to assess crashworthiness and enable comparison of steel grade performance for crumple zone applications. To this end, components may be tested as shown in Fig. 4.17 [41] in a drop tower apparatus where a weight is dropped from a prescribed height and the reduction in height after axial crash testing is assessed. Photographs of the

a) HSLA 440
22.6 cm crush distance

b) DP 590
18.4 cm crush distance

c) TRIP 780
13.6 cm crush distance

Fig. 4.18 Components after crash testing showing the response of (**a**) HSLA 440, (**b**) DP 590, and (**c**) TRIP 780 [41]. The crush distances representing the height difference between the component prior and after crash testing are also shown. (Adapted from [41])

components following interaction with the weight are shown in Fig. 4.18 where various steel grades are compared namely HSLA, DP, and TRIP steels at a variety of strength levels. The crush distance, i.e., the difference between the height of the component prior to impact and after impact, is also indicated where the lower strength HSLA material shows the greatest compaction following the weight drop, followed by DP and TRIP steels. Further analysis of this test can include assessment of crack formation in the crushed component.

4.4 TRIP-Bainitic Ferrite Steels

A further development from low-carbon TRIP steels are the so-called TRIP-bainitic ferrite (TBF) steels which again rely on the TRIP effect for improved work hardening and tensile elongation at strength levels exceeding typical levels observed for low-carbon TRIP steels. The difference in strength results from reduced or absent intercritical ferrite volume fractions. A heat-treating schematic for TBF steels is shown in Fig. 4.19, where reheating following cold rolling is in the fully austenitic rather than the intercritical region [42]. The reheating step is followed by isothermal holding and quenching to room temperature. The isothermal hold can be done at a variety of temperatures and results in austempering or the introduction of some initial martensite (depending on the temperature) followed by further austenite decomposition in low-temperature transformation products. TRIP steel compositions are suited for TBF processing, and retained austenite is present in resulting microstructures stabilized by carbon enrichment resulting predominantly from a bainitic transformation. TBF steels show attractive properties as shown in Fig. 4.20 which plots tensile properties as a function of holding temperature (referred to as annealing temperature, T_A, in Fig. 4.19) for a fixed holding time of 200 s. Increased strength levels evolve with lower holding temperatures along with reduced total and uniform elongation. TBF steels are attractive third-generation AHSS as evident from their mechanical properties.

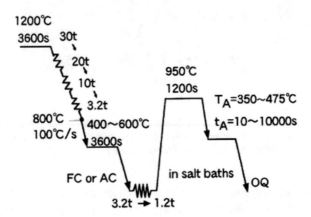

Fig. 4.19 Processing schematic detailing hot rolling, cold rolling, and annealing for a 0.20C-1.51Si-1.51Mn (wt pct) TBF steel [42]

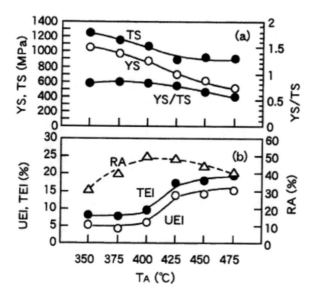

Fig. 4.20 Yield strength (YS), tensile strength (TS), their ratio (YS/TS), uniform (UEl), and total elongation (TEl) reduction in area (RA) as a function of isothermal annealing temperature (T_A) shown in Fig. 4.19 and a holding time of 200 s [42]

4.5   Quenching and Partitioning Steels

Another important family of third-generation AHSS constitutes the quenching and partitioning (Q&P) steels [43–45]. Q&P steels have a martensitic microstructure containing film-like retained austenite. Intercritical ferrite may also be present. A heat-treating and microstructural evolution schematic of Q&P is provided in Fig. 4.21. Reheating can be performed in the intercritical region or result in full austenitization as shown in the schematic and is followed by quenching at a temperature intermediate between the martensite start and finish temperatures. At this so-called quenching temperature (QT), the microstructure consists of martensite and any remaining austenite. The latter phase is stabilized by carbon enrichment during a secondary step at the same or a higher temperature. This secondary heat-treating step is referred to as the partitioning step, which aims at carbon enrichment of austenite by carbon transfer from the martensitic matrix. Quenching to room temperature may result in some additional martensite formation depending on the level of austenite stabilization and carbon enrichment.

Example micrographs of Q&P microstructures are shown in Fig. 4.22 [47, 48]. The microstructure consists of fine martensitic laths with a high dislocation density and austenite with a film-like appearance, separating the martensite laths as shown in the dark-field TEM micrograph in Fig. 4.22b illuminated using an austenite reflection. The bright-field image included in Fig. 4.22c clearly also shows transition carbides present. Quantitative analysis using Mössbauer spectroscopy

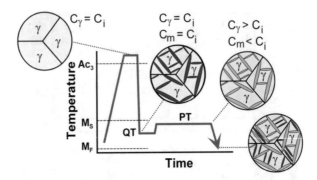

Fig. 4.21 Schematic illustrating the quenching and partitioning heat-treating process. QT is the quenching temperature, PT is the partitioning temperature and M_s and M_f are the martensite start and finish temperatures, respectively. C_i, C_γ, and C_m refer to the carbon contents of the initial alloy, austenite, and martensite, respectively [46]

[49] has indicated that lower levels of η transition carbides are observed in Q&P versus quench and tempered (Q&T) steels for the same partitioning/tempering conditions. This indicates that carbon supersaturation of martensite is relieved by carbide formation in Q&T steels whereas carbon transport to austenite and its resulting retention are predominantly operating in Q&P steels. Retained austenite levels superior to levels obtained following Q&T were also observed in this study [49]. These results suggest that both carbon partitioning from martensite into austenite and carbide precipitation are active mechanisms toward alleviating carbon supersaturation of martensite.

The effect of the quenching temperature (QT) on the fraction of austenite that can be retained assuming full carbon depletion of martensite and enrichment into the austenite is shown in Fig. 4.23. A maximum fraction is obtained for an "optimal" QT. This optimum QT is obtained by considering the fraction of martensite forming as a function of the offset with the M_s temperature and the carbon available in the formed martensite. Assumption of full carbon depletion of the martensite or "idealized" partitioning enables determination of the optimum QT and corresponding fraction [50]. Further details on the methodology and calculation can be found in [50].

Example engineering stress-strain curves are shown in Fig. 4.24 for a 0.3C-3Mn-1.6Si (wt pct) alloy processed by Q&P heat treating following full austenitization. It should be noted that this alloy chemistry is richer than traditional TRIP steel chemistries. The latter have been employed extensively to study Q&P heat-treating responses as they are suited for Q&P processing given the presence of cementite retarding alloying elements such as Si. High ultimate tensile strength levels approximating 1500 MPa and total elongation over 16 pct are apparent in the engineering stress-strain curves shown in Fig. 4.24. The curves show round-house yielding with high yield strengths and substantial amounts of strain hardening.

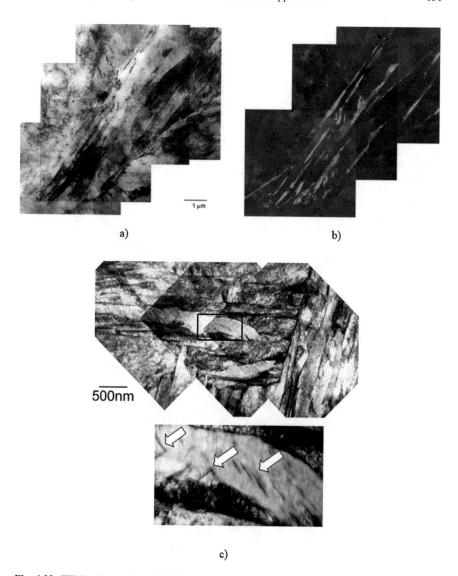

a)   b)

c)

Fig. 4.22 TEM micrographs of Q&P microstructures of a Fe-0.20C-1.63Mn-1.63Si (wt pct) composition QT: 240 °C, PT: 400 °C, Pt: 120 s. (**a**) Bright-field and (**b**) dark-field images taken using an austenite reflection, (**c**) bright-field image of a Q&P microstructure showing the presence of transition carbides (QT: 240 °C, PT: 300 °C, Pt: 120 s) [47, 48]. *QT* quenching temperature, *PT* partitioning temperature, *Pt* partitioning time

Fig. 4.23 Predicted phase fractions for a 0.19C steel containing 50 vol pct intercritical ferrite versus quenching temperature. The austenite fraction following Q&P heat treating and assuming "idealized" partitioning is shown in bold as γ_{final}. The initial martensite fraction is shown based on the Koistinen-Marburger equation as $M_{initial\ quench}$ and corresponding remaining austenite at this stage as $\gamma_{initial\ quench}$. The martensite forming during final quenching due to insufficient carbon stabilization is shown as $M_{final\ quench}$ [50]

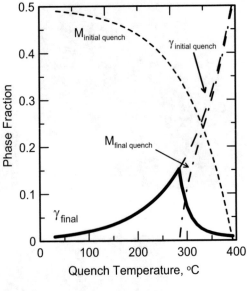

Fig. 4.24 Engineering stress-strain curves obtained in Fe-0.3C-3Mn-1.6Si (wt pct) alloy Q&P heat treated following full austenitization using a quenching temperature of 200 °C and the indicated partitioning conditions [51]

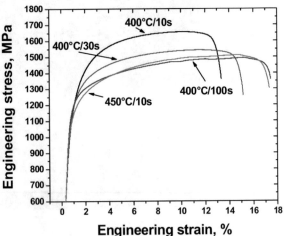

4.6 Medium Manganese Steels

Another third-generation AHSS concept are the medium manganese steels typically containing 5–12 wt pct Mn and are based on a concept proposed by Miller [52]. The heat treating for these steels consists of reheating and holding in the intercritical region to establish fractions of ferrite and austenite and to stimulate diffusion of manganese (and carbon) solute from ferrite into austenite, thereby stabilizing it to room temperature. Depending on the effectiveness of the heat treatment, all of the austenite present at high temperature can be retained to room temperature.

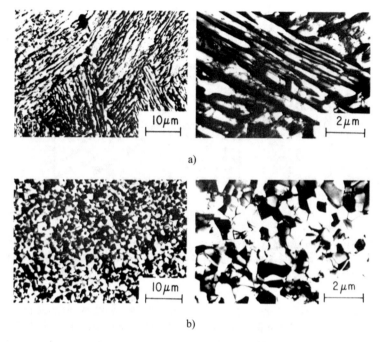

Fig. 4.25 TEM micrographs of a 0.11C-5.7Mn steel intercritically annealed at 600 °C for 1 hr. after (**a**) hot rolling, and (**b**) cold rolling [52]

An optimal intercritical annealing temperature can be calculated assuming full partitioning of solute between ferrite and austenite [53]. TEM micrographs of medium manganese steel microstructures are shown in Fig. 4.25 where intercritical annealing was done following hot rolling (Fig. 4.25a) and cold rolling (Fig. 4.25b), respectively, at a temperature of 600 °C for 1 h. Lath-like ferrite and austenite are apparent following the application of the intercritical heat treatment after hot rolling reminiscent of the martensitic hot band microstructure whereas equiaxed grains of ferrite and austenite develop following intercritical annealing of the cold rolled condition. Low dislocation densities in the ferrite are observed in both cases and far less than observed in a (cold-rolled) martensitic microstructure.

The intercritical annealing heat treatment has been explored using simulated and industrial batch annealing and continuous annealing. Thermal profiles shown in Fig. 4.26 representing batch annealing have been applied to enable effective intercritical annealing with solute partitioning approximating equilibrium levels [54]. More recently, cycles reflective of continuous annealing have been pursued, and austenite was effectively retained at room temperature resulting from solute enrichment as indicated by the TEM micrograph shown in Fig. 4.27 [55]. Energy-dispersive spectroscopy (EDS) analysis along a line scan on the micrograph clearly indicates manganese enrichment in the austenitic region following short time annealing of 180 s in the intercritical temperature range.

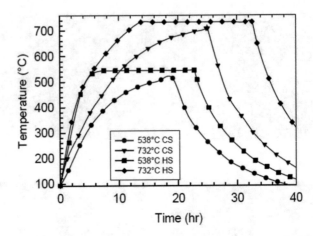

Fig. 4.26 Example annealing cycles employed in [54] to assess intercritical annealing of medium manganese steels using batch annealing. Cold-spot (CS) and hot-spot (HS) temperature-time evolution illustrated

Austenite amount and stability are important for tensile properties. Figure 4.28a shows the mechanical stability of retained austenite measured in a 0.1C-7Mn (wt pct) steel intercritically annealed for 1 week at a variety of temperatures [56]. Corresponding engineering stress-strain curves using the same color coding are included in Fig. 4.28b, and a family of stress-strain curves develops. Both mechanical stability of retained austenite and tensile behaviors are greatly dependent on the employed annealing temperature. The lowest intercritical annealing (IA) temperature of 575 °C shows a large yield point elongation (YPE) followed by limited strain hardening and intermediate elongation. At a higher IA temperature, the YPE decreases, and substantial strain hardening follows leading to very high elongation levels (e.g., 600 °C). IA at an even greater temperature of 675 °C results in a markedly different shape of the stress-strain curve where YPE is absent, yielding occurs at a much lower stress level, and very rapid strain hardening results in failure at much lower elongation levels. The corresponding retained austenite evolution with strain indicates that the mechanical stability of the austenite substantially impacts the stress-strain behavior, in particular, strain hardening where intermediate stability results in substantial strain hardening and high elongation levels. Low stability (e.g., at an IA of 650 °C) results in very pronounced strain hardening over a limited strain range resulting in the high ultimate tensile strength levels but low elongation levels.

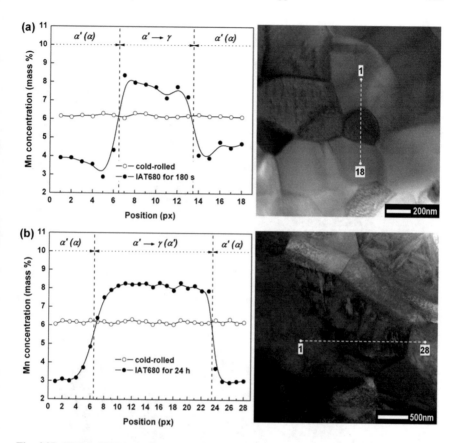

Fig. 4.27 EDS and TEM micrographs of a Fe-0.05C-6.15Mn-1.4Si-0.04Al (wt pct) steel intercritically annealed at 680 °C for (**a**) 180 s and (**b**) 24 hr. [55]

4.7   Press-Hardenable Steels

Another important industrial technology is press hardening or hot stamping that enables the manufacturing of high-strength martensitic components with complex geometries. An example BIW where press-hardenable steels (PHS) are widespread is shown in Fig. 4.29. A number of antiintrusion components such as door beams, pillars, etc., employ PHS steels. Press hardening circumvents the challenge of reduced ductility associated with higher strength levels by forming complex parts at high temperatures where high ductility exists in the austenitic phase field, followed by accelerated cooling in the dies to set the final, martensitic microstructure exhibiting high strength that will enhance vehicle performance in anti-intrusion components. Unlike the prior steels discussed, the properties of press-hardenable steels are set following shaping of the component. Hot stamping is particularly instrumental for components requiring the use of dies with tight radii that may

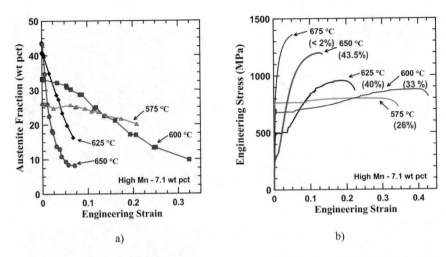

Fig. 4.28 Mechanical stability of retained austenite and tensile behavior of a 0.1C-7.1Mn (wt pct) steel intercritically annealed for 168 hours at various IA temperatures. (**a**) Austenite fraction as a function of engineering strain and (**b**) corresponding engineering stress-strain curves with initial austenite fractions (vol pct) indicated in brackets [56]

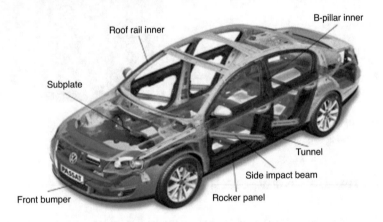

Fig. 4.29 BIW showing application of PHS steels in various locations [59]

induce failures at the edge of the die and where springback is of concern [57, 58]. PHS processing can be done with cooling taking place immediately following forming using the same die sets that remain closed to accelerate cool the part following sheet forming. Alternately, it can be done in two steps where forming is done in one set of dies, and the formed component is transferred to a secondary die set which will cool the part and where the final microstructure will be set. Schematics of both processing routes are shown in Fig. 4.30 and are referred to as direct and indirect press hardening, respectively [59].

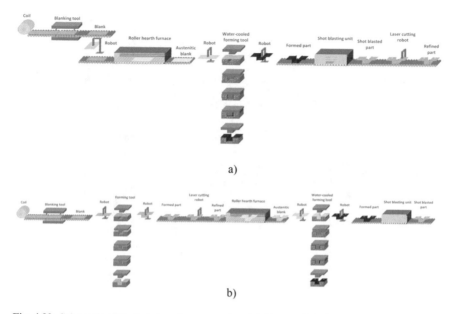

Fig. 4.30 Schematic of the hot stamping processing: (**a**) direct and (**b**) indirect [59]

In order to establish a fully martensitic high-strength microstructure, hardenability needs to be compatible with cooling rates obtained by quenching in water-cooled dies. To this end, alloying with boron is widespread, and the boron is protected from interaction with nitrogen by tying up the nitrogen into a more stable nitride such as a titanium nitride. Solute boron will segregate to prior austenite grain boundaries and poison nucleation sites for nonmartensitic transformation products. A common alloy is referred to as 22MnB5 with example chemical composition range 0.20–0.25C,1.10–1.40Mn,0.15–0.35Si,0.15–0.35Cr,0.02–0.05Ti,0.002–0.005B (wt pct) [59]. A representative continuous cooling transformation (CCT) diagram for 22MnB5 is shown in Fig. 4.31 [60, 61]. From the CCT diagram, it can be shown that the cooling rate must exceed 27 °C/s to avoid ferritic and/or bainitic transformation products.

4.8   TWIP Steels

Second-generation AHSS employ greater levels of alloying to stabilize an austenitic microstructure. As shown in Fig. 4.1, stainless steels properties fall in the same band as the second-generation AHSS properties envelope. Twinning-induced plasticity (TWIP) steels are also second-generation AHSS with a fully austenitic microstructure stabilized by additions of manganese and carbon to substitute for more expensive nickel [13, 62–68]. Engineering stress-strain curves for TWIP steels

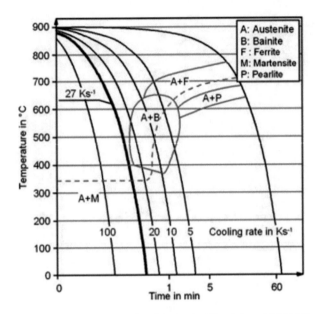

Fig. 4.31 Continuous cooling transformation diagram for 22MnB5 grade [60, 61]

are compared to first-generation AHSS, a low-carbon (low C) steel with a ferrite-pearlite microstructure, and a fully martensitic (M220) steel in Fig. 4.32 [69]. Very pronounced strain hardening results in very high tensile elongation, and the exhibited tensile properties are exceptional when compared to leaner compositions. The TWIP steels exhibit ultimate tensile strengths of approximately 1000 MPa. It should be noted that relatively low yield strengths are observed especially compared to yield strengths exhibited by third-generation AHSS such as Q&P steels. Example TWIP steel compositions are shown in Table 4.1. Heat treating following cold rolling consists of annealing to recrystallize the austenitic microstructure and set the final grain size [70, 71].

The stacking fault energy (SFE) is a critical parameter governing the stress-strain behavior, tensile properties, and strain hardening of TWIP steels. Ranges for the intrinsic stacking fault energy for various TWIP steel alloying concepts and associated deformation mechanisms, such as phase transformations, deformation twinning, and dislocation glide and cell formation, are shown in Fig. 4.33 [13, 82–85].

Transmission electron micrographs of deformed TWIP steels are shown in Fig. 4.34 [86] where dislocation cell-like structures develop in a Fe-30Mn (wt pct) steel whereas extensive twinning is observed in a Fe-22Mn-0.6C (wt pct) alloy, and the latter steel exhibits superior strain hardening and tensile behavior resulting from mechanical twin formation [86].

As shown in Fig. 4.33, phase transformations may occur depending on alloying and associated stacking fault energy. An example of α' martensite nucleated at the

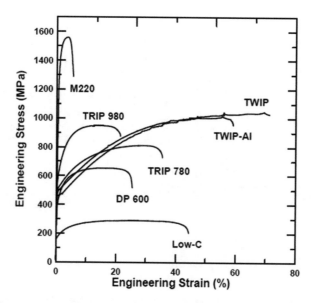

Fig. 4.32 Comparison of engineering stress-strain curves of first-generation AHSS DP and TRIP steels: a low-carbon (low-C) mild steel, a fully martensitic M220, and two TWIP steels [69]

Table 4.1 Example TWIP steel compositions

wt pct	Ref.
Fe-(18–24)Mn-(0.5–0.7)C	[72, 73]
Fe-18Mn-0.6C-(1.5–2.5)Al	[13, 74]
Fe-(15–27)Mn-(0.4–0.7)C-(1.3–3.5)Al-(0–3)Si	[75–80]
Fe-(17–22)Mn-(0.45–0.60)C-(0.1–0.3)V	[13, 81]
Fe-22Mn-0.6C-(0.1–0.3)Ti	[81]
Fe-18Mn-0.6C-0.31Nb	[81]

intersection of two ε martensite bands in a Fe–18Mn–0.25C–0.084 N (wt pct) alloy is shown in Fig. 4.35 [87].

The exploitation of phase transformations in high manganese steels has led to the investigation of TRIP/TWIP alloys [62]. Figure 4.36 shows the tensile response of Fe-xMn-3Al-3Si (wt pct) alloys with varying Mn contents ranging from 15 over 20 to 25 wt pct [62]. Reduced Mn alloying results in an inflection in the stress-strain curves associated with the activation of abundant austenite transformation. The Mn content of the alloy exhibiting this behavior is leaner than typical TWIP steel compositions.

TWIP steels have seen fairly limited application to date as the rich alloying to stabilize the austenite makes their widespread implementation cost prohibitive. Furthermore, the application of TWIP steels has been hindered in the past due to sensitivity to hydrogen embrittlement [88–90] observed following, e.g., cup drawing tests as illustrated in Fig. 4.37 [91] albeit this embrittlement sensitivity is not necessarily revealed during tensile testing in hydrogen charged specimens [69, 92].

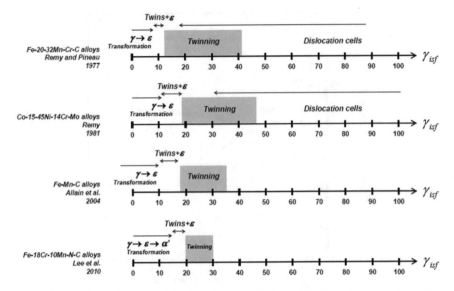

Fig. 4.33 Overview of ranges of intrinsic stacking fault energies for phase transformation, deformation twinning, and dislocation glide in austenitic Fe-(20–32)Mn-Cr-C alloys, Co-(15–45)Ni-14Cr-Mo alloys, Fe-Mn-C TWIP steels, and Fe-18Cr-10Ni-C-N alloys [13]. Based on data compiled from Remy and Pineau [82], Remy [83], Allain et al. [84] and Lee et al. [85]. (Elemental concentrations are in wt pct)

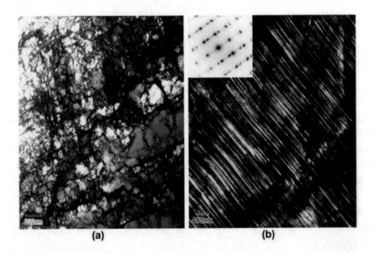

Fig. 4.34 TEM micrographs showing (**a**) a bright-field image showing well-developed dislocation cell-like structures in a Fe-30Mn (wt pct) steel deformed up to 20 pct in tension and (**b**) a dark-field image of a Fe-22Mn-0.6C (wt pct) deformed to 50 pct showing extensive mechanical twinning [86]

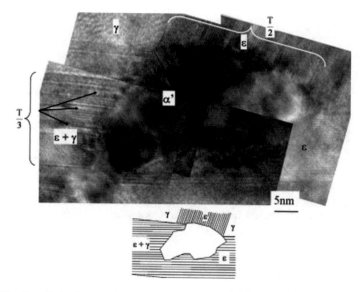

Fig. 4.35 α'martensite formed on the intersection of two ε martensite bands in a Fe–18Mn–0.25C–0.084 N (wt pct) alloy [87]

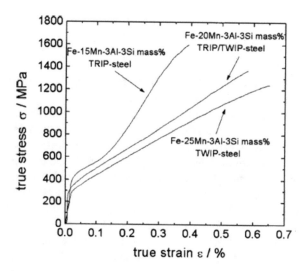

Fig. 4.36 True stress versus true strain curves showing the effect of Mn content in Fe-(15–25)Mn-3Al-3Si (wt pct) alloys [62]

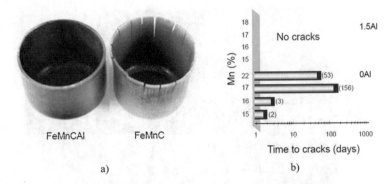

Fig. 4.37 (**a**) Delayed cracking observed following cup drawing in a Fe-Mn-C TWIP steel but not in a Fe-Mn-C-Al composition. (**b**) Time to cracks in days as a function of Mn and Al alloying [91]

Alloying with aluminum has alleviated this concern [91, 93], which is believed to be related to the suppression of ε martensite formation and due to stacking fault energy increases and reduced H segregation to mechanical twin interfaces and ε martensite [94, 95]. Figure 4.37 shows the absence of (delayed) cracking in a 1.5 wt pct Al-added TWIP steel following cup drawing.

4.9 Low-Density Steels

In order to further reduce the weight of AHSS, low-density steels are being explored that incorporate high levels of aluminum alloying in the range of 5–13 wt pct to reduce the density of steel and increase the specific strength (strength/weight ratio) [96–99]. Figure 4.38 shows the measured density of binary Fe-Al alloys as a function of Al content where a clear density reduction develops with increased Al levels [100]. Al additions have been explored in high and medium manganese ferritic, austenitic, and multiphase steels. Intermetallic formation such as Fe_3Al and FeAl has been observed to negatively impact ductility in the Fe-Al binary system, and, therefore, Al levels have typically been restricted to 10 wt pct [101, 102]. Al additions to Fe-Mn-C alloys have been studied extensively, and $(Fe,Mn)_3AlC$ carbide referred to as κ carbide develops in these alloys [101]. A summary of tensile properties of low-density steels is provided in Fig. 4.39 on an ultimate tensile strength versus total elongation diagram [96]. To date low-density steels have not seen significant application in automotive architectures.

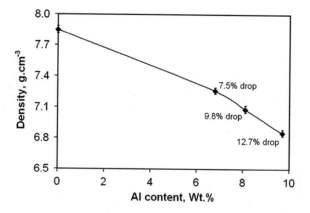

Fig. 4.38 Experimentally measured density as a function of Al content in binary Fe-Al alloys [100]

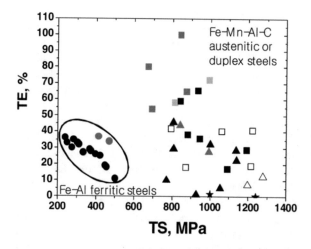

Fig. 4.39 Tensile total elongation (TE) as a function of tensile strength (TS) for various low-density steels [96]

4.10   Local Formability

The above sections have reviewed various AHSS automotive steel grades with particular emphasis on alloying, microstructure, and resulting tensile properties – an initial measure for global formability. Tensile properties measured at strain rates from quasistatic to dynamic are important in vehicle design and performance simulations and help predict achievable lightweighting levels. In order to enable implementation of these steel grades in vehicle architectures, further properties relevant to component manufacturing and in service performance assessment are required. These include formability testing, spring back, bending under tension

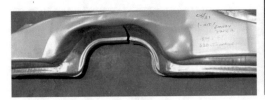

Fig. 4.40 Cracks observed in formed parts initiated at sheared edges [103]

testing, fatigue performance, bake hardening, and liquid metal and hydrogen embrittlement. A detailed review of these various aspects is beyond the scope of this chapter, and the discussion will be limited here to hole expansion testing or stretch flangeability, a measure for "local" formability. Examples of failed parts where cracks initiated at sheared edges are shown in Fig. 4.40 [103, 104].

In order to study shear edge stretching or stretch flangeability, the hole expansion test has been developed. The test consists of preparing a circular hole in a steel sheet by punching, and the hole is then expanded over a flat or conical die. The diameter expansion is measured following development of a through thickness crack. Manual measurements and automated camera systems are being used to measure the expansion. A schematic of the hole expansion test and samples expanded to failure are shown in Fig. 4.41 [105, 106]. A summary of hole expansion ratios (HER) defined as the percentage expansion of the initial diameter to failure is shown in Fig. 4.42a [107, 108]. In general, decreasing HER or reduced "local" formability is observed with increased ultimate tensile strength for most steel grades except for DP steels that deviate from the relationship developed for a variety of predominantly single-phase steel microstructures. Figure 4.42b shows the hole expansion ratio of DP steels as a function of ferrite content present in the microstructure, and an overall decreasing trend is apparent whereas uniform elongation increases with ferrite content [109]. Furthermore, it has been shown that, in general, HER increases with a reduced hardness ratio between the hardness of the martensite over the ferrite hardness [110].

The effect of the shearing process prior to forming is important and can be assessed by measuring the shear affected zone (SAZ), which is the material adjacent the sheared face work hardened during shearing [111, 112]. The degree of work hardening during shearing and, therefore, also the size of the SAZ relates to the degree of work hardening exhibited by the steel during tensile testing.

4.11 Summary

This chapter provides an overview of various advanced high-strength steels for structural applications in vehicle architectures. These AHSS enable lightweighting of vehicles via down gauging of steel components. Various families of AHSS exist

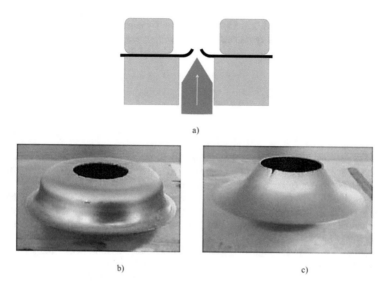

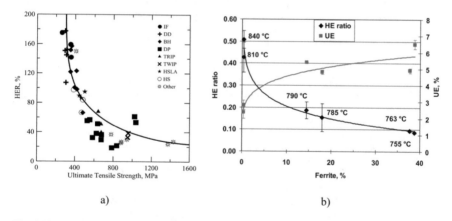

Fig. 4.41 (**a**) Schematic of hole expansion test showing the punched and expanded sample (black) and the conical die expanding the sample, and samples expanded to failure using (**b**) a flat die and (**c**) a conical die [105, 106]

Fig. 4.42 (**a**) Hole expansion ratio (HER) as a function of ultimate tensile strength for a variety of steel grades. Figure from [108] based on data from [107], (**b**) hole expansion ratio (left axis) and uniform elongation (UE-right axis) as a function of ferrite content in DP steels [109]

including DP and TRIP steels, which are examples of so-called first-generation AHSS; TWIP steels are of the second generation; and TBF, Q&P, and medium manganese steels among other concepts are being explored as third-generation AHSS. These various steels require different alloying and heat-treating strategies that were reviewed along with their microstructural characteristics. Substantial development and implementation efforts are underway toward increased usage of AHSS in vehicle architectures.

Acknowledgments The support of the sponsors of the Advanced Steel Processing and Products Research Center, an NSF industry-university cooperative research center at Colorado School of Mines, is gratefully acknowledged. Brandon Blesi is gratefully acknowledged for his work on redrawing Fig. 4.7.

References

1. Setting emission performance standards for new passenger cars as part of the Community's integrated approach to reduce CO_2 emissions from light-duty vehicles, Regulation (EC) No 443/2009 of the European parliament and of the council, April 23, 2009
2. Environmental Protection Agency (EPA), Department of Transportation (DOT), National Highway Traffic Safety Administration (NHTSA), 2017 and Later Model Year Light-Duty Vehicle Greenhouse Gas Emissions and Corporate Average Fuel Economy Standards. USA Fed. Reg. **77**, 62624–63200 (2012)
3. European Environment Agency, Average CO_2 emissions from newly registered motor vehicles, (2018)
4. Regulation (EU) No 510/2011 of the European Parliament and of the Council of 11 May 2011 setting emission performance standards for new light commercial vehicles as part of the Union's integrated approach to reduce CO_2 emissions from light-duty vehicles, Official Journal of the European Union, pp. L 145/1–18, 2011
5. J. Miller, N. Lutsey, *Consumer Benefits of Increased Efficiency in 2025–2030 Lightduty Vehicles in the U.S* (International Council on Clean Transportation, 2017)
6. "Advanced High Strength Steel (AHSS) Application Guidelines", International Iron & Steel Institute, Committee on Automotive Applications, available online at www.worldautosteel.org
7. D.K. Matlock, J.G. Speer, Design Considerations for the Next Generation of Advanced High Strength Sheet Steels, in *Proceedings of the 3rd International Conference on Structural Steels*, ed. by H. C. Lee, (The Korean Institute of Metals and Materials, Seoul, Korea, 2006), pp. 774–781
8. C.D. Horvath, J.R. Fekete, Opportunities and Challenges for Increased Usage of Advanced High Strength Steels in Automotive Applications, in *International Conference on Advanced High Strength Sheet Steels for Automotive Applications Proc*, ed. by J. G. Speer, (AIST, Warrendale, PA, 2004), pp. 3–10
9. C.D. Horvath, C.M. Enloe, J.P. Singh, J.J. Coryell, Persistent Challenges to Advanced High-Strength Steel Implementation, in *Proceedings, International Symposium on New Developments in Advanced High-Strength Sheet Steels*, ed. by E. De Moor, N. Pottore, G. Thomas, M. Merwin, J. G. Speer, (AIST, Warrendale, PA, 2017), pp. 1–10
10. W.S. Miller, L. Zhuang, J. Bottema, A.J. Wittebrood, P. De Smet, A. Haszler, A. Vieregge, Recent development in aluminium alloys for the automotive industry. Mater. Sci. Eng. A **280**, 37–49 (2000)
11. J. Hirsch, Recent development in aluminium for automotive applications. Trans. Nonferrous Met. Soc. China **24**, 1995–2002 (2014)
12. J.C. Benedyk, Aluminum alloys for lightweight automotive structures, in *Materials, Design and Manufacturing for Lightweight Vehicles*, ed. by P. K. Mallick, (Woodhead Publishing Limited, 2010), pp. 79–113
13. B.C. De Cooman, Y. Estrin, S.K. Kim, Twinning-induced plasticity (TWIP) steels. Acta Mater. **142**, 283–362 (2018)
14. R. Rana, S.B. Singh (eds.), Automotive Steels-Design, Metallurgy, Processing and Applications, Woodhead Publishing Series in Metals and Surface Engineering (2017)
15. D.K. Matlock, J.G. Speer, Third Generation of AHSS: Microstructure Design Concepts, in *Microstructure and Texture in Steels*, ed. by A. Haldar, S. Suwas, D. Bhattacharjee, (Springer,

London, 2009), pp. 185–205

16. A. T. Davenport (ed.), *Formable HSLA and Dual-Phase Steels* (TMS-AIME, Warrendale, PA, 1977)

17. R. A. Kot, J. W. Morris (eds.), *Structure and Properties of Dual-Phase Steels* (TMS-AIME, Warrendale, PA, 1979)

18. R. A. Kot, B. L. Bramfitt (eds.), *Fundamentals of Dual-Phase Steels* (TMS-AIME, Warrendale, PA, 1981)

19. A.K. De, J.G. Speer, D.K. Matlock, Color tint-etching for multiphase steels. Adv. Mat. Proc. **161**, 27–30 (2003)

20. R. Kuziak, R. Kawalla, S. Waengler, Advanced high strength steels for automotive industry. Arch. Civ. Mech. Eng. **8**, 103–117 (2008)

21. N. Fonstein, Advanced high strength sheet steels-physical metallurgy, design, processing, and properties. Springer

22. R.G. Davies, Influence of Martensite composition and content on the properties of dual phase steels. Metall. Trans. A. **9**, 671–679 (1978)

23. M.S. Rashid, B.V.N. Rao, Tempering Characteristics of a Vanadium-Containing Dual-Phase Steel, in *Fundamentals of Dual-Phase Steels*, ed. by R. A. Kott, B. Bramfitt, (TMS-AIME, Warrendale, PA, 1981), pp. 249–264

24. D.A. Korzekwa, D.K. Matlock, G. Krauss, Dislocation substructure as a function of strain in a dual-phase steel. Metall. Mater. Trans. A **15**, 1221–1228 (1984)

25. V. Zackay, E. Parker, D. Fahr, R. Busch, The enhancement of ductility in high strength steels. Trans. Am. Soc. Met. **60**, 252–259 (1967)

26. J. Mahieu, B.C. De Cooman, J. Maki, Phase transformation and mechanical properties of si-free CMnAl transformation-induced plasticity-aided steel. Metall. Mater. Trans. A **33**, 2573–2580 (2002)

27. W.S. Owen, The effect of silicon on the kinetics of tempering. Trans. Am. Soc. Met. **46**, 812–829 (1954)

28. L. Samek, E. De Moor, J. Penning, B.C. De Cooman, Influence of alloying elements on the kinetics of strain-induced martensitic nucleation in low-alloy, multiphase high-strength steels. Metall. Mater. Trans. A **37**, 109–124 (2006)

29. G.B. Olson, M. Cohen, Kinetics of strain-induced Martensite nucleation. Metall. Trans. A. **6**, 791–795 (1975)

30. P.J. Jacques, Q. Furnémont, F. Lania, T. Pardoen, F. Delannay, Multiscale mechanics of TRIP-assisted multiphase steels: I. Characterization and mechanical testing. Acta Mater. **55**, 3681–3693 (2007)

31. E. De Moor, L. Samek, J. Penning and B.C. De Cooman, Unpublished research

32. J.R. Platel, M. Cohen, Criterion for the action of applied stress in the Martensite transformation. Acta Metall. **1**, 531–538 (1953)

33. G.N. Haidemenopoulos, M. Grujicic, G.B. Olson, M. Cohen, Transformation microyielding of retained austenite. Acta Metall. **37**, 1677–1682 (1989)

34. A.N. Vasilakos, K. Papamantellos, G.N. Haidemenopoulos, W. Bleck, Experimental determination of the stability of retained austenite in low alloy TRIP steels. Steel Res. Intl. **70**, 466–471 (1999)

35. G.B. Olson, Mechanically-Induced Phase Transformation in Alloys, in *Encyclopedia of Materials: Science and Technology*, ed. by M. B. Bever, (Pergamon Press, Cambridge, MA, 1986), pp. 2929–2932

36. L. Barbé, M. De Meyer, B.C. De Cooman, *Determination of the Mσ Temperature of Dispersed Phase TRIP-Aided Steels, in: International Conference on TRIP-Aided High Strength Ferrous Alloys* (Gent, Belgium, 2002)
37. L. Barbé, M. De Meyer, B.C. De Cooman, Determination of the M_S^σ temperature of dispersed phase TRIP-aided steels, in *Proceedings International Conference on TRIP-Aided High Strength Ferrous Alloys*, ed. by B. C. De Cooman, (GRIPS, 2002), pp. 65–70
38. J. Mahieu, PhD thesis, Ghent University
39. J. Van Slycken, P. Verleysen, J. Degrieck, J. Bouquerel, B.C. De Cooman, Dynamic response of aluminium containing TRIP steel and its constituent phases. Mater. Sci. Eng. A **460–461**, 516–524 (2007)
40. L. Samek, E. De Moor, B.C. De Cooman, J. Van Slycken, P. Verleysen, J. Degrieck, Quasi-adiabatic Effects During the High Strain Rate Deformation of Dispersed-phase Ferrous Alloys with Strain Induced Martensitic Transformation, in *Intl. Conf. on Advanced High Strength Sheet Steels for Automotive Applications Proceedings*, ed. by AIST, (2004), pp. 361–374
41. T.M. Link, G. Chen, Anisotropy Effects in the Axial Crash Behavior of Advanced High-Strength Steels, in *Proceedings International Symposium on New Developments in Advanced High-Strength Sheet Steels*, (AIST, Warrendale, PA, 2013), pp. 63–70
42. K.-I. Sugimoto, T. Iida, J. Sakaguchi, T. Kashima, Retained austenite characteristics and tensile properties in a TRIP type Bainitic sheet steel. ISIJ Int. **40**, 902–908 (2000)
43. J.G. Speer, D.K. Matlock, B.C. De Cooman, J.G. Schroth, Carbon partitioning into austenite after Martensite transformation. Acta Mater. **51**, 2611–2622 (2003)
44. A.M. Streicher, J.G. Speer, D.K. Matlock, B.C. De Cooman, *Proceedings of the International Conference on Advanced High Strength Sheet Steels for Automotive Applications* (AIST, Warrendale, PA, June 6–9, Winter Park, CO, 2004), pp. 51–62
45. D.V. Edmonds, K. He, F.C. Rizzo, B.C. De Cooman, D.K. Matlock, J.G. Speer, Mater. Sci. Eng. **A438-440**, 25–34 (2006)
46. D.K. Matlock, V.E. Bräutigam, J.G. Speer, Application of the quenching and partitioning (Q&P) process to a medium-carbon, high-Si microalloyed Bar steel. Mater. Sci. For. **426–432**, 1089–1094 (2003)
47. E. De Moor, S. Lacroix, A.J. Clarke, J. Penning, J.G. Speer, Effect of retained austenite stabilized via quench and partitioning on the strain hardening of martensitic steels. Metall. Mater. Trans. A **39**, 2586–2595 (2008)
48. E. De Moor, J. Kähkönen, P. Wolfram, J.G. Speer, Current Developments in Quenched and Partitioned Steels, in *Proc. of the 5th Intl. Symposium on Steel Science, 2017, Nov. 13–16*, ed. by The Iron and Steel Institute of Japan, (Kyoto, Japan, 2017), pp. 11–17
49. D.T. Pierce, D.R. Coughlin, D.L. Williamson, K.D. Clarke, A.J. Clarke, J.G. Speer, E. De Moor, Characterization of transition carbides in quench and partitioned steel microstructures by Mössbauer spectroscopy and complementary techniques. Acta Mater. **90**, 417–430 (2015)
50. J.G. Speer, A.M. Streicher, D.K. Matlock, F. Rizzo, G. Krauss, Quenching and Partitioning: A fundamentally new process to create high strength TRIP sheet microstructures, in *Formation and Decomposition*, ed. by Austenite, (TMS/ISS, Warrendale, PA, 2003), pp. 505–522
51. E. De Moor, J.G. Speer, D.K. Matlock, J.H. Kwak, S.B. Lee, Effect of carbon and manganese on the quenching and partitioning response of CMnSi steels. ISIJ Int **51**, 137–144 (2011)
52. R.L. Miller, Ultra-fine grained microstructures and mechanical properties of alloy steels. Metall. Trans. **3**, 905–912 (1972)
53. E. De Moor, D.K. Matlock, J.G. Speer, M.J. Merwin, Austenite stabilization through manganese enrichment. Scri. Mater. **64**, 185–188 (2011)
54. M.J. Merwin, Microstructure and Properties of Cold Rolled and Annealed Low-Carbon Manganese TRIP Steels, in *Proc. of Materials Science and Technology 2007, Sept. 16–20*, (Detroit, MI, 2007), pp. 515–536
55. S. Lee, S.-J. Lee, B.C. De Cooman, Austenite stability of ultrafine-grained transformation-induced plasticity steel with Mn partitioning. Scri Mater. **65**, 225–228 (2011)

56. P.J. Gibbs, E. De Moor, M.J. Merwin, B. Clausen, J.G. Speer, D.K. Matlock, Austenite stability effects on tensile behavior of manganese-enriched-austenite transformation-induced plasticity steel. Metall. Mater. Trans. A **42**, 3691–3702 (2011)
57. W. Gan, S.S. Babu, N. Kapustka, R.H. Wagoner, Microstructural effects on the springback of advanced high-strength steel. Metall. Trans. A **37**, 3221–3231 (2006)
58. A.W. Hudgins, D.K. Matlock, J.G. Speer, C.J. Van Tyne, Predicting instability at die radii in advanced high strength steels. J. Mater. Proc. Tech. **210**, 741–750 (2010)
59. T. Taylor, A. Clough, Critical review of automotive hot-stamped sheet steel from an industrial perspective. Mater. Sci. Tech. **34**, 809–861 (2018)
60. A. Erman Tekkaya, H. Karbasian, W. Homberg, M. Kleiner, Thermo-mechanical coupled simulation of hot stamping components for process design. Prod. Eng. Res. Devel. **1**, 85–89 (2007)
61. M. Suehiro, K. Kusumi, T. Miyakoshi, J. Maki, M. Ohgami, Properties of Aluminium-coated steels for hot-forming. Nippon Steel Tech. Rep. **88**, 16–21 (2003)
62. G. Frommeyer, U. Brüx, P. Neumann, Supra-ductile and high-strength manganese-TRIP/TWIP steels for high energy absorption purposes. ISIJ Intl. **43**, 438–446 (2003)
63. O. Grässel, L. Krüger, G. Frommeyer, L.W. Meyer, High strength Fe-Mn-(Al,Si) TRIP/TWIP steels development – properties - application. Int. J. Plast. **16**, 1391–1409 (2000)
64. U. Brux, G. Frommeyer, O. Grassel, L.W. Meyer, A. Weise, Development and characterization of high strength impact resistant Fe-Mn-(Al-, Si) TRIP/TWIP steels. Steel Res. Intl. **73**, 294–298 (2002)
65. S. Martin, S. Wolf, U. Martin, L. Krüger, D. Rafaja, Deformation mechanisms in austenitic TRIP/TWIP steel as a function of temperature. Metall. Mater. Trans. A **47**, 49–58 (2016)
66. S. Martin, C. Ullrich, D. Rafaja, Deformation of austenitic CrMnNi TRIP/TWIP steels: Nature and role of the $\varepsilon-$martensite. Mat. Today Proc. **2S**, S643–S646 (2015)
67. L. Krüger, L.W. Meyer, U. Brûx, G. Frommeyer, O. Grässel, Stress-deformation behaviour of high manganese (AI, Si) TRIP and TWIP steels. J. Phys. IV France **110**, 189–194 (2003)
68. S.-M. Lee, S.-J. Lee, S. Lee, J.-H. Nam, Y.-K. Lee, Tensile properties and deformation mode of Si-added Fe-18Mn-0.6C steels. Acta Mater. **144**, 738–747 (2018)
69. J.A. Ronevich, J.G. Speer, D.K. Matlock, "Hydrogen embrittlement of commercially produced advanced high strength sheet steels," SAE technical publication #2010-01-0447. SAE Int. J. Mater. Manuf. **3**, 255–267 (2010)
70. S. Kang, Y.-S. Jung, J.-H. Jun, Y.-K. Lee, Effects of recrystallization annealing temperature on carbide precipitation, microstructure, and mechanical properties in Fe–18Mn–0.6C–1.5Al TWIP steel. Mater. Sci. Eng. A **527**, 745–751 (2010)
71. L. Bracke, K. Verbeken, L. Kestens, J. Penning, Recrystallization behaviour of an austenitic high Mn steel. Mater. Sci. For. **558–559**, 137–142 (2007)
72. J.-H. Kang, T. Ingendahl, W. Bleck, A constitutive model for the tensile behaviour of TWIP steels: Composition and temperature dependencies. Mater. Des. **90**, 340–349 (2016)
73. W.S. Choi, B.C. De Cooman, S. Sandlöbes, D. Raabe, Size and orientation effects in partial dislocation-mediated deformation of twinning-induced plasticity steel micro-pillars. Acta Mater. **98**, 391–404 (2015)
74. I.-C. Jung, B.C. De Cooman, Temperature dependence of the flow stress of Fe–18Mn–0.6C–xAl twinning-induced plasticity steel. Acta Mater. **61**, 6724–6735 (2013)
75. B. Mahatoa, S.K. Shee, T. Sahu, S. Ghosh Chowdhury, P. Sahu, D.A. Porter, L.P. Karjalainen, An effective stacking fault energy viewpoint on the formation of extended defects and their contribution to strain hardening in a Fe–Mn–Si–Al twinning-induced plasticity steel. Acta Mater. **86**, 69–79 (2015)
76. H. Idrissi, K. Renard, L. Ryelandt, D. Schryvers, P.J. Jacques, On the mechanism of twin formation in Fe–Mn–C TWIP steels. Acta Mater. **58**, 2464–2476 (2010)
77. S. Mahajan, G.Y. Chin, Formation of deformation twins in f.c.c. crystals. Acta Metall. **21**, 1353–1363 (1973)
78. J.-E. Jin, Y.-K. Lee, Effects of Al on microstructure and tensile properties of C-bearing high Mn TWIP steel. Acta Mater. **60**, 1680–1688 (2012)

79. J.S. Jeong, W. Woo, K.H. Oh, S.K. Kwon, Y.M. Koo, In situ neutron diffraction study of the microstructure and tensile deformation behavior in Al-added high manganese austenitic steels. Acta Mater. **60**, 2290–2299 (2012)
80. J. Kim, S.-J. Lee, B.C. De Cooman, Effect of Al on the stacking fault energy of Fe–18Mn–0.6C twinning-induced plasticity. Scri. Mater. **65**, 363–366 (2011)
81. C. Scott, B. Remy, J.-L. Collet, A. Cael, C. Bao, F. Danoix, B. Malar, C. Curfs, Precipitation strengthening in high manganese austenitic TWIP steels. Int. J. Mater. Res. **102**, 538–549 (2011)
82. L. Rémy, A. Pineau, Twinning and strain-induced f.c.c. → h.c.p. transformation on the mechanical properties of co-Ni-Cr-Mo alloys. Mater. Sci. Eng. **26**, 123–132 (1976)
83. L. Rémy, The interaction between slip and twinning systems and the influence of twinning on the mechanical behavior of fcc metals and alloys. Metall. Trans. A. **12**, 387–408 (1981)
84. S. Allain, J.-P. Chateau, O. Bouaziz, A physical model of the twinning-induced plasticity effect in a high manganese austenitic steel. Mater. Sci. Eng. A **387**, 143–147 (2004)
85. T.-H. Lee, E. Shin, C.-S. Oh, H.-Y. Ha, S.-J. Kim, Correlation between stacking fault energy and deformation microstructure in high-interstitial-alloyed austenitic steels. Acta Mater. **58**, 3173–3186 (2010)
86. O. Bouaziz, S. Allain, C.P. Scott, P. Cugy, D. Barbier, High manganese austenitic twinning induced plasticity steels: A reviewof the microstructure properties relationships. Curr. Opin. Solid State Mater. Sci. **15**, 141–168 (2011)
87. L. Bracke, L. Kestens, J. Penning, Transformation mechanism of α'-martensite in an austenitic Fe–Mn–C–N alloy. Scri. Mater. **57**, 385–388 (2007)
88. T. Michler, C. San Marchi, J. Naumann, S. Weber, M. Martin, Hydrogen environment embrittlement of stable austenitic steels. Intl. J. Hydrogen Energy **37**, 16231–16246 (2012)
89. M. Koyama, E. Akiyama, K. Tsuzaki, Hydrogen embrittlement in a Fe–Mn–C ternary twinning-induced plasticity steel. Corr. Sci. **54**, 1–4 (2012)
90. J.K. Jung, O.Y. Lee, Y.K. Park, D.E. Kim, K.G. Jin, Hydrogen embrittlement behavior of high Mn TRIP/TWIP steels. Korean J. Mater. Res. **18**, 394–399 (2008)
91. S.-k. Kim, G. Kim, K.-g. Chin, Development of High Manganese TWIP Steel with 980MPa Tensile Strength, in *Proceedings of the International Conference on New Developments in Advanced High-Strength Sheet Steels*, (AIST, 2008), pp. 249–258
92. J.A. Ronevich, S.K. Kim, J.G. Speer, D.K. Matlock, Hydrogen effects on cathodically charged twinning-induced plasticity steel. Scri. Mater. **66**, 956–959 (2012)
93. K.H. So, J.S. Kim, Y.S. Chun, K.T. Park, Y.G. Lee, C.S. Lee, Effect of hydrogen on fracture of austenitic Fe–Mn–Al steel. ISIJ Int. **49**, 2009 (1952–1959)
94. J.H. Ryu, S.K. Kim, C.S. Lee, D.-W. Suh, H.K.D.H. Bhadeshia, Effect of Aluminium on hydrogen-induced fracture behaviour in austenitic Fe-Mn-C steel. Proc R Soc A **469**, 20120458 (2013)
95. K.G. Chin, C.Y. Kang, S.Y. Shin, S.K. Hong, S.H. Lee, H.S. Kim, K.H. Kim, N.J. Kim, Effects of Al addition on deformation and fracture mechanisms in two high manganese TWIP steels. Mater. Sci. Eng. A **528**, 2922–2928 (2011)
96. H. Kim, D.-W. Suh, N.J. Kim, Fe–Al–Mn–C lightweight structural alloys: A review on the microstructures and mechanical properties. Sci. Technol. Adv. Mater. **14**, 1–11 (2013)
97. R. Rana, C. Lahaye, R.K. Ray, Overview of lightweight ferrous materials: Strategies and promises. JOM **66**, 1734–1746 (2014)
98. I. Zuazo, B. Hallstedt, B. Lindahl, M. Selleby, M. Soler, A. Etienne, A. Perlade, D. Hasenpouth, V. Massardier-Jourdan, S. Cazottes, X. Kleber, Low-density steels: Complex metallurgy for automotive applications. JOM **66**, 1747–1758 (2014)
99. D. Raabe, H. Springer, I. Gutierrez-Urrutia, F. Roters, M. Bausch, J.-B. Seol, M. Koyama, P.-P. Choi, K. Tsuzaki, Alloy design, combinatorial synthesis, and microstructure–property relations for low-density Fe-Mn-Al-C austenitic steels. JOM **66**, 1845–1856 (2014)
100. R. Rana, C. Liu, R.K. Ray, Low-density Low-carbon Fe–Al Ferritic Steels. Scri. Mater.**68**, 354–359 (2013)

101. O. Ikeda, I. Ohnuma, R. Kainuma, K. Ishida, Phase equilibria and stability of ordered BCC phases in the Fe-rich portion of the Fe–Al system. Intermetallics **9**, 755–761 (2001)
102. Y.-U. Heo, Y.Y. Song, S.-J. Park, H.K.D.H. Bhadeshia, D.-W. Suh, Influence of silicon in low density Fe-C-Mn-Al steel. Metall. Mater. Trans. A **43**, 1731–1735 (2012)
103. M. Chen, D.J. Zhou, "AHSS forming simulation for shear fracture and edge cracking", Great designs in steel. Am. Iron Steel Inst.
104. M. Shi, X. Chen, Prediction of Stretch Flangeability Limits of Advanced High Strength Steels Using the Hole Expansion Test, SAE Technical Paper 2007-01-1693, (2007)
105. R.J. Johnson, *Hole Expansion of Retained Austenite Containing CMnSi Bainitic/Martensitic Steels* (Colorado School of Mines, 2013)
106. C. Butcher, L. ten Kortenaar, M. Worswick, Experimental characterization of the sheared edge formability of boron steel. IDDRG Conf. Proc., 222–227 (2014)
107. S. Sadagopan, D. Urban, *Formability Characterization of a New Generation of High Strength Steels* (US Department of Energy, American Iron and Steel Institute, 2003).
108. O.R. Terrazas, K.O. Findley, C.J. Van Tyne, Influence of martensite morphology on sheared-edge formability of dual-phase steels. ISIJ Int. **57**, 937–944 (2017)
109. I. Pushkareva, C.P. Scott, M. Gouné, N. Valle, A. Redjaïmia, A. Moulin, U. De Lorraine, I. Jean, L. Crns, L. De Nancy, P. De Saurupt, F. Nancy, Distribution of carbon in martensite during quenching and tempering of dual phase steels and consequences for damage properties. ISIJ Int **53**, 1215–1223 (2013)
110. N. Pottore, N. Fonstein, I. Gupta, D. Bhattacharya, A Family of 980MPa Tensile Strength Advanced High Strength Steels with Various Mechanical Property Attributes, in *International Conference on Advanced High Strength Sheet Steels for Automotive Applications Proceedings*, ed. by J. G. Speer, (AIST, Warrendale, PA, 2004), pp. 119–129
111. S.B. Lee, J.G. Speer, D.K. Matlock, K.G. Chin, Analysis of Stretch-Flangeability Using a Ductile Fracture Model, in *3rd International Conference on Advanced Structural Steels*, (2006), pp. 841–849
112. A. Karelova, C. Krempaszky, E. Werner, P. Tsipouridis, T. Hebesberger, A. Pichler, Hole expansion of dual-phase and complex-phase AHS steels - effect of edge conditions. Steel. Res. Int. **80**, 71–77 (2009)

Chapter 5
Cast Iron–Based Alloys

Dawid Myszka

Abbreviations and Symbols

CI	Cast iron
DI	Ductile iron
ADI	Austempered ductile iron
ADI_xxx_xxx (for example: EADI_400_30)	Determination of isothermal transformation parameters of austempered ductile iron (in the example: $T_{it} = 400\,°C$, $t_{it} = 30$ min.)
CADI	Carbidic austempered ductile iron
DADI	Direct austempered ductile iron
Q&T	Quenched and tempered
HT	Heat treated
EN-GJL	Gray iron grades according to European Standard EN 1561
C_E	Carbon equivalent
Gr	Graphite
P	Pearlite
α, *F*	Ferrite
δ	High-temperature ferrite
γ, *A*,	Austenite
γ^0	Primary austenite
γ_{HC}	High-carbon austenite
α', *M*	Martensite
$\alpha+\gamma_{HC}$, *AF*	Ausferrite

D. Myszka (✉)
Department of Metal Forming and Foundry Engineering, Warsaw University of Technology,
Warszawa, Mazowsze, Poland

© Springer Nature Switzerland AG 2021
R. Rana (ed.), *High-Performance Ferrous Alloys*,
https://doi.org/10.1007/978-3-030-53825-5_5

B_U	Upper bainite
B_L	Lower bainite
B_S, B_F	Start and end of bainitic transformation,
A_S, A_F	Start and end of ausferritic transformation
M_S, M_F	Temperature of the beginning and end of martensitic transformation [^{0}C]
$T_{\gamma,A}$	Austenitizing temperature [^{0}C]
$t_{\gamma,A}$	Austenitizing time [min.]
T_{it}	Isothermal transformation temperature [^{0}C]
t_{it}	Isothermal transformation time [min.]
T_0	Temperatura pokojowa [^{0}C]
$C_\gamma{}^A$	Carbon content in austenite at austenitizing temperature [%]
C_γ	Carbon content in austenite [%]
X_γ	Austenite content in the ausferritic mixture [%]
R_{YS}	Yield strength [MPa]
R_{TS}	Tensile strength [MPa]
K_{IC}	Fracture toughness [MPa $\times$ m$^{1/2}$]
D	Critical diameter [mm]
TRIP	Transformation-induced plasticity
SEM	Scanning electron microscope
TEM	Transmission electron microscope
LM	Light microscope
XRD	X-ray diffraction

5.1 Cast Iron Vs. Steel

Among all foundry alloys for structural applications, cast iron is the best known and most frequently used material (Fig. 5.1) [1–13]. As an iron-carbon alloy, it is often confused with or compared to steel. And yet, as we shall see in this chapter, its features are completely different from steels. To discover this, cast iron needs to be studied more carefully.

5.1.1 Basics of Cast Iron

The great popularity of cast iron in both industry and research areas has survived to this day despite the fact that this casting material has been known since 1000 BC [14]. The devices for cast iron melting, i.e., shaft furnaces in China and bloomeries first, blast furnaces in Europe next, and, finally, cupolas are milestones in the development of civilization.

In addition to the practical experience of artists, i.e., foundry men, the gradual entering of science into the field of cast iron production enabled further discoveries

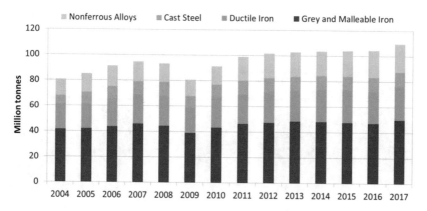

Fig. 5.1 Trends in world casting production in the years 2004–2017. (Author's own analysis based on [1–13])

that led to unflagging interest in this material lasting to the present day. It is enough to mention A.F. Meehan who in 1922 developed a method for the manufacture of inoculated cast iron, or H. Morrogh and A.P. Gagnebin who in 1947 invented the cast iron with nodular graphite known also as ductile iron. Those were the turning points in the use of cast iron as a construction material but by no means the end of spectacular discoveries in this field. Heat treatment, and especially the austempering treatment of ductile iron, developed at the end of the twentieth century as well as a multitude of scientific publications that have appeared in recent years best prove the fact that studies of this material are going on all the time [15, 16].

According to the most general definition, cast iron is an iron-carbon alloy, usually with the carbon content ranging from 2 to 4 wt.%. Typically, cast iron also contains other elements. It is produced by remelting pig iron, cast iron scrap, and steel scrap and solidifies with eutectic transformation. It contrasts with cast steel which, being also an iron-carbon alloy with other elements, has no eutectic in its composition. Depending on whether the cast iron solidifies in a stable or metastable system (Fig. 5.2), graphite eutectic or cementite eutectic (ledeburite) will form, respectively. Depending on the form of carbon, bonded or free (graphite), the cast iron is said to be white or gray, respectively (Fig. 5.2). The adjective *white* or *gray* in the name of the cast iron reflects the color of the fracture, which is bright with metallic shimmer in white cast iron and gray in gray cast iron. The gray color in the latter case is associated with the presence of graphite in the fracture.

The most spectacular properties of cast iron are obtained primarily in ductile iron, the invention of which eliminated the fundamental disadvantage of this material, i.e., the lack of plasticity. Ductile iron is the material which retains all the advantages of cast iron, and at the same time allows obtaining the strength properties comparable or superior to the properties of heat-treated steels or cast steels.

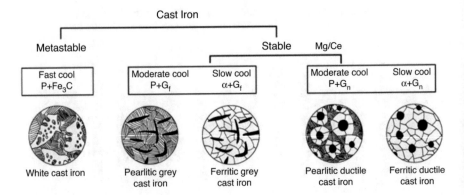

Fig. 5.2 Simplified classification of cast iron into different types depending on the type of solidification in a stable or metastable system; *P* pearlite, G_f flake shaped graphite, G_n nodule shaped graphite, α ferrite

5.1.2 Mechanical Properties

The main advantage of ductile iron is the unique combination of mechanical and plastic properties [15, 17, 18]. Comparing the minimum values of R_m (ultimate tensile strength) and A_5 (total elongation with 5 mm gauge length) obtained in conventional grades of this cast iron and in alloyed steel, carbon steel, and austempered ductile iron, the potential competitiveness of ductile iron with respect to steel deserves attention only in the case of the highest ductility (ductile iron with ferritic matrix) or highest strength (ductile iron with ausferritic matrix) (Fig. 5.3) [15]. From Fig. 5.3, it follows that austempered ductile iron (ADI) offers the properties which are a "continuation" of the cast steel properties in the direction of higher strength, while ADI grades with lower strength are competitive with the cast steel of the highest strength.

Compared to steel, another very important feature of cast iron is a 10% lower density. In combination with very high durability, it turns out that austempered ductile iron can be competitive not only to steel, but also to magnesium alloys and even aluminum alloys with respect to density (Fig. 5.4). Through small structural changes, such as the reduction of wall thickness, attempts to replace aluminum alloys with ADI have been successfully implemented [17]. The examples are available in various sources, e.g., in formation available on a website [17] about casting of a truck wheel hub of the same weight as an aluminum alloy casting (~ 15 kg) and which best proves this statement.

5.1.3 Vibration Damping Capacity

Subjected to cyclic loads, cast iron shows the ability to absorb the applied energy and convert it into heat resulting in fast damping of vibrations. The presence of

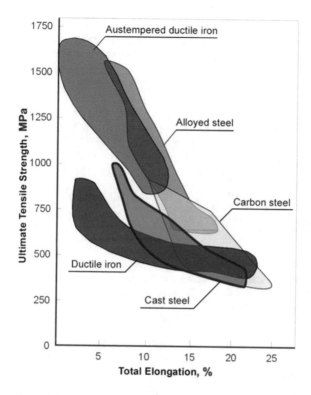

Fig. 5.3 Comparison of the achievable range of ultimate tensile strength and total elongation: austempered ductile iron, ductile iron, alloyed steel, carbon steel, and cast steel. (Author's own analysis based on [14–17])

graphite in gray cast iron results in damping of vibrations faster than in steel (Fig. 5.5) [14, 15, 17]. This is of great importance in the case of unexpected or rapidly changing loads. Cast iron offers a faster relaxation of the accumulated stresses or suppression of vibration resonance which in other cases, even for materials of much higher strength, can prove difficult and, therefore, disastrous failure can occur.

The main structural constituent that affects the ability to damp vibrations is graphite. With increasing content and size of the graphite precipitates, this ability increases too. Therefore, in gray cast iron, the ability to dampen vibrations decreases with the increasing properties and structure refinement. For this property of cast iron, the metal matrix is incomparably less important, although much more effective damping of vibrations by the bainitic matrix of cast iron compared to the pearlitic matrix has been well documented [14]. The usefulness of gray cast iron for the production of gears can be explained by the fact that stresses arising at the base of the teeth, which is the place of stress accumulation, assume values much lower in the cast iron gears due to damping of vibration than in the steel gears.

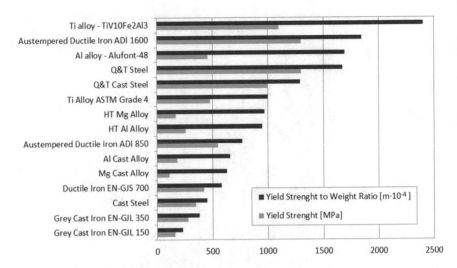

Fig. 5.4 Yield strength and the yield strength-to-weight ratio compared for selected materials from various cast irons, cast steels, and Al, Ti, and Mg alloys in different heat treatment conditions. Q&T – quenched and tempered, HT – heat treated, EN-GJL – gray iron grades according to European Standard EN 1561. (Author's own analysis based on [1–17])

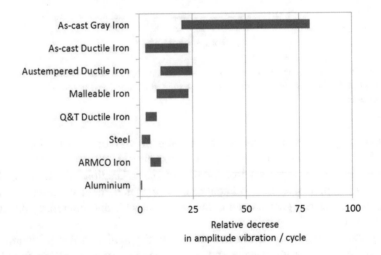

Fig. 5.5 Value of relative decrease in amplitude of vibration for different materials. (Author's own analysis based on [14, 15, 17])

5.1.4 Manufacturing Costs

Cast iron is an attractive construction or structural material also in terms of its price, and not only when the cost of a kilogram of the product at current market prices in Europe is compared (Fig. 5.6) [13], but also when the cost of producing

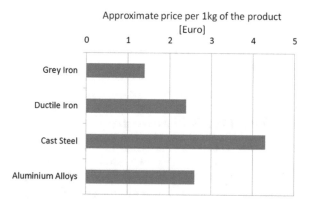

Fig. 5.6 Approximate price per 1 kg of the product for different materials (2018). (Author's own analysis based on production sources)

Table 5.1 Energy consumption in the production of gears – a comparison of ADI castings and steel forgings [17]	Operation	Energy consumption, kWh/t	
		ADI	Steel
	Making semifinished products	2500	4500
	Annealing	–	500
	Austempering	600	–
	Carburizing	–	800–1200
	Total	3100	5600–6200

complex geometries from different materials with the same properties is compared. An interesting example is the comparison of manufacturing costs of products made from carburized steel and cast from ADI which serves as parts of a machine operated under dynamic mechanical loads and friction conditions (Table 5.1) [17]. The documented almost 50% savings in costs related to energy consumption is not the only advantage of technology conversion in this particular case. Other benefits include reduction of product weight by almost 10%, reduction of noise during operation, almost three times higher durability of castings, and six times higher durability of cutting tools, resulting mainly from the ductile iron machinability before the heat treatment process [17].

5.2   Types of Cast Iron

High-quality alloys are construction materials whose production complies with all technological standards leading to a product without defects with properties that meet the minimum standards. This is also true in the case of iron castings. This study is largely devoted to the problem of the structure and structure-related properties, and not to the quality of the technological process, which should always be at the highest level. However, it should be remembered that the type of structure formed during casting solidification and cooling process will have a significant impact on the properties of this casting, also in the case of the most-advanced cast iron grades subjected to heat treatment.

5.2.1 Production – Cast Iron Structure

The structure of cast iron can be shaped in a number of different ways. The melt "refining" processes, the solidification conditions and cooling rate, or control of the chemical composition are the most important factors, which when properly selected and monitored will make the control of cast iron properties possible. Therefore, even a very brief discussion of these issues is extremely important to capture the differences between cast iron and steel in which the postcasting structure is only of minor significance.

According to the earlier definition, cast iron is composed not only of iron and carbon, but also of a number of other elements that always exist in this alloy and include silicon, manganese, phosphorus, and sulfur. Other elements such as Cu, Ni, Mo, or Cr are usually regarded as additional alloying additives. Each of them exerts a greater or lesser influence on the course of primary and eutectic crystallization, and this influence should always be considered taking into account the content of individual elements and cooling conditions. Chemical composition has a very complex effect on the structure of cast iron [20–23] although in this complex system it is carbon that plays an overriding role in shaping the cast iron microstructure.

Carbon is the main constituent determining the casting, mechanical, and functional properties of cast iron [2]. According to the binary Fe-C phase equilibrium diagram, the cast iron in a stable system contains at least 2.08 wt.% C (Fig. 5.7) [14, 16]. However, since it always contains a certain amount of Si and P, the critical carbon content can vary within a fairly wide range. Due to the strong effect of silicon and phosphorus on the carbon content at the eutectic point with a tendency to bring this content to lower values, in practice the so-called carbon equivalent C_E is used [24]. It is a measure of the deviation of the chemical composition of cast iron of its eutectic composition and is defined as follows:

$$C_E = C + \frac{Si + P}{3} \tag{5.1}$$

where C, Si, and P contents are in wt.%. Establishing this value with respect to the eutectic value of 4.26 wt.% C enables dividing the cast iron into three groups, i.e., hypoeutectic ($C_E < 4.26$), eutectic ($C_E = 4.26$), and hypereutectic ($C_E > 4.26$) [14, 19]. Among them, the most popular in industrial practice is the hypoeutectic cast iron, mainly due to the possibility of obtaining a wide range of in-service properties. From a production point of view, it is also important whether the cast iron solidifies in a stable or metastable system. The metastable system is associated

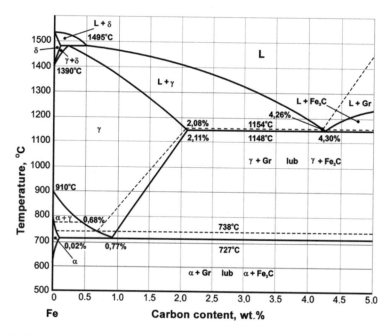

Fig. 5.7 Phase equilibrium diagram of Fe-C alloys: thin line – Fe-C graphite system; thick line – Fe-Fe$_3$C system; L – liquid solution; δ – high-temperature ferrite; γ – austenite; α – ferrite; Gr – graphite. (Based on [14, 16])

with the formation of hard and brittle cementite eutectic, while stable system ensures the formation of graphite eutectic with precipitates of free graphite. This type of cast iron is the most widespread and worldwide-appreciated casting material for structural applications. In a further part of this chapter, this material will be the main subject of discussion.

An ideal cooling curve and true cooling curve from the liquid of the hypoeutectic gray cast iron have several differences between the two runs (Fig. 5.8). First, there is a more or less visible characteristic "isothermal arrest". According to Karsay [25], its origin on the true curve is identified with undercooling below the equilibrium temperature due to which a "driving force" appears in the system to carry into effect the nucleation of the primary phase, which in the case of hypoeutectic cast iron is austenite. The rate of solidification at this point becomes quite significant, trying to compensate for some delay caused by the need to achieve sufficient undercooling of the liquid alloy. Acceleration of solidification during this period leads to an increase in temperature in the direction of the equilibrium freezing point, which results in the effect visible in Fig. 5.8. As a result of further cooling, the alloy reaches its eutectic temperature. Triggering the eutectic transformation also requires some undercooling below the equilibrium temperature. Due to the simultaneous formation of graphite and austenite, the temperature rises again, generating the phenomenon called recalescence [14, 25]. Depending on the cooling rate, the recalescence can

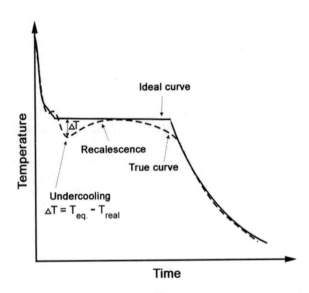

Fig. 5.8 Typical run of the ideal and true cooling curve of hypoeutectic gray iron. (Based on [25])

make the solidifying eutectic cast iron reach the equilibrium temperature or remain below this point. Further cooling is associated with the completion of solidification – first by one phase (graphite) and later by another phase (austenite) [14]. This phenomenon evokes some deviations from the ideal cooling curve with a tendency toward continuous lowering of the temperature.

The above-discussed solidification sequence of the hypoeutectic gray cast iron, which results in the formation of a primary phase in the form of austenite dendrites spread against the background of graphite eutectic, is only partly true, and this is due to the fact that solidification in the systems comprising an eutectic point takes place under the conditions of so-called competitive growth of two phases. As a consequence, in the eutectic systems, the type of microstructure depends not only on the chemical composition of the alloy but also on the rate of growth of the eutectic mixture and the primary (preeutectic) phase. This means that if the growth rate of the eutectic is higher than the growth rate of the primary phase, the obtained microstructure will be fully eutectic. The above is related to the concept of the coupled zone of eutectic growth, which is understood as a range of the alloy chemical composition, growth velocity, and temperature gradient, ensuring the formation of a completely eutectic microstructure – without primary phases, both pre- and posteutectic. Disregarding any more detailed studies of the eutectic solidification, which are described in the works of Campbell and Fraś [18, 24], it should be noted that due to the chemical composition of cast iron approaching the eutectic composition and the above-discussed conditions of coupled eutectic growth, the microstructure of cast iron is usually fully eutectic and its dendritic character is recognizable even after long-term heat treatment processes (Fig. 5.9).

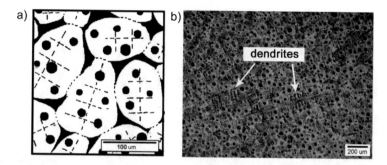

Fig. 5.9 Model of the primary structure of ductile iron: (**a**) – model of structure; black fields between cells are the last to freeze regions (LTF), dashed lines denote dendrite arms; (**b**) optical microscopic photo of ADI – Nomarsky contrast. (Author's own work)

According to the phase equilibrium diagram of Fe-C alloys (Fig. 5.7), at room temperature, two basic phases containing no alloying additions can coexist in hypoeutectic gray cast iron. These are the following phases: solid carbon solution in α iron (ferrite) and graphite (stable system). Alloyed or heat-treated cast iron may also contain solid solution of carbon in γ iron (austenite) and product of diffusion-free transformation (martensite). On the other hand, individual constituents of cast iron can give rise to the formation of two- or three-phase structural constituents, such as various types of eutectic (graphite, cementite) and pearlite or bainite. Studies of the cast iron microstructure after casting and cooling at different speeds to ambient temperature showed the possibility of the formation of different phases in the ductile iron matrix. The research presented by Rivera [26] indicates the possibility of obtaining a wide range of matrix types from martensitic, through austenitic-ferritic and pearlitic up to ferritic in nonalloyed cast iron or in cast iron with low additions of Ni < 1 wt.%, Cu < 0.7 wt.% and Mo < 0.15 wt.%.

Reducing the wall thickness is tantamount to increasing the casting solidification and cooling rate, and thus to increasing the homogeneity of the concentration of elements in cast iron [27]. However, as shown in [28, 29], as the wall thickness increases, the number of graphite precipitates in the casting decreases. This promotes the microsegregation of elements, and thus the reduction of mechanical properties. In turn, a large number of precipitates in the case of a high solidification rate (smaller wall thickness) means smaller distance between graphite precipitates, and hence higher chemical homogeneity in microregions. The mechanical properties were found to be much better in ductile iron castings with thinner walls, than in the slowly solidifying castings characterized by large and sparsely distributed graphite nodules [30, 31].

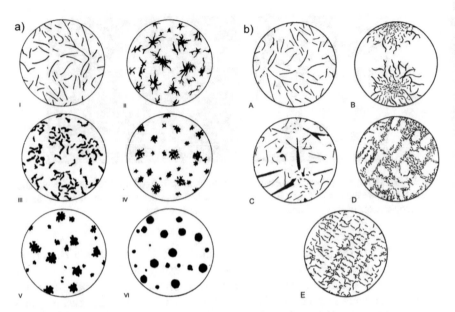

Fig. 5.10 Graphite classification by visual analysis as per International Standard ISO 945-1:2017 (E): (**a**) reference images for principal graphite forms in the cast irons from flake to nodular shape (forms: I, II, III, IV, V and VI), (**b**) reference images for graphite distribution (type A: random flake graphite in a uniform distribution, type B: rosette flake graphite, type C: kish graphite (hypereutectic compositions), type D: undercooled flake graphite, type E: interdendritic flake graphite (hypoeutectic compositions)). (Based on [14])

5.2.2 Graphite Morphology

Different forms of graphite (flake, nodular, vermicular, chunky, exploded, coral, etc.) occurring in the cast iron determine its properties (Fig. 2.4), but it was the discovery of the possibility of making the graphite grow in a spherical form during the gray cast iron solidification that has opened the way for foundry men not only to eliminate the inherent brittleness of cast iron, but also to obtain the strength properties competing with the best grades of steel (Fig. 5.10).

The following factors determine the graphite form in the cast iron: the type and structure of nuclei, physicochemical factors, and growth conditions in the directions perpendicular to the graphite crystal walls. According to many scientists [14, 18, 24], the morphology of graphite precipitates is probably most affected by variations in the speed of crystal growth in different directions, related to both the internal structure of graphite and external factors. The most important role is attributed to elements which introduced even in small amounts can clearly change the shape of graphite precipitates. Thus, Mg, Ce, and other elements promote the formation of nodular graphite, while Sb, Pb, Al, S, Ti, and Bi facilitate the crystallization of flake graphite. So far, the mechanism of interaction of these elements has not been fully explained, but it is known that they exert different effects on the growth rate of

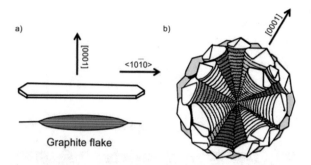

Fig. 5.11 Schematic presentation of (**a**) growth of graphite flake and (**b**) graphite nodule as a polycrystal composed of conically coiled layers. (Based on [14])

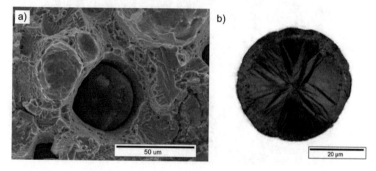

Fig. 5.12 (**a**) Scanning electron microscope (SEM) image of nodular graphite in an ausferritic matrix, (**b**) light microscope (LM) image of the cross-section through a graphite nodule. (Author's own work)

graphite crystals in [0001] and <1010> directions. It is assumed that these elements may be incorporated into the lattice of graphite or may be selectively absorbed on the walls of the growing crystal. It seems highly probable that the spheroidizing elements facilitate the formation of <1010> dislocations, which promote the growth of graphite crystals in [0001] direction but inhibit this growth in direction. From Fig. 5.11 [14] presenting the growth of graphite structure, it follows that graphite flakes in gray cast iron are monocrystals (often bent and branched), while nodular graphite precipitates are polycrystals (Fig. 5.12).

In contrast to other types of gray cast iron described in the literature [14, 24, 25], the essence of the solidification of ductile iron is in the special crystallization mode of graphite in the form of nodules. This graphite crystallizes directly from the liquid, where its nucleus in the initial stage grows freely as a specific preeutectic phase and takes on a spheroidal form (Fig. 5.11). The nuclei grow in the space between the arms of the austenite dendrites formed earlier. This location promotes the heterogeneous nucleation of the eutectic austenite grains. Austenite grows around the graphite nodule until its complete closure, contrary to flake graphite whose

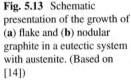

Fig. 5.13 Schematic presentation of the growth of (**a**) flake and (**b**) nodular graphite in a eutectic system with austenite. (Based on [14])

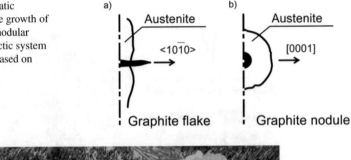

Fig. 5.14 Two neighboring eutectic cells in the ductile iron with a pearlitic-ferritic matrix. (Author's own work)

growth precedes the build-up of austenite (Fig. 5.13) [14], forming a notch stress at the tip of a sharp precipitate. After losing contact with the liquid, the growth of the nodule takes place through the diffusion of carbon from the surrounding austenite. Together with the austenite envelope, after solidification the flake or nodular graphite forms a characteristic structure commonly known as eutectic cell (Fig. 5.14). From now in this chapter, the product of eutectic transformation in the form of a mixture of phases will be referred to as a eutectic cell.

5.2.3 Influence of Chemical Composition – Microsegregation

The primary microstructure of ductile iron is one of the most important factors affecting the final properties of cast iron in both as-cast state and after heat treatment (or after other technological treatments). Therefore, its shaping through properly chosen chemical composition, and especially striving for maximum microstructural

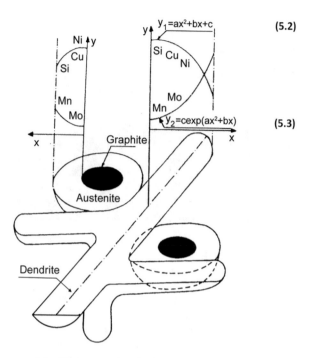

Fig. 5.15 Diagram of ductile iron structure and microsegregation of its constituents. Eq. (5.2) $y_1 = ax^2 + bx + c$; Eq. (5.3) $y_2 = c\exp(ax^2 + bx)$. (Based on [21])

homogeneity of the distribution of elements in a volume of the alloy, is of particular importance in the overall production process.

When analyzing the microstructure of ductile iron after solidification, it is convenient to use the model shown in Fig. 5.15 [21]. As demonstrated by Sękowski and Pietrowski [21, 23], a specific type of microsegregation occurs in cast iron, described by Eq. (5.2) for noncarbide forming elements and by Eq. (5.3) for carbide-forming elements (Fig. 5.15). Microsegregation raises the concentration of molybdenum, manganese, and carbon at grain boundaries and causes enrichment in silicon, nickel, and copper in areas close to graphite nodules. This has its reflection in the changes that occur when the structure becomes so complex as it starts cooling down to ambient temperature. The elements present in cast iron, having a definite influence on the formation of individual phases, also interact with each other, weakening or strengthening the corresponding transformations. For example, high silicon content promotes carbon activity, while Cu and Ni counteract the carbide-forming tendency of molybdenum. However, it should be remembered that microsegregation, which is a diffusion-controlled process and depends primarily on the cooling rate, is also a function of the size of graphite nodules and internodular spacing. In turn, the cooling rate affects not only the speed, and hence the extent of diffusion, but also the degree of dispersion of phases and structural constituents.

Fig. 5.16 Concentration model of an idealized eutectic cell and ductile iron structure: lines of equal concentration of manganese. (Based on [23])

Adopting Sękowski's idealized model of ductile iron microstructure as a set of eutectic cells [23], one can illustrate in a simplified way the microsegregation of individual elements (Fig. 5.16). It is easy to notice that microsegregation will be significantly affected by the distance between adjacent graphite nodules. Hence it follows that the refinement of nodular precipitates of graphite and an increase in their number should be the main technological guidelines in an attempt to reduce the intensity of the microsegregation of elements in ductile iron. The presence of a given element in an appropriate concentration or absence of this element also contributes to changes in the intensity of the microsegregation in cast iron. However, in terms of the microstructure homogenization, the most problematic is control of the increasing concentration of elements at the boundaries of austenite dendrites or eutectic cells [32]. Proportionally, heterogeneity occurs in the mechanical properties transferred from the microscale (e.g., different properties at the boundaries and in the centre of eutectic cells) to the macroscale, determining the properties of finished castings through different structures formed in the matrix of cast iron.

5.2.4   Importance of As-Cast Matrix Microstructure

As already discussed, the microstructure of gray cast iron is composed of graphite embedded in a metal matrix. Although the degree of spheroidization, and the size and distribution of graphite precipitates are extremely important features, the type of cast iron matrix is undoubtedly equally important. A ferritic, ferritic-pearlitic, pearlitic, martensitic, or bainitic matrix surrounding the graphite precipitate is the result of various treatments used by the foundry men to obtain the desired properties of cast iron during solidification and cooling. Appropriate control of chemical composition is the simplest method that allows obtaining practically all of the above mentioned types of microstructure.

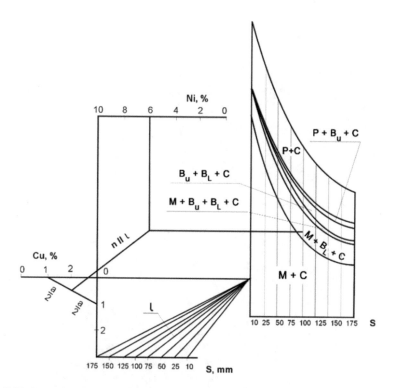

Fig. 5.17 Synergic influence of Ni, Mo, and Cu and casting wall thickness "S" on the microstructure of ductile iron in as-cast state at a constant content of 3.4 ± 0.10 wt.% C, 2.75 ± 0.10 wt.% Si, 0.35 ± 0.05 wt.% Mn, 0.045 ± 0.005 wt.% Mg. *C* molybdenum carbides, *P* pearlite, *F* ferrite, B_U upper bainite, B_L lower bainite, *M* martensite. (Based on [20])

Microstructures that provide the highest strength properties, i.e., bainitic, pearlitic, or martensitic, are easy to obtain after casting by application of special technological treatments. For example, the addition of an appropriate amount of nickel to ductile iron produces a wide range of microstructures even in the as-cast state [20, 34]. A double or triple combination of Ni, Cu, or Mo as alloying additives can produce similar effects, but it must be economically viable. Therefore, the method of obtaining the required as-cast microstructure, and thus the required properties, through the introduction of a significant amount of alloying additives is usually applied to castings massive or with intricate shapes, in the case of which the heat treatment is either difficult or even impossible. In several studies presented by Pietrowski [20, 21, 32], a detailed description of the type of microstructure obtained by adding an appropriate amount of alloying constituents to ductile iron was provided. The nomogram in Fig. 5.17, developed as the result of extensive research, perfectly reflects the possibilities of using the above-mentioned treatments. It directly indicates that for the decreasing wall thickness, the range of bainitic

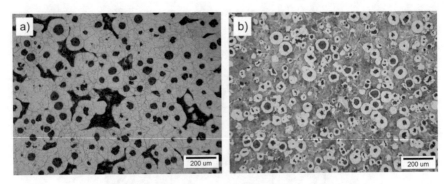

Fig. 5.18 (**a**) Ferritic-pearlitic and (**b**) pearlitic-ferritic matrix in ductile iron. (Author's own work)

Fig. 5.19 Pearlitic matrix in gray cast iron with flake graphite. (Author's own work)

microstructure with the best mechanical properties narrows, while the range of the bainitic-martensitic or martensitic microstructure becomes wider.

The functional properties of castings depend to a large extent on the type of metal matrix present in ductile iron. In terms of the matrix type, the ductile iron is divided into austenitic, ferritic, ferritic-pearlitic, or pearlitic. The matrix may also comprise bainite, ausferrite, or tempered martensite. However, controlled solidification and cooling of ductile iron usually shape its matrix as a mixture of pearlite and ferrite (Fig. 5.18). In the as-cast state, the fully pearlitic matrix is obtained either by sufficiently rapid cooling or, in the case of castings with thick walls, by introducing pearlite-forming alloying elements, such as Cu, Sn, Cr, or Ni (Fig. 5.19). Due to them, immediately after casting, the highest ductile iron grades are produced (Table 5.2) [25, 36, 37]. To obtain yield strength above 420 MPa it is necessary to either introduce a certain amount of alloying elements to the ductile iron or carry out an appropriate heat treatment [16].

Table 5.2 Designation and mechanical properties of selected cast iron grades with flake and nodular graphite according to European standards. (Based on [25, 36, 37])

Cast iron designation	Yield strength [MPa] (min.)	Tensile strength [MPa] (min.)	Total elongation [%] (min.)	Microstructure of metallic matrix
Flake graphite cast iron				
EN-GJL-150	98	150	0.3	Ferritic – pearlitic
EN-GJL-200	130	200	0.3	Pearlitic
EN-GJL-250	165	250	0.3	Pearlitic
EN-GJL-300	195	300	0.3	Pearlitic
EN-GJL-350	228	350	0.3	Pearlitic
Nodular graphite cast iron				
EN-GJS-350-22	220	350	22	Ferritic
EN-GJS-400-18	250	400	18	Ferritic
EN-GJS-450-10	310	450	10	Ferritic – pearlitic
EN-GJS-500-7	320	500	7	Pearlitic – ferritic
EN-GJS-600-3	370	600	3	Pearlitic – ferritic
EN-GJS-700-2	420	700	2	Pearlitic
EN-GJS-800-2	480	800	2	Pearlitic or martensitic
EN-GJS-900-2	600	900	2	Martensitic
Austempered ductile iron (nodular graphite)				
EN-GJS-800-10	500	800	10	Ausferritic
EN-GJS-900-8	600	900	8	Ausferritic
EN-GJS-1050-6	700	1050	6	Ausferritic
EN-GJS-1200-3	850	1200	3	Ausferritic
EN-GJS-1400-1	1100	1400	1	Ausferritic

Heat treatment can totally remodel the matrix structure without affecting the graphite precipitates in gray cast iron. It is usually applied to correct the microstructure (e.g., to dissolve carbides) or to improve the required properties of the casting (e.g., to increase hardness). However, heat treatment must always be economically justified. The correct cast iron microstructure should normally be obtained immediately after casting, but if for some reason it is not possible to provide in this way the desired properties or the customer explicitly wishes a specific type of the heat treatment to be performed on the cast iron, the heat treatment must be carried out in order to obtain a pearlitic, martensitic, or ausferritic matrix. This aspect of the ductile iron heat treatment is taken into account when the austempered ductile iron is produced. The process of making castings from this material has opened a new chapter in the production of cast iron as a technology having undoubtedly its place in the "high-tech" row. This is well illustrated by the division of cast iron grades with respect to graphite shape and matrix type according to European standards (Table 5.2).

5.3 Modern Heat Treatment of Cast Iron

The microstructure formed during solidification and cooling has a fundamental effect on the properties of cast iron. Searching for higher strength and plasticity or adding special properties to cast iron is the reason why heat treatment has become a very important factor in the development of the production of iron castings.

Due to the discovery of the nodular form of graphite, cast iron was started to be perceived as a plastic material and for a long time the heat treatment was considered unnecessary, especially that it considerably increased production costs. This is particularly well visible on the example of malleable cast iron, where long-lasting (>30 h) high-temperature annealing gives the same properties as the properties obtained in as-cast ductile iron. Heat treatment primarily affects the gray cast iron matrix, so it should be remembered that only with the nodular shape of graphite one can take full advantage of the microstructure modification. Although heat treatment is also carried out on gray iron castings with flake graphite or vermicular graphite, these are still only scientifically important processes.

5.3.1 Normalizing and Toughening

Normalizing, hardening, and toughening are carried out to increase the strength and resistance to wear. All these thermal processes have been known and used for decades but due to their existence; today modern cast iron grades are often obtained, e.g., by incorporating a conventional thermal cycle into innovative heat treatment [34]. Heat treatment is also applied for technological reasons as a means to get a homogeneous postcast structure in castings or to reduce the residual stresses to minimum.

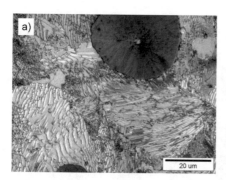

Fig. 5.20 Pearlitic (**a**) and martensitic (**b**) matrix of ductile iron. (Author's own work)

The outcome of "normalizing" is homogenization of the chemical composition and obtaining a matrix composed of highly dispersed pearlite (Fig. 5.20a). The treatment consists in annealing castings at a temperature higher by approximately 80 °C than the critical temperature (820–900 °C for gray cast iron and 870–925 °C for ductile iron) followed by rapid air cooling. The rapid air cooling treatment can be applied immediately after casting by knocking out the cast item from the foundry mold. In this way, higher strength and hardness are obtained. Normalizing is often required to dissolve carbides or to reduce the microsegregation of elements in castings with thick walls.

"Toughening" is applied when castings are expected to offer the yield strength and elongation higher than the standard grades, including grades after normalizing treatment, e.g., EN-GJS-800-2. This type of heat treatment is based on quenching followed by tempering. Tempering also removes quenching stresses and improves fracture toughness.

Rapid cooling ensuring the austenite → martensite transformation is usually carried out in an oil bath from the casting annealing temperature, i.e., from the temperature higher by 30–100 °C than the critical temperature, for a time depending on the type of cast iron. The presence of graphite distinguishes cast iron from steel and makes the holding time longer. The time taken by the austenite to get saturated with carbon amounts to about 10 min for the cast iron with ferritic matrix and flake graphite and up to 80 min for the cast iron with nodular graphite [14]. The shorter the time, the higher is the temperature of annealing and the larger is the graphite-austenite interface, i.e., the finer are the precipitates of graphite. In a pearlitic matrix, the complete transformation into austenite will proceed faster than in the ferritic matrix. For these reasons, the hardness of cast iron is much more sensitive than the hardness of steel to the time and temperature of annealing prior to quenching. The presence of graphite-austenite phase boundaries and surface development also affect the rate of matrix transformation into austenite. Therefore, gray cast iron with flake graphite shows lower hardenability than the spheroidal graphite cast iron of the same chemical composition.

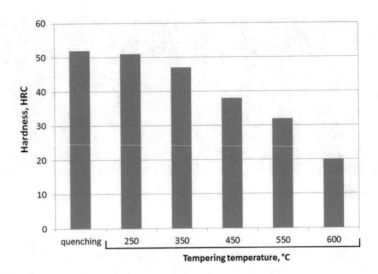

Fig. 5.21 Hardness of cast iron as a function of tempering temperature. (Based on [14, 35])

Quenching produces a martensitic matrix, providing high hardness levels to the castings (Fig. 5.20b), but it is tempering that ultimately shapes the final properties of castings, which largely depend on the time and temperature of this treatment. When high hardness and abrasion resistance are required, castings are subjected to low-temperature tempering (180–250 °C). On the other hand, high fracture toughness requires tempering carried out at high temperature (550–650 °C) (Fig. 5.21) [14, 35]. After these processes, slow cooling to ambient temperature is recommended to minimize the level of internal stresses. A special feature of cast iron tempered at a temperature above 500 °C is the occurrence in the structure of a new constituent, namely the secondary graphite [35]. It has the form of small spheroids evenly distributed in the matrix.

5.3.2 Austempering of Cast Iron

Austempering of cast iron is primarily the heat treatment of ductile iron; hence the product has the well-known name austempered ductile iron (ADI). Other types of gray cast iron, i.e., with vermicular or flake graphite, are also subjected to this treatment for better resistance to wear or fracture toughness [16, 31]. In this part of the chapter, however, the description of ADI will prevail, bearing in mind the fact that it is the cast iron most popular in today's research.

ADI is a construction material that still arouses the curiosity of scientists and the interest of practitioners. Perhaps that is why new ideas and scientific research aimed at further improvement of its properties still appear [14, 15]. This is particularly important in the aspect of the ADI implementation into practical use. It is enough to mention that since 2000 the production of this casting material in the world has been increasing at a rate of several tens of thousands of tons per year [16]. This proves the great interest in ADI of the casting users who see the opportunity to make from this material a large variety of parts of machinery and equipment operating in the automotive, railway, agricultural, and defense industries.

Austempered ductile iron is classified according to European and American standards [36, 37]. A characteristic feature of this material is the combination of high plastic and mechanical properties, comparable to many grades of steel. It has high mechanical properties owing to the properly conducted heat treatment, which consists of austenitizing and austempering operations. Both these operations are extremely important from the point of view of changes occurring in the microstructure over time and reflected in the changing properties of ADI. Austenitizing primarily determines the carbon content in austenite, while austempering following rapid cooling from the austenitizing temperature ultimately shapes the mixture of ferrite and austenite called ausferrite. The term *ausferrite* was first used in Poland at the 20th Steel Casters Conference in Raba Niżna, 1997 [16]. Ausferrite with a specific morphology and proportions of individual phases at ambient temperature is the mixture responsible for ADI properties (Fig. 5.22).

The process of heat treatment of ductile iron seems to be very simple with regard to both performance and desired effects. However, as the research shows [33], it is not always possible to meet the minimum property criteria specified by the standard. In the context of this study, it seems interesting to accurately present the kinetics of the formation of phases in ADI together with their characteristics.

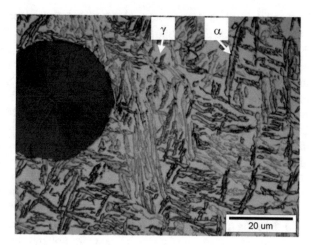

Fig. 5.22 Typical microstructure of austempered ductile iron (ADI). γ austenite, α ferrite plates. (Author's own work)

5.3.2.1 Austenitizing

Both temperature and time of the austenitizing treatment exert a significant effect on the carbon content in the austenite which forms around the graphite nodules in cast iron. The content of carbon determines the course of subsequent heat treatment and, as a consequence, affects the mechanical properties of cast iron.

By temperature setting it is possible to adjust the value of the equilibrium carbon concentration in austenite, while time decides if and when this equilibrium occurs. The higher the austenitizing temperature, the faster is the process of the carbon saturation in austenite. The solubility of carbon originating from graphite particles also increases. Austenite becomes more homogeneous and its grains start growing [38].

The relationship between austenitizing temperature and carbon content in austenite is well known [29]. For Fe-C-Si alloys, this value can be calculated from equation [29]:

$$C_\gamma^A = \frac{T_A}{420} - 0.17(Si) - 0.95 \tag{5.4}$$

where

C_γ^A – carbon content in austenite at austenitizing temperature in wt.%
T_A – austenitizing temperature [°C]
Si – content of the alloy is also in wt.%

Using this equation, a graph was developed showing influence of the austenitizing temperature T_A on carbon concentration in γ iron (Fig. 5.23). The graph is consistent with the experimental results [40].

With a sufficiently long time of austenitizing, an equilibrium carbon content can be obtained in the resulting austenite. There is no doubt that longer time of austenitizing makes the distribution of alloying elements in austenite grains more homogeneous and may contribute to the decomposition of carbides occurring in the microstructure of cast iron.

Both chemical composition and initial microstructure of cast iron are very important factors determining the effectiveness of austenitizing process. The pearlitic matrix of cast iron allows for faster C saturation in austenite than in the ferritic matrix. The factor responsible for this phenomenon is the difference in the kinetics of the transformation of pearlite and ferrite into austenite. The rate of C saturation in austenite also depends on the size and distribution of graphite nodules present in the microstructure. The larger the volume of graphite nodules and the smaller the distance between them, the shorter is the time needed by the austenite to get saturated with carbon. With the small internodular spacing, austenite will achieve the equilibrium carbon concentration in a shorter time. In addition to carbon, the most important constituent of cast iron is silicon. Its task is not only to reduce the carbon content in saturated austenite, but it also increases the temperature of

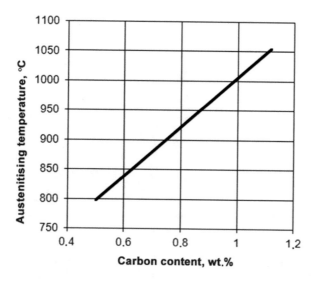

Fig. 5.23 Graph illustrating the relationship between austenitizing temperature and maximum carbon content in austenite. Carbon concentration in austenite calculated from Eq. (5.4) for the silicon content of 2.65 wt.%. (Based on [20, 40])

eutectoid transformation and extends the critical temperature range. Chromium also increases this temperature, whereas manganese and nickel reduce its value [18]. The presence of copper and nickel in cast iron has some influence on the austenitization process. Saturation of austenite with carbon depends on time and proceeds most intensely in the initial stage of this process. Cast iron containing nickel and copper takes longer time to reach the equilibrium carbon concentration in γ iron. However, studies conducted by Darwish and Eliott suggest that it is the austenitizing temperature and not the presence of alloying elements that is most critical in achieving the equilibrium carbon content in austenite [41].

In practice, austenitizing of cast iron is carried out in the temperature range of 850–930 °C, i.e., at a temperature by about 30–100 °C higher than the critical temperature (upper limit in the temperature range of eutectoid transformation). During heat treatment of ductile iron, the austenitizing temperature T_A is usually selected taking into account the microstructure of cast iron. The selected time of treatment depends on the chemical composition of cast iron and casting microstructure before austenitizing, based on the casting wall thickness, typically in the range of 48–150 s per 1 mm of the raw casting wall section [30].

5.3.2.2   Austempering

After austenitizing, austempering is carried out, i.e., rapid cooling of the casting to the temperature at which the isothermal transformation takes place (Fig. 5.24). The process starts with the nucleation and growth of ferrite plates forcing carbon

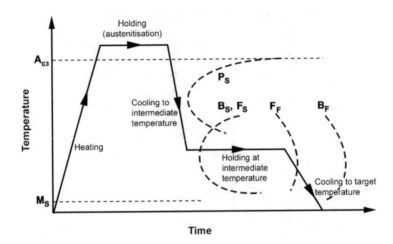

Fig. 5.24 Schematic diagram showing heat treatment steps during production of austempered ductile iron; P_S the beginning of pearlite transformation, B_S, B_F the beginning and end of bainite transformation, respectively, F_S, F_F the beginning and end of ausferrite transformation

migration to the surrounding austenite. The transport of carbon continues until a high degree of C saturation in austenite is achieved to make it stable both thermally and mechanically. According to some researchers, the process of "pumping" carbon into austenite during the growth of ferrite plates enables achieving carbon concentrations of up to 2.2 wt.% [42, 43]. This transformation stage is characterized by the achievement of an appropriate ferrite-to-austenite content ratio. Depending on the temperature of the transformation, the annealing time and the homogeneity of austenite before isothermal transformation, the resulting microstructure may have a diverse morphology.

5.3.2.3 Kinetics of Isothermal Transformation

In ductile iron, the isothermal transformation occurs in two stages illustrated in Fig. 5.25, which can be written down in the following form:

$$\text{Stage I} \Rightarrow \gamma^0 \rightarrow \gamma_{HC} + \alpha \tag{5.5}$$

$$\text{Stage II} \Rightarrow \gamma_{HC} \rightarrow \alpha + \text{carbides} \tag{5.6}$$

where:

γ^0 – primary austenite
γ_{HC} – stable (high carbon) austenite
α – lamellar ferrite

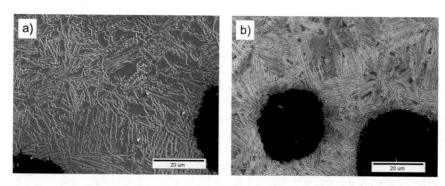

Fig. 5.25 The typical microstructures of austempered ductile iron are shown: (**a**) temperature of isothermal transformation T_{it} = 350–400 °C; (**b**) temperature of isothermal transformation T_{it} = 260–350 °C [44]. (Author's own work)

The primary austenite γ^0 with the lowest carbon content is transformed into ferrite plates and high-carbon austenite γ_{HC}. This process begins at the graphite-austenite and austenite-austenite phase boundaries and on the previously formed ferrite plates [44]. During growth of ferrite plates, carbon starts diffusing into the austenite retained in the space between the plates and the process continues until full stabilization is obtained. The end of austenite saturation with carbon to a certain level of concentration and entering the period of the structural stability $\alpha + \gamma_{HC}$ (ausferrite) opens the range of the so-called processing window (t_{it1} to t_{it2}) [44]. Annealing for a time longer than t_{it2} results in the precipitation of carbides, which is the phase undesired in the ADI microstructure. During transformation at a temperature of 350–400 °C, i.e., after the time t_{it2}, the precipitation of ε carbide or cementite starts at the α / γ_{HC} interface, leading to the formation of microstructure characteristic of the upper bainite (Fig. 5.25a). Lowering the temperature of austempering reduces the amount of austenite and widens the "processing window" [41]. For this temperature, longer time t_{it2} is associated with the formation of lower bainite with ferrite morphology (Fig. 5.25b), comprising plates with a thickness of several ten to several hundred nanometers. The low temperature of transformation reduces the diffusion rate and prevents carbon transport over long distances, which results in the formation of carbides not only at the α/γ_{HC} phase boundaries, but also inside the ferrite grains.

Studies of various types of cast iron confirm the important effect of chemical composition (especially of alloying additions) and of the austenitizing and austempering temperature on the "processing window" [44, 45]. The research described in [44] shows that increasing the content of alloying additions prolongs the time and reduces the temperature of a stable "processing window." A similar result is obtained by increasing the austenitizing temperature.

5.3.2.4   Temperature and Time of Austempering

The close dependence of mechanical properties of austempered ductile iron on the parameters of the austempering process determines the choice of temperature and duration of this treatment. Tensile strength and ductility are the most important properties of ADI. Knowing that there is a direct relationship between these properties and the content of different phases in the ADI matrix [46, 47], it is possible to control these properties through control of the content of these phases.

The time/temperature relationship between isothermal transformation and austenite content in the ADI microstructure described by Dymski [48] can be illustrated with appropriate measures of strength or elongation (Fig. 5.26) [48]. The next graph (Fig. 5.27) shows the change of carbon content in austenite depending on the temperature and time of austempering. Increasing the austempering time in the range of 15–240 min results in the increase of austenite saturation with carbon, the increase being most pronounced in the initial period of isothermal transformation. In turn, the highest saturation of austenite with carbon is observed during austempering at 300–350 °C. The degree of this saturation assumes lower values for the temperatures lower and higher than the indicated range. The lower the transformation temperature, the lower is the diffusion rate, the smaller is the size of its products, and the higher is the dislocation density and phase dispersion. All these factors strengthen the heat-treated cast iron.

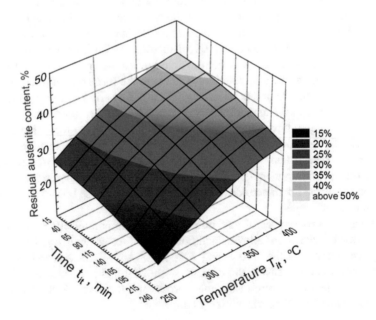

Fig. 5.26 Austenite content (in vol.%) in the ADI matrix as a function of austempering time and temperature. (Based on [48])

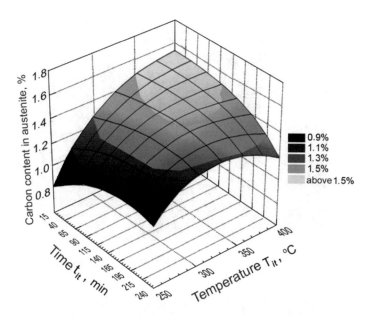

Fig. 5.27 Carbon content (in wt.%) in the ADI matrix as a function of austempering time and temperature. (Based on [48])

And yet, proper selection of the time of austempering may be difficult due to the postcasting chemical composition of the cast iron in microregions. The range of the "processing window" is not always open, which means that the same transformations may not occur simultaneously in the entire volume of the microstructure. At the temperature of the isothermal transformation, there may be permeating areas of austenite still in the process of being saturated with carbon (reaction in Eq. 5.5) and of high-carbon austenite, in which phase II of the transformation has already started (reaction in Eq. 5.6). Differentiation of chemical composition, and especially the content of strongly segregating elements, may be another reason for the modification of "processing window." The velocity of reactions (Eq. 5.5) and (Eq. 5.6) may be different due to the chemical heterogeneity of eutectic cells and individual phases. Of particular importance here is the uneven distribution of carbon on the cross-section of austenite grains.

5.3.2.5   Mechanical Instability of Austenite

Studies of ausferrite freezing processes indicate that the austenite→martensite transformation at temperatures below M_S occurs only in selected areas of the microstructure, i.e., in the block type austenite (Fig. 5.28). These places are also privileged places for the deformation-induced transformation [19, 49–52]. As a result of this transformation, hard high-carbon martensite is formed. This effect is

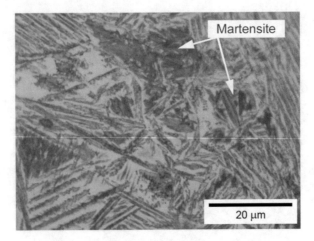

Fig. 5.28 Martensite in the fields of primary block austenite in ausferritic ductile iron. (Author's own work)

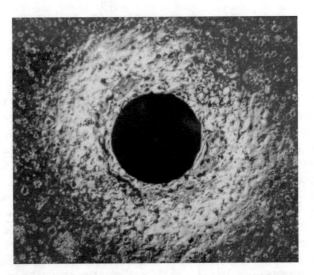

Fig. 5.29 The deformation zone formed on the surface of ausferritic ductile iron as a result of Rockwell hardness measurement; LM, x50, Nomarsky contrast. (Author's own work)

well visible under the microscope in samples subjected to hardness measurements during which the steel indenter permanently deforms the material (Figs. 5.29 and 5.30), creating a zone of plastic deformation around the impression. Martensite formed under the indenter of the hardness tester also creates problems in finding an unambiguous relationship between hardness and austenite content in ausferrite (Fig. 5.31) [33]. The same effect is also observed in other tests, e.g., in static tensile test, resulting in the occurrence of martensite on the fracture of samples which

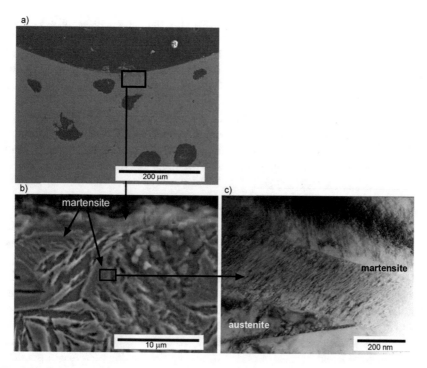

Fig. 5.30 Martensite in the ausferritic matrix of austempered ductile iron after deformation in Brinell hardness measurement: (**a**), (**b**) scanning electron microscopy, (**c**) transmission electron microscopy. (Author's own work)

Fig. 5.31 Dependence of austenite content in ausferritic ductile iron on its hardness HRC based on the data contained in [20, 21, 29, 32–34, 41, 44–52]

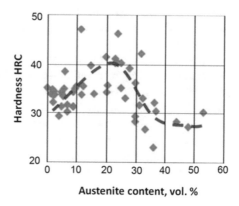

failed during the test [48]. Martensite of this type may also appear during sample preparation and during cutting and abrasion processes [50]. Therefore, based on the transmission electron microscope (TEM) images and X-ray diffractograms, it is difficult to find out whether the examined martensite is a product of transformation induced by the technological process or the result of an incorrect preparation of samples.

The phenomenon described above is related to the problem of obtaining in the material the value of stress or strain which will initiate the transformation of austenite into martensite. This phenomenon occurs in austenite-containing steels and is known as the transformation-induced plasticity (TRIP) effect.

5.3.2.6 Variations of Austempering Treatment

The tendency of austenite with low mechanical stability to deformation-induced transformation is also observed in other technological deformation processes that are not described in this chapter. Investigating the initiation of this transformation by appropriately high stresses and above all plastic deformation, it can be concluded that the transformation can occur in both surface and bulk deformation processes. The analysis of processes occurring in the entire volume of material is more difficult and requires studies of phenomena related to fracture mechanics, allowing for the microstructural heterogeneity of cast iron, casting defects, the presence of graphite, etc., during fatigue or fracture toughness testing. A general statement that can be formulated based on the analysis of the literature and the author's own research is that by minimizing the possibility of the occurrence of TRIP effect, one can obtain an improvement in the "bulk" mechanical properties of ductile iron. Any mitigation of the effect of this phenomenon has its source in the refinement of the ausferritic microstructure and in reducing the content of austenite with low mechanical stability.

As a result of studies of the ausferritic ductile iron, it was found that the value of the strain-hardening exponent n increases linearly with the increasing temperature of austempering. Based on this statement, it has been concluded that n will reach its maximum under the conditions of high austenite content, coarse ferrite, and high temperature of isothermal transformation, while the minimum will fall to the low austenite content and fine ferrite. This is confirmed by the dependence derived by Hayrynen et al., which enables the yield strength to be determined from the size of ferrite plates and the austenite content in matrix [53]:

$$\sigma = AL^{-1/2} + BX_\gamma + C \tag{5.7}$$

where:

σ – yield strength
L – size of ferrite plates

X_γ – austenite content
A, B, C – constant values

The dependence shows that it is the size of lamellar ferrite precipitates that controls the value of elastic stress in ausferritic ductile iron. This means that the finer the plates of ferrite in the cast iron matrix, the higher are the strength properties.

Studies on this issue were undertaken and reported in the literature. They were limited to special processes of the heat treatment of ductile iron. The main aim was to refine the microstructure of the cast iron matrix. The refinement was obtained through, higher than in standard processes, undercooling in the first stage of isothermal transformation during two-step austempering (Fig. 5.32) [54, 55] or by conducting low-temperature long-duration thermal processes under special conditions of chemical composition and temperature. These processes of obtaining nanostructural ausferrite is well documented (Fig. 5.33) [56].

It has been reported that compared to the common heat treatment process, undercooling of austenite to a lower temperature of austempering in the first stage of isothermal transformation results in the refinement of ausferrite grains. This reduces the thickness of ferrite plates and thus refines the ausferrite [54]. Also, it has an impact of obtaining slightly higher content of retained austenite compared to the conventional process. The higher retained austenite content and the higher degree of ferrite fragmentation in ausferrite according to relationship in Eq. 5.7 increase the value of the yield strength.

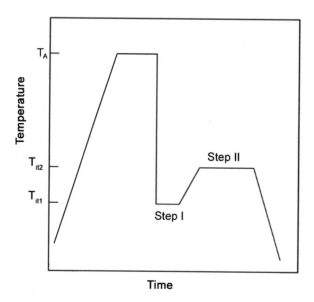

Fig. 5.32 Schematic diagram of a two-step austempering process to obtain ausferritic ductile iron [37]

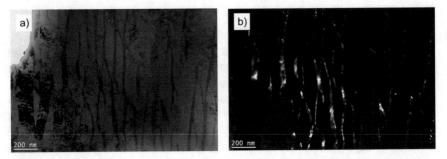

Fig. 5.33 Nanoausferritic structure of ductile iron matrix – a mixture of ferrite and retained austenite plates [56]

The microstructure obtained in the two-step process reduces the strain hardening exponent n and prevents the achievement of high elongation values (above 11%) in the quasistatic tensile test of cast iron [54]. Hence, it follows that increasing the content of stable retained austenite at the expense of mechanically less stable retained austenite will not be synonymous with improvement of the cast iron plastic properties. It can, therefore, be assumed that it is the mechanism of the deformation-induced martensitic transformation, dependent on the content and stability of retained austenite in the ausferritic microstructure, that is largely responsible for the obtained properties.

5.3.2.7 The Role of Chemical Composition

The choice of cast iron chemical composition to obtain ADI is not always similar to the composition of standard ductile iron grades. The first and one of the most important tasks is to determine the content of basic elements in such a way as to ensure the best possible spheroidization of graphite and the desired type of metal matrix. On the other hand, the quantity of these elements should be such as to counteract the formation of carbides or the tendency to form casting defects. The next task is to control the content of alloying additions, which determines the quality of the microstructure obtained in castings with different wall thicknesses, or to control different cooling rates in casting during austempering.

Austempering of gray nonalloyed cast iron in salt baths is effective only for castings with wall sections thinner than 10 mm (Fig. 5.34) [44]. A slight increase in hardenability occurs when the temperature of the austempering treatment is reduced (up to 30 mm at 250 °C). In the case of using a heating medium other than the salt bath, e.g., a fluidized bed, or in the case of castings with heavy wall sections, obtaining the correct ausferritic microstructure is possible primarily through the introduction of alloying additions [15, 21, 25, 44]. For isothermal transformation, popular and important elements are nickel, copper, and molybdenum added separately or in various combinations.

Fig. 5.34 Influence of austempering temperature on critical diameter for hardenability of ductile iron castings. (Based on [44])

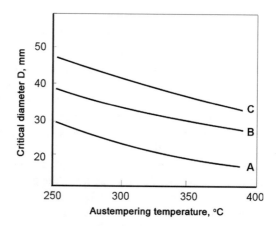

A Unalloyed ductile iron
B DI: 0.5wt.%Ni; 0.5wt.%Cu; 3.5wt.%C; 2.2wt.%Si
C DI: 0.75wt.%Ni; 0.75wt.%Cu; 3.5wt.%C; 2.2wt.%Si

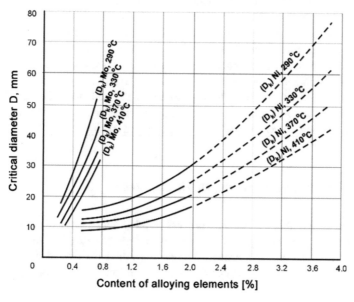

Fig. 5.35 The effect of molybdenum, nickel, and austempering temperature on critical diameter [57]

All the elements mentioned above are introduced primarily to increase the stability of austenite undercooled in the temperature range of pearlite transformation. Introduced together or separately, they increase the hardenability and critical diameter (Fig. 5.35) [57].

The strongest effect on the transformation of undercooled austenite is exerted by molybdenum or its combinations with nickel and copper. The Mo content in ductile iron is usually limited to ≤ 0.3 wt.% due to its very strong tendency to segregation and formation of $(FeMo)_3C$ carbides at the boundaries of eutectic cells [32]. Starting with the molybdenum content in cast iron of about 1 wt.% Mo, carbides of the $(FeMo)_3C_6$ and Mo_2C type can also appear [32].

The composition of ductile iron is usually enriched with 1.5% Cu, up to 4 wt.% Ni and up to 0.3 wt.% Mo. It is assumed that 1.5 wt.% of copper is equivalent to 0.3% Mo. It is, however, recommended to avoid introducing the maximum amount of copper due to its despheroidizing effect [10]. According to Pietrowski [32], copper in ductile iron changes essentially the course of pearlite and bainite transformation. In the temperature range of bainite transformation, copper increases the stability of austenite, slowing down the process of its decomposition.

Nickel added in an amount of up to 1.5% has no significant effect on the increase of austenite stability in the temperature range of pearlite transformation and in an amount of up to 1.0 wt.% in the temperature range of bainite transformation. Copper has similar effect as nickel, and hence cases are known of Ni being replaced with Cu or of both elements co-existing in the cast iron matrix to additionally counteract the formation of molybdenum or manganese carbides [29]. Like copper, nickel reduces the rate of austenite transformation and inhibits its decomposition in Stage II according to reaction in Eq. (5.6). Limitations in the use of Cu and Mo practically make nickel the choice of element that controls the microstructure of ductile iron. There are a few publications [e.g., 32] illustrating the effect of nickel (at a constant content of about Cu and Mo) and wall thickness in ductile iron castings on the microstructure obtained during cooling in the air below 750 °C. This is an example of the possibility of obtaining the ADI microstructure in cast iron of a given chemical composition when cooled in the air or in a casting mold.

5.3.3 Direct Austempering of Ductile Iron

"Direct austempering" is a specific process of alloy heat treatment, which consists of a controlled isothermal quenching carried out during casting cooling immediately after its solidification. The process has been studied for over a dozen years in numerous research and development centers for various types of foundry materials [42, 43, 58–60].

Increasing the cooling rate of casting from the solidus temperature, done by knocking out the casting from mold and cooling it down in the air or quick cooling to a predetermined temperature, is a well-known method of controlling the metal microstructure. This procedure is usually applied to aluminum alloys to simplify the heat treatment cycle and increase the hardness of the parts obtained. The solution heat treatment of aluminum alloys directly after casting is to some extent similar to the direct hardening of iron alloys. In this case, rapid cooling (e.g., in water) of an aluminum casting with a temperature below the solidus line is used immediately after removal from the casting mold (metal mold in this case). In contrast to austempering, this type of a fast cooling of Al alloys leads to a saturation of the casting with alloying constituents which, during later aging, gives rise to the effect of precipitation hardening. This process is slightly different in the case of direct austempering.

It is a well-known fact that when the cast iron cools down from the temperature of 1130–1150 °C, the casting "passes" through a temperature range of 950–815 °C, which is the range used for the austenitizing process during conventional heat treatment of ADI. So it seems logical to remove the hot casting from the mold in this temperature range and subject it to rapid cooling to a temperature at which it will be austempered. In this way it is possible to avoid cooling of casting and reheating it to the austenitizing temperature (Fig. 5.36).

It should be emphasized how important and beneficial this simplified heat treatment is. It is an indisputable fact that in addition to the appropriate properties, the attractiveness of the product also depends on its price. In the case of ADI, the cost calculation depends on several fixed factors at individual stages of the technological process. Heat treatment of ductile iron is still a very expensive element of the whole technological process. Therefore, its simplification, e.g., through the use of direct austempering, leads to significant savings, which are composed of a number of factors:

– Significant savings of electricity, which would have to be used to reheat the castings to the austenitizing temperature, and consequently reduced volume of raw materials necessary to produce such energy, less harm to the natural environment and lower environmental pollution associated with electricity generation in conventional power plants
– Elimination of additional equipment for high-temperature heat treatment, which is very important for production plants with limited resources of specialized equipment and funds for their purchase

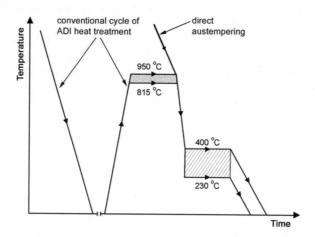

Fig. 5.36 Comparison of the conventional heat treatment cycle and direct austempering of ADI. (Author's own work based on [26, 61–63])

– Saving space, which would be occupied by devices for the austenitizing treatment
– Significant time saving of product manufacture, which should be expected especially in large-lot orders

Clearly defined benefits resulting from direct austempering are not, however, equivalent to obtaining the construction material with the desired in-service mechanical properties. Detailed analysis of the phenomena which occur during cooling and subsequent heat treatment and of the processes that occur during direct austempering suggests the possibility of some structural differences that may be reflected later in the properties of ADI castings. Although it is difficult to determine which of them and in what way (positive or negative) will affect the in-service mechanical properties of castings, from a production point of view this technology seems very interesting.

Owing to the research made in various scientific centers around the world, several ways of conducting the direct austempering process of ductile iron castings are known, to mention the use of metal molds, sand molds and the full mold process. The experience gained during their implementation allows describing the transformations and phenomena that accompany the ductile iron direct heat treatment process.

5.3.3.1 Microstructure and Properties

The description of the direct austempered ductile iron (DADI) microstructure should start with the statement that this microstructure is shaped in a different way than the microstructure of conventional ADI cast iron. The main reason is the lack of typical austenitizing treatment, i.e., absence in the direct cycle of the eutectoid $\gamma \leftrightarrow \alpha + Fe_3C$ transformation. The microstructure of DADI rather retains its original character, which means that it consists of austenite dendrites and eutectic cells on a microscale and of "macrocells" on a macroscale [26, 61]. Isothermal transformation does not exert a great influence on the structure modification, and, therefore, the primary microstructure should be expected after the casting has cooled down to ambient temperature (Fig. 5.9). The result may be greater heterogeneity in the chemical composition of ductile iron matrix, larger dimensions of products of the isothermal transformation, and directional character of the microstructure.

Rapid cooling of the casting from the temperature below the solidus line results in a specific "freezing" of the DADI primary microstructure, preventing the diffusion of its constituents, carbon in particular, at the temperature of the austenitizing treatment and lower temperature. This phenomenon results in greater homogeneity of the matrix in the vicinity of graphite precipitates than in the case of the usual cooling of cast iron during which the graphite nodules grow at the expense of carbon present in the surrounding matrix. The austenitizing treatment used in the conventional cycle makes the cast iron matrix homogeneous by redistribution of carbon, including partial dissolution of graphite nodules [32]. During the direct heat treatment of ductile iron, this effect will not occur. It should

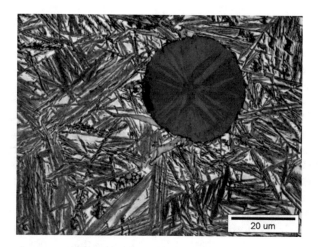

Fig. 5.37 Microstructure of DADI cast iron (austempering temperature $= 300$ °C, time $= 90$ min.). (Author's own work)

be noted that in the case of DADI, there is also absence of carbon diffusion toward the graphite nodules during slow cooling of the casting from the solidus temperature to ambient temperature. This leads to a "bull eyes" structure. It can, therefore, be assumed that in DADI the concentration of carbon in the vicinity of graphite precipitates should be comparable to ADI or even higher. Images of the microstructure of samples taken from DADI under identical conditions with ADI (austempering temperature $= 300$ °C, time $= 90$ min.), shown in Figs. 5.37, confirm these assumptions. It follows that in the case of conventional treatment, the transformation rate near the graphite/austenite interface must be high since products of this transformation show a high degree of refinement. The high rate of austenite transformation into ferrite suggests a lower carbon content in the microregions or higher silicon content allowing for the occurrence of this kinetics. Hence it follows that the extent of homogenization of the DADI matrix can be equal to that of ADI or possibly even higher.

The phenomena that occur at the boundaries of eutectic cells or dendrites may assume the course slightly different than in the case described above. Austenitization of ductile iron, carried out during conventional heat treatment aimed at obtaining ADI, may lead to a redistribution of elements grouped in these microregions. This phenomenon does not occur during direct austempering, and hence the conclusion follows that the DADI matrix will be more heterogeneous in the areas of the austenite/austenite phase boundaries. Additionally, austenitizing treatment enables dissolution of the undesired phases appearing in ductile iron, e.g. carbides. The lack of austenitizing treatment during direct austempering does not create conditions for the dissolution of carbides that may have appeared during solidification.

Analysis of individual constituents occurring in the DADI microstructure is in practice reduced to the basic phases present in the matrix, i.e., ferrite and austenite.

However, some irregularities have been found, which also suggest the need to investigate the characteristics of the precipitates of nodular graphite.

As in the case of all other types of ductile iron, nodular graphite is one of the most important constituents of the DADI microstructure. During the process of direct austempering, some differences in the formation of this phase were noted compared to the standard types of ductile iron, including ADI. The reason is that the direct heat treatment inhibits the growth of graphite nodules in the solid phase, which runs all the time during cooling of the casting, although at a constantly decreasing speed. As mentioned earlier, in the case of DADI, a specific "freezing" of the microstructure occurs, thereby inhibiting the diffusion of carbon into the graphite nodules as soon as the intensive cooling to the temperature of the austempering treatment starts. The above leads to the formation of graphite nodules characterized by a smaller volume and to milder heterogeneity of carbon distribution in the matrix surrounding the graphite and to less deformation. This is consistent with the studies presented in [62] and systematic measurements of the size of graphite particles present in the microstructure of castings made from DADI and ADI originating from the same melts. The results indicate that the average diameter of graphite nodules is approx. 31 μm in DADI, while in ADI this value is at a level of 35 μm, which means that it is by about 10% higher than in DADI.

Ferrite, present in DADI in the form of plates, is slightly different from the ferrite formed in the ADI matrix. It was found earlier that these plates are characterized by a specific orientation and form structures coarser than in ADI. The plates are longer during direct austempering at a lower temperature, but their length decreases when austempered at high temperature. Changes also occur in the shape and thickness of plates (Figs. 5.38 and 5.39). Lower austempering temperature affects the shape of the plates in such a way that they become more "sharp-pointed," resembling

Fig. 5.38 Fine lamellar ferrite in the matrix of DADI (austempering temperature = 260 °C, time = 90 min.). (Author's own work)

Fig. 5.39 Plates of "feathery" ferrite in the matrix of DADI (austempering temperature = 360 °C, time = 90 min.). (Author's own work)

martensite in their appearance, and have also smaller thickness of approx. 0.2–0.8 μm. Higher temperature of direct austempering makes the plates more ramified. They look more "feathery" and the thickness grows to approx. 1.2 μm.

Retained austenite present in the DADI matrix has also been examined in relation to the austenite found in ADI. The most important conclusion that followed this analysis is that under the same conditions of the austempering treatment, the proportional content of austenite in the microstructure of DADI is slightly higher than in ADI. Proper amount of retained austenite is usually associated with the ADI plasticity. Its presence in DADI is confirmed by a SEM image of the fracture showing numerous dimples (Fig. 5.40) [63]. The DADI matrix is mainly composed of the lamellar ferrite and retained austenite. As demonstrated by the microanalysis and TEM examinations, the matrix may also contain carbides and martensite, which are phases undesirable in the microstructure of both DADI and ADI.

The presentation of DADI mechanical properties should start with the statement that only in a very few cases of direct heat treatment, attempts to achieve the strength/ductility combination consistent with the requirements of the European Standard EN 1564 for ADI have ended in success. This fact is graphically presented in Fig. 5.41. In spite of this, the results obtained for DADI indicate that its properties far outweigh those guaranteed by other types of "ordinary" ductile iron. It is interesting to compare the properties of DADI and ADI obtained for the same conditions of the austempering treatment.

The information presented in the previous subsections reveals some basic facts concerning DADI cast iron, which would be worth emphasizing. Ausferrite in the DADI matrix differs slightly from the ausferrite obtained by conventional austenitizing and austempering. For the same heat treatment parameters, in the DADI matrix a higher content of the stable retained austenite was recorded,

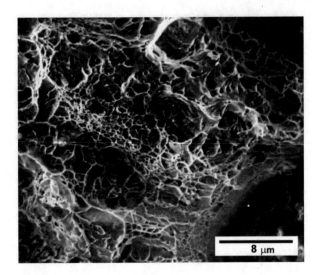

Fig. 5.40 DADI. SEM fractograph showing dimples characteristic of ductile fracture. (Author's own work based on [63])

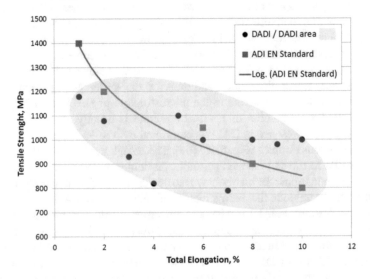

Fig. 5.41 Mechanical properties of DADI compared to the properties of ADI included in the EN 1564 Standard. (Author's own work)

accompanied by the ferrite strongly saturated with carbon. Compared to ADI, the microstructure of DADI seems to be coarser. Additionally, in DADI, some areas with increased concentration of alloying additions have been identified, especially at sites referred to as last to freeze (LTF), which were not found in the structure of ADI. Molybdenum occurring in LTF microregions can contribute to the formation of Mo_xC_y carbides (identified in the research presented in [47]), the presence of which increases DADI hardness but significantly deteriorates its plastic properties.

Based on the analysis of the thermal history of castings subjected to the direct austempering and conventional austempering, it was found that the contribution of various mechanisms of hardening the DADI matrix is slightly different than in the ADI cast iron. Closer analysis of the hardenability of both types of cast iron suggests that the most important factor responsible for the lower ductility and fracture toughness of DADI compared to ADI is a coarse metal matrix, left unrefined due to the omission of the $\gamma \leftrightarrow \alpha + Fe_3C$ eutectoid transformation, which occurs during cooling of the casting and its reheating for the austenitizing treatment. The explanation for the high yield strength $R_{0.2}$ of DADI, similar to the values obtained in ADI, should be sought in the strain hardening effect resulting from the higher density of dislocations in austenite and ferrite and from the presence of microtwins. However, the main reason for the high DADI strength is, as in ADI, the solid solution strengthening of both ferrite and austenite. Saturation of both phases with carbon is the reason for DADI in the considered variants of austempering treatment to reach a tensile strength, R_m, of 1200 MPa.

The above statements characterizing the ductile iron obtained as a result of direct heat treatment are sufficient to conclude that DADI has the mechanical properties slightly inferior to ADI. It was also noticed that extending the time of DADI austempering has increased both strength and elongation. DADI is competitive with ADI not only in terms of higher mechanical properties, hardness in particular, but also in terms of the production cost. A brief analysis of the production process shows that the cost of producing a DADI blade for the shot blasting machine is by 9% lower than in the case of ADI and by 25% lower than in the case of alloyed high-chromium cast iron with the addition of molybdenum, so far the standard material for this product [62].

5.3.4 Ausforming

The refinement of ausferrite grains in the matrix of gray cast iron can also be obtained by combined plastic forming and heat treatment, generally known under the name of ausforming. This is a very interesting issue and since it has recently gained wide recognition among scientists, special attention will be paid to this type of cast iron.

Hot or cold plastic forming of cast materials refines their structure, reducing shrinkage and gas porosity and microsegregation of alloying elements. It also provides additional nucleation sites during austenite transformation after deformation or recrystallization. These assumptions were adopted in the high-temperature deformation process before the start of austenite transformation in a ductile iron matrix. Studies show that ausforming during ausferritic structure formation significantly shortens the time necessary to trigger the austenite-ferrite transformation by introducing additional energy into the system through mechanical action [64].

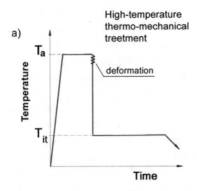
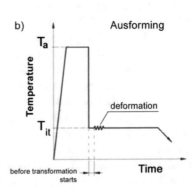

Fig. 5.42 Thermal processing of ductile iron: (**a**) high-temperature thermo-mechanical treatment, (**b**) ausforming. T_a and T_{it} are temperatures of austenitization and isothermal transformation, respectively. (Based on [65, 66])

It was also found that austenite saturation with carbon was higher than in the conventional process, which also indicates an acceleration of the reaction. The increased ferrite nucleation rate leads to fragmentation of the ausferritic structure and ensures better homogeneity of the alloy, and hence also of its component phases [65, 66]. A more homogeneous structure prevents the occurrence of martensite at ambient temperature even after a short bainite transformation time.

Processes completed with 25% deformation show that cast iron dramatically gains in mechanical properties. At least 70% increase in the yield strength and 50% increase in the tensile strength as compared to the conventional ADI were demonstrated. However, depending on the temperature and mode in which the process of deformation is carried out, castings will change their properties. For the thermoplastic treatment shown in Fig. 5.42, alloyed cast iron characterized by high hardenability often with the additions of Ni and Mo should be used. Alloying additions are also important when ausforming is carried out on products with large cross-sections and when high degrees of the deformation are applied. Ausforming carried out on nonalloyed cast iron was found to significantly reduce plasticity with only a minimum increase in strength.

For simple products, methods of forging gray iron castings have been proposed. Special preforms would be used for this purpose from which the casting would be transported to a bath ensuring thermal treatments in the temperature range of bainite transformation [64].

5.4 Mechanical Properties

The richness of the variants of cast iron structure provides this material with an extremely wide range of achievable properties. Not all of them can be discussed here, but, apart from the most important ones which include comparable strength to steels, good fracture toughness, and high abrasion resistance, it is worth paying attention to the properties less often discussed, e.g., corrosion resistance, good thermal conductivity, and resistance to the effect of low or high temperature [14, 17].

5.4.1   Strength of Cast Iron

The strength of gray cast iron is primarily a function of its structure, where the type of matrix and the content, shape, and distribution of graphite precipitates are of importance [14, 17]. In general, increasing the amount of pearlite in the matrix increases the value of R_m and reduces the plasticity, which results from the mechanical properties of pearlite and ferrite (pearlite: $R_m = 550$–850 MPa, medium plasticity: $A_5 = 2 \div 10\%$, ferrite: $R_m = 350$–500 MPa, high plasticity: $A_5 = 10 \div 22\%$). In gray cast iron with flake graphite, increasing the amount of pearlite in the matrix from 70 to 100 vol.% increases the value of the tensile strength R_m by 20–35 MPa. The low strength of cast iron with a martensitic matrix is caused by the presence of quenching stresses.

A similar relationship governs the type of ductile iron matrix and its strength. In this type of cast iron, ferrite formation leads to a reduction in the tensile strength R_m and improvement of plastic properties (Fig. 5.43) [14]. Increasing the proportion of pearlite in white cast iron (with simultaneous reduction in the content of cementite) increases not only the tensile strength, but also the ductility of white cast iron.

The parameters of the cast iron primary crystallization also exert a significant effect on the strength of this material. The tensile strength of cast iron increases with the increasing content of the dendrites of primary austenite, with the increasing degree of their branching and with the increasing number of eutectic grains (Fig. 5.44) [14].

The strength and ductility of metal matrix decrease to a large extent due to the presence of graphite precipitates in the cast iron structure. The negative influence of graphite results from the reduction of the active cross-section of the metal matrix and also from the fact that in this matrix graphite particles play the role of micronotches. With the same total volume of graphite in cast iron, the reduction of the active cross-section is the least severe in the cast iron with nodular graphite, and the most severe in the cast iron with flake graphite. For graphite flakes with elongated shapes, the degree of reduction of the active surface can reach even 50%, especially when casting walls have small cross-sections. The fact that graphite particles act as micronotches in the cast iron matrix is due to the shape of these precipitates. This effect is weakest in the nodular graphite and strongest in the graphite in the shape of sharp-ending flakes.

Summing up, it can be stated that in gray cast iron with the decreasing content of graphite precipitates and the increasing degree of their refinement, both cast iron strength and ductility tend to increase. The transition from the flake graphite to the nodular graphite significantly improves the strength and even more the ductility of the cast iron.

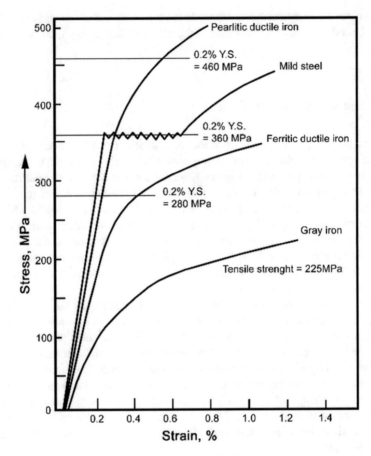

Fig. 5.43 Curves illustrating the elastic and plastic behavior of mild steel, cast iron with flake graphite, and cast iron with nodular graphite with a ferritic and pearlitic matrix. (Based on [14])

Figure 5.45 shows the effect of the content of regular nodular graphite on the tensile strength of ductile iron with different types of matrix [14]. The harmful effect of the presence of various forms of graphite is best visible in the case of purely pearlitic matrix. However, an ausferritic matrix, obtained as a result of heat treatment, mainly of ductile iron, has the strongest influence on the cast iron strength.

The advantage of ductile iron with ausferritic matrix is primarily a unique combination of mechanical properties (Fig. 5.46) [15–18]. The high temperature of austempering treatment (reaching 400 °C) enables producing the cast iron characterized by high elongation and tensile strength at a level of 850 MPa. At a relatively low transformation temperature, e.g. 260 °C, ADI grades with strengths of up to 1600 MPa are obtained. They are additionally characterized by a very high hardness and excellent abrasive wear resistance.

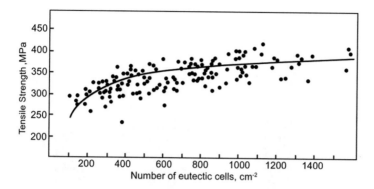

Fig. 5.44 Influence of the number of eutectic grains on the tensile strength of nonalloyed gray cast iron with flake graphite. (Based on [14])

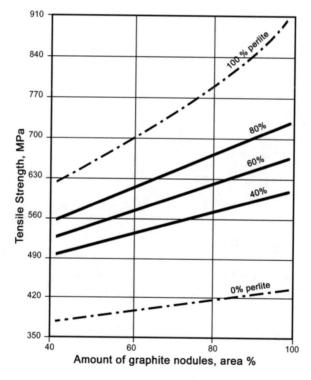

Fig. 5.45 Influence of the content of nodular graphite (lower than 100% content of the nodular graphite means its degeneration toward the vermicular or flake form) and pearlite on tensile strength. (Based on [14])

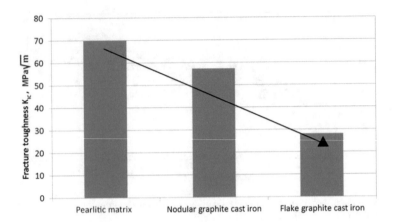

Fig. 5.46 Influence of graphite form on the fracture toughness (K_{1C} value) for eutectic cast iron with pearlitic matrix. (Author's own analysis based on [14, 17])

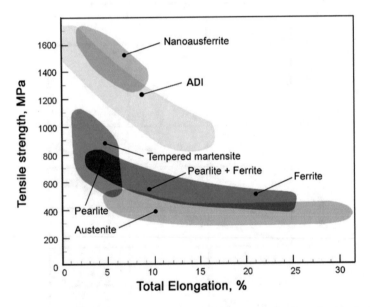

Fig. 5.47 Mechanical properties of ductile iron with various types of matrix. (Author's own analysis based on [15–18])

5.4.2 Fracture Toughness

The effect of primary structure, and especially of graphite precipitates present in this structure, on the cast iron tendency to cracking is quite significant [24]. The presence of graphite precipitates acting as micronotches drastically reduces the value of fracture toughness (K_{1C}) as shown in Fig. 5.47. for the eutectic cast iron with

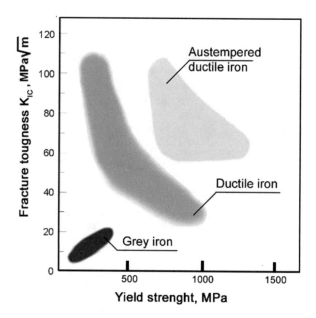

Fig. 5.48 Different types of ductile iron compared in terms of their fracture toughness (K_{IC} value). (Author's own analysis based on [14, 17, 33])

flake graphite and with nodular graphite compared to a purely pearlitic matrix [14, 17]. The most important factors that control the gray cast iron fracture toughness include the shape and dispersion of graphite precipitates and absence of carbides, nonmetallic inclusions and porosity.

Assuming that the spherical shape of graphite is the optimal one, it can be concluded that fracture toughness depends to a large extent on the type of cast iron matrix. The drawing shows the relationship between fracture toughness and various types of ductile iron matrix after heat treatment. The graph in Fig. 5.48 [14, 17, 33] indicates that the ADI grades with the highest strength have better fracture toughness than the ductile iron based on a pearlitic matrix or matrix of tempered martensite. On the other hand, compared to pearlitic ductile iron, the ADI of a lower strength has an almost double K_{IC} value. All tests in which ADI shows fracture toughness better than or comparable to other materials (e.g. ductile iron, forged steel or carburized steel) emphasize one fact, namely that along with the increase in austempering temperature, the resistance to impact loads also increases, reaching a maximum at 340–370 °C (Fig. 5.49) [17, 33]. This characteristic is a direct proof of the relationship between fracture toughness and austenite content in the ADI microstructure.

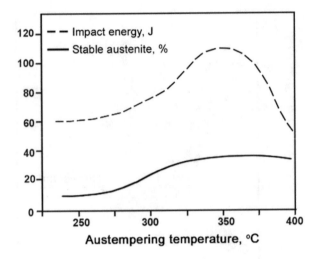

Fig. 5.49 Relationship between stabilized austenite content in ADI microstructure and elongation and impact toughness. (Author's own analysis based on [17, 33])

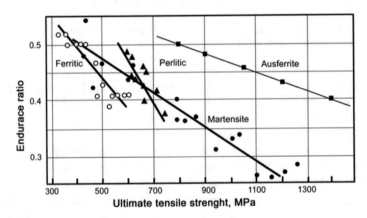

Fig. 5.50 Fatigue strength to tensile strength ratio (endurance ratio) for ductile cast iron. (Author's own analysis based on [14, 17, 25])

5.4.3 Fatigue

The structure of cast iron affects its fatigue strength primarily through the shape and size of graphite particles. Therefore the ratio of fatigue strength to tensile strength decreases from a value of about 0.5 for the fine precipitates of flake graphite to a value of about 0.3 for the cast iron with nodular graphite. Hence it follows that the change of graphite shape from flakes to nodules affects to a lesser extent the increase in fatigue strength than the corresponding increase in tensile strength (Fig. 5.50) [14, 17, 25]. The fatigue strength of cast iron decreases with the increasing size of graphite flakes.

In the case of fatigue strength, the impact of the matrix is not as spectacular as in the case of fracture toughness, but even then, it is worth paying attention to one characteristic feature of the cast iron after heat treatment. It is the possibility to mechanically harden the surface of finished ADI castings by burnishing, shot blasting or machining. This is due to the presence of carbon-saturated austenite in the deformed microstructure. Machining triggers the mechanism of the deformation-induced austenite transformation into martensite. This mechanism combined with the deformation by twinning causes additional hardening of the casting surface [52, 67]. Literature data suggest that, due to this effect, some ADI grades are in this respect comparable to or better than forged steels [14, 17].

As in the case of fracture toughness, also for fatigue strength, a relationship can be derived between this property and the heat treatment parameters or, in other words, between the content of unreacted austenite in ADI and mechanical properties of this cast iron. This is illustrated by the graph in Fig. 5.50 showing the fatigue strength-to-tensile strength ratio of ADI. For cast iron grades with lower strength, this ratio reaches 0.5 and decreases below 0.3 with an increase in tensile strength R_m.

Comparison of fatigue strength of various materials, including two classes of ADI cast iron, provides interesting information concerning the use of this material for gears [17]. From the presented results of studies it follows that surface hardening of castings made from ADI significantly increases their fatigue strength, which is higher than in the common and conventionally hardened ductile iron and cast steel grades, and is comparable to steel subjected to surface hardening treatment [24, 50–52].

5.4.4 Wear Resistance

Though abrasive wear of cast iron mainly depends on the conditions under which the wear process occurs, the structure of this material and the properties of its individual constituents also exert an important effect on the wear intensity (Table 5.3) [14, 16, 17, 19, 25, 33, 68]. Abrasive wear is usually directly related to the hardness of the material and the ratio of the cast alloy to abrasive material hardness values. Therefore, to improve the abrasion resistance of cast iron, alloying elements forming hard phases are introduced.

The material of the highest abrasion resistance under the conditions of low dynamic loads is white cast iron containing cementite [14, 17]. The chemical composition of this material is determined taking into account the carried load and the temperature at which it performs. At high load values, it is advantageous to reduce carbon and phosphorus content in spite of the related drop in hardness. Abrasion resistance of the white cast iron is observed to increase with the increasing proportional content and dispersion of cementite. In both white and gray cast irons, this property decreases with the increasing interlamellar spacing of pearlite. Wear

Table 5.3 Comparison of hardness values of cast iron structural constituents. (Based on [14, 16, 17, 19, 25, 33, 68])

Structural constituent of cast iron	Hardness HV
Unalloyed ferrite	70–200
Unalloyed austenite	170–230
Alloyed austenite	250–600
Unalloyed pearlite	250–320
Alloyed pearlite	300–460
Ausferrite	280–550
Martensite	500–1010
Cementite	840–1100
$(Cr,Fe)_7C_3$	1200–1600

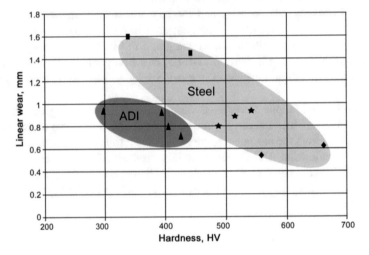

Fig. 5.51 Abrasive wear resistance compared for different materials. (Author's own analysis based on [67–69])

resistance increases with the matrix changing from pearlitic through ausferritic to martensitic.

So far, for the cast iron with an ausferritic matrix, a definite relationship between hardness and wear resistance has not been found (Fig. 5.51) [68, 69]. It turns out that slightly lower hardness of ausferritic cast iron does not prevent it from competing with steel with a hardness of 600 HV. This is due to the aforementioned surface hardenability of ADI where through mechanical action the critical stresses are overcome causing deformation-induced martensitic transformation and strain hardening by twinning. The result of this surface hardening is definite increase in hardness (Fig. 5.52). Both the surface hardening effect and the special microstructure of ADI translate into better wear resistance.

An interesting case is the new type of cast iron proposed by American scientists called carbidic austempered ductile iron (CADI) [70]. This material combines the features of ADI with the structure of cast iron containing hard chromium carbides

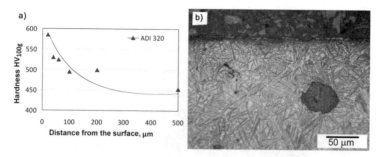

Fig. 5.52 (**a**) Hardness distribution in the surface layer and microstructure of ausferritic ductile iron (ADI 320 grade) after abrasion process carried out with loose abrasive medium, (**b**) typical microstructure after abrasion. (Author's own work)

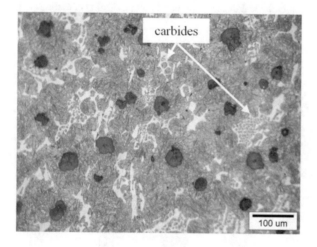

Fig. 5.53 Microstructure of carbidic austempered ductile iron (CADI). (Author's own work)

in a proportion of up to 18 vol.% (Fig. 5.53). This structure, by providing adequate strength to cast iron, enables 40% improvement in wear resistance compared to the common ADI grades [70].

5.5 Applications and Future Perspectives

Cast irons are the most popular foundry materials, also constituting the largest tonnage share. They account for 70% of world production of castings (Fig. 5.54), which is mainly due to the variety of properties offered by castings. Comparable proportions of individual foundry materials are found in all countries in which the foundry industry is highly developed (Fig. 5.50) [13] despite the growing demand for light alloys. Trends observed in global production indicate that the production

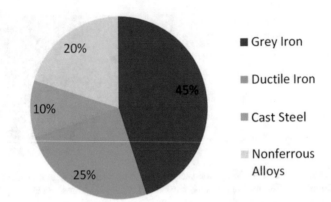

Fig. 5.54 Share of individual materials in the overall production of castings in the world (2016). (Author's own analysis based on [12])

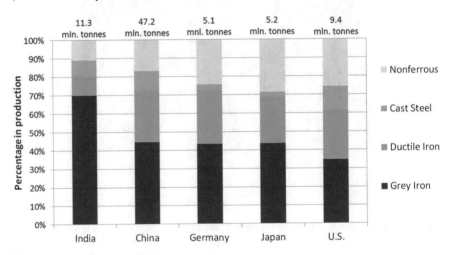

Fig. 5.55 Production of castings in top five countries producing highest amounts of cast metals (2016). (Author's own analysis based on [12])

of ductile iron or cast iron with vermicular graphite is treated as an indicator of the development of the foundry industry in a given country and of those industries for which castings are important components of end products. This is particularly well visible in the machine and automotive industries. ADI is considered the highest achievement of cast iron technology; hence it is worth quoting several possible applications of this material (Fig. 5.55):

- Automotive: gears, connecting rods, camshafts, steering axles
- Agriculture: plough ploughshares, parts of the chassis of the agricultural machines, toothed wheels
- Mining: pump bodies and casings, rotors, crankshafts, drills

- Railway applications: locomotive wheels, clutches, suspension elements of wagons
- Arms industry: missile shells, track vehicle feet, armored vehicle casings

5.6 Summary

It is not possible to describe in a few words the vast potential of all modern types of cast iron obtained in foundry processes. Each casting method (melting process, mold material, etc.), each alloying element, and each heat and surface treatment technology will have its respective share in the properties of the final product. This is true for any group of foundry materials, and this is also true for cast iron – the most popular casting alloy in the world. Good castability, the ability to exactly reproduce even most intricate configurations and at the same time the ability to shape the structure in both micro- and nanoscale to the level that provides the highest strength properties – these are the unique features of this alloy. This chapter, showing only some isolated examples, tries to convey this simple message that modern technologies, even if based on most common metal alloys, are still capable of providing the necessary properties to modern engineering constructions. Many other possibilities of shaping cast iron properties have not been mentioned here, e.g., creating unique properties by surface layer technologies, electromagnetic processing, or creating discontinuous structures with multidimensional geometry through new casting techniques or 3D printing methods [71]. The author, however, is convinced that these are areas of future development also for high-strength casting iron alloys. One can learn about the latest achievements in scientific conferences where subject matter is focused on cast iron and other important iron alloys. These are events like the International Symposium on the Science and Processing of Cast Iron and the International Conference on Modern Steels and Iron Alloys.

References

1. World Foundry Organization, 40th census of world casting production —2005. http://www.thewfo.com/census Accessed 30 Sept 2019 (2006)
2. World Foundry Organization, 41st census of world casting production —2006. http://www.thewfo.com/census Accessed 30 Sept 2019 (2007)
3. World Foundry Organization, 42nd census of world casting production —2007. http://www.thewfo.com/census Accessed 30 Sept 2019 (2008)
4. World Foundry Organization, 43rd census of world casting production —2008. http://www.thewfo.com/census Accessed 30 Sept 2019 (2009)
5. World Foundry Organization, 44th census of world casting production. http://www.thewfo.com/census Accessed 30 Sept 2019 (2010)
6. World Foundry Organization, 45th census of world casting production. http://www.thewfo.com/census Accessed 30 Sept 2019 (2011)
7. World Foundry Organization, 46th census of world casting production. http://www.thewfo.com/census Accessed 30 Sept 2019 (2012)

8. World Foundry Organization, 47th census of world casting production. Dividing up the Global Market. http://www.thewfo.com/census Accessed 30 Sept 2019 (2014)
9. World Foundry Organization, 49th census of world casting production. Modest Growth in Worldwide Casting Market. http://www.thewfo.com/census Accessed 30 Sept 2019 (2015)
10. World Foundry Organization, 50th census of world casting production. Global Casting Production Stagnant. http://www.thewfo.com/census Accessed 30 Sept 2019 (2016)
11. World Foundry Organization, 51st census of world casting production. Global Casting Production Growth Stalls. http://www.thewfo.com/census Accessed 30 Sept 2019 (2017)
12. World Foundry Organization, 52nd census of world casting production. Global Casting Production Expands. https://www.moderncasting.com/issues/december-2019 Accessed 30 Dec 2019 (2018)
13. J.J.Sobczak, E.Balcer, A.Kryczek, Polish foundry on the background of world casting. Current status and trends, Nationwide Day of the Foundryman, (Krakow 2015)
14. Podrzucki Cz, *Cast iron – structure, properties, application* (ZG STOP, Kraków, 1991)
15. Podrzucki Cz, Problems of production of castings from spheroidal cast iron ADI. Foundry J. Polish Foundrymen Assoc **10**, 260–265 (1996)
16. E. Guzik, *Processes for Refining Cast Iron - Selected Issues* (Archives of Foundry Engineering, Katowice, 2001)
17. Ductile Iron Society, Section IV. Austempered ductile iron. https://www.ductile.org/didata/Section4/4intro.htm. (1998) Accessed 30 Sept 2019
18. E. Fraś, *Crystallization of Metals and Alloys* (PWN, Warszawa, 1992)
19. H.K.D.H. Bhadeshia, *Bainite in Steels* (The Institute of Materials, Cambridge, 2001)
20. S. Pietrowski, Effects of Ni, Mo and cu on the transformation of austenite into softwoods in raw castings of nodular iron. Przegląd Odlewnictwa **5**, 157–161 (1987)
21. Pietrowski St.: ,Conversion of austenite - bainite, martensite in nodular cast iron", Inżynieria Materiałowa, 1990, nr5, 115–120
22. Rivera G., Boeri R., Sikora J., Influence of the inoculation process, the chemical composition and the cooling rate, on the solidification macro and microstructure of ductile iron, Seventh International Symposium SPCI7', (Barcelona 2002), Pre Pints
23. K. Sękowski, Heterogeneity of the chemical composition of the metal matrix of nodular cast iron. Foundry J. Polish Foundrymen Assoc **8-9**, 250–255 (1973)
24. J. Campbell, *Complete Casting Handbook. Metal Casting Processes, Metallurgy, Techniques and Design* (Elsevier, Butterworth-Heinemann, 2015)
25. S.I. Karsay, *Ductile iron I. Production* (QIT - Fer et Titane Inc, Quebec, 1992)
26. G. Rivera, R. Boeri, J. Sikora, Influence of the solidification microstructure on the mechanical properties of ductile iron. Int. J. Cast Metals Res. **11**, 533–538 (1999)
27. H. Bayati, R. Elliot, Influence of austenitising temperature on austempring kinetics in high Mn alloyed ductile cast iron. Mat. Sci. Technol. **11**(4), 776–786 (1995)
28. T. Kobayashi, H. Yamamoto, Development of high toughness in austempered type ductile cast iron and evaluation of its properties. Metall. Trans. A. **19A**, 319–327 (1988)
29. H. Bayati, R. Elliot, The concept of an austempered heat treatment processing window. Int. J. Cast Metals Res. **11**, 413–417 (1999)
30. L. Shen-Chin, H. Cheng-Hsun, C. Chao-Chia, F. Hui-Ping, Influence of casting size and graphite nodule refinement on fracture toughness of austempered ductile iron. Metall. Met. Trans. **29A**(10), 2511–2521 (1998)
31. E. Fraś, K. Wiencek, M. Górny, H.F. López, E. Olejnik, Grain count in castings: Theoretical background and experimental verification. Int. J. Cast Metals Res. **27**, 15–25 (2014)
32. S. Pietrowski, Ductile cast iron with a bainitic ferrite structure with austenite or bainitic ferrite. Archiwum Nauki o Materiałach **18**(4), 253–273 (1995)
33. P. Nawrocki, A. Kochański, D. Myszka, Statistical assessment of the impact of elevated contents of cu and Ni on the properties of Austempered ductile Iron. Arch. Metall. Mater. **61**, 2147–2150 (2016)
34. P. Nawrocki, K. Wasiluk, K. Łukasik, D. Myszka, Influence of pre-heat treatment on mechanical properties of Austempered ductile cast Iron. Arch Foundry Eng **18**, 176–180 (2018)

35. T. Szykowny, T. Giętka, The hardening and tempering of hot rolling of ductile cast iron. Arch Foundry Eng 6(19), 349–354 (2006)
36. European Standard Grades according to EN 1564
37. US Standard Grades according to ASTM A897/A897M-03
38. L. Jincheng, R. Elliot, The influence of cast structure on the austempering of ductile iron. Part 3. The role of nodule count on the kinetics, microstructure and mechanical properties of austempered Mn alloyed ductile iron. Int. J. Cast Metals Res. 12, 189–195 (1999)
39. J.M. Schissler, J. Saverna, The effect of segregation on the formation of ADI. J. Heat Treat. 4, 167–176 (1985)
40. Volgt L, Austempered ductile iron – Process control and quality assurance. Proceedings 1st International Conference on ADI, University of Michigan, Ann Arbor, USA (1986)
41. N. Darwish, R. Elliot, Austempering of low manganese ductile irons. Part 1. Processing window. Mat. Sci. Technol. 9(7), 572–586 (1993)
42. N. Varahraam, O. Yanagisawa, Properties of austempered ductile iron produced in equipment designed for consecutive instream treatment, gravity-die casting, and direct austempering. Cast Metal 3(3), 129–139 (1990)
43. S.M. Yoo, K. Moeinipour, A. Ludwig, P.R. Sahm, Numerical simulation and experimental results of in situ heat treatment austempered ductile iron. Int. J. Cast Metals Res. 11, 483–488 (1999)
44. H. Bayati, R. Elliot, Influence of austenitising temperature on austempering kinetics in high Mn alloyed ductile cast iron. Mat. Sci. Technol. 11(4), 776–786 (1995)
45. G. Cinceros, L. Perez, C. Campos, C. Valdes, The role of cu, Mo and Ni on the kinetics of the bainitic reaction during the austempering of ductile irons. Int. J. Cast Metals Res. 11, 425–430 (1999)
46. D.M. Moore, T.N. Rouns, K.B. Rundman, The relationship between microstructure and tensile properties in ADI. Am. Foundry Soc. 95, 765–774 (1987)
47. A. Trudel, E. Gangne, Effect of composition and heat treatment parameters on the characteristics of austempered ductile iron. Can. Metall. Q. 36(5), 289–298 (1997)
48. S. Dymski, Kształtowanie struktury i właściwości mechanicznych żeliwa sferoidalnego podczas izotermicznej przemiany bainitycznej (Wyd. Uczelniane ATR, Bydgoszcz, 1999)
49. J. Aranzabal, I. Guitierrez, J.M. Rodriguez-Ibabe, J.J. Urcola, Influence of the amount and morphology of retained austenite on the mechanical properties of an austempered ductile iron. Metall. Mater. Trans. A 28A, 1143–1156 (1997)
50. D. Srinivasmurthy, P. Prasad Rao, Formation of strain-induced martensite in austempered ductile iron. J. Mater. Sci. 43, 357–367 (2008)
51. J.L. Garin, R.L. Mannheim, Strain-induced martensite in ADI alloys. J. Mater. Process. Technol. 143–144, 347–351 (2003)
52. D. Myszka, Austenite-martensite transformation in austempered ductile iron. Arch. Metall. Mater. 52, 475–480 (2007)
53. D.J. Moore, J.R. Parolini, K.B. Rundman, On the kinetics of austempered gray cast iron. AFS Trans. 111, 911–930 (1990)
54. J. Yang, S.K. Putatunda, Influence of a novel two-step austempering process on the strain-hardening behaviour of austempered ductile cast iron (ADI). Mater. Sci. Eng. A 382, 265–279 (2004)
55. J. Yang, S.K. Putatunda, Effect of microstructure on abrasion wear behavior of austempered ductile cast iron (ADI) processed by a novel two-step austempering process. Mater. Sci. Eng. A 406, 217–228 (2005)
56. D. Myszka, K. Wasiluk, E. Skołek, W. Świątnicki, Nanoausferritic matrix of ductile iron. Mater. Sci. Technol. 31, 829–834 (2015)
57. M. Pachowski, Hartowanie bainityczne żeliwa sferoidalnego. Metaloznawstwo i Obróbka Cieplna 43, 16–19 (1980)
58. K. Moeinipour, P.R. Sahm, ADI - Kokillengießverfahren mit gekoppelter in-situ-Wärmebehandlung. Giessereiforschung 54(1), 22–28 (2002)

59. J. Sikora, R. Boeri, Solid state transformations in ductile iron – influence of prior austenite matrix microstructure. Int. J. Cast Metals Res. **11**, 395–400 (1999)
60. J. Massone, R. Boeri, J. Sikora, Production of ADI by hot shake out – Microstructure and mechanical properties. Int. J. Cast Metals Res. **11**, 419–424 (1999)
61. Rivera G, Boeri R, Sikora J, Influence of the inoculation process, the chemical composition and the cooling rate, on the solidification macro and microstructure of ductile iron. Pre prints of Seventh International Symposium SPCI7', Barcelona, Spain (2002)
62. J. Tybulczuk, D. Myszka, A. Pytel, A. Kowalski, M. Kaczorowski, Structural and mechanical investigations of ductile iron directly austempered from the casting mould. Arch. Foundry Eng. **2**(4), 460 (2002)
63. D. Myszka, M. Kaczorowski, Direct austempering of ductile iron produces in sand moulds. Acta Metalurgica Slovaca **3**, 271–276 (2001)
64. Nofal AA The current status of the metallurgy and processing of austempered ductile iron (ADI). Proceedings of 10th international symposium on the science and processing of cast Iron – SPCI10, Cairo, Egypt (2010)
65. M. Soliman, H. Palkowski, Nofal, Multiphase ausformed austempered ductile iron. Arch. Metall. Mater. **62**(3), 1493–1498 (2017)
66. B.N. Olson, D.J. Moore, K.B. Rundman, G.R. Simula, Potential for practical applications of ausforming austempered ductile iron. AFS Trans. **112**, 845–850 (2003)
67. D. Myszka, L. Cybula, A. Wieczorek, Influence of heat treatment conditions on microstructure and mechanical properties of austempered ductile iron after dynamic deformation test. Arch. Metall. Mater. **59**, 1181–1189 (2014)
68. A.N. Wieczorek, Operation-oriented studies on wear properties of surface-hardened alloy cast steels used in mining in the conditions of the combined action of dynamic forces and an abrasive material. Arch. Metall. Mater. **62**(1), 119–128 (2017)
69. A. Wieczorek, D. Myszka, Abrasive wear properties of Fe-based alloys designed for mining applications. Arch. Metall. Mater. **62**(3), 1521–1534 (2017)
70. K.L. Hayrynen, K.R. Brandenberg, *Carbidic Austempered Ductile Iron (CADI) – The New Wear Material* (Applied Process Technologies Division, Livonia, 2003)
71. H. Fayazfara, M. Salariana, A. Rogalskya, D. Sarkera, P. Russoa, V. Paserinb, E. Toyserkani, A critical review of powder-based additive manufacturing of ferrous alloys: Process parameters, microstructure and mechanical properties. Mater. Des. **144**, 98–128 (2018)

Chapter 6
Low-Density Steels

Shangping Chen and Radhakanta Rana

Abbreviations and Symbols

AHSS	Advanced high strength steel
DP	Dual phase
TRIP	Transformation induced plasticity
3GAHSS	3rd generation advanced high strength steel
2GAHSS	2nd generation advanced high strength steel
LDS	Low density steel
BCC	Body centred cubic
SFE	Stacking fault energy
SRO	Short range order
CALPHAD	CALPHAD is originally an abbreviation for CALculation of PHAse Diagrams, but is later expanded to refer to Computer Coupling of Phase Diagrams and Thermochemistry
BCC_A2 or A2	A solid solution FeAl of body centred cubic structure
BCC_B2 or B2	An ordered solid solution FeAl of body centred cubic structure
κ	A type of carbide with $(Fe,Mn)_3AlC$, in which Al atoms are located at corners of the unit cell, Fe and Mn at the face centred positions, and C at the body centre
κ'	A metastable κ-carbide with $(Fe,Mn)_3AlC_x$ ($x < 1$), which has the same crystal structure as the κ phase but with an incomplete occupation of the C atoms
κ^*	κ-carbides formed at grain boundaries

S. Chen · R. Rana (✉)
Tata Steel, IJmuiden, Noord-Holland, The Netherlands
e-mail: shangping.chen@tatasteeleurope.com; radhakanta.rana@tatasteeleurope.com

© Springer Nature Switzerland AG 2021
R. Rana (ed.), *High-Performance Ferrous Alloys*,
https://doi.org/10.1007/978-3-030-53825-5_6

γ	Austenite phase in Fe-Mn-Al-C steels, in which Fe or Mn and Al atoms are randomly located at cubic positions or face centred positions in an face centred cubic lattice
γ'	The solute-lean austenite decomposed during cooling from high temperature austenite in Fe-Mn-Al-C steels, having $L1_2$ crystal structure
γ''	The solute-rich austenite decomposed during cooling from high temperature austenite in Fe-Mn-Al-C steels, precursor of κ'-carbide, having $L'1_2$ crystal structure
γ_R	retained austenite phase
α	Ferrite phase forming from austenite phase
α'	Martensite phase with body centred tetragonal lattice
β-Mn	A phase with a complicated crystal structure in Fe-Mn-Al-C steels
ε	Martensite phase with hexagonal lattice
δ	Ferrite phase forming from liquid steel
DO_3	An ordered solid solution Fe_3Al of body centred cubic structure
K state	A complex short-range ordered phase in Fe-Al phase diagram
K_1	A complex short-range ordered phase in Fe-Al phase diagram
K_2	A complex short-range ordered phase in Fe-Al phase diagram
$E2_1$	A perovskite crystal structure designated in the Strukturbericht classification
$L1_2$	The crystal structure of the ordered γ phase in Fe-Mn-Al-C steels, Al atoms are located at corner positions, Fe and Mn atoms at face positions
$L'1_2$	The crystal structure of κ', which has the same crystal structure as the κ phase but with an incomplete occupation of the C atoms
a_o	The lattice parameter of austenite
λ	Oliver factor
σ_{y0}	Initial yield strength for Hall-Petch equation
σ_{uts0}	Initial ultimate tensile strength for Hall-Petch equation
r_m	Lankford parameter or normal anisotropy parameter
Δr	Planar anisotropy parameter
n	Strain hardening exponent
E_{corr}	Corrosion potential
i_{corr}	Corrosion current density
ρ	Density
$\rho_{ferritic}$	Density of ferritic phase
$\rho_{austenitic}$	Density of austenitic phase
n	An automotive part dependent index for determining specific stiffness
E_{CUN}	Impact energy measured from Charpy U-notch specimens
E_{spec}	Specific energy absorption
HSLA	High strength low alloy
DDRX	Discontinuous dynamic recrystallization

CDRX	Continuous dynamic recrystallization
r-value	Anisotropy parameter
PSN	Particle stimulated nucleation
TTT	Time-temperature-transformation
TEM	Transmission electron microscope
FCC	Face centred cubic
SEM	Scanning electron microscope
TWIP	Twinning induced plasticity
YS	Yield strength
UTS	Ultimate tensile strength
UE	Uniform elongation
TE	Total elongation
K	Hall-Petch parameter
DSBR	Dynamic slip band refinement
MBIP	Microband induced plasticity
SIP	Shear band induced plasticity
3D	Three dimensional
USFE	Unstable stacking fault energy
DC	Dislocation cells
CB	Cell blocks
ASTM E-8	Specimen geometry (tensile) conforming to the standard of ASTM International where ASTM stands for American Society for Testing and Materials
1GAHSS	1st generation advanced high strength steel
CVN	Charpy V-notch
BH	Bake hardenable
IF-HS	High strength interstitial free (IF)
DC04	A deep drawing quality according to DIN EN 10130 standard
HC300LA	A cold rolled microalloyed steel grade according to DIN EN 10268 standard
ASB	Adiabatic shear band
S-N	Alternating stress versus number of cycles to failure for fatigue test
PFZ	Precipitate free zones
FCG	Fatigue crack growth
HEC	Hole expansion capacity
DEH	Dome expansion height
BA	Bending angle
ST	Strip thickness
HAZ	Heat affected zone
GTAW	Gas tungsten arc welding
FCAW	Flux core arc welding
LBW	Laser beam welding
WA	Widmanstätten austenite
AA	Acicular austenite

M2 steel	A tungsten-molybdenum tool steel
AISI 304	An austenitic Cr-Ni stainless steel grade according to American Iron and Steel Institute (AISI) standard
AISI 316	An austenitic Cr-Ni stainless steel grade according to American Iron and Steel Institute (AISI) standard
BIW	Body-in-white
E-modulus	Elastic modulus
DP500	Dual phase steel with a minimum ultimate tensile strength of 500 MPa
DP600	Dual phase steel with a minimum ultimate tensile strength of 600 MPa
DP980	Dual phase steel with a minimum ultimate tensile strength of 980 MPa
DP1000	Dual phase steel with a minimum ultimate tensile strength of 1000 MPa
M	Martensite
B	Bainite

6.1 Introduction

The development of advanced steels with high strength, good ductility and toughness has long been pursued. Advanced high-strength steels (AHSS) including dual-phase (DP) steels, transformation-induced plasticity (TRIP) steels and retained austenite-controlled third-generation advanced high-strength steels (3GAHSS) [1–3] and high-manganese austenitic steels as second-generation AHSS (2GAHSS) [4] have been developed. Traditionally, the high level of specific strength (i.e. strength to density ratio) of advanced steels has been sought to be achieved mainly by increasing the strength of the steels. An alternative way of increasing the specific strength of steels is to develop steels with lower density by alloying with light elements such as Al (and/or Si). These alloying elements are usually added to the Fe–Mn–C based alloy systems. These steels with elevated contents of light alloying elements are referred to as 'low-density steels' (LDS).

Aluminium is expected to increase the corrosion and oxidation resistances of steels as well as to reduce the density. The Fe–Mn–Al–C steels were developed as potential substitutes for more expensive Fe–Cr–Ni-based stainless steels for different applications such as in aerospace and chemical industries for oxidation and corrosion resistance, with specific compositions recommended for each purpose. In particular, the research efforts on the development of the Fe–Al–Mn–C alloys to replace Fe–Cr–Ni–C stainless steels were active until the 1980s considering the replacement of costly Cr and Ni by less expensive Al and Mn, respectively. The chronological development of these alloys for various applications can be found in the literature, in particular, in the work of Hadfield [5], Dean and Anderson [6], Ham and Cairns [7], Kayak [8] and Kim et al. [9]. More recently [10], the possibility of adopting the Fe–Mn–C–Al alloys for automotive structural applications has attracted considerable attention due to the potential of direct mass saving in the auto bodies by using these steels.

The efforts in developing the low-density steels are summarised in comprehensive review and overview articles [11–14] and in special issues [15, 16]. The potential of lightweight steels for automotive applications [17, 18] and for the military and other transportation industries [19] has been assessed. Issues related to the processing and properties of LDS have also been examined [20, 21]. Recently, it has been reported that the B2 phase can be utilised to increase the strength of low-density steels [22]. The concept of high-entropy alloys has been applied to design multicomponent steels including Fe-Mn-Al-C steels [23]. Therefore, the field has been enriched scientifically to a great extent in the recent past. This chapter aims to summarise these researches in a concise manner highlighting the complex metallurgy of these alloys and proposing the strategies of future research directions.

6.2 Types of Low-Density Steels

Based on the matrix phase of the microstructure, different types of low-density steels have been reported such as:

1. Fully ferritic [24–30]
2. Ferrite-based duplex [31–47]
3. Austenite-based duplex [48–59]
4. Fully austenitic [60–76]

It is to be noted that for this classification 'ferrite' refers to both δ-ferrite and α-ferrite which are body-centred cubic (bcc) in crystal structure and form from liquid iron and an austenite phase, respectively, during cooling of the alloy.

Ferritic Fe–Al low-density alloys can contain up to 5% Mn but only a very low amount of C.* This type of alloys has an elongated δ-ferrite microstructure at hot working conditions and can be present as A2-disordered FeAl, B2-ordered FeAl or DO$_3$-ordered Fe$_3$Al at room temperature depending on the Al content.

Austenitic low-density steels contain a higher Mn content, typically between 12% and 30%, Al up to 12% and C between 0.6% and 2.0%. This type of alloys can have a fully equiaxed austenitic microstructure at hot working temperatures, and the austenite is metastable after fast cooling.

Ferrite-based duplex low-density steels have a microstructure of $\gamma + \delta$-ferrite with larger than 50% δ-ferrite, whereas the austenite-based duplex low-density steels have a microstructure of $\gamma + \delta$-ferrite at hot working temperatures with the austenitic phase forming the continuous matrix. The stability of austenite in

---

*All the alloy chemistries in this chapter are given in weight percentage (wt.%) and all the phase fractions are given in volume percentage (vol.%) unless otherwise mentioned.

ferrite-based duplex LDS is relatively low in comparison to the austenite-based ones, with the consequence of different and complex transformation of austenite at lower temperatures in the former type of alloys. On the other hand, in austenite-based duplex LDS the austenite phase is largely stable at room temperature due to their high contents of austenite formers and contains nanometric κ-precipitates inside the austenite grains [10] Thus, these type of LDS are also sometimes called triplex low-density steels based on their three-phase components.

6.3 Effects of Alloying Elements

As mentioned previously, besides Al, C and Mn are primarily added to Fe for creating the low-density steel alloy systems. Optionally, other common alloying elements such as Si, Cr and Ni and various microalloying elements (Nb, V, Mo, etc.) are also added to achieve their specific effects.

Carbon and manganese are gammagenic elements and, therefore, stabilise the austenite phase. Both of them also cause solid solution strengthening and increase the lattice parameters of austenite and ferrite. Furthermore, due to the presence of C the κ-carbides form, and Mn forms the β-Mn phase. In addition, Mn increases the stacking fault energy (SFE) of austenite, silicon, on the other hand, prevents formation of the β-Mn phase, stabilises the bcc phases (δ or α ferrite) and gives solid solution hardening and influences the age-hardening kinetics positively. In service, Si increases the oxidation resistance of the steel by forming a protective film of $SiO_2 + Al_2O_3$ on the steel surface, together with the presence of Al in low-density steels. The presence of Si also makes the casting easier since it increases the fluidity of liquid steel. Chromium, like Si, is a bcc stabiliser but also suppresses the κ-carbide precipitation. It increases ductility but decreases strength and gives better corrosion and oxidation resistance to the steel. Al is the primary alloying element which is present in LDS due to its positive effects on reducing the density of steels but also due to relative ease of processing over presence of Si in similar amounts. Due to the relatively large amount of Al, ferrite phases are stabilised unless countered by austenite-stabilising elements (C and Mn) as in fully austenitic LDS. At larger amounts, Al can also form ordered intermetallic phases of FeAl and Fe_3Al. Aluminium is also present in the κ-carbides and favours their formation. Aluminium also increases the lattice parameter and SFE of austenite and reduces carbon diffusivity in low-density steels.

6.4 Phase Constitutions of Fe–Mn–Al–C Alloys

Although, the broad concept of the so-called low-density steels appears to be simple, in the sense that these steels should contain elements that are lighter to iron, complex metallurgical principles come into picture to understand and manipulate the physical metallurgy of these steels to achieve a wide range of mechanical properties. To discuss these principles, to begin with, the equilibrium phase diagrams of these alloy systems are described below.

6.4.1 Fe–Al Phase Diagram

In low-density steels, the Al contents are usually limited to lower than about 12 wt.% and, therefore, it is sufficient to discuss the Fe-rich part of the Fe-Al phase diagram as shown in Fig. 6.1 [77]. It is worth noting the peculiarities in the phase diagram of the Fe-Al system. Aluminium, being a ferrite stabiliser, extends the ferritic region and limits the austenite phase field in the phase diagram to an 'austenite loop' with a maximum solubility of Al in γ-Fe of 0.65 wt.%.

The phases present in the Fe-rich part of the phase diagram are disordered A2 (α-iron) phase and ordered B2 (FeAl) and DO₃ (Fe₃Al) phases. The α-Fe contains up to ~10 wt.% Al at room temperature. The B2 FeAl phase has a superlattice and is present in the range of 10–32 wt.% Al. The DO₃ Fe₃Al phase is present between 12 and 22 wt.% Al at room temperature and also has a superlattice structure. It forms below 552 °C via a second-order phase transformation of the FeAl phase. The B2 and DO₃ phases, due to their ordered nature, reduce the ductility and cause brittle fracture at room temperature [78].

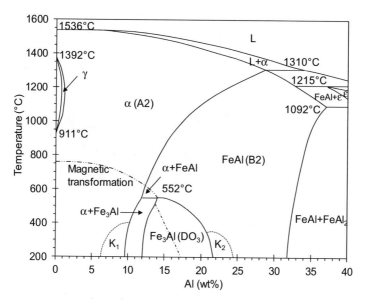

Fig. 6.1 Fe-rich part of the Fe-Al system according to Kubaschewski [77]. Phase boundaries for γ, α (A2), FeAl (B2) and Fe₃Al (DO₃) are shown together with the area in which the so-called K-state exists

In addition, there are two phase fields of the so-called K state (K_1 and K_2) where complex short-range ordering reactions occur [77–80], affecting mechanical properties [80–82]. Marcinkowski and Taylor [81] reported that considerable short-range ordering occurs during cooling of the A2 phase below 400 °C for the Al contents of 6.2–9.6 wt.% (K_1 state) with a maximum ordering at 250 °C.

6.4.2 Fe–Mn–C and Fe–Al–C Phase Diagrams

The effects of Mn and C on the phase constitution of Fe–Al alloys have been studied on the basis of the Fe–Al–Mn [83] and Fe–Al–C [84] phase equilibria separately. These effects are illustrated with the partial phase diagrams in Fig. 6.2.

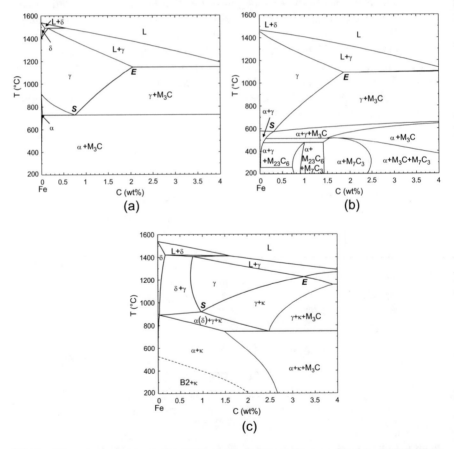

Fig. 6.2 The effects of Mn and Al addition on the phase constituent and the phase fields of the Fe–C system. (**a**) Fe–C; (**b**) Fe–15Mn–C; (**c**) Fe–7Al–C. The dashed line in Fig. 6.2c indicates the A2 to B2 transition temperature caused by Al addition. The phase diagrams are calculated using FactSage 6.4

In the Fe–C alloy system, Fig. 6.2a, the basic solid phases are ferrite (α, δ), austenite (γ) and cementite (M_3C, in which M refers to element Fe and/or Mn). When Mn is added to the Fe–C system, the high-temperature peritectic reaction gradually shifts to a lower C range with increasing Mn and disappears completely when the Mn content is higher than ~13 wt.%, as shown in Fig. 6.2b for Fe-15Mn-C. The γ single-phase field is extended to lower temperatures. The point S, the eutectoid point in the Fe-C phase diagram, indicates the lowest temperature where the γ phase exists as a single phase. Below the temperature of the S point, the γ is metastable and starts to decompose into α-ferrite and carbides. The C content and the temperature of the S point are reduced and lowered with increasing the Mn content. The temperature of the S point is 386 °C, when 30 wt.% Mn is added. However, the C concentration of the E point (the maximum solubility of C in γ in the Fe–C phase diagram) is less affected by the Mn content. The addition of Mn to the Fe–C system will form M_7C_3 and $M_{23}C_6$ types of carbides in addition to cementite (M_3C) (compare Fig. 6.2a with Fig. 6.2b) at lower temperatures. Another new phase, called β-Mn, might be introduced when the Mn content is higher than 20 wt.% [83].

The addition of Al to the Fe–C system has a large effect on the phase fields and the phase constituent (Fig. 6.2c). The ranges of composition and temperature for the high temperature peritectic transformation are enlarged, indicating that a higher level of elemental segregation (micro and macro) can occur during the solidification process. The ($\delta + \gamma$) area is extended and the single γ phase area is shifted to the right as the C concentrations of the S point and the E point are increased. A new phase, called κ-carbide, is introduced when the Al content is higher than 2 wt.%. The ($\gamma + M_3C$) area in the Fe–C system is replaced by ($\gamma + \kappa$) and ($\gamma + \kappa + M_3C$) areas.

The κ-carbide is present over a wide range of temperature. This κ-phase has a perovskite crystal structure designated as $E2_1$ in the Strukturbericht classification. It can dissolve some amount of Mn in Fe–Mn–Al–C systems. κ-carbide with an ideal stoichiometry is $(Fe,Mn)_3AlC$ [85, 86]. Figure 6.3 illustrates the crystal structure of

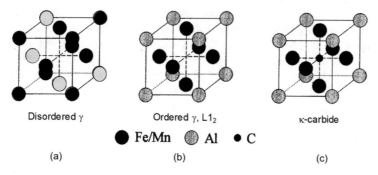

Disordered γ Ordered γ, $L1_2$ κ-carbide

● Fe/Mn   ◉ Al   • C

(a)   (b)   (c)

Fig. 6.3 Unit cell of disordered γ (**a**), ordered γ ($L1_2$) (**b**) and κ-carbide (**c**) structure in FCC alloys; κ-carbide illustrated here is with ideal stoichiometry: Al occupies each corner, Fe and Mn are located on face centres and the C atom is placed at the centre of the unit cell which is also an octahedral site made by Fe and Mn [86]

the κ-phase in comparison with that of the austenitic phase. In the austenite, Fe/Mn and Al atoms are randomly located at cubic positions or face-centred positions in a FCC lattice, as shown in Fig. 6.3a. When Al atoms are located at corner positions, Fe and Mn atoms at face positions, an ordered γ, also called $L1_2$ structure, is produced, as shown in Fig. 6.3b. Further ordering of C to the body centre octahedral (½, ½, ½) interstitial position produces the κ-carbide, as seen in Fig. 6.3c. In practice, a metastable $(Fe,Mn)_3AlC_x$ (x < 1) phase, which has the same crystal structure as the κ-phase but with an uncompleted occupation of the C atoms, may form via a spinodal decomposition [87, 88]. κ-carbides exist at a broad composition range [85]. In the literature, the ordered $L1_2$ structure is often referred as the short-range order (SRO), while the non-stoichiometric κ-$(Fe,Mn)_3AlC_x$ is also referred as κ′ or $L'1_2$ ordered crystal structure [87]. Yao et al. [89] have measured the chemical composition of the κ-carbide by atom probe tomography in a Fe–30Mn–7.7Al–1.3C (wt.%) alloy aged at 600 °C for 24 h and found depletion of both C and Al in comparison to the ideal stoichiometric bulk perovskite. The off-stoichiometric concentration of Al can, to a certain extent, be explained by the incoherency strain of κ in austenite, which facilitates occupation of Al sites in κ-carbide by Mn atoms.

In the Fe–Al–C phase diagrams, there are two important phase boundaries:

– The solvus temperature of the κ-carbide formation indicated by the line S-E in Fig. 6.2c.
– The order-disorder transition temperature of ferrite shown by the dashed line in Fig. 6.2c. The temperature of the solubility limit of the κ-carbide in the γ phase increases as the C content and/or the Al content are increased. Ferrite transforms to the ordered B2 phase at a temperature below 500 °C, and this transition temperature decreases as the C content increases.

6.4.3 Fe–Mn–Al–C Phase Diagrams

The phase equilibria of quaternary Fe–Mn–Al–C systems have been investigated for many decades [90–99]. Several researchers established the isothermal phase sections of Fe–Mn–Al–C alloys in the high temperature range of 900–1200 °C – both by experiments [90–93] as well as by theoretical calculations [94–98].

Figure 6.4 shows the isotherms of phase relations in Fe–(10,20,30)Mn–xAl–yC alloys at 900 °C including results established by experiments and calculated from thermodynamic databases. The individual points are the experimental compositions (by Kim et al. [93] in Fig. 6.4a and by Ishida et al. [90] in Fig. 6.4b and c); the red lines are the isotherms of phase relations established by Ishida et al. [91]. The dotted lines and the solid black lines are, respectively, the results calculated by Chin et al. based on a CALPHAD approach [94] and calculated by using FactSage 6.4 [96]. As seen in these diagrams, four phases, including ferrite, austenite, cementite and κ-carbide, exist in the quaternary Fe–Mn–Al–C alloys at hot working temperatures in the composition range considered. The influence of C and Al contents on

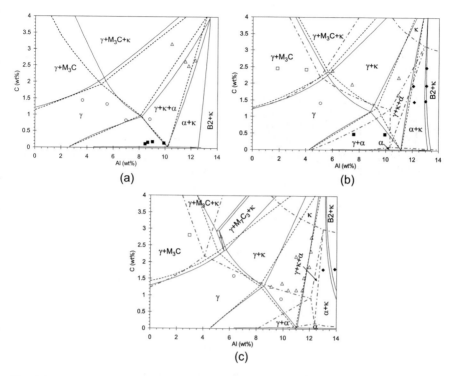

Fig. 6.4 Isothermal phase sections of Fe–Mn–Al–C alloys at 900 °C established by experiments (individual points [91, 94] and red dot-dashed lines [91]), calculated from FactSage (solid lines) and based on CALPHAD approach (dotted lines [96]): (**a**) Mn = 10%, (**b**) Mn = 20%, (**c**) Mn = 30%. The individual points labelled by the open circle, open triangle, open square, solid square and solid diamond indicate that the experimental alloy compositions locate in the γ, ($\gamma + \kappa$), ($\gamma + M_3C$), ($\alpha + \kappa$) and ($\gamma + \alpha$) phase fields, respectively

the constituent phases, in particular, the stability of the κ-carbide, is reproduced fairly well. There are differences in the composition ranges of the different phase fields between the calculations and experiments, although the calculations from the different approaches are similar. The experimental data show that the γ region is extended to the direction of the high Al concentration as Mn is increased. However, the calculations show that the γ region expands towards a higher Al concentration with increasing Mn up to 20% but shrinks with further increasing Mn. Thus, there is a derivation between the calculations from the current available databases and the experimental data as the Mn and Al contents are increased. This suggests that more accurate thermodynamic databases and models are needed for developing the phase diagrams for high Mn- and high Al-containing steels.

The experimental results and the calculations also demonstrate that the single γ region is slightly reduced as the temperature is decreased from 1200 to 900 °C [90–96]. The isothermal sections of the phase diagrams at 900 °C are important for alloy

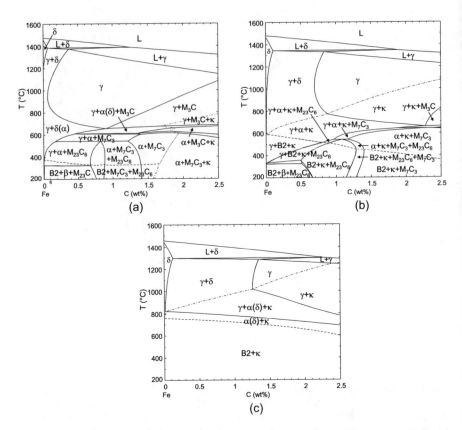

Fig. 6.5 The polythermal sections of the phase diagrams of Fe–15Mn–xAl–yC alloys calculated from FactSage as a function of C for Mn = 15% and various Al contents, (**a**) Al = 4%, (**b**) Al = 7%, (**c**) Al = 10%. The dash-dotted lines are the formation temperature of the κ-phase; the dashed lines are the disorder to order transition temperature

design because this temperature generally approximates with the final hot rolling temperature in the industrial production lines for automotive flat products.

The austenite in the Fe–Mn–Al–C alloys is not a stable phase at lower temperatures. Any heat treatment will lead to the decomposition of austenite to approach equilibrium state. The decomposition reactions and products can be predicted from polythermal phase diagrams (sections of temperature vs. Al or C content diagrams) of the quaternary Fe–C–Mn–Al alloys. Polythermal phase diagrams were experimentally established for the Fe–(20-35)Mn–10Al–xC alloy system as a function of the C content by Goretskii et al. [91]. Phase diagrams of Fe–Mn–(0–9)Al–C were also calculated by Kim et al. [93] and Chin et al. [94] based on the CAPHAD approach. Figure 6.5 shows a few phase diagrams calculated from FactSage for Fe–15Mn–xAl–yC alloys. Some important points from these experiments and phase calculations are summarised below.

- The equilibrium phases in the Fe–Mn–Al–C system at the iron-rich side may include: γ-austenite, $\delta(\alpha)$-ferrite, κ-carbide, M_3C, $M_{23}C_6$, M_7C_3 carbides and β-Mn. These phases may coexist under conditions depending on the compositions and temperature.
- The lowest temperature where γ phase exists as a stable single phase decreases to 386 °C as Mn is increased to 30% in Fe–Mn–C steels. The γ phase is not stable at temperatures lower than 386 °C in Fe–Mn–Al–C steels.
- The addition of Al in Fe–Mn–C steels enlarges the phase field of ferrite and suppresses the γ single phase region.
- In Fe–Al–C steels, the κ-phase is formed when Al is higher than 2%, while in Fe–Mn–Al–C steels, the required Al content to form the κ-phase is increased to higher than 5% when more than 2% Mn is added. The experimental evidence shown by Buckholz et al. [99] also supports this statement. This indicates that Mn restrains the formation of the κ-phase and Al restrains the formation of the other carbides such as M_3C and $M_{23}C_6$.
- The formation temperature of the κ-phase, indicating its stability, increases as the C and Al contents increase. The solvus temperature of κ-carbide decreases as Mn is increased up to 15%, and this does not change with further increase of Mn.
- As the C and Mn contents of the steels increase, the austenite region is enlarged for a given Al content. In case a single-phase austenite matrix is desired, the ferrite-stabilising effect of Al must be compensated by an increased Mn and C content.
- The predictions show that the disorder-order transition of BCC_A2 (α) to BCC_B2 occurs when the Al content is higher than 2%, which is lower than that obtained from the experiments (see Fig. 6.1). The disorder-order transition temperature increases as the Al and Mn contents increase but decreases as the C content increase. This indicates that the increases of the Al and Mn contents enhance the formation of ordered phases within the ferrite, while the increase of the C content restricts the formation of ordered phases.
- The homogeneous region of γ is enlarged by increasing Mn content up to 20%, then shrunk by further increasing the Mn content up to 30%.

The thermodynamic phase diagrams demonstrate that a variety of microstructures can be obtained in Fe–Mn–Al–C steels depending on the alloy compositions and temperature. From the microstructural point of view, the alloy design criteria should be based on the absence of κ-carbides at hot working temperatures, prevention of coarse grain boundary precipitation of κ-carbides and prevention of formation of the other carbides such as $M_{23}C_6$ and M_7C_3 at lower temperatures to achieve good mechanical properties.

6.5 Microstructure Development in Fe–Mn–Al–C Alloys

As discussed in the previous sections, the austenite in the Fe–Mn–Al–C low-density steels is not a thermally stable phase at lower temperatures. However, the stability of the austenitic γ-phase can be very high, and it can be present at room temperature if high amounts of the stabilising elements (Mn and C) are added to the alloy. The phase constituents at hot working temperatures can be predicted by applying thermodynamic calculations by assuming that the equilibrium conditions are approached. At lower temperatures, the diffusion-controlled transformations are slow and also the driving force for a martensitic transformation is too low as compared to the conventional steels. However, the formation of the phase constituents is predominantly controlled by the kinetics, and the actual microstructures may be away from those derived by thermodynamic calculations. The microstructural evolution under the industrial processing conditions is elaborated in this section.

6.5.1 Generic Processing Routes

The Fe–Mn–Al–C alloys reported in the literature have been made mostly as small ingots (~ 50 kg or lower) by using standard vacuum melting and ingot casting route on a laboratory pilot scale [24–76]. Usually, these ingots are homogenised at a temperature in the range of 1100–1250 °C for 1–3 h and then hot rolled to 2–5 mm thickness at a finish rolling temperature in the range of 850–1000 °C. In some cases, the steel strips are reheated many times between rolling passes to avoid cracking and temperature loss during hot rolling. After hot rolling, the hot rolled strips are cooled to a temperature between 500 and 650 °C and held for 1–5 h to approximately simulate the coiling step for conventional steels or directly water quenched or air cooled to room temperature.

For ferritic low-density steels and ferrite-based duplex steels, it is not practical to produce hot-rolled products by applying a hot rolling and coiling schedule similar to that used for conventional steels as the grain size of the δ-ferrite is very large and the grains are highly elongated after hot rolling. Therefore, cold rolling and annealing are needed to control their microstructures. The ferritic Fe–Al steels are generally cold rolled and heated to a temperature in the range 700–1000 °C for recrystallisation annealing to control the grain size and texture of the ferrite matrix as well as the precipitation of κ-carbides [26–30]. The continuous annealing line for producing conventional high-strength low-alloy (HSLA) steel strips can be utilised for the annealing purpose. The ferrite-based duplex steels are cold rolled and then annealed at an intercritical temperature in the range of 700–900 °C, followed by an austempering or overaging process [31–47].

As shown in Fig. 6.6, for austenite-based low-density steels, there can be many processing routes [60–76]. For hot-rolled products, after hot rolling, the strips can be directly fast cooled to a temperature in the range of 500–750 °C, then slow cooled or isothermally held, as indicated by process 1 in Fig. 6.6. Alternatively, the hot-rolled strips can be fast cooled to room temperature, followed by isothermal annealing as indicated by process 2. The cooling rate after hot rolling should be high enough to avoid the formation of the intergranular κ-carbides. The coiling temperature and the cooling rate during coiling can be manipulated to obtain fine, nanoscale κ-precipitates avoiding the coarse ones. For age-hardenable austenitic

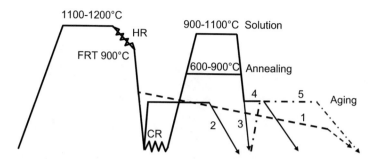

Fig. 6.6 Process variants for producing hot-rolled and cold-rolled austenitic Fe–Mn–Al–C steel strips. The numbers identify process routes as described in the text

Fe–Mn–Al–C steels, in order to avoid the uncontrolled precipitation of κ-carbides, hot-rolled or cold-rolled strips can be solution-treated at a temperature between 900 and 1100 °C in the single austenitic phase, and finally quenched in water, oil or other quench media. Subsequent annealing treatments are performed to produce precipitation hardening as indicated by processes 3, 4 and 5 in Fig. 6.6. Process 3 has no isothermal holding and processes 4 and 5 are applied to maximise the age-hardening effects. The common practice for age hardening is to hold the material isothermally for 5–20 h in the temperature range of 450–650 °C [11–16, 100]. For non-age-hardenable austenitic Fe-Mn-Al-C steels, after cold rolling, recovery or recrystallisation annealing in the temperature range of 600–900 °C for a short time (1–5 min) can be applied to restore the ductility or to refine the grain structure. In this case, the continuous annealing line for producing conventional steel strips can be used [20].

6.5.2 Microstructure Evolution in Ferritic Fe–Al Steels

As the ferritic Fe–Al steels usually contain a large amount of Al but small amount of C (< 0.03%), there is no $\alpha \rightarrow \gamma$ transformation during heating in the entire processing temperature range. Ferrite in these steels is the δ-ferrite that is directly produced from liquid during the casting process. The hot working is conducted in the single-ferrite phase range. As Al not only extends the ferrite region to high temperatures but also increases the recrystallisation temperature of ferrite, the grain size of the δ-ferrite cannot be sufficiently refined through dynamic recrystallisation during hot rolling or static recrystallisation between the hot rolling passes under the conventional hot rolling conditions. Consequently, δ-ferrite is elongated along the rolling direction and forms band-like structures although there might be some degree of recrystallisation within the δ-ferrite bands, as shown in Fig. 6.7 [28]. The as-cast δ-ferrite usually has a columnar structure with a strong cube fibre, which is inherited in the hot rolled strips. In the subsequent cooling process, coarse κ-

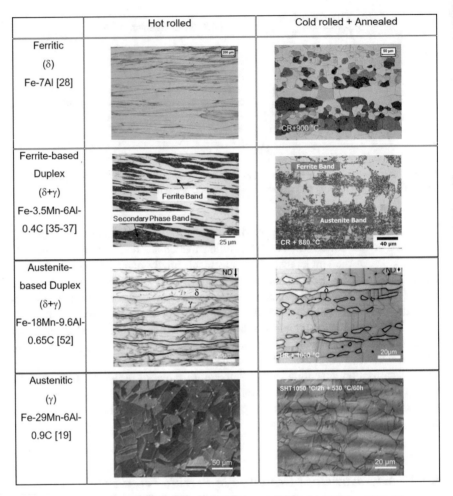

	Hot rolled	Cold rolled + Annealed
Ferritic (δ) Fe-7Al [28]		
Ferrite-based Duplex (δ+γ) Fe-3.5Mn-6Al- 0.4C [35-37]		
Austenite- based Duplex (δ+γ) Fe-18Mn-9.6Al- 0.65C [52]		
Austenitic (γ) Fe-29Mn-6Al- 0.9C [19]		

Fig. 6.7 The microstructural characteristics of low-density steels, as observed with a light optical microscope, in the hot-rolled and, cold-rolled and annealed conditions

carbides can be formed along the original δ-ferrite grain boundaries if the cooling rate is slow [25–30]. The κ-carbides in the ferritic matrix are semicoherent and in the form of thick and elongated rod-like shaped structures [46].

The hot-rolled microstructures can be modified by applying different hot rolling parameters (temperature, strain and strain rate) to activate dynamic recrystallisation [27]. At high temperatures and high strain rates, the alloy undergoes discontinuous dynamic recrystallisation (DDRX), whereas at lower temperatures and low strain rates, continuous dynamic recrystallisation (CDRX) occurs. In the case of CDRX, the original grains are elongated in the direction of the shear. Subgrains form inside the grains, preferentially close to the initial grain boundaries. Such a microstructure has a strong intensity of the cube texture, which has a relatively low formability (i.e. low Lankford parameter or r-value). While in the case of DDRX, newly recrystallised grains have no preferred orientations (very low intensity of the texture components). Therefore, it seems that a higher hot rolling temperature and higher strain rate are needed to favour DDRX in order to obtain a microstructure with recrystallized grains exhibiting random orientations which help to improve the formability.

Cold rolling and annealing are necessary steps to change the grain structure, grain size and texture in ferritic low-density steels [24–30]. A fully recrystallised microstructure can be obtained after final recrystallisation annealing at the temperature range 700–950 °C following a 50% cold rolling reduction in a Fe–7%Al steel [28]. The δ-ferrite grains become more or less equiaxed, with the grain size ranging from 40 to 90 μm, which is still much larger than that of low-carbon HSLA steels. The texture in cold rolled and annealed ferritic low-density steels is random with a relatively higher intensity of the cube fibre, {001}<110> [27, 28]. Coarse grains and the cube texture can produce high plastic anisotropy and ridging during sheet forming. The microalloying elements such as Nb, Ti, V and B can be used to refine grain size and to suppress the formation of the cube fibre in ferritic low-density steels [29]. The addition of Nb in combination with C in a Fe–8Al–5Mn–0.1C–0.1Nb steel was reported to change the microstructure and refine the grain size through enhanced recrystallisation by the particle stimulated nucleation (PSN) mechanism caused by pinning effects of κ-carbide and NbC particles [30].

6.5.3 Microstructure Evolution in Austenitic Fe–Mn–Al–C Steels

The optical microstructures of the austenitic and/or austenite-based duplex alloys in the as-cast state show a mixture of austenite (γ) and ferrite (δ) phases and present as a dendritic structure due to the high degree of micro- and macrosegregation caused by large amounts of alloying elements. The average austenite grain size in the columnar zones ranges from 100 μm at the surface to 500 μm at the ingot centre [19, 101–104]. In a series of steels containing Fe– (5-40)Mn–10Al–1.0C, transmission electron microscope (TEM) images revealed that the γ phase transforms to a mixture of γ + κ phases, and the α phase is made up of a mixture of DO_3 + coarse κ-phases at room temperature [104].

For austenitic Fe–Mn–Al–C steels, hot working is conducted in the single γ phase region. After reheating at 1100–1200 °C for a sufficient length of time, the segregation of alloying elements in the as-cast material can be largely homogenised. Recrystallisation occurs during the conventional hot rolling process, and the microstructure after hot rolling generally shows an equiaxed austenitic grain structure containing annealing twins [19] (Fig. 6.7).

When austenitic steels are slow cooled or coiled, the κ-carbides and/or α-phase tend to precipitate along the austenite grain boundaries and within the austenite matrix. These κ-carbides formed during coiling are about three to six times larger than those obtained after solutionising, quenching and aging the steel at around 550 °C [21, 102], decreasing ductility and toughness. Therefore, the austenitic steels are usually fast cooled (by water quenching) from solutionising temperatures (~900–1000 °C) [61–75] to room temperature to avoid the formation of coarse κ-carbide and/or α-phase particles.

The precipitation of the κ-carbide and the decomposition of the metastable austenite in austenitic Fe–Mn–Al–C alloys have been extensively studied [61–74, 105–130]. These studies show that, depending on the compositions of the alloys and the annealing time-temperature profiles, κ-carbide, α-ferrite (or B2, DO_3) and/or the β-Mn phase may be produced in various forms by different mechanisms. Some other carbides such as $M_{23}C_6$, M_7C_3 or M_3C are observed during the aging processes if the Al content is relatively low and/or other alloying elements are added [71, 112, 116–121]. Si has been shown to eliminate β-Mn formation but to promote the formation of B2 and/or DO_3 precipitates [105, 106, 110, 127].

Two types of κ-precipitates are observed in austenitic Fe–Mn–Al–C steels aged in the temperature range of 500–900 °C: the intragranular κ'-phase and the intergranular κ*-phase. The formation mechanisms of these two types of precipitates and their contributions to the final properties of the steel appear to be quite different. The fine intragranular κ'-phase is considered to produce age-hardening in these alloys as it raises the yield stress significantly [11–16]. However, the intergranular κ*-phase is much coarser and can result in a severe loss of ductility [10, 129].

The time-temperature-transformation (TTT) diagram for the decomposition of γ solid solution during isothermal annealing has been established for a Fe–28Mn–8.5Al–1C–1.25Si alloy [103, 105], as shown in Fig. 6.8. A series of intermediate reactions occur before equilibrium phases are reached. Chemical modulation (spinodal decomposition) occurs in the area marked as 1 followed by homogeneous κ'-carbide precipitation in the γ matrix in area 2 (γ + κ'). Area 3 corresponds to the formation of the grain boundary κ*-carbide, γ + κ* (at higher temperature) or γ + κ' + κ* (at lower temperature). Area 4 represents a discontinuous reaction product of grain boundary α(B2/DO_3) and γ + κ' + κ* + α(B2/DO_3). All equilibrium phases are present in Area 5: γ + κ + DO_3 + B2 [105]. Aging at a temperature below 500 °C greatly lengthens the time for the formation of grain boundary κ* precipitation due to slow transformation kinetics.

6.5.3.1 Intragranular κ'-Phase Precipitation

Intragranular κ'-carbides refer to the precipitates produced homogeneously within the austenitic matrix. The κ'-carbide is a metastable $(Fe,Mn)_3AlC_x$ phase (x < 1) and it is coherent with the matrix. The κ'-phase has a cube-on-cube orientation relationship with the austenite matrix: [100]κ//[100]γ and [010]κ//[010]γ and the lattice misfit is less than 3% [108–111]. The sequence of the κ' precipitation in the austenite phase has been studied in the literature [108–111] and can be described

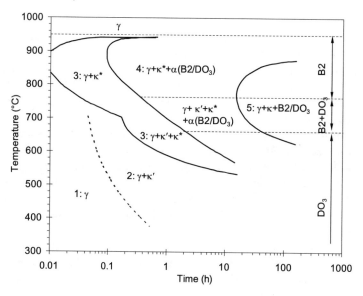

Fig. 6.8 TTT diagram of decomposition of γ-solid solution in a Fe–28Mn–8.5Al–1C–1.25Si alloy during isothermal aging [105]. Area 1: spinodal decomposition; area 2: homogeneous matrix κ′-carbide precipitation; area 3: the formation of the grain boundary κ*-carbide; area 4: discontinuous reaction product of grain boundary α(B2/DO$_3$); area 5: the formation of equilibrium phases

as $\gamma \rightarrow \gamma' + \gamma'' \rightarrow \gamma' + L1_2(SRO) \rightarrow \gamma + \kappa'$. The steps of these reactions can be described as the following:

- A spinodal reaction causes modulation of C and Al within the austenite [19, 108–115], which decomposes the high temperature γ-FCC austenite into two low-temperature FCC austenite phases: the solute-lean (C and/or Al) phase γ′ and the solute-rich phase γ″.
- A short-range ordering (SRO) reaction takes place upon further cooling to lower temperature, and the solute-rich FCC austenite γ″ transforms into the L1$_2$ phase.
- The L1$_2$ phase transforms to the E2$_1$ κ′-carbide by further ordering of carbon atoms.
- Precipitation of κ′-carbide leads to the destabilisation of the remaining austenite and ferrite is precipitated, followed by imminent β-Mn precipitation [114, 115].

The spinodal decomposition and following ordering reaction have been reported to take place during cooling (either quenching or air cooling) from solution treatment and/or during the initial stage of aging in some alloys [61–74, 105–115]. The reported observations demonstrate that the high-temperature austenite has already started to transform into low-temperature austenite and L1$_2$ phase (SRO) or L′1$_2$ (κ′) carbides during cooling after solution treatment. The decomposition products of the austenite depend on the Al and C contents and cooling rate. Alloys with a lower Al < 7% only produce SRO in the rather slow cooling rate, while alloys with high contents of Al and C form κ' (L'1$_2$)-particles directly during water quenching or even κ-particles during air cooling. The spinodal decomposition kinetics, the density of SRO and the amount of the fine L'1$_2$ within the austenite matrix increase as the C and Al contents increase in the alloys, especially in the alloys with C > 1.3% and Al > 10% [115, 116, 122, 123]. The size of SROs is reported to be less than ~2 nm and randomly distributed in solution-treated and water-quenched states.

Aging of the austenitic Fe–Mn–Al–C alloys results in a more definite precipitation of κ'-particles within austenite grains [11–15, 62, 66, 67, 107]. The L'1$_2$ phase was observed to precipitate homogeneously in the austenite as small coherent particles in the initial stage of annealing in a Fe–18Mn–7Al–0.85C alloy when the annealing temperature is below 625 °C [98]. Initially the κ'-carbide appears in cube-shaped morphology [86, 108, 114]. With increasing the annealing time, the metastable L'1$_2$ phase within the austenite matrix accumulates C and Al atoms from the surrounding carbon-lean austenite and grows upon prolonged annealing, and transforms to κ-carbides. The lattice parameter of κ'-carbide increases and the parent phase lattice parameter decreases, and this leads to a loss of coherency and a more plate-like morphology.

6.5.3.2   Intergranular κ^*-Phase Precipitation

Intergranular κ^*-phase precipitation refers to the κ-precipitates produced along austenite grain boundaries. With a prolonged holding time at higher annealing temperatures, κ-carbides are found to precipitate heterogeneously on the γ/γ boundaries in the form of coarse particles [118–120]. Grain boundary κ^* generally has a parallel orientation relationship with one of the neighbouring grains [98]. The κ^*-phase develops originally as discrete particles but soon becomes nearly continuous along the grain boundaries. After aging for a long time, the κ^*-particles grow into adjacent austenite grains and form a lamellar structure. Depending on the steel composition and the annealing temperature, the κ^*-carbides might be formed through precipitation reactions, cellular transformations or eutectoid reactions.

The precipitation reactions occur during the transformation of high-temperature austenite into distinct κ-carbide or α ferrite at temperatures in the $\gamma + \kappa$ or in the $\gamma + \alpha + \kappa$ areas [97, 98]. These precipitation reactions can be described as $\gamma \rightarrow \gamma + \kappa$ or $\gamma \rightarrow \gamma + \kappa + \alpha$. When annealing is done in the $\gamma + \kappa$ region for a long time, the κ^*-phase develops originally as discrete particles but soon becomes thin films distributed continuously along the γ grain boundaries [97]. When annealing is done in the $\gamma + \kappa + \alpha$ region, especially at a lower temperature in the phase field, both α and κ^* independently nucleate on the austenite grain boundaries. Separate α ferrite grains (in distinct form) and κ^*-carbide precipitates (as thin films) coprecipitate on the austenite grain boundaries by the precipitation reaction.

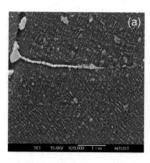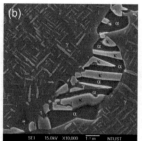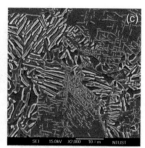

Fig. 6.9 Secondary electron images of a Fe–18Mn–7Al–0.85C austenitic steel, showing the growing feature of L'1$_2$ precipitates (κ') in the austenite and the evolution of the κ*-carbide in $\kappa + \gamma + \alpha$ lamellar colonies on the austenitic grain boundaries after isothermal holding at 700 °C for various periods of time [111]: (**a**) 1 h, (**b**) 25 h, (**c**) 100 h

The cellular transformation, which is a special precipitation reaction, is a form of discontinuous reaction, involving grain boundary migration. The cellular transformation occurs during the transformation of high-temperature austenite into lamellae of austenite, ferrite and κ-carbide or M$_{23}$C$_6$ at temperatures in the $\gamma + \alpha + \kappa$ or in the $\gamma + \alpha + \kappa + $ M$_{23}$C$_6$ phase fields [97, 98]. The cellular transformations are described by $\gamma \rightarrow \gamma + \kappa + \alpha$ or $\gamma \rightarrow \gamma + \kappa + \alpha + $ M$_{23}$C$_6$. In these cases, the κ*- and α-precipitates not only nucleate simultaneously on the grain boundaries but also grow into the austenite matrix as small nodules of the lamellar mixture. The lamellar κ*-carbide grains are always accompanied with austenite twins in the lamellar austenite grains. Figure 6.9 shows that the lamellae of $\alpha + \gamma + \kappa$ develop initially from the austenite grain boundaries as lamellar colonies and grow from the grain boundaries via the cellular transformation in a Fe–18Mn–7Al–0.85C austenitic steel [111]. Formation of coarse second-phase particles having a lamellar morphology was also observed to occur after aging at 625 °C for a longer period of time in a solution treated Fe–28Mn–9Al–0.8C alloy [62].

The eutectoid reaction involves the replacement of a high-temperature phase by a mixture of new low-temperature phases. Figure 6.10 shows the lamellae of $\alpha + \kappa + $ M$_{23}$C$_6$ formed by a eutectoid reaction in a Fe–13.5Mn–6.3Al–0.78C alloy [97]. These lamellar phases are produced from the decomposition of austenite during the eutectoid reaction of the quaternary alloy ($\gamma \rightarrow \alpha + \kappa + $ M$_{23}$C$_6$). Since the Al concentration in the steel is higher than that of the eutectoid composition, proeutectoid ferrite and κ-carbide appear in the austenite prior to the eutectoid reaction to reduce the Al content of the retained austenite. The retained austenite decomposes into ferrite, κ-carbide and M$_{23}$C$_6$ carbide during the eutectoid reaction. Lin et al. [121] have also shown that the complex carbide M$_{23}$C$_6$ may occur on γ/γ grain boundaries in Fe–26.6Mn–8.8Al–0.61C alloy during aging at 550 °C for 10–40 h.

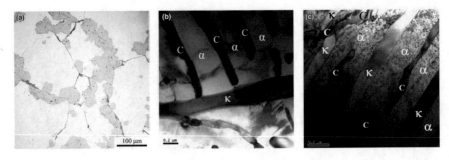

Fig. 6.10 Microstructures of a Fe–13.5Mn–6.3Al–0.78C steel after solution treatment at 1000 °C followed by aging at 600 °C for 100 h [97], (**a**) Optical microscope image showing large κ-pearlite colonies along the austenite grain boundaries; (**b**) and (**c**) TEM bright-field images revealing the internal portions of various κ-pearlites. $M_{23}C_6$ is indicated by C

6.5.3.3 Conditions for the Formation of Intra- κ′ and Intergranular κ*-Carbides

The thermodynamic calculations in Sect. 4.3 show that the κ-carbide is formed in Fe–Al–C steels when Al is higher than 2%, while in Fe–Mn–Al–C steels, an Al content of higher than 5% is required to form κ-carbides. The formation of carbides in relation with the compositions of austenitic steels has been systematically investigated in a series of alloys with Fe– (23–31)Mn– (2–10)Al– (0.4–1.0)C compositions [116, 117]. For leaner alloyed Fe–Mn–Al–C compositions even after long aging treatments at 650 °C for 360 h, no κ precipitation was observed [116]. Other carbides such as M_3C, M_7C_3 or $M_{23}C_6$ can be formed after annealing at a lower temperature in alloys with Al content less than 5.5%. Thus, it was concluded by Huang et al. [116] that the intragranular κ′-phase can precipitate only in alloys with Al and C contents higher than about 6.2% and 1.0%, and that the intergranular κ*-phase can precipitate only in alloys with Al and C contents higher than about 5.5% and 0.7%, respectively. Further, the formation of κ′-phase can be related to the lattice constant (in nm) a_o of austenitic Fe–Mn–Al–C alloys, which has been determined to be a linear function of chemical composition (in wt%) [116].

$$a_o = 0.3574 + 0.000052\,\text{Mn} + 0.00094\,\text{Al} + 0.0020\,\text{C} - 0.00098\,\text{Si} \qquad (6.1)$$

A critical lattice constant of about 0.3670 nm was found for the austenite matrix, below which the intragranular κ′ phase does not precipitate. Thus, the lattice constant of the austenite matrix was proposed to serve as a simple empirical index in determining the precipitation of intragranular κ′-phase. By use of the critical lattice constant (0.3670 nm) and Eq. (6.1), a composition boundary (in wt.%) for the precipitation of intragranular κ′ phase in Fe-Mn-Al-C alloys can be expressed as:

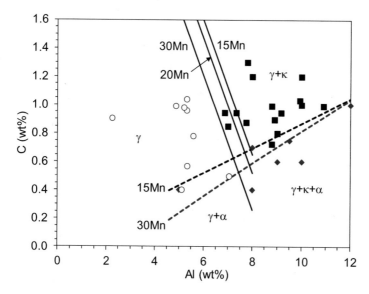

Fig. 6.11 The precipitation of intragranular κ′-phase as a function of alloy compositions. The individual points labelled by the open circles, solid squares and solid diamond symbols indicate that the experimental alloy compositions locate in the γ, γ + κ and γ + α + κ phase fields, respectively [109, 116, 130]. The solid lines are the κ′ formation boundaries predicted by Eq. (6.2) [116]. The dashed lines are the γ/α boundaries fitted from experimental data [8, 9, 109]

$$0.098 \text{ Al} + 0.208 \text{ C} = 1 - 0.0054 \text{ Mn} \qquad (6.2)$$

Figure 6.11 shows the precipitation of intragranular κ′-phase as a function of alloy compositions determined from Eq. (6.2) [116].

The effects of Al and Mn contents on the solvus of both intergranular and intragranular κ-phase were investigated for a composition range of Fe-(21–32)Mn-(2–10)Al-1C, as shown in Fig. 6.12 [117]. The results indicate that the addition of Al raises the solvus of both intergranular and intragranular κ-phases significantly, while the addition of Mn causes only a slight decrease of the solvus.

As κ-carbide has an ordered FCC structure, the character of the matrix/κ interface will be strongly dependent on the matrix structure, i.e. FCC (γ) or BCC (α). In the γ-phase, the lattice parameter of the κ-phase is 0.383–0.386 nm and γ/κ interfaces are coherent (lattice mismatch <3%) [86], whereas in the α phase, the lattice parameter of the κ-phase is 0.387–0.389 nm and α/κ interfaces are semicoherent (lattice mismatch ~6%) [46]. As a consequence, at the same aging conditions (< 100 h at 600 °C), κ-carbide is present as nanoscale cuboidal precipitates in austenitic Fe–Mn–Al–C steels [66, 69, 86], whereas in ferritic or duplex steels, these precipitates are thicker and may assume an elongated rod-type shape [46].

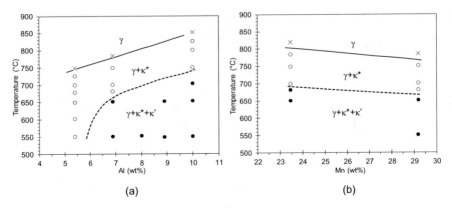

Fig. 6.12 The effects of Al and Mn contents on the solvus of κ-phase in (**a**) Fe-30Mn-xAl-1C and (**b**) Fe-xMn-7Al-1C alloys. γ: austenite; κ′: intragranular κ-phase; κ*: intergranular κ-phase [117]

The effect of the contents of the alloying elements on the formation of κ′ phase can be understood from two aspects: driving force and strain energy due to misfit between κ′ and γ. Since the κ-phase and the matrix have a parallel orientation relationship and a common {100}-type interface [109], the lattice misfit between the κ-phase and the matrix (about 2.1%) could result in considerable misfit strain energy. Increasing the Al and C contents increases the chemical driving force for the precipitation of κ′ phase, and at the same time, also increases the lattice constant of austenite (by 0.00094 nm per wt% for Al and 0.0020 nm per wt% for C), which in turn decreases the misfit between austenite and intragranular κ′-phase and thereby the coherent strain energy. Both the increase in chemical driving force and the decrease in strain energy are favourable for the precipitation of intragranular κ′-phase. Increasing the Mn content also increases the matrix lattice parameter (by 0.000065 nm per wt% Mn), and therefore it should decrease the misfit strain energy between the intragranular κ′-phase and matrix, and facilitate the intragranular κ′-phase to precipitate. However, increasing the Mn content can decrease the driving force for the κ-phase to precipitate. The net result is that increasing the Mn content is not favourable for the κ-phase to precipitate. It is to be noted that the Mn content has been shown to have only a minor effect on the range of composition within which the κ-phase can precipitate.

The effects of aging temperature on the formation of the κ-precipitates can be approximately divided into the following two ranges:

1. 450–650 °C: When the alloy is aged within this temperature range, fine κ′-carbides precipitate coherently within austenite matrix as well as on grain boundaries. The size of κ′-carbides is about 20–50 nm depending on chemical composition, aging temperature and aging time.
2. 650 ~ 800 °C: The κ-carbides precipitate not only coherently within austenite matrix, but also on grain boundaries in the form of coarser particles. The higher

the aging temperature, more is the quantity of the grain boundary κ*-carbides. Besides the precipitation of κ*-carbides, β-Mn precipitates are also observed to form on the grain boundaries through the transformation, κ → α + β-Mn, in high Mn-containing alloys when a longer annealing time is applied.

Since the intergranular κ*-carbides and β-Mn precipitates are detrimental to the ductility and toughness of austenitic low-density steels, it is desirable to avoid their formation. From a composition point of view, Mn has no significant effect in strengthening these alloys but is essential in stabilising the austenitic matrix. Therefore, it is better to increase the Mn content, possibly up to 35%. More Mn should be avoided because it may result in the formation of brittle β-Mn phase [105, 107]. The Al content should be less than about 5.5 wt% to avoid the κ*-phase [116]. However, Al is an element for strengthening and light weighting in this alloy system and should not be decreased further. In addition to being an austenite stabiliser, C is also a strong strengthening element and its content should be kept high for strength. It is suggested that C content should be more than 0.7 wt%. The aging temperature should be controlled to below the solvus of intergranular κ*-phase, and the aging time should be controlled to limit the formation of the intergranular κ*-phase and β-Mn.

6.5.3.4 Precipitation of α (B2)

In addition to the κ-precipitation, the α-precipitation may occur during annealing after solution treatment, depending on the steel composition and the annealing temperature. This is usually observed as fine stringers along the γ-austenitic grain boundaries [98, 111]. The α-precipitates can also form along the shear bands or the twin boundaries [22]. In the solution treated fully austenitic microstructure, the α-precipitates on γ grain boundaries are initially coherent with one of the austenite grains, with a Kurdjumov-Sachs orientation relationship, $(111)_γ//(110)_α$ and $[101]_γ//[111]_α$ [124]. These precipitates, however, loose coherence upon coarsening as the aging time is increased [98, 111]. In duplex δ + γ microstructures, the α-ferrite forms preferentially on the existing δ-ferrite [43]. The α-ferrite may transform to B2 or DO_3 structures at lower temperatures [15, 111] due to the high Al contents. In a Fe–8.0Al–29.0Mn–0.90C–1.5Si steel, some discrete particles having a mixture of $α + DO_3$ phases are formed along the γ grain boundaries by an ordering transition during quenching [127]. In a Ni-alloyed Fe–15Mn–10Al-0.86C–5Ni steel [22], it was observed that the B2 phase (transformed from α-ferrite) is stronger than the austenite matrix and these particles can be utilised to increase the strength of steels as long as the size of the B2 particles can be reduced to nanoscale. Figure 6.13 shows the precipitation of B2 particles during annealing of a cold-rolled Fe–15Mn–10Al–0.86C–5Ni steel [22, 48]. The strategy to be adopted to produce nanoscale B2 precipitates is not yet sufficiently clear in the literature as for κ-precipitates.

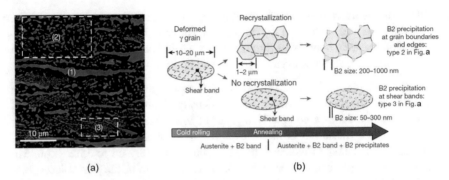

(a)   (b)

Fig. 6.13 (**a**) The microstructure of a Fe–15Mn–10Al–0.86C–5Ni steel after 70% thickness reduction in cold rolling and annealing at 900 °C for 2 min, and (**b**) schematic illustration of the formation mechanism of B2 precipitates. Area (1) in (a): B2 stringer bands transformed from δ-ferrite, which is produced from liquid; area (2) in (a): B2 particles transformed from α precipitates at γ grain boundaries and edges; area (3) in (a): B2 particles transformed from α precipitates at shear bands within a γ grain [22]

6.5.4 Microstructure Evolution in Duplex Fe–Mn–Al–C Steels

In duplex alloys, the microstructure consists of δ + γ at reheating temperatures (~1200 °C). During hot rolling, the δ-ferrite stringers are formed parallel to the rolling direction, while the γ grains are refined through recrystallisation. As a result, a banded microstructure parallel to the rolling direction, consisting of a mixture of ferrite and austenite forms, as shown in Fig. 6.7. For a given Al content, the volume fraction of austenite and the thickness of the austenitic bands increase with an increase of Mn and C contents [32–38].

During hot working (reheating and hot rolling), partitioning of elements C, Mn and Al occurs between the δ and γ phases accompanied by some amount of γ → α phase transformation. The α phase can form along δ/γ boundaries or within γ grains along dislocations, as shown in Fig. 6.14a for a Fe-5Mn-6Al-0.4C steel [35]. Due to element partitioning between two phases, the C and Mn contents in austenite are higher and the Al content in austenite is lower than the nominal composition. The stability and the transformation products of the austenite in the duplex microstructure depend on the composition of the alloy and the processing parameters.

When the amounts of C and especially Mn are relatively low, for example, in a composition with 6–8% Al, 2–6% Mn, 0.1–0.3% C [32–37], the decomposition temperature of the austenite into the α-ferrite and the κ-carbide is quite high, and κ-carbides can be formed even at hot working temperatures, at the interface between austenite and δ-ferrite, forming a shell structure surrounding the austenite [17] as shown in Fig. 6.14b. With increasing amounts of Mn, the formation of coarse κ-carbides in duplex low-density steels at hot working temperatures can be avoided.

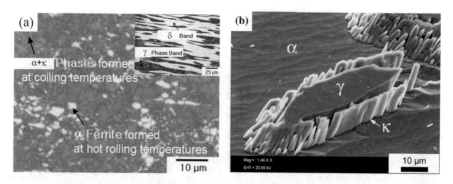

Fig. 6.14 (a) Optical microstructure showing α-ferrite formed in the γ phase band (the grey areas in the embedded microstructure) during hot rolling stage in a Fe–5Mn–6Al–0.4C steel [35]; (**b**) scanning electron microscope (SEM) image showing κ-phase formed along δ/γ boundaries during hot rolling stage in a Fe–2Mn–8Al–0.2C steel [17]

In the subsequent cooling process, if the Mn and C contents are low, such as in Fe–(3-6)Mn–(5-7)Al–(0.1–0.5)C steels [32–35], the stability of the austenite is low, and if the cooling rate is high such as in water quenching, the high temperature decomposition of γ is suppressed and the austenite may completely transform into martensite.

The transformation products within the austenite after slow cooling (air cooling or simulated coiling at 650 °C for 2 h) change with the Mn content of duplex steels. In a Fe–3Mn–5Al–0.3C alloy [32–35], it was observed that the austenite completely transforms to lamellar colonies of α + κ during a simulated coiling. In a Fe–6Mn–5Al–0.3C alloy [33], the austenite partially transforms to lamellar colonies of α + κ and partially stabilises to a lower temperature and transforms to martensite. The α + κ lamellar colonies are formed along the original austenite grain boundaries or at δ/γ boundaries. The κ-carbides may also form with a globular morphology at the grain boundaries of the δ-ferrite and/or at δ/γ interface boundaries. It was revealed that the austenite decomposes into austenite and ferrite along the grain boundaries without the formation of κ-carbides.

For duplex steels containing less than 6% Mn, the microstructure of the air cooled hot-rolled product consists of a bimodal ferrite structure, partially recrystallised bands of coarse elongated grains of δ-ferrite and some κ-carbides on the δ-ferrite grain boundaries, together with bands of α-ferrite with smaller grains and enriched with κ-carbides (produced from the original γ) and coarse κ-carbides along δ/γ interfaces. Such a microstructure cannot make best use of the κ-carbides because they are very coarse and are present as clusters.

The stability of the austenite in the hot-rolled duplex microstructure increases with increasing the Mn content to above 6%. Even by air cooling, the decomposition reaction, γ → α + κ, is suppressed and the austenite may transform into martensite, and some amount of austenite will be stabilised to room temperature (retained austenite). In a series of alloy consisting the chemistries, Fe–(9-12)Mn–6Al–(0.15–

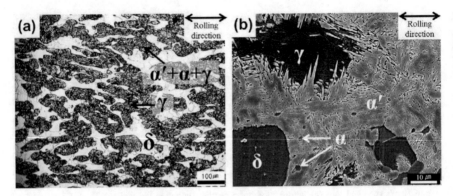

Fig. 6.15 Microstructures of the air-cooled hot-rolled specimen of the Fe–9Mn–6Al–0.15C alloy [40]: (**a**) optical microstructure composing of partially recrystallised δ-ferrite (white), austenite (γ, brown), and the other lath-like microstructure components (grey); (**b**) SEM micrograph showing the details inside the lath-like microstructure component (grey area in (a)) consisting of austenite, a lath-like martensite (α′), and small globular ferrite particles (α)

0.3)C [39, 40], the microstructure of the air-cooled hot-rolled specimen has been found to consist of partially recrystallised δ-ferrite, austenite (γ) and other minor microstructural constituents such as retained austenite (γ_R), a lath-like martensite (α′), and small globular α-ferrite particles inside the lath-like phase, as shown in Fig. 6.15. No precipitates of κ-carbide were observed in the air-cooled hot-rolled samples. The amount of martensite (α′) decreases as the amounts of Mn or C are increased.

The stability of the austenite in the hot-rolled duplex microstructure is increased further by increasing the Mn content of the alloy to 12%. In Fe–(12–18)Mn–8Al–(0.4–0.8)C alloys [49, 124], it was reported that the microstructures of the water-quenched and of the air-cooled hot-rolled specimens consist of a mixture of thin strip-like δ and γ. Fine L′1$_2$ structures were observed in the austenite, and B2 + DO$_3$ were found in the ferrite of the duplex steels [124]. It indicates that the increase in the C or Mn contents reduces the volume fraction of ferrite, restrains the precipitation of κ-carbides within the ferrite but enhances the formation of ordered phases in ferrite. The increase in the C or Al contents enhances the formation of L′1$_2$ and the precipitation of κ-carbides within the austenite matrix. When the Al content in duplex steels is increased above 7%, the presence of ordered B2 and DO$_3$ domains was observed within the ferrite grains [20, 46]. DO$_3$ usually leads to embrittlement and should be avoided. Thus, for each composition of duplex Fe–Mn–Al–C steel an upper limit for the Al content of 7% was specified for achieving good mechanical properties.

As the δ-ferrite in hot-rolled duplex steels is partially recrystallised and is present in an elongated shape, high anisotropy in mechanical properties is expected. The direct utilisation of this microstructure in hot-rolled products is not practical. To make cold-rolled products, intercritical annealing is generally applied following cold rolling to refine the grain size of the δ-ferrite and the γ-phase and to control the stability and morphology of the γ phase.

The effect of annealing temperature and time on the microstructures of duplex alloys was extensively studied to make use of the twinning-induced plasticity (TWIP) or transformation-induced plasticity (TRIP) effects in the austenite phase [37, 38] in addition to control the grains size of the δ-ferrite. According to research on Fe–3.5Mn–5.9Al–0.4C [37] and Fe–8.5Mn–5.6Al–0.3C [38] steels, the volume fraction, grain size and the SFE of the austenite phase can be controlled to obtain TRIP or TWIP effects when steels are annealed with optimised temperature-time profiles. The austenite grain size acts as a factor determining the activation of martensitic transformation and twinning.

The subsequent treatment after intercritical annealing also affects the final microstructure. The microstructural evolution in a Fe–3.5Mn–5.8Al–0.35C alloy after intercritical annealing at 780–980 °C for 50 s has been reported by Sohn et al. [36]. When a similar steel was cooled fast to a temperature at 500–600 °C followed by isothermal annealing for 60 min [46], the final microstructure composed of a polygonal ferritic matrix (recrystallised from δ-ferrite) and the lamellae formed in the prior austenitic grains. In the lamellar structure, there are three other phases: fine α-ferrite, κ-carbide and retained austenite (γ_R). The γ phase is partially decomposed by a eutectoid reaction into ferrite and needle-like κ-carbide ($\gamma \rightarrow \kappa + \alpha + \gamma_R$). The phase boundaries between γ_R and κ are coherent. The precipitation of nonstoichiometric κ-carbide has also been observed in the ferritic matrix after isothermal annealing at 500 °C in this steel. Also, this steel was cooled fast to 400 °C after intercritical annealing to avoid the decomposition of austenite followed by an austempering treatment at 400 °C for 180 s. This austempering treatment, termed also as isothermal bainitic treatment, used to enrich austenite with C by suppressing carbide formation during bainitic transformation, was applied so that the austenite can be retained at room temperature. The final microstructure is composed of a polygonal ferritic matrix (recrystallised from δ-ferrite), carbide-free bainitic ferrite and the retained austenite in the prior austenitic grains. The results showed that the thermal and mechanical stabilities of austenite decrease with the increasing austenite size and volume fraction, which can be controlled by annealing practice. When this steel was quenched to room temperature, a dual phase of $\delta + \alpha'$-martensite is obtained.

For austenite-based duplex steels or fully austenitic steels, the intercritical annealing treatment can be also applied to tune the $\alpha \leftrightarrow \gamma$ phase transformation and recrystallisation kinetics of the ferrite and austenite to obtain a composite-like mixed microstructure including recovered, non-recrystallised austenite, fine recrystallised ferrite and/or austenite. The concurrent progress of primary recrystallisation in the γ phase and $\gamma \rightarrow \alpha$ phase transformation can be utilised to produce an ultrafine-grained, duplex microstructure. Submicron-sized newly formed α-ferrite grains could effectively pin the grain boundaries of the growing recrystallised austenite grains and significantly slow down the recrystallisation kinetics [58, 59].

A partially recrystallised microstructure with fine dislocation substructures or fine grains both in ferrite (< 1 μm) and in austenite (~ 3 μm) were obtained in a Fe–

12Mn–5.5Al–0.7C steel [47] and in a Fe–0.5C–11.8Mn–2.9Al–1Si–0.32Mo–0.5 V steel [58] after annealing of the cold-rolled steels in the temperature range of 640– 750 °C, which is above the κ-carbide formation temperature. Cold-rolled sheets are generally annealed to obtain a fully recrystallised microstructure because the existence of non-recrystallised region seriously deteriorates the ductility. However, it has been shown that improved ductility and ultra-high strength can be realised by actively utilising the non-recrystallisation region in combination with the TRIP mechanism. When the mechanical stability of un-recrystallised austenite is well balanced with critical strain for the TRIP effect, stress and strain balance can be sustained in a very wide range of plastic deformation without tensile necking.

An ultrafine-grained duplex microstructure was reported to be achieved in Fe–25.7Mn–10.6Al–1.2C austenitic steel by ferrite transformation constrained, austenite partial recrystallisation treatment [59]. The hot-rolled strips were solution treated and quenched to room temperature to form a fully austenitic structure. The strips were then cold rolled and intercritically annealed at 870 °C for 30 s, followed by fast quenching (220 °C/s). A duplex microstructure including α (< 10 vol.%, size <0.5 μm), recrystallised γ (~ 50 vol.%, size <1.5 μm) and uncrystallised γ (~ 40 vol.%) was obtained during intercritical annealing. The recrystallised austenite grains were found mainly along the prior grain boundaries and were surrounded by submicron-sized ferrite grains. Nanosized intragranular κ-carbide formed during quenching preferably on the microband boundaries (4.5 ± 1.8 nm) rather than in the interior of microbands in the recovered austenite or in the recrystallised austenite grains (1.9 ± 0.7 nm) due to the segregation of the alloying elements to the defects.

As a summary, in the duplex steels, the microstructural constituents are more diverse and the microstructural evolution is more complicated than those in the ferritic or austenitic low-density steels. Because of the co-existence of the δ phase and the γ phase, the difference in recrystallisation behaviour between the δ and the γ phases, the partially reverse transformation of γ to α and the elemental partitioning between austenite and ferrite during intercritical annealing, many variants of microstructures can be obtained. The final microstructure of duplex steels may comprise of a combination from δ, γ, α, κ, martensite and bainite phases. The phase constitution, size, volume fraction and the distribution of phases as well as the substructures in each phase can be controlled by adjusting the annealing parameters in combination with the preceding cold rolling reduction and the austempering that follows the annealing. One key aspect is to control the mechanical stability of the austenite through optimising the steel composition and processing.

6.6 Strengthening Mechanisms

6.6.1 Ferritic Low-Density Steels

The solution hardening of Al in BCC iron is an important strengthening mechanism in ferritic low-density steels. Aluminium causes marked solid solution strengthening in ferrite causing ~40 MPa/wt% increase in yield strength, following Suzuki's

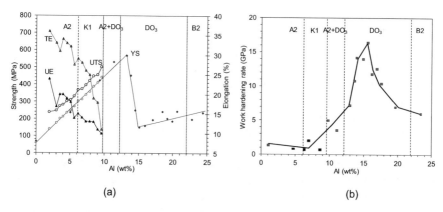

Fig. 6.16 (**a**) Tensile properties vs. Al content showing the influence of different lattice structures and the solid solution hardening effect of Al in Fe [24, 131]; (**b**) work hardening rate measured between 1% and 1.5% plastic strain as a function of Al content [25]

theory of solid solution hardening of BCC alloys [24, 25, 131]. Figure 6.16 shows the tensile properties of Fe–Al steels versus the Al concentration considering different crystal structures, the non-ordered and ordered lattices. The yield strength of a pure Fe–Al solid solution increases with increasing Al content up to 13% (22 at%), then rapidly decreases in the range of 13–15% and then remains almost constant with increasing Al content further to 25%. At room temperature, the tensile elongation decreases with increasing Al content. A transition from ductile fracture to brittle fracture without necking occurs in the range of 8–10% Al.

The formation of K_1 state, which starts from 6% Al and has a short-ordered lattice structure, causes the accelerated decrease in ductility for Al contents in the range of 6–10% and the transition from ductile to brittle fracture. The sharp drop in strength at 13–15% Al is due to the predominance of the DO_3 superlattice structure [128]. The ordered DO_3 and B2 superlattice structures are brittle in nature due to low strength and low ductility [78]. The appearances of ordering phenomena, such as DO_3 and B2 superlattice structures, have detrimental influences on the strengthening. Therefore, the Al content in ferritic Fe–Al alloys should be limited to a maximum of 9% and even down to 7% for automotive sheet applications.

Carbon (< 150 ppm), nitrogen (< 30 ppm) and silicon (< 0.3%) also increase the strength of Fe–Al steels through solid solution strengthening [131]. Grain size control can be used to strengthen Fe–Al steels further through Hall-Petch effect [29, 30, 131]. The addition of a small amount of microalloying elements (< 0.1%) such as Ti, Nb V or B can cause a remarkable grain refinement [130].

Work hardening through deformation can also be applied to increase the strength of ferritic Fe–Al steels. The work hardening rate measured over the strain range of 1.0–1.5% for Fe–Al steels as a function of the Al content is shown in Fig. 6.16b [25]. The work hardening rate is low for the disordered Fe–Al alloys with less than 7% Al, then it rises to a sharp peak at 16% Al as the ordered state DO_3 state forms, and then falls to a moderate value above 20% Al. The deformation in Fe–Al steels may involve dislocation generation and slip inside the grains. In a Fe–6Al steel, where the disordered state is obtained, dislocation glide bands have been observed, along with dislocation tangles between them after deformation [24]. In a Fe–8Al steel, where K_1 state may be produced, the deformed structure is manifested by short, straight segments of paired dislocations (super dislocations) with narrow mechanical antiphase boundaries at low strains. At high strains, ferrite exhibits dense dislocation bundles and undeformed B2 domains, indicating non-uniform deformation of ferrite [24]. It is suggested that strain hardening of the steel is dominated mainly by shearing of the ordered phases by superdislocations. It has also been observed that dislocation structures after deformation at room temperature show a transition from single, bowed dislocations for disordered materials to paired superdislocations with a strong tendency to form straight screw segments for highly ordered materials [25]. Deformation twinning occurs only at low temperatures because of insufficient dislocation mobility [24].

6.6.2 Austenitic Low-Density Steels

Unlike in ferritic low-density steels, various strengthening mechanisms are active in austenitic low-density steels. First of all, solid solution hardening plays an important role in the strengthening of Fe-Mn-Al-C steels due to the high amounts of the alloying elements C, Al and Mn present in these steels. Especially, the interstitially dissolved C increases the strength of any Fe–Mn–Al–C alloy substantially. In the solutionised and quenched condition, all the alloying elements Mn, C and Al dissolve in the FCC austenite. The effect of C on the yield strength of austenite in Fe–Mn–Al–C steels has been reported to be 187–300 MPa/wt% C (187 MPa/wt% C in Mn-TWIP steels) [70]. The presence of Al in solid solution of austenite slightly increases the yield strength (< 10 MPa/wt% Al) [21, 103].

Grain refinement is another strengthening mechanism that can be applied for non-age-hardenable Fe–Mn–Al–C alloys. The austenitic grain size can be refined by a combination of cold rolling and recrystallisation annealing. The effects of austenitic grain size on the tensile properties have been determined on the basis of the Hall-Petch relationship, using the following values: an initial strength of $\sigma_{y0} = 288$ MPa and Hall-Petch parameter $K = 461$ MPaμm$^{1/2}$ for yield strength, and an initial strength of $\sigma_{uts0} = 742$ MPa and $K = 351$ MPaμm$^{1/2}$ for ultimate tensile strength (UTS) [57].

Precipitation hardening is a significant strengthening mechanism in highly alloyed Fe–Mn–Al–C compositions. The first form of precipitation is that of fine (nanoscaled) and homogeneously distributed κ'-carbides. The presence of the κ'-carbide has a significant influence on the movement and arrangement of dislocations during deformation [11–15]. The second form of precipitation is that of the α-ferrite, which can be present as fine stringers in the γ matrix of the high Al-alloyed compositions [98, 111]. The α-ferrite will transform to B2 or DO$_3$ (intermetallic phases), which may also lead to strengthening of Fe–Mn–Al–

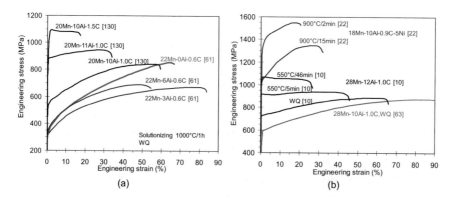

Fig. 6.17 Typical engineering stress-strain curves of austenitic and austenite-based duplex steels: (**a**) effects of Al and C contents; (**b**) effect of κ′-carbide and B2 intermetallic phases

C steels. The formation of controlled amount of α-ferrite grains also influences the recrystallisation behaviour of austenite and helps to obtain an ultrafine-grained, duplex microstructure to increase the strength [58, 59].

Strengthening through work hardening is a very important mechanism for austenitic Fe–Mn–Al–C alloys. Intensive studies [60–76] have demonstrated that the tensile deformation behaviour and strain hardening rate in austenitic Fe-Mn-Al-C steels are different from those in high Mn-TWIP steels. The difference becomes larger as the Al and C contents are increased and the trend can be seen from a series of typical engineering stress-strain curves of Fe–Mn–Al–C steels, as shown in Fig. 6.17. In Fig. 6.17a, all the steels with different Al and C contents are solution treated at 1000 °C and water quenched and have an optical microstructure of a single γ phase [61, 130]. In Fig. 6.17b, the Fe–28Mn–10Al–1C steel consists of a single γ phase after solutionising at 1000 °C and water quenching [63]. The Fe–28Mn– 12Al–1C alloy is an austenite-based duplex steel (with 8 vol.% δ-ferrite) [10]. The stress-strain curves of this steel in the water quenched and two aged conditions (aged at 550 °C for 5 min or 46 min after solution treatment at 1050 °C for 25 min) are given. The Fe–28Mn–10Al–1C steel [63] has the highest total elongation (~ 100%) of all the Fe–Mn–Al–C steels reported in the literature. Compared to it, the Fe–28Mn–12Al–1C alloy has a similar UTS, higher YS but a lower TE (~ 65%). Here the higher YS is related to the nanosized κ′-precipitates, and the lower total elongation (TE), which is attributed to the presence of the δ-phase and the strengthening effect of κ′-precipitates. The Fe–18Mn–10Al–0.9C–5Ni alloy [22] is an austenite-based duplex steel with ~20 vol.% δ-ferrite with a process histories of 70% cold rolling reduction followed by annealing at 900 °C for 2 min or 15 min. A large ductility and a high strain hardening capacity even at ultrahigh yield strength levels of over 1 GPa are observed (see Fig. 6.17b). The B2 intermetallic phase has been used as a strengthening element in this steel.

As can be seen, the strain hardening rate of the Al-containing austenitic steels is always lower than that of the Al-free Mn-TWIP steel (Fig. 6.17a). For steels containing less than 6% Al, the strength is lower than that of the Al-free Mn-TWIP steel. For steels containing high contents of Al and C, as Al or C content is increased, the YS and the ratio between the YS and the UTS (yield ratio) increase and the TE decreases. As the aging time is increased, similar trends can be observed for the Fe–28Mn–12Al–1.0C steel with κ'-precipitation (Fig. 6.17b). The effects of the cooling rate after solution treatment are similar to those of aging, i.e. an increase in strength but degradation of the strain hardenability and elongation caused by a decrease in the cooling rate [130]. The engineering stress-strain curves become nearly horizontal as the Al or C content and the aging time are increased or the cooling rate is decreased. Even negative work hardening rate (work softening) is observed in the highly alloyed Fe–20Mn–10Al–1.5C steel in the water quenched condition (Fig. 6.17a) and in the Fe–28Mn–12Al–1C steel after a longer aging time (Fig. 6.17b). In such cases, the steels practically do not reveal any strain hardening, but instead even have a minor stress drop at the beginning of yielding. However, the steels do not seem to become unstable but can rather be deformed to a higher elongation prior to fracture. Typically, conventional metallic alloys, which do not undergo any substantial strain hardening, do not reach such a huge ductility at these high strength levels. Such an unusual behaviour is unique for Fe–Mn–Al–C steels bearing high contents of Al and C. The Fe–18Mn–10Al–0.9C–5Ni steel shows a different strain hardening behaviour.

It, therefore, appears that there are two scenarios for austenite-based low-density steels. For alloys containing less than 6% Al, there is no κ' precipitation. The dislocation-based strain hardening, which is followed at higher loads by mechanical twinning, is the primary strain hardening mechanism. The lower work hardening rate in alloys containing less than 6% Al is attributed to the suppression of the mechanical twinning. With a higher Al content (> 7%), twinning is fully inhibited and lower strain hardening rates are essentially associated with dislocation planar gliding, which is affected by highly concentrated alloying elements, SRO and κ'-carbides. Microband-induced plasticity (MBIP) [62, 63], dynamic slip band refinement (DSBR) [74] and shear-band-induced plasticity (SIP) [10] are the mechanisms, newly discovered, to describe the plastic deformation of Fe–Mn–Al–C austenitic steels with a high SFE. The different characteristics of these deformation mechanisms, together with the steel compositions, process parameters and microstructures are summarised in Table 6.1. The variations of the true stress and strain hardening rate with the true strain of the alloys, which were used for studying these mechanisms [10, 63, 74], are shown in Fig. 6.18, in comparison with those for a typical TWIP steel [83]. All reported research results in this respect are based on SFE calculations and TEM observations with different deformation strains. For details of mechanisms, readers are referred to the relevant literature. Typically, structural metallic alloys reveal a monotonic decay of the strain hardening rate as a function of strain. The strain hardening of the TWIP steel (e.g. Fe–22Mn–0.6C alloy from [72]) exhibits the typical 'hump' associated with the strong microstructure refinement by deformation twinning. This is ascribed to the role of twin interfaces on plasticity and the activation of deformation twinning at low stress levels, close to the yield stress. The Fe–30Mn–2Al–1.2C steel with a low

Table 6.1 Comparison of the experimental conditions for three deformation mechanisms

Mechanism	SIP [10]	MBIP [63]	DSBR [74]
Characteristics	Homogenous shear band formation, sustained by nanosized κ'-carbides	Grain subdivision by microband intersections	Grain refinement of slip band structure through formation of Taylor lattices, cell blocks and dislocation cells
Alloy composition	Fe–27Mn–12Al–0.9C	Fe–28Mn–10Al–1.0C	Fe–30.4Mn–8Al–1.2C
Solution treatment	1050 °C/25 min	1000 °C/60 min	1100 °C/120 min
Cooling	Water quenching	Water quenching	Oil quenching
As-quenched microstructure	$\gamma + 8$ vol% $\delta + \kappa'$ (20–30 nm)	γ + SRO (< 2 nm)	γ + SRO (< 2 nm)
γ grain size	~ 30 μm	64 μm	40 μm
SFE	110 mJ/m^2	120 mJ/m^2	85 mJ/m^2
Tensile sample	Geometry not reported; 1 mm thickness	25.4 mm × 6 mm × 2 mm	40 mm × 6 mm diameter
Strain rate	10^{-4}/s	10^{-3}/s	5×10^{-4}/s
UTS/TE	875 MPa/58%	873 MPa/98.9%	900 MPa/68%

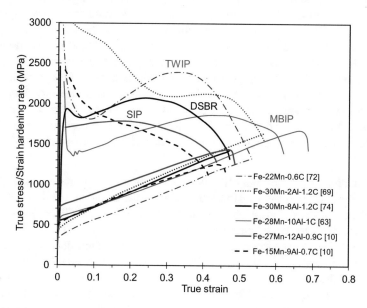

Fig. 6.18 True stress and strain hardening rate vs. true strain curves measured by room temperature tensile tests of the steels used for studying various deformation mechanisms [10, 63, 69, 72, 74]

Al content shows a multistage strain hardening behaviour, which are linked to two different deformation mechanisms: refinement of dislocation substructure at lower strains and refinement of deformation twinning structure at higher strains [69]. As seen, the strain hardening rate of the three alloys selected for displaying MBIP (Fe–28Mn–10Al–1.0C), DSBR (Fe–30.4Mn–8Al–1.2C) and SIP (Fe–27Mn–12Al– 0.9C) mechanisms show qualitatively the same strain dependence as observed in TWIP steels although the microstructural mechanisms are different. No changes in deformation mechanisms are observed. The application of these mechanisms depends on the presence of SRO or κ'-carbides. The MBIP and DSBR effects were observed in the austenite where SRO was present while SIP was observed in the austenite where the κ'-carbide already existed. The 'hump' in the strain hardening rate vs. true strain curves decreases and becomes flat as the aging time is increased. These mechanisms can explain the strain hardening behaviours in the solution-treated and underaged states in the specific austenitic Fe–Mn–Al–C steels studied but cannot be generalised. For example, the alloy Fe–15Mn–9Al– 0.7C shows a different behaviour, a monotonic decay of strain hardening with increasing strain (black dashed curves [10] in Fig. 6.18). Above all, none of these mechanisms can completely explain the strain hardening behaviours in the highly alloyed steels or peak aged steels, where a negative strain hardening rate is observed. Further investigations are required to elucidate the dislocation–particle interactions, the role of κ'-carbides in the formation of planar dislocation substructures and the contribution of these dislocation substructures to strain hardening.

Previous researches have demonstrated that the excellent mechanical properties in high Mn-TWIP steels are caused by massive mechanical twinning instead of dislocation and cell hardening [72]. However, the deformation of the austenite based Fe–Mn–Al–C steels with different Al levels was found to be dominated by planar glide before occurrence of mechanical twinning regardless of the SFE value and aging conditions. The addition of Al to Mn-TWIP steels leads to the following changes which control the mechanical properties of density-reduced Fe–Mn–Al–C steels.

- The addition of Al to high Mn-TWIP steels significantly increases the SFE of austenite and suppresses the formation of deformation twins, which results in the transition of the deformation mechanisms from TRIP and TWIP to dislocation gliding, characterised by planar slip.
- The addition of Al suppresses the cementite precipitation due to the decrease in both activity and diffusivity of C in austenite and promotes the formation of the SRO through ordering of Al and C in the solid solution. The SRO produces the strain energy required for spinodal decomposition and the eventual precipitation of nanosized κ'-carbide. The interaction between SRO or κ'-carbides and dislocations leads to the development of planar dislocation structures.

The deformation mechanisms of austenitic steels are strongly influenced by SFE [61–76]. For twinning to occur, it is generally admitted that the SFE of the steel must be in the range of 18 mJ/m^2 < SFE < 50–80 mJ/m^2 [131, 132]. If the SFE is lower, twinning is replaced by ε-martensite transformation. When the SFE is higher than

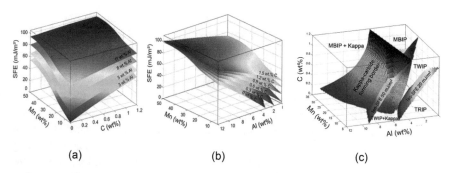

Fig. 6.19 Effect of (a) Al and (b) C on SFE, and (c) composition-dependent SFE map of Fe–Mn–Al–C steels at room temperature presented in 3D [134]

30–40 mJ/m^2, the formation of a dislocation cell is observed during deformation of high-Mn alloys. Thus, with decreasing SFE, the plasticity mechanisms change from dislocation glide (slip) to dislocation glide and mechanical twinning (TWIP) and finally to dislocation glide and martensitic transformations (TRIP: ε or α') [131–134]. Dislocation slip occurs at any SFE, but it predominates at lower degrees of deformation and/or in high SFE alloys.

The effects of Al, C and Mn on the SFE of austenitic steels are shown in 3D plots in Fig. 6.19a, b [132–134]. The addition of Al and C linearly increases the SFE of austenite by 10 mJ/m^2 per wt% of Al and 40 mJ/m^2 per wt% of C, respectively. Thus, Al and C suppress the mechanical twinning. Manganese decreases the SFE in the range of 0 to15 ~ 20% Mn, and then increases the SFE by 1.2–3.0 mJ/m^2 per wt% of Mn with further increase of the Mn content [132]. Some sources indicate that Si decreases the SFE of the austenite [135] by 2.4–2.8 mJ/m^2/wt%, while others suggest that the dependence of the SFE on Si alloying is non-monotonic [136, 137]. Addition of Cr results in a decrease of the SFE [136]. The SFE decreases as a result of increase in the austenitic grain size [132].

The deformation mechanisms of the austenite-based Fe–Mn–Al–C steels, i.e. TRIP, TWIP and MBIP, can be predicted based on the thermodynamic calculation of SFE maps, as shown in Fig. 6.19c [134]. With increasing the content of Al in steels, which significantly increases the SFE, the deformation mechanism shifts from TRIP to TWIP and further to MBIP and MBIP coupled with κ-carbide precipitates in Fe–Mn–Al–C steels.

However, the mechanisms based only on the SFE cannot completely explain the strain hardening behaviour in Fe–Mn–Al–C steels. In spite of its high SFE with increasing the Al content, Fe–Mn–Al–C steels exhibit planar glide characteristics and no cell formation occurs even up to failure. This fact indicates that the SFE is not a critical material parameter determining the slip mode, i.e. planar glide or wavy slip.

In general, for FCC materials, planar glide is expected to occur if they have a low SFE and so their extended partial dislocations are widely separated. If the SFE is high enough, cross-slip of the screw component of extended partials is easy to occur, tending to wavy slip forming a cell structure [138]. However, planar glide has often been observed in concentrated FCC solid-solution alloys with a high SFE [139]. Gerold and Karnthaler [139] explained planar glide at high SFE in the term 'glide plane softening', associated with the interaction between dislocations and short-range orders (or clustering). High concentrations of solute elements produce SRO after solution treatment. The initial dislocation that moves through the FCC lattice destroys the SRO and as such, this dislocation will face a higher resistance to slip than dislocations that follow. Thus, plastic deformation of subsequent dislocations becomes localised into slip bands, and this effect is called glide plane softening [139]. SRO in Fe–Mn–Al–C steels of high Al contents has been well reported in the solid solution state, manifested by the presence of weak superlattice reflections [61–65, 71–74], which is similar to that of κ'-carbide. However, no other experimental proofs for SRO have been published so far. Medvedeva et al. [140] have studied the Mn, Al and C distributions and their effect on the SFE in austenite using ab initio simulation. They demonstrated that Al and C as well as Mn at concentrations higher than 15 at.% increase the SFE, while the formation of Mn–C pairs and Al-ordering restrain the SFE growth. Short-range Al-ordering strongly decreases the unstable stacking fault energy (USFE) making the formation of stacking faults much easier but does not affect the SFE that can explain the observed planar glide deformation before the occurrence of mechanical twinning regardless of the SFE. Thus, addition of Al promotes SRO or κ'-precipitation and a phenomenon of glide plane softening, which contribute to planar slip and increase the ductility but decrease the work hardening rate. Buckholz et al. [99] have demonstrated that SRO produces the glide plane softening responsible for increasing elongation where the TE increases but the work hardening rate decreases with the Al/C ratio for solid solution in Fe–30Mn–(3, 6, 9)Al–(0.8–1.6)C alloys. However, it seems that SRO or κ'-carbide has different contributions to deformation as they lead to different deformation mechanisms (MBIP and DSBR vs. SIP). Further studies on the nature of SRO and its effect on the deformation behaviour in Fe–Mn–Al–C steels are required.

6.6.3 Austenite-Based Duplex Low-Density Steels

The deformation of austenite-based duplex Fe–Mn–Al–C steels is dominated by the behaviour of the austenitic matrix if the δ-ferrite content is less than 10 vol.%. The tensile response of these duplex steels is similar to that of the fully austenitic steels, i.e. exhibition of an inflection in the strain hardening rate curve [10, 51]. The deformation of the steels is dominated by the austenite, and the superior plasticity has been attributed to the homogeneous shear deformation of austenite.

As the amount of ferrite increases, the austenite–ferrite duplex steels exhibit multi-stage strain hardening characteristics due to the contribution of the δ-ferrite [47–50, 52, 54]. A Fe–18Mn–9.6Al–0.65C steel, where the volume fraction of δ-ferrite is about 21 vol.%, has been reported [52, 54] to exhibit distinct three-stage strain hardening behaviour. In stage A, an early rapid decrease in strain hardening rate was observed which is attributed mainly to the deformation of ferrite. In stage B, an approximately linear work hardening rate was dominated by the response of austenite. In stage C, a sharp decline in the strain hardening rate was ascribed to severe strain localisation in both austenite and ferrite. Austenite possesses strain hardenability greater than ferrite, while ferrite is easier to deform. The deformation mode of austenite (SFE ~ 66 mJ/m^2) is dominated by the MBIP mechanism. However, wavy glide in ferrite leads to the formation of dislocation nodes, dislocation cells (DC) and cell blocks (CB). Neither secondary twin nor strain-induced martensite was found during tensile deformation.

The ordered phases in the ferrite in duplex steels [55] have been found to affect the deformation behaviour. The B2 domains bound by swirled thermal antiphase boundaries are formed inside the disordered ferrite matrix. In addition to the B2 domains, fine DO_3 phases are evenly distributed throughout both B2 domains and the disordered ferrite matrix. The deformed structure of ferrite is manifested by short, straight segments of paired dislocations (superdislocations) with narrow mechanical antiphase boundaries. In austenite, the single planar dislocation glide seems to be dominant at low strains and multiple planar slip occurs at high strains. It is suggested that strain hardening of the steel is dominated mainly by shearing of the ordered phases by superdislocations in ferrite and planar gliding dislocations in austenite. The steel exhibits relatively high yield strength and low initial strain hardening rate, leading to a moderate elongation. A Ni-alloyed Fe–15Mn–10Al–0.86C–5Ni austenite-based duplex steel (with ~20 vol.% δ-ferrite) was reported [22] to consist of γ as the matrix and two types of ferrite phases: the δ (B2) stringer bands and the α (B2) nanosized particles along the γ grain boundaries or at shear bands within the γ grains. The high strain hardening rate was attributed to the interaction of dislocations with α (B2) particles. Dislocations were observed to be piled up or bowing out at α (B2) phase boundaries. The α (B2) particles are not sheared by gliding dislocations.

6.6.4 Ferrite-Based Duplex Low-Density Steels

Ferrite-based duplex steels possess a bimodal structure consisting of clustered austenite particles (10–40 vol.%) in a coarse grained δ-ferrite matrix owing to their low C (0.1–0.4%), medium Mn (3–9%) and medium Al (3–6%) contents. In these steels, most of the deformation is accommodated by the ferrite, while some refined austenite grains homogeneously distributed in ferrite matrix undergo a deformation-induced martensitic transformation [27–47]. A multi-stage strain hardening behaviour was observed in a duplex Fe–8.5Mn–5.6Al–0.3C steel, annealed at 900 °C for 30 min, containing 40 vol.% austenite [38]. In Stage 1, the work hardening was dominated by deformation in the ferrite phase, and the austenite grains were deformed by dislocation slip. In Stage 2, deformation twins and α′-martensite were independently formed inside individual austenite grains. Finally in Stage 3, a number of thick deformation twins formed inside the austenite grains. The enhanced tensile properties of ferrite-based duplex steels are ascribed to the simultaneous formation of strain-induced martensite and deformation twins. Deformation twinning having extremely small (about 5 nm) thickness and spacing are activated in these steels [47]. The TRIP effect in these steels is mainly dependent on the crystal orientation of the metastable austenite and its mechanical stability against the formation of strain-induced martensite. The martensitic transformation preferentially occurs in regions having high resolved shear stresses, i.e. with high Schmid factors. The C, Mn and Al contents are not varied much with annealing conditions of these steels and have little effect on mechanical stability. The austenite grain size also acts as a factor determining the activation of martensitic transformation and twinning [37]. Under optimal conditions, TRIP and TWIP mechanisms can simultaneously be operative in austenite grains. Beside the deformation mechanisms, partial recrystallisation in the austenite and the fraction and size of ferrite also influence the tensile properties of these alloys [47].

6.7 Tensile Properties of Fe–Mn–Al–C Alloys

Plots of total elongation (TE) against ultimate tensile strength (UTS) are a key tool in claiming success in alloy design and property differentiation. There is a large variation in the TE of Fe–Mn–Al–C steels reported in literature as tensile specimens with various geometries were tested. The tensile strength and yield strength are reasonably independent of specimen geometry, but the elongation, especially TE, is dependent on the gauge length and other geometrical parameters of the tested specimen, such as the ratio of the gauge length to parallel length and the width/thickness ratio. The tensile properties also depend on the surface condition of the specimens, especially for hot-rolled strips. For a fair comparison, in this chapter, the total elongations tested with various specimen geometries in the literature are converted to those for ASTM E-8 specimen geometry (gauge length 50 mm, width 12.5 mm, and thickness 1 mm) using an expression proposed by Oliver [141] with an Oliver factor $\lambda = 1.14$.

Depending on the alloy composition and the process route, Fe–Mn–Al–C steels provide a wide range of tensile properties. Figure 6.20 is a property diagram showing the TE vs. the UTS of Fe–Mn–Al–C lightweight alloys in the solution-treated and quenched conditions. The positions of the conventional steels are indicated in dotted areas for comparison. The tensile properties collected from the literature were measured from specimens with tensile direction parallel to the strip rolling direction. It has been reported that the tensile properties of duplex and austenitic Fe–Mn–Al–C steels have very low planar anisotropy [20, 142]. As seen, the tensile properties (UTS, TE) of Fe–Al ferritic alloys are similar to those of conventional low-alloyed C–Mn and HSLA steels. The ferrite-based duplex steels are located on the upper bound of the first-generation advanced high-strength steels (1GAHSS). The single-phase austenitic steels cover the location of Mn-TWIP steels in the property map but with a larger area. The austenite-based duplex steels show a

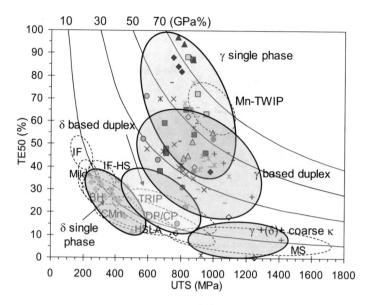

Fig. 6.20 The TE-UTS property map of Fe–Mn–Al–C alloys for a uniform specimen geometry corresponding to ASTM E-8 standard (50 mm × 12.5 mm × 1 mm). The tensile data were taken for the solution-treated and water-quenched conditions [10, 18, 23, 24, 28, 29, 38, 47–76, 142]

lower TE than the single-phase austenitic steels, while the strength levels are similar. With the formation of coarse κ-carbide in the austenitic matrix, the duplex alloys display very poor tensile elongation values irrespective of the strength.

6.7.1 Ferrite-Based Low-Density Steels

Fe–Al alloys with an Al content up to 9% and with additions of microalloying elements such as Ti, Nb and V (< 0.1%) exhibit an equivalent strength-elongation balance to the conventional bake hardenable (BH), low-alloyed C–Mn and HSLA steels at room temperature [24, 28, 29, 131]. Typical engineering stress-strain curves of a Fe–6.8Al and a Fe–8.1Al steel [28] are shown in Fig. 6.21. These steels achieve an UTS of 450–500 MPa and a TE of 15–30%. The strain hardening exponent (n) of BCC iron decreases slightly with the addition of Al. The n value has been reported to be in the range of 0.19–0.21 for Al contents in the range of 6–9% [28, 131]. The average normal anisotropy value or Lankford parameter (r_m) decreases as Al is increased, from 1.4 for 1.35% Al to 1.0 for 8.5% Al [28, 29, 131]. The planar anisotropy (Δr) is about 0.03–0.77 for microalloyed Fe-6Al steels [29]. Note that some of the reported r_m values in literature were calculated from the measured texture [28] and are higher than those measured from tensile tests. The actual r value obtained from tensile tests for a Fe-7%Al steel might be equivalent

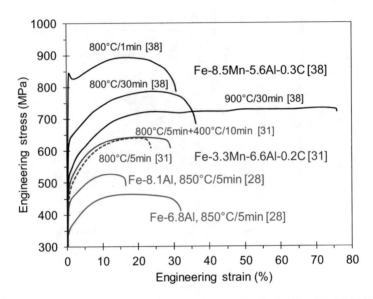

Fig. 6.21 Engineering stress-strain curves of two ferritic steels (Fe-6.8Al and Fe-8.1Al [28]) and two ferrite-based duplex steels (Fe-3.3Mn-6.6Al-0.2C [31] and Fe-8.5Mn-5.6Al-0.3C [38]) after different heat treatments

to that of the dual phase steels (about 0.6 to 1.1). The inferior r_m values of the Fe–Al alloys, with respect to BH steels, can be correlated with the much less intense and discontinuous γ fibre formation in the former. Optimal combinations of microalloying elements such as Ti, Nb V or B can cause a grain refinement (from ~90 to ~40 μm) and provide about 30 MPa increase in the tensile strength and also suppress the formation of the {001}<110 > sheet texture and, therefore, improve the deep drawing and stretch forming properties as well as the strength and ductility [29]. When Al is higher than 7%, the formation of the short-range ordering (K1 state) can cause brittle fracture in the sheet forming, and even during the process step of cold rolling. To achieve good formability, the Al content should be limited to a maximum of 7%. Ferritic low-density steels, from dent resistance considerations, have the potential to replace the conventional BH steels in the door outer panels of automobiles. However, the weight-saving potential might be compromised by the reduced Young's modulus caused by Al addition.

The ferrite-based duplex Fe–Mn–Al–C steels can provide a wide range of tensile properties. The engineering stress-strain curves of two ferrite-based duplex steels, Fe–3.3Mn–6.6Al–0.2C [31] and Fe–8.5Mn–5.6Al–0.3C [38], after different annealing treatments are shown in Fig. 6.21. The Fe–3.3Mn–6.6Al–0.2C steel has a dual phase (δ-ferrite + Martensite) microstructure after annealing at 850 °C and water quenching but has a mixture of δ-ferrite + bainite + martensite + retained austenite after quenching to 400 °C and austempering there for 10 min. The austempering provides a higher elongation. The Fe–8.5Mn–5.6Al–0.3C steel [38] has a bimodal structure consisting of clustered austenite particles and a coarse δ-ferrite matrix after different annealing and water quenching. The TRIP and/or TWIP mechanisms are activated during tensile deformation depending on the mechanical stability, the SFE and the size and orientation of the austenite grains. The TRIP is operative in the samples annealed at 800 °C for 1 min and 800 °C for 30 min, while the TRIP and TWIP take place simultaneously in the sample annealed at 900 °C for 30 min. By adjusting the annealing parameters, the ferrite based duplex steels can provide a high tensile strength of 600–950 MPa and a high total elongation of 30–70%. There are no reports in literature on the anisotropy parameters of these steels.

6.7.2 Duplex and Austenitic Fe–Mn–Al–C Steels

6.7.2.1 Solution-Treated and Quenched Conditions

Figure 6.22 shows the tensile properties of two series of Fe–Al–Mn–C steels as a function of the Al content [10, 18, 71]. For the Fe–30Mn–1.2C–xAl series in Fig. 6.22a, the microstructure of the alloys in the range of 0–9% Al after solution treatment is a single-phase austenite; δ-ferrite is introduced when the Al content is higher than 10%. Compared with the tensile properties of an Al-free TWIP steel (Fe–30Mn–1.2C), the addition of less than 7% of Al to the steel leads to a slight increase in the YS but a decrease in the UTS and TE. The increase in the YS is due

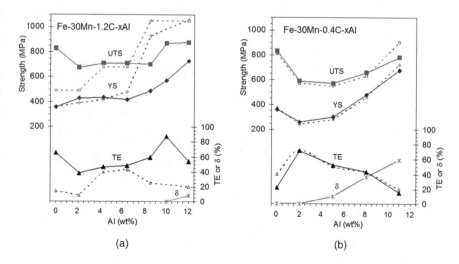

Fig. 6.22 Tensile properties of two series of Fe–Mn–Al–C steels showing the effect of Al on the tensile properties through changes in microstructure. (**a**) Fe–30Mn–1.2C–xAl, (**b**) Fe–30Mn–0.4C–xAl. The steels were solutionised at 1100 °C for 2 h. Solid curves are properties from the water-quenched samples; dotted curves are properties from the water-quenched and aged (at 550 °C for 24 h in air atmosphere) samples. (Data are taken from refs. [10, 18, 71])

to the solid solution strengthening of Al. The decrease in the UTS is attributed to a suppression of mechanical twinning caused by the increase of the SFE. However, with increasing the Al content further beyond 8%, the YS, UTS and TE all increase again, and the tensile response is similar to that of the Al-free alloy. In alloys containing high Al (> 7%) and high C (> 1.0%), metastable κ-carbide (L'1$_2$) may form during quenching after solution treatment, which increases the strength and does not impair the elongation of the steels. With further increasing Al to above 10%, the increase of strength is accompanied by a severe deterioration of ductility. This is due to the fact that the δ phase and/or coarse κ-carbide are introduced in the austenite grain boundaries. The presence of grain boundary δ-ferrite decreases the plasticity of such alloys and enhances the anisotropy of mechanical properties.

For Fe–20Mn–0.4C–xAl steels (Fig. 6.22b), the effect of Al on the tensile properties is found to be different. The Fe–30Mn–0.4C steel in an as-solutionised condition contains a very small amount of ε-martensite, which leads to a rather low elongation. The addition of 2% Al to the Fe–30Mn–0.4C steel produces a fully austenitic microstructure and causes an increase in the SFE, resulting in a decrease both in the YS and UTS but an increase in the work hardening rate and TE. With a further increase in the Al content from 2% to 11%, the microstructure changes from a single phase austenite to a duplex microstructure consisting of austenite and δ-ferrite, and the amount of δ-ferrite increases as Al is increased. The SFE of the steels is also increased. The strength increases as the amount of the ferrite is increased with increasing Al. This is due to the fact that the solution hardening effect of Al in the ferrite (40 MPa/wt% of Al) [24] is stronger than that in austenite (10 MPa/wt% of Al) [21]. The elongation drop is due to the increase in the amount of ferrite and the suppression of twinning in the austenite. The δ-ferrite in the high Al-containing alloys transforms to DO$_3$ or B2 type lattice, which becomes brittle.

The effect of C on the tensile properties of low-density steels is shown in Fig. 6.23 for steels containing similar Al contents but with two significantly different Mn levels [62, 66, 67, 142]. With lower C levels, Fe–Mn–Al–C steels have duplex microstructures. As the C content is increased, the amount of the austenite increases, and the steels will have a single-phase austenite when C reaches a certain level (~ 0.7% for Fe–30Mn–9Al–xC steels and 1.0% for Fe–20Mn–10Al–xC steels). For duplex steels, the YS decreases slightly and UTS does not change significantly as the amount of austenite is increased with increasing the C content. This is due to the fact that the solution hardening of Al in the austenite is decreased. The Al content in austenite is reduced as a result of the elemental partitioning between two phases, and it decreases as the C content increases. On the other hand, the deformation capacity of the austenitic phase cannot fully unfold because of early failure from the δ/γ interfaces. The TE of duplex steels increases as the amount of austenite increases and reaches a maximum value until the matrix becomes single-phase austenite. For single-phase austenitic steels, the strength increases but the elongation decreases with increasing C content, which is related to the increased amount of κ-carbide precipitation (or ordering) in the austenite upon cooling after solution treatment.

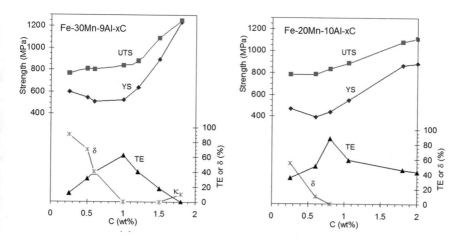

Fig. 6.23 Tensile properties of two series of Fe–Mn–Al–C steels showing the effect of C on the tensile properties through changes in microstructure. (**a**) Fe–30Mn–9Al–xC, (**b**) Fe–20Mn–10Al–xC. (Data are taken from refs. [62, 66, 67, 142])

The effect of Mn on the tensile properties of Fe–Mn–Al–C steels is related to the Al and C contents of the steels [11, 12]. Manganese, forming an almost perfect substitutional solid solution with iron, displays almost no solid solution hardening in austenite. Manganese can indirectly affect the properties by increasing the volume fraction of austenite as well as the solubility of Al and C in the γ-solid solution. Manganese content in the range of 22–30% show similar tensile properties [11, 12]. When the Mn concentration exceeds 30%, it gives rise to the β-Mn precipitation, causing extreme brittleness [108, 109, 130].

Adding Si has also been shown to increase the strength and hardness during aging but to decrease the work hardening rate [19, 106, 110]. Silicon has been reported to increase the kinetics of κ-carbide precipitation by increasing the activity of C in austenite and thus increasing the partitioning of C into the κ-carbide during aging. Silicon can prevent or suppress the formation of brittle β-Mn during extended aging treatment. Chromium addition to Fe–Mn–Al–C alloys results in an increase of ductility and a decrease of strength due to the suppression of κ-carbide precipitation [130, 147, 148]. The effects of other elements such as Mo, V, and Nb have also been studied and summarised in the literature [11, 12], indicating that the effects of these elements on the mechanical properties of Fe–Mn–Al–C steels are rather limited.

The solution temperature and cooling rate after solution treatment have also their effects on the tensile properties of austenite-based low-density steels. The strength decreases but the elongation increases as the solution temperature is increased due to a larger austenitic grain size [61, 63]. The strength increases as the cooling rate is decreased due to the formation of a higher density of SRO or κ′-precipitates [130].

6.7.2.2 Age-Hardenable Austenitic Steels

A. *Effect of aging*

The dotted curves in Fig. 6.22 show the tensile properties of these alloys after aging at 550 °C for 24 h [10, 18, 71]. Aging greatly reduces both the UTS and TE of the Al-free Fe–30Mn–1.2C steel but very weakly affects the tensile properties of the alloys, Fe–30Mn–1.2C with Al additions of 2–7%. The strong hardening effect can only be seen in the Fe–30Mn–1.2C–xAl steels when Al content is higher than 7%. The increase in strength is attributed to the precipitation hardening effect of κ'-carbides. However, aging of Fe–20Mn–0.4C–xAl duplex alloys at 550 °C does not introduce a significant change in tensile properties although it may slightly increase the amount of α-ferrite or introduce κ'-carbides in the alloys with increasing the Al content.

Kalashnikov et al. [129] have also studied the effect of Al and C contents on the tensile properties and impact toughness of the austenite-based Fe–Mn–Al–C alloys in the aged condition. The tensile properties and the impact toughness as a function of Al or C contents are given in Fig. 6.24. Again, alloys with intermediate Al concentrations (about 2–7%) do not offer improved tensile properties with increasing the Al content. However, the austenitic Fe–Mn–Al–C alloys containing Al > 7% and C > 0.7% can produce pronounced strengthening after aging treatment through the formation of fine and dispersed κ'-carbides.

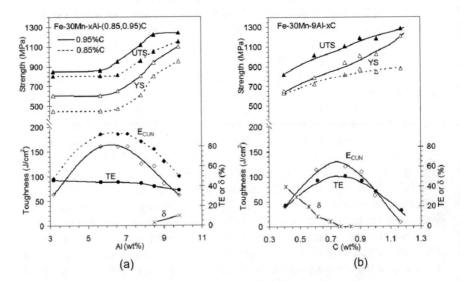

Fig. 6.24 (a) Effect of Al on mechanical properties of Fe–30Mn–xAl–(0.85, 0.95)C steels. The alloys were quenched from 1150 °C and aged at 550 °C for 16 h. E_{CUN} is the impact energy measured from Charpy U-notch specimens (1 mm notch radius). (b) Effect of C on mechanical properties of Fe–30Mn–9Al–xC steels. The alloys were water quenched from 1050 °C (dashed curve) and aged at 550 °C for 16 h (solid curves). (Data are taken from ref. [129])

B. *Aging kinetics*

Age hardening is associated with the homogenous precipitation of κ′-carbides within the austenite. The strengthening potential of the κ′-phase is between 100 and 350 MPa for a volume fraction of approximately 0.05–0.35 [133].

As discussed in the previous sections, in Fe–Mn–Al–C austenitic steels containing high amounts of Al and C, SRO zones or nanosized κ′-carbides form uniformly in the water-quenched steel and the coarse cuboidal κ′-carbides precipitate in the air-cooled steel. This indicates the very fast precipitation kinetics of κ′-carbides in the early stage of precipitation.

Figure 6.25 illustrates the age hardening behaviour of austenitic Fe–Mn–Al–C steels as functions of aging temperature and the steel compositions. Figure 6.25a shows the age hardening response of a Fe–30.4Mn–8Al–1C–0.35Si steel at various aging temperatures [109]. As can be seen, the Fe–Mn–Al–C alloys follow classic

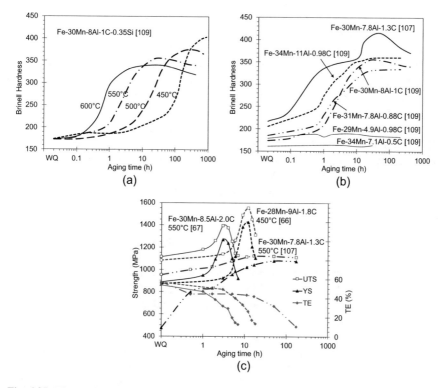

Fig. 6.25 The age hardening of austenitic Fe–Mn–Al–C alloys established using hardness measurements or tensile tests: (**a**) the aging curves of a Fe–30Mn–8Al–1C–0.35Si steel [109] showing the temperature dependence of the aging kinetics; (**b**) the aging curves of several alloys aged at 550 °C, showing the effect of the Al and C contents. Data are extracted from refs. [107, 109] and converted to Brinell hardness for comparison; (**c**) the tensile properties of several alloys as a function of aging time, showing the effect of the C content and temperature. Data are taken from refs. [66, 67, 107]

age hardening trends; the peak hardness increases for longer times at lower aging temperatures. Figure 6.25b summarises the age hardening response of six Fe–Mn–Al–C alloys aged at 550 °C [107, 109]. Two alloys, Fe–29Mn–4.9Al–0.98C and Fe–34Mn–7.8Al–0.5C, do not show any age hardening response because the Al or C content is too low to form the κ-carbides. For the other alloys, the hardness increases as the C content is increased in the alloys [88, 107]. Figure 6.25c demonstrates the age hardening response of three Fe–Mn–Al–C alloys aged at 550 °C or 450 °C with a higher C content (> 1.3%), showing the effect of the aging temperature and the C content on the aging kinetics. The solution treatment temperatures were 980, 1200 and 1200 °C for the Fe–30Mn–7.8Al–1.3C [107], Fe–28Mn–9Al–1.8C [66] and Fe–30Mn–8.5Al–2C [67] alloys, respectively, to obtain full austenite during solution treatment. These tensile data clearly show that the peak strength increases and the aging kinetics decrease as the aging temperature is decreased, and that the peak strength increases and the aging kinetics increase as the C content increases. The increment of the peak strength is about 350 MPa when C content increases from 1.3% to 1.8%. After the peak aging time, both the strength and ductility gradually drop due to the formation of the coarse intergranular κ*-carbide, the grain boundary lamellar colonies composed of discontinuously coarsened κ-carbides and transformed α-ferrite phase and/or β-Mn precipitates on the γ/γ grain boundaries.

The increase in the Al content from 7% to 11% gives an increase in the age hardening effect, as can be seen in Fig. 6.22 and Fig. 6.23, but has less influence on the age hardening kinetics, as seen in Fig. 6.25b. The effect of Mn content in the range from 24% to 34% on the age hardening behaviours was examined by Kalashinikov et al. [129] using fixed Al and C concentrations of 9% and 0.9%, respectively. The YS and UTS reached a maximum at 26% Mn, while ductility and impact toughness peak at 31% Mn.

From Fig. 6.25, it can be seen that it takes a rather long time for age hardenable Fe–Mn–Al–C steels to reach the peak strength and the TE decreases monotonously with increasing aging time. However, the YS and TE of Fe–Mn–Al–C steels are sensitive to the cooling rate after solution treatment or short aging. That is, a wide range of mechanical properties can be realised by processing control without altering the steel chemistry: (a) water quenching can produce a single austenite phase with SRO that exhibits a high tensile strength and high ductility; and (b) either controlled cooling or short aging time can produce a single austenite phase with κ′-carbide that shows a high yield strength and moderate ductility. This fact provides engineering significance of this steel grade for automotive applications.

Cold rolling prior to aging has been shown to retard the subsequent aging process but to produce a higher strength [143, 144]. At the same time, the $\gamma \rightarrow \alpha + \beta$-Mn reaction is accelerated making the alloy extremely brittle. However, warm rolling before aging seems to suppress this reaction, and this might be a promising approach to produce appropriate combinations of mechanical strength and fracture toughness [9, 145]. But, understanding on the effects of deformation on the κ′ precipitation in the literature is limited and warrants more research in this direction.

6.7.2.3 Non-age-Hardenable Austenitic Steels

For non-age-hardenable austenitic steels with the Al content less than 7%, it is possible to combine cold rolling with annealing to obtain high strength and formability simultaneously. After cold rolling, annealing in the temperature range of 500–1000 °C for a short period of time above the formation temperature of M_7C_3 can be applied to partially recover the ductility or to obtain fine recrystallised austenitic grains. An effective grain refinement is caused by the generation of a large number of nanoscale twin boundaries during cold rolling. By subsequent recovery annealing, the deformation twins exhibit thermal stability while the dislocation density is reduced, providing the opportunity for cold formability on the one hand and high strength by extreme grain refinement on the other hand. Figure 6.26a shows the tensile properties of a Fe–22Mn–1.2Al–0.3C steel after annealing at various temperatures following different amounts of cold rolling [146]. Figure 6.26b shows the tensile properties of a Fe–22Mn–7Al–0.8C steel after annealing at various temperatures following a 50% thickness reduction by cold rolling [20]. The microstructure with a lower degree of recrystallisation will provide a relatively high strength but a lower elongation. With increasing both the annealing time and/or temperature, the yield behaviour changes from continuous to discontinuous yielding due to the increasing recrystallised volume fraction of the microstructure. Strong increases in yield strength and tensile strength are observed due to the austenitic grain refinement when a lower annealing temperature and a shorter annealing time are used. When the annealing is conducted at an intercritical temperature, the controlled amount of α-ferrite grains can be introduced to influence the recrystallisation behaviour of austenite and to obtain an ultrafine-grained, duplex microstructure consisting of recrystallised austenite grains surrounded by submicron-sized ferrite grains, and recovered austenite regions which might be strengthened with preferential nano-κ′-

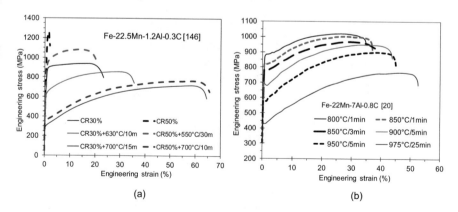

Fig. 6.26 Engineering stress-strain curves of two Fe–Mn–Al–C steels in annealed condition: (**a**) Fe–22.5Mn–1.2Al–0.3C steel with different cold rolling and annealing [146]; (**b**) Fe–22Mn–(7–10)Al–(0.7–1.0)C steels with different recrystallisation annealing parameters applied after a 50% cold rolling reduction [20]

carbide precipitation [58, 59]. This partially recrystallised duplex microstructure demonstrates excellent strength-ductility combinations, e.g. a yield strength of 1251 MPa, an ultimate tensile strength of 1387 MPa and a total elongation of 43%, arising from the composite response by virtue of diverging constituent strength and strain hardening behaviours [59].

6.8 Application Properties of Fe–Mn–Al–C Alloys

6.8.1 Impact Toughness

For structural applications, a good combination of mechanical strength and fracture toughness is desirable. The impact toughness as a function of temperature in some austenitic Fe–Mn–Al–C steels has been established [19, 147–151], as shown in Fig. 6.27, in comparison with that of the conventional steels and the various types of stainless steels. The Fe–Mn–Al–C steels in the as-solutionised condition show high impact energy of 200 J/cm^2 at room temperature, which is equivalent to that of a typical austenitic stainless steel and is higher than that of the conventional steels. However, these Fe–Mn–Al–C(Si) alloys exhibit a ductile-to-brittle transition at temperatures lower than -100 °C, as indicated by curves 1–3 in Fig. 6.27. The impact energy decreases as the aging time is increased. After peak aging, the Charpy V-notch (CVN) impact energy of the Fe–28.3Mn–5.38Al–1.04C steel at

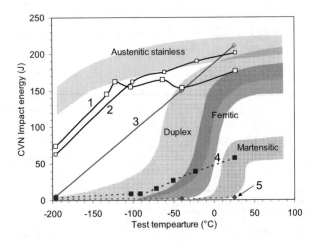

Fig. 6.27 CVN impact energy as a function of test temperature for two austenitic steels [148, 149]: curve 1: Fe–28Mn–5Al–1C solutionised at 1050 °C for 75 min (grain size d = 41 μm); curve 2: Fe–28Mn–5Al–1C solutionised at 1050 °C for 8 h (d = 135 μm); curve 3: Fe–30Mn–10Al–1C–1Si solutionised at 1050 °C for 1 h (d = 120 μm); curve 4: Fe–28Mn–5Al–1C solutionised at 1050 °C for 8 h and aged at 550 °C for 16 h; curve 5: Fe–30Mn–10Al–1C–1Si solutionised at 1050 °C for 1 h and aged at 550 °C for 16 h

room temperature is much reduced, and the energy value falls continuously with decreasing temperature (curve 4), while the impact energy of the Fe–30Mn–10Al–1C–1Si alloy is reduced to almost zero value (curve 5). Microvoid coalescence was observed for fracture of the solution-treated specimen, and the aged microstructure produced brittle grain boundary or cleavage fracture.

The CVN impact toughness of duplex steels at room temperature is similar to the values obtained for the duplex stainless steels. The impact toughness can be optimised by adjusting the annealing parameters to change the size and the distribution of the δ phase and the γ grain size in duplex steels [54]. Ultrahigh CVN toughness levels of about 450 J/cm^2 (measured from specimen where the V-notch direction is perpendicular to the rolling direction) was obtained in Fe–5Mn–(3–4)Al–(0.05–0.2C) steels [151]. The steel had a laminated dual phase structure of ferrite and martensite after hot rolling and has a strength of 0.8–1.2 GPa in the rolling direction. This impact toughness is much higher when compared with that of the conventional AHSS (usually <300 J/cm^2) with a similar strength.

The dynamic fracture toughness of Fe–30Mn–9Al–0.9C–xSi steels, measured from specimens with a sharp crack, is much higher in the solution-treated condition (71.5 J/cm^2) than in age-hardened conditions (9.5–37.5 J/cm^2) [19], which is comparable with that of a cast 4130 steel in quenched and tempered conditions (9.4–13.6 J/cm^2).

The impact toughness of Fe–Mn–Al–C steels depends on the phase constituents and their distributions, in particular, the formation of κ-carbide and the presence of δ-ferrite. The impact toughness as a function of Al and C contents is shown in Fig. 6.24. When the Al content is below 7%, there is no κ-carbide formed but instead, other carbides such as Fe_3C or M_7C_3 may be produced along the grain boundaries during aging. The distribution of these carbides has little effect on the strength but reduces the ductility and toughness significantly. As Al is increased above 7%, the formation of Fe_3C or M_7C_3 carbides is retarded and κ-carbides form. The strength increases but ductility and impact toughness decrease due to an increasing volume fraction of the hardening κ-carbide particles. The strengthening effect of the intragranular κ'-precipitates is associated with a decrease of ductility and can lead to brittle fracture before yielding at subambient temperatures [147]. On the other hand, the intergranular κ^*-phase alone can result in a severe loss in impact energy at both room and subambient temperatures [148]. As the Al content exceeds 10%, δ-ferrite is introduced in the microstructure as a banded phase, which further decreases the plasticity and toughness of such alloys and enhances the anisotropy of mechanical properties. The main crack source derives from the γ/δ interfaces in duplex steels because of the incompatibility in deformation between the two phases [129].

The effect of C on the impact toughness is given in Fig. 6.24b. An increase in C content increases strength, ductility and the impact toughness up to ~0.7% C in aged conditions due to the increase in the volume fraction of the austenitic phase, solid solution hardening by C and κ'-precipitation hardening in austenite. Above ~0.7% C, strength continues to increase but ductility and impact toughness decrease due to the increased amount of the κ*-precipitates. Silicon and phosphorus have been reported to decrease the fracture toughness of the austenitic lightweight steels [19].

In summary, the impact toughness of the austenitic steels in the solution-treated and water-quenched states is higher than that of the conventional AHSS but is lower than that of the austenitic stainless steels. The impact toughness of austenitic low-density steels is greatly reduced in the age-hardening conditions and also significantly influenced by the cooling rate after solution treatment due to the formation of SRO and κ-carbide. There is not enough information in the literature to draw conclusions on the toughness of the other types of low-density steels and needs future research attention.

6.8.2 High Strain Rate Properties

A very important parameter which characterises the impact behaviour of deep drawing steels for automotive bodies and frame structures is the specific energy absorption, E_{spec}, defined as the deformation energy per unit volume at a given temperature and strain rate in the order of 10^2–10^3/s. Figure 6.28 represents the E_{spec} value of the Fe–Mn–Al–C steels in comparison with some conventional deep drawing steels, such as DC04, HC300LA and high-strength IF (IF-HS) steels [10, 18, 75]. It is shown that the E_{spec} value of Fe–Mn–Al–C steels is about 0.5 J/mm^3, when the Al content is higher than 7%, which is similar to that of Mn-TWIP steels and about 50% higher than that of the conventional deep drawing steels (between 0.16 and 0.25 J/mm^3). The enhanced energy absorption of Fe–Mn–Al–C steels is due to the severe shear band formation at high strain rates.

The high strain rate behaviour of Fe–Mn–Al–C alloys has been investigated for solution-treated [26, 45] and age-hardened [26, 142] conditions utilising a split

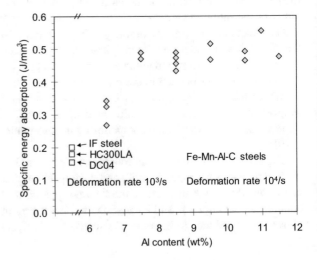

Fig. 6.28 Specific energy absorption values of Fe–Mn–Al–C steels in comparison with different conventional deep-drawing steels at the crash-relevant strain rate of 10^3/s [10, 18]

Hopkinson bar compression test for strain rates from 10^3/s to 10^4/s [45]. The Fe–Mn–Al–C alloys in both solutionised and age-hardened conditions show increasing yield strengths and decreasing failure strains with increasing deformation rate. A Fe–28Mn–12Al–1C steel in the aged condition [10] has shown compressive true strengths greater than 1.5 GPa and true fracture strains exceeding 40% at strain rates between 10^3/s and 10^4/s. At higher strain rates the work hardening exponent also increases, the activation volume for dislocation mobility decreases, and the degree of dislocation entanglement increases. Fracture in this loading regime occurs by highly localised deformation bands or shear bands [130]. The high deformation rate causes momentary localised high temperatures to occur within these bands. Low thermal-conducting materials cannot dissipate the heat causing lower stress within the band. This phenomenon has, therefore, been designated as adiabatic shear band (ASB) formation. Solution-treated Fe–Mn–Al–C alloys have demonstrated a resistance to fracture by adiabatic shear bands as work hardening occurs prior to shear localisation [152].

Another study shows that in fully austenitic Fe–22Mn–(0, 3, 6)Al–0.6C steels, which were dynamically compressed with nearly constant strain rate of ~3200/s at room temperature [153], the strain-hardening rate of the Al-free steel was the highest and that of other two steels was nearly the same. This suggests that dynamic flow of the Al-free steel was associated with both TRIP and TWIP effects, and that of other two steels was dominated by dislocation gliding – mainly, planar glide for the 3Al steel and the combination of both planar glide and wavy glide for the 6Al steel.

6.8.3 Fatigue Behaviour

The S–N fatigue strength of the austenitic Fe–Mn–Al–C steels in the as-quenched condition is equivalent to high Mn-TWIP steels [154] and is higher than that of austenitic Cr–Ni steels, but is inferior to martensitic chromium steels, provided that cyclic stresses do not promote plastic deformation at low temperature [155]. The fatigue strength of Fe–Mn–Al–C alloys in the temperature range of 250 to 550 °C is higher than that at room temperature. The behaviour of Fe–Mn–Al–C alloys (Fe–29Mn–9Al–0.9C) at the higher temperature ranges under the action of cyclic loading, including elasto-plastic deformation and cyclic temperatures, is better than the martensitic Cr-containing steels (e.g. Fe-12Cr-1.5Ni-0.2 V-1.8 W-0.5Mo-0.15C) [164].

The duplex Fe–Mn–Al–C steels exhibit a similar S–N cyclic life response to that of the austenitic Fe–Mn–Al–C steels under equal stress amplitude despite differences in microstructure and strength [142, 160]. The presence of the soft and stringed configuration of the ferrite phase and the grain size play an important role on the fatigue life.

Fig. 6.29 The cyclic
response of an austenitic
Fe–30Mn–9Al–1C steel after
three aging treatments
(550 °C for 2 h, 550 °C for
24 h, or 710 °C for 2 h) under
a strain amplitude of 0.8%
(dotted curves) [161], and a
duplex Fe–30Mn–10Al–0.4C
steel containing (65 vol.%
γ + 35 vol.% δ) in the
solution-treated condition
under strain amplitudes
ranging from 0.2 to 2.0%
(solid curves) [162]

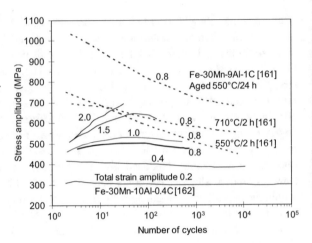

The intergranular, fine κ' precipitates and intragranular coarse κ^* precipitates affect the fatigue properties [156, 157]. The behaviour of cyclic softening is observed without any early stages of cyclic hardening for the age hardened Fe–Mn–Al–C alloys containing coherent κ'-precipitates [156, 161, 162]. Fine κ'-carbides within the grains are effective barriers to slip dislocations during fatigue. As a result, planar arrays of dislocations and incomplete cell structure tend to form within the grains. Thus, an inhomogeneous distribution of a high density of intergranular carbide and the formation of planar arrays of dislocations led to an observed two-slope behaviour in the Coffin-Manson relationship.

Figure 6.29 shows the controlled strain amplitude testing results in an solutionised and aged austenitic Fe–29.7Mn–8.7Al–1C alloy and in a solution-treated Fe–30Mn–10Al–0.4C alloy with a duplex microstructure of austenite (~ 65 vol.%) and ferrite (the solid curves) [162]. In the austenitic steel, cyclic softening was observed during fatigue for all three aging practices under strain amplitude of 0.8%, as shown by dashed curves in Fig. 6.29. Cyclic softening was associated with planar slip, shearing of κ-carbides, mechanical dissolution of the precipitates and formation of persistent slip bands. Tjong and Ho [161] postulated that the formation of precipitate-free zones (PFZ) in the specimen aged at 710 °C was responsible for the cyclic softening, which facilitated slip band formation. The precipitates in the specimen aged at 550 °C sheared early in the test due to their small size, thus producing persistent slip bands and cyclic softening behaviour similar to solid solution alloys. The specimen aged for 24 h endured higher stress amplitudes (from ~1000 MPa to ~750 MPa) and softened cyclically because of planar slip, dissolution of the precipitates and formation of persistent slip bands. In duplex steels, as shown by solid curves in Fig. 6.29, cyclic softening was observed below strain amplitudes of 0.6%; the formation of persistent slip bands produced cyclic softening in the austenite, and the applied strain was accommodated by these slip bands. At strain amplitudes above 0.6%, cyclic strengthening was achieved before critical stress amplitude, followed by cyclic softening. The fatigue crack initiation occurred in ferrite. Cyclic softening was not observed at strain amplitudes greater than 2.0%.

The addition of Al to high Mn-TWIP steels is reported to increase the resistance to fatigue crack growth (FCG) [158]. The austenitic Fe–Mn–Al–C alloys in the solution-treated state offers a higher fatigue crack growth resistance, compared with the conventional 304 L austenitic stainless steel [159, 163]. The nature of κ'-carbide has a significant effect on the FCG rate [159]. The aging treatment for the austenitic steels causes the decrease in the fatigue crack propagation resistance. The FCG rates measured in the solution treated Fe–9.33Al–25.9Mn–1.45C steel are about two times higher than those in the solution-treated Fe–8.25Al–29.9Mn–0.85C steel.

In summary, the S–N fatigue properties of the Fe–Mn–Al–C steels in the solution-treated conditions are better than those of the stainless steels and conventional high-strength steels. However, the nature of κ'-carbides has drastic effects on notched fatigue properties, i.e. the fatigue growth rate. Further, the mechanisms for improved fatigue properties are not clearly understood.

6.8.4 Formability

Formability, which is important for automotive and also other strip applications, is the ability of a material to undergo the desired shape changes without necking failures. The values of normal anisotropy factor and the strain hardening exponent are usually considered the most important indicators for sheet formability. In general, the full austenitic Fe–Mn–Al–C steels display a lower work-hardening rate compared with high Mn steels although there is no report about the anisotropy value. A strain-hardening exponent (n value) of 0.58 was measured for a Fe–28Mn–12Al–1C steel [10]. This value is higher than the reported n values of 0.33–0.45 for the austenitic 304 stainless steel. Other tests such as bending and hole expansion are used to characterise the stretch-forming and deep-drawing abilities. Limited reports about the bendability and hole expansion capacity (HEC) of Fe–Mn–Al–C steels [20, 56] are found for a few duplex steels in the literature, and these results are given in Table 6.2. The HEC values of Fe–Mn–Al–C steels are comparable to that of DP600 steels. The bending angle and the dome expansion height (DEH) of Fe–Mn–Al–C steels are lower compared with conventional AHSS of similar strength levels. The low bending angle in duplex steels might be related to the formation and the presence of δ-ferrite stringers within the γ-austenitic matrix. It is known that ordered ferrite possesses brittle behaviour [79]. In view of the limited amount of experimental data, these conclusions cannot be generalised with confidence.

The effects of Al addition on the deformation mechanisms in tensile and cup-forming tests in a high Mn-TWIP steel (22Mn–0.6C) and an Al-added TWIP steel (18Mn–1.2Al–0.6C) were studied by Chin et al. [165]. The cracking or delayed fracture takes place in the 22Mn steel but not in the 18Mn–1.2Al steel during the cup-forming test where the strain rate is considerably high (0.02/s), although the 22Mn steel had higher tensile strength, elongation and strain-hardening rate than

Table 6.2 Formability of duplex Fe–Mn–Al–C steels in comparison with dual-phase (DP) steels

Steel	Annealing treatment	Microstructure	UTS (MPa)	TE [a] (%)	HEC[a] (%)	BA[a] (°)	DEH[a] (mm)	ST (mm)
Fe–26Mn–9.7Al–0.47C [47]	830 °C1 min	γ + δ (21.8 vol.%)	945	22	26	135	8.5	1
	900 °C1 min	γ + δ (28.3 vol.%)	882	30	37	142	10	1
Fe–22Mn–(7–11)Al–(0.7–1.0)C [20]	850 °C3 min	γ + δ (~7 vol.%)	974	51	43	152	17.2	0.7
Fe–25Mn–(9–13)Al–(0.9–1.3)C [20]	850 °C3 min	γ + δ (~5 vol.%) + κ	999	35	34	97	20.7	1.3
DP500–DP1000		α + M	500–1000	15–29	20–40	150–180	25–35	1

[a] *HEC*: hole expansion capacity, *BA*: bending angle, *DEH*: dome expansion height, *ST*: strip thickness

the 18Mn–1.2Al steel (tensile strain rate $= 6.7 \times 10^{-3}$/s). This is explained by the fact that in the 18Mn–1.2Al steel where Al is added and the Mn content is reduced, twinning occurs more homogeneously and with less intensity during the cup-forming process than in the 22Mn steel. This indicates that a small amount of Al addition might be an effective way to improve the formability of TWIP steels. Furthermore, the resistance to hydrogen embrittlement in a Fe–18Mn–0.6C–1.5Al steel was reported to be improved [166], but a solution-treated and aged austenitic Fe–26Mn–11Al–1.2C alloy was found to be still susceptible to hydrogen embrittlement [167–169]. As the cracking during cup forming or the delayed fracture after cup forming is one of the critical problems limiting the application of Mn-TWIP steels, intensive studies should be conducted on the effect of Al and Mn on deformation mechanisms occurring during cup forming.

6.8.5 Weldability

The weldability of Fe–Mn–Al–C alloys has not been extensively investigated in the literature. It has been shown that the properties of weldment in these steels are related to welding methods. Fe–28Mn–(5, 6)Al–1C alloys were welded by electron-beam [170, 171] and continuous-wave CO_2 laser techniques [170]. The microstructure of the welded metals consisted mainly of the columnar dendrites with a preferred orientation and some equiaxed austenitic structures in the central portion of the weld pool. The growth and convergence of these columnar dendrites led to the formation of an apparent 'parting' in the weld centreline. The parting exhibited a lower hardness value while peak hardness was observed in the heat-affected zone (HAZ) of the weld.

The tensile and impact tests indicated that the weld materials exhibited lower tensile strength, lower elongation and lower impact energy than those of the base alloy. Creep rupture tests of the weldment and the base metal were conducted at 600 °C over a stress range of 225–350 MPa. Creep fracture of the weldments was observed to occur in the region of the parting. The rupture life and rupture ductility of the weldments were considerably lower than those of the base metal. On the other hand, Fe–30Mn–xAl–xC alloys were welded with gas tungsten arc welding (GTAW) process [172–174]. It was observed that all the welds presented satisfactory properties; relatively low properties were observed only for the particular composition with the highest carbon content. In a study with flux core arc welding (FCAW), the ductility as well as the strength of the Fe–Mn–Al–C welds showed superior behaviour (except the alloy with the highest Mn content due to the brittle β-manganese phase formation) in comparison with the 310 and 316 stainless steels welds [175].

Laser beam welding (LBW) is the most widely used technique by automotive manufacturers to join similar and dissimilar parts. However, when LBW is applied for high Fe–Mn–Al–C steels, a drastic evaporation of the Mn in the fusion zone (due to an increase in Mn vapour pressure as the temperature is increased) could decrease the SFE, thus changing locally the deformation mechanisms [177].

The welding between medium manganese Fe–Mn–Al–C alloys and other high-strength steels (e.g. TRIP steels, HSLA steels, and DP980 steels) has shown good joint efficiency [178]. In contrast, in high-manganese Fe–Mn–Al–C steels, the welds have shown a decrease in the mechanical properties for laser welding but not for beam welding [179]. Moreover, in the welds of high-manganese duplex Fe–Mn–Al–C steels a new region, called incompletely melted zone, has been found to form during the laser welding process [180]; in this zone the austenite matrix was melted completely, and the ferrite was melted partially with formation of precipitates.

The microstructural change of a Fe–Mn–Al–C weldment during aging treatment [176] affects the weld properties. For a duplex steel consisting of $\gamma + \alpha$ phases, the initial disordered α phase transformed to an ordered DO_3 phase in the as-welded condition. As the aging temperature increased from 550 to 650 °C, the ferrite in the weld metal changed from DO_3 to B2, and the ferrite base metal changed from $\alpha + DO_3$ to B2 + DO_3 and finally to B2. The formation of these ordered precipitates drastically decreased the elongation and toughness of the weldment.

The hot crack susceptibility was evaluated for Fe–29Mn–8Al–(0.53, 0.81, 1.17)C alloys [181, 182]. The alloy with 0.53% C is a duplex steel containing about 4 vol.% δ-ferrite, while the other two alloys have a fully austenitic structure. After welding, the ferrite content in the weld metal was found to be higher than in the base metal. The hot-crack susceptibility is affected by the characteristic morphologies of welds. Compared with the vermicular ferrite or lacy ferrite morphology types, the Widmanstätten austenite (WA) and acicular austenite (AA) morphology types exhibited greater hot-crack susceptibility. Both intergranular cracks and transgranular cracks were observed in WA and AA morphology types. A DO_3 type superlattice was found in the ferritic matrix of weld metal. After heat treatment at 1050 °C, the amount of ferrite in the weld metal is reduced and annealing twins were found. Tensile strengths in the weld metal were equivalent to those of the base metal but elongations in the weld metal were lower.

6.8.6 Oxidation Resistance

The oxidation resistance of Fe–Mn–Al–C steels is higher than that of the conventional steels over the temperature range between 450 and 900 °C in water vapour, and in wet and dry oxygen. This is attributed to the formation of a continuous $FeAl_2O_4$ and Al_2O_3 layer, which is promoted by the presence of Si and by a fine grain size [183–187]. At 700 °C, stainless steel resists oxidation better than the Fe–Mn–Al–C alloys. However, at 500 °C the Fe–Mn–Al–C steel specimens show improved oxidation resistance over the 304 stainless steel [183]. The most oxidation resistant alloys, having compositions within the range Fe–(5–10)Mn–(6–10)Al, develop continuous protective alumina scales and are totally ferritic [184, 185]. Austenite is detrimental to the oxidation resistance of duplex alloys as it promotes the breakdown of pre-existing alumina scales and the growth of bulky Mn-rich oxides. Sauer et al. [186] reported that the Fe–32Mn–7.5Al–0.6C alloys containing 1–2% Si have good oxidation resistance up to 850 °C. The austenitic Fe–8.7Al–29.7Mn–1.04C alloy was found to exhibit much higher oxidation rates compared with the duplex Fe–10Al–29Mn–0.4C alloy [187]. It is explained that the higher Al concentration of the duplex alloy enables a more effective Al_2O_3 layer to develop more rapidly. The small addition of elements Cr and Ni are found to further improve the oxidation resistance and corrosion resistance of Fe–Al–Mn–C alloys.

6.8.7 Corrosion Resistance

The Fe–Al–Mn–C alloys are anticipated to possess good corrosion resistance in both aqueous and gaseous environments, which are expected to derive from the high Al content. Al is expected to increase the corrosion resistance of the Fe–Mn–Al–C steels as Al with a high passivity coefficient forms a stable Al_2O_3 film, increases the corrosion potential (E_{corr}) and decreases the corrosion current density (i_{corr}). The effects of Mn on corrosion resistance are contrary to that of Al. However, the corrosion resistance of Fe–Mn–Al–C alloys in aqueous environments (3.5% sodium chloride, acetic acid, ammonia, and sodium sulphide solutions) is far inferior to that of 304 stainless steel [188–191, 194] and is comparable to that of the conventional steels [20]. Many researchers tried to improve the corrosion resistance of the austenitic Fe–Mn–Al–C alloys by adding Cr and decreasing C [190, 191]. However, since Cr and C are ferrite and austenite formers, respectively, Fe–Mn–Al–Cr–C alloys with higher Cr and lower C are dual-phase (austenite + ferrite) steels and their mechanical properties are far inferior to those of the austenitic steels [191, 192]. The corrosion resistance of the duplex Fe–Mn–Al–C alloys is worse than that of the austenitic Fe–Mn–Al–C alloys because in the duplex Fe–Mn–Al–C alloys pitting is the primary type of corrosion, and it takes place preferentially within the α grains and on the α/γ grain boundaries [193]. In order to obtain a fully austenitic structure, a proper combination of Cr and C contents is needed. The stress corrosion cracking behaviour of Fe–Mn–Al–C alloys is not improved due to the addition of Al, which is expected to provide active protective function [193, 194].

6.8.8 Wear Resistance

The abrasive wear resistance of Fe–Mn–Al–C steels is generally lower than that of M2 and AISI 304 stainless steels [195–197] and that of the Hadfield steel [198]. Several important results have been concluded [196]:

1. The nature of κ-carbides affects the abrasive wear resistance, coarsened grain boundary κ*-carbides (in overaged conditions) promoting excessive brittleness and reducing the wear resistance, fine-sized intergranular κ′ carbides (in under aged conditions) acting as obstacles for the grooving abrasives and leading to an increase in the wear resistance [197]
2. An increase in aluminium content reduces the wear resistance
3. An increase in carbon content increases the wear resistance
4. The wear resistance of austenitic steels increases as the SFE of steels is increased [198]
5. There is no clear correlation between the wear resistance and the initial or worn hardness of the steels

The cavitation wear resistance of Fe–Mn–Al–C steels in the solution-treated state was shown to be higher than that of AISI 304 stainless steel [199]. Aging treatment does not improve the resistance to cavitation. The alloys with higher carbon and aluminium content showed the best resistance to cavitation. The erosion wear resistance of Fe–Mn–Al–C steels is generally lower than that of the AISI 316 stainless steel. The erosion wear resistance of Fe–Mn–Al–C steels increases as the Mn and Al contents are increased due to the high solid solution strengthening [200].

In summary, the understanding of the wear behaviour of Fe–Mn–Al–C steels is a growing research topic, and there are several phenomena that are not fully understood as such, for instance the role of the SFE or the role of aging treatment on the wear resistance. The effect of microstructure on the different types of wear response is still unclear.

6.9 Physical Properties of Fe–Mn–Al–C Low-Density Steels

6.9.1 Density

Weight reduction is the main driving force for developing Fe–Mn–Al–C steels for automotive applications. Alloying elements with a lower density than Fe (7.8 g/cm^3) such as Al (2.7 g/cm^3), Si (2.3 g/cm^3), Mn (7.21 g/cm^3) and Cr (7.19 g/cm^3) are often added to Fe–C steels to reduce the density as well as to control the phase constitution. The lower density results from the fact that these light elements change the lattice parameter of steels and at the same time reduce density by virtue of their low atomic masses [131, 135, 201]. For example, an addition of 12% aluminium

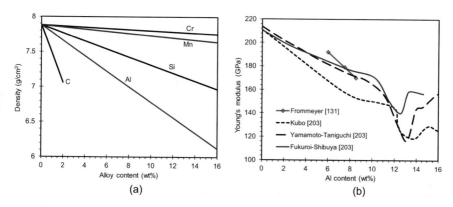

Fig. 6.30 Effect of alloy elements on the physical properties of iron; (**a**) density reduction of austenitic iron by elements lighter than Fe [131, 202]; (**b**) the reduction of Young's modulus of Fe–Al steels in the annealed state as a function of Al content [131, 203]

will reduce the density of iron by 17% of which lattice dilatation contributes 10% and atomic mass reduction contributes an additional 7%.

Figure 6.30a shows the effects of alloying elements on density reduction in austenitic steels up to a maximum of 16% alloy content. The data for Al were obtained directly from the literature [131], while those for other elements were calculated using different formulae derived by Bohnenkamp and Sandström [202]. The density of steel decreases linearly with increasing addition of the elements Al, C, Si and Mn. Considering its strong effect on density reduction and the engineering aspects such as alloy making and workability, Al has emerged as the chief alloying element in low-density bulk steels. Sometimes, Si is added in combination with Al.

Austenitic steels have higher density (8.15 g/cm^3 for γ-Fe vs. 7.87 g/cm^3 for α-Fe) and lower elastic modulus (than ferritic steels 195 GPa vs. 207 GPa). The increase in the Al content and the ferrite fraction will decrease the mass density due to the smaller atomic weight of Al compared to Fe and the difference in atomic density between the austenite (FCC) and ferrite (BCC) structures in steels [131, 135, 201]. The overall density reduction of the coexisting austenitic and ferritic Fe(Mn, Al) solid solutions was analysed based upon the combined effect of the lattice dilatation and the average molar mass of the alloys. By a linear combination of the influences from each element, the density (ρ) of Fe–Al ferritic steels and Fe–Mn–Al–C austenitic alloys can be expressed respectively as [12, 131]:

$$\rho_{\text{ferritic}} \left(\text{g/cm}^3 \right) = 7.874 - 0.098 \,(\text{wt\%Al}) \tag{6.3}$$

$$\rho_{\text{austenitic}} \left(\text{g/cm}^3 \right) = 8.15 - 0.101 \,(\text{wt\%Al}) - 0.41 \,(\text{wt\%C}) - 0.0085 \,(\text{wt\%Mn}) \tag{6.4}$$

The effectiveness of Al in density reduction is almost the same in both the ferritic and austenitic alloys since the coefficients for Al in Eqs. (6.3) and (6.4) are nearly identical (0.098 vs. 0.101). This indicates 1.3% reduction in density per 1% Al addition. The addition of C is very effective in density reduction for austenitic low-density steels. The effectiveness of C is about four times higher than that of Al.

6.9.2 Young's Modulus

An increase in the Young's modulus (E modulus) improves the stiffness of structural parts, e.g. the body-in-white (BIW) in automobiles. One of the critical disadvantages of low-density steels is that the addition of Al decreases the Young's modulus. The elastic moduli of polycrystalline Fe–Al alloys in the annealed state at room temperature are shown in Fig. 6.30b as a function of the Al content [131, 203]. The collected data of the Young's modulus were measured with dynamic measurements such as the resonance method or the ultrasonic method, which are more precise than those determined in quasistatic tensile tests. A 2.0–2.5% reduction in the Young's modulus per 1% Al addition is observed. The decrease of the Young's modulus with increasing the Al content is caused by a reduction of lattice energy of the Fe–Al solid solution and the larger distance between coexisting Fe and Al atoms in the lattice [135]. Silicon and chromium are reported to slightly increase the E modulus of steels containing high amounts of Al while Mn slightly decreases it [204].

Young's moduli of Fe–Mn–Al–C steels also depend on the processing condition [20, 104] as shown in Fig. 6.31. A relatively low E modulus is observed in the as-cast state, and it is increased by hot rolling. After cold rolling and annealing, the E modulus is slightly decreased again from the as-hot rolled value. The Young's modulus in the as-cast condition is lower due to the casting defects such as

Fig. 6.31 Young's moduli of Fe–(15–28)Mn–(7–10)Al–(0.7-1.2)C steels after different processing steps (solid lines) [20] and in the as-cast state of Fe–(5–40)Mn–10Al–1C steels (dashed band) [104]. The E moduli were measured using the resonance method

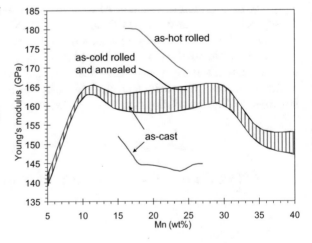

shrinkage cavities and segregations among dendritic arms. The presence of the microcavities decreases the Young's modulus of a material [205]. The addition of Al widens the composition range and temperature range of the primary δ-phase during solidification, which promotes the formation of cavities. The high amount of Al makes the steel-making complicated, and the defects in the cast structures also depend on the atmosphere used in the melting and the dimension of the ingots [206, 207]. This may explain the large difference in the measured Young's modulus in the as-cast condition between the two steels from different sources [20, 104]. The Young's modulus in the hot-rolled condition is higher because the microcavities are reduced during reheating and hot rolling. Cold deformation does have a reducing influence on the Young's modulus [208, 209]. Annealing may change the texture, which also influences the E modulus [210]. Compared with the HSLA grades (E ~ 210 GPa), decrease of around 20% in the Young's modulus is observed for cold-rolled strips of Fe–20Mn–(7–10)Al–(0.7–1.2)C steels.

As mentioned earlier, the weight saving potential of a low-density steel in structural applications depends on both the Young's modulus and density. For such purposes, materials with the highest 'specific stiffness', E^n/ρ, should be considered, where E is the Young's modulus, ρ is the density and n is a part dependent index. Depending on the shape of a structural part, the index varies: $n = 1$ for a tie, $n = 1/2$ for a beam and $n = 1/3$ for a panel. A comparison of the specific stiffness of a HSLA steel, a Fe–8Al ferritic steel [131], a Fe–22Mn–8Al–0.8C [20] austenitic steel and a typical 6xxx series aluminium alloy is shown in Fig. 6.32. It clearly shows that the specific stiffness of the low-density steels is not higher than that of HSLA steels and aluminium alloys. Therefore, to save weight, low-density steels have to be used for parts which are not strictly limited by the value of stiffness.

6.10 Challenges in Scalability of Low-Density Steels

All the properties presented and discussed in the previous sections were obtained from small-scale laboratory experiments. The next logical step should consist essentially in elaborating on the efforts to scale-up these steels from the laboratory to industrial manufacturing. From the technological point of view, processing of these steels on a large scale remains a great challenge. High Mn-TWIP steels are still not widely used in automotive applications. The production of these alloys is very difficult owing to their very high Mn content and its associated vapour pressure which makes the liquid metal processing very challenging. This will be also true for Fe–Mn–Al–C light-weight steels.

As illustrated in Fig. 6.33, many new problems (given in boxes) arise in steelmaking and in upstream and downstream processing due to the high amount of Al addition in Fe–Mn–Al–C steels [20, 21, 102, 206, 211, 212], although it

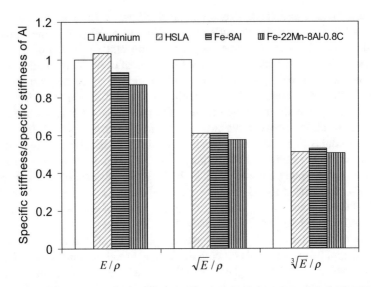

Fig. 6.32 Comparison of the specific stiffness of Fe–8Al [131], Fe–22Mn–8Al–0.8C [20], HSLA and aluminium 6xxx alloys

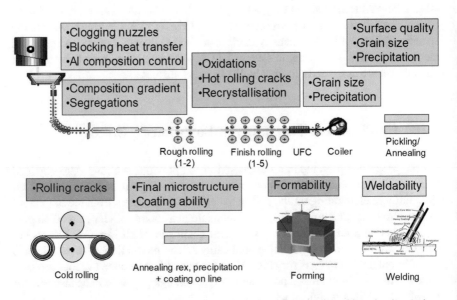

Fig. 6.33 Schematic illustration of the problems for industrial production of low-density steels

is reported that solidification cracking resistance of Fe–18Mn–0.6C alloys can be improved by the addition of Al [213]. Aluminium is well known as a deoxidiser in steel production. However, when a high amount of Al is used as an alloying element, it becomes a sign of low-quality steel in steelmaking practice. Intensive chemical reactions can occur between the melt and the refractory materials, casting powders and covering materials, which can result in deviation from the aimed chemical compositions. The formation of alumina in the liquid state can lead to clogging of the nozzles during continuous casting. Heavy and dense Al oxides can form on the surfaces during the strip casting stage as a result of the chemical reactions between the as-cast strip and the atmosphere. A very adherent and chemically stable layer of (Al,Mn,Fe)-oxides can be formed during the hot working stage, which is very difficult to eliminate. Surface defects, decarburisation, brittle phases and even cracks can occur during the downstream processing of these higher Al-alloyed compositions. Cracking can occur during hot forming as a result of the reduced solidus temperatures caused by heavy segregation of elements and their lower thermal conductivity. Cracking can occur during cold forming due to the presence of the grain boundary oxides, α-phase and/or κ-carbides as well the disorder to order transition. All these aspects indicate that the current processing routes for producing the conventional automotive steels could not be applied to produce Fe–Mn–Al–C steels without additional processing difficulties.

As a first-generation advanced high-strength steel, the DP1000 steel can be produced in the current available production lines as cold-rolled products providing typical tensile properties of UTS = 1000 MPa and TE = 13%. Stainless steels based on Cr and Ni–Cr alloying have been under development for many decades and offer very desirable combinations of properties and fabricability. The austenitic Cr–Ni stainless steels have good strength, ductility and toughness from cryogenic to elevated temperatures and have adequate weldability and machinability, superior formability and excellent oxidation and corrosion resistance. However, Cr–Ni stainless steels are used as special purposes due to the high costs of alloying elements and melting and processing. Where do Fe–Mn–Al–C alloys stand? The factors deciding if the Fe–Mn–Al–C light-weight steels can have a market niche include properties and processability. In Table 6.3, these factors of the Fe–Mn–Al–C steels, the high Mn-TWIP steels and the 304 stainless steel are scored to compare with DP1000 steel, which is used as a base line. The factor of the reference steel is defined as '0'. A higher positive score indicates more advantage or easier productivity. A higher negative score indicates less advantage or less productivity. As can be seen, the Fe–Mn–Al–C alloys show promise as lighter and stronger oxidation-resistance alloys of lower cost. However, the advantage of low density has not yet compensated for the existing disadvantages of producing Fe–Mn–Al–C steels on an industrial scale for automotive applications.

Table 6.3 Comparison of Fe–Mn–Al–C steels vs. DP1000 steel and Fe–Cr–Ni stainless steel

Process/property	DP1000	Fe–Cr--Ni	Mn–TWIP	Fe–Mn–Al–C
Melting and liquid processing	0	−1	−1	−3
Casting	0	−1	−1	−3
Hot rolling	0	−2	−2	−3
Cold rolling	0	−1	−1	−2
Coating	0	+4	−2	−3
Welding	0	0	−2	−2
Room temperature				
Strength	0	+1	+1	+1
Ductility	0	+1	+2	+2
Toughness	0	+1	−1	−1
Fatigue	0	+1	−1	−1
Cryogenic temperature				
Strength	0	+1	+1	+2
Ductility	0	+1	+1	+1
Toughness	0	+1	−1	−1
Elevated temperature				
Strength	0	+2	+1	+1
Ductility	0	+2	+1	+1
Formability	0	−1	−2	−3
Oxidation resistance	0	+2	0	+2
Corrosion resistance	0	+1	0	0
Density	0	0	0	+4
Alloy cost	0	−4	−1	−2

0 baseline as reference; + better than baseline; − worse than baseline

6.11 Summary

Aluminium is a main alloying element in Fe–Mn–Al–C steels for low-density purposes. The addition of 1% Al leads to 1.3% reduction in density, which is an advantage, and at the same time brings 2% reduction in Young's modulus, which is a disadvantage. The weight-saving potential can be balanced by the loss of the stiffness to some extent due to the reduced Young's modulus.

The addition of Al to steels enlarges the δ-phase area, widens the temperature range and composition range of the primary δ-phase and liquid phase area, creating a new phase κ. As such the physical metallurgy of the Fe–Mn–Al–C system is largely different from that of conventional steels.

The Fe–Mn–Al–C low-density alloys can produce a variety of microstructures and achieve a very wide range of mechanical properties. The most important group of steels in this category are the Fe–Mn–Al–C austenitic steels, which include precipitation-hardenable ($\gamma + \kappa$) and non-precipitation-hardenable steels (γ), followed by the ferrite-based Fe–Mn–Al–C duplex steels ($\delta + \gamma$), the austenite-based Fe–Mn–Al–C duplex or triplex steels ($\gamma + \delta$ or $\gamma + \delta + \kappa$), and the Fe–Al ferritic steels.

The tensile properties (product of UTS x TE) of Fe–Mn–Al–C steels have been well established. The austenite-based low-density steels can be placed in the space provided for the second-generation AHSS in the traditional strength vs. elongation property map, very often quite similar to those of Mn-TRIP and Mn-TWIP steels. The ferrite-based duplex low-density steels can be located at the upper bound of the first-generation AHSS in the property map while the ferritic Fe–Al steels show behaviours typical of HSLA steels of 400–500 MPa strength level. Thus, there are no significant gains in the combination of strength and ductility when a high amount of Al is added to steels within the corresponding groups.

In ferritic Fe–Al steels, the microstructure is a single δ-phase; the solid solution strengthening by Al is the major strengthening mechanism. To impart a lower density, high Al concentration is required. As Al is increased, the strength increases and density decreases. When the Al content is increased to more than 7%, the alloys suffer from serious embrittlement due to the BCC_A2 to K1 or BCC_B2 or DO_3 transition, which leads to poor workability and limited formability. Also, coarse grains and a strong cube texture formed in the strips may create hot shortness, cold rolling cracks, high plastic anisotropy and ridging. As the grain size of the δ-phase cannot be effectively controlled during hot rolling process, Fe–Al steels can only be produced as cold-rolled products. The cold rolling and annealing operations are necessary steps to modify the grain structure, grain size and the texture. Grain size control through the addition of small amounts of the microalloying elements, Ti, Nb or V, can only give limited increase in strength. Considering the low strength level, low formability and low weight-saving potential, the application prospects of this kind of steels for automotive parts are limited.

In austenitic Fe–Mn–Al–C steels, the microstructure in solution-treated condition is a single γ-phase, with some SRO or nanosized κ′-carbides uniformly distributed in the matrix, depending on the composition and the cooling rate after solution treatment. Solid solution hardening, grain size refinement, κ′-carbide precipitation and strain hardening are the possible strength mechanisms in these steels.

Strain hardening is a very important strengthening mechanism in austenite-based Fe–Mn–Al–C alloys. Aluminium brings two main effects in the solution-treated austenitic steels: increasing the SFE, producing short-range orders (SRO) or κ′-carbides. It can be generalised that dislocation planar gliding is a dominant deformation mechanism in austenitic Fe–Mn–Al–C alloys regardless of the SFE and aging conditions. In austenitic Fe–Mn–Al–C steels with a lower Al content, TWIP is still an active deformation mechanism in addition to the dislocation planar glide. Thus, strain-hardening behaviour very similar to high Mn-TWIP steels can be obtained. When an increasing amount of Al is added, the TWIP effect is suppressed due to the increase in the SFE. The existence of SRO or nanosized κ′-carbides and its interaction with dislocations play an important role on the planar glide mode of deformation of the high Al austenitic steels. Thus, the high solute elements contribute to the deformation gliding and the high SFE supresses the TWIP effect. New deformation mechanisms such as the microband-induced plasticity (MBIP), the dynamic slip band refinement (DSBR) and the shear-band-induced plasticity (SIP) have been invoked to describe plastic deformation of Fe–Mn–Al–C austenitic steels with a high SFE. Work hardening through these mechanisms contributes to the exceptionally high ductility of these alloys, sometimes to the tune of about 100% total elongation. The above deformation and strengthening mechanisms are effective in case of the austenite-based duplex Fe–Mn–Al–C alloys also. These new mechanisms can successfully explain the strain hardening behaviour in the austenitic steels in the solution-treated state and under-aged state. However, they cannot completely explain the negative work hardening rate in the alloys bearing higher Al and/or in peak aged condition, where the engineering stress-strain curve is almost horizontal or shows a downward trend after initial yielding. Therefore, more investigations are needed on this aspect.

A very significant strengthening mechanism in austenitic low-density alloys is the precipitation hardening provided by the uniform distribution of SRO or nanosized κ'-precipitates. It appears that Al > 7% and C > 0.7% are necessary to form intragranular κ'-precipitates in austenitic Fe–Mn–Al–C alloys. However, when Al is higher than 10%, δ-ferrite is introduced. Therefore, the effective Al composition for austenitic Fe–Mn–Al–C alloys for the best combination of mechanical properties is quite narrow and limited between 7% and 10%. Typical compositions for austenitic low-density steels should be in the range of 15–30% Mn, 7–10% Al and 0.7–1.2% C, with the balance being Fe. Aging below 650 °C causes κ'-carbide precipitates in a homogeneous fashion throughout austenite, and overaging produces heterogeneous grain boundary phases deleterious to mechanical properties. Another form of precipitation may occur by way of formation of α-ferrite by transformation from the γ matrix. These α-ferrite particles may transform into the B2 or DO$_3$ intermetallic phases, leading to significant strengthening. The strategy to be adopted to produce nanosized B2 precipitates is not yet very clear in contrast to that for κ-carbides. Strengthening in the non-precipitation hardenable austenitic Fe–Mn–Al–C alloys can be effected through solid solution hardening and by grain refinement through cold rolling and annealing.

In austenite-based duplex low-density steels, small amount of δ-ferrite is present as elongated bands covering many austenite grains in the hot-rolled state and transforms to the brittle ordered B2 or DO$_3$ phase in the final microstructure as a result of the high Al content in the δ-ferrite. The ductility and toughness are significantly reduced although the strength is not affected.

Ferrite-based duplex low-density steels have a mixture of $\delta + \gamma$ microstructure in the hot-rolled condition but develop a complex mixture from $\delta + \gamma + M + B$ phases after cold rolling and annealing, depending on the alloy composition and the annealing parameters. This type of steel usually shows a bimodal microstructure, consisting of clustered austenite or martensite particles in a coarse-grained δ-ferrite matrix and provides many microstructural variants so that plastic deformation of the δ-matrix and the TRIP and/or TWIP effect of the austenite can be combined to achieve a super combination of strength and ductility.

As the Fe–Mn–Al–C low-density steels are primarily meant for automotive structural applications, properties such as impact toughness, fatigue, formability, weldability and coatability are important and must be looked into and evaluated.

The impact toughness of the austenitic low-density steels in the solution-treated and quenched-state is better than that of the conventional AHSS but is lower than that of the austenite stainless steels. However, the impact toughness of austenitic low-density steels is greatly reduced in the aged hardening conditions and also significantly influenced by the cooling rate after solution treatment.

The specific energy absorption of austenite-based low-density steels during high strain rate tests is equivalent to that of high Mn-TWIP steels and is higher than that of the conventional high-strength steels. Excellent high strain rate deformation behaviour in addition to the low density makes this alloy family an interesting candidate possibly for armour components.

The S-N fatigue behaviour of Fe–Mn–Al–C steels in the solution treated and quenched state is better than that of the austenite stainless steels at room temperature and at elevated temperatures. The fatigue growth rate da/dN is sensitive to the nature of the κ'-carbides. Therefore, it will be premature to compare the formability of Fe–Mn–Al–C steels vis-à-vis existing AHSS of comparable strength.

The strain-hardening rate of austenitic Fe–Mn–Al–C steels is not better than that of high Mn-TWIP steels, which indicates a worse sheet-forming capacity. The hole expansion ratio of the austenite-based duplex low-density steels is comparable with that of DP600 steels. The bending angle and the dome expansion are lower compared with those of the AHSS of similar strength level. No formability tests have been reported on the austenitic steels and on the ferrite-based duplex steels.

The oxidation resistance of the low-density steels at elevated temperature is increased due to the formation of Al_2O_3 films. However, the corrosion resistance of the Fe–Mn–Al–C alloys does not show any significant improvement due to addition of Al, which is expected to provide an active protective function. Therefore, coating of these steels is very much needed for use as automotive parts. The Fe–Mn–Al–C alloys are susceptible to stress corrosion cracking and hydrogen embrittlement.

6.12   Future Developments

Future work on low-density steels should be carried out on two fronts. On the research front, more work is needed in the gap areas of the physical metallurgy of the Fe–Mn–Al–C steels, areas such as mechanisms of deformation and annealing, and the effect of microstructures on the formability, coatability and weldability in Fe–Mn–C–Al alloy systems. On the technology front, new and innovative processing and manufacturing techniques must be developed.

The following points may be mentioned as areas where attention in future is needed:

- For making use of κ'-carbide precipitation, it is critical to control the nature of κ-carbide precipitation. Solution treatment, fast cooling and aging are need for this purpose, which are not compatible with the conventional industrial installations for automotive products.
- For making use of nanosized B2 and/or DO_3 precipitates, the mechanisms and conditions to control the formation of these precipitates are not clear. Measures

to effectively utilise BCC_B2 and/or DO_3 intermetallics as the second phases should be explored.

For making use of grain refinement and TWIP effects, a combined process of cold deformation and recovery or recrystallisation processing needs to be explored to adhere to the industrial conditions.

- For duplex low-density steels, the laminae of the δ grain structures, the amount of the austenite phase and the stability of the retained austenite should be controlled through alloy design or process for utilising the TWIP or TRIP options to the fullest possible extent.
- Intensive application tests, such as formability, fatigue, coatability and weldability, are needed to understand the relationship between the application properties and the governing microstructural parameters and provide proper justification for the use as automotive parts.
- Investigations on processing problems during steel making, hot rolling and pickling should be conducted to look for proper processing parameters in industrial production lines.
- The alloying and process strategies to improve the Young's modulus of low-density steels should be explored.
- New processing routes have to be explored to overcome the problems that can occur in processing the low-density steels in the thermomechanical treatment lines at present used for producing conventional steels. Thin strip casting and 3D printing could be two promising new technologies for producing new Fe–Mn–Al–C steels.

References

1. D. Matlock, J.G. Speer, E. De Moor, P.J. Gibbs, Recent developments in advanced high strength sheet steels for automotive applications: An overview. JESTECH **15**, 1–12 (2012)
2. L. Galán, L. Samek, P. Verleysen, K. Verbeken, Y. Houbaert, Advanced high strength steels for automotive industry. Rev. Metal. **48**, 118–131 (2012)
3. H.K.D.H. Bhadeshia, Nanostructured bainite. Proc. R. Soc. A **466**, 3–18 (2010)
4. B.C. De Cooman, O. Kwon, K.G. Chin, State-of-the-knowledge on TWIP steel. Mater. Sci. Technol. **28**(5), 513–527 (2012)
5. R. Hadfield, T.H. Burnham, *Special Steels*, 2nd edn. (The Pitman Press, New York, 1933)
6. Dean RS, Anderson CT. Alloys. US Patent 2329186, 1943
7. J.L. Ham, R.E. Cairns Jr., Manganese joins aluminium to give strong stainless. Prod. Eng. **29**, 51–52 (1958)
8. J.L. Kayak, Fe-Mn-Al precipitation-hardening austenitic alloys. Met. Sci. Heat Treat. **22**, 95–97 (1969)
9. Y.G. Kim, Y.S. Park, J.K. Han, Low temperature mechanical behaviour of microalloyed and controlled-rolled Fe-Mn-Al-C-X alloys. Metall. Trans. A. **16A**, 1689–1693 (1985)
10. G. Frommeyer, U. Brüx, Microstructures and mechanical properties of high-strength Fe–Mn–Al–C light-weight triplex steels. Steel Res. Int. **77**, 627–633 (2006)
11. R.A. Howell, D.C. Van Aken, A literature review of age hardening Fe-Mn-Al-C alloys. Iron Steel Technol. **6**, 193–212 (2009)

12. H. Kim, D.W. Suh, N.J. Kim, Fe–Al–Mn–C lightweight structural alloys: A review on the microstructures and mechanical properties. Sci. Technol. Adv. Mater. **14**, 1–11 (2013)
13. R. Rana, C. Lahaye, R.K. Ray, Overview of lightweight ferrous materials: Strategies and promises. JOM **66**, 1734–1745 (2014)
14. S. Chen, R. Rana, A. Haldar, R.K. Ray, Current state of Fe-Mn-Al-C low density steels. Prog. Mater. Sci. **89**, 345–391 (2017)
15. S.D. Woo, N.J. Kim, Viewpoint set 53: Low density steels. Scr. Mater. **68**, 337–442 (2013)
16. R. Rana, Special topic: Low density steels. JOM **66**, 1730–1876 (2014)
17. I. Zuazo, B. Hallstedt, B. Lindahl, M. Selleby, M. Soler, A. Etienne, et al., Low-density steels: Complex metallurgy for automotive applications. JOM **66**, 1747–1758 (2014)
18. D. Raabe, H. Springer, I. Gutierrez-Urrutia, F. Roters, M. Bausch, J.B. Seol, et al., Alloy design, combinatorial synthesis, and microstructure-property relations for low-density Fe-Mn-Al-C austenitic steels. JOM **66**, 1845–1855 (2014)
19. L. Bartlett, D. Van Aken, High manganese and aluminium steels for the military and transportation industry. JOM **66**, 1770–1784 (2014)
20. Bausch M, Frommeyer G, Hofmann H, Balichev E, Soler M, Didier M et al. Ultra high-strength and ductile FeMnAlC light-weight steels. Final report RFCS Grant No. RFSR-CT-2006-00027, 2009
21. C.J. Altstetter, A.P. Bentley, J.W. Fourie, A.N. Kirkbride, Processing and properties of Fe-Mn-Al alloys. Mater. Sci. Eng. **82**, 13–25 (1986)
22. S.H. Kim, H. Kim, N.J. Kim, Brittle intermetallic compound makes ultrastrong low-density steel with large ductility. Nature **518**, 77–80 (2015)
23. D. Raabe, C.C. Tasan, H. Springer, M. Bausch, From high-entropy alloys to high-entropy steels. Steel Res. Int. **86**, 1–10 (2015)
24. J. Herrmann, G. Inden, G. Sauthoff, Deformation behaviour of iron-rich iron-aluminium alloys at low temperatures. Acta Mater. **51**, 2847–2857 (2003)
25. D.G. Morris, M.G. Munoz-Morris, L.M. Requejo, Work hardening in Fe–Al alloys. Mater. Sci. Eng. A **460–461**, 163–173 (2007)
26. L. Falat, A. Schneider, G. Sauthoff, G. Frommeyer, Constitution and microstructures of Fe–Al–M–C (M= Ti, V, Nb, Ta) alloys with carbides and Laves phase. Intermetallics **13**, 1256–1262 (2005)
27. C. Castan, F. Montheillet, A. Perlade, Dynamic recrystallization mechanisms of an Fe–8% Al low density steel under hot rolling conditions. Scr. Mater. **68**, 360–364 (2013)
28. R. Rana, C. Liu, R.K. Ray, Low-density low-carbon Fe–Al ferritic steels. Scr. Mater. **68**, 354–359 (2013)
29. U. Brüx, G. Frommeyer, J. Jimenez, Light-weight steels based on iron-aluminium: Influence of micro alloying elements (B, Ti, Nb) on microstructures, textures and mechanical properties. Steel Res. Int. **73**, 543–548 (2012)
30. A. Zargaran, H. Kim, J.H. Kwak, N.J. Kim, Effects of Nb and C additions on the microstructure and tensile properties of lightweight ferritic Fe–8Al–5Mn alloy. Scr. Mater. **89**, 37–40 (2014)
31. R. Rana, C. Liu, R.K. Ray, Evolution of microstructure and mechanical properties during thermomechanical processing of a low-density multiphase steel for automotive application. Acta Mater. **75**, 227–245 (2014)
32. S.Y. Han, S.Y. Shin, S. Lee, N.J. Kim, J.H. Kwak, K.G. Chin, Effect of carbon content on cracking phenomenon occurring during cold rolling of three light-weight steel plates. Metall. Mater. Trans. A **42**, 138–146 (2010)
33. S.S. Sohn, B.J. Lee, S. Lee, J.H. Kwak, Effect of Mn addition on microstructural modification and cracking behaviour of ferritic light-weight steels. Metall. Mater. Trans. A **45**, 5469–5485 (2014)
34. S.S. Sohn, B.J. Lee, S. Lee, J.H. Kwak, Effects of aluminium content on cracking phenomenon occurring during cold rolling of three ferrite-based lightweight steel. Acta Mater. **61**, 5626–5635 (2013)

35. S.Y. Shin, H. Lee, S.Y. Han, C.H. Seo, K. Choi, S. Lee, et al., Correlation of microstructure and cracking phenomenon occurring during hot rolling of lightweight steel plates. Metall. Mater. Trans. A **41**, 138–148 (2010)

36. S.S. Sohn, B.J. Lee, S. Lee, N.J. Kim, J.H. Kwak, Effect of annealing temperature on microstructural modification and tensile properties in 0.35C–3.5Mn–5.8Al lightweight steel. Acta Mater. **61**, 5050–5066 (2013)

37. C.H. Seo, K.H. Kwon, K. Choi, K.H. Kim, J.H. Kwak, S. Leed, et al., Deformation behaviour of ferrite–austenite duplex lightweight Fe–Mn–Al–C steel. Scr. Mater. **66**, 519–522 (2012)

38. S.S. Sohn, K. Choi, J.H. Kwak, N.J. Kim, S. Lee, Novel ferrite–austenite duplex lightweight steel with 77% ductility by transformation induced plasticity and twinning induced plasticity mechanisms. Acta Mater. **78**, 181–189 (2014)

39. S. Lee, J. Jeong, Y.K. Lee, Precipitation and dissolution behaviour of κ-carbide during continuous heating in Fe-9.3Mn-5.6Al-0.16C lightweight steel. J. Alloys Compd. **648**, 149–153 (2015)

40. J. Jeong, C.Y. Lee, I.J. Park, Y.K. Lee, Isothermal precipitation behaviour of κ-carbide in the Fe–9Mn–6Al–0.15C lightweight steel with a multiphase microstructure. J. Alloys Compd. **574**, 299–304 (2013)

41. S.J. Park, B. Hwang, K.H. Lee, T.H. Lee, D.W. Suh, H.N. Han, Microstructure and tensile behaviour of duplex low-density steel containing 5 mass% aluminium. Scr. Mater. **68**, 365–369 (2013)

42. V. Rigaut, D. Daloz, J. Drillet, A. Perlade, P. Maugis, G. Lesoult, Phase equilibrium study in quaternary iron-rich Fe–Al–Mn–C alloys. ISIJ Int. **47**, 898–906 (2007)

43. Z. Cai, H. Ding, Z. Ying, R.D.K. Misra, Microstructural evolution and deformation behaviour of a hot-rolled and heat treated Fe-8Mn-4Al-0.2C steel. J. Mater. Eng. Perform. **23**, 1131–1137 (2014)

44. S.S. Sohn, H. Song, J.G. Kim, J.H. Kwak, H.S. Kim, S. Lee, Effects of annealing treatment prior to cold rolling on delayed fracture properties in ferrite-austenite duplex lightweight steels. Metall. Mater. Trans. A **47**, 706–717 (2015)

45. S.Y. Han, S.Y. Shin, H.J. Lee, B.J. Lee, S. Lee, N.J. Kim, et al., Effects of Annealing temperature on microstructure and tensile properties in ferritic lightweight steels. Metall. Mater. Trans. A **43A**, 843–853 (2012)

46. J.B. Seol, D. Raabe, P. Choi, H.S. Park, J.H. Kwak, C.G. Park, Direct evidence for the formation of ordered carbides in a ferrite-based low-density Fe–Mn–Al–C alloy studied by transmission electron microscopy and atom probe tomography. Scr. Mater. **68**, 348–353 (2013)

47. S.S. Sohn, H. Song, B.C. Suh, J.H. Kwak, B.J. Lee, N.J. Kim, et al., Novel ultra-high-strength (ferrite + austenite) duplex light weight steels achieved by fine dislocation substructures (Taylor lattices), grain refinement, and partial recrystallization. Acta Mater. **96**, 301–310 (2015)

48. M.X. Yang, F.P. Yuan, Q.G. Xie, Y.D. Wang, E. Ma, X.L. Wu, Strain hardening in Fe-16Mn-10Al-0.86C-5Ni high specific strength steel. Acta Mater. **109**, 213–222 (2016)

49. Z.Q. Wu, H. Ding, H.Y. Li, M.L. Huang, F.R. Cao, Microstructural evolution and strain hardening behaviour during plastic deformation of Fe–12Mn–8Al–0.8C steel. Mater. Sci. Eng. A **584**, 150–155 (2013)

50. S.W. Hwang, J.H. Ji, E.G. Lee, K.T. Park, Tensile deformation of a duplex Fe–20Mn–9Al–0.6C steel having the reduced specific weight. Mater. Sci. Eng. A **528**, 196–203 (2011)

51. F. Yang, R. Song, Y. Li, T. Sun, K. Wang, Tensile deformation of low density duplex Fe–Mn–Al–C steel. Mater. Des. **76**, 32–39 (2015)

52. L. Zhang, R. Song, C. Zhao, F. Yang, Work hardening behaviour involving the substructural evolution of an austenite–ferrite Fe–Mn–Al–C steel. Mater. Sci. Eng. A **640**, 225–234 (2015)

53. C. Zhao, R. Song, L. Zhang, F. Yang, T. Kang, Effect of annealing temperature on the microstructure and tensile properties of Fe–10Mn–10Al–0.7C low-density steel. Mater. Des. **91**, 348–360 (2016)

54. L. Zhang, R. Song, C. Zhao, F. Yang, Y. Xu, S. Peng, Evolution of the microstructure and mechanical properties of an austenite–ferrite Fe–Mn–Al–C steel. Mater. Sci. Eng. A **643**, 183–193 (2015)
55. M.C. Ha, J.M. Koo, J.K. Lee, S.W. Hwang, K.T. Park, Tensile deformation of a low density Fe–27Mn–12Al–0.8C duplex steel in association with ordered phases at ambient temperature. Mater. Sci. Eng. A **586**, 276–283 (2013)
56. R. Rana, J. Loiseaux, C. Lahaye, Microstructure, mechanical properties and formability of a duplex steel. Mater. Sci. Forum **706–709**, 2271–2717 (2012)
57. A. Etienne, V. Massardier-Jourdan, S. Cazottes, X. Garat, M. Soler, I. Zuazo, et al., Ferrite effects in Fe-Mn-Al-C triplex steels. Metall. Mater. Trans. A **45**, 324–334 (2014)
58. S.S. Sohn, H. Song, M.C. Jo, T. Song, H.S. Kim, S. Lee, Novel 1.5 GPa-strength with 50%-ductility by transformation-induced plasticity of non-recrystallized austenite in duplex steels. Sci. Rep. **7**, 1225 (2017)
59. J. Zhang, D. Raabe, C.C. Tasan, Designing duplex, ultrafine-grained Fe-Mn-Al-C steels by tuning phase transformation and recrystallization kinetics. Acta Mater. **141**, 374–387 (2017)
60. I.C. Jung, L. Cho, B.C. De Cooman, In situ observation of the influence of Al on deformation-induced twinning in TWIP steel. ISIJ Int. **55**, 870–876 (2015)
61. K.T. Park, K.G. Jin, S.H. Han, S.W. Hwang, K. Choi, C.S. Lee, Stacking fault energy and plastic deformation of fully austenitic high manganese steels: Effect of Al addition. Mater. Sci. Eng. A **527**, 3651–3661 (2010)
62. J.D. Yoo, K.T. Park, Microband-induced plasticity in a high Mn–Al–C light steel. Mater. Sci. Eng. **A496**, 417–424 (2008)
63. J.D. Yoo, S.W. Hwang, K.T. Park, Origin of extended tensile ductility of a Fe-28Mn-10Al-1C steel. Metall. Mater. Trans. A **40**, 1520–1523 (2009)
64. K.T. Park, Tensile deformation of low-density Fe–Mn–Al–C austenitic steels at ambient temperature. Scr. Mater. **68**, 375–379 (2013)
65. K. Choi, C.H. Seo, H. Lee, S.K. Kim, J.H. Kwak, K.G. Chin, et al., Effect of aging on the microstructure and deformation behaviour of austenite base lightweight Fe–28Mn–9Al–0.8C steel. Scr. Mater. **63**, 1028–1031 (2010)
66. K.M. Chang, C.G. Chao, T.F. Liu, Excellent combination of strength and ductility in a Fe–9Al–28Mn–1.8C alloy. Scr. Mater. **63**, 162–165 (2010)
67. C.L. Lin, C.G. Chao, H.Y. Bor, T.F. Liu, Relationship between microstructures and tensile properties of a Fe-30Mn-8.5Al-2.0C alloy. Mater. Trans. **51**, 1084–1088 (2010)
68. C.L. Lin, C.G. Chao, J.Y. Juang, J.M. Yang, T.F. Liu, Deformation mechanisms in ultrahigh-strength and high-ductility nanostructured FeMnAlC alloy. J. Alloys Compd. **586**, 616–620 (2014)
69. I. Gutierrez-Urrutia, D. Raabe, Multistage strain hardening through dislocation substructure and twinning in a high strength and ductile weight-reduced Fe–Mn–Al–C steel. Acta Mater. **60**, 5791–5802 (2012)
70. I. Gutierrez-Urrutia, D. Raabe, High strength and ductile low density austenitic FeMnAlC steels: Simplex and alloys strengthened by nanoscale ordered carbides. Mater. Sci. Technol. **30**, 1099–1104 (2014)
71. H. Springer, D. Raabe, Rapid alloy prototyping: Compositional and thermo-mechanical high throughput bulk combinatorial design of structural materials based on the example of 30Mn–1.2C–xAl triplex steels. Acta Mater. **60**, 4950–4959 (2012)
72. I. Gutierrez-Urrutia, D. Raabe, Dislocation and twin substructure evolution during strain hardening of an Fe–22 wt.% Mn–0.6 wt.% C TWIP steel observed by electron channeling contrast imaging. Acta Mater. **59**, 6449–6462 (2011)
73. I. Gutierrez-Urrutia, D. Raabe, Influence of Al content and precipitation state on the mechanical behaviour of austenitic high-Mn low-density steels. Scr. Mater. **68**, 343–347 (2013)
74. E. Welsch, D. Ponge, S.M. Hafez Haghighat, S. Sandlöbes, P. Choi, M. Herbig, S. Zaefferer, D. Raabe, Strain hardening by dynamic slip band refinement in a high-Mn lightweight steel. Acta Mater. **116**, 188–199 (2016)

75. G. Frommeyer, U. Brüx, P. Neumann, Supra-ductile and high-strength manganese-TRIP/TWIP steels for high energy absorption purposes. ISIJ Int. **43**, 438–446 (2003)
76. H.J. Lai, C.M. Wan, The study of work hardening in Fe-Mn-Al-C alloys. J. Mater. Sci. **24**, 2449–2453 (1989)
77. O. Kubaschewski, *Iron—Binary Phase Diagrams*, 1st edn. (Springer-Verlag, Berlin, 1982)
78. N.S. Stoloff, Iron aluminides: Present status and future prospects. Mater. Sci. Eng. **A258**, 1–14 (1998)
79. R.G. Davies, An X-ray and dilatometric study of order and the "K-state" in iron-aluminium alloys. J. Phys. Chem. Solids **24**, 985–988 (1963)
80. R.K.W. Marceau, A.V. Ceguerra, A.J. Breen, M. Palm, F. Stein, S.P. Ringer, et al., Atom probe tomography investigation of heterogeneous short-range ordering in the 'komplex' phase state (K-state) of Fe-18Al (at.%). Intermetallics **64**, 23–31 (2015)
81. M.J. Marcinkowski, M.E. Taylor, Relationship between atomic ordering and fracture in Fe-Al alloys. J. Mater. Sci. **10**, 406–414 (1975)
82. E. Frutos, D.G. Morris, M.A. Muñoz-Morris, Evaluation of elastic modulus and hardness of Fe–Al base intermetallics by nano-indentation techniques. Intermetallics **38**, 1–3 (2013)
83. W. Koster, W. Tonn, The iron corner of the iron-manganese-aluminum system. Arch. Eisenhüttenwes **7**, 365–366 (1933)
84. A. Schneider, L. Falat, G. Sauthoff, G. Frommeyer, Microstructures and mechanical properties of Fe3Al-based Fe–Al–C alloys. Intermetallics **13**, 1322–1331 (2005)
85. M. Palm, G. Inden, Experimental determination of phase equilibria in the FeAlC system. Intermetallics **3**, 443–454 (1995)
86. Y. Kimura, K. Handa, K. Hayashi, Y. Mishima, Microstructure control and ductility improvement of the two-phase γ-Fe/κ-(Fe, Mn)₃AlC alloys in the Fe–Mn–Al–C quaternary system. Intermetallics **12**, 607–617 (2004)
87. K.H. Han, J.C. Yoon, W.K. Choo, TEM evidence of modulated structure in Fe-Mn-Al-C austenitic alloys. Scr. Metall. **20**, 33–36 (1986)
88. K. Sato, K. Tagawa, Y. Inoue, Age hardening of an Fe–30Mn–9Al–0.9C alloy by spinodal decomposition. Scr. Metall. **22**, 899–902 (1988)
89. M.J. Yao, P. Dey, J.B. Seol, P. Choi, M. Herbig, R.K.W. Marceau, T. Hickel, J. Neugebauer, D. Raabe, Combined atom probe tomography and density functional theory investigation of the Al off-stoichiometry of κ-carbides in an austenitic Fe-Mn-Al-C low density steel. Acta Mater. **106**, 229–238 (2016)
90. K. Ishida, H. Ohtani, N. Satoh, R. Kainuma, T. Nishizawa, Phase equilibria in Fe-Mn-Al-C alloys. ISIJ Int. **30**, 680–686 (1990)
91. G.P. Goretskii, K.V. Gorev, Phase equilibria in Fe-Mn-Al-C alloys. Russ. Metall. **2**, 217–221 (1990)
92. G.B. Krivonogov, M.F. Alekseyenko, G.G. Solov'yeva, Phase transformation kinetics in steel 9G28Yu9MVB. Phys. Met. Metallogr. **4**, 86–92 (1975)
93. M.S. Kim, Y.B. Kang, Development of thermodynamic database for high Mn–high Al steels: Phase equilibria in the Fe–Mn–Al–C system by experiment and thermodynamic modelling. CALPHAD **51**, 89–103 (2015)
94. K.G. Chin, K.J. Lee, J.H. Kwak, J.Y. Kang, B.J. Lee, Thermodynamic calculation on the stability of (Fe, Mn)₃AlC carbide in high aluminum steels. J. Alloys Compd. **505**, 217–223 (2010)
95. D. Connétable, J. Lacaze, P. Maugis, B. Sundman, Calphad **32**, 361–370 (2008)
96. C.W. Bale, E. Belisle, P. Chartrand, S.A. Decterov, G. Eriksson, K. Hack, et al., FactSage thermochemical software and databases — Recent developments. Calphad **33**, 295–311 (2009)
97. W.C. Cheng, Y.S. Song, Y.S. Lin, K.F. Chen, P.C. Pistorius, On the eutectoid reaction in a quaternary Fe-C-Mn-Al alloy: Austenite → ferrite + kappa-carbide + M₂₃C₆ carbide. Metall. Mater. Trans. A **45A**, 1199–1216 (2014)
98. W.C. Cheng, Phase transformations of an Fe-17.9Mn-7.1Al-0.85C austenitic steel after quenching and annealing. JOM **66**, 1809–1819 (2014)

99. S.A. Buckholz, D.C. Van Aken, L.N. Bartlett, On the influence of aluminum and carbon on abrasion resistance of high manganese steels. AFS Trans. **121**, 495–510 (2013)
100. I.S. Kalashnikov, O. Acselrad, A. Shalkevich, L.D. Chumakova, L.C. Pereira, Heat treatment and thermal stability of Fe-Mn-Al-C alloys. Int. J. Mater. Prod. Technol. **136**, 72–79 (2003)
101. H. Ishii, K. Ohkubo, S. Miura, T. Mohri, Mechanical properties of α+γ two-phase lamellar structure in Fe–Mn–Al–C alloy. Mater. Trans. **44**, 1679–1681 (2003)
102. J.C. Benz, H.W. Leavenworth Jr., An assessment of Fe-Mn-Al alloys as substitutes for stainless steel. J. Meteorol. **37**(3), 36–39 (1985)
103. J.C. Garcia, N. Rosas, R.J. Rioja, Development of oxidation resistant Fe-Mn-Al alloys. Met. Prog. **122**(3), 47–50 (1982)
104. C.Y. Chao, C.H. Liu, Effects of Mn contents on the microstructure and mechanical properties of the Fe–10Al–xMn–1.0C alloy. Mater. Trans. **43**, 2635–2642 (2002)
105. O. Acselrad, I. Kalashnikov, E. Silva, M. Khadyev, R. Simao, Diagram of phase transformations in the austenite of hardened alloy Fe-28% Mn-8.5% Al-1% C-1.25% Si as a result of aging due to isothermal heating. Met. Sci. Heat Treat. **48**, 543–553 (2006)
106. O. Acselrad, I. Kalashnikov, E. Silva, R. Simao, C. Achete, L. Pereira, Phase transformations in FeMnAlC austenitic steels with Si addition. Metall. Mater. Trans. A **33A**, 3569–3573 (2002)
107. W.K. Choo, J.H. Kim, J.C. Yoon, Microstructural change in austenitic Fe-30Mn-7.8Al-1.3C (in wt%) initiated by spinodal decomposition and its influence on mechanical properties. Acta Mater. **45**, 4877–4885 (1997)
108. K. Sato, K. Tagawa, Y. Inoue, Spinodal decomposition and mechanical properties of an austenitic Fe-30wt.%Mn-9wt.%Al-0.9wt.%C alloy. Mater. Sci. Eng. A **111**, 45–50 (1989)
109. K. Sato, K. Tagawa, Y. Inoue, Modulated structure and magnetic properties of age-hardenable Fe-Mn-Al-C alloys. Metall. Trans. A. **21**, 5–11 (1990)
110. L. Bartlett, D. Van Aken, J. Medvedeva, D. Isheim, N.I. Medvedeva, K. Song, An atom probe study of kappa carbide precipitation and the effect of silicon addition. Metall. Mater. Trans. A **45**, 2421–2435 (2014)
111. W.C. Cheng, C.Y. Cheng, C.W. Hsu, D.E. Laughlin, Phase transformation of the $L1_2$ phase to kappa-carbide after spinodal decomposition and ordering in a Fe–C–Mn–Al austenitic steel. Mater. Sci. Eng. A **642**, 128–135 (2015)
112. Cheng WC, Cheng CY, Hsu CW. Phase transformations, spinodal decomposition, precipitation reaction, and eutectoid reaction, of an Fe-12.5Mn-6.53 Al-1.28C austenitic steel. In: Proceeding of international conference on solid-solid phase transformation in inorganic materials (PTM 2015), 2015 June 28–July 3, Whistler, BC, Canada
113. A.P. Bentley, Ordering in Fe-Mn-Al-C austenite. J. Mater. Sci. Lett. **5**, 907–908 (1986)
114. K.H. Han, W.K. Choo, Phase decomposition of rapidly solidified Fe-Mn-Al-C austenitic alloys. Metall. Trans. A. **20A**, 205–213 (1989)
115. C.Y. Chao, L.K. Hwang, T.F. Liu, Spinodal decomposition in Fe-9.0Al-30.5Mn-xC alloys. Scr. Metall. **29**, 647–650 (1993)
116. H. Huang, D. Gan, P.W. Kao, Effect of alloying additions on the κ phase precipitation in austenitic Fe-Mn-Al-C alloys. Scr. Metall. Mater. **30**, 499–504 (1994)
117. M.C. Li, H. Chang, P.W. Kao, D. Gan, The effect of Mn and Al contents on the solvus of κ phase in austenitic Fe-Mn-Al-C alloys. Mater. Chem. Phys. **59**, 96–99 (1999)
118. Y.H. Tuan, C.L. Lin, C.G. Chao, T.F. Liu, Grain boundary precipitation in Fe-30Mn-9Al-5Cr-0.7C alloy. Mater. Trans. **49**, 1589–1593 (2008)
119. C.Y. Chao, C.N. Hwang, T.F. Liu, Grain boundary precipitation in an Fe-7.8Al-31.7Mn-0.54C alloy. Scr. Metall. Mater. **28**, 109–114 (1993)
120. C.Y. Chao, C.N. Hwang, T.F. Liu, Grain boundary precipitation behaviours in an Fe-9.8Al-28.6Mn-0.8Si-1.0C alloy. Scr. Mater. **34**, 75–81 (1996)
121. Y.L. Lin, C.P. Chou, $M_{23}C_6$ carbide in a Fe-26.6Mn-8.8Al-0.61C alloy. Scr. Metall. Mater. **27**, 67–70 (1992)
122. C.S. Wang, C.N. Hwang, C.G. Chao, T.F. Liu, Phase transitions in a Fe–9Al–30Mn–2.0C alloy. Scr. Mater. **57**, 809–812 (2007)

123. G.D. Tsay, Y.H. Tuan, C.L. Lin, C.G. Chao, T.F. Liu, Effect of C on spinodal decomposition in Fe-26Mn-20Al-C alloys. Mater. Trans. **52**, 521–525 (2011)
124. Wang CA. The observations and analyses of nano-scale $(Fe,Mn)_3AlCx$ Carbides and Ordered Phases in the Fe-15Mn-8Al-0.7C alloy. PhD thesis, Southern Taiwan University of Science and Technology, 1999
125. W.C. Cheng, H.Y. Lin, The precipitation of FCC phase from BCC matrix in a Fe-Mn-Al alloy. Mater. Sci. Eng. **323**, 462–466 (2002)
126. S.C. Tjong, Electron microscope observations of phase decompositions in an austenitic Fe-8.7Al-29.7Mn-1.04C alloy. Mater. Charact. **24**(3), 275–292 (1990)
127. T.F. Liu, J.S. Chou, C.C. Wu, Effect of Si addition on the microstructure of an Fe–8.0Al–29.0Mn–0.90C alloy. Metall. Mater. Trans. **21A**, 1891–1899 (1990)
128. R.G. Baligidad, K.S. Prasad, Effect of Al and C on structure and mechanical properties of Fe–Al–C alloys. Mater. Sci. Technol. **23**, 38–44 (2007)
129. I.S. Kalashnikov, O. Acselrad, L.C. Pereira, Chemical composition optimization for austenitic steels of the Fe-Mn-Al-C system. J. Mater. Eng. Perform. **9**, 597–602 (2000)
130. Y. Sutou, N. Kamiya, R. Umino, I. Ohnuma, K. Ishida, High-strength Fe–20Mn–Al–C-based alloys with low density. ISIJ Int. **50**, 893–899 (2010)
131. G. Frommeyer, E.J. Drewes, B. Engl, Physical and mechanical properties of iron-aluminium-(Mn, Si) lightweight steels. Rev. Met. Paris **97**, 1245–1253 (2000)
132. A. Saeed-Akbari, J. Imlau, U. Prahl, W. Bleck, Derivation and variation in composition-dependent stacking fault energy maps based on subregular solution model in high-manganese steels. Metall. Mater. Trans. A **40**, 3076–3090 (2009)
133. W. Song, W. Zhang, J. von Appen, R. Dronskowski, W. Bleck, κ-Phase formation in Fe–Mn–Al–C austenitic steels. Steel Res. Int. **86**, 1161–1169 (2015)
134. W. Song, T. Ingendahl, W. Bleck, Control of strain hardening behaviours in high-Mn austenitic steels. Acta Metall. Sin. (Engl. Lett.) **27**, 546–556 (2014)
135. G.R. Lehnhoff, K.O. Findley, B.C. De Cooman, The influence of Si and Al alloying on the lattice parameter and stacking fault energy of austenitic steel. Scr. Mater. **92**, 19–22 (2014)
136. A. Dumay, J.P. Chateau, S. Allain, S. Migot, O. Bouaziz, Influence of addition elements on the stacking-fault energy and mechanical properties of an austenitic Fe–Mn–C steel. Mater. Sci. Eng. A **483–484**, 184–187 (2008)
137. K.R. Limmer, J.E. Medvedeva, D.C. Van Aken, N.I. Medvedeva, Ab-initio simulation of alloying effect on stacking fault energy in fcc Fe. Comput. Mater. Sci. **99**, 253–255 (2015)
138. D. Khulmann-Wilsdorf, Q: Dislocations structures - how far from equilibrium? A: Very close indeed. Mater. Sci. Eng. A **315**, 211–216 (2001)
139. V. Gerold, H.P. Karnthaler, On the origin of planar slip in F.C.C. alloys. Acta Metall. **37**, 2177–2183 (1989)
140. N.I. Medvedeva, M.S. Park, D.C. Van Aken, J.E. Medvedeva, First-principles study of the Mn, Al and C distribution and their effect on the stacking fault energies in austenite. J. Alloys Compd. **582**, 475–482 (2013)
141. D.A. Oliver, Proposed new criteria of ductility from a new law connecting the percentage elongation with size of test-piece. Proc. Inst. Mech. Eng. **2**, 827–864 (1928)
142. M.T. Jahn, S.C. Chang, Y.H. Hsiao, Transverse tensile and fatigue properties of Fe-Mn-Al-C alloys. J. Mater. Sci. Lett. **8**, 723–724 (1989)
143. J.E. Krzanowski, The effects of heat-treatment and cold working on the room-temperature and cryogenic mechanical properties of Fe–30Mn–9Al–0.9C steel. Metall. Trans. **19**, 1873–1876 (1988)
144. K.H. Han, T.S. Kang, D.E. Laughlin, Thermomechanical treatment of FeMnAlC sideband alloy, in *Proceedings of the International Conference on Precipitation Phenomena*, (Chicago, USA, 1988), pp. 69–75
145. K.H. Han, The microstructures and mechanical properties of an austenitic Nb-bearing FeMnAlC alloy processed by controlled rolling. Mater. Sci. Eng. A **279**, 1–9 (2000)

146. C. Haase, L.A. Barrales-Mora, F. Roters, D.A. Molodov, G. Gottstein, Applying the texture analysis for optimizing thermomechanical treatment of high manganese twinning-induced plasticity steel. Acta Mater. **80**, 327–340 (2014)
147. R.K. You, P.W. Kao, D. Gan, Mechanical properties of Fe-30Mn-10Al-1C-1Si alloy. Mater. Sci. Eng. A **117**, 141–148 (1989)
148. K.T. Luo, P.W. Kao, D. Gan, Low temperature mechanical properties of Fe-28Mn-5Al-1C alloy. Mater. Sci. Eng. A **51**, L15–L18 (1992)
149. O. Acselrad, J. Dille, L.C. Pereira, J.L. Delplancke, Room-temperature cleavage fracture of Fe-Mn-Al-C steels. Metall. Trans. A. **35A**, 3863–3866 (2004)
150. N.I. Medvedeva, R.A. Howell, D.C. Van Aken, J.E. Medvedeva, Effect of phosphorus on cleavage fracture in κ-carbide. Phys. Rev. **2010, 81B**, 012105-1-4
151. W. Cao, M. Zhang, C. Huang, S. Xiao, H. Dong, Y. Weng, Ultrahigh Charpy impact toughness (~450J) achieved in high strength ferrite/martensite laminated steels. Sci. Rep. **7**, 41459: 1–8 (2017)
152. S.T. Chiou, W.C. Cheng, W.S. Lee, Strain rate effects on the mechanical properties of a Fe-Mn-Al alloy under dynamic impact deformations. Mater. Sci. Eng. A **392**, 156–162 (2005)
153. S.W. Hwang, J.H. Ji, K.T. Park, Effects of Al addition on high strain rate deformation of fully austenitic high Mn steels. Mater. Sci. Eng. **528**, 7267–7275 (2011)
154. A.S. Hamada, L.P. Karjalainen, J. Puustinen, Fatigue behaviour of high-Mn TWIP steels. Mater. Sci. Eng. A **517**(1–2), 68–77 (2009)
155. I.S. Kalashnikov, G.V. Chudakova, M.F. Alekseyenko, V.A. Konstantinov, I.M. Efimova, Structure and properties of nonmagnetic steels. Nauka Moscow, 155–158 (1982). (in Russian)
156. N.J. Ho, C.M. Chen, S.C. Tjong, Cyclic softening of age hardened Fe-Mn-Al-C alloys containing coherent precipitates. Scr. Metall. **21**(10), 1319–1322 (1987)
157. S.C. Tjong, Low cycle fatigue behavior of an austenitic Fe-25Mn-5Al-1C alloy containing grain boundary precipitates. Mater. Sci. Eng. A **203**, L13–LI6 (1995)
158. H.K. Yang, D. Doquet, Z.F. Zhang, Fatigue crack growth in two TWIP steels with different stacking fault energies. Int. J. Fatigue **98**, 247–258 (2017)
159. J.B. Duh, W.T. Tsai, J.T. Lee, Fatigue crack growth in FeAlMn alloys. Scr. Metall. **21**, 95–98 (1987)
160. S.C. Chang, Y.H. Hsiau, M.T. Jahn, Tensile and fatigue properties of Fe-Mn-Al-C alloys. J. Mater. Sci. **24**, 1117–1120 (1989)
161. N.J. Ho, S.C. Tjong, Cyclic stress-strain behaviour of austenitic Fe-29.7Mn-8.7Al-1.04C alloy at room temperature. Mater. Sci. Eng. **94**, 195–202 (1987)
162. N.J. Ho, L.T. Wu, S.C. Tjong, Cyclic deformation of duplex Fe-30Mn-10Al-0.4C alloy at room temperature. Mater. Sci. Eng. A **102**, 49–55 (1988)
163. V.K. Saxena, M.S.G. Krishna, P.S. Chhaunker, V.M. Radhakrishnan, Fatigue and fracture behaviour of a Ni–Cr free austenitic steel. Int. J. Press. Vessel. Pip. **60**, 151–157 (1994)
164. I.S. Kalashnikov, O. Acselrad, L.C. Pereira, T. Kalichak, M.S. Khadyyev, Behaviour of Fe-Mn-Al-C steels during cyclic tests. J. Mater. Eng. **9**, 334–337 (2000)
165. K.G. Chin, C.Y. Kang, S.Y. Shin, S. Hong, S. Lee, H.S. Kim, et al., Effects of Al addition on deformation and fracture mechanisms in two high manganese TWIP steels. Mater. Sci. Eng. **A528**, 2922–2928 (2011)
166. J.H. Ryu, S.K. Kim, C.S. Lee, D.W. Suh, H.K.D.H. Bhadeshia, Effect of aluminium on hydrogen-induced fracture behavior in austenitic Fe–Mn–C steel. Proc. R. Soc. A **469**, 1–14 (2013)
167. M. Koyama, H. Springer, S.V. Merzlikin, K. Tsuzaki, E. Akiyama, D. Raabe, Hydrogen embrittlement associated with strain localization in a precipitation-hardened Fe-Mn-Al-C light weight austenitic steel. Int. J. Hydrog. Energy **39**, 4634–4646 (2014)
168. C.A. Pierpoint, T.S. Sudarshan, M.R. Loutham, T.A. Place, Hydrogen susceptibility of a Mn-Al austenitic stainless steel. Weld Fail Anal. Met. **14**, 423–435 (1985)
169. T.S. Sudarshan, D.P. Harvey, T.A. Place, Mechanistic similarities between hydrogen and temperature effects on the ductile-to-brittle transition of a stainless steel. Metall. Trans. A. **19**, 1547–1553 (1988)

170. J.S. Ku, N.J. Ho, S.C. Tjong, Properties of electron-beam-welded and laser-welded austenitic Fe-28Mn-5Al-1C alloy. J. Mater. Sci. **28**(10), 2808–2814 (1993)
171. S.C. Tjong, S.M. Zhu, N.J. Ho, J.S. Ku, Solidification microstructure and creep rupture behaviour of electron beam welded austenitic Fe-28M-6Al-1C alloy. Mater. Sci. Technol. **13**(3), 251–256 (1997)
172. C.P. Chou, C.H. Lee, Effect of carbon on the weldability of Fe–Mn–Al alloys. J. Mater. Sci. **25**(2b), 1491–1496 (1990)
173. C.P. Chou, C.H. Lee, The influence of carbon content on austenite–ferrite morphology in Fe–Mn–Al weld metals. Metall. Trans. A Phys. Metall. Mater. Sci. **20**(11), 2559–2561 (1989)
174. C.-P. Chou, C.-H. Lee, Weld metal characteristics of duplex Fe–30wt.%Mn–10wt.%Al–xC alloys. Mater. Sci. Eng. A **118**, 137–146 (1989)
175. K. Makhamreh, D. Aidun, Mechanical properties of flux cored iron–manganese–aluminium weld metal. Weld. J. (USA) **71**(3), 104–114 (1991)
176. Y.-L. Lin, C.-P. Chou, D03–B2-a phase transition in an Fe–Mn–Al–C weldment. Scr. Metall. Mater. **28**(10), 1261–1266 (1993)
177. L. Mujica Roncery, S. Weber, W. Theisen, Welding of twinning-induced plasticity steels. Scr. Mater. **66**(12), 997–1001 (2012)
178. N. Lun, D.C. Saha, A. Macwan, H. Pan, L. Wang, F. Goodwin, Y. Zhou, Microstructure and mechanical properties of fibre laser welded medium manganese TRIP steel. Mater. Des. **131**, 450–459 (2017)
179. J.S. Ku, N.J. Ho, S.C. Tjong, Properties of electronbeam-welded and laser-welded austenitic Fe–28Mn–5Al–1C alloy. J. Mater. Sci. **28**(10), 2808–2814 (1993)
180. C.H. Chao, N.J. Ho, Fusion boundary structures in a laser welded duplex Fe–Mn–Al–C alloy. J. Mater. Sci. **27**(15), 4139–4144 (1992)
181. C.P. Chou, C.H. Lee, Weld metal characteristics of duplex Fe-30wt.%Mn-10wt.%Al-xC alloys. Mater. Sci. Eng. A **118**, 137–146 (1989)
182. C.P. Chou, C.H. Lee, Effect of carbon on the weldability of Fe-Mn-Al alloys. J. Mater. Sci. **25**(2), 1491–1496 (1990)
183. J.C. Garciak, N. Rosas, R.J. Rioja, Development of oxidation resistant Fe-Mn-Al alloys. Metals Prog. **122**, 47–50 (1982)
184. P.R.S. Jackson, G.R. Wallwork, High temperature oxidation of iron-manganese-aluminium based alloys. Oxid. Met. **21**, 135–170 (1984)
185. C.H. Kao, C.M. Wan, Effect of temperature on the oxidation of Fe-7.5Al-0.65C alloy. J. Mater. Sci. **23**, 1943–1947 (1988)
186. J.P. Sauer, R.A. Rapp, J.P. Hirth, Oxidation of iron-manganese-aluminium alloys at 850 and 1000°C. Oxid. Met. **18**, 285–294 (1982)
187. S.C. Tjong, High temperature oxidation of the austenitic Fe-9Al-30Mn-1.0C and duplex Fe-10Al-29Mn-0.4C alloys. Trans. Jpn. Inst. Metals **28**, 671–678 (1987)
188. Y.J. Gau, J.K. Wu, The influence of alloying elements on the corrosion behaviour of Fe-Mn-Al alloys. Corros. Prev. Control **44**, 55–60 (1997)
189. Y.H. Tuan, C.S. Wang, C.Y. Tsai, C.G. Chao, T.F. Liu, Corrosion behaviours of austenitic Fe–30Mn–7Al–xCr–1C alloys in 3.5% NaCl solution. Mater. Chem. Phys. **114**, 595–598 (2009)
190. C.S. Wang, C.Y. Tsai, C.G. Chao, T.F. Liu, Effect of chromium content on corrosion behaviours of Fe-9Al-30Mn-(3,5,6.5,8)Cr-1C alloys. Mater. Trans. **48**, 2973–2977 (2007)
191. J.W. Lee, C.C. Wu, T.F. Liu, The influence of Cr alloying on microstructures of Fe–Al–Mn–Cr alloys. Scr. Mater. **50**, 1389–1393 (2005)
192. C.F. Huang, K.L. Ou, C.S. Chen, C.H. Wang, Research of phase transformation on Fe–8.7Al–28.3Mn–1C–5.5Cr alloy. J. Alloys Compd. **488**, 246–249 (2009)
193. S.C. Tjong, C.S. Wu, The microstructure and stress corrosion cracking behaviour of precipitation-hardened Fe-8.7Al-29.7Mn-1.04C alloy in 20% NaCl solution. Mater. Sci. Eng. **80**, 203–209 (1986)
194. S.C. Chang, J.Y. Liu, H.K. Juang, Environment-assisted cracking of Fe-32%Mn-9%Al alloys in 3.5% sodium chloride solution. Corrosion **51**, 399–406 (1995)

195. O. Acselrad, A.R. de Souza, I.S. Kalashnikov, S.S. Camargo, A first evaluation of the abrasive wear of an austenitic FeMnAlC steel. Wear 257(9–10), 999–1005 (2004)
196. D. Van Aken, L.N. Bartlett, S.A. Buckholz, On the influence of aluminium and carbon on abrasion resistance of high manganese steels, in *117th metal casting congress*, (2013)
197. S.-G. Peng, R.-B. Song, T. Sun, Z.-Z. Pei, C.-H. Cai, Y.-F. Feng, Z.-D. Tan, Wear behaviour and hardening mechanism of novel lightweight Fe–25.1Mn–6.6Al–1.3C steel under impact abrasion conditions. Tribol. Lett. 64(1), 13 (2016)
198. O.A. Zambrano, Y. Aguilar, J. Valdés, S.A. Rodríguez, J.J. Coronado, Effect of normal load on abrasive wear resistance and wear micromechanisms in FeMnAlC alloy and other austenitic steels. Wear 348–349, 61–68 (2016)
199. S.C. Chang, W.H. Weng, H.C. Chen, S.J. Lin, P.C.K. Chung, The cavitation erosion of Fe-Mn-Al alloys. Wear 181–183, 511–515 (1995)
200. W. Aperador, J.H. Bautista, J.D. Betancur, Resistencia Al desgaste erosivo-corrosivo de aceros austeníticos fermanal. Rev. EIA 18(18), 49–59 (2012)
201. C.M. Chu, H. Huang, P.W. Kao, D. Gan, Effect of alloying chemistry on the lattice constant of austenitic Fe-Mn-Al-C alloys. Scr. Metall. Mater. 30, 505–508 (1994)
202. U. Bohnenkamp, R. Sandström, Evaluation of the density of steels. Steel Res. 71, 88–93 (2000)
203. M. Yamamoto, S. Taniguchi, The density, magnetic properties, Young's modulus, and ΔE-effect, and their change due to quenching in ferromagnetic iron-aluminium Alloys. II: Young's modulus and the ΔE-effect. Sci. Rep. Res. Inst. Tohuku Univ. Ser. A 8, 193–204 (1956)
204. U. Bohnenkamp, R. Sandström, Evaluation of the elastic modulus of steels. Steel Res. 71, 94–99 (2000)
205. A. Salak, *Ferrous Powder Metallurgy* (Cambridge International Science Publishing, Cambridge, 1995)
206. V.V. Satya Prasad, S. Khaple, R.G. Baligidad, Melting, processing, and properties of disordered Fe-Al and Fe-Al-C based alloys. JOM 66, 1785–1793 (2014)
207. S.L. Case, K.R. Van Horn, *Aluminum in Iron and Steel* (Wiley, New York, 1953), p. 4
208. M. Yang, Y. Akiyama, T. Sasaki, Evaluation of change in material properties due to plastic deformation. J. Mater. Process. Technol. 151, 232–236 (2004)
209. C.L. Davis, P. Mukhopadhyay, M. Strangwood, M. Potter, S. Dixon, P.F. Morris, Comparison between elastic modulus and ultrasonic velocity anisotropy with respect to rolling direction in steels. Ironmak. Steelmak. 35, 359–366 (2008)
210. A. Granato, A. Hikata, K. Lücke, Recovery of damping and modulus changes following plastic deformation. Acta Metall. 6, 470–480 (1958)
211. E. Mazancova, I. Ruziak, I. Schindler, Influence of rolling conditions and aging process on mechanical properties of high Mn steels. Arch. Civil Mech. Eng. 12, 142–147 (2012)
212. A. Mohamadizadeh, A. Zarei-Hanzaki, H.R. Abedi, S. Mehtonen, D. Porter, Hot deformation characterization of duplex low-density steel through 3D processing map development. Mater. Charact. 107, 293–301 (2015)
213. J. Yoo, B. Kim, Y. Park, C. Lee, Microstructural evolution and solidification cracking susceptibility of Fe-18Mn-0.6C-xAl steel welds. J. Mater. Sci. 50, 279–286 (2015)

Chapter 7
High-Modulus Steels

Hauke Springer and Christian Baron

Abbreviations and Symbols

Ar	Argon
B	Bulk modulus
E	Young's modulus
E/ρ	Specific modulus
G	Shear modulus
HIP	Hot isostatic pressed
HMS	High-modulus steels
LMD	Laser metal deposition
ODS	Oxide dispersion strengthened steels
PCD	Polycrystalline diamond
PM	Powder metallurgy
SiC	Silicon carbide
TE	Total elongation
TiB_2	Titanium diboride
TiC	Titanium carbide
TMT	Thermo-mechanical treatments
wt.%	Weight percent
YS	Yield strength
YS/ρ	Specific yield strength
θ	Contact angle
ρ	Mass density/density
σ	Surface energy

H. Springer · C. Baron (✉)
Department Microstructure Physics and Alloy Design, Max-Planck-Institut für Eisenforschung GmbH, Düsseldorf, Nordrhein-Westfalen, Germany
e-mail: h.springer@mpie.de; c.baron@mpie.de

© Springer Nature Switzerland AG 2021
R. Rana (ed.), *High-Performance Ferrous Alloys*,
https://doi.org/10.1007/978-3-030-53825-5_7

7.1 Motivation and Approach

Saving weight in engineering constructions allows reduction of inertia of accelerated components, thus decreasing their energy consumption and improving performance. Bridges may span farther distances, trains and aeroplanes can transport heavier payloads, machines and robots are able to move quicker and weight-reduced body constructions can compensate for the yet heavy battery packs of electric cars. Lightweight design is, therefore, a critical technological frontier, and a strong driving force for the structural materials for engineering constructions. Despite the recent advances in the development of polymers, metals remain the backbone of structural applications due to their wide range of properties, which can be achieved at comparatively low cost, as well as their excellent suitability for recycling. As illustrated throughout this book, steels are especially attractive, as they are non-toxic, require relatively little energy to be produced, and can utilise multiple equilibrium and non-equilibrium phase transformations and an extensive solubility for alloying elements. These result in a vast range of various attractive properties at low costs, rendering steels as the predominant structural materials of today [1–3].

With regards to lightweight design, the most obvious approach relies on increasing the strength of steels, thus enabling thinner parts to carry the equivalent mechanical load. Most important is the yield strength (YS), as the majority of parts for engineering constructions are supposed to transfer mechanical forces without permanent deformation. However, strength needs to be balanced with ductility, indicated by the total elongation (TE) and toughness: The material's ability to undergo plastic deformation beyond the elastic regime is required for forming the desired parts and acts as a safety margin in a crash situation. In order to overcome this trade-off, tremendous advances have been realised by carefully fine-tuning steel microstructures and the associated deformation mechanisms (Fig. 7.1a). However, the greatest shortcoming of steels in competition with other metallic materials for lightweight design is its comparatively high density (ρ). While it can be partly compensated for by higher strength, only few designated concepts such as maraging steels can match the specific yield strength (YS/ρ) of Al and Ti alloys [1]. As described in a Chap. 6 of this book, most promising are the recently developed low-density steels as pioneered by Frommeyer et al. [4], which are based on significant density-reduction through additions of up to 12 wt.% Al (Fig. 7.1b).

However, focussing exclusively on the improvement of strength, ductility and density, neglects the stiffness of materials, expressed by their Young's modulus (E): Most parts of engineering constructions interact with others, such as the gear teeth in a transmission or parts in a suspension system, and they consequently rely on distinct shapes and tolerances. Even though they may not plastically deform when designed correctly, the respective parts will still deform elastically under load, affecting their performance during interaction. Stiffness is also the only material (and not geometrical) property relevant for the stability of a part, and thus critical for the resistance against buckling or bending, for example of a bridge beam or an aeroplane wing. Furthermore, a high stiffness translates into an elevated natural or

Fig. 7.1 Materials property maps of illustrating the key factors for lightweight design: (**a**) Areas of YS vs. TE for different material systems. (**b**) YS per ρ vs. TE of highest performing alloys. (**c**) Unique performance of HMS over conventional structural materials. The images are with permission based on previously published work [68]

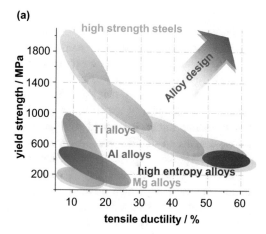

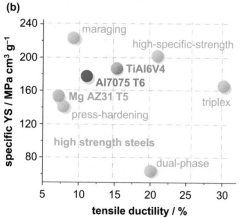

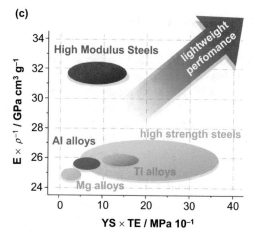

eigen-frequency, relevant for avoiding resonance-related phenomena. However, this most critical material property has been for a long time overlooked or simply taken as a given intrinsic property. Interestingly, the "heavy" steels are favourable again in this view, as their elevated density ($\rho_{Fe} = 7.85$ g cm^{-3}) is compensated by their relatively high stiffness ($E_{Fe} \sim 208$ GPa).

High-modulus steels (HMS) address all those key criteria for superior performance in weight-critical applications (Fig. 7.1c). This new class of high performance ferrous alloys is created by blending low-density and stiff particles into strong, tough and ductile steel matrices, resulting in composite steels. However, aiming to improve the physical properties (E, ρ) without sacrificing the mechanical properties – especially ductility – adds substantial complexity to the already challenging material design process. Especially important is the integration of the synthesis and processing techniques into the alloy development. In order to establish HMS as commercially successful products, which does not allow for disregarding the associated costs, a comprehensive approach to design these novel steels is required. Consequently, this chapter covers the underlying aspects of physical metallurgy and derived strategies for alloy and process design of HMS. Based on these fundamental aspects, the current state of the art of HMS development is described, culminating in an outlook on critical challenges as well as future perspectives and opportunities.

HMS open new dimensions for structural material design, and their development has only just begun to receive significant attention in the last two decades. Accordingly, the associated terminology has not yet been universally standardised, but the authors believe that "high modulus steels" as coined by Tanaka et al. [5], highlights the most important aspect – an increase in specific stiffness – and is therefore an appropriate nomenclature.

7.2 Physical Metallurgy

As shown in Fig. 7.2a, Fe exhibits with about 26 GPa cm^3 g^{-1} a similar specific modulus (E/ρ ratio) as all other widely used metals, from Mg, Al and Ti to Ni and even W. The same trend applies to derived structural materials (e.g. Fe-based steels vs. Al or Ti alloys), as the effects of alloying elements and microstructure are limited. In order to achieve a substantial increase in the specific modulus, steels have to be blended with compounds of increased stiffness and lower density. However, the scope of HMS is wider than one specific alloy concept, but instead has the potential to form the basis for a new class of steels: Decreasing the mass of parts by raising the specific modulus is of interest across the whole range of steel applications, e.g. from high-strength to stainless and hard tool steels. The differing requirements determine the optimum matrix microstructure as well as the suitable particle phase, morphology and size, to name but a few. In this context it will become clear that matrix, particles and their mechanical interaction and synthesis technique are all closely interlinked. Depending on the targeted application, a balance must be struck

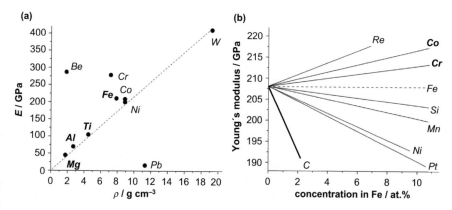

Fig. 7.2 (a) Young's modulus and density of metallic elements [1], and (b) literature data for the effect of alloying additions on the stiffness of Fe. The images are with permission based on previously published work [1, 86]

between mechanical, chemical and physical properties, as well as the feasibility and associated costs to synthesise the desired composite. The aim is to provide in this section the metallurgical basis for deriving the necessary guidelines and strategies for alloy and processing design of HMS alloys.

7.2.1 Matrix

Throughout this book, it is shown how the vast range of mechanical properties that steels are capable of can be realised with various microstructures, achievable by fine-tuning the alloy composition and processing parameters. For HMS the "physical" properties E and ρ need to be considered too, adding a new level of complexity.

The density of metals relies on their crystallographic lattice. Their stiffness is additionally depending on the electronic configuration, affecting the inter-atomic bonding strength [6]. Hence, ferrite exhibits higher stiffness than austenite despite being less closely packed. E values are, therefore, also orientation dependent, with closely packed planes such as {111} yielding the highest stiffness values. Steels are most commonly used as polycrystalline materials, unlike in special applications such as single-crystalline turbine blades. Unless excessive texture is present, E values are, therefore, quasi-isotropic. Lattice defects such as grain boundaries, stacking faults or dislocations – induced for example by martensitic transformation – disrupt the stacking of atoms and atomic planes, and consequently lower E [7]. Decreasing ρ of steel is achievable by implementing substitutional atoms of light elements such as Al or Si. The associated lattice distortion (as atoms of these elements are of different size than Fe atoms) typically lowers the density even further, but this crystallographic defect also decreases the stiffness [7, 8]. Furthermore, in order to overcome embrittlement induced by ordering and intermetallic phases observed for Al- and Si-containing steels, large quantities of austenite-stabilising elements such as Mn need to be added [9]. Together, these phenomena can severely limit the weight saving potential of low density steels, of course depending on the chosen application: While ρ values of less than 6.8 g cm^{-3} can be achieved, the parallel drop in E to about 170 GPa renders the specific modulus – as an indicator for the lightweight performance – virtually unaffected [10]. Certain elements like Cr, Co or Re raise E of steel [8] due to electronic effects without increasing the density unduly, but in most cases only to a limited extent as, for example, of about 0.5 GPa/wt.% in case of Cr (Fig. 7.2b).

It becomes clear that the alloying elements and crystallographic defects chosen for the constitution and phase balance, and thus the desired mechanical properties of the matrix, can be expected to have a strong – and mostly detrimental – effect on the physical properties they need to be balanced with. Furthermore, the matrix must be able to curtail the embrittling effect of the stiff and hard particles embedded into it, which act as stress and strain concentrators. From a physical metallurgy standpoint, the optimal matrix for a maximised E/ρ ratio should have a ferritic, Cr- and or Co-containing microstructure, with large grains free of defects, and little texture. The defects are necessary though to achieve sufficient strength, and the most suitable pathway (e.g. precipitations, martensite, or reduction in grain size) needs to be chosen carefully with respect to the targeted application. If possible, localised treatments such as surface hardening are preferable to those affecting the entire bulk of the material. Balancing the effects of elevated strength is even more complex as, for example, inherently ductile austenite (especially if obtained through effective but comparatively heavy Ni) decreases the stiffness of the matrix (Fig. 7.2b). As with other high-performance steels, multiphase microstructures appear as a promising pathway, but their realisation within the composite material can be become even more complex to achieve.

7.2.2 Particles

Key factor for HMS is the improvement of the specific modulus which is based on composite structures containing stiffer constituents of lower density. As they exceed the role of reinforcements (i.e. improving more than the strength) and are typically applied blended throughout the steel matrix as more or less equiaxed compounds of μm size, we refer to them as particles throughout this chapter.

As a general rule, particle phases exhibit a lower density by being constituted of elements lighter than steel, while an electronic structure that leads to predominantly ionic and/or covalent bonding, which is much stronger than the metallic type, ensures superior stiffness. Consequently, possible particle phases are numerous and range from oxides to nitrides, carbides, borides to intermetallic phases. The main criterion for the selection of particle phases is their efficiency, i.e. their E/ρ ratio, as it determines the fraction of particles necessary for improving the specific modulus of the composite. This is important, as such ceramics and metalloids are

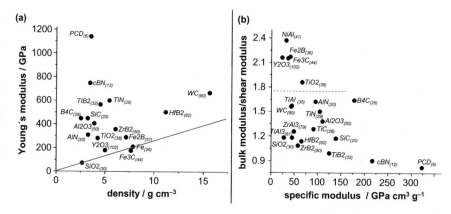

Fig. 7.3 Particle selection criteria: (**a**) The specific modulus E/ρ of most particle phases is higher than those of metals (red line). (**b**) Particles with a B/G ratio above 1.75 are considered being more ductile. The images are with permission based on previously published work [14, 15]

typically stiff and of low density, but also hard and brittle, and limiting their required fraction is generally beneficial for the mechanical performance of HMS. Figure 7.3a shows E and ρ values of various particle phases. Polycrystalline diamond (PCD), for example, exhibits a superior E/ρ ratio of about 350 GPa cm^3 g^{-1}. The majority of phases does not reach that extreme level, but even more commonly used and cost-effective compounds such as SiC readily exceed the specific modulus of metals (indicated by the red line) by several times. Their contribution to the E/ρ ratio of the composite can be approximated with a simple rule of mixture [11], while more complex considerations also factor in the particle's morphology influencing stiffness, i.e. from fibrous to ellipsoid to spherical particles [12, 13]. The particle morphology can be estimated based on existing knowledge, for example, concerning the typical shape in which carbides precipitate in steel microstructures, or even be controlled beforehand if the particle phase is synthesised ex situ prior to be blended into the matrix as detailed in Sect. 7.3 of this chapter.

While the plethora of particle phases can be narrowed down for their suitability in HMS design based on their effectiveness on the specific modulus, additional factors need to be considered. The elevated stiffness of particles typically translates into satisfactory hardness and strength, but their ductility and/or toughness follows an inverse trend in a similar manner as with metals. The ratio between bulk (B) and shear modulus (G) has been suggested as a suitable criterion (see Fig. 7.3b), with values above 1.75 indicating potentially elevated ductility [14]. Recent investigations based on ab initio calculations are aimed at overcoming this contradicting behaviour [15], and phases such as Mo$_2$BC might become candidates for developing future HMS concepts. Furthermore, the price of the constituting elements of particles becomes more and more relevant with increasing particle fraction and targeted production volume of a respective HMS concept. In this light, phases such as Al$_2$O$_3$ appear especially interesting. However, following these isolated views on particles and matrices, it is just as important to consider phenomena governing their interaction as a composite in order to derive meaningful guidelines for alloy and processing design.

It should be noted that the literature information regarding the physical – let alone mechanical – properties of particle phases is far from complete, which can be expected in view of the multitude of oxides, carbides, borides, intermetallics, MAX phases, etc. Even if available, E values for example are often extrapolated from crystallographic parameters and theoretical predictions, leaving open a wide field for experimental investigations, preferably with high-throughput techniques [16]. A compilation of available data can be found in [17].

7.2.3 Particle-Matrix Interaction

As with all composite materials, the chosen constituents need to interact success-fully in order to ensure the desired performance. Rooted in the differences between metallic matrix and ceramic particles, several specific factors govern the respective phenomena for HMS concepts. Due to the vast range of possible combination depending on the respective requirements, they need to be carefully evaluated once a combination of particle and matrix has been selected.

Differences in thermal expansion coefficients of particle and matrices lead to localised stresses during any thermal cycle, e.g. during solidification, heat treatments, welding or changes in operational temperature, which is mainly caused by their different nature. Differences in their electro-chemical properties might lead to localised corrosion taking place, as observed with changes in Cr content related to carbide formation in stainless steels [18]. Similar considerations apply to other properties affected by the constitution of the composite, such as conductivity for heat and electricity, magnetism or others.

For lightweight design though, most critical is the interaction between matrix and particles during mechanical deformation, both during the production and subsequent application of parts. Linked to the respective bonding types and crystallographic structure, their deformation mechanisms are different as well, rendering particles strong and brittle, and the metallic matrix soft and ductile in comparison. However, their interface determines whether both can be utilised to their full potential. Ideally, the interface can transfer load up to the respective limits in strength or ductility of the weakest partner, as any premature failure serves as an additional stress concentration, and decreases the stiffness of the composite. The interface is governed by multiple factors such as interfacial energy, lattice misfit and coherency [6, 19]. It is therefore possible to pre-evaluate based on crystallographic information of both components, but as discussed later due to the complexity in synthesis, the actual interfacial quality of a HMS concept needs to be assessed experimentally. Once the interface is sufficiently strong and ductile, the behaviour of the bulk composite during deformation relies on fraction, size and especially morphology of the particles. Small particles in the order of 50 nm are most effective as dislocation obstacles and thereby strengthening constituents, as smaller particles are readily cut through whereas bigger particles become circumvented [20]. Larger particles, on the other hand, can be advantageous from a tribological perspective, were the depth of an abrasive attack should be smaller than the particles, important when its wear resistance is a critical property of a HMS concept [21]. Dependent on additional desired properties, the above-mentioned characteristics have to be tailored individually. The desired morphology of particles depends on the loading conditions the fabricated part is subjected to during its application. Sharp, stress concentrating edges should be generally avoided, especially with oscillating load (fatigue). Accordingly, the surrounding matrix should exhibit maximum toughness to limit propagation of cracks. This is controlled mainly by its constitution (e.g. austenite is typically more tough than ferrite) and grain size (i.e. a large density of grain boundaries) and therefore contrary to the desired effect on the E/ρ ratio. If the loading direction is well known and mostly constant, the performance of HMS can be tuned accordingly, by creating elongated particles along the loading direction until they reach an aspect ratio of whiskers or even fibres. However, such laminar structures are difficult to produce in steel matrices, and materials with an isotropic behaviour are better suited for established design and production techniques of metallic parts [22].

   In summary, in steel matrices most desirable are highly effective particles, constituted of cost-effective elements and of spherical shape, exhibiting a sufficiently strong and ductile interface with a surrounding matrix of maximum toughness. For elevated strength and non-critical wear applications, a small particle size is preferable, together with a matrix ensuring sufficient ductility for the targeted application. These considerations represent the basis for the alloy and processing design guidelines. The following section describes how the desired composite structures of HMS concepts can be obtained.

7.3 Alloy Design

The design of metal-based composites is more complex than the design of polymer based composites, where matrix and reinforcement can typically be chosen solely based on their respective property profile. With carbon-fibre-reinforced polymers for example, the adhesive matrix (e.g. polyurethane) can be readily blended with carbon fibres via injection moulding or other fabrication techniques [23]. With metallic materials, however, the associated temperatures during casting and heat treatments are much higher (more than 1500 °C for steels opposed to less than 300 °C for polymers). Most importantly, the interfacial phenomena with the ceramic particles are complex, and it is more challenging to achieve a suitable composite structure with metals and, especially, steels than with polymers as a matrix material. Consequently, designing HMS requires a broader view, taking into account the physical metallurgy of matrix, particles and their mechanical and chemical interactions. While this allows narrowing down the vast range of possible choices, it does not include the critical steps of synthesis and processing. All these factors mutually influence each other and determine whether the desired microstructure and, thus, property profile of a HMS concept can be achieved. Only a comprehensive approach allows identifying the alloy composition together with the suitable production technique and processing parameters. This section is therefore concerned with outlining how the fundamental metallurgical considerations can be turned into alloy design strategies, and how they can be implemented.

7.3.1 Strategy

Once a promising matrix and particle combination has been selected, the following factors should be considered to validate whether they can be combined successfully and transferred into a HMS.

1. Stability of Particles Within the Matrix

The thermodynamic stability of the particle phase within the chosen matrix determines its tendency to decompose, transform or even dissolve during synthesis and processing. Diamond, for example, exhibits outstanding physical properties ($E > 1000$ GPa; ρ 3.51 g cm^{-3} [24]) but will dissolve rapidly in Fe-based matrices even at only slightly elevated temperatures and is, therefore, extremely difficult to utilise in HMS. Oxides such as Al_2O_3, on the other hand, are typically stable even above the liquidus temperature of most steels. Stability in liquid steel is not a requirement though, or even always beneficial as detailed further below. If a particle phase is not stable down to room temperature, it should be possible to retain it by rapid cooling (i.e. quenching). This metastability is relevant for intermetallic phases, whose appearance typically relies on crystallographic ordering in a specific composition and temperature range. It also applies to certain carbides such as cementite (Fe_3C) or low-alloy carbides (M_7C_3), which can be dissolved and precipitated again in a controlled manner with solid-state heat treatments [25]. However, associated ripening and growth limits the applicable time/temperature parameter window during synthesis and processing of HMS concepts with metastable particle phases even further, more than with composites relying on particles stemming from solidification, such as most boride-based HMS. In this context it should be noted that the prediction of stability critically relies on the available thermodynamic data. Especially for the most commonly researched borides, though, the validity of the published databases is still limited. The recent interest in HMS and the growing amount of research into high–modulus alloys – which venture into yet only sparsely covered composition ranges of complex alloy systems – is promising to overcome this limitation in the near future.

2. Bonding Between Particles and Matrix

A particle phase does not only require being stable in the chosen steel matrix, it must also be possible to form a sufficiently strong and ductile interface with it.

Fig. 7.4 An example of
decohesion between particles
and matrix: Al_2O_3 particle in
a ferritic matrix after tensile
testing of the composite
structure. (Reproduced from
own work Ref. [74])

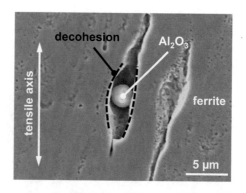

Insufficient interfacial bonding between particle and matrix leads to decohesion
(Fig. 7.4), which negates the intended purpose of the composite structure. In
solid-state synthesis techniques such as sintering, this is more easily achieved
as the interfacial bonding is formed by interdiffusion processes once the thin
barrier layers on powder sphere surfaces (such as oxide layers) are broken down
with sufficient pressure, temperature and time. In liquid processing, however, the
interface formation is mainly controlled by wetting. The degree of wetting is
described by the interplay of surface energies (σ) [26], the contact angle (θ) and the
work of adhesion (w_a) representing the adhesion strength [27]. Sufficient wetting is
typically described by θ values below 90° (Fig. 7.5a). While θ can be influenced
by processing conditions such as temperature or thin film coatings [28–32], the
general rule applies that particles with a stronger intramolecular bond interact less
intensive with a metallic liquid, decreasing their wettability [27]. This agrees well
with trends shown in Fig. 7.5b, where θ values of various particle phases with liquid
Fe (filled symbols) and Ni (empty symbols) in vacuum are plotted as a function
of their formation enthalpy. Particles of phases with a more covalent bond character
(e.g. carbides and borides) are wetted more easily than more stable ionic compounds
(such as oxides). Especially Ti- and Zr carbides, borides and nitrides combine
adequate stability with high wettability [27, 28, 33–36].

3. Particle Dispersion

Lastly, the performance of a HMS does not only depend on whether an individual
particle is stable and bonded to the matrix. In order to achieve a suitable isotropy
and favourable mechanical interaction during deformation, it is also important that a
multitude of them can be obtained in a sufficiently even dispersion in the composite
structure.

Solid-state synthesis such as sintering renders this to be more readily controllable
but requires respective efforts in preparation (i.e. powder mixing). Precipitation
reactions typically ensure a very even macroscopic, global dispersion of particles,
but the microstructural, local distribution can be disadvantageous. Particles precipitating preferably on grain boundaries, for example, can lead to severe embrittlement,
and it can be difficult to alter their precipitation location. Liquid synthesis is

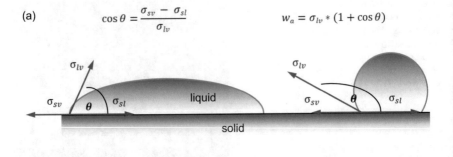

(a)

$$\cos \theta = \frac{\sigma_{sv} - \sigma_{sl}}{\sigma_{lv}}$$

$$w_a = \sigma_{lv} * (1 + \cos \theta)$$

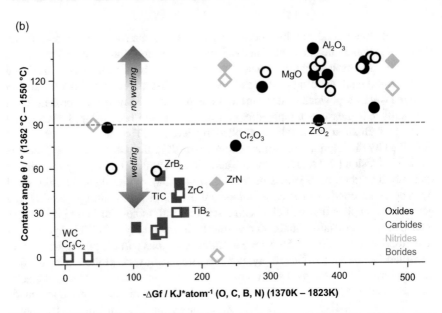

(b)

Fig. 7.5 Wetting of particle phases by liquid metal: (**a**) equations (top) and schematic illustration (bottom) of the contact angle (θ) between solid and liquid; here specific surface tensions (σ) given for each interface: σ_{sl} solid-liquid, σ_{sv} solid-vapour, σ_{lv} liquid-vapour. By θ and the work of adhesion (w_a) the adhesion can be described. (**b**) Overview plot of the contact angle θ (y-axis) of particles in vacuum or inert atmosphere as a function of the formation enthalpy ΔGf (x-axis) in Fe (filled symbols) and Ni (unfilled symbols) (Reproduced from own work Ref. [74])

typically faster and easier to perform from an engineering standpoint, but here the particle dispersion is more complicated to control: Particles with a higher density than the liquid or solidifying matrix will sink to the bottom whereas lower density phases – most relevant for HMS design – float to the top. Primary TiB$_2$ particles (ρ 4.52 g cm^{-3}), for example, readily agglomerate and even form a closed layer at the top of a cast ingot once the solidification rate is below about 5 K s^{-1} [37]. It, therefore, represents an interesting route to obtain self-assembling hard surfaces for wear-resistant materials, which could be further amplified, for example, by casting assisted by centrifugal forces. Particles with an even more pronounced difference in density to the matrix, such as many oxides or nitrides, can only be kept incorporated when the turbulence of the melt is very high and the solidification accordingly fast; otherwise they form a slag on top of the meltpool.

Thermo-mechanical treatments (TMT; e.g. hot rolling, forging or extrusion) do not allow to fully remedy pronounced particle agglomerations but still can be utilised to remove certain artefacts from the previous synthesis. Networks of ledeburitic carbides or Fe_2B, for example, can be effectively broken up and turned into well dispersed elliptical particles by hot rolling [38]. The dispersion of TiB_2 on the other hand is very difficult to affect by TMT without inducing severe damage to the brittle particles. Experimental investigations regarding the effect and extent of TMT post-processing are, therefore, inevitable.

7.3.2 Implementation

The last step in the alloy design process is concerned with transferring the theoretical considerations into bulk HMS materials. When evaluating suitable synthesis techniques, it is necessary to consider whether it is desirable, and if so indeed possible, to obtain the chosen phases during the composites synthesis (in situ), or if the particles have to be created beforehand (ex situ) and subsequently blended into the composite structure (non-reactive synthesis) (Fig. 7.6). In situ formation requires that the respective alloy system (i.e. the sum of alloying elements required for matrix and particles) allows that a homogenous melt or solid decomposes into

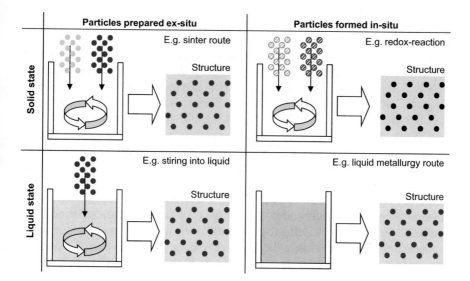

Fig. 7.6 Schematic visualisation of synthesis routes for metal matrix composites

the desired particles and matrix, for example, in a eutectic or eutectoid reaction. It, therefore, relies strictly on suitable thermodynamic stabilities and formation kinetics to achieve the targeted HMS structure. In turn, the interface formation and bonding is usually not hindered by any barrier layers such as oxides or hydrides. Furthermore, in situ synthesis typically eliminates the need for complex pre-fabrication and mixing procedures as required for ex situ processes, which are utilised, for example, in the production of oxide dispersion strengthened (ODS) steels [19, 39].

Synthesis techniques are a key element in the design and development of HMS as they determine whether the targeted in situ or ex situ approach can be followed. Furthermore, their specific characteristics, for example, regarding peak temperature or solidification kinetics, determine whether the desired HMS structure can be achieved. The multitude of available techniques is typically grouped into solid- and liquid-state processes (Fig. 7.6). However, this distinction is not always clear, as, for example, solid particles can be injected into melts, or components mixed in the solid state can be sintered in a partially liquid temperature regime. This holds also true for the laser-powder-based processes used in additive manufacturing. In the following we give a short overview of the most commonly applied techniques for HMS production to this point with future developments and opportunities being discussed in Sect. 7.5.

The most commonly used approach for producing metal-based composites, and consequently the starting point for HMS development [40, 41], is utilising solid-state powder metallurgy (PM). It is based on substantial experience with the production of tool steels and ODS alloys [19, 42–44]. The constituents, i.e. particles and matrix, are produced as powders (with various techniques), mixed (with or without additional agents and/or mechanical milling) and pre-pressed. These "green" raw samples are then subsequently consolidated into a composite at elevated temperature by either sintering or hot isostatic pressed (HIP), which remove remaining porosity and bond the powder particles via diffusion processes. Afterwards a secondary processing step such as extrusion, rolling or forging is often applied to achieve the final microstructure (Fig. 7.7). For HMS this established approach has been followed to incorporate TiB_2 into stainless steel [40, 45] or Fe-Cr alloys [41, 46, 47] as well as in process variations such as self-propagating synthesis utilising redox reactions between Al and TiO_2 in both solid [48] and liquid state [38, 49]. These synthesis procedures following a PM route result in a complex chain of productions with comparatively large efforts regarding machinery, time and quality control steps: Obtaining powders with a small particle size, for example, requires time-consuming ball milling. Consolidation via HIP relies on applying uniform (isostatic) pressure by an inert gas, typically Ar, so the mixed powders have to be filled into precisely manufactured containers. They are subsequently individually evacuated and sealed by welding to prevent oxidation within the pressure chamber – which is limited in size – where they are annealed for several hours, and need to be removed by machining afterwards [19, 43]. As a consequence, PM HMS can be designed with a high degree of freedom, especially regarding matrix and particle

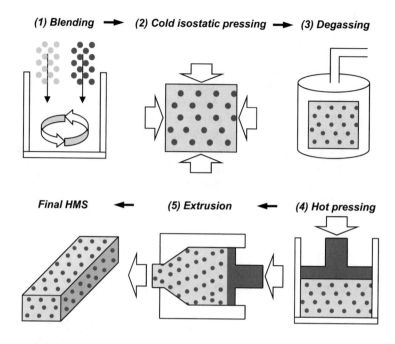

(1) Blending ➡ **(2) Cold isostatic pressing** ➡ **(3) Degassing**

Final HMS ⬅ **(5) Extrusion** ⬅ **(4) Hot pressing**

Fig. 7.7 Sketch of the typical powder metallurgy synthesis route: A complex sequence of processing steps is required to obtain the desired particle distribution in HMS

combinations, but the large associated costs limit them to niche applications rather than mass production.

Liquid metallurgical synthesis techniques therefore appear more appealing from a production point of view, ideally if HMS production can be performed on existing, industrially established machinery such as block or strip casting. Ex situ injection techniques have not been found favourable, tough, as they still require prefabrication of particles (with the drawbacks detailed above), and can be problematic to allow for achieving a suitable particle dispersion (tendency for particle agglomeration and floatation) and sufficient wetting. Therefore, most developmental work has been performed on HMS systems which allow for in situ reactions (Fig. 7.6). Especially the Fe–Ti–C [50–52] and Fe–Ti–B [37, 38, 47, 53–74] systems have received considerable attention. They are eutectic systems, where TiC and TiB_2 are precipitated from a homogenous liquid, yielding a ferritic matrix. Challenges are related to particles precipitating in the liquid already at elevated temperatures, as they dramatically decrease the viscosity of the melt, rendering casting processes difficult [51]. Furthermore, they are often causing a density-induced demixing, especially as large-scale industrial casting processes solidify relatively slow. They are therefore limiting the particle fraction – and thus the gain in the specific modulus – below the eutectic concentration. This favours alloy systems where this concentration is located at large fraction of particles, which should be as efficient as possible (maximum E/ρ ratio). These considerations, especially the superb physical properties of TiB_2 ($E = 545GPa$, $\rho = 4.52 \text{ g cm}^{-3}$) render Fe–TiB_2 based alloys the most prominent HMS concepts at the current stage.

At this point it should be pointed out that the relevance of experimental investigations to evaluate the feasibility of a chosen HMS concept cannot be understated. This is caused by the increased complexity of composite materials, for example, regarding the lack of precise thermodynamic data and particle properties as well as the difficulties in reliably predicting matrix/particle interactions. Basic microstructure information, mechanical properties and density of bulk HMS can be efficiently probed with high-throughput techniques, but stiffness measurements are notoriously difficult to perform reliably and reproducibly. While a multitude of techniques exist to measure bulk E values, extra care needs to be taken to pay attention to their specific artefacts and to avoid possible uncertainties in derived values [7].

7.4 Current State of the Art: Fe–TiB$_2$-Based HMS

The basis for HMS development was laid in the 1970s when Fe-based boride systems were researched mainly focussing on its application for wear-resistant steels [53, 75, 76]. Around the year 2000 the first publications recognized their potential for a new generation of structural materials for lightweight design [40, 46]. Following the alloy design strategy outlined above, especially the Fe–Ti–B system has since been then the subject of intense research due to the outstanding effectiveness of TiB_2 and the possibility of in situ liquid metallurgy synthesis, and respective HMS concepts have started to spark industrial interest [57, 58, 77]. In the following we present an overview of the current understanding regarding fundamental aspects of the ternary Fe–Ti–B system, recent efforts in preparing them for industrial production and technical applications as well as first insights into alloying and processing strategies to improve the property profile of Fe–TiB$_2$-based HMS.

7.4.1 Ternary Fe–Ti–B Materials

Figure 7.8 shows the pseudo-binary Fe–TiB$_2$ section of the Fe–Ti–B phase diagram. The eutectic point is located at about 12 vol.% of TiB_2 and 1380 °C. Figure 7.9 shows respective microstructures with increasing particle fractions, stemming from small arc-melting ingots subjected with "intermediate" solidification kinetics (about 5 K s^{-1}) [63], similar to those found in industrially established casting processes. The eutectic morphology is not perfectly lamellar, and the particles stemming from primary solidification (TiB_2 fraction above the eutectic concentration) are large (diameter in the order of 10 μm) and of sharp-edged, polygonal morphology. It

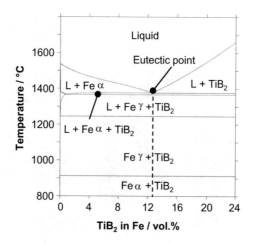

Fig. 7.8 Pseudo-binary Fe–TiB_2 section of the Fe–Ti–B phase diagram, calculated with Thermocalc 2015a. Database provided by Thermocalc Software AB

should be noted that the thermodynamic databases are still under refinement [46, 53, 55], for example, regarding effects of the Ti/B ratio. Experimental results shown in Fig. 7.10 [78] reveal that Ti amounts below or at the exact stoichiometric concentration (left, middle) required for the formation of TiB_2 lead to the formation of Fe_2B, in agreement with simulation results [46]. For over-stoichiometric Ti concentrations (right), the predicted Fe_2Ti Laves-type intermetallics – which would lead to additional embrittlement – cannot be observed. As the effectiveness of Fe_2B (about 40 GPa cm^3 g^{-1}) is much lower than TiB_2 (about 125 GPa cm^3 g^{-1}) [17], most Fe–TiB_2-based HMS rely on slightly elevated Ti concentrations (about 20% over-stoichiometric).

Increasing the particle fraction effectively increases the specific modulus (Fig. 7.11a), but in turn leads to a deterioration of the mechanical performance (Fig. 7.11b) [78, 79]. For a HMS with 20 vol.% TiB_2 in a ferritic matrix produced by established casting and hot rolling procedures, a specific modulus of about 30 to 32 GPa cm^3 g^{-1} can be expected, representing an increase of about 20% over conventional high-strength steels, Al and Ti alloys. The ultimate tensile strength is increased from about 300 (pure Fe) to about 600 MPa, but the ductility is decreased to less than 10%, along with a respective decrease in toughness. This property profile renders such basic HMS alloys not competitive with other structural materials for lightweight design. It has to be pointed out though that surprisingly little mechanical testing results – even basic tensile testing data – of various HMS concepts are published [37, 38, 51, 66, 68, 69].

More information is available on microstructural phenomena: transmission electron microscopy investigations showed that the Fe – TiB_2 interface is semi-coherent indicating elevated strength (Fig. 7.12) [61]. In situ scanning electron microscopy tensile tests revealed fracture of large particles as the dominant failure mode rather than interface delamination (Fig. 7.13) [59]. This highlights the importance of avoiding or at least limiting the number of coarse TiB_2 particles, which can be achieved by closely controlling the synthesis conditions. Any carbon present in the

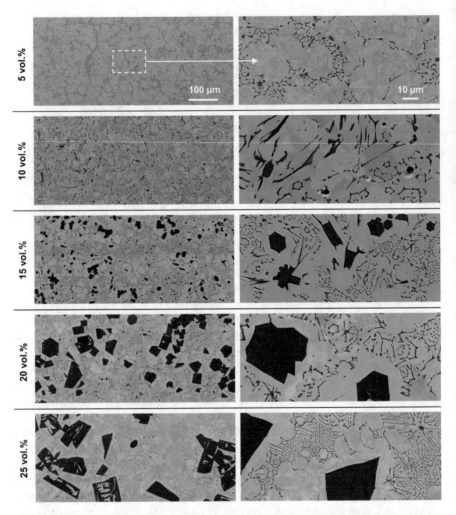

Fig. 7.9 Micrographs (backscatter electron contrast) of HMS microstructures in as-cast condition with increasing fractions of TiB_2 (dark grey) in Fe (light grey, intermediate solidification rate): Eutectic TiB_2 particles of irregular morphology prevail until a fraction of 10 vol.% above coarse primary particles can be observed. (Reproduced from own work Ref. [78])

melt leads to the formation of TiC particles, which seem to act as nucleation sites for TiB_2 and may lead to its coarsening [63, 74, 80]. Furthermore, it was found that the particle microstructure (especially size and dispersion) can be dramatically changed by altering the solidification kinetics [37, 63]. As shown in Fig. 7.14, the primary particle size (by area) remains roughly constant (about 50 μm^2) down to a cooling rate of about 50 K s^{-1} (10 mm mould thickness) and then begins to strongly decrease with increased solidification velocity. As the primary TiB_2 particles effectively hinder grain growth of the surrounding ferrite, both particle

Fe – 7.04Ti – 3.86B   Fe – 8.57Ti – 3.86B   Fe – 10.10Ti – 3.86B

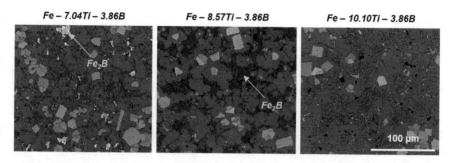

Fig. 7.10 Effect of Ti/B stoichiometry in Fe on the phases formed: colour-coded electron backscatter diffraction (EBSD) phase maps with image quality superimposed in grey scale; lean and stoichiometric Ti concentrations (left, middle) reveal TiB$_2$ (green) additional Fe$_2$B (yellow). Overstoichiometric Ti concentrations (right) suppress the Fe$_2$B formation. (Reproduced from own work Ref. [78]

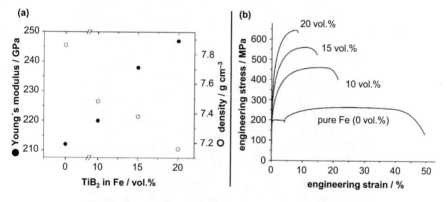

Fig. 7.11 Inverse relationship between physical and mechanical performance: While the E/ρ ratio of the HMS is improving with increasing TiB$_2$ particle fraction (**a**), it leads to severe embrittlement at only intermediate strength (**b**). (Reproduced from own work Ref. [78])

and matrix grain sizes are closely linked. This mechanism is especially important as it has the advantage of improving the materials strength without lowering its impact toughness [81]. The effects on the microstructure are shown in more detail in Fig. 7.15. Very slow solidification close to equilibrium conditions (top) at 0.016 K s^{-1} leads to extremely coarse primary particles which have floated to the top of the solidifying melt. Underneath the eutectic material remains, which is free of any primary TiB$_2$ particles or ferrite grains. With intermediate cooling rates (middle) the particle size remains constant as already mentioned, but the tendency of the light TiB$_2$ particles to de-mix and agglomerate is clearly visible. At extreme cooling rates achievable with techniques such as meltspinning or splat cooling (reaching up to 10^7 K s^{-1}), both particles and matrix can be refined to the order of tens of nanometres. The HMS alloy can even be super-cooled into an amorphous, glass-like

(a)   (b)

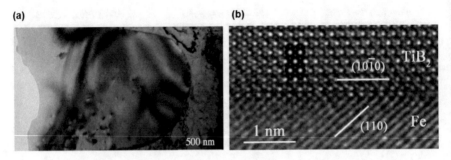

Fig. 7.12 Fe–TiB$_2$ interface: Transmission electron microscopy (TEM) bright-field image (**a**) and high-resolution transmission electron microscopy (HRTEM) image of the interface parallel to the prismatic plane (1010), showing semi-coherency (**b**) [61]

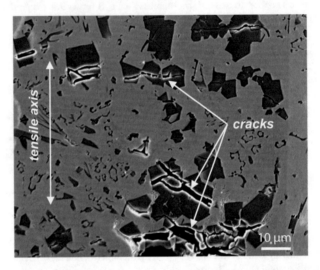

Fig. 7.13 Fractured particles after tensile testing: Deformation of Fe–TiB$_2$ HMS is typically accompanied by cracking of coarse particles along the loading direction. The images are with permission based on previously published work [37]

state, from which nano-scaled ferrite and TiB$_2$ can be precipitated by subsequent annealing [70].

One of the most attractive attributes of steels is their extremely competitive price. Only when it can be kept at a minimum even for high-performance alloys, innovative materials such as HMS have a chance to be used broadly across the entire range of weight critical applications and not remain in a niche. The elevated alloying costs of HMS caused by the large Ti additions – 20 vol.% TiB$_2$ require more than 10 wt.% of Ti – can be overcome by utilising redox-reactions of cheaper Ti and B oxides [48, 82]. More problematic is that industrial steel production with its large volumes, obtained mainly by strip or slab casting, results in solidification kinetics in the critical intermediate regime. To avoid the uncontrolled floatation of

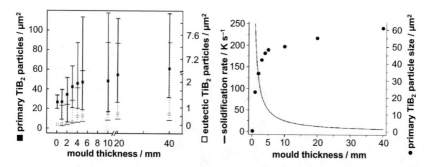

Fig. 7.14 Effect of the solidification rate on TiB$_2$ particle size: By decreasing the mould thickness, increasing cooling rates are achieved (left), but only above about 50 K s^{-1} a substantial reduction of the particle rate is observed (right). The images are with permission based on previously published work [63]

primary particles and ensure satisfactory mechanical properties, published industrial interest has centred on eutectic Fe–TiB$_2$ HMS [57, 58, 77]. However, it is difficult to precisely target the ternary eutectic composition without hyper and hypo-eutectic constituents, i.e. purely ferritic grains and primary TiB$_2$ particles, respectively. This can be overcome by utilising the density-induced separation in hyper-eutectic alloys described above: By switching to large scale discontinuous block casting with very slow solidification through appropriate moulds, the head containing the risen primary particles can be easily removed afterwards, leaving behind a large volume of HMS material of regular eutectic microstructure with relatively simple and straightforward processing. Laboratory-scale trials revealed that it is thereby possible to achieve HMS with an increased fracture toughness compared to faster solidified materials of identical particle content [37]. Processes such as centrifugal casting can be utilised to enforce this separation even further and steer the coarse primary particles to form a self-assembling hard surface layer, ideally suited for wear-critical applications such as break disks or mining drills [37].

As the TiB$_2$ particles – both primary and eutectic – stem from solidification, their microstructure can only be influenced to a minor extent by subsequent TMT processing. A slight spheroidisation (increasing with temperature and time) of the eutectic lamellae can be achieved during annealing, for example, during soaking [64]. During rolling or forging, the desired recrystallisation of the matrix can be severely affected by the particles microstructure, especially if they are closely positioned to each other. Such areas, often as a consequence of agglomerations during casting, can also lead to particles fracturing. The induced damage leads to fractures at the edges, which can progress rapidly to destroy the entire slab, and are another factor why eutectic composition is favourable from an industrial point of view. Respective hot-rolled, thin-sheet HMS have been produced, and first evaluations of critical engineering processes to form automotive body parts out them, such as deep drawing or welding, have been performed [56]. The processing of thicker gauge materials has not yet been investigated, where especially the possibility of surface hardening, relevant for example for drive train components is of critical importance.

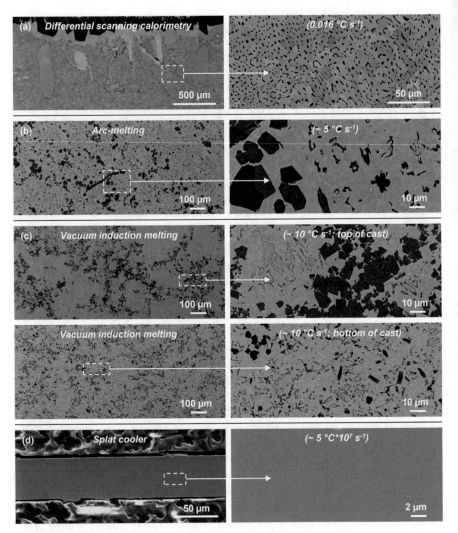

Fig. 7.15 Micrographs (backscatter electron contrast) showing the effect of cooling rates on the as-cast microstructures of Fe–TiB$_2$ HMS: Solidification rates close to equilibrium conditions (**a**) lead to floatation of extremely coarse primary particles to the top of the melt, while the eutectic material remains without primary precipitates underneath. For intermediate cooling rates (**b**, **c**) particle size remains constant, but light de-mixing and agglomeration emerges. Utilising extreme cooling rates achieved by meltspinning or splat cooling, resulting in amorphisation (**d**). The images are with permission based on previously published work [63]

7.4.2 Alloying Effects

Ternary Fe–Ti–B HMS are hindered by an inherent trade-off: they either allow for a strong increase in the specific modulus but at the same time they are rather brittle (hyper-eutectic TiB$_2$ fraction), or they yield satisfactory mechanical properties but at a limited E/ρ ratio improvement (hypo-eutectic TiB_2 fraction). Both mechanical and physical performance need to be obtained simultaneously on a suitable level for HMS, to be not only competitive with, but even superior to other structural materials for lightweight design. While it is possible to overcome this mutual exclusivity by adapting the solidification rate (to either very slow or very fast kinetics), this is not always feasible in view of production volume. Meltspun ribbons, for example, are with maximum 100 μm too thin for most structural applications, and most steel production plants are set up for continuous strip casting rather than block casting. The promising new pathways opened by additive manufacturing are described in the following section.

An alternative route to obtain HMS whose mechanical and physical performance are satisfactory, and which can be produced with established steel production techniques, is to utilise additional alloying elements. This route has not yet left the stage of fundamental scientific investigations. The goal is to counteract the embrittlement induced by the stiff, hard and sharp-edged particles and simultaneously increase the strength at least to the level of high-strength steels. The alloying additions – together with appropriate TMT parameters of course – allow modifying the constitution of the matrix away from pure ferrite but are expected to also interact with the morphology, structure and properties of the TiB_2 particles. Elevated dimensions in chemical composition also increase the thermodynamic complexity, highlighting the need of experimental investigations.

The effect of various alloying elements on the particle microstructure is shown in Fig. 7.16. Additions of 5 wt.% Cr, Ni, Co, Mo, W, Mn, Al, Si, V, Ta, Nb and Zr, respectively, were found to slightly increase the size of the primary TiB_2 particles, and Co additions led to the most even size distribution. The eutectic particles were decreased in size by all elements except Ni, while their aspect ratio was only marginally affected [64]. The most common alloying elements used in steel design and relevant for achieving a modified matrix microstructure and constitution can therefore be considered mostly independent from the particle microstructure, with the exception of carbon, which leads to the formation of TiC in case of excessive Ti. This induces coarsening of TiB_2, and any remaining B is incorporated into Fe_2B borides. As a result, no C is left available in solution or in cementite and therefore cannot be utilised for associated heat treatment pathways such as martensitic hardening as shown in Fig. 7.17 [74].

Alternatively, Mn appears favourable as it is cost effective and allows for a wide range of matrix constitutions. Concentrations below ~5 wt.% lead to solid-solution strengthened ferrite, while intermediate concentrations (~ 10 wt.% Mn) result in α' and ε martensite, which may be coupled with reverted austenite formed during intercritical annealing. Concentrations above ~20 wt.% generate austenitic matrices with decreasing stacking-fault energy (SFE) can be obtained as utilised for TRIP and TWIP phenomena in high Mn steels. However, while such microstructures allow achieving elevated strength, toughness and ductility, the altered constitution is typically associated with a decrease in stiffness. First investigations showed that predominantly austenitic matrices with 20 and 30 wt.% Mn, respectively, did not translate into a mechanical performance which could overcome the significantly

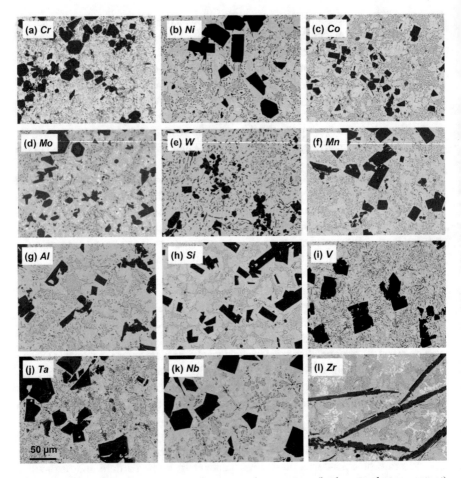

Fig. 7.16 Effect of alloying additions on the as-cast microstructures (backscatter electron contrast) of Fe–TiB$_2$ HMS: Additions of 5 wt.% Cr, Ni, Co, Mo, W, Mn, Al, Si, V, Ta, Nb and Zr, respectively, slightly increase the size of the primary TiB$_2$ particles, and Co additions lead to the most even size distribution. The eutectic particles were decreased in size by all elements except Ni, while their aspect ratio was only marginally affected. The images are with permission based on previously published work [64]

reduced E values. Martensitic/austenitic matrices of 10 wt.% Mn concentrations, on the other hand, were found to improve the matrix/particle co-deformation without sacrificing too much of the stiffness of the composite, especially for hypo-eutectic Fe–TiB$_2$-based HMS, where the mechanical performance of dual-phase (ferrite/martensite) steel could be matched with a specific modulus of about 29 GPa cm^3 g^{-1} [66].

The options to further improve the strength of HMS are limited as C-based martensite is not available, plastic deformation involves the risks of inducing internal damage and solid-solution strengthening requires large alloying concentrations.

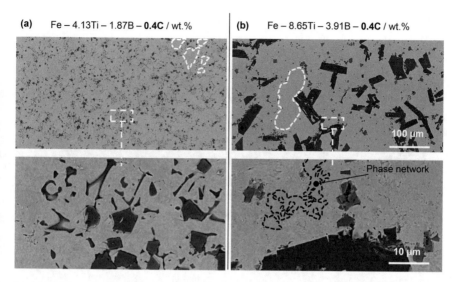

Fig. 7.17 Micrographs (backscatter electron contrast) showing changes of the as-cast Fe–TiB$_2$ HMS microstructures caused by C additions: Both (**a**) hypo- and (**b**) hyper-eutectic Fe–TiB$_2$ alloys reveal particle-free areas (areas within dashed white lines) strongly changed eutectic particles. Primary particles of the hyper-eutectic alloy are vastly enlarged and partially broken. Phase networks (Fe$_2$B) are marked with dashed black lines (Reproduced from own work Ref.[74])

It was, therefore, investigated whether precipitation strengthening of the inherently soft and ductile ferritic matrix can be utilised in a similar manner as deployed in interstitial free or electrical steels [83–85]. Alloying with Si, Mn and Ni led to the formation of G-phase (base composition Ti$_6$Si$_7$Ni$_{16}$) in aged materials and elevated the strength of hyper-eutectic HMS to more than 1100 MPa, but its preferential formation on grain boundaries strongly decreased the ductility. Additions of 1 and 2 wt.% Cu led to lower strength after quenching (less solid-solution strengthening) but also to significant strengthening via ageing with a parallel drop in ductility. Both Cu and G-phase additions lowered the specific modulus, though, most notably for Cu additions with about 3 GPa cm^3 g^{-1} per wt.% [69].

Figure 7.18 summarises the mechanical (tensile testing data) and physical properties (E/ρ) of these various alloying strategies for Fe–TiB$_2$-based HMS [66, 69]. It becomes clear that while a promising start has been made and suitable alloy design strategies have been identified, there remains substantial room for improvement. The goal in these complex developments is to further optimise the co-deformation processes without sacrificing the desired gain in specific modulus for maximum lightweight performance. Most of the underlying phenomena for the encountered problems are inherent to the Fe–Ti–B base alloy system and the established synthesis methodologies, though. It is therefore of strong interest to re-evaluate the current state of the art and to investigate the possibilities offered by alternative alloy systems and new production routes.

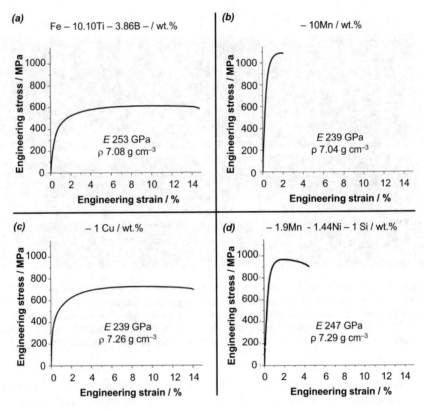

Fig. 7.18 Summary of tensile testing results, E and ρ values from hyper-eutectic Fe–TiB$_2$ HMS with alloying additions, in hot-rolled and quenched states: (**a**) reference alloy, (**b**) 10 wt.% Mn, (**c**) 1 wt.% Cu, (**d**) G-Phase. The images are with permission based on previously published work [66, 69]

7.5   Recent Developments and Outlook

The currently most researched HMS concepts, i.e. Fe–TiB$_2$-based alloys, are limited by the trade-off between mechanical and physical performance when synthesised by standard liquid metallurgy techniques, linked to their intermediate solidification kinetics. Furthermore, they lack the strong fundamental advantage associated with steels, namely the possibility to alter the mechanical properties over a wide range by simple and straightforward heat treatments. These limitations have led to research into novel alloy systems and production techniques.

7.5.1   Alternative Alloy Systems

Fe–boride systems are advantageous as detailed in the previous sections concerned with the fundamentals of alloy design for HMS. Investigations into alternatives for Ti as a boride forming element, though, are accordingly complex, as the thermodynamics of the multitude of ternary Fe–B–X systems for example are not fully described yet, and the properties of formed phases and their interaction towards bulk properties are difficult to predict. This motivated property-driven combinatorial investigations, adding 5 at.% of various boride forming elements to an Fe–10 at.% B base alloy, and measure the resulting E, ρ and mechanical properties [67]. As shown in the selected results of Fig. 7.19a, this high-throughput approach allowed to efficiently identify Cr as an attractive element for a number of reasons: Cr is not only much cheaper as an alloying element than Ti, but also beneficial for the matrix stiffness if kept in solution [86]. It leads to the in situ formation of M(Fe,Cr)$_2$B type borides from the melt. These particles are increasing in stiffness with their Cr concentration, from about 290 to 450 GPa from Fe$_2$B to Cr$_2$B. The elevated density (about 7 g cm^{-3}) is beneficial for liquid metallurgy as it limits the tendency to float and agglomerate (when precipitated from liquid) as observed with the less dense TiB$_2$ (about 4.5 g cm^{-3}), rendering the solidification rate a much less critical parameter. On the other hand this appears disadvantageous for the efficient improvement of lightweight performance but the specific modulus of the bulk HMS is important: The drop in effectiveness of the M$_2$B particles can be readily compensated for by an inherently stiffer matrix when excessive Cr is utilised,

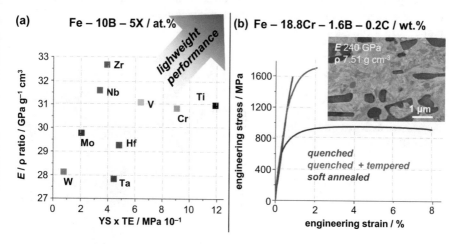

Fig. 7.19 Property-driven approaches establishing alternative HMS systems: (**a**) Alloying addition (5 at.%) of various boride-forming elements to an Fe–10 at.% B base alloys reveals potential candidates for HMS design. (**b**) C alloying of Fe–Cr–B-based HMS allows to change the mechanical performance by simple heat treatments. The images are with permission based on previously published work [67, 72]

which also raises the matrix's strength and also its corrosion resistance. It can be boosted further by elevated particle fractions without causing production drawbacks even when standard casting procedures are used as the eutectic concentration in Fe is higher than those of TiB_2. Furthermore, the hypo-eutectic M_2B borides are of favourable elliptic morphology, greatly improving the co-deformation with the matrix. This beneficial effect on ductility and toughness also translates into a strongly decreased susceptibility for fracturing during TMT procedures. Most importantly, though, due to the limited stability of Cr carbides compared to Ti carbides, Fe–Cr–B-based alloys allow to keep substantial amounts of C into solution within austenite at elevated temperatures. This free C allows utilising the full range of established steel heat treatments. As exemplified in Fig. 7.19b, soft annealing or quenching and tempering allows for varying the mechanical performance over a huge range, which can also be readily applied locally (e.g. surface hardening) [72]. Together with the possibility to also achieve stainless steel matrices, this demonstrates the potential of Fe–Cr–B-based alloys to spread the HMS concept to a dramatically wider range of applications, to more than the commonly referred design goal of high-strength sheets for automotive parts.

Other Fe–boride systems appear just as interesting though. The elongated, whisker-like morphology of highly effective Zr-borides for example might be utilised for achieving highly anisotropic stiffness [67], ideally suited for applications where the loading direction in a specific part is unidirectional, such as bolts. Moving on, the vast range of possible particle phases offers much more yet untapped potential for future lightweight materials. Even the combination of different particle phases might be a feasible route to follow, offering the potential advantage that each phase can be kept lower than its critical fraction (e.g. their eutectic concentration). An example of a combination of both in situ TiC and TiB_2 particles in a ferritic matrix is shown in Fig. 7.20 [74]. However, the associated

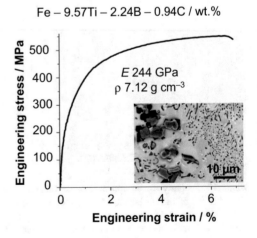

Fig. 7.20 An example of a HMS with more than one particle phase: TiC and TiB_2 formed in situ in Fe. (Reproduced from own work Ref. [74])

thermodynamic stabilities, which are already complex for ternary boride systems, get even more complicated to predict. As already mentioned, HMS are, therefore, a prime field for combinatorial search missions, both with bulk techniques such as rapid alloy prototyping [87] and as thin film methods based on ab initio calculations [15].

7.5.2   Novel Synthesis and Processing Techniques

As made obvious in the previous sections, the design of HMS is strongly interrelated with their respective synthesis techniques. While this limits them somewhat in not always being readily produced on established equipment, it opens the possibility to find innovative materials by utilising the specific artefacts of adapted as well as completely novel techniques. As already mentioned, extra slow solidifications in discontinuous block casting can be exploited for highly controlled eutectic Fe–TiB$_2$ HMS, and centrifugal casting can lead to self-assembling hard surfaces for wear-critical applications. The development of synthesis and processing is progressing rapidly though, with multiple processes that can be tailored for HMS production, which exceeds the scope of this chapter. Just to give a few examples, friction stir processing could be ideally suited to refine the size and dispersion of particles [88], while advanced sintering techniques might increase the effectiveness in PM-based production routes. Particularly interesting for HMS development are laser-based additive manufacturing techniques. They have gained significant attention recently as they allow for superb freedom for geometrical design, near net-shape synthesis and on-demand production without being economically limited in the number of required parts. Despite all these advantages, the production route already is comparatively expensive, and it becomes even more relevant to achieve a true gain in terms of achievable properties. Currently, the specific processing artefacts make it difficult to be competitive when materials are used that have been optimised over decades (if not centuries) for a completely different processing route (i.e. casting and forging or hot rolling). The main characteristics of additive manufacturing processes are a very high heating rate, a small meltpool and a high cooling rate in the upper temperature regime. These render them ideally suited for realising completely new material solutions, especially for HMS design.

As a first example, the rapid solidification of a small liquid zone can be exploited to effectively refine the microstructure of Fe–Ti–B alloys down to the nanometric range: the TiB$_2$ particles have very limited time to grow and agglomerate and subsequently pin the matrix grain boundaries, preventing their movement and thus grain growth in the solid state. Subsequent TMT can therefore be utilised – if possible – to remove any remaining porosity. The basic feasibility of this approach was demonstrated by spray forming experiments [68]. While not being a true additive manufacturing technique as it lacks their precise geometrical control, it is similar regarding the relevant physical metallurgical phenomena. As shown in Fig. 7.21, the drastic refinement of the microstructure results in doubling the strength to the level of advanced dual-phase steels (ultimate tensile strength of more than

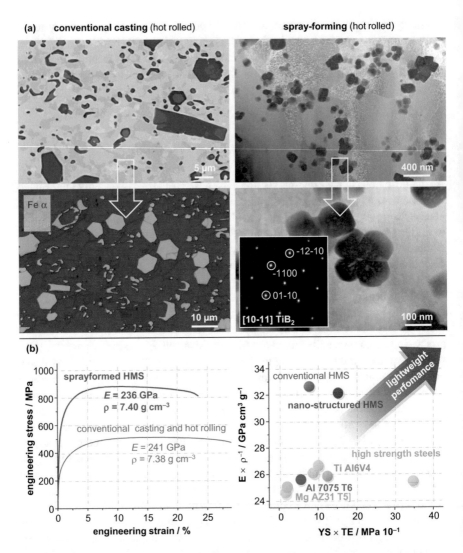

Fig. 7.21 Implementation of novel synthesis techniques: (**a**) Microstructure (top, backscatter electron contrast; bottom, image quality map superimposed by colour coded backscatter diffraction map) obtained after conventional casting can be successfully refined utilising the spray compaction technique from micrometre to nanometre range (right, backscatter electron contrast). (**b**) Effect of nano-structuring on mechanical performance. E and ρ remain virtually unchanged. The images are with permission based on previously published work [68]

850 MPa) without sacrificing ductility (total elongation of more than 20%) at a specific modulus of over 32 GPa cm^3 g^{-1}. It is thereby a prime example how innovative lightweight design solutions can be achieved through novel synthesis techniques. The transfer to 3D printing processes is currently under investigation.

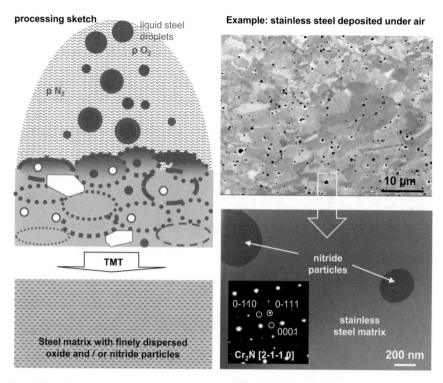

Fig. 7.22 Atmospheric reactions in liquid metal deposition: The schematic sketch on the left illustrates the straightforward synthesis and processing approach utilising distinct atmospheric reactions. The microstructures (top, secondary electron contrast; bottom, scanning transmission electron microscopy – high-angle annular dark-field imaging) on the right show the successfully synthesised material inhering about 2 vol.% nitrides in a stainless steel matrix. The images are with permission based on previously published work [82]

Another new opportunity offered by additive manufacturing techniques is the possibility to incorporate oxides and nitride particles into composite structures as the rapidly solidifying liquid zone allows overcoming their difficult wetting and tendency to quickly demix in the melt. They can be brought into the process synthesised ex situ, but this of course requires separate preparation steps, and the interaction of small particles with the high intensity laser beam can be critical as even stable oxides can vapourise. Itis therefore more appealing to save the efforts of particle preparation and rather utilise their in situ formation by reactions of metallic powder particles with nitrogen and/or oxygen being present in the process atmosphere. As shown in Fig. 7.22 [82], preliminary trials with various liquid metal deposition techniques were successful in incorporating more than 2 vol.% of Cr-nitrides into a stainless steel matrix [82]. For further developing this approach, powder-spray techniques (such as laser metal deposition; LMD) appear to be better suited than powder-bed processes (selective laser melting; SLM) as flying liquid powder particles have a more intense interaction with the atmospheric gases.

Opposed to oxygen-dispersed strengthened (ODS) steels, the particle size is not so critical for HMS, but much larger fractions are required for the desired improvement of the specific modulus. Typical values for oxides and nitrides are in the order of 50–100 GPa cm^3 g^{-1} [17].

The examples described in this chapter showcase the dynamic state of alloy and processing design for high-modulus steels. Being merely at the start of their development, numerous new insights and discoveries are to be expected on the way towards the next generation of structural materials for lightweight constructions.

References

1. M.F. Ashby, *Materials Selection in Mechanical Design* (Butterworth-Heinemann, Burlington, 2005)
2. H. Berns, W. Theisen, *Eisenwerkstoffe – Stahl und Gusseisen* (Springer, Berlin/Heidelberg, 2008)
3. G. Krauss, *Steels—Processing, Structure, and Performance*, 2nd edn. (ASM International, Materials Park, 2015)
4. G. Frommeyer, U. Brux, Microstructures and mechanical properties of high-strength Fe-Mn-Al-C light-weight TRIPLEX steels. Steel Res. Int. **77**(9–10), 627–633 (2006)
5. K. Tanaka, Ultra high modulus steel reinforced with Ti Boride particles, R&D Review of Toyota CRDL **35**(1), 1 (2000)
6. G. Gottstein, *Physikalische Grundlagen der Materialkunde* (Springer, Berlin/Heidelberg, 2007)
7. S. Münstermann, W. Bleck, Einflussgrößen auf den Elastizitätsmodul von Stählen für den Karosseriebau. Mater. Test. **47**(6), 337–344 (2005)
8. W.C. Leslie, Iron and its dilute substitutional solid solutions. Metall. Trans. **3**(1), 5–26 (1972)
9. C.N. Hwang, T.F. Liu, Grain boundary precipitation reactions in a wrought Fe-8Al-5Ni-2C alloy prepared by the conventional ingot process. Metall. Mater. Trans. A **29**(2), 693–696 (1998)
10. M. Bausch, G. Frommeyer, H. Hofmann, E. Balichev, M. Soler, M. Didier, L. Samek, Final Report – Ultra high-strength and ductile FeMnAlC light-weight steels (MnAl-steel) (2013)
11. J.C.H. Affdl, J.L. Kardos, The Halpin-Tsai equations: a review. Poly. Eng. Sci. **16**(5), 344–352 (1976)
12. N. Chawla, R.S. Sidhu, V.V. Ganesh, Three-dimensional visualization and microstructure-based modeling of deformation in particle-reinforced composites. Acta Mater. **54**(6), 1541–1548 (2006)
13. N. Chawla, K.K. Chawla, Microstructure-based modeling of the deformation behavior of particle reinforced metal matrix composites. J. Mater. Sci. **41**(3), 913–925 (2006)
14. S.F. Pugh, XCII. Relations between the elastic moduli and the plastic properties of polycrystalline pure metals. Lond. Edinb. Dubl. Phil. Mag. **45**(367), 823–843 (1954)
15. J. Emmerlich, D. Music, P. Braun, P. Fayek, F. Munnik, J.M. Schneider, A proposal for an unusually stiff and moderately ductile hard coating material: Mo2BC. J. Phys. D: Appl. Phys. **42**(18), 185406 (2009)
16. J. Cui, Y.S. Chu, O.O. Famodu, Y. Furuya, J. Hattrick-Simpers, R.D. James, A. Ludwig, S. Thienhaus, M. Wuttig, Z. Zhang, I. Takeuchi, Combinatorial search of thermoelastic shape-memory alloys with extremely small hysteresis width. Nat. Mater. **5**, 286 (2006)
17. C. Baron, H. Springer, Properties of particle phases for metal-matrix-composite design. Data Brief **12**, 692–708 (2017)

18. Q. Wu, W. Li, N. Zhong, Corrosion behavior of TiC particle-reinforced 304 stainless steel. Corros. Sci. **53**(12), 4258–4264 (2011)
19. N. Chawla, K.K. Chawla, *Metal Matrix Composites* (Springer US, New York, 2006)
20. C.W. Nan, D.R. Clarke, The influence of particle size and particle fracture on the elastic/plastic deformation of metal matrix composites. Acta Mater. **44**(9), 3801–3811 (1996)
21. K.H.Z. Gahr, Wear by hard particles. Tribol. Int. **31**(10), 587–596 (1998)
22. K.U. Kainer, Grundlagen der Metallmatrix-Verbundwerkstoffe (2004)
23. G. Streukens, E. Langkabel, M. Ortelt, Polyurethane prepregs – a new concept for CFRP processing. JEC Compos. Mag. (87), 102–104 (2014)
24. G.E. Spriggs, Properties of diamond and cubic boron nitride; Figs. 13.1–13.91, in *Materials · Powder Metallurgy Data. Refractory, Hard and Intermetallic Materials, Landolt-Börnstein - Group VIII Advanced Materials and Technologies*, ed. by P. Beiss, R. Ruthardt, H. Warlimont, (Springer, Berlin/Heidelberg, 2002)
25. J.H. Kang, P.E.J. Rivera-Díaz-del-Castillo, Carbide dissolution in bearing steels. Comp. Mater. Sci. **67**, 364–372 (2013)
26. T. Young, An essay on the cohesion of fluids. Philos. Trans. R. Soc. Lond. **95**(0), 65–87 (1805)
27. J.V. Naidich, The Wettability of Solids by Liquid Metals, in *Progress in Surface and Membrane Science*, ed. by D. A. Canhead, J. F. Danielli, (Elsevier, Saint Louis, 1981), pp. 353–484
28. G.V. Samsonov, The Oxide Handbook (1973)
29. N. Eustathopoulos, N. Sobczak, A. Passerone, K. Nogi, Measurement of contact angle and work of adhesion at high temperature. J. Mater. Sci. **40**(9–10), 2271–2280 (2005)
30. N.Y. Taranets, Y.V. Naidich, Wettability of aluminum nitride by molten metals. Powder Metall. Met. Ceram. **35**(5–6), 282–285 (1996)
31. M. Kivio, L. Holappa, T. Yoshikawa, T. Tanaka, Interfacial phenomena in Fe-TiC systems and the effect of Cr and Ni. High Temp. Mater. Processes (Lond.) **31**(4–5), 645–656 (2012)
32. R.M. Aikin, The mechanical properties of in-situ composites. JOM **49**(8), 35–39 (1997)
33. M. Humenik, W.D. Kingery, Metal-ceramic interactions: III, surface tension and wettability of metal-ceramic systems. J. Am. Ceram. Soc. **37**(1), 18–23 (1954)
34. J.-G. Li, Wetting and interfacial bonding of metals with ionocovalent oxides. J. Am. Ceram. Soc. **75**(11), 3118–3126 (1992)
35. G.A. Yasinskaya, The wetting of refractory carbides, borides, and nitrides by molten metals. Poroshkovaya Metallurgiya (translated) **7**(43), 53–56 (1966)
36. G. Arth, A. Samoilov, Metall-Matrix-Verbundwerkstoffe auf Eisenbasis. BHM Berg- und Hüttenmännische Monatshefte **157**(8–9), 306–312 (2012)
37. H. Zhang, H. Springer, R. Aparicio-Fernández, D. Raabe, Improving the mechanical properties of Fe – TiB2 high modulus steels through controlled solidification processes. Acta Mater. **118**, 187–195 (2016)
38. C. Baron, H. Springer, D. Raabe, Efficient liquid metallurgy synthesis of Fe–TiB2 high modulus steels via in-situ reduction of titanium oxides. Mater. Design **97**, 357–363 (2016)
39. Z. Oksiuta, P. Olier, Y. de Carlan, N. Baluc, Development and characterisation of a new ODS ferritic steel for fusion reactor application. J. Nucl. Mater. **393**(1), 114–119 (2009)
40. Z. Kulikowski, A. Wisbey, T.M.T. Godfrey, P.S. Goodwin, H.M. Flower, Mechanical properties of high performance lightweight steels. Mater. Sci. Technol. **16**(7–8), 925–928 (2000)
41. Z. Kulikowski, T.M.T. Godfrey, A. Wisbey, P.S. Goodwin, F. Langlais, H.M. Flower, J.G. Zheng, D.P. Davies, Mechanical and microstructural behaviour of a particulate reinforced steel for structural applications. Mater. Sci. Technol. **16**(11–12), 1453–1464 (2000)
42. E. Pagounis, V.K. Lindroos, M. Talvitie, Influence of matrix structure on the abrasion wear resistance and toughness of a hot isostatic pressed white iron matrix composite. Metall. Mater. Trans. A **27**(12), 4183–4191 (1996)
43. E. Pagounis, V.K. Lindroos, Processing and properties of particulate reinforced steel matrix composites. Mater. Sci. Eng, A-Struct. **246**(1–2), 221–234 (1998)
44. R. Lindau, A. Moslang, M. Rieth, M. Klimiankou, E. Materna-Morris, A. Alamo, A.A.F. Tavassoli, C. Cayron, A.M. Lancha, P. Fernandez, N. Baluc, R. Schaublin, E. Diegele, G. Filacchioni, J.W. Rensman, B. van der Schaaf, E. Lucon, W. Dietz, Present development status

of EUROFER and ODS-EUROFER for application in blanket concepts. Fusion Eng. Des. **75–79**, 989–996 (2005)

45. D.H. Bacon, L. Edwards, J.E. Moffatt, M.E. Fitzpatrick, Fatigue and fracture of a 316 stainless steel metal matrix composite reinforced with 25% titanium diboride. Int. J. Fatigue **48**, 39–47 (2013)

46. K. Tanaka, T. Saito, Phase equilibria in TiB2-reinforced high modulus steel. JPE **20**(3), 207–214 (1999)

47. T. Saito, K. Tanaka, High-modulus iron-based alloy with a dispersed boride, Japan (1998), p. 14

48. A. Anal, T.K. Bandyopadhyay, K. Das, Synthesis and characterization of TiB2-reinforced iron-based composites. J. Mater. Process. Technol. **172**(1), 70–76 (2006)

49. C.C. Degnan, P.H. Shipway, A comparison of the reciprocating sliding wear behaviour of steel based metal matrix composites processed from self-propagating high-temperature synthesised Fe–TiC and Fe–TiB2 masteralloys. Wear **252**(9–10), 832–841 (2002)

50. P.H. Booker, *Ternary phase equilibria in the systems Ti-Fe-C, Ti-Co-C and Ti-Ni-c: phase equilibria of the type metal carbonitride + graphite + nitrogen in the systems Ti-C-N, Zr-C-N, and Hf-C-N* (Department of Materials Science, University of Oregon, Oregon Graduate Center, 1979)

51. Y. Feng, *Strengthening of steels by ceramic phases, Fakultät für Georessourcen und Material-tenik* (RWTH Aachen, 2013)

52. H. Ohtani, T. Tanaka, M. Hasebe, T. Nishizawa, Calculation of the Fe-C-Ti ternary phase diagram. Calphad **12**(3), 225–246 (1988)

53. A.K. Shurin, V.E. Panarin, Phase equilibria and structure of Fe-TiB2, Fe-ZrB2 and Fe-HfB2 alloys. Izvestiya Akademii Nauk SSSR **5**, 235–239 (1974)

54. A. Antoni-Zdziobek, M. Gospodinova, F. Bonnet, F. Hodaj, Experimental determination of solid-liquid equilibria with reactive components: example of the Fe-Ti-B ternary system. J. Phase Equilib. Diffus. **35**(6), 701–710 (2014)

55. A. Antoni-Zdziobek, M. Gospodinova, F. Bonnet, F. Hodaj, Solidification paths in the iron-rich part of the Fe–Ti–B ternary system. J. Alloys Compd. **657**, 302–312 (2016)

56. E. Bayraktar, F. Ayari, D. Katundi, J.P. Chevalier, F. Bonnet, *Weldability and Toughness Evaluation of the Ceramic Reinforced Steel Matrix Composites (TIB2-RSMC)* (2011), pp. 85–92

57. F. Bonnet, O. Bouaziz, J.-C. Chevallot, in *Tole d'acier pour la fabrication de structures allegees et procede de fabrication de cette tole*, ed. by A. France, (France, 2008), p. 20

58. F. Bonnet, V. Daeschler, G. Petitgand, High modulus steels: new requirement of automotive market. How to take up challenge? Can. Metall. Q. **53**(3), 243–252 (2014)

59. Z. Hadjem-Hamouche, J.P. Chevalier, Y.T. Cui, F. Bonnet, Deformation behavior and damage evaluation in a new titanium diboride (TiB2) steel-based composite. Steel Res. Int. **83**(6), 538–545 (2012)

60. M. Dammak, I. Ksaeir, O. Brinza, M. Gasperini, Experimental analysis of damage of Fe-TiB2 metal matrix composites under complex loading, 21ème Congrès Français de Mécanique, Bordeaux, 2013

61. S. Lartigue-Korinek, M. Walls, N. Haneche, L. Cha, L. Mazerolles, F. Bonnet, Interfaces and defects in a successfully hot-rolled steel-based composite Fe–TiB2. Acta Mater. **98**, 297–305 (2015)

62. Y.Z. Li, M.X. Huang, *Interfacial Strength Characterization in a High-Modulus Low-Density Steel-Based Fe-TiB2 Composite* (2015), pp. 453–460

63. H. Springer, R. Aparicio Fernandez, M.J. Duarte, A. Kostka, D. Raabe, Microstructure refinement for high modulus in-situ metal matrix composite steels via controlled solidification of the system Fe–TiB2. Acta Mater. **96**(0), 47–56 (2015)

64. R. Aparicio-Fernández, H. Springer, A. Szczepaniak, H. Zhang, D. Raabe, In-situ metal matrix composite steels: effect of alloying and annealing on morphology, structure and mechanical properties of TiB 2 particle containing high modulus steels. Acta Mater. **107**, 38–48 (2016)

65. H. Springer, M. Belde, D. Raabe, Combinatorial design of transitory constitution steels: coupling high strength with inherent formability and weldability through sequenced austenite stability. Mater. Des. **90**, 1100–1109 (2016)
66. C. Baron, H. Springer, D. Raabe, Effects of Mn additions on microstructure and properties of Fe–TiB2 based high modulus steels. Mater. Des. **111**, 185–191 (2016)
67. C. Baron, H. Springer, D. Raabe, Combinatorial screening of the microstructure–property relationships for Fe–B–X stiff, light, strong and ductile steels. Mater. Des. **112**, 131–139 (2016)
68. H. Springer, C. Baron, A. Szczepaniak, V. Uhlenwinkel, D. Raabe, Stiff, light, strong and ductile: nano-structured high modulus steel. Sci. Rep. **7**(1) (2017)
69. A. Szczepaniak, H. Springer, R. Aparicio-Fernández, C. Baron, D. Raabe, Strengthening Fe – TiB2 based high modulus steels by precipitations. Mater. Des. **124**, 183–193 (2017)
70. R. Aparicio-Fernández, A. Szczepaniak, H. Springer, D. Raabe, Crystallisation of amorphous Fe – Ti – B alloys as a design pathway for nano-structured high modulus steels. J. Alloys Compd. **704**, 565–573 (2017)
71. Y.Z. Li, M.X. Huang, Interfacial strength characterization in a high-modulus low-density steel-based Fe-TiB2 Composite, in *Characterization of Minerals, Metals, and Materials 2017*, ed. by S. Ikhmayies, B. Li, J. S. Carpenter, J. Li, J.-Y. Hwang, S. N. Monteiro, D. Firrao, M. Zhang, Z. Peng, J. P. Escobedo-Diaz, C. Bai, Y. E. Kalay, R. Goswami, J. Kim, (Springer International Publishing, Cham, 2017), pp. 453–460
72. C. Baron, H. Springer, D. Raabe, Development of high modulus steels based on the Fe – Cr – B system. Mater. Sci. Eng., A **724**, 142–147 (2018)
73. Z.C. Luo, B.B. He, Y.Z. Li, M.X. Huang, Growth mechanism of primary and eutectic TiB2 particles in a hypereutectic steel matrix composite. Metall. Mater. Trans. A **48**(4), 1981–1989 (2017)
74. C. Baron, On the design of alloys and synthesis for composite steels, Institut für Metallkunde und Metallphysik, RWTH Aachen University, Veröffentlicht auf dem Publikationsserver der RWTH Aachen University; Dissertation, RWTH Aachen University, 2017, 2017
75. A.K. Shurin, G.P. Dmitrieva, Phase diagrams of iron alloys with zirconium and hafnium carbides. Met. Sci. Heat Treat. **16**(8), 665–667 (1974)
76. A.K. Shurin, V.E. Panarin, Phase diagrams for iron with interstitial phases as a basis for developing wear-resistant and eutectic steels. Met. Sci. Heat Treat. **26**(2), 166–169 (1984)
77. R. Rana, C. Liu, Effects of ceramic particles and composition on elastic modulus of low density steels for automotive applications. Can. Metall. Q. **53**(3), 300–316 (2014)
78. H. Springer, *Integrated Alloy and Processing Design of High Modulus Steels, Fakultät für Georessourcen und Materialtechnik* (RWTH Aachen, 2018)
79. F.B. Pickering, *Physical Metallurgy and the Design of Steels* (Applied Science Publishers, London, 1978)
80. L. Cha, S. Lartigue-Korinek, M. Walls, L. Mazerolles, Interface structure and chemistry in a novel steel-based composite Fe–TiB2 obtained by eutectic solidification. Acta Mater. **60**(18), 6382–6389 (2012)
81. R. Valiev, Nanostructuring of metals by severe plastic deformation for advanced properties. Nat. Mater. **3**(8), 511–516 (2004)
82. H. Springer, C. Baron, A. Szczepaniak, E.A. Jägle, M.B. Wilms, A. Weisheit, D. Raabe, Efficient additive manufacturing production of oxide- and nitride-dispersion-strengthened materials through atmospheric reactions in liquid metal deposition. Mater. Des. **111**, 60–69 (2016)
83. J. Marandel, H. Michel, M. Gantois, Coherent precipitation and age hardening mechanisms in ferritic and martensitic alloys. Metall. Trans. A **6**(3), 449 (1975)
84. R. Rana, W. Bleck, S.B. Singh, O.N. Mohanty, Development of high strength interstitial free steel by copper precipitation hardening. Mater. Lett. **61**(14–15), 2919–2922 (2007)
85. Y. Kimura, S. Takaki, Phase transformation mechanism of Fe-Cu alloys. ISIJ Int. **37**(3), 290–295 (1997)

86. G.R. Speich, A.J. Schwoeble, W.C. Leslie, Elastic constants of binary iron-base alloys. Metall. Trans. **3**(8), 2031–2037 (1972)
87. H. Springer, D. Raabe, Rapid alloy prototyping: compositional and thermo-mechanical high throughput bulk combinatorial design of structural materials based on the example of 30Mn–1.2C–xAl triplex steels. Acta Mater. **60**(12), 4950–4959 (2012)
88. Y.X. Gan, D. Solomon, M. Reinbolt, Friction stir processing of particle reinforced composite materials. Materials **3**(1), 329–350 (2010)

Chapter 8
Nanostructured Steels

Rosalia Rementeria, Carlos Capdevila, and Francisca G. Caballero

Abbreviations

APT	Atom probe tomography
ARB	Accumulative roll bonding
CCGT	Combined cycle gas turbine
CVD	Chemical vapor deposition
DED	Direct energy deposition
DLP	Direct in-line patenting
DP	Direct patenting
ECAP	Equal channel angular pressing
ED	Easy drawable
GAR	Grain aspect ratio
GARS	Gas atomization reactive synthesis
HAGB	High-angle grain boundaries
HPT	High-pressure torsion
LAGB	Low-angle grain boundary
LM	Liquid metal
LMP	Larson-Miller parameter
MA	Mechanical alloying
MBE	Molecular beam epitaxy
MF	Multiaxial forging

R. Rementeria (✉)
Additive Manufacturing – New Frontier, ArcelorMittal Global R&D, Avilés (Asturies), Spain
e-mail: rosalia.rementeria@arcelormittal.com

C. Capdevila · F. G. Caballero
Physical Metallurgy Department, National Centre for Metallurgical Research (CENIM-CSIC),
Madrid, Spain
e-mail: ccm@cenim.csic.es; fgc@cenim.csic.es

© Springer Nature Switzerland AG 2021
R. Rana (ed.), *High-Performance Ferrous Alloys*,
https://doi.org/10.1007/978-3-030-53825-5_8

NFA   Nanostructured ferritic alloy
ODS   Oxide dispersion strengthened
PM   Powder metallurgy
PVD   Physical vapor deposition
RFCS   Research fund for coal and steel
SEM   Scanning electron microscopy
SFR   Sodium-cooled fast reactor
SLM   Selective laser melting
SPD   Severe plastic deformation
TE   Tensile elongation
TEM   Transmission electron microscope
TMT   Thermomechanical treatment
UFG   Ultrafine-grained materials
UTS   Ultimate tensile strength
XRD   X-ray diffraction

Symbols

A_{c3} Temperature of the $\gamma/(\alpha + \gamma)$ phase boundary detected upon heating
A_{cm} Temperature of the $\gamma/(\gamma + \theta)$ phase boundary detected upon heating
Ae'_3 Temperature of the $(\alpha + \gamma)/\gamma$ paraequilibrium phase boundary
Ae''_3 Temperature of the $(\alpha + \gamma)/\gamma$ paraequilibrium phase boundary allowing for
             the stored energy in ferrite
B_S Bainite start temperature
d_p Size of precipitates
G Shear modulus of the material
k Constant
$\overline{L}$ Mean lineal intercept as a measure of the effective grain size
M_S Martensite start temperature
N_p Number density of precipitates
P Average die pressure
Q Heat of formation per unit mass
So Interlamellar spacing
t Bainitic ferrite plate thickness
T Temperature
T_e Eutectoid temperature
T_M Melting temperature
T_0 Temperature at which austenite and ferrite of the same chemical composition
             have identical free energy
T'_0 Temperature at which austenite and ferrite of the same chemical composition
             have identical free energy and the strain term for ferrite is incorporated
V_α Volume fraction of ferrite
$V_{\gamma,0}$ Initial volume fraction of retained austenite

V_γ	Volume fraction of retained austenite
$x'_{\gamma\alpha}$	Carbon content of austenite given by $(\alpha + \gamma)/\gamma$ paraequilibrium phase boundary
x'_{T_0}	Carbon content of austenite at the T'_0 curve
α	Ferrite
ΔG	Gibbs free energy
ε	Strain
ε_p	True plastic strain
γ	Austenite
μ	Average coefficient of friction
ρ	Density
σ_ρ	Dislocation forest strengthening
σ_d	Drawing stress
σ_{HP}	Hall-Petch strengthening
σ_p	Nano-oxide precipitation strengthening
σ_Y	Yield strength

8.1 Introduction and Definitions

Since Herbert Gleiter (the founding father of nanotechnology) in the 1980s [170, 190], the field of nanomaterials has flourished over the last three decades due to its scientific and technological importance. Gleiter's basic idea was formulated and explored experimentally in a 1981 paper on "materials with ultra-fine grain sizes" [84], where he announces a new class of materials referred to as "interfacial" or "microcrystalline", not using yet the term "nanocrystalline materials." Nevertheless, Gleiter's description of these new materials meets the definition of nanotechnology owing to nanoscale-dependent material properties. According to his first ideas, the atoms in the boundary or interface region of a nanometer scale domain can adjust their positions in order to increase the strength and decrease the energy of the boundary, irrespective of the usual constraints from the volume or bulk of a material. In addition, Gleiter foresees that the structure and properties of the material having a volume of interfaces comparable or larger than the volume of crystals may be different from the structure and properties of the crystalline state of the same material.

The mechanical and physical properties of steels are determined by several parameters, namely the intrinsic strength of pure annealed iron, solid solution strengthening and various microstructural components including particle or precipitation contributions, dislocation strengthening and grain size effects. Among them, the average grain size of the material generally plays a very significant and, often, dominant, role. The dependence of strength on grain size is expressed in terms of the Hall-Petch equation for equiaxed structures [92, 188], strength depending on $\overline{L}^{-1/2}$, where $\overline{L}$ is the mean lineal intercept as a measure of the *effective grain size*. In the case of lath or plate-shaped grains, such as those of bainite and martensite, strength depends on $\overline{L}^{-1}$ [163, 169]. It was predicted that in nanostructured steels resistance to plastic deformation by dislocation motion would steadily increase as grain size is reduced [82]. This proved to be true, except that nanocrystalline grain sizes of certain metals produce often "negative or inverse Hall-Petch effect", not observed in the particular case of iron-based materials [37, 45].

The term "nanostructured materials", with "nanocrystalline materials" and "nanophase materials" as backups, has become a generic reference to a wide range of grain and precipitate structures. In order to provide a rationalized and simple classification, *nanostructured materials* are here defined to represent cases where the governing lengthscale $\overline{L}$ is below 100 nm, as opposed to *nanocrystalline materials*, which are those with crystallite sizes smaller than 100 nm. Materials whose governing lengthscale lies within the sub-micrometer scale (100 nm–1 μm) are termed *submicron materials*, while more generically, the term *ultrafine-grained materials* (UFG) is used to refer to both nanostructured and submicron materials. In order to qualify as *bulk nanostructured materials*, the condition to fulfill is that they can be manufactured in parts which are *large* in all three dimensions, with uniform properties throughout. The definition of large depends on the eye of the beholder; for this chapter, large is big enough to produce a technologically relevant component.

The techniques used nowadays to produce bulk nanostructured materials are usually divided into two categories, i.e., the *top-down* and the *bottom-up* approaches [255], to which a third *middle-out* approach can be added. In the *top-down* approach, a bulk solid with a relatively coarse grain size is processed to produce an UFG microstructure, often submicron and sometimes nanoscale, through severe plastic deformation. In the *bottom-up* approach, nanostructured materials are fabricated by assembling individual atoms or by consolidating nanoscale solids. The *middle-out* approach does not involve any mechanical operation or chemical reaction; nanostructures are obtained by solid-solid phase transformations through controlled heat treatment.

This chapter provides an overview of the most relevant and promising processing strategies to produce nanostructured steels and the structures and related properties thus obtained. These are nanostructured pearlitic wires obtained by severe plastic deformation, nanostructured ferritic steels produced by mechanical alloying and nanostructured pearlitic and bainitic steels formed by solid reaction.

8.2   Processing and Design of Bulk Nanostructured Steels

8.2.1   Nanostructured Steels Produced by Severe Plastic Deformation

The benefits of heavily deformed steels have been known to humankind for more than three millennia. It is a fact that technological progress in metallurgy has first benefitted the development of weapons, from daggers to cannons, and that this development has been led by the know-how and fighting techniques across the ages and civilizations. The earliest examples of steels subjected to severe plastic deformation can be found in slashing and stabbing weapons, such as swords, glaives, sabers and daggers, discovered in archeological excavations. For instance, the Persian Damascus blades made from wootz steel were known to be the finest weapons produced in Eurasia and were reputed to cut even silk [221]. The exact processing technique of Damascus steel is still shrouded in mystery, but it is thought that small hockey puck-sized high-carbon steel ingots were forge-welded together by hammering at a temperature close to or above Ac_m, producing a flat strip, which is repeatedly folded and re-forged, each fold doubling the number of layers contained through-thickness [258]. Although the final microstructure has been found to contain cementite nanowires encapsulated by carbon nanotubes [196], shown in Fig. 8.1, the structure of the matrix remains micrometric. An overview of the metallurgy and microstructures of antique weapons is given in [58].

Nowadays, hypereutectoid nanopearlitic steel wires produced by drawing have the highest strength of all mass-produced steel materials, reaching tensile strengths above 6 GPa [135]. These are used for a wide variety of applications including, but not limited to, steel cords for reinforcing automobile tires, galvanized wires for suspension bridges and piano wires.

In the case of wire drawing, grain size reduction is accumulated through the reduction in the cross-section area. Alternative processes have been developed in which the overall dimensions of the workpiece remain practically unchanged after each deformation cycle, so that nanostructures are obtained by repetitive cyclic plastic deformation. The term *severe plastic deformation* (SPD) is frequently reserved to refer to processes which do not involve a net change of shape. However, the steel industry is conservative to adapt any of the net-shape SPD processes, primarily due to cost and scalability factors.

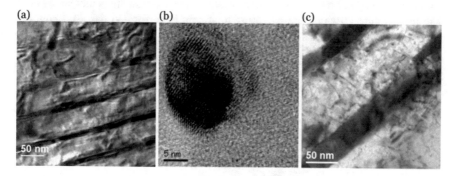

(a)   (b)   (c)

Fig. 8.1 (a) Bright-field transmission electron microscope (TEM) image of cementite nanowires in a Damascus sabre; the dark stripes indicate wires of several hundreds of nanometers in length, (b) high-resolution TEM image of the cross section of a cementite nanowire piercing the image plane and (c) bright-field TEM image of dislocation lines tangled at cementite nanowires. Reprinted with permission from Springer Nature: Springer Physics and Engineering of New Materials, Discovery of Nanotubes in Ancient Damascus Steel, Reibold et al. [196]

8.2.1.1 Heavily Deformed Pearlite Wires

Drawing operations involve pulling a wire, rod or bar (usually having a circular cross-section) through a die or converging channel to decrease its cross-sectional area and increase length. Wire drawing is produced by the combination of a pulling force and a pressure force from the die, which extends the wire and reduces its cross-sectional area while passing through the die, as schematized in Fig. 8.2. The combined effect of this drawing force should be less than the force that would cause the wire to stretch, neck, and break downstream from the die. Also, a too large reduction in cross-sectional area might break the wire. In industrial practice, pulling loads are rarely above 60% of the as-drawn strength, and the area reduction in a single drawing pass is rarely above 30% or 35% and is often much lower [270].

The reason why drawing cannot be achieved by simply stretching the wire with a pulling force is the necking phenomenon. Essentially, after a certain amount of uniform reduction in cross-sectional area, all further elongation concentrates at a single location or neck, which rapidly stretches and breaks. A heavily drawn wire has little or no work-hardening capability and immediately necks when subjected to simple stretching. Die-less drawing systems based on simple stretching are of limited application because of necking vulnerability of heavily drawn steels, and heating of the system is required [109].

In pearlitic steels, the starting microstructure for wire production is critical, and a previous heat treatment, termed patenting, is needed. Patenting consists of heating above Ac_3 or A_{cm}, followed by either continuous cooling or isothermal holding to produce a uniform and fine pearlite microstructure. In this sense, the work hardening rate during drawing and the delamination resistance are enhanced through elimination of upper bainite in the microstructure [171]. Figure 8.3 shows an isothermal transformation diagram for an eutectoid steel and the relevant transformation products as a function of the temperature, where the range to produce the desired fine pearlitic microstructure for wire drawing is indicated [172]. The temperature at which the patenting treatment is performed has a pronounced effect on the interlamellar spacing [248]. Coarse microstructures would give rise to

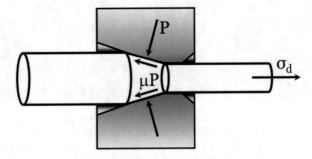

Fig. 8.2 Schematic illustration of the main forces in drawing, P is the average die pressure, μP is average frictional stress, where μ is the average coefficient of friction, and σ_d is the drawing stress

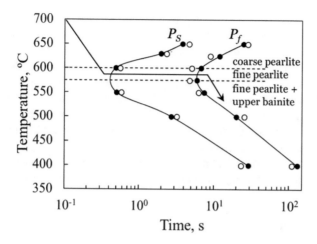

Fig. 8.3 Isothermal time-temperature-transformation diagram of a 0.8C-0.2Si-0.6Mn (wt.%) steel (hollow dots) and a 0.9C-0.2Si-0.3Mn-0.2Cr (wt.%) steel (filled dots) showing the transformation temperature range needed to produce fine pearlite by patenting. (Adapted from [126, 172])

brittleness if the material was left in the heat-treated condition, but this effect is not noticed after a few drawing passes. The patented wire is heavily strained by wet wire drawing to avoid heat generation [86].

Patenting may be applied to hot-rolled rods at the start of wire drawing or to cold-drawn wire as an intermediate heat treatment prior to further wire drawing. Fine pearlite has been traditionally produced in wires by isothermal transformation in molten lead baths (lead patenting). Alternative processing, such as Stelmor® or direct patenting (DP), easy drawable (ED) or direct in-line patenting (DLP) systems, have been developed to produce fine pearlite directly after hot rolling to rod in high-speed bar mills by controlled cooling processes [60, 176].

All carbon steel wire microstructures are prone to both static and dynamic strain aging. Static strain aging refers to the transient stress peaks observed in dilute alloys when a pre-strained specimen is unloaded and aged (even at room temperature) for a time and then reloaded with the same strain rate as in pre-straining. It is commonly accepted that the effect is related to the formation of Cottrell atmospheres around dislocations by diffusing solute atoms ageing [127, 263]. Dynamic strain ageing is the consequence of recurrent pinning of dislocations repeatedly arrested at obstacles to their motion in the process of straining. After a carbon steel wire has been in service or storage for a few months at ambient temperature, it is assumed to be fully strain aged. Static strain ageing may lead to the appearance of yield point phenomena, paneling and discontinuous yielding during further forming operations. Besides, dynamic strain aging is a major consideration during steel wire drawing, since the adiabatic heat produced by plastic deformation and the frictional heat generated between the dies and the wire rod result in deformation temperatures above 200 °C. The temperatures at which dynamic strain aging occurs during steel wire drawing are dependent on both the microstructure and the strain rate [81], and they can be reasonably reached by the wire surface. In practice, the drawing speeds should be selected so as to minimize immediate static strain aging and avoid dynamic strain aging for the bulk of the wire cross section during drawing [126].

It is not surprising that large residual stresses are retained in the wires after heavy cold drawing. These depend strongly on the position in the wire; compressive stresses are found at the center and tensile stresses arise near the surface [210]. In an industrial context, stress relieving of cold drawn steel wires is performed at temperatures as low as 200 °C, in order to maximize strength and minimize stress relaxation. Alternatively, residual stresses can be relieved to a great extent by a stabilization thermomechanical procedure, i.e., by applying a plastic deformation on the wire surface, thus turning the tensile residual stresses into compressive ones [6].

8.2.1.2 Net-Shape Severe Plastic Deformation

Deforming steels to very large plastic strains without introducing any cracks is not a trivial task. Conventional processing of steel, such as forging or rolling, is carried out at relatively high temperatures where large strains are needed. In net-shape SPD processes, compressive hydrostatic stresses are present along with the shear stresses required for plastic deformation. Shear stresses impart the required plastic strains, while hydrostatic stresses prevent the samples from cracking. Given that SPD processes are cyclic, all of them should be such that at the end of any cycle the overall shape remains the same, ensuring that the process can be repeated again, thereby accumulating further plastic strain [116].

The most relevant net-shaped SPD processes for steels are equal channel angular pressing (ECAP) [5, 55, 68, 107], high pressure torsion (HPT) [67, 158, 202, 280] and multiaxial forging (MF) [141, 178, 236, 245, 276]. Steels produced by SPD processing such as ECAP, HPT, or MF processes usually have an average grain size between 0.2 and 0.5 µm, with a highly distorted crystal lattice. However, the crystallite size, or the size of the coherent domains, as determined from X-ray diffraction (XRD) analyses is of about 50 nm. Although not truly nanostructured materials, this technicality lets these SPD materials to be classified as nanocrystalline materials.

Accumulative roll bonding (ARB) [47, 48, 249] is a process suitable for large-scale production which involves the repeated rolling and folding of sheet material in order to accomplish strain increments without thinning the sample entering the rolls. However, the process does not lead to particularly fine grains, which tend to be closer to micrometers than nanometers in size.

Net-shape SPD has not demonstrated to be a relevant technology to produce bulk nanostructured steels and is left out of this chapter. Needless to say, this does not mean that further improvements in the processes, a finer control of the strain and temperature to which steels are subjected, and tailored alloy compositions could not lead to a new class of competitive bulk nanostructured steels. However, scalability and processing costs remain an unresolved issue.

8.2.2 Nanostructured Steels Produced by Mechanical Alloying

An alloy can be created without melting, by violently deforming mixtures of different powders [10, 91, 219]. This technique was developed around 1966 by Benjamin and his co-workers at the Inco Paul D. Merica Research Laboratory as a part of the program to produce oxide dispersion strengthened (ODS) Ni-based superalloys for gas turbine applications [13]. Nowadays, there are two main classes of mechanical alloys which are of commercial significance, the ODS iron-base superalloys and the ODS nickel-base superalloys. They all contain chromium and/or aluminum for corrosion and oxidation resistance, and yttrium or titanium oxides for creep strength. Yttrium oxide cannot be introduced into either iron or nickel by any method other than mechanical alloying; indeed, this was the motivation for the original work by Benjamin [13]. Since the topic of this book is on ferrous materials, the focus is put on iron-base ODS, and, hereafter, the term ODS will refer only to those with iron as solution matrix. Nanostructured ferritic alloys (NFAs) are a subcategory of ODS steels, characterized by ultrafine matrix grain size (200–400 nm) and an extremely high number density ($>10^{24}$ m^{-3}) of nanoclusters/nanoprecipitates (2–4 nm diameter) in the grain interiors and precipitates decorating the grain boundaries [153, 174].

ODS alloys are manufactured by mechanical alloying techniques involving powder metallurgy. This concept is understood to mean the refining of elementary or alloyed metal powders by high-energy milling. This results in alloy powders of extremely fine-grained structure, in which inert oxides, the dispersion particles responsible for boosting the strength of the material, and the alloying elements themselves are introduced uniformly into the microstructure. By means of hot compaction of the mechanically alloyed powder, the fully dense material is derived and this is then worked to create the semi-finished product. Semi-finished ODS superalloys are further processed to finished components by hot-forming methods, machining, or chip-less metal forming. Heat treatment then induces recrystallization, either into a coarse columnar grain structure or into a fine, equiaxed set of grains. The parts are then assembled by mechanical means (bolts or rivets) or by welding or brazing. A scheme of the typical processing route of ODS steels is presented in Fig. 8.4.

There are several potential advantages in employing ODS ferritic steels for high-temperature power plant applications: in addition to the lower raw material cost, the alloys have a higher melting point, lower density, and lower coefficient of thermal expansion than the current nickel- or cobalt-base alloys. However, the mechanical strengths of the alloys in the cast and wrought condition at temperatures in excess of about 600 °C were too low for them to be considered for critical structural applications. Dispersion strengthening with stable oxide particles is an ideal method for improving high temperature strength without sacrificing the excellent surface stability of the matrix alloy. Fe-based ODS alloys, such as MA956 and PM2000, have a composition and microstructure designed to impart creep and oxidation resistance in components operating at temperatures from ~1050 °C to 1200 °C and

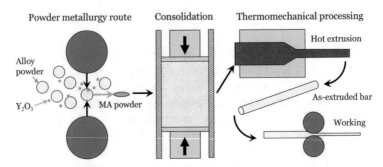

Fig. 8.4 Schematics of the processing route of ODS alloys

Table 8.1 Chemical composition (wt.%) of ODS steels and NFAs

Steel	Cr	Mo	W	Ti	Al	Dispersoid	Ref.
MA956	20	–	–	0.5	4.5	$0.50Y_2O_3$	[42]
MA957	14	0.3	–	1	–	$0.25 Y_2O_3$	[42]
PM2000	19	–	–	0.5	5.5	$0.50 Y_2O_3$	[122]
ODM 751	16.5	1.5		0.6	4.5	$0.50Y_2O_3$	[118]
12Y1	12					$0.25 Y_2O_3$	[1]
12YWT	12	–	2.5	0.4	–	$0.25 Y_2O_3$	[1]
14Y1	14					$0.25 Y_2O_3$	[1]
14YWT	14	–	3	0.4	–	$0.25 Y_2O_3$	[1]
14Cr CEA	14	–	1	0.3	–	$0.25 Y_2O_3$	[63]

above. These alloys achieve their creep resistance from a combination of factors, i.e., the presence of a very coarse, highly textured, high grain aspect ratio (GAR) structure that results from and is sensitive to the alloy thermomechanical processing history [218], and the dispersion of fine scale (20–50 nm diameter) Y_2O_3 particles introduced during mechanical alloying that are highly stable to Ostwald ripening. NFAs typically contain ≥ 12 wt.% Cr along with tungsten and/or molybdenum for solid solution strengthening and ferrite stabilization, as well as small amounts of yttrium, titanium, and oxygen. NFAs are usually designated by their weight percentage of Cr content followed by YWT, as in 12YWT and 14YWT [29, 173]. Table 8.1 collects typical chemical compositions of ODS steels and NFAs.

8.2.2.1 Alternative Routes for Producing ODS Steels

The standard powder metallurgy (PM) route for the fabrication of ODS steels (including NFAs) involves several steps, such as gas atomization to produce a pre-alloyed powder, mechanical alloying (MA) with fine oxide powders, consolidation, and finally thermal/thermomechanical treatment (TMT). This fabrication route is complex and expensive, and the scale up for industrial production is very limited. The suitability of this family of steels for high temperature applications in harsh

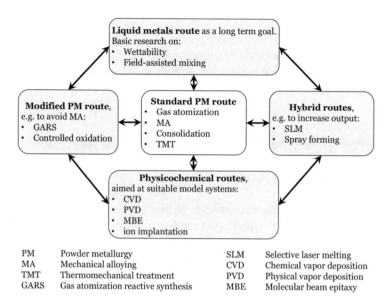

Fig. 8.5 and legend:

PM	Powder metallurgy	SLM	Selective laser melting
MA	Mechanical alloying	CVD	Chemical vapor deposition
TMT	Thermomechanical treatment	PVD	Physical vapor deposition
GARS	Gas atomization reactive synthesis	MBE	Molecular beam epitaxy

Fig. 8.5 Schematic classification of fabrication routes toward ODS steels conceptualized from [14]

environments [4, 88, 146, 177, 194, 235, 252] has led to a growing interest in complementary or alternative fabrication routes able for industrial scale-up offering a reasonable balance of cost, convenience, and properties. Bergner et al. [14] recently reviewed the most promising new routes toward ODS steels. Assuming the conventional PM/AM route as the benchmark for the fabrication of ODS steels, the focus is mainly put in hybrid routes that comprise aspects of both the PM route and more radical liquid metal (LM) routes as promising approaches for larger volumes and higher throughput of fabricated material. A summary of the alternative routes from the above research is shown in Fig. 8.5. The well-established and default processes for the production of ODS steels based on a PM route that depend on an extended MA step to reach an intimate mixture of oxides in a steel matrix requires quite extended alloying times (typically 60 h). The precipitation of the necessary high number density of oxide nanoparticles then occurs during consolidation, heat treatment, and any subsequent thermomechanical processing. In an attempt to avoid the extended and usually costly MA step, and to facilitate greater volumes of ODS material, a range of LM and hybrid routes have been developed nowadays, including ultrasonic dispersion of particles in the melt and in situ oxidation approaches based on atomization, spray forming, and melt spinning. Although none of the processes have reached a maturity where the competitiveness with PM/MA can be assessed, there is sufficient encouragement in early results to pursue alternatively manufactured ODSsteels. Regarding scalability, in terms of consolidation of powders along with hot isostatic pressing plus hot extrusion, spark plasma sintering and other similar field-assisted techniques have now convincingly shown promise for scaling-up to an industrial frame. Given the importance of ODS alloys, particularly to the niche of energy-generating industry, and civil nuclear power in particular, it is necessary that alternative manufacturing techniques for small components, such as additive manufacturing technologies [56], are standardized.

8.2.3 Nanostructured Steels Produced by Solid Reaction

8.2.3.1 Nanostructured Pearlite

Pearlite, as the very first microstructure to be ever observed under a microscope, is probably the best understood among the solid-solid phase transformations in steels. However, quoting Robert's [150] paper on the structure and rate of formation of pearlite [150, 151], "no one could readily believe that the subject is exhausted, and there can be no doubt but that it will continue to interest the investigator for many years to come."

The course of the isothermal transformation of a steel after austenitization and further quenching to a temperature level above the martensite start temperature is conveniently represented by the so-called S curve or time-temperature-transformation (TTT) diagram. Figure 8.3 schematizes transformations taking place above the martensite start temperature (M_S) in low alloy steels, where two transformation products are distinguished: pearlite forming above the knee of the S curve and bainite forming below. It is well-known that the interlamellar spacing of pearlite decreases as the reaction temperature is lowered [184]. Back in the beginning of the twentieth century, when the resolution power of microscopes was limited, transformation products at large undercoolings near the knee of the S curve were not resolvable. Metallurgists thought that the pearlite nodules thus produced were indeed lamellar pearlite, but of an interlamellar spacing too small for the resolution of the metallurgical microscope [123, 124]. Zener's simple but yet useful thermodynamic approach can be used to illustrate the relationship between the interlamellar spacing and the transformation temperature in pearlitic steels. Under Zener's treatment, a pearlite nodule of interlamellar spacing So advances into the parent austenite, as illustrated in Fig. 8.6. As the nodule grows, the free energy remains unchanged in a region that includes one cementite and one ferrite plate (or two halves of ferrite plates at both sides of the cementite plate) with an interlamellar distance of S_0 and depth of W, as indicated by dotted lines in Fig. 8.6. As the nodule advances a distance dx, the volume of austenite transformed in the region under consideration is $S_0 W dx$, and the mass of the austenite transformed is $\rho S_0 W dx$, where ρ is the density. The free energy that is available at temperature T for the formation of new interfaces is given by

$$\Delta G = Q \frac{T_e - T}{T} \rho S_0 W dx \tag{8.1}$$

Fig. 8.6 Illustration of an advancing pearlite nodule

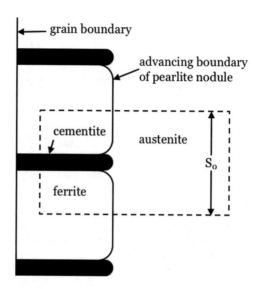

where Q is the heat of formation of pearlite per unit mass and T_e is the eutectoid temperature. Growth of the lamellae is possible only if the increase in the surface energy is outweighed by the increase of free energy resulting from the transformation. The increase in the total interface energy is $2SWdx$, where S is the surface energy per unit area. Equating the available free energy to the increase in interface energy, one obtains an expression for the interlamellar spacing as a function of the transformation temperature:

$$S_0 = \frac{2T_e S}{\rho Q \left(T_e - T\right)} \tag{8.2}$$

Thus, the interlamellar spacing is inversely proportional to the degree of under-cooling and the heat of formation of pearlite per unit mass, which of course is a function of the composition.

In this regard, recent work was devoted to assess whether a nanostructured pearlite could be obtained by increasing the driving force for pearlite transformation using tailored compositions [271, 272]. Cobalt has been long known to decrease the interlamellar spacing by increasing the driving force for the formation of pearlite [97], and the same effect is observed through aluminum additions [112, 137]. The interlamellar spacing that can be obtained in a eutectoid pearlitic steel with cobalt and aluminum additions by relatively slow continuous cooling at 0.1 °C/s can be as small as 50 nm [271, 272]. It could be speculated that tailored pearlite isothermal transformation at low temperatures would lead to even finer interlamellar spacings.

In general terms, the effect of alloying elements upon pearlite transformation is two-fold: (1) they modify the eutectoid transformation temperature, thereby changing the degree of undercooling at a fixed transformation temperature, and

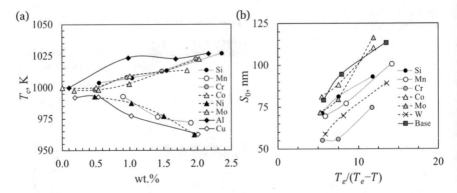

Fig. 8.7 (a) Effect of alloying elements on the eutectoid transformation temperature, T_e and (b) relationship between the reciprocal of undercooling, $T_e/(T_e - T)$ and interlamellar spacing, S_0, in a 0.8C-0.3Si-0.5Mn (wt.%) steel with different alloying additions of 1 wt.%, except for Mo which is 0.5 wt.%. (Data from [242])

(2) they change the chemical driving force for pearlite formation at a fixed degree of undercooling. For instance, chromium refines the interlamellar spacing at a fixed pearlite transformation temperature by increasing the undercooling, while at the same undercooling, the lamellar spacing is refined by increasing the chemical driving force [104].

Figure 8.7a shows the effect of alloying element additions in a 0.8C-0.3Si-0.5Mn (wt.%) base steel on the eutectoid transformation temperature [242]. It is shown that ferrite stabilizers, silicon, chromium, molybdenum, and aluminum, increase the eutectoid transformation temperature whereas austenite stabilizers, manganese, nickel, and copper, decrease it. At the same time, Figure 8.7b shows the relationship between the reciprocal of undercooling, $T_e/(T_e - T)$ and interlamellar spacing in a 0.8C-0.3Si-0.5Mn (wt.%) steel with 1 wt.% additions of the aforementioned alloying elements (except for Mo, where the addition is of the 0.5 wt.%). In all cases, the interlamellar spacing decreases with the undercooling and reaches values below 100 nm for reciprocal undercoolings below 10, qualifying all these pearlitic structures as nanoscaled. It should be noted that the literature dedicated to basic research on pearlite transformation does not pay much attention to the nanoscaled nature of the structures formed at large undercoolings in alloyed steels, but several examples can be found if the results are carefully examined [104, 137, 193, 261].

Besides, in hypereutectoid pearlitic steels, a detrimental continuous or semi-continuous cementite on the prior austenite grain boundaries may develop during cooling. To avoid this, the steel is alloyed with silicon and vanadium; silicon suppresses grain boundary cementite formation and microadditions of vanadium assists in the suppression process by forming submicron vanadium carbide particles taking the excess carbon from the austenite grain boundary areas [26].

The nanostructure in pearlitic steels is usually achieved at the end of the hot-rolling step by controlled cooling to the desired transformation temperatures. The heat treating schedule presented in Fig. 8.3 to produce the pearlite structure during patenting can be taken as a reference frame for the production of nanostructured pearlite following large undercoolings. However, forced cooling followed by isothermal treatments is not a common practice in long products, where these steels have their most relevant applicability. Instead, controlled continuous cooling is applied so that transformations occur in a temperature range where the interlamellar spacing is kept to a minimum. Further details will be given in Sect. 5.3.

8.2.3.2 Nanostructured Bainite

As for the case of pearlite, the general trend in bainitic steels is that the plate thickness decreases when the transformation temperature is decreased [220]. The thickness of the bainitic ferrite plates depends primarily on the "strength" of the austenite at the transformation temperature, and the driving force for the transformation, when these variables are treated independently. Strong austenite offers more resistance to interface motion (growth), while a large driving force increases the nucleation rate, both leading to microstructural refinement [46, 220]. The effect of the temperature is implicitly included in both terms, lower temperatures providing austenite strength and increased chemical driving forces. The theory to estimate the bainite and martensite start temperatures, B_S and M_S, respectively [16, 80, 117], was used in the early 2000s to estimate the lowest temperature at which bainite can form.

Figure 8.8a shows the calculated B_S and M_S temperatures whereas Fig. 8.8b shows the time required for bainite transformation to start in a Fe-2Si-3Mn (wt.%) steel as a function of the carbon content [72, 73]. There seems to be no lower limit to the temperature at which bainite can be generated as long as the carbon concentration is increased. In parallel, the rate at which bainite forms slows down

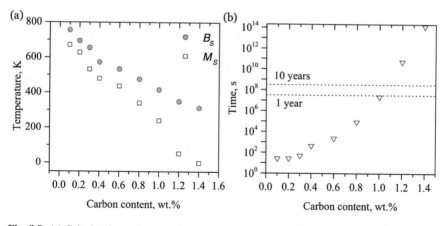

Fig. 8.8 (a) Calculated transformation start temperatures in a Fe-2Si-3Mn (wt.%) steel as a function of the carbon concentration, and (b) calculated time required to initiate bainite transformation after [72, 73]

exponentially as the transformation temperature is reduced. It may take hundreds or thousands of years to generate bainite at room temperature.[1] For practical purposes, transformation times are nowadays kept up to a maximum of days, corresponding to transformation temperatures between 200 and 350 °C and carbon concentrations from 0.7 to 1.0 wt.% with silicon additions from 1.5 to 3.0 wt.%. The presence of other alloying elements is adjusted to allow hardenability without compromising transformation kinetics. Silicon, as a carbide inhibitor, is added to avoid the precipitation of cementite from austenite during isothermal bainitic transformation. The carbon that is rejected from the bainitic ferrite enriches the residual austenite, thereby stabilizing it (partially or totally) at ambient temperature. However, carbide inhibitors are capable of preventing the precipitation of cementite between the subunits of bainitic ferrite, but they cannot avoid the precipitation of carbides within the ferrite plates in lower bainite [125]. Indeed, during bainite transformation, there are one or more transient intermediate states that have a short lifetime [149, 269], and the relevant precipitation reactions can be best studied in silicon steels since this element retards the precipitation of cementite without influencing the formation of transient carbides [98]. Before nanostructured bainite arose as a metallurgical concept, Sandvik [207, 208] studied the evolution of the bainitic structures formed in high-carbon high-silicon steels transformed at temperatures between 290 and 380 °C, taking advantage of the slow reaction kinetics that permitted a detailed tracking of the different reaction products.

The overall thermodynamics of the diffusionless growth of bainitic ferrite are represented by the T_0 curve, which is the locus of all points on a temperature versus carbon concentration plot, where austenite and ferrite of the same composition have the same free energy, $\Delta G^{\gamma \to \alpha} = 0$ [16, 282]. In Fig. 8.9a, growth without diffusion can only occur when the free energy of ferrite becomes less than that of austenite of the same composition, i.e., when the concentration of austenite lies to the left of the intersection between the two Gibbs free energy curves. Assuming that a plate of bainitic ferrite forms without diffusion, the excess carbon would afterward be rejected into the surrounding austenite, and the presence of silicon will avoid any precipitation from the carbon-enriched austenite. The next plate then has to grow from carbon-enriched austenite in a process that would stop when the austenite carbon concentration reaches the T_0 curve. This effect is known as the *incomplete reaction phenomenon* [98] since austenite does not reach its equilibrium composition, $x'_{\gamma\alpha}$ given by the Ae'_3 curve, i.e., the $(\alpha + \gamma)/\gamma$ paraequilibrium phase boundary. Considering the stored energy associated with bainite transformation ($\Delta G_s = 400$ J/mol^{-1}) in the Gibbs free energy of ferrite, the condition $\Delta G^{\gamma \to \alpha} = 0$ is fulfilled at lower carbon content values than in the latter case, as depicted in Figure

---

[1]To test this theory, two samples of a steel that would take 100 years to transform to bainite at room temperature were manufactured in 2004, one of which can be seen at the Science Museum of London while the other one is stored at Sir Harry Bhadeshia's office in Cambridge University. The samples, sealed in an inert atmosphere, are mirror-polished so that phase changes will be evident through the surface rumples caused by the shear transformation, hopefully in 2104.

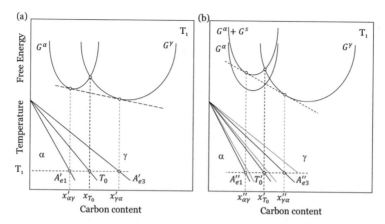

Fig. 8.9 Schematic illustration of the origin of the (**a**) T_0 and (**b**) T'_0 curves on the phase diagram. The T'_0 curve incorporates the strain energy term for ferrite, illustrated on the diagram by raising the free energy curve of ferrite by the term ΔGs. Ae'_1 and Ae'_3 refer to the paraequilibrium phase boundaries $\alpha/(\alpha + \gamma)$ and $(\alpha + \gamma)/\gamma$, respectively. Ae''_1 and Ae''_3 are the corresponding phase boundaries allowing for the stored energy

8.9b. Thus, the T'_0 and Ae''_3 curves are obtained when the T_0 and Ae'_3 curves are modified to account for the stored energy of transformation, respectively [20]. Under these circumstances, the process will cease when the austenite carbon concentration reaches the T'_0 curve (x'_{T_0}).

Therefore, nanostructured bainite, also termed nanobainite, superbainite, or low-temperature bainite, is the reaction product of the austenite obtained in high-carbon, high-silicon steels after isothermal transformation at temperatures below 350 °C. It is a form of lower bainite consisting of a mixture of bainitic ferrite plates with nanoscale precipitates and retained austenite. In practice, the nanostructures are obtained outside a continuous production line through proper heat treating, usually requiring the use of salt baths, with the exception of smaller dimensions components amenable to dry heat treatments. The temperature control during the isothermal step is key to obtain the structures, given that transformation times grow exponentially with decreasing the transformation temperature. Deviations from the target temperature to lower temperatures may lead to too short times available for the steel to be fully bainitic, with large portions of detrimental untransformed retained austenite. The block morphology of untransformed austenite is deleterious to toughness when it transforms into martensite at an early stage of deformation.

8.3   Microstructure Description at the Multiscale

8.3.1   Nanostructures in Steels Produced by Wire Drawing

The carbon content in cold-drawn pearlitic steel wires is in the range of 0.75
and 1.00 wt.%, with small additions of other alloying elements, and little or no
addition of carbide formers, i.e., they are fully pearlitic after proper patenting
treatment. The initial material usually consists of a patented wire with a diameter of
1.70 mm, as described in [241]. Figure 8.10 shows the cross-section microstructure
of a 0.82 wt.% C patented steel wire [103]. Figure 8.10a presents a bright-
field transmission electron microscope (TEM) image containing two micron-sized
pearlite colonies, with a uniform interlamellar spacing within each colony (about
130 nm). The bright-field TEM micrograph in Fig. 8.10b was taken from the
cementite reflection encircled in Fig. 8.10a and shows planar defects (indicated
by arrows) that can be attributed to stacking faults or slip traces resulting from
internal strain [110, 214]. Both the patenting temperature and the composition of
the steel have a remarkable effect on the interlamellar spacing, the interlamellar
spacing decreasing with the increase in the carbon composition of the steel and the
decrease in the patenting temperature [171, 248].

Successive cold drawing reduces the width of both ferrite and cementite lamellae
due to increasing strain and also aligns the originally randomly oriented lamellae belonging to different colonies parallel to the wire axis. In pearlite wires,
the microstructural architecture is not changed during moderate deformations, but
soaring strains cause morphological changes, namely fiber curling [105, 238, 241].
Figure 8.11 presents the cross-section microstructure of a 0.82 wt.% C patented
steel wire drawn to a true strain of 4.22 [103]. Figure 8.11a shows a bright-field
TEM micrograph where the pearlite lamellae exhibit bending around the wire axis

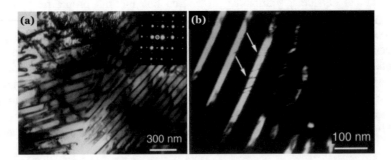

Fig. 8.10 Cross-sectional (**a**) bright-field and (**b**) dark-field TEM images of patented 0.82C-
0.5Mn-0.25Si (wt.%) pearlitic steel wire. The beam direction is close to the [001] zone of
ferrite. The image in (**b**) was obtained using the cementite reflection circled in (**a**). Reprinted by
permission from Springer Nature: Springer Metals and Materials Transactions A, Atom Probe and
Transmission Electron Microscopy Investigations of Heavily Drawn Pearlitic Steel Wire, Hong et
al. [103]

(a)   (b)   (c)

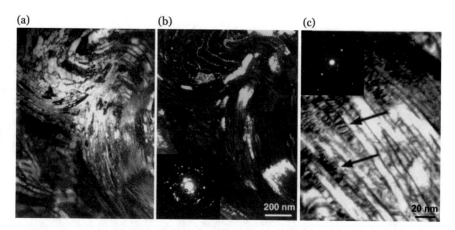

Fig. 8.11 Cross-sectional TEM images of a 0.82C-0.5Mn-0.25Si (wt.%) pearlitic steel wire drawn to a true strain of 4.22 (**a**) bright field and (**b**) corresponding dark field where the cementite reflection used is encircled in the selected-area diffraction pattern (inset), and (**c**) bright-field image showing the contrast arising from regions of local strain (arrowed) found throughout the ferrite lamellae. Reprinted by permission from Springer Nature: Springer Metals and Materials Transactions A, Atom Probe and Transmission Electron Microscopy Investigations of Heavily Drawn Pearlitic Steel Wire, Hong et al. [103]

with a curled structure, tending to fragment into a ribbon-like morphology. The interlamellar spacing varies significantly from one ferrite to another, and it appears that adjacent ferrite ribbons have different orientations, given the darker or lighter contrast of the ferrite. Figure 8.11b shows a dark-field TEM micrograph taken from the cementite reflection of the inset. After heavy straining, cementite lamellae are fragmented into small grains, and individual dislocations, as shown in Fig. 8.10b, are no longer resolvable. The bright-field TEM micrograph in Fig. 8.11b reveals strain contrast throughout the ferrite lamellae, which is thought to arise from highly-dislocated regions. Furthermore, selected-area diffraction patterns in the insets of Fig. 8.11b and c include both ferrite reflections and rings corresponding to cementite interplanar spacings, the latter resulting from randomly oriented fragmented cementite grains, not revealed in the dark-field in Fig. 8.11b.

Figure 8.12 shows three-dimensional carbon atom maps of a 0.98C-0.31Mn-0.20Si-0.20Cr (wt.%) pearlitic steel wire in longitudinal (left) and cross-sectional (right) views relative to the drawing direction in samples drawn to low (1.96), medium (4.19), and extremely high strains (6.52) [136]. The 7 at.% C green isosurfaces are drawn to separate the carbon-enriched regions identified as cementite from the carbon-depleted ferrite regions. At low drawing strains, the lamellae align in parallel to the drawing directions without a significant change in the morphology but with certain fragmentation of the cementite lamellae. As the drawing strain increases, the volume fraction of cementite continuously decreases by mechanically driven chemical decomposition thereby releasing carbon into the ferrite matrix, which gets carbon-supersaturated [79, 94, 110, 111, 130, 157, 211, 214]. Successive

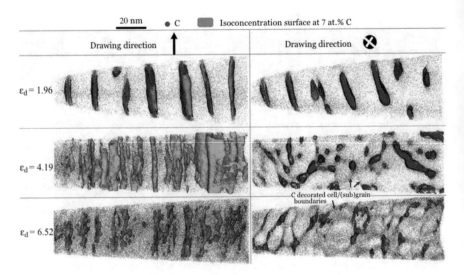

Fig. 8.12 Three-dimensional carbon atom maps with the 7 at.% C isosurfaces superimposed in both longitudinal (parallel to the drawing direction, left) and cross-section (perpendicular to the drawing direction, right) of a 0.98C-0.31Mn-0.20Si-0.20Cr (wt.%) pearlitic steel wire cold drawn to different drawing strains. Blue arrows indicate some of the subgrain boundaries decorated with carbon atoms. ε_d is the true drawing strain. Reprinted by permission from American Physics Society: American Physics Society and Physical Review Letters, Segregation Stabilizes Nanocrystalline Bulk Steel with Near Theoretical Strength, Li et al. [136]

cold-drawing results in a reduction of the cementite volume fraction to values of about 6 vol.% or less, with a concomitant decrease of its carbon content from 25 at.% (6.67 wt.%) to values around 12.5 at.% (3.0 wt.%), depending on the thickness of the individual lamellae [23]. At extremely high strains, the lamellar structure evolves into a nearly equiaxed ferrite subgrain structure with carbon atoms segregating at the boundaries (indicated by the blue arrows).

The stability of the nanoscaled multiphase structure can be easily tested by annealing treatments at different temperatures. Figure 8.13 shows the effect of various annealing treatments on the microstructure of a 0.82C-0.5Mn-0.25Si (wt.%) pearlitic steel wire drawn to a true strain of 4.22 in longitudinal view [103]. The microstructure of the as-drawn wire in Fig. 8.13a shows little or no difference with the microstructure after annealing at 200 °C for 1 h in Fig. 8.13b. Indeed, the strain contrast in ferrite remains after annealing at 200 °C, while the cementite preserves the fragmented nanocrystalline structure. Annealing at higher temperatures leads to significant microstructural changes. After annealing at 400 °C for 1 h (Fig. 8.13c), the interlamellar spacing is coarser than in the as-drawn state, and cementite grains of about 30 nm and above are observed all throughout the microstructure, even within the ferrite lamellae. The lamellar structure is lost after annealing at 500 °C for 1 h, originated by spheroidization of cementite and recovery and recrystallization of ferrite, as Fig. 8.13d illustrates. In general, there are no substantial changes in the

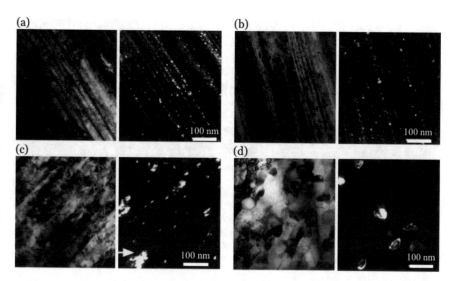

Fig. 8.13 The effect of annealing on the microstructure of a 0.82C-0.5Mn-0.25Si pearlitic steel wire drawn to a true strain of 4.22. Bright-field/dark-field TEM image pairs are taken perpendicular to the wire axis (longitudinal view): (**a**) as-drawn wire; (**b**) annealed at 200 °C for 1 h; (**c**) annealed at 400 °C for 1 h; and (**d**) annealed at 500 °C for 1 h. Reprinted by permission from Springer Nature: Springer Metals and Materials Transactions A, Atom Probe and Transmission Electron Microscopy Investigations of Heavily Drawn Pearlitic Steel Wire, Hong et al. [103]

nanostructures after annealing at temperatures below 250 °C, whereas annealing at higher temperatures results in time-dependent spheroidization of both ferrite and cementite lamellae [23, 101, 103, 135, 156, 182, 239].

8.3.2 Nanostructures in Steels Produced by Mechanical Alloying

8.3.2.1 Structure of the Matrix in NFAs

Immediately after the mechanical alloying process, the powders have a grain size that can be as fine as 1–2 nm locally [113]. This is a consequence of the extent of the deformation during mechanical alloying, with true strains of the order of 9, equivalent to stretching a unit length by a factor of 8000. The consolidation process involves hot extrusion and rolling at temperatures of about 1000 °C, which leads to microstructure of as-extruded NFAs consisting on elongated grains along the extrusion direction with a high grain aspect ratio.

Figure 8.14 shows the grain structure in a 14 wt.% Cr ODS steel hot extruded into bars at 1100 °C, in both the longitudinal and transverse directions [39]. The grain structure in the longitudinal direction is composed of large elongated grains,

Fig. 8.14 TEM bright-field micrographs from as-received ODS Fe-14 wt.% Cr steel (ferritic). (**a**) Longitudinal direction revealing elongated as well as equiaxed grains. (**b**) Transverse direction revealing small equiaxed grains. In both directions, regions with different dislocation density are apparent. Reprinted with permission from Elsevier: Journal of Nuclear Materials, Microstructure characterization and strengthening mechanisms of oxide dispersion strengthened (ODS) 9 wt.% Cr and 14 wt.% Cr extruded bars, Chauhan et al. [39]

together with regions of small equiaxed grains, while in the transverse direction the grains are smaller and equiaxed. The average dislocation density is about 5×10^{14} m^{-2}, with the presence of regions with different dislocation density.

The typical grain size in NFAs is below 300 nm, usually showing a bimodal distribution of the grain size in different regions of the transverse direction. This bimodal grain structure is also observed in conventional ODS alloys and is likely to be due to differences in the amount of milling locally imparted during mechanical alloying and the resulting grain refinement [39, 154, 155, 174].

It is known that during the course of consolidation, the material may dynamically recrystallize several times [35, 36, 93]. It should be emphasized that the submicron grains are not low-misorientation cell structures, but true grains with large relative misorientations [17].

The resulting crystallographic texture in ODS steels and NFAs is a consequence of the manufacturing route [8, 38, 54, 145, 179]. For instance, the PM2000 alloy supplied in the form of a hot-rolled tube (finish rolling temperature of ~1050 °C) of 100 mm diameter and 7.9 mm thickness air cooled to room temperature shows an incomplete α-fiber texture (RD∥<110>) with a dominant {001}<110> component, i.e., the {100} crystallographic planes parallel to the tube surface and the <110> crystallographic directions parallel to the rolling direction. On the other hand, the MA956 alloy having a similar composition to that of the PM2000 alloy, but supplied in the form of a hot-extruded bar (extrusion temperature of ~1050 °C) of 60 mm diameter air cooled to room temperature, presents an α-fiber (ED∥ < 110>) texture with a strong {111}<100 > γ-fiber component. Figure 8.15a and b show the orientation distribution function (ODF) at $\varphi_2 = 45°$ for as-hot-rolled PM2000 alloy and as-hot-extruded MA956 alloy, respectively. To ease the interpretation, Fig. 8.15c shows the ODF section with $\varphi_2 = 45°$ indicating the position of the major components of texture in ODS steels.

Upon subsequent cold-rolling of ODS steels and NFAs, the crystallographic texture produced consists of both α-fiber components such as {001}<110> and

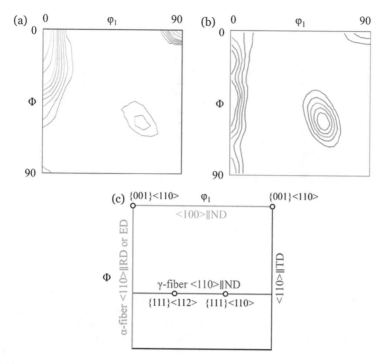

Fig. 8.15 Orientation distribution function (ODF) at $\varphi_2 = 45°$ for (**a**) as-hot-rolled PM2000 alloy, (**b**) as-hot-extruded MA956 alloy, and (**c**) the major components of texture in ODS steels. (Adapted from [38])

{112}<110> and γ-fiber components such as {111}<110> and {111}<112> [9, 132, 133, 216], depending on the cold rolling direction relative to the original hot-rolling or -extruding direction. In this sense, investigations on cold-rolling processes have been intended to change the original texture after fabrication and its effect upon recrystallization.

8.3.2.2 Structure of the Oxides in NFAs

During intensive milling of the system, the powder particles get work-hardened and their grain structures are refined. At the same time, yttrium and titanium atoms are forced into the matrix, either forming solid solution or amorphous sub-nanometric fragments [100]. The lattice distortions are intensified by the addition of large solute elements (yttrium, titanium) at substitutional positions and oxygen atoms at interstitial sites, which can promote the build-up of a large dislocation density by reducing the level of dynamic recovery. Since the solubility of both titanium and yttrium in alpha iron is low, the driving force for the precipitation of titanium and yttrium oxides in the matrix is very high. Precipitation being a

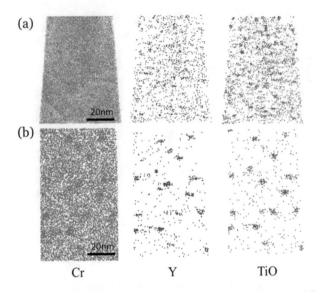

Fig. 8.16 Five-nanometer-thick slice of an atom map showing the distribution of Cr, Y, and Ti-O ions in a 14Cr–2 W–0.3Ti–0.3Y$_2$O$_3$ (nominal, wt.%) ferritic steel (**a**) powder immediately after MA and (**b**) after degassing; HIP consolidation at 1150 °C for 4 h and further annealing at 1150 °C for 100 h. Reprinted with permission from Elsevier: Acta Materialia, The formation and evolution of oxide particles in oxide-dispersion-strengthened ferritic steels during processing, Williams et al. [268]

diffusion-activated mechanism, it is assumed that they form only during the high-temperature consolidation process [24]. Yttrium is added under the form of yttrium oxides (Y$_2$O$_3$), and the oxygen content is, therefore, directly related to the content of these oxides. In addition to this direct oxygen source, there is a second oxygen source referred to as "excess oxygen" resulting from contamination during milling, although the oxygen quantity brought by this second source is far from negligible [268].

Clustering of yttrium and titanium is observed in the powder after a certain milling time [268], as shown in Fig. 8.16a. Perfect solid solution with a random distribution of solute atoms is difficult to achieve, especially for elements with low solubility in bcc iron. However, there is general agreement with the fact that these sub-nanometric clusters are homogeneously distributed in the powder. Also, after annealing or consolidation at high temperature, the clusters tend to crystallize and form nano-oxide precipitates [271, 272]. The number densities and volume fractions of the nano-oxides decrease and their radii increase with increasing consolidation temperature [2].

The importance of dissolved titanium relies on the refinement of the Y$_2$O$_3$ precipitates [250] due to the reactions leading to the formation of complex oxides having typical sizes of a few nanometers [175, 251, 275]. The stoichiometry and crystal structure of the oxide nanoparticles have been long debated, partly due to

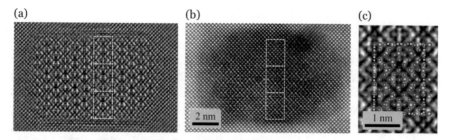

Fig. 8.17 (**a**) Phase of the reconstructed exit wave from focal series (under, near, and over focus) in a high-resolution TEM image of a large nano-oxide, (**b**) corresponding high-angle annular dark-field (HAADF) scanning transmission electron microscope (STEM) image where the dashed lines highlight the periodicity of the Moiré pattern in the overlap region, (**c**) magnified views of the periodic structure in the exit-wave reconstruction image. The dashed lines show the periodic repeated pattern of 5 × 7 Fe unit cells, while the colored balls represent lateral positions of Y (green) and Ti (blue) columns relative to the Fe (yellow) matrix. Reprinted with permission from Elsevier: Acta Materialia, The crystal structure, orientation relationships and interfaces of the nanoscale oxides in nanostructured ferritic alloys, Wu et al. [273]

the use of different techniques, each of them having different limitations: XRD, TEM, atom probe tomography (APT), and small-angle neutron and X-ray scattering (SANS and SAXS) [173, 175]. Nowadays, there is general agreement that the smallest cubic and cuboidal oxide nanoparticles are the face-centered cubic (fcc) $Y_2Ti_2O_7$ pyrochlore phase [174], while larger oxides in NFAs are observed to be Y_2TiO_5 oxides [271, 272]. As an example, Figure 8.17a and b show a periodic image pattern in the central region of a small nano-oxide, reflecting the overlap the $Y_2Ti_2O_7$ pyrochlore and the ferrite matrix crystal structures formed in the MA957 NFA. Figure 8.17c shows the corresponding overlays of an ideal $Y_2Ti_2O_7$ pyrochlore atomic model with the observed orientation relationship. The dashed lines correspond to a periodic array of 5 × 7 Fe periodic cells, while the colored balls represent lateral positions of Y (green) and Ti (blue) atoms relative to the Fe matrix (yellow), based on a visual best fit position adjustment of the image intensities [273]. Thus, the smallest nano-oxides have a coherent cube-on-edge interface, where the oxides are under compressive stress while the matrix is under tension [273]. The interface of larger oxides is semicoherent [274], with misfit dislocation structures in the ferrite matrix. At the same time, there is evidence of oxide nanoparticles exhibiting core-shell structures with oxygen, chromium and/or titanium enrichments at the interface [140]. The relevance of the oxide-ferrite interface in NFAs relies on its ability to trap helium.

Besides, the $Y_2Ti_2O_7$ have a remarkable thermal stability, remaining essentially stable for thousands of hours at temperatures below 900 °C [50, 175]. Such stability derives from the low solubility of yttrium in local equilibrium with $Y_2Ti_2O_7$ [50]. At higher temperatures, the coarsening mechanism is pipe diffusion along dislocations, where the yttrium is more soluble than in the matrix and migrates more rapidly. In parallel, grain sizes and dislocation densities are stable at the same temperatures, at least for some tens of thousands of hours [49, 50].

8.3.2.3 Recrystallization Behavior of NFAs

Recrystallization in ODS steels occurs at exceptionally high homologous temperatures, of the order of 0.9 of the melting temperature (T_M in Kelvin). This contrasts with ordinary cold-deformed metals that recrystallize readily at about 0.6 T_M. The reason for such intriguing behavior remains unclear; some authors have speculated [114] that recrystallization occurs when the grain boundaries overcome solute drag and the mobility rises suddenly at high temperatures. This is inconsistent with some experimental evidences that demonstrate that the recrystallization temperature can be reduced by many hundreds of Kelvin by a slight additional inhomogeneous deformation [41, 133, 195]. Conversely, other authors suggested that the fine particles of yttrium oxide may offer a hard pinning for moving boundaries during recrystallization but this does not explain the reason for the enormous limiting grain size following recrystallization. In any case, recrystallization is found to be insensitive to the overall pinning force [165].

The recrystallization process of ODS steels and NFAs consists of two well-defined stages: extended recovery and abnormal grain growth [24, 189]. During the whole recrystallization process the submicron elongated grains along the rolling direction in the as-rolled microstructure evolve to assemble a coarse microstructure with the millimeter-sized grains presenting a preferential <112> orientation parallel to the rolling direction. At the first stage, the alloy undergoes an extended recovery process characterized by a geometrical change in the grain morphology from an elongated to an equiaxed structure. Here, new grains are not nucleated, and no significant change in material texture is observed [189], although there is a strengthening in the texture component towards the (001)<110> (lower Taylor factor) due to grain rotation driven by dislocation glide.

The uniform microstructure with the α-fiber texture component prior to coarse grain microstructure is strengthened by the increase of low-angle grain boundaries (LAGBs) due to subgrain rotation driven by dislocation glide. This promotes the orientation pinning mechanism of high-angle grain boundaries (HAGBs), which require an enormous energy to unpin the grain boundaries. This fact explains the reason for recrystallization (generation of coarse grained microstructure) requiring elevated temperatures of 0.9 of the absolute melting temperature. The orientation pinning term was used for the first time by Juul Jensen et al. [115] to describe the fact that the grain boundary mobility depends on the orientation relationship between the growing grains and the surrounding matrix material. If the growing grain is separated from the matrix by a HAGB, and the misorientation between this grain and the matrix is reduced because of the grain rotation, the HAGB evolves to a LAGB which presents a substantially reduced mobility. Therefore, the growth of the grain is stopped after a certain time. This is consistent with the increasing number density of LAGBs with annealing time detected during the recrystallization of PM2000 [35]. Besides, the increased number of LAGBs is also responsible for the pinning of the HAGBs between grains with a strong texture [115].

In this context, weak texture grains of <112> present a substantially higher mobility than those of the matrix with strong texture (RD || <110>). Therefore,

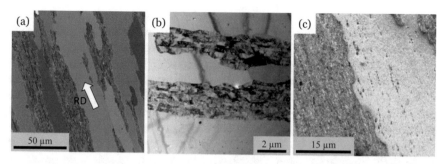

Fig. 8.18 Microstructure of the 14YWT NFA after 60% compression and further annealing at 1100 °C for 7.5 h showing recrystallized, abnormally grown, and unrecrystallized regions. (**a**) Electron backscatter diffraction (EBSD) orientation distribution (inverse pole figure) map, where the arrow indicates the rolling direction (RD) perpendicular to the compression direction and the low-angle grain boundaries having misorientation of 0.5°–2° are the red lines, (**b**) bright-field TEM image, and (**c**) EBSD band contrast gray scale map showing the low-angle grain boundaries having misorientation of 0.5°–2° (pink lines) in abnormally grown and recovered and unrecrystallized grains. Reprinted with permission from Elsevier: Acta Materialia, High temperature microstructural stability and recrystallization mechanisms in 14YWT alloys, Aydogan et al. [9]

the <112> oriented grains, present in the as-hot-rolled microstructure with their associated relatively high-mobility HAGBs, provide the seeds for the nucleation of abnormal grain growth in the second stage of the recrystallization process.

Figure 8.18 illustrates the structure of the 14YWT NFA after 60% compression and further annealing at 1100 °C for 7.5 h, where recrystallized, abnormally grown (size >50 μm), and unrecrystallized/recovered regions coexist, with no texture change in the latter regions [9]. These results corroborate that in NFAs, abnormal grain growth occurs with no incubation time, and recrystallization and abnormal grain growth occur simultaneously above 850 °C [206].

As aforementioned, the increase of misorientation between neighboring grains due to cold deformation weakens the orientation pinning, triggering the recrystallization [9, 132, 133, 216]. When NFA products require cold rolling after hot extrusion, such as in clad tubes,[2] recrystallization control is essential to obtain the required strength and ductility in the hoop direction. Here, thermomechanical processing aims to produce unrecrystallized cold-rolled forms at an intermediate process followed by a recrystallization only at final heat treatment [168]. Recrystallization in NFAs after cold deformation remains a key issue particular to each case study. In general terms, the temperature of the onset of recovery and recrystallization is closely related with deformation texture and dissolution-reprecipitation mechanisms of the nano-oxides [9, 132, 133, 216].

---

[2]Claddings are tubes with an outer diameter of ~10 mm and a wall thickness of 1 mm. Seamless tubes are produced according to well-established procedures leading to a pronounced texture after heat treatment.

The improved high-temperature strength of NFAs is the result of the effect of oxide nanoparticles in pinning movable dislocations and inhibiting grain growth at elevated temperatures. During the extended recovery stage, particles play a crucial role in suppressing grain boundary mobility by the well-known Zener pinning effect. The presence of nanoscale oxide particles in the microstructure may exert an anisotropic pinning force on grain boundaries, and this explain the differences between transverse and longitudinal grain boundary velocities reported by Capdevila et al. [36]. Nevertheless, the equiaxed grain morphology during the extended recovery stage is closely related to the coarsening of the oxides located at the grain boundaries. After the extended recovery stage, the grain growth is only suppressed by the orientation pinning mechanism described previously [189]. Only those grains which are not likely to undergo orientation pinning will evolve to form the recrystallized coarse grained microstructure characteristic of this class of material.

The coarsening rate of the oxide particles and their link to recrystallization of the matrix are altered as a function of the annealing temperature and differ from one alloy to another [216]. As an example, unrecrystallized grains in Fig. 8.18b have particles with a large aspect ratio while the nanoparticles in the recrystallized and abnormally grown grains are round and coarser. This is explained on the basis of a dissolution-reprecipitation process of the nanoparticles [206]. It is speculated that elongated and undissolved particles in the unrecrystallized areas pin the dislocations during annealing, while dissolution and reprecipitation of nano-oxides occur on the dislocation boundaries along the deformation direction. As a consequence, particles aligned in the deformation direction prevent recrystallization (Fig. 8.18b), whereas in the regions lacking these particles, after deformation, undergo recovery before reprecipitation occurs (Fig. 8.18c).

8.3.3   Nanostructures in Steels Produced by Solid Reaction

8.3.3.1   Nanostructured Pearlite

The basic structure of pearlite has been introduced in Sect. 3.1 for patented wires and is not repeated here for the sake of brevity. Figure 8.19 shows TEM images of nanostructured pearlite produced in a hypereutectoid steel containing vanadium microadditions. Two pearlite nodules are observed, where the Moiré fringes between the ferrite and cementite lamellae indicate that the structure is not parallelly aligned with the beam, and thus the interlamellar spacing here measured is greater than the true spacing. The dark-field image in Fig. 8.19 reveals a uniform distribution of tiny vanadium carbides within the ferrite lamellae.

Measurements of the interlamellar spacing of pearlite can be conducted in a simple manner by using properly prepared and etched metallographic specimens using circular test grids to determine the mean random spacing using either scanning electron microscopy (SEM) replicas or SEM images. A number of randomly chosen

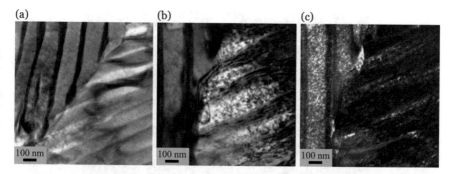

Fig. 8.19 TEM images of pearlite produced in a hypereutectoid steel micro-alloyed with vanadium. (**a**) Bright-field image of ferrite and cementite in two pearlite nodules, (**b**) bright-field image of vanadium carbides in interlamellar ferrite, and (**c**) dark-field image of vanadium carbides in ferrite locations of the left nodule (ArcelorMittal Global R&D internal report)

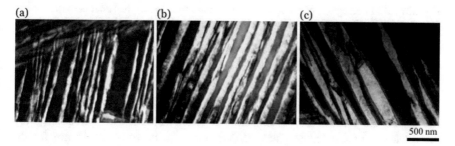

Fig. 8.20 Bright-field TEM images of the structures formed in a 0.98C-1.46Si-1.89Mn-0.26Mo-1.26Cr-0.09 V (wt.%) steel isothermally transformed at (**a**) 200 °C for 15 days, (**b**) 250 °C for 30 h and, (**c**) 300 °C for 9 h. (Authors' own unpublished work)

fields should be measured to obtain adequate statistics, and the mean true spacing is calculated as half the mean random spacing [253], meaning that direct rough estimations without stereological correction are overestimated. Directed spacing measurements in TEM thin foils by tilting the specimen take much more effort, and the correlation between the directed spacing measurement and the true interlamellar spacing is empirical in nature [256, 257]. The interlamellar spacing in the steel shown in Fig. 8.19 measured using SEM micrographs is about 85 nm.

8.3.3.2 Nanostructured Bainite

Figure 8.20 presents bright-field TEM images of the structures obtained in a 0.98C-1.46Si-1.89Mn-0.26Mo-1.26Cr-0.09 V (wt.%) steel isothermally transformed at temperatures between 200 and 300 °C. The bright contrast corresponds to the bainitic ferrite plates, while the dark contrast corresponds to retained austenite

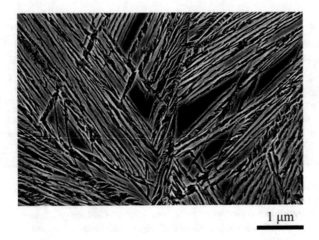

1 μm

Fig. 8.21 SEM micrograph of a 1.0C-1.47Si-0.74Mn-0.97Cr (wt.%) steel after incomplete transformation at 200 °C for 22 h showing large blocks of retained austenite in between bainite sheaves
formed at a prior austenite grain boundary and thin films between bainitic ferrite plates within the
sheaves. Lower relief etched regions correspond to bainitic ferrite, while upper relief regions are
retained austenite. (Authors' own unpublished work)

regions. It is easy to note that the scale of the bainitic ferrite plates is refined when
decreasing the transformation temperature. After stereological correction [143],
the thickness of the ferrite plates is determined to be of about 35 nm for the
steel isothermally transformed at 200 °C (Fig. 8.20a), 55 nm for transformation
at 250 °C (Fig. 8.20b), and 125 nm for transformation at 300 °C (Fig. 8.20c) [72,
73]. Thus, nanostructured bainite mainly consists of bainitic ferrite plates embedded
in a network of retained austenite.

Retained austenite presents two distinguishable morphologies, i.e., thin films
between platelets of bainitic ferrite and blocks between sheaves of bainite, as
presented in Fig. 8.21. The carbon content distribution in the austenite goes hand
in hand with the austenitic feature size distribution, thin films being more enriched
in carbon than blocks [31, 75, 199, 200]. Austenite transformation into bainitic
ferrite is regarded as a division process, where ferrite plates divide the austenite
into films, and the sheaves (or aggregates of plates) divide the austenite into blocks.
In the early stages of bainite transformation in silicon-containing steels there are two
populations of austenite: one having a higher carbon content in the surrounding of
the ferrite plates and the other, having a carbon content close to nominal in areas far
from the ferrite [247]. As the transformation progresses, with the subsequent carbon
partitioning from the ferrite towards the austenite, the austenite blocks diminish in
scale getting enriched in carbon and evolving into films as bainitic ferrite plates
breach them. The carbon content of the retained austenite after transformation is
complete is slightly above the value given by the T_0' curve.

Figure 8.22a exemplifies an APT needle obtained from a 0.66C-1.45Si-1.35Mn-
1.02Cr (wt.%) steel transformed at 220 °C for 168 h, where carbon isoconcentration

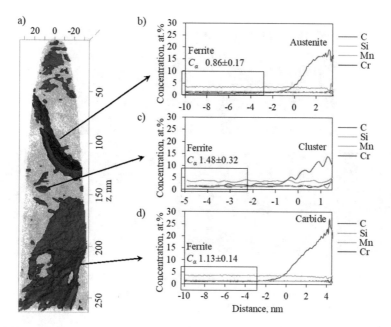

Fig. 8.22 APT measurements on a 0.66C-1.45Si-1.35Mn-1.02Cr (wt.%) steel transformed at 220 °C for 168 h showing carbon isoconcentration surfaces at 6 at.% superimposed with the carbon atom map and proximity histograms across the interfaces indicated by arrows which correspond to (**a**) ferrite/austenite interface, (**b**) ferrite/cluster interface, and (**c**) ferrite/carbide interface. (Authors' own unpublished work)

surfaces at 6 at.% C are superimposed with the carbon atom map. As no crystallographic information is available through this technique, the low carbon regions are usually assumed to be the ferrite phase, while high carbon regions are identified according to their carbon content and morphology [32, 246]. Figure 8.22b, c and d show the proximity histograms obtained across a ferrite/austenite interface, a carbon cluster, and a ferrite/carbide interface, respectively. As here observed, nanostructured bainitic ferrite exhibits a complex nonhomogeneous distribution of carbon atoms in arrangements with a specific composition, a few nanometers in size in most cases. These carbon-enriched regions are identified from their carbon content as Cottrell atmospheres ($\sim$8 at.% C), carbon clusters ($\sim$11 at.% C), the $Fe_{32}C_4$ carbide (20 at.% C), and cementite or η-carbide precipitates (25 at.% C) [32, 33, 186, 199, 200, 246].

Besides, APT observations show that large amounts of excess carbon remain in the baintic ferrite matrix [30, 199, 200, 246], justified by a body-centered tetragonal, rather than body-centered cubic, symmetry of the ferrite lattice [77, 108, 247]. It is suggested that the tetragonality detected would be the result of carbon clusters, with a locally increased tetragonality, surrounded by a depleted matrix [199, 200], which is representative of the early stages of decomposition of ferrous martensites [244]. The reasons for such carbon-supersaturation still remain a matter of discussion

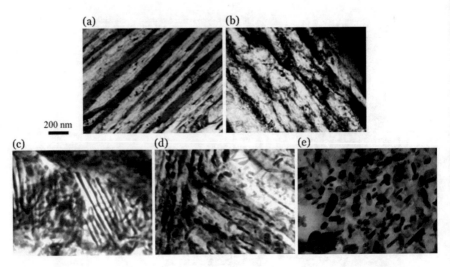

Fig. 8.23 Bright-field TEM images of a 0.98C-1.46Si-1.89Mn-0.26Mo-1.26Cr-0.09V (wt.%) steel isothermally transformed at 200 °C for 45 days and tempered at (**a**) 400 °C for 1 h, (**b**) 450 °C for 1 h, (**c**) 550 °C for 1 h, (**d**) 600 °C for 30 min, and (**e**) 600 °C for 1 day

[186], but it is sensible thinking that it is the result of the diffusionless, or partially diffusionless, transformation of nanostructured bainite, where vacancies would be playing a role [201].

The thermal stability of nanostructured bainite is tested by tempering experiments presented in Fig. 8.23. Tempering at 400 °C for 1 h does not introduce any perceptible change in the original microstructure, as shown in Fig. 8.23a, while the same treatment in nanostructured pearlite leads to spheroidization and coarsening of both ferrite and cementite (see Fig. 8.13d). Tempering at 450 °C for 1 h leads to decomposition of the austenite films by the precipitation of fine carbides (Fig. 8.23b), the length scale of the bainitic ferrite plates remaining intact. With increasing the tempering temperature to 550 °C, retained austenite decomposes completely, with larger austenite regions leading to cooperative growth of ferrite and cementite in the form of pearlite, which nucleates at cementite particles located at ferrite/austenite interfaces (Fig. 8.23c) [95]. Thus, carbon-rich austenite films are less thermally stable than austenite blocks due to both their higher amount of carbon in solid solution and the presence of higher amounts of carbon-containing defects [209]. Tempering of the original structure at 600 °C for 1 h produces major changes involving general coarsening, presented in Fig. 8.23d, and recrystallization when the heat treatment is prolonged to 1 day (Fig. 8.23e). It is thought that the resistance to tempering of nanostructured bainite is a consequence of carbide precipitation at the ferrite/austenite interface, which hinders the plate coarsening process [40, 74, 183].

8.4   Mechanical Performance of Nanostructured Steels

The ideal slip resistance of metals is about $G/30$, where G is the shear modulus of the material [27]. This level of strength is only achieved when the size of the crystal becomes sufficiently small. This was first demonstrated by Taylor, with the preparation of a 30 μm antimony wire whose tensile strength resulted in values about 30 times greater than those of 4 mm antimony crystals [243]. In the case of iron, crystals in the form of whiskers 1.6 μm in diameter have a yield strength of about 13.4 GPa and an ultimate tensile strength (UTS) of 16–23 GPa before fracture occurs [27]. The ideal strength of pure iron is thus computed to be around 13 GPa [43], and this value is dramatically decreased as materials scale-up [21]. It is in this context that the relationship between strength and nanoscale of current bulk nanostructured steels is examined.

8.4.1   Strength and Ductility of Nanostructured Pearlite

The highest tensile strengths that can be achieved in bulk material are found with eutectoid steels wires, where strength is routinely above 1.5 GPa but can reach more than 6 GPa [135]. In pearlitic steels, there are fundamentally four strengthening mechanisms: (1) Hall-Petch strengthening, (2) strain hardening, (3) solid solution hardening, and (4) dispersion hardening (only when cementite turns amorphous after large strains). In microstructural terms, the mechanical properties of pearlitic steels are mainly governed by the thickness of the cementite lamellae. In the case of cold-drawn pearlitic wires, stress-strain data presented in Fig. 8.24 indicate that two regimes are present: at true drawing strain, $\varepsilon \leq 3$ the strength increment as a function

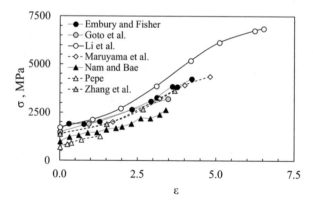

Fig. 8.24 Tensile strength, σ, for different true drawing strains, ε, as evaluated by different authors: Embury and Fisher [59], Goto et al. [86], Li et al. [136], Maruyama et al. [148], Nam and Bae [166], Pepe [185], and Zhang et al. [283]

of strain is lower, and increases when increasing strain presumably due to alignment of the cementite lamellae parallel to the drawing direction [167]. Therefore, the Hall-Petch strengthening is considered the most relevant mechanism in pearlitic steels, following the equation:

$$\Delta\sigma_{HP} = k\overline{L}^{-1/2} \tag{8.3}$$

where $k = 0.422$ MPa·m$^{-1/2}$ is the Hall-Petch constant [265] and $\overline{L}$ is the effective grain size, corresponding to the interlamellar spacing S_0 in pearlitic steels. The concomitant strengthening effects of strain hardening, solid solution hardening, and dispersion hardening are remarkable for $\varepsilon \leq 2 - 3$ and are reviewed elsewhere [23]. The same strengthening mechanisms should apply to nanostructured pearlite produced by both cold drawing and solid reaction.

The fracture mechanism in pearlitic wires varies depending on the drawing strain. For $\varepsilon = 1 - 3$, the material undergoes ductile failure characterized by a cup and cone fracture accompanied by necking. This form of fracture occurs by the initial formation of microvoids in the interior of the material followed by microvoid enlargement and formation of a crack. As deformation continues, the crack grows rapidly spreading laterally toward the edges of the specimen forming an angle of 45° to the loading axis. The final shearing produces a cup type shape on one fracture surface and a matching cone shape on the other [23]. For wires cold drawn to $\varepsilon = 4 - 5$, observations indicate that fracture is ductile with the typical dimples indicating microvoid coalescence with a transition from a necking to the nonnecking mode between $\varepsilon = 3$ and $\varepsilon = 4$ [23, 86, 136].

8.4.2 Temperature Dependence of Strength and Ductility of Nanostructured Ferritic Alloys

Three dominant contributions are considered in the flow stress of ODS steels, namely, (1) Hall-Petch hardening due to the ferrite matrix, (2) hardening from dislocation forests, and (3) precipitation strengthening resulting from the oxide dispersoids. However, the superposition of these terms has been suggested to adopt a Pythagorean (rather a linear) superposition in the form [39]:

$$\sigma_Y = \sigma_{HP} + \sqrt{\sigma_\rho + \sigma_p} \tag{8.4}$$

where σ_Y is the yield strength, and σ_{HP}, σ_ρ, and σ_p are the Hall-Petch, dislocation forest, and nano-oxide precipitation strengthening contributions, respectively. For the first term, Hall-Petch hardening is considered to follow Eq. 8.3, with $k = 0.268$ MPa·m$^{-1/2}$ [39, 191]. The contribution from dislocation strengthening is estimated by the Bailey-Hirsch relationship in terms of the forest intersection mechanism at the boundaries [39, 191]. Last but not least, nanoparticle strengthening adopts

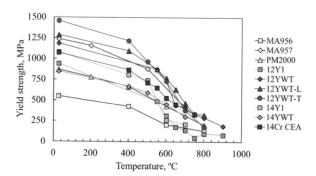

Fig. 8.25 Yield strength as a function of the testing temperature for various ODS steels and NFAs. The compositions of the alloys are given in Table 8.1, and the suffixes L and T indicate that samples were tested in the longitudinal and transversal directions, respectively. (Data adapted from [7, 57, 102, 121, 223])

different equations according to the particular dislocation-particle interaction mechanisms observed, which are strongly dependent on the type, size, shape, and coherency of the oxide. Under a general assumption [120, 262], oxide dispersion strengthening depends on the number density (N_p) and size (d_p) of the nanoparticles such that $\sigma_p \propto \sqrt{N_p d_p}$.

Figure 8.25 shows the yield strength as a function of the testing temperature for various conventional ODS steels (MA956 and PM2000) and NFAs, whose chemical composition is given in Table 8.1. Significant variations in the yield strength of the different steels are evidenced for testing temperatures below 800 °C; the lower the temperature the larger the difference. There is a general collapse of the tensile properties above 400 °C that can be explained by a change in the deformation mechanism. However, this behavior is interpreted in two different ways; some authors find that at the lowest temperatures, precipitates play a major role by pinning dislocation movement [39, 191], while others argue that the low-temperature properties are largely affected by the grain size and fraction of unrecrystallized material, where small atom clusters do not play a major role [7, 121]. At higher temperatures, the hardening role of the precipitates is still observed, until strength of all the alloys merges for extrapolations to 900 °C. The breakdown of the strength at elevated temperatures is due to control by other mechanisms. The strength depends in different ways on temperature, grain size, and strain rate [212]. Hence, the specific role of the nano-oxides in the strengthening of NFAs is complex and still needs to clarified from the point of view of both Hall-Petch strengthening by refining the ferrite matrix and precipitation strengthening itself.

Total elongation values presented in Fig. 8.26 as a function of the testing temperature for various conventional ODS steels and NFAs given in Table 8.1 indicate that ductility in these materials is relatively poor as compared to conventional steels. Most of the alloys show a peak in the elongation versus temperature at about 600–700 °C. At this temperature range, the deformation mechanism is modified, with an intragranular character turning into intergranular for higher temperatures. Damage

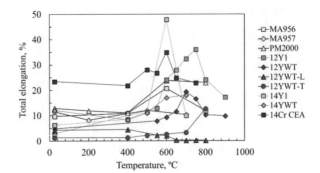

Fig. 8.26 Total elongation as a function of the testing temperature for various ODS steels and NFAs. The compositions of the alloys are given in Table 8.1, and the suffixes L and T indicate that samples were tested in the longitudinal and transversal directions, respectively. Data adapted from [7, 57, 102, 121, 223]

is more severe as the temperature is increased, particularly when the formation of cavities leading to decohesion along the grain boundaries starts [191].

8.4.3   Strength and Ductility of Nanostructured Bainite

Most of the strength in nanostructured bainite is the result of the bainitic ferrite plate thickness, t. As mentioned in the introduction, there is an inverse dependence of the strength on the effective grain size, which in this case corresponds to the true plate thickness as measured by mean linear intercept and follows the equation [129]:

$$\Delta\sigma = k\overline{L}^{-1} \tag{8.5}$$

where $k = 115$ MPa·μm and $\overline{L} = 2t$ [143]. The fact that strength does not obey the Hall-Petch relationship is because the transmission of slip across the effective grain boundaries is determined by the energy required to expand dislocation loops rather than by dislocation pile-up at the boundaries [129].

The total strength of ferrite alone is the result of (1) the plate thickness, (2) dislocation forests, (3) solid solution strengthening (including the carbon excess), and (4) carbide precipitates [20]. The strengthening contributions of these factors are minute as compared to that due to the bainitic ferrite plate size. This is held as a reason for the properties of the nanostructure to be insensitive to tempering, until the onset of plate coarsening [96].

Considering austenite alone, strength is simply described as a function of the composition and the temperature [278]. The strength of the whole ferrite plus austenite mixture in nanostructured bainite has been found to be mainly controlled

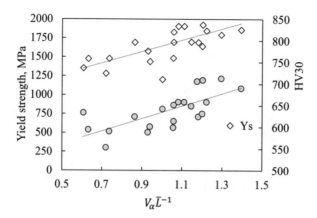

Fig. 8.27 Yield strength and hardness (HV30) of different bainite nanostructures as a function of $V_\alpha \overline{L}^{-1}$. V_α is the volume fraction of bainitic ferrite and $\overline{L} = 2t$, where t is the bainitic ferrite plate thickness. Adapted from [226–228]

by the inverse of the effective grain size weighted by the amount of the bainitic ferrite [71, 76], i.e., $V_\alpha \overline{L}^{-1}$, where V_α is the volume fraction of ferrite (Fig. 8.27).

Ductility in nanostructured bainite remains an unpredictable property, given that there is no correlation between the initial volume fraction and/or composition of austenite and the tensile ductility [159]. On the one hand, a large amount of unstable austenite is not appropriate for total elongation. On the other hand, a highly stable austenite, unable to undergo mechanically induced martensitic transformation, or implying a low rate of this transformation, is neither beneficial [161]. In this sense, microstructures exhibiting similar retained austenite contents and bainitic ferrite plate thickness, even produced from the same steel, show largely different tensile ductility [159, 226–228].

It has long been suggested that failure is a consequence of austenite isolation, i.e., failure occurs when austenite loses percolation. Modeling of this phenomenon, considering ellipsoidal objects placed in a matrix, predicts that the percolation threshold, and thus failure, occurs when $V_\alpha \approx 0.1$ [18]. Recently, a simple linear relationship has been recently proposed [229, 230] in an attempt to quantify, but not predict, the relationship between retained austenite stability and tensile ductility:

$$\ln \left(V_{\gamma,0}\right) - \ln \left(V_\gamma\right) \propto k\varepsilon_p \tag{8.6}$$

where $V_{\gamma,0}$ is the initial retained austenite content, V_γ is the retained austenite in the gauge length after applying a true plastic strain of ε_p, and k is a constant. Fittings to Eq. 8.6 in Fig. 8.28 indicate that values of k above 0.2 indicate rapid mechanical destabilization of austenite with increasing strain, leading to brittle fracture. Values of k below 0.2 are related to ductile fracture, with tensile elongation increasing as k is decreased. These results are conflicting with the suggested existence of a percolation threshold below which ductile deformation is no longer possible.

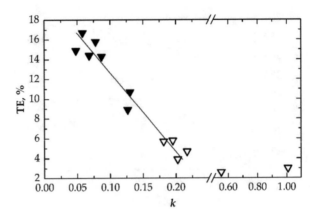

Fig. 8.28 Tensile elongation (TE) as a function of the value of k for different bainite nanostructures. Hollow symbols are for brittle ruptures; full symbols are for ductile ruptures, after [229, 230]

A general observation indicates that poor ductility is largely associated with the lowest transformation temperatures (about 220 °C), whereas transformation at slightly higher temperatures (about 250 °C) provide ductile behavior [229, 230]. The reduction of the mechanical mismatch between austenite and bainitic ferrite is speculated to be the reason behind the better performance of samples treated at higher temperature [160, 161].

Furthermore, recent studies show that mechanically induced martensitic transformation in nanostructured bainite is stress assisted, where mechanical twinning of austenite is not a necessary event prior to martensitic nucleation [162]. However, extensive twinning has been observed to occur in austenite blocks with a crystal orientation highly stable against martensitic transformation. Thus, both mechanically induced martensitic transformation and twinning contribute to plasticity and strain hardening in nanostructured bainite [162].

8.5 In-Use Properties and Industrial Applications of Nanostructured Steels

8.5.1 Applications and Failure Mechanisms in Nanostructured Pearlite Wire Ropes

Nanostructured pearlite wires are employed in a wide range of applications where a combination of ultra-high strength and sound ductility is required. The most relevant usages of drawn wires are tire cords, lifting ropes, and bridge cables [181], where individual wires are wither twisted or braided to form a rope. For both applications, the design requirements rely on tensile and torsion strength, stiffness, ductility, corrosion resistance, and fatigue endurance of the drawn wires.

A wire rope or cable under tensile or bending loads places the individual wire elements in tension, torsion, and bending [65, 66]. The geometry of a strand of wire rope and its stress state under loading can acquire high levels of complexity, and they are of primary consideration for metallurgical development and tailoring of the properties. Effective design should minimize bending and torsion stresses and maximize axial stresses. Bending and torsion stresses can lead to splitting and delamination of the wire by inducing normal tensile stresses in the subsurface of the wire [28]. Furthermore, contact and cyclic stresses exist within each wire due to manufacturing and service loading [232]. For a proper design, torsion tests and fretting and fatigue behavior of the wires need to be effectively evaluated. The testing methodologies and properties of wire ropes and commercial cold drawn pearlitic wires are detailed elsewhere [65, 66]. Therefore, under normal circumstances, where no corrosion, excessive heat, severe overloading, or chemical damage is involved, failure of ropes occurs by either fatigue at the zones subjected to the greatest cyclic loading or abrasion [259]. This type of failure is predictable under proper inspection, given that individual wires fail first before the whole rope breaks down [260].

Besides, nanostructured pearlite wires are low alloyed, and corrosion resistance is thus poor. Wires exposed to environments where protection is needed are normally zinc coated by either hot zincing or galvanizing process. Only for exceptional cases, stainless steel wires are used as rope wires. Corrosion is a frequent failure in ropes when this occurs internally, i.e., when individual wires at the core of the rope do not receive proper protection preventing the inspector to foresee the problem [11]. Less common failure mechanism in ropes are those that cannot be predicted and usually lead to catastrophic failure, such as ropes "jumping the sheaves," lightning strikes, hydrogen embrittlement, and kinks [259].

For the reader to envision the magnitude of the requirements of nanostructured pearlite wires, a rousing example is given by Borchers and Kirchheim [23]: The Strait of Messina bridge is a suspension bridge projected to connect Torre Faro in Sicily and Villa San Giovanni in mainland Italy and designed to have the largest span in the world at 3.3 km. The suspension system of the bridge relies on two pairs of pearlitic steel cables, each of them having a diameter of 1.24 m and a total length of 5.3 km, corresponding to about 100 tons of steel cable alone. When considering the dynamic effects of traffic and an earthquake of 6.9 in the moment magnitude scale, the cables would be subjected to total deformations of more than 1 m or unitary strains of $3 \cdot 10^{-4}$. This value is far less than the elastic strain of nanostructured pearlite wires, which reaches a saturation value of 1.6% with increasing drawing strain. Thus, failure due to overloading is unlikely to occur in this class of material when design is safely performed.

8.5.2 Industrial Applications and In-Use Properties of Nanostructured Ferritic Alloys

The efficiency of plants that produce electric energy or heat is usually increased when the operation temperature is raised. Some examples requiring increased operation temperatures for improved efficiencies are gas and steam turbines, ultra-high temperature coal gasification, solar thermal applications, and advanced nuclear fusion and fission power plants. These latter are currently being studied within the framework of the International Generation IV initiative, where the sodium-cooled fast reactor (SFR) has received the greatest share of funding over the years. For the safe and efficient operation of the SFR, it is necessary to develop a clad material that meets the core specifications in terms of dimensions and surface finish, ductility, compatibility with helium coolant, and irradiation conditions at temperatures of 1000 °C during normal operation [12]. Besides, technologies and means for developing biomass plants with higher energy conversion efficiencies are essential to commit to renewable biomass energy generation in the future. Advanced, indirect combined cycle gas turbine (CCGT) systems offer overall biomass energy conversion efficiencies of 45% and above in comparison with the 35% efficiency of a conventional biomass steam plant. However, to attain this efficiency in CCGT operations, it is necessary to develop a heat exchanger capable of gas-operating temperatures of approximately 1100 °C and pressures of 15–30 bar. Current structural steels and superalloys are at the limits of their applicability under such severe environments, and the development of components with improved performance remains a strong driving force for further progress. ODS steels are candidate materials for use in these next-generation high-temperature applications.

8.5.2.1 Oxidation and Corrosion Resistance of Nanostructured Ferritic Alloys

In the case of ODS ferritic steels containing ~5 wt.% Al, such as PM2000 and MA956, an α-alumina scale forms upon high-temperature exposure, which acts as a good barrier against oxidation [51, 147, 152, 267]. In cases of exposure to environments with low-oxygen content, formation of the protective layer is not guaranteed, but a preoxidation treatment, i.e., oxidation of the alloy prior to exposure, can be applied [69, 90]. Besides their good oxidation resistance, PM2000 and MA956 ODS steels present an outstanding creep performance [231]. On the contrary, the absence of aluminum in the 12YWT and 14YWT NFAs makes them susceptible to corrosion under certain media, such as liquid lead-bismuth eutectic cooled reactors and spallation sources, where aluminum-containing ODS steels show an excellent performance [106]. It is worth mentioning that a decreased grain size plays a positive role against localized oxidation, given that grain boundaries act as fast diffusion paths inside the material preventing deep localized grain boundary oxidation [61, 106].

8.5.2.2   Creep Resistance of Nanostructured Ferritic Alloys

Besides the homogeneous dispersion of nanosized oxides in the ferritic matrix, ODS ferritic steels and NFAs usually exhibit high-strength and creep resistance properties as a consequence of being far from the equilibrium state [3, 83]. Many features, such as a large proportion of interfaces and triple junctions, irregular distributions of alloying elements, the occurrence of nonequilibrium phases and supersaturated solutions, residual stresses, and excess concentrations of lattice defects, increase the Gibbs free energy. All these features are closely connected with the nonequilibrium conditions of the ODS fabrication methods by powder technology [87, 144]. The excellent creep properties of the NFAs are due to an attractive interaction of dislocations with oxides described in the well-known model by Rösler and Arzt [203]. Creep usually exhibits the threshold stress, which correlates well with the Orowan theory according to which, at a given temperature, the threshold stress is inversely proportional to the distance of the oxides. Thus, any coarsening of the oxides causes a degradation of the creep properties.

Figure 8.29 illustrates the superior creep performance of different ODS steels, and specially of the MA957, 12YWT, and 14YWT NFAs, as compared with the 9Cr TMP steel having similar chemical composition but not reinforced with oxide particles. Creep resistance is here expressed by means of the Larson-Miller parameter (LMP), which allows predicting the lifetime of material vs. time and temperature using a correlative approach based on the Arrhenius rate equation. The value of the parameter is usually expressed as $LMP = T(c + \log t)$, where c is a material specific constant, approximated as values between 20 and 30, t is the time in hours, and T is the temperature in Kelvin.

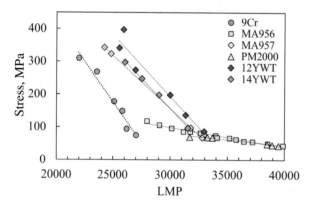

Fig. 8.29 Larson Miller parameter (LMP) for various conventional ODS steels and NFAs. (Data adapted from [102, 121])

8.5.2.3 Irradiation Resistance of Nanostructured Ferritic Alloys

Irradiation resistance in NFAs is obtained two ways. First, they contain highly stable dislocation sinks with a large number density of thermally stable, nanometer-scale precipitates that trap helium in bubbles to avoid swelling and helium migration to the matrix grain boundaries with the subsequent intergranular fracture. Second, they present high creep strength, permitting operation at temperatures above the displacement damage regime. These two characteristics are responsible for the superior irradiation resistance of ODS steels and NFAs as compared to thermomechanically processed heat resistant steel that do not contain nano-oxide precipitates. As long as NFAs are operated at temperatures above the displacement damage-swelling regime, they are able to manage very high helium levels, reaffirming their suitability for fusion reactor structures. An extensive overview of the irradiation damage in these and other heat-resistant steels can be found in [99, 175].

8.5.2.4 Scalability of Nanostructured Ferritic Alloys

The key point for the rather feeble industrial application may be that producing NFAs and ODS by the current PM route is certainly expensive. Even if the high costs could be justified by the absence of materials with comparable properties, the potential applications need to be established for manufacturers[3] to emerge and promote the development of this class of alloys. For this reason, applications cannot be limited to the niche of nuclear fusion and fission power plants but to systems where there is room for increases in the operation temperatures, as previously introduced. Anyhow, NFAs are used in applications requiring small amounts of material in complex multimaterial hybrid systems. A perspective of cost is given by Odette [174] for a first wall and divertor hybrid structure in a large demonstration fusion reactor requiring 10 tons of NFA at a target price of $50/kg for consolidated billets. The cost of the material alone would be of five million dollars, remaining a small fraction of the total cost of a 10,000 million dollar fusion power reactor. Even though the cost of construction does not reflect the global cost of electric power economics (including the environmental cost), it is thought that the use of either ODS steels or NFAs is a feasibility enabling issue [174].

---

[3]The production of MA957 was discontinued by INCO, and so it was the production of PM2000 by Plansee, while 12YWT was produced only once as a small heat by Kobe steel.

8.5.3 Industrial Applications and Failure Mechanisms of Nanostructured Pearlite Produced by Solid Reactions

The main industrial use of fully pearlitic structures is in rail steels. The first steel rail to be ever used is said to be the one laid in Derby station on the Midland Railway in 1857 [277] when no one even knew what pearlite is. The essential microstructure of those rails was similar to that of the rail steels used today, pearlitic structures based on nearly eutectoid compositions. Of course improvements have been made in the last 150 years in order to enhance the basic features around the fundamental structure by optimizing the chemical composition and impurity/cleanliness levels and by controlling the austenitization temperatures and cooling paths that lead to microstructural refinement [233].

The main failure mechanism in rails is wear caused by rolling contact fatigue arising from cyclic loading at the railhead. The main stress components at the rails are the rolling contact pressure, shear, and bending forces due to certain sliding in the rail/wheel contact, and residual stresses arising from manufacturing and welding. Less relevant failure considerations are related to overloads and harsh environmental conditions [222].

In eutectoid pearlitic steels, fatigue strength is found to be insensitive to the interlamellar spacing [52]. However, reduced pearlite interlamellar spacings have a greater flow stress and work-hardening rate [237], both of which lead to a reduction in the wear rate [44, 187, 237]. The strong correlation between the wear resistance and the interlamellar spacing (also reflected an increase in hardness) has led to processing and alloying approaches to produce fine fully pearlitic structures.

Figure 8.30 shows the specific-wear resistance as a function of the hardness of the nanostructured pearlite as compared to somewhat softer pearlitic steels and a variety of bainitic steels, some of which are much harder [53, 226–228]. The results emphasize the fact that wear performance cannot be simply estimated on the basis of phase fractions or hardness when the mechanisms of wear are dependent on microstructural parameters.

An effective and widely used processing approach to produce refined pearlitic structure on the surface of the rails is the *head hardening* heat treatment [204]. Head hardening is applied by accelerated cooling with forced air, water sprays, or oil or aqueous polymer quenching either in the production line while the steel is still austenitic immediately after hot rolling [26] or separately by reheating of as-rolled rails [205]. Accelerated cooling is applied so that a refined pearlitic microstructure is obtained at the railhead where the rolling contact and sliding occur. By careful adaption of both the steel composition and the thermal history during head hardening, it is currently feasible to obtain pearlite nanostructures in an industrial environment [25].

Nevertheless, it should be emphasized that the steel rail industry is strongly cost driven, and while microalloying with expensive elements such as vanadium or molybdenum is admitted, relatively high cobalt additions as those proposed in [53, 271, 272] are simply inconceivable. Also, pearlite formed under isothermal

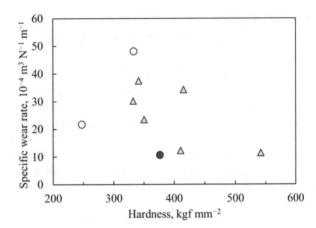

Fig. 8.30 Specific wear rates as a function of the initial hardness measured in rolling/sliding tests. The triangles represent bainitic steels, circles correspond to pearlitic steels, and the filled point is nanostructured pearlite. Data estimated from [53, 226–228]

conditions near the B_S temperature exhibits the finest achievable interlamellar spacing for a given composition (illustrated in Fig. 8.3 for patented wires). Unlike for sheets, for which the rolling and coiling operations ease isothermal transformations, long product manufacturing routes are not typically suited for controlled cooling followed by isothermal paths.

With the development of heavy-load high-speed railways, the in-use demands for rails are increasingly pushing current metallurgical approaches to their limits. Safety and cost of railway transportations are partly related to the quality of rail steel, and improved nanostructures with reliable damage tolerance design and effective maintenance methods are under continuous development.

8.5.4 In-Use Properties and Industrial Applications of Nanostructured Bainite

Nanostructured bainite has attracted considerable attention over the past 15 years, far from but getting closer to industrial applications. Research efforts have focused on evaluating the in-use performance of these materials for various applications implying mainly fatigue and wear resistance, beyond looking at basic tensile strength. This implies understanding the microstructure-properties relationships and also comparing the results to currently used steel grades and other manufacturing methods for such applications as well as evaluating the costs of the different alternatives.

Until today, the Research Fund for Coal and Steel (RFCS) of the European Commission has financially supported three projects aimed at the industrialization of nanostructured bainite, namely Nanobain [226–228], Mecbain [229, 230], and more recently Bainwear [192]. These were consortia gathering scientists, steel makers, and end users sharing a common intrigue and interest for the bainitic nanostructures and ended up with unique developments in understanding both the physical metallurgy and properties of these steels.

The evaluation of the notched fatigue performance for a variety of load ratios and notch severity of nanostructured bainite isothermally transformed at 220 or 250 °C to a UTS over 2 GPa shows that the materials behave 10–20% better than 100Cr6 isothermally transformed to a UTS of 2.35 GPa [76, 226–228]. The fatigue response of nanostructured bainite is strongly dependent on cleanliness (as for most high-strength steels) and the crystallographic grain size as the critical microstructural parameter [164, 197] (Table 8.2).

Wear resistance of nanostructured bainite is far more promising with significant improvements under specific dry sliding, dry rolling-sliding, and rolling contact fatigue conditions [131, 198]. As a general conclusion, for each wear system the nanostructures can be optimized by adjusting the amount, size, distribution, and stability of phases. To this end, the specific and synergistic role of each phase in the wear mechanisms must be first understood.

Field tests have been recently carried out to assess the suitability of bainite nanostructures in rollers, punches, bearings, and reducing elements [192], and the results are summarized as follows: Guide rollers for rolling mills made of nanostructured materials show better performance in warm applications than current grades (see Fig. 8.31), although they do not offer advantages in Calow straightener rollers in cold applications. In the case of sheet metal cutting punches, nanostructured bainite performs favorably when used in extremely harsh conditions, outperforming

Table 8.2 Tensile properties and fatigue data for different loading and notch conditions for various nanostructured bainitic steels and reference materials (50CrMo4 and 100Cr6). Results underline the excellent performance of the 0.6C-1.5Si under both isothermal treatment conditions. YS is the 0.2% offset yield strength, UTS is the ultimate tensile strength, UE is the uniform elongation, TE is the total elongation, $S_{a, 50\%}$ is the fatigue strength for a probability of failure of 50% (mean value) at 10^7 cycles, R is the maximum to minimum stress ratio, and K_t is the stress concentration factor. Adapted from [229, 230]

Variant	Isothermal treatment	YS, MPa	UTS, MPa	UE, %	TE, %	$S_{a, 50\%}$, MPa $K_t = 2$ R = 0.1	R = -1	$K_t = 4$ R = 0.1
0.6C-1.5Si	220 °C 114 h	1643	2102		4.5	350	605	240
0.6C-1.5Si	250 °C 16 h	1448	1990	8.2	14.3	365	605	230
0.6C-2.5Si	250 °C 16 h	1483	1950	8.1	8.6	255	–	–
1C-1.5Si	250 °C 16 h	1740	2170	8.8	10.7	310	–	–
1C-2.5Si	250 °C 16 h	1738	2106	11.7	16.8	310	445	210
1C-2.5Si	250 °C 40 h	1785	2101	4.3	15.8	335	420	200
50CrMo4	To 37 HRC	1118	1183	5.1	14.7	305	333	
100Cr6		2150	2350			355	566	200

Fig. 8.31 (a) Set-up of the vertical guide rolls located at the entrance of the hot rolling stage in order to facilitate the passage of the bars, (b) nanostructured bainite guide roll, and (c) undisclosed pearlitic roll, which is the current main alternative. Measurements with a template, evaluation of weight loss, and comparison of flank profiles all show an advantage of nanostructured bainite rollers. (Adapted from [192])

Fig. 8.32 Images of selected punches: (a) Reference DIN 1.2379 in quench and tempered conditions, showing severe chipping at the tip and failure after 10 cycles, (b) nanostructured bainitic steel uncoated showing wear and blunting of the cutting surface but no failure before 850 cycles, and (c) coated nanostructured bainitic steels showing almost no wear and no failure after 850 cycles. (Adapted from [192])

the reference DIN1.2379 tool steel, as illustrated in Fig. 8.32. Nanostructured bainite bearing components tested under artificial pollution outperform the reference 100Cr6 material, with a significant margin of benefit. Finally, trials on reducing elements were not conclusive, and under the studied conditions, nanostructured bainite grades did not offer advantages over the conventional 1.2379 material, where the wear mechanisms proved to be very different; this was considered interesting for further applications.

In contrast with its intense academic research activity, nanostructured bainite is still far from mass production for two main reasons [225]. First, the heat-treatment remains complex and commonly requires the use of salt baths with the finest temperature control. Such installations are far less frequent than conventional and simpler quenching and tempering furnaces. Furthermore, existing lines are often designed to handle treatments significantly shorter than those required to produce nanostructured bainite. Second, the benefits remain to be undoubtedly established. In particular, for components where fatigue is the main damage mechanism, nanostructured bainite should bring a sufficiently large advantage to justify a change from existing solutions. For components where wear is the main damage mechanism, it is firmly established that bainite nanostructures have exceptional potential, but further work is required before they can be transferred to actual applications.

With regard to heat treatment durations needed for isothermal transformation, several attempts have been made to improve the reaction kinetics. These have been focused in modifications of the composition [72, 73, 226–228] and heat-treatment schedules, e.g., by producing partial martensite transformation before subsequent isothermal bainite transformation [85] or by multistep isothermal heat treatments [138, 139]. A new approach is currently being explored under the auspices of the RFCS-funded project, Tianobain [70], where leaner medium-carbon alloys and shorter-processing times via thermomechanical ausforming are expected to lead to at least ultrafine bainitic steels with tensile strengths above 1600 MPa [78]. Trial products will be produced and tested using laboratory rolled materials, and recommendations for full-scale production parameters will be made.

Independent of commercial developments, nanostructured bainite can be considered as the upper limit of nanostructures obtained by solid-solid phase transformations, and their study is key for further improvements and understanding of this class of materials.

8.6 Future Trends

The set of *enabling technologies* that will supposedly lead us to the so-called *Industry 4.0* or the fourth Industrial Revolution, enabling the complete digitization of most industrial processes include cognitive and cloud computing, cyber-physical systems, virtual reality, the Internet of Things, big data, and last but not least, additive manufacturing [213]. Additive manufacturing is attracting much attention in the field of metallurgy, owing to the potential for direct manufacturing of complex parts which would not be possible to machine using current methods. Relevant additive manufacturing processes for metals consist in successively depositing wire or powder material that is locally melted on the top surface of the sample to add a new layer. This is achieved by selectively melting the material deposited using an energetic electron or laser beam. The technology is currently being used in rapid prototyping, cladding, coating, tooling and parts repair, and screening of novel alloys or functionally graded materials.

The main additive manufacturing processes for steels, both selective laser melting (SLM) and direct energy deposition (DED), consist in local melting and solidification of the steel powder. In principle, the structures obtained can be controlled through the process parameters, i.e., the laser power, the speed, and the hatch distance or the distance in between two laser passes. To fully understand the correlation between the microstructure and the thermal history, efforts are put to measure the temperatures of the melted zones and the cooling rates in situ [15, 62], which are often said to be in the overwhelming range of 10^4–10^6 °C/s for SLM.

The very particular, powder-metallurgical and nonequilibrium characteristics of additive manufacturing processing leaves room to alloys and micro/nanostructures that are not accessible through classical metallurgical routes. This includes super-saturated solid solutions, suppression of unwanted phase precipitation, composite synthesis, and in situ metal-gas reactions. Indeed, SLM-processed parts have completely different microstructures to those found in cast or wrought steel, which are not even recognizable for an experienced metallurgist. Phase distribution and dislocation network structures arising from the rapid solidification process are quite unique and different from the microstructures and dislocation cells observed in deformed metals. The resulting tensile properties in bulk steel parts additively produced are usually superior than those of wrought or cast material [119, 138, 139, 224].

The benefits of rapid solidification can be exploited by careful alloying as a method to promote massive heterogeneous nucleation and lower the energy of grain boundaries, bringing stable nanocrystalline steel structures. The opportunities that additive manufacturing bring in producing not only nanostructured but a new range of steels have not yet been exploited. Actually, the materials portfolio for metal additive manufacturing only includes a few stainless and tool steels having the standard composition of the wrought or cast grades. The race for new steel compositions adapted to the process is running for both scientist and the main powder manufacturers. Until now, there are no innovative solutions available in the market. The most remarkable work on steels specifically designed for additive manufacturing being that developed in the Max Planck Institute in conjunction with the Fraunhofer Institute for Laser Technology [128].

Besides, the myth that of 3-D printing in metal will allow producers to replace mass manufacturing with mass customization needs to be dispelled [22]. Additive manufacturing will be a disruptive technology in the sense that it will allow to produce complex steel parts with graded mechanical properties [142, 215] which cannot be achieved by conventional means, but replacing of traditional manufacturing techniques is not expected at least in the upcoming decades. If any, hybrid manufacturing approaches combining both mass conventional processing and additive manufacturing are a prospect [234]. Thus, the development and understanding of bulk nanostructured steels that can be produced by the classical routes is expected to keep in vogue.

8.7 Sources of Further Information

All the findings and developments carried out in nanostructured steels since their appearance do not fit in this chapter, and each of the metallurgical concepts well deserves its own book. Here there are some useful references where the reader can taste the complexity and beauty of each of the topics: bulk nanostructured materials (general) [254, 266, 281]; pearlitic steel wires [23, 64, 270]; pearlite and rails [34, 89, 134, 180, 193, 240]; nanostructured bainite [19, 226–230]; and NFAs and heat-resistant steels [154, 155, 217, 264, 279].

Acknowledgements The authors acknowledge financial support from the Spanish Ministerio de Economia y Competitividad (MINECO) in the form of a coordinate project (MAT2016-80875-C3-1-R) and the Research Fund for Coal and Steel of the European Commission under the contract SuperHigh (RFSR-CT-2014-00019).

References

1. M. J. Alinger, On the Formation and Stability of Nanometer Scale Precipitates in Ferritic Alloys during Processing and High Temperature Service. Doctor of Philosophy, University of California (2004)
2. M.J. Alinger, G.R. Odette, D.T. Hoelzer, On the role of alloy composition and processing parameters in nanocluster formation and dispersion strengthening in nanostuctured ferritic alloys. Acta Mater. **57**(2), 392–406 (2009)
3. R.A. Andrievskii, A.M. Glezer, Size effects in nanocrystalline materials: I. structure characteristics, thermodynamics, phase equilibria, and transport phenomena. Phys. Met. Metallogr. **88**(1), 45–66 (1999)
4. B.M. Arkhurst, J.H. Kim, Evolution of microstructure and mechanical properties of oxide dispersion strengthened steels made from water-atomized Ferritic powder. Met. Mater. Int. **24**(3), 464–480 (2018)
5. E.G. Astafurova, G.G. Zakharova, E.V. Naydenkin, S.V. Dobatkin, G.I. Raab, Influence of equal-channel angular pressing on the structure and mechanical properties of low-carbon steel 10G2FT. Phys. Met. Metallogr. **110**(3), 260–268 (2010)
6. J.M. Atienza, M. Elices, J. Ruiz-Hervias, L. Caballero, A. Valiente, Residual stresses and durability in cold drawn eutectoid steel wires. Met. Mater. Int. **13**(2), 139–143 (2007)
7. M.A. Auger, V. de Castro, T. Leguey, M.A. Monge, A. Muñoz, R. Pareja, Microstructure and tensile properties of oxide dispersion strengthened Fe–14Cr–0.3Y2O3 and Fe–14Cr–2W–0.3Ti–0.3Y2O3. J. Nucl. Mater. **442**(Supplement 1), S142–S147 (2013)
8. E. Aydogan, S.A. Maloy, O. Anderoglu, C. Sun, J.G. Gigax, L. Shao, F.A. Garner, I.E. Anderson, J.J. Lewandowski, Effect of tube processing methods on microstructure, mechanical properties and irradiation response of 14YWT nanostructured ferritic alloys. Acta Mater. **134**, 116–127 (2017)
9. E. Aydogan, O. El-Atwani, S. Takajo, S.C. Vogel, S.A. Maloy, High temperature microstructural stability and recrystallization mechanisms in 14YWT alloys. Acta Mater. **148**, 467–481 (2018)
10. M.M. Baloch, H.K.D.H. Bhadeshia, Directional recrystallisation in Inconel MA 6000 nickel base oxide dispersion strengthened superalloy. Mater. Sci. Technol. UK **6**(12), 1236–1246 (1990)
11. P. Barnes, T. McLaughlin, Corrosion fatigue behaviour of high strength steel wire in various aqueous environments. NACE - International Corrosion Conference Series. 2015-January (2015)
12. C. Behar, Technology roadmap update for generation IV nuclear energy systems. E. N. E. A. f. t. G. I. I. Forum. **17**, 2014–2003 (2014)
13. J.S. Benjamin, Dispersion strengthened superalloys by mechanical alloying. Metall. Trans. **1**(10), 2943–2951 (1970)
14. F. Bergner, I. Hilger, J. Virta, J. Lagerbom, G. Gerbeth, S. Connolly, Z. Hong, P.S. Grant, T. Weissgärber, Alternative fabrication routes toward oxide-dispersion-strengthened steels and model alloys. Metall. Mater. Trans. A Phys. Metall. Mater. Sci. **47**(11), 5313–5324 (2016)

15. U.S. Bertoli, G. Guss, S. Wu, M.J. Matthews, J.M. Schoenung, In-situ characterization of laser-powder interaction and cooling rates through high-speed imaging of powder bed fusion additive manufacturing. Mater. Des. **135**, 385–396 (2017)

16. H.K.D.H. Bhadeshia, A rationalisation of shear transformations in steels. Acta Metall. **29**(6), 1117–1130 (1981)

17. H. Bhadeshia, Recrystallisation of practical mechanically alloyed iron-base and nickel-base superalloys. Mater. Sci. Eng. Struct. Mater. Prop. Microstruct. Proces. **223**(1–2), 64–77 (1997)

18. H. Bhadeshia, Properties of fine-grained steels generated by displacive transformation. Mater. Sci. Eng. A **481**, 36–39 (2008)

19. H.K.D.H. Bhadeshia, *Bainite in Steels: Theory and Practice* (Maney Publishing, 2015)

20. H.K.D.H. Bhadeshia, J.W. Christian, Bainite in steels. Metall. Trans. **A 21 A**(4), 767–797 (1990)

21. H.K.D.H. Bhadeshia, H. Harada, High-strength (5 GPa) steel wire: An atom-probe study. Appl. Surf. Sci. **67**(1), 328–333 (1993)

22. J. Bonnín-Roca, P. Vaishnav, J. Mendonça, G. Morgan, Getting past the hype about 3-D printing. MIT Sloan Manag. Rev. **58**(3), 57 (2017)

23. C. Borchers, R. Kirchheim, Cold-drawn pearlitic steel wires. Prog. Mater. Sci. **82**, 405–444 (2016)

24. X. Boulnat, N. Sallez, M. Dadé, A. Borbély, J.L. Béchade, Y. De Carlan, J. Malaplate, Y. Bréchet, F. De Geuser, A. Deschamps, Influence of oxide volume fraction on abnormal growth of nanostructured ferritic steels during non-isothermal treatments: An in situ study. Acta Mater. **97**, 124–130 (2015)

25. B. Bramfitt, F. Fletcher, A Perspective on the Manufacture of Modern-Day High-Strength Steel Rail. AREMA 2013 Annual Conference, Indianapolis (2013)

26. B.L. Bramfitt, R.L. Cross, D.P. Wirick, Advanced in-line head hardening of rail. Iron Steelmak. **22**(1), 17–21 (1995)

27. S.S. Brenner, Tensile strength of whiskers. J. Appl. Phys. **27**(12), 1484–1491 (1956)

28. A. Brownrigg, R. Boelen, M. Toyama, Delamination of hard drawn eutectoid steel, in *Fracture 84*, ed. by S. R. Valluri, D. M. R. Taplin, P. R. Rao, J. F. Knott, R. Dubey, (Pergamon, 1984), pp. 1431–1438

29. T.S. Byun, J.H. Yoon, D.T. Hoelzer, Y.B. Lee, S.H. Kang, S.A. Maloy, Process development for 9Cr nanostructured ferritic alloy (NFA) with high fracture toughness. J. Nucl. Mater. **449**(1), 290–299 (2014)

30. F.G. Caballero, M.K. Miller, S.S. Babu, C. Garcia-Mateo, Atomic scale observations of bainite transformation in a high carbon high silicon steel. Acta Mater. **55**(1), 381–390 (2007)

31. F.G. Caballero, C. Garcia-Mateo, M.J. Santofimia, M.K. Miller, C. García de Andrés, New experimental evidence on the incomplete transformation phenomenon in steel. Acta Mater. **57**(1), 8–17 (2009)

32. F.G. Caballero, M.K. Miller, C. Garcia-Mateo, Carbon supersaturation of ferrite in a nanocrystalline bainitic steel. Acta Mater. **58**(7), 2338–2343 (2010)

33. F.G. Caballero, M.K. Miller, C. Garcia-Mateo, J. Cornide, M.J. Santofimia, Temperature dependence of carbon supersaturation of ferrite in bainitic steels. Scr. Mater. **67**(10), 846–849 (2012)

34. J.W. Cahn, W.C. Hagel, Theory of the Pearlite Reaction, in *The Selected Works of John W. Cahn*, (1998), pp. 133–198

35. C. Capdevila, Y.L. Chen, N.C.K. Lassen, A.R. Jones, H. Bhadeshia, Heterogeneous deformation and recrystallisation of iron base oxide dispersion strengthened PM2000 alloy. Mater. Sci. Technol. **17**(6), 693–699 (2001)

36. C. Capdevila, Y.L. Chen, A.R. Jones, H. Bhadeshia, Grain boundary mobility in Fe-base oxide dispersion strengthened PM2000 alloy. ISIJ Int. **43**(5), 777–783 (2003)

37. C.E. Carlton, P.J. Ferreira, What is behind the inverse Hall–Petch effect in nanocrystalline materials? Acta Mater. **55**(11), 3749–3756 (2007)

38. J. Chao, R. Rementeria, M. Aranda, C. Capdevila, J.L. Gonzalez-Carrasco, Comparison of ductile-to-brittle transition behavior in two similar ferritic oxide dispersion strengthened alloys. Materials **9**(8), 637 (2016)
39. A. Chauhan, F. Bergner, A. Etienne, J. Aktaa, Y. de Carlan, C. Heintze, D. Litvinov, M. Hernandez-Mayoral, E. Oñorbe, B. Radiguet, A. Ulbricht, Microstructure characterization and strengthening mechanisms of oxide dispersion strengthened (ODS) Fe-9%Cr and Fe-14%Cr extruded bars. J. Nucl. Mater. **495**, 6–19 (2017)
40. C.-Y. Chen, Microstructure characterization of nanocrystalline bainitic steel during tempering. J. Alloys Compd. **762**, 340–346 (2018)
41. T.S. Chou, H. Bhadeshia, Grain control in mechanically alloyed oxide dispersion-strengthened Ma-957 steel. Mater. Sci. Technol. **9**(10), 890–897 (1993)
42. T.S. Chou, H.K.D.H. Bhadeshia, Recrystallization temperatures in mechanically alloyed oxide-dispersion-strengthened MA956 and MA957 steels. Mater. Sci. Eng. A **189**(1), 229–233 (1994)
43. D.M. Clatterbuck, D.C. Chrzan, J.W. Morris, The ideal strength of iron in tension and shear. Acta Mater. **51**(8), 2271–2283 (2003)
44. P. Clayton, D. Danks, Effect of interlamellar spacing on the wear resistance of eutectoid steels under rolling-sliding conditions. Wear **135**(2), 369–389 (1990)
45. H. Conrad, J. Narayan, On the grain size softening in nanocrystalline materials. Scr. Mater. **42**(11), 1025–1030 (2000)
46. J. Cornide, C. Garcia-Mateo, C. Capdevila, F.G. Caballero, An assessment of the contributing factors to the nanoscale structural refinement of advanced bainitic steels. J. Alloys Compd. **577**, S43–S47 (2013)
47. A.L.M. Costa, A.C.C. Reis, L. Kestens, M.S. Andrade, Ultra grain refinement and hardening of IF-steel during accumulative roll-bonding. Mater. Sci. Eng. A **406**(1), 279–285 (2005)
48. F. Cruz-Gandarilla, A.M. Salcedo-Garrido, M. Avalos, R. Bolmaro, T. Baudin, J.G. Cabañas-Moreno, H.J. Dorantes-Rosales, *EBSD Characterization of an IF Steel Processed by Accumulative Roll Bonding* (IOP Publishing, n.d.)
49. N. Cunningham, Y. Wu, D. Klingensmith, G.R. Odette, On the remarkable thermal stability of nanostructured ferritic alloys. Mater. Sci. Eng. A **613**, 296–305 (2014)
50. N.J. Cunningham, M.J. Alinger, D. Klingensmith, Y. Wu, G.R. Odette, On nano-oxide coarsening kinetics in the nanostructured ferritic alloy MA957: A mechanism based predictive model. Mater. Sci. Eng. A **655**, 355–362 (2016)
51. A. Czyrska-Filemonowicz, D. Clemens, W.J. Quadakkers, The effect of high temperature exposure on the structure and oxidation behaviour of mechanically alloyed ferritic ODS alloys. J. Mater. Process. Tech. **53**(1–2), 93–100 (1995)
52. M.d.G.M. da Fonseca Gomes, L.H. de Almeida, L.C.F.C. Gomes, I. Le May, Effects of microstructural parameters on the mechanical properties of eutectoid rail steels. Mater. Charact. **39**(1), 1–14 (1997)
53. S. Das Bakshi, A. Leiro, B. Prakash, H.K.D.H. Bhadeshia, Dry rolling/sliding wear of nanostructured pearlite. Mater. Sci. Technol. **31**(14), 1735–1744 (2015)
54. A. Das, H.-W. Viehrig, E. Altstadt, F. Bergner, J. Hoffmann, Why do secondary cracks preferentially form in hot-rolled ODS steels in comparison with hot-extruded ODS steels? Crystals **8**(8), 306 (2018)
55. J. De Messemaeker, B. Verlinden, J. Van Humbeeck, Texture of IF steel after equal channel angular pressing (ECAP). Acta Mater. **53**(15), 4245–4257 (2005)
56. C. Doñate-Buendía, F. Frömel, M.B. Wilms, R. Streubel, J. Tenkamp, T. Hupfeld, M. Nachev, E. Gökce, A. Weisheit, S. Barcikowski, F. Walther, J.H. Schleifenbaum, B. Gökce, Oxide dispersion-strengthened alloys generated by laser metal deposition of laser-generated nanoparticle-metal powder composites. Mater. Des. **154**, 360–369 (2018)
57. P. Dubuisson, Y.d. Carlan, V. Garat, M. Blat, ODS Ferritic/martensitic alloys for sodium fast reactor fuel pin cladding. J. Nucl. Mater. **428**(1), 6–12 (2012)
58. M. Durand-Charre, *Of Swords and Swordmaking. Microstructure of Steels and Cast Irons* (Springer, 2004), pp. 13–34

59. J.D. Embury, R.M. Fisher, The structure and properties of drawn pearlite. Acta Metall. **14**(2), 147–159 (1966)
60. M. Enomoto, W. Huang, H. Ma, Modeling pearlite transformation in super-high strength wire rods: II. Simulation in Fe–C Base multi-component alloys. ISIJ Int. **52**(4), 632–637 (2012)
61. J. Farmer, B. El-dasher, J. Ferreira, M. S. d. Caro, A. Kimura, Coolant Compatibility Studies for Fusion and Fusion-Fission Hybrid Reactor Concepts: Corrosion of Oxide Dispersion Strengthened Iron-Chromium Steels and Tantalum in High Temperature Molten Fluoride Salts, Lawrence Livermore National Lab.(LLNL), Livermore, CA (United States) (2010)
62. M.H. Farshidianfar, A. Khajepour, A.P. Gerlich, Effect of real-time cooling rate on microstructure in laser additive manufacturing. J. Mater. Process. Technol. **231**, 468–478 (2016)
63. C. Fazio, A. Alamo, A. Almazouzi, S. De Grandis, D. Gomez-Briceno, J. Henry, L. Malerba, M. Rieth, European cross-cutting research on structural materials for generation IV and transmutation systems. J. Nucl. Mater. **392**(2), 316–323 (2009)
64. K. Feyrer, *Wire Ropes* (Springer, 2007)
65. K. Feyrer, Wire Ropes Under Tensile Load, in *Wire Ropes: Tension, Endurance, Reliability*, ed. by K. Feyrer, (Springer, Berlin/Heidelberg, 2015a), pp. 59–177
66. K. Feyrer, Wire Ropes, Elements and Definitions, in *Wire Ropes: Tension, Endurance, Reliability*, ed. by K. Feyrer, (Springer, Berlin/Heidelberg, 2015b), pp. 1–57
67. R.B. Figueiredo, F.L. Sicupira, L.R.C. Malheiros, M. Kawasaki, D.B. Santos, T.G. Langdon, Formation of epsilon martensite by high-pressure torsion in a TRIP steel. Mater. Sci. Eng. A **625**, 114–118 (2015)
68. Y. Fukuda, K. Oh-ishi, Z. Horita, T.G. Langdon, Processing of a low-carbon steel by equal-channel angular pressing. Acta Mater. **50**(6), 1359–1368 (2002)
69. M.C. García-Alonso, J.L. González-Carrasco, P. Pérez, V.A.C. Haanappel, M.L. Escudero, J. Chao, M.F. Stroosnijder, A surface modified ODS superalloy by thermal oxidation for potential implant applications. J. Mater. Sci. Mater. Med. **12**(7), 589–596 (2001)
70. C. Garcia-Mateo, TIANOBAIN-towards industrial applicability of (medium C) nanostructured bainitic steels-RFCS. Impact **2018**(1), 94–96 (2018)
71. C. Garcia-Mateo, F.G. Caballero, Ultra-high-strength bainitic steels. ISIJ Int. **45**(11), 1736–1740 (2005)
72. C. Garcia-Mateo, F.G. Caballero, H.K.D.H. Bhadeshia, Development of hard Bainite. ISIJ Int. **43**(8), 1238–1243 (2003a)
73. C. Garcia-Mateo, C. Fg, B. Hkdh, Acceleration of low-temperature bainite. ISIJ Int. **43**(11), 1821–1825 (2003b)
74. C. Garcia-Mateo, M. Peet, F.G. Caballero, H. Bhadeshia, Tempering of hard mixture of bainitic ferrite and austenite. Mater. Sci. Technol. **20**(7), 814–818 (2004)
75. C. Garcia-Mateo, F.G. Caballero, M.K. Miller, J.A. Jimenez, On measurement of carbon content in retained austenite in a nanostructured bainitic steel. J. Mater. Sci. **47**(2), 1004–1010 (2012)
76. C. Garcia-Mateo, T. Sourmail, F.G. Caballero, V. Smanio, M. Kuntz, C. Ziegler, A. Leiro, E. Vuorinen, R. Elvira, T. Teeri, Nanostructured steel industrialisation: Plausible reality. Mater. Sci. Technol. **30**(9), 1071–1078 (2014)
77. C. Garcia-Mateo, J.A. Jimenez, H.W. Yen, M.K. Miller, L. Morales-Rivas, M. Kuntz, S.P. Ringer, J.R. Yang, F.G. Caballero, Low temperature bainitic ferrite: Evidence of carbon super-saturation and tetragonality. Acta Mater. **91**, 162–173 (2015)
78. C. Garcia-Mateo, G. Paul, M.C. Somani, D.A. Porter, L. Bracke, A. Latz, C. Garcia De Andres, F.G. Caballero, Transferring nanoscale Bainite concept to lower C contents: A perspective. Metals **7**(5), 159 (2017)
79. V.G. Gavriljuk, Decomposition of cementite in pearlitic steel due to plastic deformation. Mater. Sci. Eng. A **345**(1–2), 81–89 (2003)
80. G. Ghosh, G.B. Olson, Computational thermodynamics and the kinetics of martensitic transformation. J. Phase Equilibria **22**(3), 199–207 (2001)

81. C. Ghosh, M. Shome, Dynamic strain aging during wire drawing and its effect on electro-chemical behaviour. Ironmak. Steelmak. **44**(10), 789–795 (2017)
82. H. Gleiter, Chapter 9: Microstructure, in *Physical Metallurgy*, ed. by R. W. Cahn, P. Haasen, 4th edn., (Oxford, North-Holland, 1996), pp. 843–942
83. H. Gleiter, Nanostructured materials: Basic concepts and microstructure. Acta Mater. **48**(1), 1–29 (2000)
84. H. Gleiter, N. Hansen, A. Horsewell, T. Leffers, H. Lilholt, Deformation of polycrystals: Mechanisms and microstructures
85. W. Gong, Y. Tomota, S. Harjo, Y.H. Su, K. Aizawa, Effect of prior martensite on bainite transformation in nanobainite steel. Acta Mater. **85**, 243–249 (2015)
86. S. Goto, R. Kirchheim, T. Al-Kassab, C. Borchers, Application of cold drawn lamellar microstructure for developing ultra-high strength wires. Trans. Nonfer. Metal Soc. China (English Edition) **17**(6), 1129–1138 (2007)
87. G. Gottstein, L.S. Shvindlerman, Triple junction drag and grain growth in 2D polycrystals. Acta Mater. **50**(4), 703–713 (2002)
88. T. Gräning, M. Rieth, A. Möslang, A. Kuzmin, A. Anspoks, J. Timoshenko, A. Cintins, J. Purans, Investigation of precipitate in an austenitic ODS steel containing a carbon-rich process control agent. Nucl. Mater. Energy **15**, 237–243 (2018)
89. S.L. Grassie, *Mechanics and Fatigue in Wheel/Rail Contact* (Elsevier, 2012)
90. V. Guttmann, A. Mediavilla, O. Ruano, Preoxidized ma 956 in an S-O-C-bearing atmosphere. Mater. High Temp. **11**(1–4), 42–50 (1993)
91. G.A.J. Hack, Developments in the production of oxide dispersion strengthened superalloys. Powder Metall. **27**(2), 73–79 (1984)
92. E.O. Hall, The deformation and ageing of mild steel: III discussion of results. Proc. Phys. Soc. Sect. B **64**(9), 747 (1951)
93. R.G. Hamerton, D.M. Jaeger, A.R. Jones, New enhanced performance stainless steel. Mater. World **1**(1), 9–10 (1993)
94. K. Han, G.D.W. Smith, D.V. Edmonds, Pearlite phase transformation in Si and V steel. Metall. Mater. Trans. A **26**(7), 1617–1631 (1995)
95. H.S. Hasan, M.J. Peet, H.K.D.H. Bhadeshia, Severe tempering of bainite generated at low transformation temperatures. Int. J. Mater. Res. **103**(11), 1319–1324 (2012)
96. H.S. Hasan, M.J. Peet, M.N. Avettand-Fènoël, H.K.D.H. Bhadeshia, Effect of tempering upon the tensile properties of a nanostructured bainitic steel. Mater. Sci. Eng. A **615**, 340–347 (2014)
97. M.F. Hawkes, R.F. Mehl, The effect of cobalt on the rate of nucleation and the rate of growth of pearlite. Trans. Am. Inst. Min. Metall. Petrol. Eng. **172**, 467–492 (1947)
98. R.F. Hehemann, K.R. Kinsman, H.I. Aaronson, A debate on the bainite reaction. Metall. Trans. **3**(5), 1077–1094 (1972)
99. J. Henry, S.A. Maloy, *9 - Irradiation-Resistant Ferritic and Martensitic Steels as Core Materials for Generation IV Nuclear Reactors. Structural Materials for Generation IV Nuclear Reactors. P. Yvon* (Woodhead Publishing, 2017), pp. 329–355
100. I. Hilger, M. Tegel, M.J. Gorley, P.S. Grant, T. Weißgärber, B. Kieback, The structural changes of Y2O3 in ferritic ODS alloys during milling. J. Nucl. Mater. **447**(1–3), 242–247 (2014)
101. C.E. Hinchliffe, G.D.W. Smith, Strain aging of pearlitic steel wire during post-drawing heat treatments. Mater. Sci. Technol. **17**(2), 148–154 (2001)
102. D. T. Hoelzer, J. Bentley, M. K. Miller, M. K. Sokolov, T. S. Byun, M. Li, Development of High-Strength ODS Steels for Nuclear Energy Applications. ODS 2010 Materials Workshop. ODS 2010 Materials Workshop, Qualcomm Conference Center Jacobs Hall, University of California, San Diego (2010)
103. M.H. Hong, K. Hono, W.T. Reynolds, T. Tarui, Atom probe and transmission electron microscopy investigations of heavily drawn pearlitic steel wire. Metall. Mater. Trans. A **30**(3), 717–727 (1999)
104. M. Honjo, T. Kimura, K. Hase, Effect of Cr on lamellar spacing and high-temperature stability in eutectoid steels. ISIJ Int. **56**(1), 161–167 (2016)

105. K. Hono, M. Ohnuma, M. Murayama, S. Nishida, A. Yoshie, T. Takahashi, Cementite decomposition in heavily drawn pearlite steel wire. Scr. Mater. **44**(6), 977–983 (2001)

106. P. Hosemann, H.T. Thau, A.L. Johnson, S.A. Maloy, N. Li, Corrosion of ODS steels in lead–bismuth eutectic. J. Nucl. Mater. **373**(1), 246–253 (2008)

107. C.X. Huang, G. Yang, B. Deng, S.D. Wu, S.X. Li, Z.F. Zhang, Formation mechanism of nanostructures in austenitic stainless steel during equal channel angular pressing. Philos. Mag. **87**(31), 4949–4971 (2007)

108. C.N. Hulme-Smith, I. Lonardelli, A.C. Dippel, H.K.D.H. Bhadeshia, Experimental evidence for non-cubic bainitic ferrite. Scr. Mater. **69**(5), 409–412 (2013)

109. Y.-M. Hwang, T.-Y. Kuo, Dieless drawing of stainless steel tubes. Int. J. Adv. Manuf. Technol. **68**(5), 1311–1316 (2013)

110. A. Inoue, T. Ogura, T. Masumoto, Burgers vectors of dislocations in cementite crystal. Scr. Metall. **11**(1), 1–5 (1977)

111. Y.V. Ivanisenko, W. Lojkowski, R.Z. Valiev, H.J. Fecht, The strain induced cementite dissolution in carbon steel - experimental facts and theoretical approach. Solid State Phenom. **94**, 45–50 (2003)

112. Y.S. Jang, M.P. Phaniraj, D.-I. Kim, J.-H. Shim, M.-Y. Huh, Effect of Aluminum content on the microstructure and mechanical properties of hypereutectoid steels. Metall. Mater. Trans. A **41**(8), 2078–2084 (2010)

113. A.R. Jones, J. Ritherdon, Reduction in defect content of oxide dispersion strengthened alloys. Mater. High Temp. **16**(4), 181–188 (1999)

114. C.P. Jongenburger, R.F. Singer, *Recrystallization of Ods Superalloys* (Dgm Metallurgy Information, New York, 1989)

115. D. Juul Jensen, Growth rates and misorientation relationships between growing nuclei/grains and the surrounding deformed matrix during recrystallization. Acta Metall. Mater. **43**(11), 4117–4129 (1995)

116. R. Kapoor, *Severe Plastic Deformation of Materials. Materials Under Extreme Conditions* (Elsevier, 2017), pp. 717–754

117. L. Kaufman, M. Cohen, Thermodynamics and kinetics of martensitic transformations. Prog. Met. Phys. **7**, 165–246 (1958)

118. B. Kazimierzak, J.M. Prignon, R.I. Fromont, An ODS material with outstanding creep and oxidation resistance above 1100°C. Mater. Des. **13**(2), 67–70 (1992)

119. K. Kempen, E. Yasa, L. Thijs, J.P. Kruth, J. Van Humbeeck, Microstructure and mechanical properties of selective laser melted 18Ni-300 steel. Phys. Procedia **12**, 255–263 (2011)

120. J.H. Kim, T.S. Byun, D.T. Hoelzer, C.H. Park, J.T. Yeom, J.K. Hong, Temperature dependence of strengthening mechanisms in the nanostructured ferritic alloy 14YWT: Part II—Mechanistic models and predictions. Mater. Sci. Eng. A **559**, 111–118 (2013)

121. R.L. Klueh, J.P. Shingledecker, R.W. Swindeman, D.T. Hoelzer, Oxide dispersion-strengthened steels: A comparison of some commercial and experimental alloys. J. Nucl. Mater. **341**(2), 103–114 (2005)

122. G. Korb, M. Rühle, H.P. Martinz, *New Iron-Based ODS-Superalloys for High Demanding Applications* (International Gas Turbine and Aeroengine Congress and Exposition, American Society of Mechanical Engineers Digital Collection, 1991)

123. Kourbatoff, Contribution à l'étude métallographique des aciers trempés. Rev. Met. Paris **2**(1), 169–186 (1905)

124. Kourbatoff, Contribution à l'étude métallographique des aciers trempés(1). Rev. Met. Paris **5**(10), 704–710 (1908)

125. E. Kozeschnik, H.K.D.H. Bhadeshia, Influence of silicon on cementite precipitation in steels. Mater. Sci. Technol. **24**(3), 343–347 (2008)

126. G. Krauss, High-carbon steels: Fully pearlitic microstructures and applications. Steels Process. Struct. Perform., 281–295 (2005)

127. L.P. Kubin, Y. Estrin, C. Perrier, On static strain ageing. Acta Metall. Mater. **40**(5), 1037–1044 (1992)

128. P. Kürnsteiner, M.B. Wilms, A. Weisheit, P. Barriobero-Vila, E.A. Jägle, D. Raabe, Massive nanoprecipitation in an Fe-19Ni-xAl maraging steel triggered by the intrinsic heat treatment during laser metal deposition. Acta Mater. **129**, 52–60 (2017)
129. G. Langford, M. Cohen, Calculation of cell-size strengthening of wire-drawn iron. Metall. Mater. Trans. B **1**(5), 1478–1480 (1970)
130. J. Languillaume, G. Kapelski, B. Baudelet, Cementite dissolution in heavily cold drawn pearlitic steel wires. Acta Mater. **45**(3), 1201–1212 (1997)
131. A. Leiro, E. Vuorinen, K.G. Sundin, B. Prakash, T. Sourmail, V. Smanio, F.G. Caballero, C. Garcia-Mateo, R. Elvira, Wear of nano-structured carbide-free bainitic steels under dry rolling–sliding conditions. Wear **298-299**, 42–47 (2013)
132. B. Leng, S. Ukai, Y. Sugino, Q. Tang, T. Narita, S. Hayashi, F. Wan, S. Ohtsuka, T. Kaito, Recrystallization texture of cold-rolled oxide dispersion strengthened Ferritic steel. ISIJ Int. **51**(6), 951–957 (2011)
133. B. Leng, S. Ukai, T. Narita, Y. Sugino, Q. Tang, N. Oono, S. Hayashi, F. Wan, S. Ohtsuka, T. Kaito, Effects of two-step cold rolling on recrystallization Behaviors in ODS Ferritic steel. Mater. Trans. **53**(4), 652–657 (2012)
134. R. Lewis, U. Olofsson, *Wheel-Rail Interface Handbook* (Elsevier, 2009)
135. Y.J. Li, P. Choi, S. Goto, C. Borchers, D. Raabe, R. Kirchheim, Evolution of strength and microstructure during annealing of heavily cold-drawn 6.3GPa hypereutectoid pearlitic steel wire. Acta Mater. **60**(9), 4005–4016 (2012)
136. Y. Li, D. Raabe, M. Herbig, P.-P. Choi, S. Goto, A. Kostka, H. Yarita, C. Borchers, R. Kirchheim, Segregation stabilizes Nanocrystalline bulk steel with near theoretical strength. Phys. Rev. Lett. **113**(10), 106104 (2014)
137. S. Liu, F. Zhang, Z. Yang, M. Wang, C. Zheng, Effects of Al and Mn on the formation and properties of nanostructured pearlite in high-carbon steels. Mater. Des. **93**, 73–80 (2016)
138. L. Liu, Q. Ding, Y. Zhong, J. Zou, J. Wu, Y.-L. Chiu, J. Li, Z. Zhang, Q. Yu, Z. Shen, Dislocation network in additive manufactured steel breaks strength–ductility trade-off. Mater. Today **21**(4), 354–361 (2018a)
139. N. Liu, X. Zhang, J. Ding, J. He, F.-x. Yin, Microstructure and mechanical properties of nanobainitic steel subjected to multiple isothermal heat treatments. J. Iron Steel Res. Int. (2018b)
140. A.J. London, S. Lozano-Perez, M.P. Moody, S. Amirthapandian, B.K. Panigrahi, C.S. Sundar, C.R.M. Grovenor, Quantification of oxide particle composition in model oxide dispersion strengthened steel alloys. Ultramicroscopy **159**, 360–367 (2015)
141. R. Łyszkowski, T. Czujko, R.A. Varin, Multi-axial forging of Fe3Al-base intermetallic alloy and its mechanical properties. J. Mater. Sci. **52**(5), 2902–2914 (2017)
142. E. Ma, T. Zhu, Towards strength–ductility synergy through the design of heterogeneous nanostructures in metals. Mater. Today **20**(6), 323–331 (2017)
143. C. Mack, M.S. Bartlett, *On Clumps Formed When Convex Laminae or Bodies Are Placed at Random in Two or Three Dimensions* (Cambridge University Press, 1956)
144. T.R. Malow, C.C. Koch, Grain growth in nanocrystalline iron prepared by mechanical attrition. Acta Mater. **45**(5), 2177–2186 (1997)
145. S. A. Maloy, E. Aydogan, O. Anderoglu, C. Lavender, I. Anderson, J. Rieken, J. Lewandowski, D. Hoelzer, G. R. Odette, Characterization of Tubing from Advanced ODS alloy (FCRD-NFA1), Los Alamos National Lab.(LANL), Los Alamos, NM (United States); Pacific Northwest National Lab.(PNNL), Richland, WA (United States); Ames Lab., Ames, IA (United States); Oak Ridge National Lab.(ORNL), Oak Ridge, TN (United States) (2016)
146. X. Mao, S.H. Kang, T.K. Kim, S.C. Kim, K.H. Oh, J. Jang, Microstructure and mechanical properties of ultrafine-grained austenitic oxide dispersion strengthened steel. Metall. Mater. Trans. A Phys. Metall. Mater. Sci. **47**(11), 5334–5343 (2016)
147. L. Marechal, B. Lesage, A.M. Huntz, R. Molins, Oxidation behavior of ODS Fe-Cr-Al alloys: Aluminum depletion and lifetime. Oxid. Met. **60**(1–2), 1–28 (2003)
148. N. Maruyama, T. Tarui, H. Tashiro, Atom probe study on the ductility of drawn pearlitic steels. Scr. Mater. **46**(8), 599–603 (2002)

149. S.J. Matas, R.F. Hehemann, The structure of bainite in hypoeutectoid steels. Trans. Metall. Soc. AIME **221**(1), 179–185 (1961)
150. R.F. Mehl, The structure and rate of formation of pearlite. Trans. Am. Soc. Metal. **29**(4), 813–862 (1941)
151. R.F. Mehl, The structure and rate of formation of pearlite. Metall. Microstruct. Analy. **4**(5), 423–443 (2015)
152. G. Merceron, R. Molins, J.L. Strudel, Oxidation behaviour and microstructural evolution of FeCrAl ODS alloys at high temperature. 4th Int. Conf. Micros. Oxid. **17**(1), 149–157 (2000)
153. M.K. Miller, C.M. Parish, Role of alloying elements in nanostructured ferritic steels. Mater. Sci. Technol. **27**(4), 729–734 (2011)
154. M.K. Miller, C.M. Parish, Q. Li, *Advanced Oxide Dispersion Strengthened and Nanostructured Ferritic Alloys* (Taylor & Francis, 2013a)
155. M.K. Miller, C.M. Parish, Q. Li, Advanced oxide dispersion strengthened and nanostructured ferritic alloys. Mater. Sci. Technol. **29**(10), 1174–1178 (2013b)
156. N. Min, W. Li, X. Jin, X. Wang, T. Yang, C. Zhang, Influence of aging on the mechanical property of a cold drawn pearlite steel. Jinshu Xuebao/Acta Metallurgica Sinica **42**(10), 1009–1013 (2006)
157. N. Min, W. Li, H. Li, X. Jin, Atom probe and mössbauer spectroscopy investigations of cementite dissolution in a cold drawn eutectoid steel. J. Mater. Sci. Technol. **26**(9), 776–782 (2010)
158. Y. Mine, D. Haraguchi, Z. Horita, K. Takashima, High-pressure torsion of metastable austenitic stainless steel at moderate temperatures. Philos. Mag. Lett. **95**(5), 269–276 (2015)
159. L. Morales-Rivas, *Microstructure and Mechanical Response of Nanostructured Bainitic Steels* (Universidad Carlos III de Madrid, 2016)
160. L. Morales-Rivas, A. González-Orive, C. Garcia-Mateo, A. Hernández-Creus, F.G. Caballero, L. Vázquez, Nanomechanical characterization of nanostructured bainitic steel: Peak force microscopy and Nanoindentation with AFM. Sci. Rep. **5**, 17164 (2015)
161. L. Morales-Rivas, C. Garcia-Mateo, T. Sourmail, M. Kuntz, R. Rementeria, F.G. Caballero, Ductility of nanostructured bainite. Metals **6**(12), 302 (2016)
162. L. Morales-Rivas, F. Archie, S. Zaefferer, M. Benito-Alfonso, S.-P. Tsai, J.-R. Yang, D. Raabe, C. Garcia-Mateo, F.G. Caballero, Crystallographic examination of the interaction between texture evolution, mechanically induced martensitic transformation and twinning in nanostructured bainite. J. Alloys Compd. **752**, 505–519 (2018)
163. S. Morito, H. Yoshida, T. Maki, X. Huang, Effect of block size on the strength of lath martensite in low carbon steels. Mater. Sci. Eng. A **438-440**, 237–240 (2006)
164. I. Mueller, R. Rementeria, F.G. Caballero, M. Kuntz, T. Sourmail, E. Kerscher, A constitutive relationship between fatigue limit and microstructure in nanostructured bainitic steels. Materials **9**(10), 831 (2016)
165. K. Murakami, K. Mino, H. Harada, H. Bhadeshia, Nonuniform recrystallization in a mechanically alloyed nickel-base superalloy. Metall. Trans.Phys. Metall. Mater. Sci. **24**(5), 1049–1055 (1993)
166. W.J. Nam, C.M. Bae, Void initiation and microstructural changes during wire drawing of pearlitic steels. Mater. Sci. Eng. A **203**(1–2), 278–285 (1995)
167. W.J. Nam, H.R. Song, C.M. Bae, Effect of microstructural features on ductility of drawn Pearlitic carbon steels. ISIJ Int. **45**(8), 1205–1210 (2005)
168. T. Narita, S. Ukai, B. Leng, S. Ohtsuka, T. Kaito, Characterization of recrystallization of 12Cr and 15Cr ODS ferritic steels. J. Nucl. Sci. Technol. **50**(3), 314–320 (2013)
169. J.P. Naylor, The influence of the lath morphology on the yield stress and transition temperature of martensitic- bainitic steels. Metall. Trans. A. **10**(7), 861–873 (1979)
170. A. Nordmann, Invisible origins of nanotechnology: Herbert Gleiter, materials science, and questions of prestige. Perspect. Sci. **17**(2), 123–143 (2009)
171. I. Ochiai, S. Nishida, H. Ohba, A. Kawana, Application of hypereutectoid steel for development of high strength steel wire. Tetsu-to-Hagane **79**(9), 1101–1107 (1993)

172. I. Ochial, Effect of metallurgical factors on strengthening of steel tire cord. Wire J. Int. (USA) **26**(12), 50–61 (1993)

173. G.R. Odette, Recent Progress in developing and qualifying nanostructured Ferritic alloys for advanced fission and fusion applications. JOM **66**(12), 2427–2441 (2014)

174. G.R. Odette, On the status and prospects for nanostructured ferritic alloys for nuclear fission and fusion application with emphasis on the underlying science. Scr. Mater. **143**, 142–148 (2018)

175. G.R. Odette, M.J. Alinger, B.D. Wirth, Recent developments in irradiation-resistant steels. Annu. Rev. Mater. Res. **38**, 471–503 (2008)

176. H. Ohba, S. Nishida, T. Tarui, K. Yoshimua, M. Sugimoto, K. Matsuoka, N. Hikita, M. Toda, High-performance wire rods produced with DLP. Nippon Steel Tech. Rep. **96**(6), 50–56 (2007)

177. H. Oka, T. Tanno, S. Ohtsuka, Y. Yano, T. Kaito, Effect of nitrogen concentration on nano-structure and high-temperature strength of 9Cr-ODS steel. Nucl. Mater. Ener. **16**, 230–237 (2018)

178. A.K. Padap, G.P. Chaudhari, V. Pancholi, S.K. Nath, Warm multiaxial forging of AISI 1016 steel. Mater. Des. **31**(8), 3816–3824 (2010)

179. S. Pal, M.E. Alam, G.R. Odette, J. Lewandowski, D.T. Hoelzer, S.A. Maloy, Characterization of microstructure and texture of NFA-1 for two deformation-processing routes. Fusion Mater. Semiannu. Prog. Rep. **58**, 29–41 (2015)

180. A. S. Pandit, Theory of the pearlite transformation in steels (2011)

181. H.G. Paris, D.K. Kim, Metallurgy, processing and applications of metal wires. TMS (1996)

182. D.B. Park, J.W. Lee, Y.S. Lee, K.T. Park, W.J. Nam, Effects of the annealing temperature and time on the microstructural evolution and corresponding the mechanical properties of cold-drawn steel wires. Met. Mater. Int. **14**(1), 59–64 (2008)

183. M.J. Peet, S.S. Babu, M.K. Miller, H. Bhadeshia, Tempering of low-temperature bainite. Metall. Mater. Trans. A **48**(7), 3410–3418 (2017)

184. G.E. Pellisier, M.F. Hawkes, W.A. Johnson, R.F. Mehl, The interlamellar spacing of pearlite. Trans. Am. Soc. Metal. **30**, 1049–1086 (1942)

185. J.J. Pepe, Deformation structure and the tensile fracture characteristics of a cold worked 1080 pearlitic steel. Metall. Trans. **4**(10), 2455–2460 (1973)

186. E.V. Pereloma, Critical assessment 20: On carbon excess in bainitic ferrite. Mater. Sci. Technol. **32**(2), 99–103 (2016)

187. A.J. Perez-Unzueta, J.H. Beynon, Microstructure and wear resistance of pearlitic rail steels. Wear **162**, 173–182 (1993)

188. N.J. Petch, The cleavage strength of polycrystals. J. Iron Steel Inst. **174**, 25–28 (1953)

189. G. Pimentel, J. Chao, C. Capdevila, Recrystallization process in Fe-Cr-Al oxide dispersion-strengthened alloy: Microstructural evolution and recrystallization mechanism. JOM **66**(5), 780–792 (2014)

190. J.-P. Poirier, The Coming of Materials Science. R.W. Cahn. Pergamon, 2001 (568 pages). 130 NL guilders, hardback, ISBN 0-08-042679-4." Eur. Rev. 9(4): 517–522 (2001)

191. M. Praud, F. Mompiou, J. Malaplate, D. Caillard, J. Garnier, A. Steckmeyer, B. Fournier, Study of the deformation mechanisms in a Fe–14% Cr ODS alloy. J. Nucl. Mater. **428**(1), 90–97 (2012)

192. J. Pujante, D. Casellas, T. Sourmail, F. G. Caballero, A. Soto, J. M. Llanos, E. Vuorinen, B. Prakash, J. Hardell, P. V. Moghaddam, Novel Nano-structured Bainitic steels for enhanced durability of Wear resistant components: Microstructural optimisation through simulative Wear and field tests, BAINWEAR (2018)

193. M.P. Puls, J.S. Kirkaldy, The pearlite reaction. Metall. Trans. **3**(11), 2777–2796 (1972)

194. K.G. Raghavendra, A. Dasgupta, C. Ghosh, K. Jayasankar, V. Srihari, S. Saroja, Development of a novel ZrO2 dispersion strengthened 9Cr ferritic steel: Characterization of milled powder and subsequent annealing behavior. Powder Technol. **327**, 267–274 (2018)

195. H. Regle, A. Alamo, Secondary recrystallization of oxide dispersion strengthened ferritic alloys. Journal De Physique, Paris, Fr, Publ by Editions de Physique (1993)

196. M. Reibold, P. Paufler, A.A. Levin, W. Kochmann, N. Pätzke, D.C. Meyer, Materials: Carbon nanotubes in an ancient Damascus sabre. Nature **444**(7117), 286 (2006)
197. R. Rementeria, L. Morales-Rivas, M. Kuntz, C. Garcia-Mateo, E. Kerscher, T. Sourmail, F.G. Caballero, On the role of microstructure in governing the fatigue behaviour of nanostructured bainitic steels. Mater. Sci. Eng. A **630**, 71–77 (2015)
198. R. Rementeria, M.M. Aranda, C. Garcia-Mateo, F.G. Caballero, Improving wear resistance of steels through nanocrystalline structures obtained by bainitic transformation. Mater. Sci. Technol. **32**(4), 308–312 (2016)
199. R. Rementeria, J.A. Jimenez, S.Y.P. Allain, G. Geandier, J.D. Poplawsky, W. Guo, E. Urones-Garrote, C. Garcia-Mateo, F.G. Caballero, Quantitative assessment of carbon allocation anomalies in low temperature bainite. Acta Mater. **133**, 333–345 (2017a)
200. R. Rementeria, J.D. Poplawsky, M.M. Aranda, W. Guo, J.A. Jimenez, C. Garcia-Mateo, F.G. Caballero, Carbon concentration measurements by atom probe tomography in the ferritic phase of high-silicon steels. Acta Mater. **125**, 359–368 (2017b)
201. R. Rementeria, C. Garcia-Mateo, F.G. Caballero, New insights into carbon distribution in Bainitic ferrite. HTM J. Heat Treat. Mater. **73**(2), 68–79 (2018)
202. S.O. Rogachev, V.M. Khatkevich, R.O. Kaibyshev, M.S. Tikhonova, S.V. Dobatkin, Nitrided 08Kh17T steel after high-pressure torsion. Russ. Metall. (Metally) **2015**(11), 861–867 (2015)
203. J. Rösler, E. Arzt, A new model-based creep equation for dispersion strengthened materials. Acta Metall. Mater. **38**(4), 671–683 (1990)
204. D. A. Rutherford, R. J. McWilliams, Surface Hardening of Rails, Google Patents (1970)
205. K. Saeki, K. Iwano, Progress and prospects of rail for railroads. Nippon Steel Sumitomo Met. Tech. Rep. **105**, 19–25 (2013)
206. N. Sallez, C. Hatzoglou, F. Delabrouille, D. Sornin, L. Chaffron, M. Blat-Yrieix, B. Radiguet, P. Pareige, P. Donnadieu, Y. Bréchet, Precipitates and boundaries interaction in ferritic ODS steels. J. Nucl. Mater. **472**, 118–126 (2016)
207. B.P.J. Sandvik, The bainite reaction in Fe– Si– C alloys: The primary stage. Metall. Trans. A. **13**(5), 777–787 (1982a)
208. B.P.J. Sandvik, The Bainite reaction in Fe– Si– C alloys: The secondary stage. Metall. Trans. A. **13**(5), 789–800 (1982b)
209. M.A. Santajuana, R. Rementeria, M. Kuntz, J.A. Jimenez, F.G. Caballero, C. Garcia-Mateo, Low-temperature Bainite: A thermal stability study. Metall. Mater. Trans. A **49**(6), 2026–2036 (2018)
210. S. Sato, K. Wagatsuma, S. Suzuki, M. Kumagai, M. Imafuku, H. Tashiro, K. Kajiwara, T. Shobu, Relationship between dislocations and residual stresses in cold-drawn pearlitic steel analyzed by energy-dispersive X-ray diffraction. Mater. Charact. **83**, 152–160 (2013)
211. X. Sauvage, J. Copreaux, F. Danoix, D. Blavette, Atomic-scale observation and modelling of cementite dissolution in heavily deformed pearlitic steels. Philos. Mag. A Phys. Cond. Matt. Struct. Defect. Mech. Prop. **80**(4), 781–796 (2000)
212. J.H. Schneibel, M. Heilmaier, Hall-Petch breakdown at elevated temperatures. Mater. Trans. **55**(1), 44–51 (2014)
213. K. Schwab, *The Fourth Industrial Revolution* (Crown Business, 2017)
214. J.G. Sevillano, Room temperature plastic deformation of pearlitic cementite. Mater. Sci. Eng. **21**(C), 221–225 (1975)
215. C.W. Shao, P. Zhang, Y.K. Zhu, Z.J. Zhang, Y.Z. Tian, Z.F. Zhang, Simultaneous improvement of strength and plasticity: Additional work-hardening from gradient microstructure. Acta Mater. **145**, 413–428 (2018)
216. J. Shen, H. Yang, Z. Zhao, J. McGrady, S. Kano, H. Abe, Effects of pre-deformation on microstructural evolution of 12Cr ODS steel under 1473–1673 K annealing. Nucl. Mater. Ener **16**, 137–144 (2018)
217. A. Shirzadi, S. Jackson, *Structural Alloys for Power Plants: Operational Challenges and High-temperature Materials* (Elsevier, 2014)
218. C.T. Sims, N.S. Stoloff, W.C. Hagel, *Superalloys II* (Wiley, 1987)

219. R.F. Singer, G.H. Gessinger, Hot isostatic pressing of oxide dispersion strengthened superalloy parts. Powder Metall. Int. **15**(3), 119–121 (1983)

220. S.B. Singh, H.K.D.H. Bhadeshia, Estimation of bainite plate-thickness in low-alloy steels. Mater. Sci. Eng. A **245**(1), 72–79 (1998)

221. C.S. Smith, *A History of Metallography: The Development of Ideas on the Structure of Metals before 1890* (University of Chicago Press, Chicago, 1960)

222. R.A. Smith, Fatigue in transport: Problems, solutions and future threats. Process Saf. Environ. Prot. **76**(3), 217–223 (1998)

223. M.A. Sokolov, D.T. Hoelzer, R.E. Stoller, D.A. McClintock, Fracture toughness and tensile properties of nano-structured ferritic steel 12YWT. J. Nucl. Mater. **367-370**, 213–216 (2007)

224. B. Song, X. Zhao, S. Li, C. Han, Q. Wei, S. Wen, J. Liu, Y. Shi, Differences in microstructure and properties between selective laser melting and traditional manufacturing for fabrication of metal parts: A review. Front. Mech. Eng. **10**(2), 111–125 (2015)

225. T. Sourmail, Bainite and Superbainite in long products and forged applications. HTM J. Heat Treat. Mat. **72**(6), 371–378 (2017)

226. T. Sourmail, V. Smanio, Low temperature kinetics of bainite formation in high carbon steels. Acta Mater. **61**(7), 2639–2648 (2013)

227. T. Sourmail, F.G. Caballero, C. Garcia-Mateo, V. Smanio, C. Ziegler, M. Kuntz, R. Elvira, A. Leiro, E. Vuorinen, T. Teeri, Evaluation of potential of high Si high C steel nanostructured bainite for wear and fatigue applications. Mater. Sci. Technol. **29**(10), 1166–1173 (2013a)

228. T. Sourmail, V. Smanio, C. Ziegler, V. Heuer, M. Kuntz, F.G. Caballero, C. Garcia-Mateo, J. Cornide, R. Elvira, A. Leiro, *Novel Nanostructured Bainitic Steel Grades to Answer the Need for High-Performance Steel Components (Nanobain)* (D.-G. f. R. a. I. D. G. I. T. U. G. R. F. f. C. a. Steel, Brussels, European Comission, 2013b)

229. T. Sourmail, F. Danoix, C. Garcia-Mateo, F.G. Caballero, R. Rementeria, L. Morales-Rivas, R. Pizzaro, R. Janisch, S. Sampath, I. Mueller, E. Kerscher, M. Kuntz, *Understanding Basic Mechanism to Optimize and Predict in Service Properties of Nanobainitic Steels (MECBAIN)* (E. C. D.-G. f. R. a. I. D. D. I. T. U. D. C. a. Steel, Brussels, European Comission, 2017a)

230. T. Sourmail, C. Garcia-Mateo, F.G. Caballero, L. Morales-Rivas, R. Rementeria, M. Kuntz, Tensile ductility of nanostructured bainitic steels: Influence of retained austenite stability. Metals **7**(1), 31 (2017b)

231. D. Sporer, K. Lempenauer, PM ODS materials for high temperature applications, in *13th International Plansee Seminars*, (Freund Publ. House, Reutte, Austria, 1994)

232. T. S. Srivatsan, C. Daniels, A. Prakash, High cycle fatigue behavior of high carbon steel wires. Proceedings of the Annual Convention of the Wire Association International (1997)

233. D. H. Stone, G. G. Knupp, T. American Society for, S. S. Materials. Committee A-1 on Steel and A. Related, Rail steels, developments, processing, and use: a symposium sponsored by ASTM Committee A-1 on Steel, Stainless Steel, and Related Alloys, American Society for Testing and Materials, Denver, Colo., 17–18 Nov. 1976, American Society for Testing and Materials (1978)

234. D. Strong, I. Sirichakwal, G.P. Manogharan, T. Wakefield, Current state and potential of additive–hybrid manufacturing for metal parts. Rapid Prototyp. J. **23**(3), 577–588 (2017)

235. K. Suresh, M. Nagini, R. Vijay, M. Ramakrishna, R.C. Gundakaram, A.V. Reddy, G. Sundararajan, Microstructural studies of oxide dispersion strengthened austenitic steels. Mater. Des. **110**, 519–525 (2016)

236. P.J. Szabó, P. Bereczki, B. Verő, The effect of multiaxial forging on the grain refinement of low alloyed steel. Period. Polytech. Mech. Eng. **55**(1), 63–66 (2011)

237. T. Takahashi, M. Nagumo, Flow stress and work-hardening of Pearlitic steel. Trans. Jpn. Inst. Metals **11**(2), 113–119 (1970)

238. J. Takahashi, T. Tarui, K. Kawakami, Three-dimensional atom probe analysis of heavily drawn steel wires by probing perpendicular to the pearlitic lamellae. Ultramicroscopy **109**(2), 193–199 (2009)

239. J. Takahashi, M. Kosaka, K. Kawakami, T. Tarui, Change in carbon state by low-temperature aging in heavily drawn pearlitic steel wires. Acta Mater. **60**(1), 387–395 (2012)

240. E.M. Taleff, J.J. Lewandowski, B. Pourladian, Microstructure-property relationships in pearlitic eutectoid and hypereutectoid carbon steels. JOM **54**(7), 25–30 (2002)
241. T. Tarui, N. Maruyama, J. Takahashi, S. Nishida, H. Tashiro, Microstructure control and strengthening of high-carbon steel wires. Nippon Steel Tech. Rep **91**, 56–61 (2005)
242. H. Tashiro, H. Sato, Effect of alloying elements on the lamellar spacing and the degree of regularity of pearlite in eutectoid steel. J. Jpn. Inst. Metals **55**(10), 1078–1085 (1991)
243. G.F. Taylor, A method of drawing metallic filaments and a discussion of their properties and uses. Phys. Rev. **23**(5), 655 (1924)
244. K.A. Taylor, M. Cohen, Aging of ferrous martensites. Prog. Mater. Sci. **36**, 151–272 (1992)
245. M. Tikhonova, R. Kaibyshev, X. Fang, W. Wang, A. Belyakov, Grain boundary assembles developed in an austenitic stainless steel during large strain warm working. Mater. Charact. **70**, 14–20 (2012)
246. I.B. Timokhina, H. Beladi, X.Y. Xiong, Y. Adachi, P.D. Hodgson, Nanoscale microstructural characterization of a nanobainitic steel. Acta Mater. **59**(14), 5511–5522 (2011)
247. I.B. Timokhina, K.-D. Liss, D. Raabe, K. Rakha, H. Beladi, X.Y. Xiong, P.D. Hodgson, Growth of bainitic ferrite and carbon partitioning during the early stages of bainite transformation in a 2 mass% silicon steel studied by in situ neutron diffraction, TEM and APT. J. Appl. Crystallogr. **49**(2), 399–414 (2016)
248. J. Toribio, E. Ovejero, Effect of cumulative cold drawing on the pearlite interlamellar spacing in eutectoid steel. Scr. Mater. **39**(3), 323–328 (1998)
249. N. Tsuji, Y. Saito, H. Utsunomiya, S. Tanigawa, Ultra-fine grained bulk steel produced by accumulative roll-bonding (ARB) process. Scr. Mater. **40**(7), 795–800 (1999)
250. S. Ukai, M. Harada, H. Okada, M. Inoue, S. Nomura, S. Shikakura, K. Asabe, T. Nishida, M. Fujiwara, Alloying design of oxide dispersion strengthened ferritic steel for long life FBRs core materials. J. Nucl. Mater. **204**, 65–73 (1993)
251. S. Ukai, T. Okuda, M. Fujiwara, T. Kobayashi, S. Mizuta, H. Nakashima, Characterization of high temperature creep properties in recrystallized 12Cr-ODS ferritic steel claddings. J. Nucl. Sci. Technol. **39**(8), 872–879 (2002)
252. S. Ukai, R. Miyata, S. Kasai, N. Oono, S. Hayashi, T. Azuma, R. Kayano, E. Maeda, S. Ohtsuka, Super high-temperature strength in hot rolled steels dispersing nanosized oxide particles. Mater. Lett. **209**, 581–584 (2017)
253. E.E. Underwood, *The Mathematical Foundations of Quantitative Stereology. Stereology and Quantitative Metallography* (ASTM International, 1972)
254. R.Z. Valiev, R.K. Islamgaliev, I.V. Alexandrov, Bulk nanostructured materials from severe plastic deformation. Prog. Mater. Sci. **45**(2), 103–189 (2000)
255. R. Z. Valiev, A. P. Zhilyaev, T. G. Langdon, Bulk Nanostructured Materials: Fundamentals and Applications, Wiley (2013)
256. G.F. Vander Voort, The Interlamellar spacing of pearlite. Prac. Metall. **52**(8), 419–436 (2015)
257. G.F. Vander Voort, A. Roósz, Measurement of the interlamellar spacing of pearlite. Metallography **17**(1), 1–17 (1984)
258. J.D. Verhoeven, A.H. Pendray, W.E. Dauksch, S.R. Wagstaff, Damascus steel revisited. JOM **70**(7), 1331–1336 (2018)
259. R. Verreet, What we Can Learn from Wire Rope Failures: Predictable and Unpredictable Rope Failures. Safe Use of Ropes: Proceedings of the OIPEEC Conference 2011. College Station, Texas (2011)
260. R. Verret, W. Lindsey, Wire rope inspection and examination. Aachen, PR GmbH Werbeagentur und Verlag: 31 (1996)
261. M. Wang, J. Shan, C. Zheng, M. Zhang, Z. Yang, F. Zhang, Effects of deformation and addition of aluminium on spheroidisation of high-carbon-bearing steel. Mater. Sci. Technol. **34**(2), 161–171 (2018)
262. G.S. Was, *Irradiation Hardening and Deformation. Fundamentals of Radiation Materials Science: Metals and Alloys. G. S. Was* (Springer, Berlin/Heidelberg, 2007), pp. 581–642
263. P. Watté, J. Van Humbeeck, E. Aernoudt, I. Lefever, Strain ageing in heavily drawn eutectoid steel wires. Scr. Mater. **34**(1), 89–95 (1996)

264. M. K. West, Processing and characterization of oxide dispersion strengthened 14YWT ferritic alloys. PhD, University of Tennessee (2006)
265. S. Westerkamp, Electrical Resistance, Critical Shear Strength, and Microstructure of Pearlitic Steel Cold Drawn to Varying Degrees of Deformation (2010)
266. S.H. Whang, *Nanostructured Metals and Alloys: Processing, Microstructure, Mechanical Properties and Applications* (Elsevier, 2011)
267. J.D. Whittenberger, Effect of strain rate on the fracture behavior at 1366 K of the bcc iron base oxide dispersion strengthened alloy MA 956. Metall. Trans. A. **10**(9), 1285–1295 (1979)
268. C.A. Williams, P. Unifantowicz, N. Baluc, G.D.W. Smith, E.A. Marquis, The formation and evolution of oxide particles in oxide-dispersion-strengthened ferritic steels during processing. Acta Mater. **61**(6), 2219–2235 (2013)
269. E.A. Wilson, The $\gamma \rightarrow \alpha$ transformation in iron and its dilute alloys. Scr. Metall. **4**(4), 309–311 (1970)
270. R.N. Wright, *Wire Technology: Process Engineering and Metallurgy* (Butterworth-Heinemann, 2016)
271. K.M. Wu, H.K.D.H. Bhadeshia, Extremely fine pearlite by continuous cooling transformation. Scr. Mater. **67**(1), 53–56 (2012)
272. Y. Wu, E.M. Haney, N.J. Cunningham, G.R. Odette, Transmission electron microscopy characterization of the nanofeatures in nanostructured ferritic alloy MA957. Acta Mater. **60**(8), 3456–3468 (2012)
273. Y. Wu, J. Ciston, S. Kräemer, N. Bailey, G.R. Odette, P. Hosemann, The crystal structure, orientation relationships and interfaces of the nanoscale oxides in nanostructured ferritic alloys. Acta Mater. **111**, 108–115 (2016)
274. Y. Wu, T. Stan, J. Ciston, G. R. Odette, DOE Fusion Materials Semiannual Progress Report, DOE/ER-0313/61 (2017)
275. S. Yamashita, S. Ohtsuka, N. Akasaka, S. Ukai, S. Ohnuki, Formation of nanoscale complex oxide particles in mechanically alloyed ferritic steel. Philos. Mag. Lett. **84**(8), 525–529 (2004)
276. Z. Yanushkevich, A. Mogucheva, M. Tikhonova, A. Belyakov, R. Kaibyshev, Structural strengthening of an austenitic stainless steel subjected to warm-to-hot working. Mater. Charact. **62**(4), 432–437 (2011)
277. J.K. Yates, British steel: Innovation in rail steel. Sci. Parliament **53**, 2–3 (1996)
278. c.H. Young, H. Bhadeshia, Strength of mixtures of bainite and martensite. Mater. Sci. Technol. **10**(3), 209–214 (1994)
279. P. Yvon, *Structural Materials for Generation IV Nuclear Reactors* (Woodhead Publishing, 2016)
280. G.G. Zakharova, E.G. Astafurova, The influence of severe plastic deformation by high pressure torsion on structure and mechanical properties of Hadfield steel single crystals. J. Phys. Conf. Ser. **240**(1), 012139 (2010)
281. M.J. Zehetbauer, Y.T. Zhu, *Bulk Nanostructured Materials* (Wiley, 2009)
282. C. Zener, Kinetics of the decomposition of austenite. Trans. Am. Inst. Min. Metall. Eng. **167**, 550–595 (1946)
283. X. Zhang, A. Godfrey, X. Huang, N. Hansen, Q. Liu, Microstructure and strengthening mechanisms in cold-drawn pearlitic steel wire. Acta Mater. **59**(9), 3422–3430 (2011)

Chapter 9
Iron-rich High Entropy Alloys

Fritz Körmann, Zhiming Li, Dierk Raabe, and Marcel H. F. Sluiter

Abbreviations

AFM	Atomic force microscopy
AIMD	Ab initio molecular dynamics
APT	Atom probe tomography
BCC	Body-centered cubic
CALPHAD	Calculation of phase diagrams
CE	Cluster expansion
CPA	Coherent potential approximation
DFT	Density functional theory
DP	Dual phase
EBSD	Electron back scatter diffraction
ECC	Electron channeling contrast

F. Körmann
Department Computational Materials Design, Max-Planck-Institut für Eisenforschung GmbH, Düsseldorf, Nordrhein-Westfalen, Germany
e-mail: koermann@mpie.de

Z. Li
School of Materials Science and Engineering, Central South University, Changsha, Hunan, China
e-mail: lizhiming@csu.edu.cn

D. Raabe
Department Microstructure Physics and Alloy Design, Max-Planck-Institut für Eisenforschung GmbH, Düsseldorf, Nordrhein-Westfalen, Germany
e-mail: raabe@mpie.de

M. H. F. Sluiter (✉)
Department of Materials Science and Engineering, Delft University of Technology, Delft, Zuid Holland, The Netherlands
e-mail: M.H.F.Sluiter@TUDelft.NL

© Springer Nature Switzerland AG 2021
R. Rana (ed.), *High-Performance Ferrous Alloys*,
https://doi.org/10.1007/978-3-030-53825-5_9

EDS	Energy-dispersive X-ray spectroscopy
EMTO	Exact muffin-tin orbital method
EPMA	Electron probe microanalysis
FCC	Face-centered cubic
FIB	Focused ion beam
HCP	Hexagonal-close packed
HEA	High-entropy alloy
iHEA	Interstitial high-entropy alloy
KKR	Korringa-Kohn-Rostoker
LAM	Laser additive manufacturing
LMD	Laser metal deposition
RAP	Rapid alloy prototyping
SEM	Scanning electron microscope
SLM	Selective laser melting
SPR	Spin-polarized relativistic
SQS	Special quasi-random structures
TCFE-9	Thermo-Calc iron database, version 9
TCHEA1	Thermo-Calc high-entropy alloy database version 1
TCHEA2	Thermo-Calc high-entropy alloy database version 2
TEM	Transmission electron microscopy
TRIP	Transformation-induced plasticity
TRIP-DP	Transformation-induced plasticity-assisted dual-phase
TWIP	Twinning-induced plasticity
XRD	X-ray diffraction

9.1 Introduction

Widely considered a breakthrough in conventional alloy design, high-entropy alloys (HEAs) have drawn great attention because the alloys offer numerous opportunities in the huge unexplored compositional space of multicomponent alloys [1–6]. The practical significance is evidenced also by the large number of patent applications that specifically mention "high-entropy alloy", about 500 so far, of which a quarter in the current year according to Google patents. A large number of studies in this field have been motivated by the original HEA concept which suggested that achieving maximized configurational entropy, through equiatomic ratios of multiple principal elements, could stabilize single-phase massive solid solution phases [1]. However, an increasing number of studies have revealed that formation of single-phase solid solutions in HEAs shows only a weak dependence on maximization of the configurational entropy through equiatomic ratios of elements [7–10]. Rather, it was found that maximum entropy is not the most essential parameter when aiming at the design of multicomponent alloys with superior properties [11, 12]. These findings have encouraged efforts to relax the unnecessary restrictions on (1) the equiatomic ratio of multiple principal elements and also on (2) the requirement to arrive at a strictly single-phase solid solution. In this context, reaching beyond the original concept of single-phase solid solutions, several variants of this new design approach have been suggested including nonequiatomic, multiphase, interstitial, duplex, precipitate containing, and metastable HEAs [4, 13–21]. As shown in Fig. 9.1, several typical HEA systems have been developed, including face-centered cubic (FCC) structure-based strong and ductile HEAs, body-centered cubic (BCC) structure-based refractory HEAs, hexagonal-close packed (HCP) structure-based HEAs, light-weight HEAs, and precious-metal functional HEAs. The number of HEA compositions explored is increasing rapidly due to the growing research efforts in this field. Drivers for this research are the promising properties observed for some of these materials such as for instance good cryogenic toughness and high strain hardening [3–6].

Many of the alloys developed so far contain elements that are costly and often have limited availability. Besides the platinum group metals, constituents such as Co, Zr, Hf, Ta, Nb, and to a somewhat lesser extent W, Mo, Ni, and Cr are problematic in this regard. For HEAs to find applications beyond narrow niches, it is thus necessary to formulate HEAs using widely available low cost constituents such as Fe, Al, Mn, Mg, and Si. Perhaps surprisingly, several of the rare earth elements such as La and Ce are actually not as prohibitive as their classification would suggest. Considering cost and availability, Fe is an ideal ingredient, and therefore, in this chapter the focus will be on such HEAs in which Fe is a significant elemental constituent with a more than equiatomic concentration.

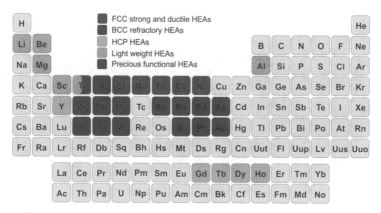

Fig. 9.1 Element groupings highlighting the main ingredients used to synthesize alloys pertaining to the five most typical HEA families, i.e., CALPHAD-based strong and ductile HEAs, BCC-based refractory HEAs, HCP-structured HEAs, light-weight HEAs, and precious functional HEAs [22]. For simplicity, other HEA systems or variants of the alloy classes mentioned above including other substitutional elements or interstitial ones such as B, C, N, O, etc., are not color coded here. Image reproduced from Ref. [22] with permission

9.2   Substitutional Fe-rich High-Entropy Alloys

Thermodynamic investigations of substitutional Fe-rich HEAs showed that the configurational entropy curve of these alloys is rather flat, indicating that a wide range of compositions, aside from the equiatomic ones, assume similar configurational entropy values; see fig. 1 in ref. [23]. Compared to the equiatomic substitutional HEAs with equimolar ratios of all alloying elements, nonequiatomic HEA variants widen the accessible compositional space substantially. Indeed, recent studies have revealed that outstanding mechanical properties exceeding those of substitutional equiatomic HEAs can be achieved by nonequiatomic Fe-rich HEAs [4, 13]. As an example shown in Fig. 9.2, when switching from the equiatomic $Fe_{20}Mn_{20}Ni_{20}Co_{20}Cr_{20}$ (at.%) system to the nonequiatomic $Fe_{80-x}Mn_xCo_{10}Cr_{10}$ (at.%) system, a novel type of transformation-induced plasticity-assisted dual-phase (TRIP-DP) HEAs was developed [4, 24]. The two constituent phases in the Fe-rich HEA, i.e., FCC matrix and the HCP phase, are compositionally equivalent and thus can be both referred to as high-entropy phases [4]. The nonequiatomic HEA features a significantly improved strength-ductility combination compared

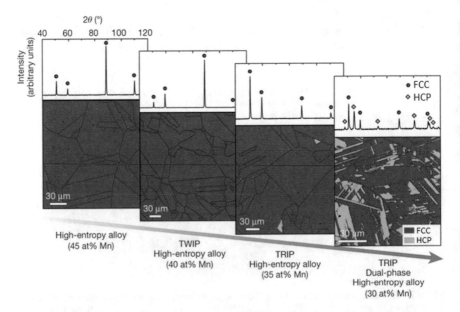

Fig. 9.2 X-ray diffraction (XRD) patterns and electron back-scattered diffraction (EBSD) phase maps of $Fe_{80-x}Mn_xCo_{10}Cr_{10}$ (x = 45 at.%, 40 at.%, 35 at.% and 30 at.%) HEAs [4]. The results reveal the capability for seamless continuous compositional tuning from the single phase into the two-phase HEA regime, once the strict equimolar composition rule is relaxed. Images reproduced from Ref. [4] with permission

to the corresponding equiatomic HEAs. This is mainly due to the combination of massive solid-solution strengthening, providing increased yield strength, and the transformation-induced plasticity (TRIP) effect, providing strain hardening increase particularly in those loading regimes where dislocation hardening gets exhausted [4, 13].

9.3 Interstitially Enhanced Fe-rich High-Entropy Alloys

Besides the substitutional HEA, substantial interstitial element fractions can also be introduced into the strong and ductile Fe-rich HEAs to further improve their mechanical properties. HEAs with functional and significant interstitial content will be referred to by the acronym iHEA. As an example, carbon was added as interstitial element into the TRIP-DP-HEA in the pursuit of two main trends [25]:

(i) The addition of interstitial carbon leads to a slight increase in stacking fault energy and hence phase stability, enabling the tuning of the FCC matrix phase stability to a critical point so as to trigger the twinning-induced plasticity (TWIP) effect while maintaining the TRIP effect, thereby further improving the strain-hardening ability of the alloy.
(ii) HEAs can benefit profoundly from interstitial solid solution strengthening with its huge local distortions instead of only the established massive substitutional solid solution strengthening provided by its multiple principal elements.

The iHEA is indeed characterized by the combination of various strengthening mechanisms originating from interstitial and substitutional solid solution, TWIP, TRIP, nanoprecipitates, dislocation interactions, stacking faults, and grain boundaries, leading to twice the tensile strength compared to the equiatomic $Co_{20}Cr_{20}Fe_{20}Mn_{20}Ni_{20}$ reference HEA while maintaining identical ductility [25]. Adding interstitial carbon has other microstructural benefits as well. In $Co_{20}Cr_{20}Fe_{20}Mn_{20}Ni_{20}$ alloy, an increase of C content leads to a significantly higher energy barrier to recrystallization during annealing. The fine-grained microstructure in partially recrystallized ($\sim$20 vol %) interstitial HEA with a C content of 0.8 at. % shows a more than five times higher yield strength compared to the as-homogenized coarse-grained ($\sim$200 μm) carbon-free alloy [26]. In other alloys too, such as in $Fe_{40.4}Ni_{11.3}Mn_{34.8}Al_{7.5}Cr_6$ alloy, interstitials such as C and B are shown to significantly enhance strength, ductility, and work-hardening rate simultaneously [27]. Furthermore, carbon has been shown to reduce corrosion rates in austenitic high-Mn steels under high pH conditions [28]. However, interstitial carbon content might be limited if strong carbide formers are present in the alloy [29].

9.4 Simulation Techniques for High-Entropy Alloys

To explore the large composition space spanned by HEAs, experimental techniques alone, which are discussed in Sect. 9.5 below, are not sufficient. Different simulation techniques can provide complementary insights for these alloys in particular for determining phase stabilities (e.g., for identifying single-phase alloys) and for characterizing their mechanical as well as other (e.g., magnetic or electrical) materials properties.

The most common simulation techniques in this respect for HEAs are currently atomistic simulations (typically realized by first-principles calculations see, e.g., [30]) and thermodynamic approaches based on the CALPHAD (acronym for calculation of phase diagrams) method in combination with empirical databases (see, e.g., [31]).

9.4.1 The CALPHAD Approach for Inspecting Phase Diagrams and Phase Stabilities

The CALPHAD method [31] is one of the most efficient approaches for evaluating phase stabilities. It is based on advanced parametrization techniques of empirical databases of assessed experimental data (mostly unaries, binaries, and selected ternaries). It provides reasonable predictions of phase stability of multicomponent alloys (see, e.g., Refs. [32, 33] and references therein).

A number of reviews have been devoted recently to related CALPHAD studies for multicomponent alloys [32, 34–39]. The main underlying concept of the CALPHAD approach is to minimize the Gibbs energy at a given temperature, composition, and pressure. The main strength of the method is that, based on assessed binary and ternary alloys, it allows to extrapolate toward multicomponent phase space. Even though the predictability of this extrapolation cannot be estimated a priori, it often turns out to be in excellent quantitative agreement with experiment [32, 35].

The main conceptual steps within the CALPHAD method is to first collect a comprehensive collection on phase equilibria and thermodynamic information of available experiments and simulations. The thermodynamic information is then quantified by fitting an analytic Gibbs energy expression with adjustable parameters to the available data including coexisting phases. The assessed data are often bundled into usually commercial databases. The power of the approach is that this procedure also allows description of phase diagrams and stable phases in regions which were not included in the assessment, in particular, the combination of subsystems to describe multicomponent alloys such as HEAs.

For HEAs, the most popular employed realizations of the CALPHAD approach in practice are the Gibbs energy minimizing codes such as PANDAT [40] or the Thermo-Calc package [41]. Many of the investigations referred to in Sects. 9.2 and 9.3 are guided by CALPHAD simulations.

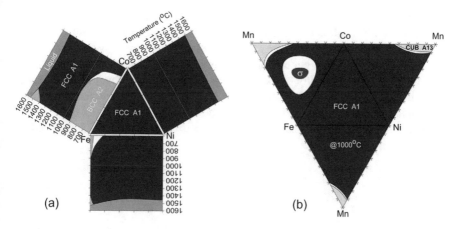

Fig. 9.3 (a) Subcomponent binaries and isothermal sections at 1000 °C for FeCoNi system. (b) Coupled isothermal section diagrams at 1000 °C for a quaternary FeCoNiMn system. (Images reproduced from [33] with permission)

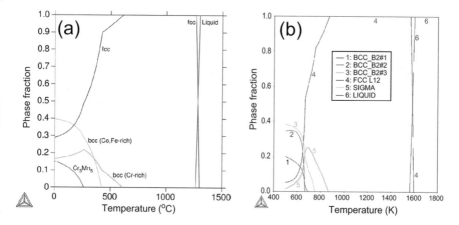

Fig. 9.4 Phase diagram of FeCoNiCrMn system (a) reproduced from [42] and (b) reproduced from [38] with permission. Both databases reveal a BCC Cr-rich phase at ~500 °C. The calculations have both been performed using Thermo-Calc together with a (a) self-developed database [42] and (b) based on the TCHEA1 database [43]

As an example, in Fig. 9.3a the binary phase isothermal section for FeCoNi system at 1000 °C (middle) and the binary phase diagrams of the constitutive binaries are shown [33]. At 1000 °C there is a perfect solid solution irrespective of the compositional ratio. This is a direct consequence of the continuous solid solution formation of the involved binaries. In Fig. 9.3b the isothermal section at 1000 °C for the five component FeCoNiCrMn (Cantor) alloy is shown. The addition of Mn results in the formation of a sigma phase in the FeCoMn ternary phase space. As another application, in Fig. 9.4a, b phase diagrams of the five-component

FeCoNiCrMn alloy constructed from two independent databases are shown. In both calculations a Cr-rich sigma phase appears at lower temperatures, which has also been reported in recent experimental studies [8].

These phase diagram simulations can thus give an important insight into the single-phase forming ability of high-entropy alloys [44]. However, when sufficient experimental/computational input data is lacking the predictions become rather uncertain and various CALPHAD databases may give conflicting results. This is shown in Fig. 9.5 where phase fractions have been determined for some alloys

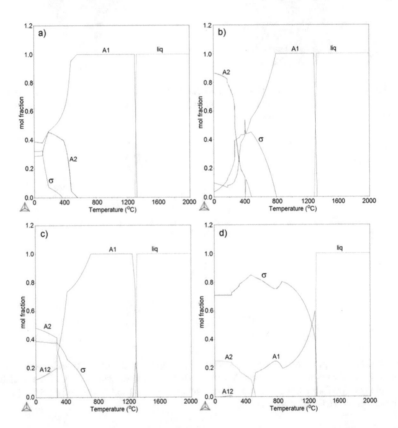

Fig. 9.5 Phase fractions as function of temperature for the alloys (**a**, **b**) $Co_{20}Cr_{20}Fe_{20}Mn_{20}Ni_{20}$ (**c**, **d**) $Co_{10}Cr_{10}Fe_{40}Mn_{40}$ (**e**, **f**) $Co_{10}Cr_{10}Fe_{50}Mn_{30}$ (**g**, **h**) $Co_{10}Cr_{10}Fe_{49.5}Mn_{30}Co_{0.5}$. The TCFE-9 database was used for panels on the left (**a**, **c**, **e**, **g**) and TCHEA2 was used for panels on the right (**b**, **d**, **f**, **h**). Experiments indicate that $Co_{20}Cr_{20}Fe_{20}Mn_{20}Ni_{20}$ and $Co_{10}Cr_{10}Fe_{40}Mn_{40}$ alloys form A1 (FCC) solid solutions at ambient, while $Co_{10}Cr_{10}Fe_{50}Mn_{30}$ and $Co_{10}Cr_{10}Fe_{49.5}Mn_{30}Co_{0.5}$ alloys form mixtures of A1 (FCC) and A3 (HCP) solid solutions, with the latter, $Co_{10}Cr_{10}Fe_{49.5}Mn_{30}Co_{0.5}$ alloy, also containing carbides such as $M_{23}C_6$. Although outside its stated range of validity, it appears that TCFE-9 gives results that are closer to the experimentally observed phases than TCHEA2. The sigma phase particularly is much more abundantly present in the TCHEA2 results than in the actual alloys. It should be noted that neither database predicts the presence of A3 (HCP) solid solutions

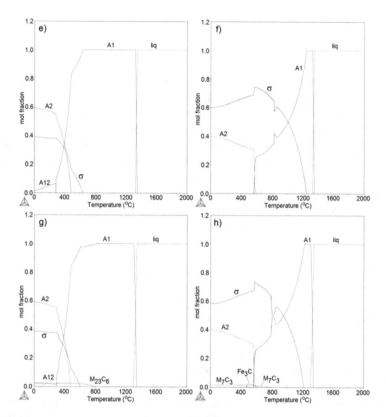

Fig. 9.5 (continued)

according to two thermodynamic databases, the Thermo-Calc iron database (TCFE-9) and the Thermo-Calc high-entropy alloy database (TCHEA2) [45]. Furthermore, the CALPHAD approach addresses questions pertaining to phase stability only; other important properties such as mechanical properties and atomistic information are lacking.

In general, therefore, the often-successful extrapolation of assessed subsystems to describe multicomponent alloys is the most powerful advantage of the CAL-PHAD approach but it is also its strongest limitation at the same time because it is a priori not evident, how well thermodynamic properties of alloys can be extrapolated. An alternative simulation technique is provided by parameter-free first-principles calculations and allows the unbiased determination of the alloy properties [24, 30, 46].

9.4.2 Atomistic and Electronic Structure Simulation Techniques for HEAs

First-principles calculations offer an alternative, complementary approach to study thermodynamic and materials properties of HEAs. Such calculations are solely based on quantum mechanical laws and natural constants and thus allow for unbiased simulations of materials without empirical input. The most common realization of such first-principles-based approaches is density functional theory (DFT) [47, 48]. In this theory the quantum-mechanical many-body Schrödinger equation for the electrons is mapped onto an effective one-electron problem in which the electron density plays the key role. DFT is nowadays one of the most often employed techniques in theoretical solid state physics and also recently gained increasing popularity for HEAs; see [30] and references therein.

To simulate chemical disorder for the description of solid solutions, two DFT-based techniques are commonly employed, the supercell approach (usually in combination with the concept of special quasi-random structures (SQS) [49]) and the coherent potential approximation (CPA) (see, e.g., [50–53]).

The supercell-based approaches sketched in Fig. 9.6a employ large (periodic) simulation cells to mimic the chemical disorder. To optimize the latter, the method is usually combined with the special quasi-random structures (SQS) technique [49]. The SQS technique is based on mimicking the correlation functions of the true random state for a small number of clusters such as pairs in near-neighbor shells and compact many-body clusters in order to obtain an improved description of chemical disorder in a finite supercell. The main advantage of this approach is that local effects (e.g., local lattice distortions) and interatomic forces (for computing lattice vibrations or carrying out molecular dynamics simulations) are readably accessible. The main disadvantage of this approach is the computational cost associated with large simulation cells. In principle it is also possible to mimic

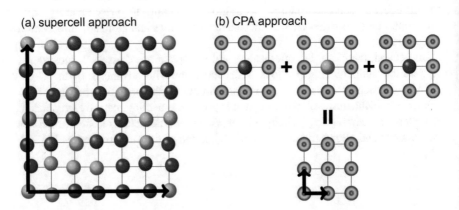

Fig. 9.6 Schematics of the (a) supercell approach and (b) CPA to simulate chemical disorder. (Figure reproduced from [30] with permission)

magnetically disordered configurations within this approach. It is, however, often rather difficult to stabilize such configurations in the self-consistent iterations, i.e., supercell calculations initialized with a specific defined initial magnetic configuration could converge into energetically close but different magnetic states. This often complicates simulation of paramagnetic HEAs in practice. Most of such calculations are carried out with established DFT codes such as WIEN2K [54], QUANTUM ESPRESSO [55], ABINIT [56], or VASP [57], to name a few.

An alternative, mean-field based approach is provided by the coherent potential approximation (CPA). This approach is sketched in Fig. 9.6b. Here the ideal random solid solution is approximated as follows: Each constituting element individually is embedded into an effective medium which is determined in a self-consistent manner from all constituting elements under the condition that the electronic scattering from each of the embedded elements gives rise to a stationary state [58]. In this way the CPA can mimic ideal mixing of chemical elements for any composition in a single primitive cell. It is thus possible to compute some properties of FCC or BCC alloys based on 1-atom unit cells, and this feature makes this approach computationally exceptionally attractive [46, 59–61]. The inherent mean field approximation of CPA allows one to vary the compositional ratio continuously in contrast to supercell approaches in which usually, depending on the chosen supercell size, only discrete and limited alloying concentrations can be computed. Magnetic disorder can also be efficiently included based on the disordered local-moment approach [62, 63]. The disordered local-moment approach is nowadays routinely being used to investigate the paramagnetic state of magnetic materials. The disadvantage of the CPA is, however, that it does not capture interatomic forces and thus lacks the ability of taking proper atomic relaxations into account. In particular for large size-mismatched alloys and/or interstitially alloyed alloys as discussed in Sect. 9.3 such relaxation effects can be crucial. The CPA is implemented in many DFT codes such as, e.g., the SPR-KKR (spin polarized relativistic Korringa-Kohn-Rostoker) code [64] or the exact muffin-tin orbital method (EMTO) [51, 65] and had been recently applied to study various properties of HEAs such as phase stability, elastic properties, or stacking fault energies [30].

As the application of DFT for HEAs has been recently comprehensively reviewed [30], here some of the potential applications are summarized, and the readers are referred to [30] for further details. As also discussed in Sects. 9.2 and 9.3, most DFT simulations for HEAs are so far devoted to study thermodynamic phase stability, thermally induced excitations such as, e.g., due to lattice vibrations (phonons), magnetic properties such as, e.g., Curie temperatures, order-disorder transitions, chemical short-range order, elastic properties, lattice distortions, and stacking fault energies.

Alloy phase stability are usually addressed by cluster expansion techniques [66–70] or perturbational approaches such as the generalized perturbation method [70–74] or the concentration wave approach [75]. In the cluster expansion (CE) technique the alloy under consideration is mapped onto an Ising-type Hamiltonian. The so-called effective cluster interactions can be obtained via the Connolly-Williams method (also sometimes referred to as structure inversion method) [66] in which a pool of DFT energies is used to fit cluster interactions. The cluster expanded Hamiltonian can then be solved, e.g., by Monte Carlo techniques. Several CE codes have been developed which can be used in conjunction with DFT codes like VASP, such as, e.g., the ATAT (the alloy theoretic automated toolkit) program [67] or the UNCLE (universal cluster expansion) code [76]. Recent applications include, e.g., stability calculations for AlNiCoFeTi alloys [77] or MoNbTaVW [78]. An alternative to this approach are perturbation-based schemes such as the generalized perturbation method [70–73] or the concentration wave approach [75]. Both schemes are based on the CPA formalism introduced above and are often implemented in KKR and EMTO codes. The generalized perturbation method (GPM) allows one to compute effective interactions based on the CPA effective medium. These interactions can be used similar as the CE Hamiltonian in combination with Monte Carlo simulations to investigate phase stability as, e.g., recently done for BCC MoNbTaW [46]. In the concentration wave approach, chemical fluctuations are derived from chemical concentration variations of the Gibbs energy functional [61, 74] and can be directly related to Warren-Cowley short-range order parameters [79]. Recent applications include, e.g., AlCoCrFeMn alloys [80]. The main advantage of the perturbation-based approaches is their computational efficiency similar as for the CPA compared to the supercell approach. The main limitations are that relaxation effects are difficult to include as well as that the calculations rely on the solid solution medium (CPA). The supercell-based cluster expansion approaches can include relaxation effects but are therefore also typically less computationally efficient. New advances in machine learning potentials (see [81–83]) have recently provided an alternative approach. In this way a potential is trained by machine learning methods which could be used in Monte Carlo simulations to study the phase stability. This approach has recently been applied to study the phase stability of BCC NbMoTaW [84]. A main advantage of this approach is that it converges faster as a comparable CE while keeping the advantages of the supercell approach (e.g., taking local relaxations into account) [82].

For computing finite-temperature excitations such as lattice vibrations, typically three approaches are used for HEAs. These are with increasing accuracy but also increasing computational expenses analytic approximations based on the elastic properties such as the Debye model [24, 85, 86], harmonic approximations based on phonon calculations [87, 88], and finally ab initio molecular dynamics simulations (AIMD). The Debye model approximations are the by far computationally most efficient ones. In this approach an analytic Debye model is used to express the vibrational Gibbs energy in terms of the Debye temperature which in turn is derived from an equation of state. The latter can be computed based on total energy calculations of the supercell or primitive cell (in combination with the CPA approach). In this way the thermodynamic properties due to vibrations and corresponding phase stabilities can be computed very efficiently [86]. The model is implemented in, e.g., the GIBBS code [89]. The more accurate approach is based on the harmonic approximation for which vibrations are explicitly computed by, e.g., the finite-displacement method (the so-called frozen phonon approach) [90].

This approach is implemented in several program codes such as, e.g., PHONOPY [91] (an overview is also given in Tab. 2 in [92]) which are linked to existing DFT codes such as, e.g., VASP. The harmonic free energy can directly be used to compute, e.g., the Gibbs energy contributions of ordered and disordered alloys [87]. The most accurate method to compute the vibrational Gibbs energy is the AIMD approach [93, 94]. This approach does not make any assumption on the harmonic potential and includes by construction also anharmonic effects. It can be used to study the impact of vibrations on thermal displacements [95–97]. In combination with the thermodynamic integration, the approach can be used to compute the full vibrational Gibbs energy including anharmonicity. The main challenge is to construct an efficient reference potential. The construction of empirical potentials, such as based on the embedded atom method, are challenging for multicomponent alloys [98]. Very recent advances in machine learning potentials have revealed that robust potentials can be constructed with almost DFT-accuracy that can speed up the AIMD simulations for free energies by orders of magnitude [98].

As many Fe-based HEAs are magnetic, magnetic contributions are one of the key properties to be addressed by simulations. The most common technique for multicomponent alloys is based on an effective localized Heisenberg Hamiltonian. The effective magnetic pair interactions can be extracted, e.g., via perturbation approaches typically based on the magnetic force theorem [99], implemented in many DFT codes such as, e.g., the SPR-KKR code [64]. The interactions can be used for example in combinations with analytic mean field expressions for the Curie temperature [100] to compute the magnetic transition temperature or in combination with Monte Carlo simulations [101]. For an overview of methods capturing the magnetic free energies from ab initio the readers are referred to [30, 102–104].

The elastic tensor can be computed straightforwardly either through supercell or CPA approaches by applying various strain tensors [105]. Such computations are now routinely done for HEAs (see [30] and refs. therein).

For the mechanical performance, the deformation mechanisms are important. A key property linking atomistic simulations and materials behavior is the stacking fault energy. Controlling the stacking fault energies by tuning the composition for instance is important strategy for developing alloys showing, e.g., TWIP or TRIP effects [24].

9.5 Experimental Techniques for Synthesis and Processing of High-Entropy Alloys

Experimental techniques for HEAs mainly concern their synthesis, processing, and characterization. For HEAs based on 3d transition metals, particularly the Fe-rich ones, well-established bulk metallurgical processes are available to synthesize high-quality alloy sheets provided that the thermomechanical processing parameters are controlled properly. Except for the traditional casting and melting setups, recently developed combinatorial approaches can be employed also to achieve a rapid screening of suitable alloy compositions. In the following, several combinatorial methods and related examples of their use in the field of HEA development are discussed. The related characterization methods for samples produced by various combinatorial synthesis techniques are discussed briefly as well.

9.5.1 Combinatorial Rapid Alloy Prototyping of Bulk High-Entropy Alloys

Bulk combinatorial material probing for identifying novel HEAs has been conducted using various types of methods, particularly the so-called rapid alloy prototyping (RAP) approach [106]. This is a combinatorial metallurgical screening method consisting of a semicontinuous high-throughput bulk casting, rolling, heat treatment, and sample preparation sequence. It has been successfully applied to the combinatorial screening of lightweight TWIP-type steels [106, 107], high-strength martensitic steels [108], and HEAs [4, 14]. Since this approach involves synthesizing bulk samples of specific compositions, it is particularly suited for screening structural alloys, where the strength, toughness, and ductility are strongly dependent on the entire thermomechanical history and microstructural evolution with strong effects associated with internal length scales such as grain size, phase dispersion, elemental partitioning, and the resulting phase stability.

As schematically illustrated in Fig. 9.7a, the RAP approach enables casting of five different alloys with tuned compositions of a material system in one operational approach. This is achieved by using a set of five copper molds which can be moved stepwise inside the furnace [106]. Although the solidification rate in the RAP casting setup is deliberately high to avoid macrosegregations, in as-cast conditions the distributions of multiple principal elements are typically not fully homogeneous in the bulk HEAs with their coarse dendritic microstructure owing to classical Scheil segregation [109]. Following casting, the alloy blocks with a thickness of 10 mm and varying compositions are subsequently hot-rolled in air (Fig. 9.7b). This step is required for removing the dendritic microstructure and possible inherited casting porosity. The hot-rolling temperature can be adjusted for the specific HEA compositions, and the rolling reduction is generally above 50%. Sometimes, even the hot-rolled HEAs still carry some inherited compositional inhomogeneity. Therefore, after the last hot-rolling pass, the HEA plates are homogenized at high temperatures ($\sim$1200 °C) for several hours ($\sim$2 h) under Ar atmosphere, followed by water quenching (Fig. 9.7c). For HEAs, appropriate homogenization is critically important for achieving uniform distributions of the multiple principal elements in the solid solution structure owing to the sometimes low substitutional diffusion constants in these materials. Therefore, homogenization is often conducted on deformed samples to exploit the higher interfacial diffusion coefficients. The homogenized HEA plates are cut perpendicular to the hot-rolling direction to obtain some segments with smaller size for the subsequent processes and characterization

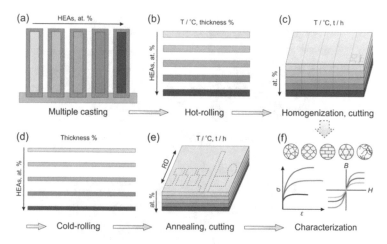

Fig. 9.7 Illustration of the RAP approach for combinatorial synthesis and processing of high-entropy alloys. The RAP approach can also be applied to many other material systems including advanced high-strength steels. (Images reproduced from Ref. [22] with permission)

(Fig. 9.7c). Since the homogenized HEA plates generally exhibit a large grain size, cold-rolling (Fig. 9.7d) and annealing (Fig. 9.7e) processes are required to refine the microstructure toward better mechanical properties. The thickness reduction during cold rolling is generally higher than 50% to obtain a fully deformed microstructure and sufficient stored strain energy in the alloys. Annealing is conducted for recrystallization of the microstructure and for the control of the grain sizes; hence, times and temperatures are in each case adjusted according to the targeted grain sizes and crystallographic textures. After these microstructure-tuning processing steps, samples for microstructure investigation and performance testing are machined from the alloy segments by spark erosion. Following the sample preparation, various characterizations (Fig. 9.7f) can be efficiently performed to determine the composition-microstructure-properties relationships in the HEA systems.

Since bulk samples can be produced by the RAP technique, all characterization methods used in conventional one-alloy-at-a-time practice, such as X-ray diffraction (XRD), electron backscatter diffraction (EBSD), energy-dispersive X-ray spectroscopy (EDS), tensile testing etc., can also be applied to explore the composition-microstructure-properties relationships in the combinatorial RAP processed HEAs.

9.5.2 Combinatorial Diffusion-Multiple Probing of High-Entropy Alloys

As an extension of diffusion-couples, diffusion-multiples allow three or more metal blocks to be placed in diffusional contact, which enables the probing of ternary and higher-order alloy systems. Generally, such diffusion-multiple setups are subjected

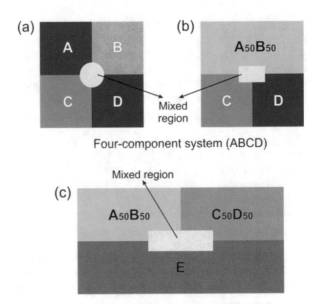

Fig. 9.8 Schematic sketches showing some examples of diffusion-multiple setups for combinational HEA research [22]. (**a–b**) Two typical setups for four-component alloy systems. (**c**) A typical setup for five-component HEA systems. (Images reproduced from Ref. [22] with permission)

to a high temperature to induce thermal interdiffusion to continuously form compositionally graded solid solutions and/or intermetallic compounds [110, 111]. The primary investigations on the combinatorial design of alloys by diffusion-multiples have been conducted by Zhao et al. [110–114]. It was demonstrated that for binary and ternary systems, complete composition libraries of all single-phase regions (including intermetallic compounds) can be achieved by using diffusion-multiples [110–113]. Currently this technique is also being extended to the high-throughput development of multicomponent HEAs [115].

Figure 9.8 presents some typical examples of diffusion-multiple setups for combinatorial HEA research and discovery. For four-component systems (ABCD), four pure metal blocks can be used to build the assembly, and the quaternary mixed region is located at the center of the whole assembly (Fig. 9.8a). Also, setup made by one block of an equiatomic binary alloy (e.g., $A_{50}B_{50}$) and two pure metal blocks (e.g., C and D, respectively) can be employed to investigate the four-component systems (Fig. 9.8b). For five-component systems (ABCDE), two different binary alloy blocks (e.g., $A_{50}B_{50}$ and $C_{50}D_{50}$, respectively) together with one pure metal block (e.g., E) are suitable for building a diffusion-multiple assembly (Fig. 9.8c).

In a diffusion-multiple assembly, the contacted surfaces of different metal blocks are required to be polished and cleaned without contamination. Generally, to avoid contamination from the environment the assembly needs to be evacuated and welded in vacuum along the peripheral junction using methods such as electron beam welding [113]. The whole assembly is then treated in a hot isostatic pressing (HIP) condition at a high temperature (e.g., 1200 °C) for several hours to achieve close, pore-free and dense interface contact of all constituent blocks. Subsequently, the diffusion multiple is heat treated at a high temperature (e.g., 1200 °C) for a long period (generally more than 24 h). The long heat treatment time is a consequence of the requirement of forming sufficiently spatially extended diffusion profiles, so that the solid solution phases and/or intermetallic compounds formed in the diffusion-multiple are large enough to be characterized without overlapping influence from neighboring phases.

Besides the sample-making process, the diffusion-multiple technique also involves localized property measurements using multiple microscale probing methods. Special microscale characterization is necessary because the phases formed in a diffusion-multiple have a very small length scale. Therefore, many of the characterization techniques commonly used in conventional one-alloy-at-a-time practice cannot be applied directly.

After the high-temperature heat-treatment, the diffusion-multiple can be cut, ground, and polished for various local measurements. This enables efficient probing of corresponding composition-structure-property relationships for the various solid-solution phases and/or intermetallic compounds. This is significantly more effective than the conventional one-composition-at-a-time approach. There are a number of probing methods applicable to localized measurements for diffusion-multiples, including electron probe microanalysis (EPMA), EBSD, instrumented nanoinden-tation, and microscale thermal conductivity evaluation. With the capability of detecting most elements of the periodic table (from Be to U), EPMA [116] can provide accurate compositional information of the interdiffusion regions. EBSD [117, 118] can be used to identify the crystal structure of the solid solution phases and/or intermetallic compounds formed in the diffusion-multiples. In an experimental EBSD pattern, since phase identification is conducted by a direct match of the diffraction bands with simulated patterns generated using known structures and lattice parameters, all known crystal structures in the sample are required to be considered by EBSD probing. Therefore, EBSD is not applicable for readily identifying unknown phases or to distinguish between phases with closely related space group relationships (e.g., an ordered phase from its disordered parent). Such ordered phases – disordered parent phase pairs – are quite commonly observed in diffusion-multiple samples of HEAs. In this regard, the combination of focused ion beam (FIB) and transmission electron microscopy (TEM) is an applicable approach to overcome this limitation of EBSD characterization. The FIB technique enables the preparation of site-specific thin foil samples from the regions of interest in the diffusion-multiple while TEM allows the precise determination of crystal structures and even some physical and chemical properties of the targeted region of interest [119, 120]. Therefore, by using EBSD to identify known phases and a FIB-TEM combination to investigate the unknown phases, most structures in diffusion-multiples can be identified effectively.

Mechanical, physical, and chemical properties of various solid-solution phases and/or intermetallic compounds in the diffusion-couples can also be probed by microscale techniques such as instrumented nanoindentation, dynamic force modulation atomic force microscopy (AFM), or the time-domain thermoreflectance technique. Nanoindentation [121–123] enables efficient hardness measurements of microscale phases. The composition-structure-hardness relationships obtained for various phases in the diffusion-couples reflect the solution hardening/strengthening behavior and, thus, are very useful for the design of novel high-strength HEAs. Nanoindentation is also suited to measure the reduced elastic modulus of the materials [121, 124], where "reduced" refers to the fact that a mixed measure of the intrinsic elastic modulus normalized by the Poisson ratio is retrieved by indentation. Dynamic force modulation AFM using a nanoindenter tip with a suitable sensing system is applicable for micro- and nanoscale measurements of the Young's modulus with rather high accuracy [125]. Thermal conductivity measurements at the microscale for identifying and revealing specific phases in such diffusion-couples can be conducted by using a time-domain thermoreflectance technique [110, 126]. The probed thermal conductivity data of metallic phases and/or intermetallic compounds is strongly associated with elemental partitioning, point defects and ordering, and can be used to guide alloy design toward or away from specific phase regimes. Additionally, by using the FIB technique mentioned above, micro- and nanopillars can be prepared and mechanically tested in situ in scanning electron microscopes (SEM) to obtain compressive strength and approximate ductility measures of the solid-solution phases and/or intermetallic compounds in the diffusion-couples, although this is not in all cases very efficient compared to the RAP technique that uses bulk samples as discussed above. Yet, pillar probing can be an essential way to identify individual properties of novel high-component phases that cannot be measured by other means.

The diffusion-multiple technique has proven to be a powerful tool for combinatorial research and development of multicomponent HEAs. For instance, Wilson et al. [115] employed the Co-Cr-Fe-Mn-Ni quinary diffusion multiples with a setup shown in Fig. 9.8c to accelerate alloy development in this HEA system. They confirmed that a disordered quinary region with FCC structure can be formed in the diffusion multiple, and the disordered quinary region was examined by using EDS and nanoindentation. The results revealed that the highest hardness did not correlate well to the maximum in atomic mismatch, suggesting that the severe lattice distortion hypothesis is not a predominant factor for the strengthening of the single-phase solid solution in the Co-Cr-Fe-Mn-Ni HEA system [115].

In principle, all possible compositional variants (i.e., complete composition libraries) of single-phase regions of a given HEA system could be produced in a properly constructed diffusion-multiple, although the size of a certain compositional variant might be too small to be efficiently characterized. It is relatively inexpensive to make diffusion-multiple samples. Therefore, the diffusion-multiple technique has some unique advantages. Especially when applied to HEA development when screening wider composition ranges where multiple phases can occur and where various properties must be probed, it can be advantageous that small samples suffice.

9.5.3 Combinatorial Laser Additive Manufacturing of High-Entropy Alloys

Laser additive manufacturing (LAM) has been employed as an efficient rapid solidification technique for synthesizing bulk HEAs [42, 127–129]. An important advantage of LAM is that it also provides the possibility to explore materials that cannot be cast conventionally, e.g., alloys containing solute ingredients above the solubility limit or elements that are metallurgically hard to blend for instance due to vapor pressure limits. This inherent advantage of LAM is due to its rapid melting and solidification kinetics applied to confined volume regions of the melt pool exposed to the laser [42, 127–130]. There are mainly two types of LAM approaches that can be employed in that context, i.e., laser metal deposition (LMD) and selective laser melting (SLM) [131]. The LMD process has also been referred to as laser engineered net shaping (LENS). During LMD (Fig. 9.9a), the alloy or premixed

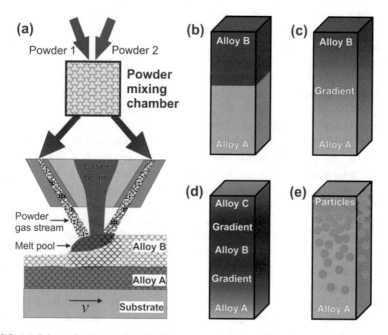

Fig. 9.9 (**a**) Schematic sketch of the LMD process and its capability for producing multilayered bulk alloys. During the LMD process, the premixed powders are transferred into the interaction zone of the laser beam and onto the substrate by injecting them through nozzles with a carrier plasma gas. All blends can be obtained by fractionized contribution from the feeding powders with various compositions. Powders are mixed in a separate chamber prior to injection. (**b–e**) Schematic sketches show examples of compositionally-graded alloys including HEAs that can be achieved by the LMD process. The colors in (**b–e**) refer to alloys of various compositions and blended colors indicate a transition from one composition to another [22]. (Images reproduced from Ref. [22] with permission)

powder blends are transferred into the interaction zone of the laser beam and onto the substrate by injecting them through nozzles with a carrier gas (e.g., Ar). In the SLM process [129, 132], a laser beam is scanning over a powder layer, melting it locally. A fresh powder layer bed is applied after each areal laser scan, and the process is then repeated after the respective 2D layered portion of the shape of the part has been built stepwise in the preceding layers. Both the various types of LMD and SLM methods have been used to produce bulk HEAs [128, 129]. For instance, Joseph et al. [128] fabricated bulk $Al_x CoCrFeNi$ alloys (x = 0.3, 0.6 and 0.85 M fraction) HEAs by utilizing an LMD process, in this case referred to as direct laser fabrication in Ref. [128], from simple elemental powder blends. The resultant alloys showed indeed similar mechanical properties as the reference specimens that had been prepared via conventional arc-melting and solidification. Brif et al. [129] investigated the four-component equiatomic FeCoCrNi HEA prepared by SLM from a prealloyed and gas-atomized powder. The produced alloy showed significantly enhanced yield and ultimate tensile strengths compared to reference material that had been produced by conventional casting. This effect was attributed to the fine microstructure of the single-phase FCC solid solution obtained by laser melting and the associated rapid solidification.

Other than the ability of efficiently synthesizing bulk HEAs with uniform chemical composition across a single part as discussed above, LAM can also be used to produce compositionally graded metal alloys, which renders it a powerful combinatorial approach for rapid alloy development [130, 132–134]. In regard to the above-introduced two types of LAM methods, LMD is an ideal technique for building up compositionally graded materials and in situ synthesis of new materials, while the SLM is generally less suited to fabricate samples with compositional gradients as the process requires a homogeneous powder bed [130, 132]. The current commercial LMD systems can be installed with four or more different feedstock nozzles to deliver powders with varying compositions to the laser, which allows for a nearly infinite combination of compositional gradients [130]. Figure 9.9b–e shows the schematic sketches of four different compositionally graded alloys including HEAs that can be efficiently synthesized with the aid of the LMD process. The compositionally graded part can be designed to include large compositional steps (Fig. 9.9b), and it can also contain smoothly graded transitions from one composition to another (Fig. 9.9c–d) [130]. Metal-matrix-composites, e.g., nanoparticle-reinforced HEA-based materials, with a gradient in particle density and with a composition gradient in the matrix can be fabricated by LMD, as schematically illustrated in Fig. 9.9e. This can be realized when one of the powders being transferred into the laser has a significantly higher melting temperature than the others. Then, the high melting particles are prevented from prealloying with lower melting point powders [130].

Fabricating compositionally graded parts via laser deposition has been shown as an attractive approach to systematically screen the influence of compositional changes on microstructural evolution and associated physical and mechanical properties of HEAs. For instance, Borkar et al. [133] studied compositionally graded $Al_x CrCuFeNi_2$ (0 < x < 1.5) HEAs (referred to as complex concentrated alloys in

[133]) produced by laser deposition (referred to as laser engineered net shaping process in [133]) from elemental powder blends. They found that the microstructure of the HEAs gradually changes from a FCC matrix containing an ordered $L1_2$ phase to a BCC matrix containing an ordered B2 phase [133]. Moreover, the microhardness gradually increases with the increase of the Al content, while the saturation magnetization and coercivity increases and reaches a maximum value when $x = 1.3$ [133]. These findings demonstrated that compositional gradients prepared by laser deposition synthesis methods provide a powerful combinatorial approach for probing composition-microstructure-properties relationships in HEAs.

The number of compositional variants that is accessible in compositionally graded HEA samples produced by combinatorial laser deposition process is huge. The process is associated with relatively high costs, particularly due to the requirement of feeding large amounts of expensive powders. Many of these powders need to be carefully and cleanly prealloyed and produced by established power metallurgical methods. The characterization techniques applicable to diffusion-multiple samples discussed above can be employed also for laser-deposited graded samples. The combinatorial laser deposition process can be used to tap novel HEAs that cannot be cast conventionally, such as those containing solute contents far above the solubility limit, as the laser process involves rapid solidification. This is particularly relevant to HEAs since rapid solidification reduces segregation in such multiple principal element systems.

9.5.4   Combinatorial Thin-Film Synthesis of High-Entropy Alloys

Alloy design can be accelerated by combinatorial thin-film synthesis for the efficient and fast exploration of wide compositional ranges. Functional materials that are in some cases less dependent on details of the microstructure and texture such as certain shape memory alloys can be developed by this method by searching for suitable crystal structures [135]. This high efficiency is due to the fact that for several functional materials, the intrinsic material characteristics such as reversibility of phase transformations [135] are essentially dependent on the chemical composition, with much less significant dependence on grain size, dislocation content, and sample dimensions. Whereas the exploration of compositional and crystal structure phase space is well addressed by the thin-film combinatorial approach, it is currently investigated how this approach can be extended to include also effects of microstructure, e.g., by using step heaters for the systematic variation of thin film microstructures [136].

The combinatorial methodology applied to thin-film growth consists of forming controlled composition gradients from multiple physical deposition sources, most often magnetron sputter sources because of their broad applicability [137]. The result is a materials library comprised of smoothly varying mixtures of the source materials that are characterized by high-throughput methods to identify the optimum composition or composition ranges for one or multiple properties of interest. Because all compositions in a library have been formed at the same time, with the same conditions and can be further processed together (e.g., annealing, environmental exposure), many sources of sample-to-sample variability can be eliminated, clarifying investigations of cause and effect. Additionally, properties that may reach a narrow maximum or only occur in a very restricted composition range can be investigated.

Several approaches to constructing combinatorial libraries by independently varying the individual constituent elements can be used, all based on magnetron sputtering. (1) Codeposition from cathodes that are tilted with respect to the substrate plane results in wedge-shaped nanoscale thin films that are thickest from the geometrically closest edge of the substrate to thinnest at the farthest edge [138]. Multiple cathodes evenly distributed around the center of a substrate (3 cathodes: 120° separation, 4 cathodes: 90° separation, 5 cathodes: 72° separation) each produce such a wedge, with the resulting composition at any point on the substrate being the sum of the material arriving from each cathode. Cosputtered films are atom-scale mixtures of the materials that are often subsequently annealed to form thermodynamically stable phases and reduce defects. (2) When the sputtering cathode is directly opposite to the substrate, a blade shutter positioned parallel to the substrate surface can be moved during the deposition to produce a thickness wedge. Next by rotating the substrate and moving a different target material into position, a controlled series of wedges can be grown. The wedges are typically 10 or 20 nm at the thickest end, and the complete multilayer stack can be repeated many times. With suitable subsequent annealing the multilayers can be compositionally mixed to produce a combinatorial library. (3) A hybrid of these two approaches can be made when the substrate is rotated past several opposed targets, with additional 90° substrate rotation steps in combination with deposition profile-shaping apertures [139]. With typical deposition rates, the net result is a film built up of multilayer gradients that are individually generally less than 1 nm thick through much of their length. Thus, while material is not being cosputtered, neither is the material deposited by a single rotation past a single target truly a well-defined layer, resulting in a hybrid between the cosputtered and multilayer approaches.

For HEAs with multiple principal elements, it eventually happens that the number of constituent elements desired exceeds the number of cathodes available. Increasing the number of confocal cathodes requires increasing the diameter of the circle on which they are situated and increasing the cathode center to substrate center distance, with the consequent problems of larger vacuum chamber sizes and longer target to substrate distances. Alternatively, decreasing the individual cathode's diameter leads to physically smaller combinatorial libraries having steeper composition gradients in turn making it often difficult to measure and characterize distinct, individual alloy compositions. These issues can be addressed by using homogeneously mixed, multiple-element targets that are confocally cosputtered with other mixed or elemental targets. Or, a single target can be made of separate material segments [140], yielding a combinatorial library after sputtering. However, in order to deposit different composition-spread ranges, custom targets of the appropriate composition need to be fabricated. Since the as-deposited element compositions are in general not identical to the target composition and can change as the target is used, the added complexity limits the versatility of this approach.

The combinatorial thin-film materials library approach is not only dependent upon subsequent high-throughput measurement and characterization of individual composition points; it is also highly conducive to this by having regular, designed gradients organized on a common substrate. As long as the library is kept together on this substrate, all sample areas experience the same environments during postdeposition handling as they similarly did during any pretreatments and during the thin film growth itself. Additionally, characterization instruments need only similarly sized programmable x-y translation tables and a common measurement grid to yield data sets that can be directly compared, combined, and archived. Physical properties such as thickness measurement by scanning needle profilometry or confocal laser microscopy, elemental composition mapping by EDS, phase mapping by XRD, surface structures characterization by SEM, and scanning probe microscopy (SPM) are all automated for data collection. More recently, thin film chemical properties can be measured by automated scanning X-ray photoelectron spectroscopy (XPS) [141] and (photo)electrochemical scanning droplet cell techniques [142]. Further examples are mechanical property screening by automated nanoindentation [143], magnetic properties by the magneto-optical Kerr effect (MOKE) [144], and optical properties by photostand [145].

Combinatorial libraries are particularly useful in refining and verifying predictions made by theoretical calculations and are suitable for designing novel HEAs [146]. Sputtered thin films, particularly when codeposited from elemental sources with the arriving elements being mixed on the atomic scale, provide a time- and effort-efficient alternative to the homogenization steps needed for bulk HEA manufacture such as by arc-melting [147].

9.6 Microstructure and Properties of Fe-rich High-Entropy Alloys

As introduced above, various Fe-rich HEAs can be prepared and characterized by multiple techniques [21]. Similar to the single-phase equiatomic $Co_{20}Cr_{20}Fe_{20}Mn_{20}Ni_{20}$ alloy, the Fe-rich single-phase nonequiatomic $Fe_{40}Mn_{27}Ni_{26}Co_5Cr_2$ [11] and $Fe_{40}Mn_{40}Co_{10}Cr_{10}$ [12] alloys can also form a fully recrystallized microstructure containing a high amount of annealing twins in equiaxed matrix grains with uniformly distributed elements via appropriate processing routes. When shifting the single-phase nonequiatomic HEAs to dual-phase (DP) nonequiatomic HEAs and further to iHEAs, the complexity of the microstructures gradually increases.

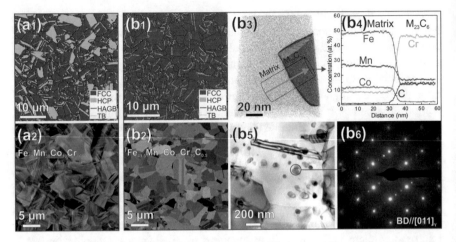

Fig. 9.10 Typical microstructures of the Fe-rich $Fe_{50}Mn_{30}Co_{10}Cr_{10}$ and $Fe_{49.5}Mn_{30}Co_{10}Cr_{10}C_{0.5}$ alloys after recrystallization annealing of 3 min [21]. (a_1) is an EBSD phase map and (a_2) is an ECC image of the DP $Fe_{50}Mn_{30}Co_{10}Cr_{10}$ alloy. (b_1), (b_2), (b_3), (b_4), (b_5), and (b_6) are an EBSD phase map, ECC image, APT tip reconstruction, elemental profiles across an interface of matrix and carbide, TEM bright-field image, and selected area diffraction pattern of the interstitial $Fe_{49.5}Mn_{30}Co_{10}Cr_{10}C_{0.5}$ alloy, respectively. The diffraction spots marked by red circles in (b_6) show the FCC structure of the $M_{23}C_6$ carbides. (Images reproduced from Ref. [21] with permission)

Figure 9.10 shows the typical microstructures of the Fe-rich nonequiatomic $Fe_{50}Mn_{30}Co_{10}Cr_{10}$ TRIP-DP-HEA [4, 13] and $Fe_{49.5}Mn_{30}Co_{10}Cr_{10}C_{0.5}$ TRIP-TWIP-HEA [25] revealed by EBSD, electron channeling contrast (ECC) imaging, atom probe tomography (APT), and TEM techniques. The TRIP-DP-HEA consists of two phases, namely, FCC γ matrix and HCP ε phase (Fig. 9.10a_1). The HCP ε phase is formed within the FCC γ matrix and mainly exhibits laminate morphology (Fig. 9.10a_{1-2}). Annealing twins, stacking faults, and dislocations are presented also in the FCC γ matrix in a recrystallized state [4, 13]. With the addition of C, the fraction of HCP ε phase in the iHEA is significantly reduced after annealing (Fig. 9.10b_1) compared to the reference alloy without C (Fig. 9.10a_1). This is due to the slight increase of the stacking fault energy and the corresponding higher FCC phase stability through the addition of C. A high density of annealing twins can also be observed in the interstitially alloyed HEA in the annealed state (Fig. 9.10b_1). Furthermore, particles with an average size of 50 ~ 100 nm and a volume fraction of ~1.5 vol. % are observed (Fig. 9.10b_{2-6}). The particles are determined to be $M_{23}C_6$ carbides with M = Cr, Mn, Fe, and Co (APT: Fig. 9.10b_{3-4}) with an $M_{23}C_6$ FCC lattice parameter that is three times that of the FCC γ matrix (TEM: Fig. 9.10b_{5-6}) with a strong orientation relationship.

Figure 9.11 summarizes the ultimate tensile strength and total elongation data obtained from the various Fe-rich transition metal HEAs. The homogenized single-phase nonequiatomic $Fe_{40}Mn_{27}Ni_{26}Co_5Cr_2$ (#1) [11] alloy shows a slightly lower

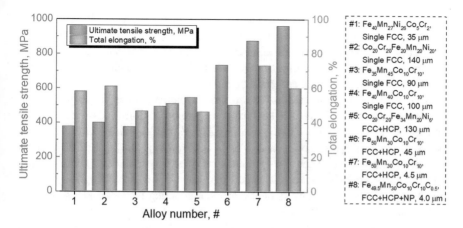

Fig. 9.11 Overview of the ultimate tensile strength and total engineering elongation obtained from the various nonequiatomic high-entropy alloys. For comparison, the data of the equiatomic $Co_{20}Cr_{20}Fe_{20}Mn_{20}Ni_{20}$ alloy (#2) are also shown here. All the alloys were produced with similar processing routes shown in Fig. 9.7 to have full control of the experimental setup. All these data stem from uniaxial tensile tests conducted on bulk samples with identical dimensions at room temperature with a strain rate of 1×10^{-3} s^{-1} [21]. For CALPHAD-predicted phase fractions for alloys #2, #4, #7, and #8, see Fig. 9.5. (Image reproduced from Ref. [21] with permission)

(by ~20 MPa) ultimate strength and slightly lower (by ~3%) total elongation compared to the homogenized single-phase equiatomic $Co_{20}Cr_{20}Fe_{20}Mn_{20}Ni_{20}$ alloy (#2) [24] even though the former one has a finer grain size (35 μm vs. 140 μm). The trend is similar for the homogenized nonequiatomic $Fe_{35}Mn_{45}Co_{10}Cr_{10}$ alloy (#3). One of the key mechanisms behind this trend is that twinning can occur in the single-phase equiatomic $Co_{20}Cr_{20}Fe_{20}Mn_{20}Ni_{20}$ alloy to a certain extent while mere dislocation hardening prevails in the nonequiatomic $Fe_{40}Mn_{27}Ni_{26}Co_{5}Cr_{2}$ (#1) [11] and $Fe_{35}Mn_{45}Co_{10}Cr_{10}$ (#3) alloys upon tensile deformation at room temperature. Interestingly, the homogenized single-phase nonequiatomic $Fe_{40}Mn_{40}Co_{10}Cr_{10}$ alloy (#4) shows a significantly higher (by ~95 MPa) ultimate strength compared to the equiatomic alloy (#2) due to more intense twinning activity [12]. Along with the alloy design strategies explained above, the newly designed quinary TRIP-DP $Co_{20}Cr_{20}Fe_{34}Mn_{20}Ni_{6}$ alloy (#5) [24] has a further enhanced ultimate strength compared to the TWIP-assisted $Fe_{40}Mn_{40}Co_{10}Cr_{10}$ alloy (#4).

Furthermore, the homogenized DP nonequiatomic $Fe_{50}Mn_{30}Co_{10}Cr_{10}$ alloy (Fig. 9.11; #6) shows a much higher ultimate strength compared to the various single-phase alloys (#1–4) and the DP $Co_{20}Cr_{20}Fe_{34}Mn_{20}Ni_{6}$ alloy (#5), while maintaining a total elongation above 50%. Interestingly, the alloy with the same composition but refined FCC matrix grains (#7) exhibits further significant increase of both strength and ductility. This is ascribed to the substantially improved work-hardening ability of the alloy due to the well-tuned phase stability via adjustment of grain size and phase fractions [4, 13]. With the addition of the interstitial element carbon into the DP microstructure, the grain-refined $Fe_{49.5}Mn_{30}Co_{10}Cr_{10}C_{0.5}$ alloy

(#8) shows a further increased ultimate strength up to nearly 1 GPa with a total elongation of ~60%. These superior mechanical properties are attributed to the joint activity of various strengthening mechanisms including interstitial and substitutional solid solution strengthening, TWIP, TRIP, nanoprecipitates, dislocation interactions, stacking faults, and grain boundaries [25].

The mechanisms responsible for the above microstructure-property relations in various iron-rich HEAs can be further explained as follows. Comparing the single-phase $Fe_{40}Mn_{27}Ni_{26}Co_5Cr_2$ and $Fe_{35}Mn_{45}Co_{10}Cr_{10}$ alloys with the single-phase $Co_{20}Cr_{20}Fe_{20}Mn_{20}Ni_{20}$ and $Fe_{40}Mn_{40}Co_{10}Cr_{10}$ alloys, a TWIP effect has been introduced. Then the presence of phase boundaries (DP structure) and the TRIP effect have been included into the DP $Co_{20}Cr_{20}Fe_{34}Mn_{20}Ni_6$ and $Fe_{50}Mn_{30}Co_{10}Cr_{10}$ alloys. Next, in the carbon-containing $Fe_{49.5}Mn_{30}Co_{10}Cr_{10}C_{0.5}$ alloy, interstitial solid solution and nanoprecipitate strengthening effects have been additionally utilized, thereby unifying all known strengthening mechanisms in one material. Indeed, the multiple deformation mechanisms enable a significant improvement of the strain-hardening capacity and strength-ductility combinations. This clearly shows that tuning deformation mechanisms via the adjustment of composition is key to the design of strong and ductile HEAs. Therefore, it is suggested that the introduction of multiple deformation mechanisms that are activated gradually or, respectively, sequentially during mechanical and/or thermal loading is a future key route for the development of new HEAs with superior mechanical properties.

9.7 Summary and Outlook

The most extensively studied HEA family, those based on the 3d transition metals including Fe-rich alloys, exhibits a broad spectrum of microstructures and mechanical behaviors. To design strong and ductile HEAs, the intrinsic stacking fault energies and/or the free energy differences between the FCC and BCC phases for various compositions can be probed in an effort to tune the phase stabilities by adjusting the compositions. Advances in parameter-free ab initio thermodynamic calculations allow efficient screening of the large compositional space for such promising alloys. Various techniques have been discussed, and it has been shown that a combined strategy of employing state-of-the-art simulation methods with advanced experimental processing and characterization techniques is a powerful approach for designing and realizing new alloys. Proper processing routes toward homogeneous solid solutions with multiple principal elements and with appropriate grain sizes are important also for achieving the targeted strength-ductility combinations.

Although outstanding strength and ductility have been achieved already in several transition metal HEAs, e.g., in the TWIP-TRIP-HEAs where an ultimate tensile strength of ~1 GPa combined with a total elongation of ~60% was reached at room temperature, further research on simultaneously strong and ductile HEAs still has a high potential. A number of research and development opportunities, which can be addressed by a similar combined simulation/experimental approach, are discussed below.

1. The strength and ductility of many HEAs at low and at elevated temperatures remain unknown. There is great potential for developing new HEAs with excellent strength-ductility combinations at low and elevated temperatures.
2. For widely studied transition metal HEAs with good strength-ductility combinations, other properties such as the resistance to hydrogen-induced degradation [148, 149], corrosion resistance, fatigue behavior, and magnetic performance can be explored also in an effort to find superior combinations of properties, i.e., multifunctionalities, to justify their relatively high cost compared to established high-strength and austenitic stainless steels. Moreover, following the design approach associated with the stacking fault energies and/or the near equivalence of FCC-BCC free energies, a shape memory effect could be introduced into nonequiatomic HEAs.
3. Other than the 3d transition metal family, HEAs containing refractory elements such as Ti, Nb, Ta, Zr, Hf, V, Mo, and W [150] can be developed for high-performance refractory HEAs applied as high-temperature load-bearing structures in the aerospace industry.
4. Considerable research work can also be conducted to screen the effects associated with different minor interstitial element fractions (e.g., C, N, B, O, etc.) to further improve the performance of the various HEAs and understand the corresponding mechanisms.
5. The current iron-rich high-performance TWIP-TRIP-HEAs contain significant amounts of Co. If Co can be substituted for lower cost alloying elements, the range of applications might be much expanded.

Acknowledgments One of the authors (MHFS) wishes to acknowledge helpful discussions with Prof. V. Soare of INMR (Romania), Prof. B. Podgornik of IMT (Slovenia), and dr. T.P.C. Klaver of TU Delft (Netherlands). This research received funding through the ERA-NET integrated computational materials engineering (ICME) program under project 4316 "HEAMODELL" as financed by NWO "domein Exacte en Natuurwetenschappen", project number 732.017.107. Fundings from the Deutsche Forschungsgemeinschaft (SPP 2006) and from NWO/STW (VIDI grant 15707) are gratefully acknowledged.

References

1. J.W. Yeh et al., Nanostructured high-entropy alloys with multiple principal elements: Novel alloy design concepts and outcomes. Adv. Eng. Mater. **6**(5), 299–303 (2004)
2. B. Cantor et al., Microstructural development in equiatomic multicomponent alloys. Mater. Sci. Eng. A **375–377**, 213–218 (2004)
3. Y. Zhang et al., Microstructures and properties of high-entropy alloys. Prog. Mater. Sci. **61**, 1–93 (2014)

4. Z. Li et al., Metastable high-entropy dual-phase alloys overcome the strength–ductility trade-off. Nature **534**, 227–230 (2016)
5. F. Otto et al., The influences of temperature and microstructure on the tensile properties of a CoCrFeMnNi high-entropy alloy. Acta Mater. **61**(15), 5743–5755 (2013)
6. D.B. Miracle, O.N. Senkov, A critical review of high entropy alloys and related concepts. Acta Mater. **122**, 448–511 (2017)
7. Y.P. Wang, B.S. Li, H.Z. Fu, Solid solution or Intermetallics in a high-entropy alloy. Adv. Eng. Mater. **11**(8), 641–644 (2009)
8. F. Otto et al., Decomposition of the single-phase high-entropy alloy CrMnFeCoNi after prolonged anneals at intermediate temperatures. Acta Mater. **112**, 40–52 (2016)
9. C.C. Tasan et al., Composition dependence of phase stability, deformation mechanisms, and mechanical properties of the CoCrFeMnNi high-entropy alloy system. JOM **66**(10), 1993–2001 (2014)
10. F. Otto et al., Relative effects of enthalpy and entropy on the phase stability of equiatomic high-entropy alloys. Acta Mater. **61**(7), 2628–2638 (2013)
11. M.J. Yao et al., A novel, single phase, non-equiatomic FeMnNiCoCr high-entropy alloy with exceptional phase stability and tensile ductility. Scr. Mater. **72–73**, 5–8 (2014)
12. Y. Deng et al., Design of a twinning-induced plasticity high entropy alloy. Acta Mater. **94**, 124–133 (2015)
13. Z. Li et al., A TRIP-assisted dual-phase high-entropy alloy: Grain size and phase fraction effects on deformation behavior. Acta Mater. **131**, 323–335 (2017)
14. K.G. Pradeep et al., Non-equiatomic high entropy alloys: Approach towards rapid alloy screening and property-oriented design. Mater. Sci. Eng. A **648**, 183–192 (2015)
15. T. Niendorf et al., Unexpected cyclic stress-strain response of dual-phase high-entropy alloys induced by partial reversibility of deformation. Scr. Mater. **143**, 63–67 (2018)
16. S.S. Nene et al., Enhanced strength and ductility in a friction stir processing engineered dual phase high entropy alloy. Sci. Rep. **7**(1), 16167 (2017)
17. M.G. Pini, P. Politi, R.L. Stamps, Anisotropy effects on the magnetic excitations of a ferromagnetic monolayer below and above the Curie temperature. Phys. Rev. B **72**, 014454 (2005)
18. F.G. Coury et al., High throughput discovery and design of strong multicomponent metallic solid solutions. Sci. Rep. **8**(1), 8600 (2018)
19. T. Yang et al., L12-strengthened high-entropy alloys for advanced structural applications. J. Mater. Res. **33**(19), 2983–2997 (2018)
20. T. Yang et al., Multicomponent intermetallic nanoparticles and superb mechanical behaviors of complex alloys. Science **362**(6417), 933–937 (2018)
21. Z. Li, D. Raabe, Strong and ductile non-equiatomic high-entropy alloys: Design, processing, microstructure, and mechanical properties. JOM J. Miner. Met. Mater. Soc. **69**(11), 2099–2106 (2017)
22. Z. Li et al., Combinatorial metallurgical synthesis and processing of high-entropy alloys. J. Mater. Res. **33**(19), 3156–3169 (2018)
23. D. Ma et al., Phase stability of non-equiatomic CoCrFeMnNi high entropy alloys. Acta Mater. **98**, 288–296 (2015)
24. Z. Li et al., Ab initio assisted design of quinary dual-phase high-entropy alloys with transformation-induced plasticity. Acta Mater. **136**, 262–270 (2017)
25. Z. Li et al., Interstitial atoms enable joint twinning and transformation induced plasticity in strong and ductile high-entropy alloys. Sci. Rep. **7**, 40704 (2017)
26. Z. Li, Interstitial equiatomic CoCrFeMnNi high-entropy alloys: Carbon content, microstructure, and compositional homogeneity effects on deformation behavior. Acta Mater. **164**, 400–412 (2019)
27. Z. Wang, I. Baker, Interstitial strengthening of a f.c.c. FeNiMnAlCr high entropy alloy. Mater. Lett. **180**, 153–156 (2016)
28. A. Chiba et al., Interstitial carbon enhanced corrosion resistance of Fe-33Mn-xC austenitic steels: Inhibition of anodic dissolution. J. Electrochem. Soc. **165**(2), C19–C26 (2018)

29. M. Beyramali Kivy, C.S. Kriewall, M.A. Zaeem, Formation of chromium-iron carbide by carbon diffusion in AlXCoCrFeNiCu high-entropy alloys. Mater. Res. Lett. **6**(6), 321–326 (2018)
30. Y. Ikeda, B. Grabowski, F. Körmann, Ab initio phase stabilities and mechanical properties of multicomponent alloys: A comprehensive review for high entropy alloys and compositionally complex alloys. Mater. Charact. **147**, 464–511 (2018)
31. H.L. Lukas, S.G. Fries, B. Sundman, *Computational Thermodynamics, the Calphad Method* (Cambridge University Press, Cambridge, 2007)
32. M.C. Gao et al., Computational modeling of high-entropy alloys: Structures, thermodynamics and elasticity. J. Mater. Res. **32**(19), 3627–3641 (2017)
33. S. Gorsse, F. Tancret, Current and emerging practices of CALPHAD toward the development of high entropy alloys and complex concentrated alloys. J. Mater. Res. **33**(19), 2899–2923 (2018)
34. C. Zhang et al., Computational thermodynamics aided high-entropy alloy design. JOM **64**(7), 839–845 (2012)
35. M. Gao, D. Alman, Searching for next single-phase high-entropy alloy compositions. Entropy **15**(10), 4504–4519 (2013)
36. F. Zhang et al., An understanding of high entropy alloys from phase diagram calculations. Calphad **45**, 1–10 (2014)
37. O.N. Senkov et al., Accelerated exploration of multi-principal element alloys for structural applications. Calphad **50**, 32–48 (2015)
38. A. Abu-Odeh et al., Efficient exploration of the high entropy alloy composition-phase space. Acta Mater. **152**, 41–57 (2018)
39. C. Zhang, M.C. Gao, CALPHAD modeling of high-entropy alloys, in *High-Entropy Alloys*, (Springer, Cham, 2016), pp. 399–444
40. S.L. Chen et al., The PANDAT software package and its applications. Calphad **26**(2), 175–188 (2002)
41. J.O. Andersson et al., Thermo-Calc & DICTRA, computational tools for materials science. Calphad **26**(2), 273–312 (2002)
42. C. Haase et al., Combining thermodynamic modeling and 3D printing of elemental powder blends for high-throughput investigation of high-entropy alloys – Towards rapid alloy screening and design. Mater. Sci. Eng. A **688**, 180–189 (2017)
43. H. Mao, H.-L. Chen, Q. Chen, TCHEA1: A thermodynamic database not limited for "high entropy" alloys. J. Phase Equilib. Diffus. **38**(4), 353–368 (2017)
44. T. Klaver, D. Simonovic, M. Sluiter, Brute force composition scanning with a CALPHAD database to find low temperature body centered cubic high entropy alloys. Entropy **20**(12), 911 (2018)
45. H.-L. Chen, H. Mao, Q. Chen, Database development and Calphad calculations for high entropy alloys: Challenges, strategies, and tips. Mater. Chem. Phys. **210**, 279–290 (2018)
46. F. Körmann, A.V. Ruban, M.H.F. Sluiter, Long-ranged interactions in bcc NbMoTaW high-entropy alloys. Mater. Res. Lett. **5**(1), 35–40 (2017)
47. P. Hohenberg, W. Kohn, Inhomogeneous electron gas. Phys. Rev. **136**(3B), B864–B871 (1964)
48. E. Engel, R.M. Dreizler, *Density Functional Theory* (Springer-Verlag, Berlin/Heidelberg, 2011)
49. A. Zunger et al., Special quasirandom structures. Phys. Rev. Lett. **65**(3), 353–356 (1990)
50. B.L. Gyorffy, Coherent-potential approximation for a nonoverlapping-muffin-tin-potential model of random substitutional alloys. Phys. Rev. B **5**(6), 2382–2384 (1972)
51. L. Vitos, Total-energy method based on the exact muffin-tin orbitals theory. Phys. Rev. B **64**(1), 014107 (2001)
52. L. Vitos, I.A. Abrikosov, B. Johansson, Anisotropic lattice distortions in random alloys from first-principles theory. Phys. Rev. Lett. **87**(15), 156401 (2001)
53. P.A. Korzhavyi et al., Madelung energy for random metallic alloys in the coherent potential approximation. Phys. Rev. B **51**(9), 5773–5780 (1995)

54. K. Schwarz, P. Blaha, G.K.H. Madsen, Electronic structure calculations of solids using the WIEN2k package for material sciences. Comput. Phys. Commun. **147**(1–2), 71–76 (2002)
55. P. Giannozzi et al., QUANTUM ESPRESSO: A modular and open-source software project for quantum simulations of materials. J. Phys. Condens. Matter **21**(39), 395502 (2009)
56. X. Gonze et al., First-principles computation of material properties: The ABINIT software project. Comput. Mater. Sci. **25**(3), 478–492 (2002)
57. J. Hafner, Ab-initio simulations of materials using VASP: Density-functional theory and beyond. J. Comput. Chem. **29**(13), 2044–2078 (2008)
58. F. Yonezawa, K. Morigaki, Coherent potential approximation. Basic concepts and applications. Prog. Theor. Phys. Suppl. **53**, 1–76 (1973)
59. Y. Zhang et al., Influence of chemical disorder on energy dissipation and defect evolution in concentrated solid solution alloys. Nat. Commun. **6**, 8736 (2015)
60. K. Jin et al., Tailoring the physical properties of Ni-based single-phase equiatomic alloys by modifying the chemical complexity. Sci. Rep. **6**, 20159 (2016)
61. P. Singh, A.V. Smirnov, D.D. Johnson, Atomic short-range order and incipient long-range order in high-entropy alloys. Phys. Rev. B **91**(22), 224204 (2015)
62. B.L. Gyorffy et al., A first-principles theory of ferromagnetic phase transitions in metals. J. Phys. F **15**(6), 1337–1386 (1985)
63. V.P. Antropov et al., Spin dynamics in magnets: Equation of motion and finite temperature effects. Phys. Rev. B **54**(2), 1019–1035 (1996)
64. H. Ebert, D. Ködderitzsch, J. Minár, Calculating condensed matter properties using the KKR-Green's function method—recent developments and applications. Rep. Prog. Phys. **74**(9), 096501 (2011)
65. L. Vitos, *Computational Quantum Mechanics for Materials Engineers: The EMTO Method and Applications* (Springer Science & Business Media, London, 2007)
66. J.W.D. Connolly, A.R. Williams, Density-functional theory applied to phase transformations in transition-metal alloys. Phys. Rev. B **27**(8), 5169–5172 (1983)
67. A. Van De Walle, M. Asta, G. Ceder, The alloy theoretic automated toolkit: A user guide. Calphad **26**(4), 539–553 (2002)
68. D.D. Fontaine, Cluster approach to order-disorder transformations in alloys, in *Solid State Physics*, ed. by H. Ehrenreich, D. Turnbull, (Academic, New York, 1994), pp. 33–176
69. K. Terakura et al., Electronic theory of the alloy phase stability of Cu-Ag, Cu-Au, and Ag-Au systems. Phys. Rev. B **35**(5), 2169–2173 (1987)
70. M. Sluiter, P. Turchi, Electronic theory of phase stability in substitutional alloys: The generalized perturbation method versus the Connolly-Williams method. Phys. Rev. B **40**(16), 11215–11228 (1989)
71. A.V. Ruban et al., Atomic and magnetic configurational energetics by the generalized perturbation method. Phys. Rev. B **70**(12), 125115 (2004)
72. P. Turchi et al., First-principles study of ordering properties of substitutional alloys using the generalized perturbation method. Phys. Rev. B **37**(10), 5982 (1988)
73. F. Ducastelle, *Order and Phase Stability in Alloys* (North – Holland, Amsterdam, 1991)
74. P.E.A. Turchi et al., First-principles study of phase stability in Cu-Zn substitutional alloys. Phys. Rev. Lett. **67**(13), 1779–1782 (1991)
75. B. Gyorffy, G. Stocks, Concentration waves and Fermi surfaces in random metallic alloys. Phys. Rev. Lett. **50**(5), 374 (1983)
76. D. Lerch et al., UNCLE: A code for constructing cluster expansions for arbitrary lattices with minimal user-input. Model. Simul. Mater. Sci. Eng. **17**(5), 055003 (2009)
77. M.C. Nguyen et al., Cluster-expansion model for complex quinary alloys: Application to alnico permanent magnets. Phys. Rev.Appl. **8**(5), 054016 (2017)
78. A. Fernandez-Caballero et al., Short-range order in high entropy alloys: Theoretical formulation and application to Mo-Nb-Ta-VW system. J. Phase Equilib. Diffus. **38**(4), 391–403 (2017)
79. J. Cowley, An approximate theory of order in alloys. Phys. Rev. **77**(5), 669 (1950)

80. P. Singh et al., Tuning phase-stability and short-range order through Al-doping in FeMn-CoCrAlx (x<= 20 at.%) high entropy alloys. arXiv preprint **arXiv**, 1803.06771 (2018)
81. K. Gubaev et al., Accelerating high-throughput searches for new alloys with active learning of interatomic potentials. arXiv preprint **arXiv**, 1806.10567 (2018)
82. A. Shapeev, Accurate representation of formation energies of crystalline alloys with many components. Comput. Mater. Sci. **139**, 26–30 (2017)
83. E.V. Podryabinkin, A.V. Shapeev, Active learning of linearly parametrized interatomic potentials. Comput. Mater. Sci. **140**, 171–180 (2017)
84. F.K. Tatiana Kostiuchenko, J. Neugebauer, A. Shapeev, Impact of local lattice relaxations on phase stability and chemical ordering in bcc NbMoTaW high-entropy alloys explored by ab initio based machine-learning potentials. npj Comput. Mater. **5**(1), 1–7 (2018)
85. V. Moruzzi, J. Janak, K. Schwarz, Calculated thermal properties of metals. Phys. Rev. B **37**(2), 790 (1988)
86. D. Ma et al., Ab initio thermodynamics of the CoCrFeMnNi high entropy alloy: Importance of entropy contributions beyond the configurational one. Acta Mater. **100**, 90–97 (2015)
87. Y. Wang et al., Computation of entropies and phase equilibria in refractory V-Nb-Mo-Ta-W high-entropy alloys. Acta Mater. **143**, 88–101 (2018)
88. F. Körmann et al., Phonon broadening in high entropy alloys. npj Comput. Mater. **3**(1), 36 (2017)
89. M.A. Blanco, E. Francisco, V. Luaña, GIBBS: Isothermal-isobaric thermodynamics of solids from energy curves using a quasi-harmonic Debye model. Comput. Phys. Commun. **158**(1), 57–72 (2004)
90. G. Kresse, J. Furthmüller, J. Hafner, Ab initio force constant approach to phonon dispersion relations of diamond and graphite. EPL (Europhys. Lett.) **32**(9), 729 (1995)
91. A. Togo, I. Tanaka, First principles phonon calculations in materials science. Scr. Mater. **108**, 1–5 (2015)
92. Y. Wang et al., First-principles calculations of lattice dynamics and thermal properties of polar solids. npj Comput. Mater. **2**, 16006 (2016)
93. B. Grabowski et al., Ab initio up to the melting point: Anharmonicity and vacancies in aluminum. Phys. Rev. B **79**(13), 134106 (2009)
94. A.I. Duff et al., Improved method of calculatingab initiohigh-temperature thermodynamic properties with application to ZrC. Phys. Rev. B **91**(21), 214311 (2015)
95. B.F.A.M. Widom, Elastic stability and lattice distortion of refractory high entropy alloys. Mater. Chem. Phys. **210**, 309–314 (2017)
96. M. Widom, Entropy and diffuse scattering: Comparison of NbTiVZr and CrMoNbV. Metall. Mater. Trans. A **47**(7), 3306–3311 (2016)
97. M. Widom et al., Hybrid monte carlo/molecular dynamics simulation of a refractory metal high entropy alloy. Metall. Mater. Trans. A **45**(1), 196–200 (2013)
98. P. Srinivasan et al., The effectiveness of reference-free modified embedded atom method potentials demonstrated for NiTi and NbMoTaW. Model. Simul. Mater. Sci. Eng. **27**(6), 065013 (2019)
99. A.I. Liechtenstein et al., Local spin density functional approach to the theory of exchange interactions in ferromagnetic metals and alloys. J. Magn. Magn. Mater. **67**(1), 65–74 (1987)
100. F. Körmann et al., "Treasure maps" for magnetic high-entropy-alloys from theory and experiment. Appl. Phys. Lett. **107**(14), 142404 (2015)
101. S. Huang et al., Mechanism of magnetic transition in FeCrCoNi-based high entropy alloys. Mater. Des. **103**, 71–74 (2016)
102. F. Körmann, T. Hickel, J. Neugebauer, Influence of magnetic excitations on the phase stability of metals and steels. Curr. Opinion Solid State Mater. Sci. **20**(2), 77–84 (2016)
103. I.A. Abrikosov et al., Recent progress in simulations of the paramagnetic state of magnetic materials. Curr. Opinion Solid State Mater. Sci. **20**(2), 85–106 (2016)
104. F. Körmann et al., Lambda transitions in materials science: Recent advances in CALPHAD and first-principles modelling. Phys. Status Solidi B **251**(1), 53–80 (2014)

105. M. de Jong et al., Charting the complete elastic properties of inorganic crystalline compounds. Sci. Data **2**, 150009 (2015)
106. H. Springer, D. Raabe, Rapid alloy prototyping: Compositional and thermo-mechanical high throughput bulk combinatorial design of structural materials based on the example of 30Mn–1.2C–xAl triplex steels. Acta Mater. **60**(12), 4950–4959 (2012)
107. D. Raabe et al., From high-entropy alloys to high-entropy steels. Steel Res. Int. **86**(10), 1127–1138 (2015)
108. H. Springer, M. Belde, D. Raabe, Combinatorial design of transitory constitution steels: Coupling high strength with inherent formability and weldability through sequenced austenite stability. Mater. Des. **90**, 1100–1109 (2016)
109. Z. Li, D. Raabe, Influence of compositional inhomogeneity on mechanical behavior of an interstitial dual-phase high-entropy alloy. Mater. Chem. Phys. **210**(1), 29–36 (2018)
110. J.-C. Zhao, X. Zheng, D.G. Cahill, High-throughput diffusion multiples. Mater. Today **8**(10), 28–37 (2005)
111. J.-C. Zhao, Combinatorial approaches as effective tools in the study of phase diagrams and composition–structure–property relationships. Prog. Mater. Sci. **51**(5), 557–631 (2006)
112. J.-C. Zhao et al., A diffusion multiple approach for the accelerated design of structural materials. MRS Bull. **27**(4), 324–329 (2002)
113. J.-C. Zhao, Reliability of the diffusion-multiple approach for phase diagram mapping. J. Mater. Sci. **39**(12), 3913–3925 (2004)
114. J.-C. Zhao, A combinatorial approach for structural materials. Adv. Eng. Mater. **3**(3), 143–147 (2001)
115. P. Wilson, R. Field, M. Kaufman, The use of diffusion multiples to examine the compositional dependence of phase stability and hardness of the Co-Cr-Fe-Mn-Ni high entropy alloy system. Intermetallics **75**, 15–24 (2016)
116. D. Misell, C. Stolinski, *Scanning Electron Microscopy and X-ray Microanalysis. A Text for Biologists, Material Scientists and Geologists* (Pergamon, New York, 1983)
117. A.J. Schwartz et al., *Electron Backscatter Diffraction in Materials Science* (Springer, Boston, 2000)
118. D. Dingley, Progressive steps in the development of electron backscatter diffraction and orientation imaging microscopy. J. Microsc. **213**(3), 214–224 (2004)
119. D.B. Williams, C.B. Carter, The transmission electron microscope, in *Transmission Electron Microscopy*, (Springer, Boston, 1996), pp. 3–17
120. B. Fultz, J.M. Howe, *Transmission Electron Microscopy and Diffractometry of Materials* (Springer Science & Business Media, Berlin, 2012)
121. A.C. Fischer-Cripps, *Nanoindentation* (Springer, New York, 2011)
122. M.F. Doerner, W.D. Nix, A method for interpreting the data from depth-sensing indentation instruments. J. Mater. Res. **1**(4), 601–609 (1986)
123. W.C. Oliver, G.M. Pharr, An improved technique for determining hardness and elastic modulus using load and displacement sensing indentation experiments. J. Mater. Res. **7**(6), 1564–1583 (1992)
124. J.J. Vlassak, W.D. Nix, A new bulge test technique for the determination of Young's modulus and Poisson's ratio of thin films. J. Mater. Res. **7**(12), 3242–3249 (1992)
125. S.A.S. Asif et al., Quantitative imaging of nanoscale mechanical properties using hybrid nanoindentation and force modulation. J. Appl. Phys. **90**(3), 1192–1200 (2001)
126. S. Huxtable et al., Thermal conductivity imaging at micrometre-scale resolution for combinatorial studies of materials. Nat. Mater. **3**(5), 298 (2004)
127. V. Ocelík et al., Additive manufacturing of high-entropy alloys by laser processing. JOM **68**(7), 1810–1818 (2016)
128. J. Joseph et al., Comparative study of the microstructures and mechanical properties of direct laser fabricated and arc-melted AlxCoCrFeNi high entropy alloys. Mater. Sci. Eng. A **633**, 184–193 (2015)
129. Y. Brif, M. Thomas, I. Todd, The use of high-entropy alloys in additive manufacturing. Scr. Mater. **99**, 93–96 (2015)

130. D.C. Hofmann et al., Compositionally graded metals: A new frontier of additive manufacturing. J. Mater. Res. **29**(17), 1899–1910 (2014)
131. M. Rombouts et al., Fundamentals of selective laser melting of alloyed steel powders. CIRP Ann. Manuf. Technol. **55**(1), 187–192 (2006)
132. H. Knoll et al., Combinatorial alloy design by laser additive manufacturing. Steel Res. Int. **88**(8), 1600416 (2017)
133. T. Borkar et al., A combinatorial assessment of AlxCrCuFeNi2 (0 < x < 1.5) complex concentrated alloys: Microstructure, microhardness, and magnetic properties. Acta Mater. **116**, 63–76 (2016)
134. B.A. Welk, M.A. Gibson, H.L. Fraser, A combinatorial approach to the investigation of metal systems that form both bulk metallic glasses and high entropy alloys. JOM **68**(3), 1021–1026 (2016)
135. J. Cui et al., Combinatorial search of thermoelastic shape-memory alloys with extremely small hysteresis width. Nat. Mater. **5**, 286 (2006)
136. H. Stein et al., A structure zone diagram obtained by simultaneous deposition on a novel step heater: A case study for Cu2O thin films. Phys. Status Solidi A **212**(12), 2798–2804 (2015)
137. A. Ludwig et al., Development of multifunctional thin films using high-throughput experimentation methods. Int. J. Mater. Res. **99**(10), 1144–1149 (2008)
138. Y. Li et al., Combinatorial strategies for synthesis and characterization of alloy microstructures over large compositional ranges. ACS Comb. Sci. **18**(10), 630–637 (2016)
139. V. Chevrier, J. Dahn, Production and visualization of quaternary combinatorial thin films. Meas. Sci. Technol. **17**(6), 1399 (2006)
140. A. Kauffmann et al., Combinatorial exploration of the high entropy alloy system Co-Cr-Fe-Mn-Ni. Surf. Coat. Technol. **325**, 174–180 (2017)
141. C. Brundle, G. Conti, P. Mack, XPS and angle resolved XPS, in the semiconductor industry: Characterization and metrology control of ultra-thin films. J. Electron Spectrosc. Relat. Phenom. **178**, 433–448 (2010)
142. H.S. Stein et al., New materials for the light-induced hydrogen evolution reaction from the Cu–Si–Ti–O system. J. Mater. Chem. A **4**(8), 3148–3152 (2016)
143. O.L. Warren, T.J. Wyrobek, Nanomechanical property screening of combinatorial thin-film libraries by nanoindentation. Meas. Sci. Technol. **16**(1), 100 (2004)
144. S.W. Fackler et al., Combinatorial study of Fe-Co-V hard magnetic thin films. Sci. Technol. Adv. Mater. **18**(1), 231–238 (2017)
145. S. Thienhaus et al., Rapid identification of areas of interest in thin film materials libraries by combining electrical, optical, X-ray diffraction, and mechanical high-throughput measurements: A case study for the system Ni–Al. ACS Comb. Sci. **16**(12), 686–694 (2014)
146. Y. Lederer et al., The search for high entropy alloys: A high-throughput ab-initio approach. arXiv preprint **arXiv**, 1711.03426 (2017)
147. N. Gurao, K. Biswas, In the quest of single phase multi-component multiprincipal high entropy alloys. J. Alloys Compd. **697**, 434–442 (2017)
148. H. Luo et al., Hydrogen effects on microstructural evolution and passive film characteristics of a duplex stainless steel. Electrochem. Commun. **79**, 28–32 (2017)
149. H. Luo, Z. Li, D. Raabe, Hydrogen enhances strength and ductility of an equiatomic high-entropy alloy. Sci. Rep. **7**(1), 9892 (2017)
150. M.C. Gao et al., Design of refractory high-entropy alloys. JOM **67**(11), 2653–2669 (2015)

Chapter 10
Iron-Based Intermetallics

Martin Palm and Frank Stein

Abbreviations and Symbols

a_0	Lattice constant a
A2	Strukturbericht symbol
AIM	Air induction melting
at.%	Atom percent
B2	Strukturbericht symbol
B20	Strukturbericht symbol
BDTT	Brittle-to-ductile transition temperature
b.c.c.	Body-centered cubic
°C	Degrees centigrade
cm	Centimeter
cP8	Pearson symbol
CVD	Chemical vapor deposition
$D8_8$	Strukturbericht symbol
$D0_3$	Strukturbericht symbol
$\dot{\varepsilon}$	Strain rate
E_{Corr}	corrosion potential
ESR	Electroslag remelting
f.c.c.	Face-centered cubic
g	Gram
GPa	Gigapascal
h	Hour
hcp	Hexagonal close-packed

M. Palm (✉) · F. Stein
Department Structure and Nano-/Micromechanics of Materials, Max-Planck-Institut für
Eisenforschung GmbH, Düsseldorf, Nordrhein-Westfalen, Germany
e-mail: palm@mpie.de; stein@mpie.de

© Springer Nature Switzerland AG 2021
R. Rana (ed.), *High-Performance Ferrous Alloys*,
https://doi.org/10.1007/978-3-030-53825-5_10

hP6	Pearson symbol
hP16	Pearson symbol
HCF	High-cycle fatigue
k_p	Parabolic rate constant
$L2_1$	Strukturbericht symbol
LMD	Laser metal deposition
m	Meter
μB	Bohr magneton
MPa	Megapascal
μm	Micron meter
oC48	Pearson symbol
Ω	Ohm
ORNL	Oak Ridge National Laboratories
s	Second
$\sigma_{0.2}$	0.2% yield stress
SCE	Saturated calomel electrode
SOFC	Solid oxide fuel cell
T_C	Curie temperature
TCP	Topological close-packed
tP3	Pearson symbol
TRIP	Transformation-induced plasticity
TRW Inc.	Thompson Ramo Wooldridge Incorporated
US	United States
V	Voltage
VIM	Vacuum induction melting
vol.%	Volume percent
wt.%	Weight percent
YSA	Yield stress anomaly

10.1   Introduction

Intermetallic phases – commonly abbreviated as intermetallics – are phases which have different crystallographic structures than the elements they constitute of [1]. They can appear as precipitates, e.g., Laves, μ, or σ phase in steels, or form the base material like FeAl and Fe_3Al in iron aluminide-based alloys or Fe_3Si in Fe–Si alloys. The ordered intermetallic phases are usually rather strong, and many of them are stable up to high temperatures. Therefore, they have been considered as strengthening precipitates since long.

The most prominent example of precipitation hardening of steels by intermetallic phases are classical maraging steels. In these low-carbon steels with Ni contents of about 12–25 wt.% and additions of Ti, Al, Nb, Mo, Co etc., finely distributed intermetallic phases are precipitated when martensite transforms to austenite during aging (maraging) at 400–500 °C [2, 3]. Which intermetallic phases form depends on alloy composition, temperature, and time of maraging. Besides (Ni,Fe)Al, Ni_3Ti, Ni_3Mo, and NiTi, the σ-phase FeMo and the Laves phases Fe_2Ti and Fe_2Mo are frequently observed intermetallic precipitates. The fine intermetallic precipitates effectively hinder the movement of dislocations. The resulting high-strength steels are, e.g., used as tool steels for application at high temperature. More recently, the concept of maraging has been transferred to other steels and combined with the transformation-induced plasticity (TRIP) effect [4]. On the other hand, Laves and σ-phases have been considered as detrimental in steels and superalloys because they may cause grain boundary embrittlement [5–7] and a loss in creep strength after prolonged service [8].

This chapter focuses on ferrous materials that are based on intermetallics, i.e., iron aluminides and iron silicides, as well as ferrous materials which gain there properties from specific microstructures achieved by intermetallic phases.

10.2   Iron Aluminides

Alloys based on the phases Fe_3Al and FeAl are lightweight (densities 5.7–6.7 g/cm^3) Fe-based materials with high wear resistance and excellent corrosion resistance. For applications, alloys in the composition range between about 15 and 40 at.% Al (8–24.5 wt.% Al) are of specific interest. The lower value corresponds to the Al content necessary to form protective Al_2O_3 scales; above the upper value the alloys become inherently brittle. Already around 1900 it was recognized that Fe becomes wear and oxidation resistant by alloying with Al. From the 1940s onward, numerous alloy developments were carried out, however usually with the outcome that brittleness and insufficient strength at high temperatures precluded their industrial application. New alloy concepts and economic pressure to minimize the use of strategic elements have revived industrial interest in these alloys in recent years [9].

10.2.1   Phases and Phase Diagram

Figure 10.1 shows the Fe–Al phase diagram. The liquidus temperatures in the Fe–Al system were established in 1908 and the outline of the phase equilibria by 1930, and then the first compendium on binary phase diagrams already contained a version, which was close to the present one [10]. The system contains six intermetallic phases in all. The three Al-rich phases $FeAl_2$, Fe_2Al_5, and Fe_4Al_{13} all have low-symmetrical crystal structures, which make them rather brittle. They have only limited homogeneity ranges and melting temperatures between 1150 and 1158 °C. The phase Fe_5Al_8 is only stable above 1095 °C and decomposes eutectoidally even during quenching.

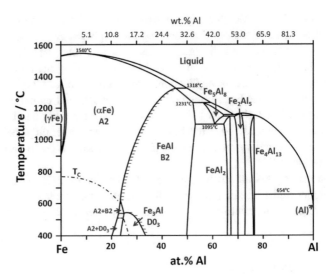

Fig. 10.1 Fe–Al phase diagram constructed on the basis of [11] with changes in the Al-rich part according to [12]

In view of structural applications, the two Fe-rich phases Fe₃Al and FeAl are of interest. The cubic D0₃-ordered phase Fe₃Al is stable between about 23 and 36 at.% Al and up to a temperature of 545 °C. B2-ordered FeAl is also cubic. It has a wide homogeneity range between 23 and 53 at.% Al and is stable up to 1318 °C. On heating, Fe₃Al transforms to FeAl by a second-order transition. Alloys of up to 45 at.% Al transform from FeAl to disordered cubic (αFe), and the transition temperatures increase markedly with increasing Al content. On cooling both transformations cannot be suppressed by quenching.

10.2.2 Alloy Developments

The first reports about "aluminum steels" and Al in cast iron date from 1890 [13, 14]. By the 1930s patented alloys exited in Russia (Cugal) and in Britain [15, 16]. The first better documented alloy developments are Alfenol and Thermenol by the US Naval Ordnance Laboratory and Pyroferal in former Czechoslovakia in the 1950s. Alfenol was originally developed for magnetic applications. The binary alloy contained 28.3 at.% Al (16 wt.%) of which strips and tapes were produced by hot and cold rolling [17]. Thermenol is a Fe–Al–Mo-based alloy designed for high-temperature use, e.g., for jet engine compressor blades [18]. Both alloy developments were continued at the Ford Motor Company in conjunction with the Wright Air Development Center and resulted in the production and testing of a number of parts, e.g., heat treatment boxes and a turbine exhaust cone [19–21].

Other alloy developments at this time in the USA to note are "DB-2" a Fe–Al–Cr–Nb–Zr-based alloy at The Martin Company and Fe–Cr–Al alloys at the Battelle Memorial Institute [22, 23]. Pyroferal is a Fe–Al–C-based alloy, with an Al content of 44.5–46.5 at.%, from which various cast parts for use in industrial furnaces were produced on a larger scale [24, 25]. Thermagal denotes a series of Fe–Al–C-based alloys developed in the 1960s in France and was used to produce a large variety of parts, e.g., crucibles, apparatuses for chemical industries, parts for mineral crushers, heating elements, permanent magnets, etc. [26]. Activities on Alfenol and specifically Thermenol were continued until the early 1970s under contract of the Office for Saline Water in search for corrosion-resistant tubes for seawater desalinization [27]. Pratt & Whitney and TRW Inc. under contract of the Air Force Wright Aeronautical Laboratories investigated the potential of Fe–Al alloys for application in aircraft engines in the 1980s [28, 29].

The alloy developments pursued at Oak Ridge National Laboratories (ORNL) are most prominent. Already involved in the study of dispersion-strengthened iron aluminides in 1960 [30], ORNL started large-scale activities on the development of iron aluminide-based alloys in the mid-1980s within the Fossil Energy Materials Program of the US Department of Energy. Series of alloys termed FA, FAL, FAS, etc. were developed and extensively characterized, and processing was studied in detail involving numerous industries. The work yielded a substantial understanding of iron aluminide alloys and is well documented in the Proceedings of the Annual Conference on Fossil Energy Materials and a number of review papers [31–35].

Other more recent activities to note are the development of the Al-rich oxide dispersion-strengthened alloy Grade 3 in France [36] and Fe–Al–C-based alloys in India at the Defense Metallurgical Research Laboratory [37, 38]. Also at the authors' institution, research on iron aluminides has a long tradition. Currently, at authors' institution alloys are developed using a variety of alloying concepts, and their processing and behavior under application conditions is studied in cooperation with industry [39].

10.2.3 Peculiar Features of Iron Aluminides

Iron aluminides show three peculiar features: unusual high vacancy concentrations, a yield stress anomaly (YSA) and environmental embrittlement. As all of them strongly influence the mechanical behavior, understanding these features is essential to assess the mechanical behavior of iron aluminide alloys.

FeAl and Fe_3Al both have low enthalpies for the formation of constitutional vacancies and therefore can contain up to 2 vol.% of vacancies at room temperature [40, 41]. The quenched-in vacancies have a strong hardening effect by lowering the mobility of dislocations, and therefore the yield strength at room temperature for a given alloy can vary by two to five times in dependence on the vacancy concentration [42, 43]. As the amount of quenched-in vacancies increases with increasing cooling rate, processing has a marked influence on the yield strength.

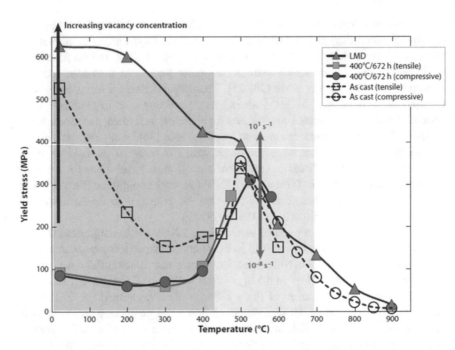

Fig. 10.2 Tensile or compressive $\sigma_{0.2}$ yield stress of binary Fe–Al with 26–28 at.% Al at a strain rate of $\dot{\varepsilon} = 1 \times 10^{-4}$ s^{-1}. In the orange shaded area, the yield strength is dominated by quenched-in vacancies: high vacancy concentration after rapid cooling after laser metal deposition (LMD), intermediate vacancy concentration after moderate cooling during casting, and minimum vacancy concentration after annealing at 400 °C for 672 h. The blue shaded area denotes the temperature range where the yield stress is affected by the yield stress anomaly (YSA), where the blue arrow indicates the dependence of the YSA on deformation rate. (Figure taken from Ref. [9])

A minimum of quenched-in vacancies can be attained by annealing at 400 °C for 100 h [44]. Figure 10.2 shows the influence of varying vacancy concentrations on the $\sigma_{0.2}$ yield stress of binary Fe–Al.

Between about 400 °C and 600 °C, iron aluminide-based alloys show an increase of the yield strength with increasing temperature, a feature not unusual for ordered intermetallic phases (Fig. 10.2). As the yield strength is expected to decrease with increasing temperature, this phenomenon has been termed yield stress anomaly (YSA). In the case of the iron aluminides, the YSA is caused by a change in the dislocation structure at these temperatures, decomposition of superdislocations creating local pinning points, and interaction of the debris with quenched-in vacancies [45–50]. The YSA is strongly strain rate dependent, decreasing with decreasing strain rate and vanishing at strain rates below 1×10^{-7} s^{-1} [51, 52] (Fig. 10.2). The practical implications are that the YSA makes hot forming between 400 and 600 °C difficult while it does not contribute to strengthening during creep. It is also noted that the actual yield strength at the maximum of the YSA and the temperature where this maximum occurs can be markedly influenced by alloying.

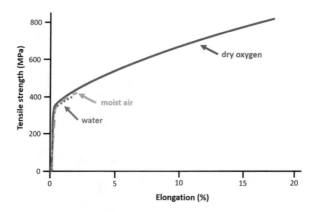

Fig. 10.3 Tensile yield stress of binary Fe–Al in different test environments. (Reproduction of the classic experiments by C.T. Liu et al. as shown in Ref. [31])

Fe$_3$Al- and FeAl-based alloys usually show a limited ductility of 2–4% total elongation at room temperature, which is still not bad compared to other intermetallic phases. Both phases are ductile when tested in hydrogen-free environments but fail in a brittle manner in the presence of moisture or hydrogen (Fig. 10.3). This phenomenon is termed environmental embrittlement and is caused by an embrittlement of crack tips by hydrogen, which is produced by the reaction $3H_2O + 2Al^{3+} \leftrightarrow 6H^+ + Al_2O_3$ [53, 54]. A detailed review on the topic is given by Zamanzade and Barnoush [55].

10.2.4 Strength, Ductility, Fatigue, Wear, and Erosion

Lack in strength at high temperatures has been the main obstacle for a wider use of iron aluminides. Therefore a substantial part of the research on iron aluminides has been devoted to evaluate different strengthening mechanisms [56–61].

Binary Fe–Al alloys lose their strength when Fe$_3$Al transforms to FeAl, i.e., at 545 °C at the latest. Solid solution hardening by a third element is well possible as many elements have a large solid solubility in Fe$_3$Al and FeAl. By solid solution hardening, the yield strength can be markedly raised, e.g., by alloying with 4 at.% V, Ti, or Mo, the yield strength can be increased from about 20 MPa to about 95–115 MPa at 800 °C [61] (Fig. 10.4).

In analogy to steels, much work has been devoted to strengthen iron aluminide-based alloys by carbide precipitates. However, no substantial strengthening has been achieved. For Fe–Al–C alloys containing Fe$_3$AlC precipitates, the yield strength drops markedly above 600 °C and equals that of a precipitate-free alloy of same Al content at 800 °C [64, 65] (Fig. 10.4). Fe–Al–X–C alloys with different carbide precipitates and additional solid solution hardening of the matrix show a less

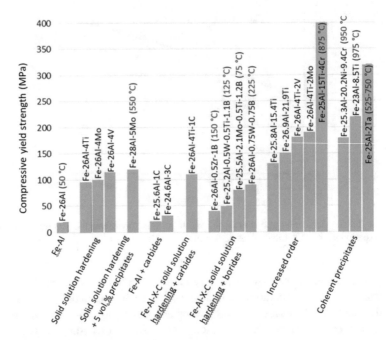

Fig. 10.4 Compressive yield strength at 800 °C of iron aluminide alloys containing 23–28 at.% Al and strengthened by different mechanisms (strain rate $\dot{\varepsilon} = 1 \times 10^{-4} \, \mathrm{s}^{-1}$). Temperatures given in red indicate brittle-to-ductile transition temperatures. (Data from Refs. [61–74])

pronounced decrease of the yield strength above 600 °C, but with about 110 MPa, they only show about the same strength as solid solution hardened alloys [65] (Fig. 10.4). Analysis of published data shows that carbides have a limited strengthening effect above 600 °C and no effect at 800 °C. Reasons are difficulties to attain an even distribution of fine carbide precipitates, which effectively hinder dislocation movement and rapid coarsening of carbides at high temperatures.

Borides have about the same limited strengthening effect as carbides, but they may form finer precipitates, which do not coarsen at high temperatures [75] (Fig. 10.4). Precipitation hardening by another intermetallic phase is also possible, specifically by Laves phases, as in many Fe–Al–X systems, respective phase equilibria between Fe_3Al or FeAl and a Laves phase exist [61]. As Laves phases have a much higher strength than the iron aluminides, strengthening will depend on the volume fraction of the precipitates. Though this would yield in principle very strong alloys, again generation of fine and evenly distributed precipitates is very difficult, and respective alloys are therefore rather brittle [76]. However, in a few Fe–Al–X systems, the precipitated Laves phase forms a film on the iron aluminide grain boundaries, and such alloys show an appreciable strength [77]; see Sect. 10.5. For strengthening by the formation of coherent microstructures involving α-Fe, FeAl, and Fe_3Al/Fe_2XAl, see Sect. 10.4.

Iron aluminides lose their strength when f.c.c. Fe_3Al transforms to b.c.c. FeAl; therefore stabilizing the f.c.c. structure at higher temperatures by alloying is another viable method to increase their strength. Actually the transition temperature can be raised substantially by a number of elements, most prominently by Ti up to 1212 °C [78]. Respective alloys can have a yield strength of about 400 MPa at 800 °C but are also rather brittle [68] (Fig. 10.4).

With respect to creep resistance, the various strengthening concepts are as effective as for increasing the yield strength. Compared to the advanced Cr-containing steel P92, iron aluminide-based alloys containing carbide or boride precipitates show an inferior creep resistance at 650 °C, while those strengthened by coherent precipitates or those with Fe_2XAl matrix can have substantially higher creep resistance [39].

As for other materials, strengthening of iron aluminide-based materials in most cases reduces the limited ductility even more [79]. While the change from brittle to ductile behavior for binary Fe_3Al and FeAl gradually takes place until about 100 °C where the alloys are ductile, ternary and higher-order alloys usually show a marked brittle-to-ductile transition temperature (BDTT), which can be higher than 800 °C [39] (Fig. 10.4). Until today, no general principle for increasing the ductility of iron aluminide-based materials has been found. Reduction of the grain size below about 10 μm, optimizing the Al content, and alloying with boron for a better cohesion of the grain boundaries and with chromium to reduce the effect of environmental embrittlement have been suggested for maintaining or even improving ductility [31, 60, 80, 81]. However, none of these methods is without exemption, and their effects are limited.

As iron aluminides show rather quick coarsening at elevated temperatures, control of the grain size, e.g., by precipitates at grain boundaries, is important. It has also been shown that thermomechanical processing can increase the ductility compared to their as-cast counterparts. Minimizing the amount of quenched-in vacancies may also enhance ductility. Still ductility may be limited to 2–4%, but this has proven to be sufficient for many applications.

Comparable few data are available for fatigue of iron aluminides, and as these data have been obtained on a variety of alloys and by different tests, general trends are difficult to identify [82]. Studies of the fatigue crack growth in Fe_3Al- and FeAl-based alloys revealed that environmental embrittlement by moist air again has a major effect, though no straightforward relation between ductility under static loading and cyclic loading has been found [83, 84]. Other observations were that alloys with coherent (αFe) + Fe_3Al microstructures showed a higher resistance against high-cycle fatigue (HCF) than single-phase Fe_3Al, and it was found that the anomalous strengthening in the temperature range of the YSA does not necessarily result in an increased fatigue life [85, 86].

Iron aluminides possess a good to excellent wear resistance, and, therefore, quite a number of investigations have been performed [87, 88]. In view of the variety of test methods, test conditions, and tested alloys, it is again difficult to find general trends.

Because of their good wear resistance, iron aluminides have not only been explored as bulk alloys but also as coatings, as binder for cemented hard phases, or as hard phase in a ductile binder. Wear resistance increases with increasing hardness, which increases with the Al content in binary Fe–Al alloys, by the formation of an Al_2O_3 oxide scale, through alloying, or by addition of a hard second phase. The beneficial effect of an oxide scale in improving wear by reducing scuffing has been attributed to its higher hardness and the ceramic nature of the scale [89]. Environmental embrittlement again may have some influence, because a decrease in wear resistance has been observed during wet abrasion [90].

The erosion behavior of iron aluminide alloys and cermets shows a ductile behavior [91]. The erosion resistance increases with increasing Al content, which is explained by an increase of strain hardening rates, leading to rapid work hardening of the surface, which limits deformation to a shallow region [91]. Specifically at high temperatures in oxidizing atmospheres, the erosion behavior of an iron aluminide can be favorable compared to steel [88].

10.2.5 Corrosion

The outstanding corrosion resistance of Fe_3Al and FeAl is related to their ability to readily form passive layers or thin and adherent oxides scales, for which, depending on environmental conditions, a minimum of 15–18 at.% Al is necessary (Fig. 10.5). Specifically in oxidizing atmospheres, they form Al_2O_3 scales which have parabolic rate constants up to two magnitudes lower than for Cr_2O_3-forming alloys, and, therefore, Fe_3Al and FeAl show a better oxidation resistance in many environments [56, 92–94]. Al_2O_3 scales are also protective in many other corrosive environments. For example, resistance against metal dusting, i.e., disintegration of Fe-based materials due to carburization, can be markedly improved [95, 96]. For protection in nonoxidizing environments, the scale can be generated by pre-oxidation.

The aqueous corrosion resistance of iron aluminides is considered not to be good [97]. Electrochemical studies of the aqueous corrosion resistance of Fe–Al have already been reviewed some time ago [98]. Fe_3Al and FeAl passivate by forming films consisting of Al- and Fe-hydroxides or oxides. The aqueous corrosion behavior of Fe–Al depends on the electrolyte. While they show passivation in neutral and alkaline electrolytes, corrosion rates are moderate to high in strong acids, specifically in the presence of Cl^-. Passivation can be improved by alloying with Cr and/or Mo [99]. Recently it has been shown that the aqueous corrosion resistance can be markedly improved through generating an Al_2O_3 scale by pre-oxidation [100]. Immersion tests in seawater and neutral salt spray tests showed that the corrosion resistance improved with increasing Al content and that dense layers of Fe_3O_4 (magnetite) form under these conditions [27, 101].

Corrosion by salt deposits or salt melts at high temperatures (hot corrosion) has been investigated as well [35]. Also under these conditions, iron aluminides show

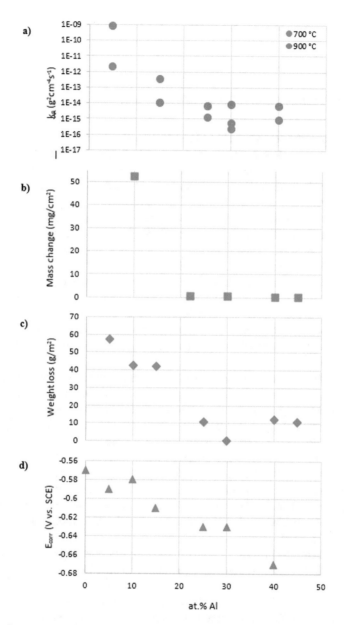

Fig. 10.5 Corrosion resistance of binary Fe–Al in dependence on Al content; (**a**) parabolic rate constants k_p for oxidation in synthetic air at 700 °C and 900 °C [104]; (**b**) mass change after exposure in steam at 700 °C for 672 h [105]; (**c**) weight loss after neutral salt spray testing for 168 h [101]; (**d**) corrosion potential in deaerated 0.0126 M H_2SO_4 at 25 °C [100]

an excellent corrosion resistance if oxidation is the prevailing corrosion mechanism [102, 103].

10.2.6 Synthesis and Processing

Alloys based on the phases Fe_3Al and FeAl can be processed with equipment readily available in steel industry, and process parameters may be similar to those for Cr steels [31, 39, 106]. However, conditions will depend on individual alloys, and thermomechanical processing of alloys with Al contents above about 30 at.% is difficult [107].

Air induction melting (AIM), electroslag remelting (ESR), and the Exo-Melt™ process have been employed for the production of iron aluminide-based alloys on an industrial scale, while a variety of techniques including vacuum induction melting (VIM), arc melting, directional solidification, etc. are used on laboratory scale [31, 108, 109]. Besides producing the alloys from the elements or pre-alloys, production from scrap has also been widely investigated [110–113]. Iron aluminide-based alloys have also been synthesized by various powder metallurgical methods. The strong exothermic reaction between Fe and Al or Al_2O_3 also makes preparation by self-propagating synthesis of alloys and oxide dispersion-strengthened materials feasible [114–116].

Iron aluminides are very well suited for casting as they show good form filling, low porosity, and smooth surfaces, and parts have been produced on an industrial scale by sand casting, centrifugal casting, and investment casting, while strip casting has been employed for rolled products [25, 109, 117, 118] (Fig. 10.6). Careful drying of molds is important to avoid hydrogen porosity, and, as iron aluminide-based alloys tend to form coarse-grained microstructures which may easily crack, respective measures by alloying and/or during casting have to be taken.

Cast or powder metallurgically produced precursors are used for forging, which can be performed in air. Forging of iron aluminides is frequently employed for refining the microstructure, but also parts have been produced on an industrial scale [28, 119–121].

Plates, sheets, and tapes have been produced by conventional hot and cold rolling of iron aluminides with Al contents up to about 30 at.% [107, 122, 123]. At higher Al contents, additional measures have to be taken such as using spacers, canning of the alloys, or use of powder metallurgical precursors, as otherwise severe cracking may occur [124–126]. Rolled products often show increased ductility, and various parts have been produced [122, 127].

As for other intermetallic materials, powder metallurgical processing has been studied intensively. Consolidation of pre-alloyed powders by hot extrusion, often with additions of oxides or borides for increasing strength, has been frequently performed, and steel tubes with an iron aluminum cladding have been produced by co-extrusion of both materials [29, 128–130].

Fig. 10.6 Sand casting of Fe–Al. The melt was produced by AIM, and the part has a diameter of about 600 mm and weighs about 160 kg. (Photograph courtesy of Otto Junker GmbH, Simmerath, Germany)

In order to avoid costly machining, near net shape production by additive manufacturing is currently widely studied. Samples and parts have been successfully produced by all currently available techniques, and additive manufacturing has also been employed for the production of chemically graded materials, e.g., iron aluminide/steel composites [131–134].

Though there have been numerous investigations on processing-property relationships, no final conclusion can be drawn. Regarding strength, quenched-in vacancies have a major influence as detailed above. Therefore, variations in cooling rate after processing at higher temperatures may have a larger effect on strength than variations in processing parameters or even use of different processing routes. For ductility there is some agreement that it should increase with decreasing grain size. However, converse observations have been reported [135]. Any type of further processing usually increases ductility, i.e., decreases BDTT, of as-cast iron aluminide alloys, though the mechanisms are not always clear [136, 137].

Because of their excellent corrosion and wear resistance, iron aluminide coatings have been widely studied, and these activities have been recently reviewed [138, 139]. Machining of iron aluminide-based alloys by all standard techniques has been demonstrated. But as they may show strong work hardening, use of specific tools, shortened tool lifetime, and slower machining speeds may be expected, though not necessarily so. Nevertheless, economical machining of iron aluminides is a critical issue [140, 141]. Welding of iron aluminides and making iron aluminide/steel bonds has also been evaluated. Cracking due to thermal stresses or by residual hydrogen is an issue and has to be taken care of [142–144].

10.2.7 Applications

Though iron aluminide-based alloys have a great potential and though there have been quite some efforts to make use of this potential for various applications, little has been achieved up to now. Because of their excellent corrosion resistance, iron aluminide-based alloys have been successfully employed as hardware within various industrial furnaces, e.g., for pyrite roasting, carburizing, and glass melting, or in the aluminum industry [21, 25, 109, 117, 145]. Parts for turbines or turbochargers, i.e., where additional benefit is gained from the lower density, are also applications that have been looked at intensively [24, 121, 146]. Other parts for automotive applications that have been produced are exhaust valves and brake discs though published evidence is not in favor of the latter [33, 147]. Different parts for application in marine engines have been tested recently as well [148]. Regarding corrosion- and wear-resistant coatings based on iron aluminides, numerous patents exist, and sintered iron aluminide filter elements for hot gas cleaning are commercially available [149, 150].

10.3 Iron Silicides

Iron alloyed with silicon is a material of tremendous industrial importance since more than 100 years. This is especially because of the excellent soft magnetic properties of electrical steels, which are Fe–Si alloys with silicon contents up to 6.5 wt.% (corresponding to 12.1 at% Si) applied in electric motors, transformers, generators, and static induction devices. However, there are more applications of Fe–Si alloys, for example, making use of the excellent corrosion resistance of Fe_3Si in even very aggressive environments or taking advantage of the magnetic behavior of this intermetallic phase for possible applications in spintronic devices. Below these diverse applications are briefly introduced after a discussion of the binary phase diagram and some remarks on the properties and processing of this type of material.

It should also be mentioned that, as silicon and iron belong to the most frequent elements on earth, iron silicon alloys play a role not only in engineering applications but also in the geological science of the earth core and in the chemistry of terrestrial and extraterrestrial minerals. These aspects will also be briefly described in the following section dealing with the phases and phase diagram of the binary system.

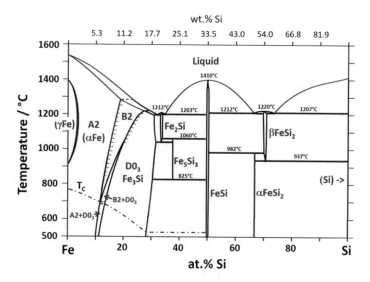

Fig. 10.7 Fe–Si phase diagram constructed on the basis of [151] with changes in the Fe-rich part according to [153]

10.3.1   Phases and Phase Diagram

There is general agreement in the literature about the number and type of inter-metallic phases in the Fe–Si system, and the phase diagram version presented in Kubaschewski's classical compilation of Fe binary systems from 1982 [151] remained unchanged for many years and is identical to the phase diagram shown in 2010 in Okamoto's handbook on binary phase diagrams [152]. Only more recently, Ohnuma et al. [153] performed a detailed reinvestigation of the Fe-rich part of the system combining experimental work and thermodynamic calculations, and Cui and Jung included all these data into their thermodynamic reassessment of the system [154]. The phase diagram shown in Fig. 10.7 follows these studies especially including the new results of Ohnuma et al. for the Fe-rich part.

Similar to the Fe–Al system, large amounts of Si can be dissolved in the ferritic α-Fe solid solution, and the ordered B2 and D0$_3$ (Fe$_3$Si) superstructures of the A2 b.c.c. lattice can form depending on Si content and temperature. Compared to the Fe–Al system, there is a stronger tendency to ordering occurring already for Si contents as low as 10 at.% (corresponding to about 5 wt.%). Early work by Schlatte, Inden, and Pitsch [155–158] indicated a second-order transition from disordered A2 to ordered B2 with increasing Si content and proved the existence of a two-phase B2 + D0$_3$ phase field. Later, results from neutron diffraction experiments led to some doubt about the occurrence of B2 at temperatures below 700 °C [159]. This was also discussed in a more recent review on the ordering phenomena in Fe–Si alloys [160]. Following the conclusions of Ohnuma's experimental and theoretical work [153], the phase diagram in Fig. 10.7 does not contain a B2 phase region below 700 °C, but shows the sequence A2 to A2 + D0$_3$ to D0$_3$ with increasing Si content.

The D0$_3$ ordered Fe$_3$Si phase exists over a wide composition range up to approximately 30 at.% Si. For Si contents below about 28 at.%, Fe$_3$Si transforms from D0$_3$ to B2 ordered structure with increasing temperature. The transformation temperature increases with increasing Si content. On further heating, Fe$_3$Si alloys with less than about 19 at.% Si undergo an additional transformation from B2 to disordered A2 before melting.

At higher Si concentrations, four more intermetallic phases exist in the Fe–Si system, all of them having only very small homogeneity ranges. Fe$_2$Si is a high-temperature phase melting congruently at 1212 °C and crystallizing with a hexagonal structure (Pearson symbol hP6) [161]. Another high-temperature phase is Fe$_5$Si$_3$, which forms on heating in a eutectoid reaction and decomposes in a peritectoid solid-state reaction. Its crystal structure is of the hexagonal D8$_8$ type (hP16) [162]. FeSi melts congruently and exhibits a cubic B20-type structure (cP8) [163], while FeSi$_2$ occurs with two structural variants (low-temperature α-FeSi$_2$ with an orthorhombic (oC48) and high-temperature β-FeSi$_2$ with tetragonal (tP3) structure) [164]. The low-temperature phase α-FeSi$_2$ no longer has a metallic character but shows semiconducting behavior, which is why today it is intensively discussed as an eco-friendly and cheap thermoelectric material [165]. The final phase in the binary system is the Si solid solution, which has an extremely low solubility for Fe, reported to be as low as 3×10^{-5} at.% Fe at 1200 °C [166].

Fe silicides were also found as mineral phases of extraterrestrial origin at various places on the earth [167]. Fe$_3$Si occurs with its D0$_3$ ordered structure (gupeiite [168]) as well as in a disordered b.c.c. state (suessite [169, 170]), but also the other intermetallic phases Fe$_2$Si (hapkeite) [171], Fe$_5$Si$_3$ (xifengite) [168], FeSi (naquite) [172], and FeSi$_2$ (linzhiite) [173] were detected in meteorites.

High pressures in the range of several 10–100 GPa strongly affect the phase relations in the Fe-rich part of the Fe–Si system. Neither the disordered ferritic solid solution nor any of the intermetallic phases Fe$_3$Si, Fe$_2$Si, Fe$_5$Si$_3$, and FeSi are stable under very high pressure. Instead, an Fe-16.4 at.% Si (9 wt.% Si) alloy transforms to a close-packed hexagonal (hcp) structure (similar as does pure Fe). An alloy with 27.5 at.% Si (16 wt.% Si) was found to become two-phase hcp + B2, and Fe-50 at.% Si (33.5 wt.%) adopts a single-phase B2 structure [174, 175]. This is of high importance for the geoscience of the interior of the earth as geochemical models based on cosmochemical arguments suggest that the core of the earth mainly consist of iron and could contain up to 33 at.% Si (20 wt.% Si). Exact composition, temperature, and pressure of the core of the earth are still under discussion [176–178], but an Fe–Si alloy in the inner core most likely is a mixture of hcp and B2 phases as is stated in [174].

10.3.2   *Properties*

Density and lattice parameter of the disordered Fe solid solution decrease approximately linearly with addition of silicon; the room temperature density decreases from 7.87 g/cm^3 for pure Fe to about 7.55 g/cm^3 for 10 at.% Si [179–181]. The occurrence of ordering results in a slight shrinkage of the lattice corresponding to a small but clearly measurable increase of the density. As for example, an 1100 °C annealed and slowly cooled alloy with 12.1 at.% Si (6.5 wt.%) has an about 0.15% higher room-temperature density than the same alloy after water-quenching from 750 °C [181].

The vacancy concentration in thermal equilibrium in Fe$_3$Si reaches high values; an extrapolation to the melting temperature yields a vacancy concentration of 4 vol.% for stoichiometric Fe$_3$Si [182]. The vacancies are located mainly on the Fe sublattice, and their concentration strongly decreases when deviating from the stoichiometric composition to the Fe-rich side. At 427 °C, the thermal vacancy concentration in Fe-21 at.% Si is found to be by a factor of 30 lower than for 25 at.% Si [182]. Fe atom diffusion occurs by nearest-neighbor jumps on the Fe sublattice and is mediated by thermal vacancies. This explains the observation that Fe diffusion in the D0$_3$ ordered Fe$_3$Si is much faster and the corresponding activation enthalpy is considerably lower than in b.c.c. Fe or in disordered Fe-rich Fe–Si alloys [183].

With respect to the mechanical behavior of Fe–Si alloys, the embrittling effect of the occurrence of ordered intermetallic phases is known since long. Additions of up to about 10 at.% Si result in strengthening of the ferritic A2 material while still allowing plastic deformation at room temperature. Further increasing the Si content results in the occurrence of ordered B2 or D0$_3$ phase giving rise to an upward jump of the yield stress and a loss of ductility; see, e.g., Fig. 10.8, showing room-temperature compressive yield stress data taken from the systematic work of Lakso and Marcinkowski on Fe–Si alloys with up to 25 at.% Si (Fe$_3$Si) [184, 185]. D0$_3$ ordered Fe$_3$Si shows no ductility at room temperature and a low fracture strength of 400 MPa in tensile tests. On heating, the material shows a brittle-to-ductile transformation between 500 and 550 °C, and at 600 °C, a fair combination of strength (525 MPa) and ductility (4.5% strain to fracture) was observed [186].

The electrical resistivity of Fe approximately linearly increases with addition of Si in the disordered Fe solid solution. The occurrence of ordering results in a decrease with local resistivity minima for the compositions of the intermetallic phases Fe$_3$Si and Fe$_5$Si$_3$. The composition dependence of the room temperature resistivity is shown in Fig. 10.9 [187–189]. It should be mentioned that the data measured by Varga et al. [189] (15–34 at.% Si) were obtained from melt-spun material, which instead of being amorphous contained the ordered intermetallic phases Fe$_3$Si and Fe$_5$Si$_3$.

As the magnetic properties of Fe–Si alloys are of special importance for applications (see the respective section below), they were well investigated in the past with the first investigations dating back to the year 1900 [190]. Both the disordered α-Fe solid solution and the D0$_3$ ordered intermetallic phase Fe$_3$Si are

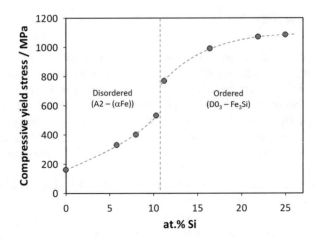

Fig. 10.8 Room-temperature compressive yield stresses of as-cast Fe–Si alloys. (Data taken form a figure shown in Ref. [179])

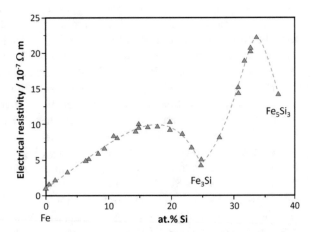

Fig. 10.9 Room-temperature electrical resistivity of Fe–Si alloys as a function of composition [187–189]. (Data taken from a figure shown in Ref. [189])

ferromagnetic with their Curie temperature and saturation magnetization decreasing continuously with increasing Si content; see, e.g., [191]. The Curie temperature T_C of stoichiometric Fe_3Si is 550 °C, and its mean magnetic moment μ amounts to 1.67 μB [192]. The high-temperature Fe-rich intermetallic phase Fe_5Si_3 can easily be metastably retained at room temperature, for example, as nanoparticles also showing ferromagnetic behavior [193]. For the Curie temperature of Fe_5Si_3, values of 102–108 °C were reported [162, 193, 194]. In the Fe solid solution, Si addition effects an increase of magnetic permeability and electrical resistance as well as a decrease of the magnetic anisotropy, coercivity, and magnetostriction, where the latter approaches zero for a Si content of 12.1 at.% (6.5 wt.%); see, e.g., [195, 196].

The corrosion behavior of Fe_3Si is extraordinarily good; Fe_3Si is resistant even to boiling sulfuric and nitric acid [197, 198]. The good corrosion resistance of Fe_3Si is a result of the formation of a protecting SiO_2 layer. Studying the corrosion behavior in sulfuric acid, Yamaguchi [198] found that firstly the acid removes iron from the surface until a pure silicon layer remains, which then is oxidized to SiO_2. According to Wasmuht [199], who studied the corrosion resistance in various acidic media for different Si contents, the existence of the single-phase, ordered intermetallic structure is a precondition for the excellent corrosion behavior with formation of protecting SiO_2.

10.3.3   Processing and Applications

Huge amounts of silicon steels with typical Si contents of about 6 at.% (~3 wt.%) are produced every year by conventional casting and rolling procedures for application as electrical steels especially in electric motors and transformers. The beneficial effect of Si additions to Fe for improving the soft magnetic properties is well-known already since the seminal work of Hadfield and co-workers at the start of the last century [190]. Optimum soft magnetic behavior is reached for a Si content of 12.1 at.% (6.5 wt.%), where the magnetostriction decreases to zero resulting in noise-free power transmission and minimum core losses during cyclic magnetization processes; see, e.g., [195, 196]. Unfortunately, such an improvement of the magnetic properties is accompanied by a dramatic decrease in ductility starting at about 8 at.% Si (~4 wt.%), which makes conventional processing and specifically the required cold rolling of such alloys impossible. Besides segregation of Si to grain boundaries, the decisive factor is the occurrence of ordering leading to embrittlement of the material.

As the financial losses due to the application of material with non-optimized properties and high power losses are enormous, great efforts were put into finding suitable methods to produce Si-enriched Fe-6.5 wt.% Si electrical steel sheets either by trying to avoid ordering or by applying alternative processing methods for the ordered material. It should be noted that the occurrence of B2 or D0$_3$ (Fe$_3$Si) ordered regions per se is not the disqualifying criterion, but instead can be even beneficial for the magnetic properties, e.g., resulting in reduction of the coercivity [200]. The crucial point rather is the brittleness of these intermetallic phases. As commercial-scale production of Fe-6.5 wt.% Si sheets via a conventional cold rolling process is not possible, alternative production routes were explored including siliconizing of cold-rolled Fe-3 wt.% sheets by CVD (chemical vapor deposition) [201–203] or hot-dipping [204], spray-forming (based on atomization of a stream of the liquid alloy by Ar inert gas) [205, 206], direct powder rolling of iron and silicon powder mixtures with subsequent heat treatments [207], rapid quenching methods such as melt spinning [195, 200, 208], or ductilizing by microalloying with boron [209].

Today, Fe–Si alloys are still the most important soft magnetic material for electrical power conversion. Besides the conventional Fe-3 wt.% Si sheet cores, so-called super cores with 6.5 wt.% Si produced via the CVD route were patented under the brand name JNEX® by JFE Steel Corp. [210, 211]. These non-oriented electrical steel sheet cores exhibit virtually zero magnetostriction resulting in low core losses and low noise in high-frequency applications. Competitors of sheet cores are compacted powder cores as the so-called Mega Flux core that was patented by Chang Sung Corp. [212, 213]. As the lack of ductility of Fe-6.5 wt.% Si prevents powder from being compacted, the Si content is adjusted to 4.5 wt.% in Mega Flux [213]. The properties and characteristics of the different core materials were compared in many studies revealing that the optimum choice strongly depends on the particular application [212–217].

The first Si-rich Fe–Si-based soft magnetic material, which already long time ago entered successfully into industrial applications, was so-called Sendust invented by Hakaru Masumoto and Tatsuji Yamamoto in 1937 [218]. Sendust, the name of which is a combination of Sendai and dust, is a ternary Fe-based alloy containing 9.6 wt.% Si and 5.4 wt.% Al and having the $D0_3$ Fe_3Si crystal structure. Due to its extreme brittleness, it is usually applied as sintered, partially or completely amorphous powder or rapidly quenched ribbons, but was also produced in sheet form through metal powder rolling techniques [219]. Sendust cores can have extremely high magnetic permeability with simultaneously low coercivity and as Sendust at the same time shows very good wear properties, it is used for magnetic recording heads [220, 221]. Composites consisting of Sendust flakes embedded in a polymer were produced for high-frequency applications in, e.g., mobile phones or personal computers to act as electromagnetic noise suppressor [222, 223]. Today, various composite powder cores are available with different standardized sizes and permeabilities, which can be combined to get new effective permeabilities [224]. Compared to the Fe–Si Mega Flux cores, Sendust cores show superior performance in transformers and inductors with respect to acoustic noise emission [225].

Due to its excellent corrosion resistance, castings of Fe_3Si are used since long as industrial components in contact with chemically aggressive media. An alloy with 25.2 at.% Si (14.5 wt.%) was developed and patented as "Duriron" by P.D. Schenck, W.E. Hall, and J.R. Pitman, who soon after (in 1912) incorporated the "During Casting Company" [226–228]. Duriron is not only corrosion resistant in inorganic and organic acidic environments but also in various alkaline solutions [199]. Its excellent corrosion behavior is comparable to that of noble metals, but as it is much cheaper, this material is extensively used in chemical industry since more than 100 years. The major drawback for production of parts from Fe_3Si is its extreme brittleness and hardness, which is why such parts can be only produced by casting.

With its good corrosion resistance, high hardness, and high electrical resistivity, Fe_3Si is also a candidate as protective surface coating, for example, on Fe–Si electrical steels. This was proven by Schneeweiss et al. [229] who successfully prepared an Fe_3Si coating by firstly CVD of a 1 μm Si layer on Fe-3wt.%Si steel followed by some heat treatments that transformed the Si layer to a protective, 3–4-μm-thick Fe_3Si coating.

Fe$_3$Si thin films are also good candidates for applications in microelectronics, e.g., as ferromagnetic electrode in spintronic devices utilizing the high spin polarization and high Curie temperature of Fe$_3$Si. Thin, epitaxial films were grown on semiconductor substrates such as GaAs [230, 231] or on SiO$_2$ for application as magnetic tunnel junctions [232, 233]. As a possible starting material for preparation of such thin layers, Fe$_3$Si nanoparticles (size approximately 5 nm) were successfully produced by a high-temperature chemical reduction method employing a reaction of silicon tetrachloride with iron pentacarbonyl [234].

10.4 Iron-Based Ferritic Superalloys

This chapter deals with ferritic iron-based alloys, where a disordered b.c.c. phase forms a coherent microstructure with an intermetallic phase. As the microstructures resemble those of the Ni-based superalloys, they have been termed ferritic superalloys. These alloys usually show high strength at elevated temperature and, therefore, recently have attracted much attention.

Because of lower costs, lower thermal expansion, and higher thermal conductivity, ferritic alloys are an interesting alternative to austenitic steels for high-temperature applications [235]. However, their poor creep resistance above 600 °C has limited their usage so far. Strengthening by coherent precipitates is an efficient method to obtain appreciable strength at high temperatures, as realized by the γ/γ' microstructures in superalloys. These superalloys were originally iron-based, but shifted to Ni and Co base already in the 1940s, and these alloys are still the backbone for applications at high temperatures [236, 237]. Strengthening the matrix by coherent NiAl precipitates is another method that has been successfully employed for maraging steels [238–240]. This concept has been recently refined to strengthening by coherent nanoprecipitates [241]. Also more recently f.c.c. + NiAl alloys with lamellar microstructures gained some interest because they show high strength and good ductility at room temperature [242]. In the following, ferrous materials other than steels with coherent microstructures are dealt with.

Specifically in many Fe–Al–X systems (X = Ni, Co, V, Ti . . .), miscibility gaps exist in which on cooling coherent microstructures of (α-Fe) + B2 (Fe,X)Al, (α-Fe) + L2$_1$ Heusler-type Fe$_2$AlX, or (Fe,X)Al + Fe$_2$AlX form. The three phases are crystallographically closely related to each other and have lattice constants quite close to each other (a$_0$ (α-Fe) $\approx$ a$_0$ (Fe,X)Al $\approx$ ½ a$_0$ Fe$_2$AlX), and it is therefore that they readily form coherent microstructures. It has been shown that increased strength does not depend on whether the disordered or the intermetallic phase forms the precipitates [243], though alloy developments are preferentially aiming at precipitating the intermetallic phase. Microstructures can vary in dependence on the volume fractions from isolated precipitates to chessboard-like to maze-like [70]. The strength of the alloys will be determined by the size of the precipitates, width of the channels between them, and the coherency stresses. At least in some of the investigated systems, the precipitates form by spinodal decomposition, which gives rise to second and third generations of precipitates. These fine-scaled precipitates have an additional strengthening effect but may also cause severe embrittlement at lower temperatures [70].

10.4.1 Alloy Developments

Initial alloy developments were based on the Fe–Al–X systems with Ni and Ti. Currently the focus is on Fe–Al–Ni–Ti–Cr alloys with minor additions of other elements [244]. By microstructural engineering the misfit between matrix and precipitates, i.e., the coherency stress, is optimized, and formation of sub-precipitates within the primary precipitates is employed for further strength increase [245]. These measures also help to avoid rapid coarsening which otherwise can be an issue [246]. A different approach to generate coherent microstructures is pursued in the Fe–Al–X systems with Nb and Ta. In these systems precipitation of the stable Laves phase is kinetically retarded, and a metastable Heusler phase forms instead [57]. The Heusler phase forms coherent microstructures with the matrix, which, dependent on temperature and composition, can be (α-Fe) or (Fe,X)Al. Though metastable, the coherent microstructures are stable for long time at elevated temperatures. Above 700–750 °C, the stable Laves phase will form, but as precipitation is primarily along grain boundaries, strength at high temperature can be maintained [73].

10.4.2 Properties

The strength of ferritic superalloys outperforms that of advanced Cr steels and matches that of Ni-based superalloys (Fig. 10.10). The yield strength at 800 °C varies between 100 and 300 MPa [70, 72, 73]. At 700 °C, secondary creep rates of 10^{-8} s^{-1} are observed at applied stresses between 90 and 200 MPa [72, 73, 247, 248], and the rate-determining deformation mechanisms have been intensively studied [243, 244, 247–250]. As the alloys are rather strong, they show only limited ductility at room temperature. It has been shown that ductility mainly depends on the volume fraction of precipitates [251, 252]. Specifically alloys with fine-scaled second- and third-generation precipitates can be rather brittle, and ductility may enhance after coarsening the fine precipitates through annealing [251, 253]. For sufficient corrosion resistance, the constituting phases must have an adequate content of Al plus Cr. Then they will show an excellent corrosion resistance like the iron aluminide alloys discussed in Sect. 10.2.

Fig. 10.10 Double logarithmic graph showing applied tensile stress against rupture life for advanced commercial steels (P92, P122, T91, T122, Cr12) in comparison to novel ferritic superalloys. (Figure adopted from Ref. [247])

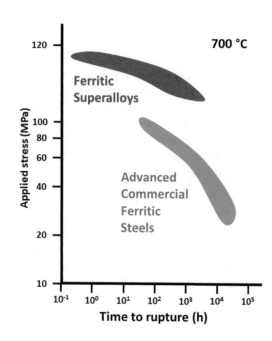

10.4.3   Processing

Casting, forging, and additive manufacturing of ferritic superalloys have been demonstrated [121, 133]. Because of their high strength, forging must be performed at temperatures above the solvus. Forged and additive processed alloys showed comparable strength as their as-cast counterparts but partly increased ductility [133, 136]. As these alloys are still under development, their processing/property relationships must be still evaluated, and no applications have been realized yet.

10.5   Iron-Based Alloys with TCP Phases

Laves phases and other topological close-packed (TCP) phases are common precipitates in steels. They are very strong phases, and therefore they have been considered for strengthening steels already long time ago, either as additional precipitates to carbides or in C-free alloys and low-carbon steels, e.g., maraging steels [3, 254–256]. However, it has been found that they precipitate preferentially at grain boundaries, where they do not contribute much to strengthening but instead induce formation of cavities and cause embrittlement [6, 257]. They are also unwanted, because after formation they tend to rapidly coarsen, thereby extracting the elements responsible for solid solution hardening like W and Mo from the matrix [8, 257–260]. However, reports are varying, and if compositions are carefully adjusted, Laves and other TCP phases may be beneficial in attaining strength at high temperature [261, 262].

The key for gaining strength at high temperatures by TCP precipitates is their distribution. Early attempts used fine precipitates of spheroidized Laves phase [254]. A new concept is to precipitate the TCP phases as more or less continuous film along the grain boundaries. The resulting steels are very strong, and it has been demonstrated that though the TCP phases are rather brittle as monolithic phase, the films show good mechanical behavior, i.e., they do not show cracking after plastic deformation [263, 264]. Both ferritic and austenitic steels are currently under development.

10.5.1   Ferritic(–Martensitic) TCP Steels

Strengthening ferritic 9–12 wt.% Cr steels by Laves phase and other TCP precipitates has been investigated for quite some time. Usually elements such as Nb, W, and Mo are added to provoke their formation. In general it is found that the Laves phase, which precipitates during creep, initially increases the creep resistance. However, due to rapid coarsening, the creep rate may increase again after some time [258].

More recently, specifically Cr-rich steels are developed which gain their strength from TCP phases. 15 wt.% Cr steels containing Laves phase and μ phase (and χ phase) show a superior creep resistance due to fine precipitation of the TCP phases in the matrix [265].

Crofer 22H is a ferritic steel strengthened by Laves precipitates within the α(Fe,Cr) matrix [266]. It was developed as a high strength steel for interconnections in solid oxide fuel cells (SOFCs). Compared to the precipitate-free Cr steel JS-3, it shows a much better creep resistance due to solid solution hardening and the precipitation of micron-sized $Fe_2Nb(W)$ Laves phase precipitates. The oxidation behavior does not deteriorate by the addition of Nb, W, and Si, and this steel has a total elongation of >20% at room temperature [266]. Further optimization of the microstructure showed that distributing the Laves phase at the grain boundary is very effective to prevent coarsening of the ferritic matrix as the Laves phase itself coarsens only very slowly [267, 268]. The concept is further elaborated in the development of HiperFer, another high chromium-containing steel for highly efficient steam power plants that has a better steam oxidation resistance and higher strength than P92 [263, 269].

10.5.2   Austenitic TCP Steels

Based on the design concept of grain boundary precipitation strengthening, Takeyama et al. developed a highly creep-resistant but still sufficiently ductile steel of composition Fe-30Cr-20Ni-2Nb (in at.%) [270, 271]. This steel has a three-dimensional network of the Fe_2Nb Laves phase precipitated as a film of <1 μm thickness on the austenite grain boundaries and additional fine nanometer-sized plates of Ni_3Nb inside the austenite grains [264]. It has been shown that the properties crucially depend on the area fraction of the grain boundaries covered by Laves phase, i.e., that strength increases and ductility decreases with increasing fraction of Laves phase. An alloy with about 90% coverage of the grain boundaries attained a creep life of 880 h at 700 °C and 140 MPa [270]. However, crack formation neither inside the Laves phase nor at the Laves/γ-Fe interface was observed [264]. The steel has a good corrosion resistance in steam in that it readily forms a protective Cr_2O_3 scale [272]. Main application of the steel could be in advanced thermal power plants.

10.6   Summary and Future Outlook

In view of their potential to replace high-alloyed steels or even Co- or Ni-based superalloys, iron-based intermetallic materials are an inexpensive alternative. As they are iron based, they can be produced using equipment readily available in iron and steel companies. Though these materials are not "another steel," at least part of the processing can be based on experience gained from the processing of cast iron or steels. Their limited ductility is still adequate for many applications. It may cause a bit more time-consuming machining, but near net shape processing, which has been successfully demonstrated, can substantially reduce costs for machining.

Their excellent strength-to-weight ratio, i.e., their specific strength, their partly outstanding corrosion resistance, and high wear resistance, opens up new possibilities to use these iron-based materials under demanding conditions, otherwise only bearable by Co- and Ni-based superalloys. In view of these properties, iron-based intermetallic materials offer an increased lifetime combined with reduced energy consumption. Iron, aluminum, and silicon are the most abundant elements, and the materials discussed in this chapter need no or little additional alloying. Therefore, besides economic considerations, with increasing lack of strategic elements, these materials for sure will get more attention in the future.

References

1. G.E.R. Schulze, *Metallphysik* (Akademie-Verlag, Berlin, 1967), pp. 1–76
2. S. Floreen, The physical metallurgy of maraging steels. Metall. Rev. **13**, 115–128 (1968)
3. V.K. Vasudevan, S.J. Kim, C.M. Wayman, Precipitation reactions and strengthening behavior in 18 wt pct nickel maraging steels. Metal. Trans. **21A**(10), 2655–2668 (1990)
4. D. Raabe et al., Designing ultrahigh strength steels with good ductility by combining transformation induced plasticity and martensite aging. Adv. Eng. Mater. **11**(7), 547–555 (2009)

5. S. Floreen, An examination of chromium substitution in stainless steels. Metal. Trans. **13A**(11), 2003–2013 (1982)
6. K. Asakura, Y. Kobuchi, T. Fujita, Embrittlement factors in high-Cr ferritic heat resisting steels. Tetsu-to-hagane **75**(7), 1209–1216 (1989)
7. D.J. Thoma, Intermetallics: Laves phases, in *Encyclopedia of Materials: Science and Technology*, ed. by K. H. J. Buschow et al., (Elsevier, Amsterdam, 2001), pp. 4205–4213
8. P.J. Ennis et al., Microstructural stability and creep rupture strength of the martensitic steel P92 for advanced power plant. Acta Mater. **45**(12), 4901–4907 (1997)
9. M. Palm, F. Stein, G. Dehm, Iron aluminides. Ann. Rev. Mater. Res. **49**, 297–326 (2019)
10. M. Hansen, K. Anderko, Aluminium-iron, in *Constitution of Binary Alloys*, ed. by M. Hansen, K. Anderko, (McGraw Hill Book Company, New York, 1958), pp. 90–95
11. F. Stein, M. Palm, Re-determination of transition temperatures in the Fe-Al system by differential thermal analysis. Int. J. Mat. Res. **98**(7), 580–588 (2007)
12. X. Li et al., The Al-rich part of the Fe-Al phase diagram. J. Phase Equil. Diffus. **37**(2), 162–173 (2016)
13. R.A. Hadfield, Aluminium-steel. J. Iron Steel Inst. **2**, 161–230 (1890)
14. W.J. Keep, Aluminum in cast iron. Trans. AIME **18**, 102–122 (1890)
15. H.F. Durnenko, Aluminevy cugun (Cugal). Liteiscik **7** (1934)
16. J.W. Bampfylde, *B.C.I.R.A. Improvements in and Relating to the Manufacture of Cast Iron Alloys* (UK, 1938)
17. J.F. Nachman, W.J. Buehler, *The Fabrication and Properties of 16-ALFENOL – A Non-strategic Aluminum-Iron Alloy* (Naval Ordnance Laboratory, 1953), pp. 1–23
18. W.J. Buehler, C.G. Dalrymple, Coming: better thermenol alloys. Met. Prog. **73**(5), 78–81 (1958)
19. E.R. Morgan, V.F. Zackay, Ductile iron aluminum alloys. Met. Prog. **68**(10), 126–128 (1955)
20. W.J. Lepkowski, J.W. Holladay, *The Present State of Development of Iron-Aluminum-Base Alloys* (Battelle Memorial Institute, 1957), pp. 1–39
21. R. Brooks, A. Volio, *Iron-Aluminum Alloy Systems, Part 14 Welding of Iron Aluminum Alloys* (Wright Air Development Center; Wright Patterson Airforce Base, OH, USA, 1959), pp. 1–24
22. W. Chubb et al., *Constitution, Metallurgy, and Oxidation Resistance of Iron-Chromium-Aluminum Alloys* (Battelle Memorial Institute, 1958), pp. 1–104
23. J.W. Holladay, *Review of Developments in Iron-Aluminum-Base Alloys* (Battelle Memorial Institute, 1961), pp. 1–57
24. M.A. Plesinger, Les nouveaux alliages refractaires developpes en Tchecoslovaquie. Fonderie **157**(2), 75–88 (1959)
25. P. Kratochvil, The history of the search and use of heat resistant Pyroferal alloys based on FeAl. Intermetallics **16**(4), 587–591 (2008)
26. Z. Tyszko, Fontes a haute teneur en aluminium. Fonderie **278**, 221–233 (1969)
27. J.F. Nachman, E.R. Duffy, Effect of alloying additions on sea water corrosion resistance of iron-aluminum base alloys. Corrosion **30**(10), 357–365 (1974)
28. G. Culbertson, C.S. Kortovich, *Development of Iron Aluminides* (AF Wright Aeronautical Laboratories, 1986), pp. 1–149
29. R.G. Bordeau, *Development of Iron Aluminides* (AF Wright Aeronautical Laboratories, 1987), pp. 1–264
30. B. King, *Dispersion Strengthening of Iron Aluminium Base Alloys: A Feasability Study* (Oak Ridge National Laboratory, 1960), pp. 1–45
31. C.G. McKamey, Iron aluminides, in *Physical Metallurgy and Processing of Intermetallic Compounds*, ed. by N. S. Stoloff, V. K. Sikka, (Chapman & Hall, New York, 1996), pp. 351–391
32. C.G. McKamey et al., A review of recent developments in Fe$_3$Al-based alloys. J. Mater. Res. **6**(8), 1779–1805 (1991)
33. S.C. Deevi, V.K. Sikka, Nickel and iron aluminides: an overview on properties, processing, and applications. Intermetallics **4**(5), 357–375 (1996)

34. C.T. Liu et al., Recent advances in B2 iron aluminide alloys: deformation, fracture and alloy design. Mater. Sci. Eng. A **258**, 84–98 (1998)
35. P.F. Tortorelli, K. Natesan, Critical factors affecting the high-temperature corrosion performance of iron aluminides. Mater. Sci. Eng. A **258**, 115–125 (1998)
36. S. Revol, R. Baccino, F. Moret, *Industrial Applications of FeAl40Grade3, A High Specific Properties Iron Aluminides. in EUROMAT99* (Wiley-VCH, Weinheim, 2000)
37. R.G. Baligidad, A. Radhakrishna, Effect of hot rolling and heat treatment on structure and properties of high carbon Fe-Al alloys. Mater. Sci. Eng. A **308**, 136–142 (2001)
38. R.G. Baligidad, V.V. Satya Prasad, A. Sambasiva Rao, Effect of Ti, W, Mn, Mo and Si on microstructure and mechanical properties of high carbon Fe-10.5 wt-% Al alloy. Mater. Sci. Technol. **23**(5), 613–619 (2007)
39. M. Palm, Fe-Al materials for structural applications at high temperatures: current research at MPIE. Int. J. Mater. Res. **100**(3), 277–287 (2009)
40. D. Paris, P. Lesbats, J. Levy, An investigation of the distribution of vacancies in an ordered Fe-Al alloy by field ion microscopy. Scr. Metall. **9**, 1373–1378 (1975)
41. M. Kogachi, T. Haraguchi, Quenched-in vacancies in B2-structured intermetallic compound FeAl. Mater. Sci. Eng. **A230**(1-2), 124–131 (1997)
42. Y. Yang, I. Baker, The influence of vacancy concentration on the mechanical behavior of Fe-40Al. Intermetallics **6**, 167–175 (1998)
43. G. Hasemann, J.H. Schneibel, E.P. George, Dependence of the yield stress of Fe_3Al on heat treatment. Intermetallics **21**(9), 56–61 (2012)
44. G. Dlubek, O. Brümmer, B. Möser, The recovery of quenched-in vacancies in Fe-Al (6.3 to 28.3 at.%) alloys studied by positron annihilation. Cryst. Res. Technol. **17**(8), 951–961 (1982)
45. I. Baker, D.J. Gaydosh, Flow and fracture of Fe-Al. Mater. Sci. Eng. **96**, 147–158 (1987)
46. J.T. Guo et al., Discovery and study of anomalous yield strength peak in FeAl alloy. Scr. Metall. Mater. **29**(6), 783–785 (1993)
47. K. Yoshimi, S. Hanada, M.H. Yoo, Yielding and plastic flow behavior of B2-type Fe-39.5 mol.% single crystals in compression. Acta Metall. Mater. **43**(11), 4141–4151 (1995)
48. D.G. Morris, M.A. Munoz-Morris, The stress anomaly in FeAl-Fe_3Al alloys. Intermetallics **13**(12), 1269–1274 (2005)
49. D.G. Morris, M.A. Munoz-Morris, A re-examination of the pinning mechanisms responsible for the stress anomaly in FeAl intermetallics. Intermetallics **18**(7), 1279–1284 (2010)
50. E.P. George, I. Baker, Thermal vacancies and the yield strength anomaly of FeAl. Intermetallics **6**, 759–763 (1998)
51. D.J. Schmatz, R.H. Bush, Elevated temperature yield effect in iron-aluminum. Acta Metall. **16**(2), 207–217 (1968)
52. J.H. Song, T.K. Ha, Y.W. Chang, Anomalous temperature dependence of flow stress in a Fe_3Al alloy. Scr. Mater. **42**(3), 271–276 (2000)
53. C.T. Liu, E.H. Lee, C.G. McKamey, An environmental effect as the major cause for room-temperature embrittlement in FeAl. Scr. Metall. **23**, 875–880 (1989)
54. N.S. Stoloff et al., Environmental embrittlement of Fe_3Al alloys under monotonic and cyclic loading, in *Processing, Properties, and Applications of Iron Aluminides*, (TMS, Warrendale, 1994)
55. M. Zamanzade, A. Barnoush, An overview of the hydrogen embrittlement of iron aluminides. Proc. Mater. Sci. **3**, 2016–2023 (2014)
56. D. Hardwick, G. Wallwork, Iron-aluminium alloys: a review of their feasibility as high-temperature materials. Rev. High-Temp. Mater. **4**(1), 47–74 (1978)
57. M.G. Mendiratta et al., A review of recent developments in iron aluminides. Mater. Res. Soc. Symp. Proc. **81**, 393–404 (1987)
58. D.G. Morris, M.A. Morris, Strengthening at intermediate temperatures in iron aluminides. Mater. Sci. Eng. A **239**, 23–38 (1997)
59. D.G. Morris, Possibilities for high-temperature strengthening in iron aluminides. Intermetallics **6**, 753–758 (1998)

60. A. Bahadur, Enhancement of high temperature strength and room temperature ductility of iron aluminides by alloying. Mater. Sci. Technol. **19**, 1627–1634 (2003)
61. M. Palm, Concepts derived from phase diagram studies for the strengthening of Fe-Al-based alloys. Intermetallics **13**(12), 1286–1295 (2005)
62. F. Stein, A. Schneider, G. Frommeyer, Flow stress anomaly and order-disorder transitions in Fe_3Al-based Fe-Al-Ti-X alloys with X = V, Cr, Nb, or Mo. Intermetallics **11**, 71–82 (2003)
63. D. Risanti et al., Dependence of the brittle-to-ductile transition temperature (BDTT) on the Al content of Fe-Al alloys. Intermetallics **13**(12), 1337–1342 (2005)
64. A. Schneider et al., Microstructures and mechanical properties of Fe_3Al-based Fe-Al-C alloys. Intermetallics **13**(12), 1322–1331 (2005)
65. L. Falat et al., Mechanical properties of Fe-Al-M-C (M = Ti, V, Nb, Ta) alloys with strengthening carbides and Laves phase. Intermetallics **13**(12), 1256–1262 (2005)
66. R. Krein et al., Microstructure and mechanical properties of Fe_3Al-based alloys with strengthening boride precipitates. Intermetallics **15**(9), 1172–1182 (2007)
67. M. Palm, G. Sauthoff, Deformation behaviour and oxidation resistance of single-phase and two-phase $L2_1$-ordered Fe-Al-Ti alloys. Intermetallics **12**, 1345–1359 (2004)
68. R. Krein, M. Palm, The influence of Cr and B additions on the mechanical properties and oxidation behaviour of $L2_1$-ordered Fe-Al-Ti-based alloys at high temperatures. Acta Mater. **56**(10), 2400–2405 (2008)
69. X. Li, P. Prokopcakova, M. Palm, Microstructure and mechanical properties of Fe–Al–Ti–B alloys with additions of Mo and W. Mater. Sci. Eng. A **611**, 234–241 (2014)
70. C. Stallybrass, G. Sauthoff, Ferritic Fe-Al-Ni-Cr alloys with coherent precipitates for high-temperature application. Mater. Sci. Eng. A **387–389**, 985–990 (2004)
71. C. Stallybrass, *The Precipitation Behaviour and Mechanical Properties of Novel Fe-Al-Ni-Cr Alloys with Coherent Precipitates* (Shaker Verlag, Aachen, 2008), pp. 1–194
72. R. Krein, M. Palm, M. Heilmaier, Characterisation of microstructures, mechanical properties, and oxidation behaviour of coherent A2 + $L2_1$ Fe-Al-Ti alloys. J. Mater. Res. **24**(11), 3412–3421 (2009)
73. D.D. Risanti, G. Sauthoff, Microstructures and mechanical properties of Fe-Al-Ta alloys with strengthening Laves phase. Intermetallics **19**, 1727–1736 (2011)
74. M. Eumann, M. Palm, G. Sauthoff, Alloys based on Fe_3Al or FeAl with strengthening Mo_3Al precipitates. Intermetallics **12**(6), 625–633 (2004)
75. T. Doucakis, K.S. Kumar, Formation and stability of refractory metal diborides in an Fe_3Al matrix. Intermetallics **7**, 765–777 (1999)
76. F. Stein, M. Palm, G. Sauthoff, Mechanical properties and oxidation behaviour of two-phase iron aluminium alloys with $Zr(Fe,Al)_2$ Laves phase or $Zr(Fe,Al)_{12}$ τ_1 phase. Intermetallics **13**(12), 1275–1285 (2005)
77. P. Prokopcakova, M. Svec, M. Palm, Microstructural evolution and creep of Fe-Al-Ta alloys. Int. J. Mater. Res. **107**(5), 396–405 (2016)
78. I. Ohnuma et al., Ordering and phase separation in the b.c.c. phase of the Fe-Al-Ti system. Acta Mater. **46**(6), 2083–2094 (1998)
79. C.G. McKamey et al., Effects of alloying additions on the microstructures, mechanical properties and weldability of Fe_3Al-based alloys. Mater. Sci. Eng. A **174**(1), 59–70 (1994)
80. R. Balasubramaniam, Alloy development to minimize room temperature hydrogen embrittlement in iron aluminides. J. Alloys Compd. **253–254**, 148–151 (1997)
81. D.G. Morris, M.A. Morris-Munoz, The influence of microstructure on the ductility of iron aluminides. Intermetallics **7**(10), 1121–1129 (1999)
82. K. Vedula, FeAl and Fe3Al, in *Intermetallic Compounds*, Practice, ed. by J. H. Westbrook, R. L. Fleischer, vol. 2, (Wiley, Chichester, 1995), pp. 199–209
83. N.S. Stoloff, D.A. Alven, C.G. McKamey, An overview of Fe_3Al alloy development with emphasis on creep and fatigue, in *International Symposium on Nickel and Iron Aluminides: Processing, Properties, and Applications*, (ASM International, 1997)
84. A. Tonneau, M. Gerland, G. Henaff, Environment-sensitive fracture of iron aluminides during cyclic crack growth. Metall. Mater. Trans. A **32**(9), 2345–2356 (2001)

85. G.E. Fuchs, N.S. Stoloff, Effects of temperature, ordering and composition on high cycle fatigue of polycrystalline Fe₃Al. Acta Metall. **36**(5), 1381–1387 (1988)
86. H.Y. Yasuda, A. Behgozin, Y. Umakoshi, Fatigue behavior of Fe-48.at% Al polycrystals with B2 structure at high temperatures. Scr. Mater. **40**(2), 203–207 (1999)
87. M. Johnson et al., The resistance of nickel and iron aluminides to cavitation erosion and abrasive wear. Wear **140**, 279–289 (1990)
88. A. Magnee, Generalized law of erosion: application to various alloys and intermetallics. Wear **181–183**, 500–510 (1995)
89. J. Xia, Thermal oxidation treatment of iron aluminide for improved tribological properties. Surf. Eng. **21**(1), 6–11 (2005)
90. J.P. Tu, M.S. Liu, Wet abrasive wear of ordered Fe₃Al alloys. Wear **209**, 31–36 (1997)
91. Y.S. Kim, J.H. Song, Y.W. Chang, Erosion behavior of Fe-Al intermetallic alloys. Scr. Mat. **36**(7), 829–834 (1997)
92. H. Hindam, D.P. Whittle, Microstructure, adhesion and growth kinetics of protective scales on metals and alloys. Oxid. Met. **18**, 245–284 (1982)
93. P. Tomaszewicz, G. Wallwork, Iron-aluminium alloys: a review of their oxidation behaviour. Rev. High-Temp. Mater. **4**(1), 75–104 (1978)
94. F.H. Stott, K.T. Chuah, L.B. Bradley, Oxidation-sulphidation of iron aluminides at high temperature. Mater. Corros. **47**, 695–700 (1996)
95. J. Klöwer, High-temperature corrosion behavior of iron aluminides and iron-aluminum-chromium alloys. Mater. Corros. **47**, 685–694 (1996)
96. A. Schneider, J. Zhang, Metal dusting of ferritic Fe-Al-M-C (M = Ti, V, Nb, Ta) alloys in CO-H₂-H₂O gas mixtures at 650 °C. Mater. Corros. **54**(10), 778–784 (2003)
97. D.J. Duquette, Corrosion of intermetallic compounds, in *Intermetallic Compounds*, Principles, ed. by J. H. Westbrook, R. L. Fleischer, vol. 1, (Wiley, Chichester, 1995), pp. 965–975
98. V. Shankar Rao, A review of the electrochemical corrosion behaviour of iron aluminides. Electrochim. Acta **49**, 4533–4542 (2004)
99. J.G. Kim, R.A. Buchanan, Pitting and crevice corrosion of iron aluminides in a mild acid-chloride solution. Corrosion **50**(9), 658–668 (1994)
100. J. Peng et al., Influence of Al content and pre-oxidation on the aqueous corrosion resistance of binary Fe-Al alloys in sulphuric acid. Corros. Sci. **149**, 123–132 (2019)
101. M. Palm, R. Krieg, Neutral salt spray tests on Fe–Al and Fe–Al–X. Corros. Sci. **64**, 74–81 (2012)
102. S. Frangini, Corrosion behavior of AISI 316L stainless steel and ODS FeAl aluminide in eutectic Li₂CO₃–K₂CO₃ molten carbonates under flowing CO₂–O₂ gas mixtures. Oxid. Met. **53**(1/2), 139–156 (2000)
103. M. Amaya et al., High temperature corrosion performance of FeAl intermetallic alloys in molten salts. Mater. Sci. Eng. A **349**, 12–19 (2003)
104. V. Marx, M. Palm, Oxidation of Fe-Al Alloys (5-40 at.% Al) at 700 and 900 °C. Mat. Sci. Forum **879**, 1245–1250 (2017)
105. D. Vogel et al., Corrosion behaviour of Fe-Al(-Ti) alloys in steam. Intermetallics **18**, 1375–1378 (2010)
106. W.J. Lepkowski, J.W. Holladay, *The Present State of Development of Iron-Aluminum-Base Alloys* (Battelle Memorial Institute, 1957), pp. 1–39
107. C. Sykes, J.W. Bampfylde, The physical properties of iron-aluminium alloys. J. Iron Steel Inst. **130**, 389–418 (1934)
108. Y.D. Huang et al., Preparation and mechanical properties of large-ingot Fe₃Al-based alloys. J. Mater. Proc. Technol. **146**, 175–180 (2004)
109. V.K. Sikka et al., Melting and casting of FeAl-based alloy. Mater. Sci. Eng. A **258**, 229–235 (1998)
110. A. Radhakrishna, R.G. Baligidad, D.S. Sarma, Effect of carbon on structure and properties of FeAl based intermetallic alloy. Scr. Mater. **45**, 1077–1082 (2001)
111. T. Itoi et al., Preparation of recycle-typed Fe₃Al alloy and its application for cutting tool materials. Intermetallics **18**(7), 1396–1400 (2010)

112. D.F.L. Borges, D.C.R. Espinosa, C.G. Schön, Making iron aluminides out of scrap. J. Mater. Res. Technol. **3**(2), 101–106 (2014)
113. K. Matsuura, Y. Watanabe, Y. Hirashima, Use of recycled steel machining chips and aluminum can shreds for synthesizing iron aluminide intermetallic alloys. ISIJ Int. **44**(7), 1258–1262 (2004)
114. B.H. Rabin, R.N. Wright, Synthesis of iron aluminides from elemental powders: reaction mechanisms and densification behavior. Metall. Trans. A **22**(2), 277–286 (1991)
115. E. Godlewska et al., FeAl materials from intermetallic powders. Intermetallics **11**(4), 307–312 (2003)
116. S. Paris et al., Spark plasma synthesis from mechanically activated powders: a versatile route for producing dense nanostructured iron aluminides. Scr. Mater. **50**(5), 691–696 (2004)
117. D.G. Morris, M.A. Munoz-Morris, Recent developments toward the application of iron aluminides in fossil fuel technologies. Adv. Eng. Mater. **13**(1-2), 43–47 (2011)
118. J.R. Blackford et al., Production of iron aluminides by strip casting followed by cold rolling at room temperature. Scr. Mat. **34**(10), 1595–1600 (1996)
119. O. Flores et al., Forging of FeAl intermetallic compounds, in *Processing, Properties and Applications of Iron Aluminides*, ed. by J. H. Schneibel, M. A. Crimp, (TMS, 1994), pp. 31–38
120. D.G. Morris, M.A. Munoz-Morris, High creep strength, dispersion-strengthened iron aluminide prepared by multidirectional high-strain forging. Acta Mater. **58**(10), 6080–6089 (2010)
121. P. Janschek et al., Forging of steam turbine blades with an Fe₃Al-based alloy. Mater. Res. Soc. Symp. Proc. **1128**, 47–52 (2009)
122. J.F. Nachman, W.J. Buehler, 16 percent aluminum-iron alloy cold rolled in the order-disorder temperature range. J. Appl. Phys. **25**(3), 307–313 (1954)
123. J.R. Blackford et al., Effect of process variables on tensile properties of ingot processed versus strip cast iron aluminides. Mater. Sci. Technol. **14**(11), 1132–1138 (1998)
124. C. Testani et al., FeAl intermetallics and applications: an overview, in *International Symposium on Nickel and Iron Aluminides: Processing, Properties, and Applications*, ed. by S. C. Deevi et al., (ASM International, Cincinnati, 1997), pp. 213–222
125. D. Kuc, G. Niewielski, I. Bednarczyk, The influence of thermomechanical treatment on structure of FeAl intermetallic phase-based alloys. JAMME **29**(2), 123–130 (2008)
126. I. Schindler et al., Forming of cast Fe – 45 at.% Al alloy with high content of carbon. Intermetallics **18**, 745–747 (2010)
127. S.C. Deevi et al., Processing and properties of FeAl sheets obtained by roll compaction and sintering of water atomized powders, in *High-temperature Ordered Intermetallic Alloys VIII*, (MRS, Warrendale, Boston, 1999)
128. S. Strothers, K. Vedula, Hot extrusion of B2 iron aluminide powders. Prog. Powder Metall. **43**, 597–610 (1987)
129. B. Kad et al., Optimization of high temperature hoop creep response in ODS-Fe₃Al tubes, in *17th Annual Conference on Fossil Energy Materials*, ed. by R. R. Judkins, (United States. Dept. of Energy. Office of Scientific and Technical Information, Baltimore, 2003), pp. 1–10
130. V.K. Sikka, S. Viswanathan, C.G. McKamey, Development and commercialization status of Fe₃Al-based intermetallic alloys, in *Structural Intermetallics*, (TMS, Seven Springs, 1993)
131. T. Durejko, M. Lazinska, W. Przetakiewicz, Manufacturing of Fe₃Al based materials using LENS method. Inzyniera Materialowa (Inz. Mater.) **35**(5), 353 (2012)
132. B. Song et al., Fabrication and microstructure characterization of selective laser-melted FeAl intermetallic parts. Surf. Coat. Technol. **206**(22), 4704–4709 (2012)
133. A. Michalcová et al., Laser additive manufacturing of iron aluminides strengthened by ordering, borides or coherent Heusler phase. Mater. Des. **116**, 481–494 (2016)
134. T. Durejko et al., Thin wall tubes with Fe₃Al/SS316L graded structure obtained by using laser engineered net shaping technology. Mater. Des. **63**, 766–774 (2014)
135. G. Rolink et al., Laser metal deposition and selective laser melting of Fe–28 at.% Al. J. Mater. Res. **29**(17), 2036–2043 (2014)

136. P. Hanus et al., Mechanical properties of a forged Fe–25Al–2Ta steam turbine blade. Intermetallics **18**, 1379–1384 (2010)
137. J. Konrad et al., Hot deformation behavior of a Fe3Al-binary alloy in the A2 and B2-order regimes. Intermetallics **13**(12), 1304–1312 (2005)
138. N. Cinca, J.M. Guilemany, Thermal spraying of transition metal aluminides: an overview. Intermetallics **24**, 60–72 (2012)
139. N. Cinca, C.R.C. Lima, J.M. Guilemany, An overview of intermetallics research and application: status of thermal spray coatings. J. Mater. Res. Technol. **2**(1), 75–86 (2013)
140. B. Denkena, H.K. Tönshoff, D. Boehnke, An assessment of the machinability of iron-rich iron-aluminium alloys. Steel Res. Int. **76**(2/3), 261–264 (2005)
141. J. Köhler, A. Moral, B. Denkena, Grinding of iron-aluminides. Proc. CIRP **9**, 2–7 (2013)
142. S.A. David et al., Welding of iron aluminides. Weld. J. **68**(9), 372s–381s (1989)
143. M.L. Santella, An overview of the welding of Ni3Al and Fe3Al alloys, in *International Symposium on Nickel and Iron Aluminides: Processing, Properties, and Applications*, ed. by S. C. Deevi et al., (ASM Int.: Proc. Materials Week, Cincinnati, 1997), pp. 321–327
144. P.D. Sketchley, P.L. Threadgill, I.G. Wright, Rotary friction welding of an Fe3Al based ODS alloy. Mater. Sci. Eng. A **329–331**, 756–762 (2002)
145. Z. Eminger, Beitrag zur Frage der Herstellung von Gußstücken aus der Eisen-Aluminium-Legierung "Pyroferal". Freiberger Forschungshefte **B 24-I**, 121–144 (1957)
146. J. Cebulski, Application of FeAl intermetallic phase matrix based alloys in the turbine components of a turbo charger. Metalurgija **54**(1), 154–156 (2015)
147. P.J. Blau, H.M. Meyer III, Characteristics of wear particles produced during friction tests of conventional and unconventional disc brake materials. Wear **255**, 1261–1269 (2003)
148. F. Moszner et al., Application of iron aluminides in the combustion chamber of large bore 2-stroke marine engines. Metals **2019**(9), 847–857 (2019)
149. S Series PSS® Filter Elements Data Sheet, in Element Data Sheet E20c, P. Corporation, pp. 1–3.
150. C.G. McKamey et al., Characterization of field-exposed iron aluminide hot gas filters, in *5th International Symposium on Gas Cleaning at High Temperature*, (Morgantown, 2002), pp. 18–20
151. O. Kubaschewski, *Iron – Binary Phase Diagrams* (Springer, Berlin, 1982), p. 185
152. H. Okamoto (ed.), *Desk Handbook Phase Diagrams for Binary Alloys*, 2nd edn. (Materials Park, ASM International, 2010), p. 855
153. I. Ohnuma et al., Experimental and thermodynamic studies of the Fe-Si binary system. ISIJ Int. **52**(4), 540–548 (2012)
154. S. Cui, I.-H. Jung, Critical reassessment of the Fe-Si system. Calphad **56**, 108–125 (2017)
155. G. Inden, W. Pitsch, Ordering transitions in body centered cubic iron silicon alloys – II. Experimental determination of the atomic configurations. Z. Metallkd. **63**, 253–258 (1972)
156. G. Schlatte, G. Inden, W. Pitsch, Ordering transitions in body centered cubic iron silicon alloys – IV. Theory with simultaneous consideration of chemical and magnetic interactions. Z. Metallkd. **65**, 94–100 (1974)
157. G. Schlatte, W. Pitsch, Ordering transitions in body centered cubic iron silicon alloys – V. Electronmicroscopic observations. Z. Metallkd. **66**, 660–668 (1975)
158. G. Schlatte, W. Pitsch, Ordering transitions in body centered cubic iron silicon alloys – VI. Comparison between theory and experiment. Z. Metallkd. **67**, 462–466 (1976)
159. K. Hilfrich et al., Revision of the Fe-Si-phase diagram: no B2-phase for 7.6 At.% < c_{Si} < 10.2 At.%. Scr. Metall. Mater. **24**(1), 39–44 (1990)
160. F. González, Y. Houbaert, A review of ordering phenomena in iron-silicon alloys. Rev. Metal. (Madrid, Spain) **49**(3), 178–199 (2013)
161. H. Kudielka, Die Kristallstruktur von Fe2Si, ihre Verwandtschaft zu den Ordnungsstrukturen des α-(Fe,Si)-Mischkristalls und zur Fe5Si3-Struktur. Z. Kristallogr. Crystal. Mater. **145**(3–4), 177–189 (1977)
162. P. Lecocq, A. Michel, Etude magnétique et structurale de phases semi-métalliques. Les composés ferromagnétiques de structure D85, Mn5Ge3 et Fe5Si3. Bull. Soc. Chim. Fr. **1965**, 307–310 (1965)

163. F. Wever, H. Möller, Über den Kristallbau des Eisensilizides FeSi. Z. Kristallogr. **75**, 362–365 (1930)
164. C. Le Corre, J.M. Genin, Transformation mechanisms of the $\alpha \leftrightarrows \beta$ Transition in $FeSi_2$. Phys. Status Solidi B **51**(1), K85–K88 (1972)
165. W.A. Jensen et al., Eutectoid transformations in Fe-Si Alloys for thermoelectric applications. J. Alloys Compd. **721**, 705–711 (2017)
166. E. Weber, H.G. Riotte, The solution of iron in silicon. J. Appl. Phys. **51**(3), 1484–1488 (1980)
167. A.E. Rubin, C. Ma, Meteoritic minerals and their origins. Chem. Erde Geochem. **77**(3), 325–385 (2017)
168. Z. Yu, Two new minerals Gupeiite and Xifengite in cosmic dusts from Yanshan. Acta Petrol. Mineral. Anal. **3**, 231–238 (1984)
169. K. Keil, J.L. Berkley, L.H. Fuchs, Suessite, Fe_3Si: a new mineral in the North Haig ureilite. Am. Mineral. **67**, 126–131 (1982)
170. M.I. Novgorodova et al., First occurrence of Suessite on the Earth. Int. Geol. Rev. **26**(1), 98–101 (1984)
171. M. Anand et al., New lunar mineral HAPKEITE*: product of impact-induced vapor-phase deposition in the regolith? in *Proc. Lunar and Planetary Science Conference XXXIV*, (2003), p. 1818
172. N. Shi et al., Naquite, FeSi, a new mineral species from Luobusha, Tibet, Western China. Acta Geol. Sin. (Engl. Ed.) **86**(3), 533–538 (2012)
173. G. Li et al., Linzhiite, $FeSi_2$, a redefined and revalidated new mineral species from Luobusha, Tibet, China. Eur. J. Mineral. **24**(6), 1047–1052 (2012)
174. R.A. Fischer et al., Phase relations in the Fe–FeSi system at high pressures and temperatures. Earth Planet. Sci. Lett. **373**, 54–64 (2013)
175. R.A. Fischer et al., Equation of state and phase diagram of Fe–16Si alloy as a candidate component of Earth's core. Earth Planet. Sci. Lett. **357–358**, 268–276 (2012)
176. C.J. Allègre et al., The chemical composition of the Earth. Earth Planet. Sci. Lett. **134**(3), 515–526 (1995)
177. J. Wade, B.J. Wood, Core formation and the oxidation state of the Earth. Earth Planet. Sci. Lett. **236**(1), 78–95 (2005)
178. B. Martorell et al., The elastic properties of hcp-$Fe_{1-x}Si_x$ at Earth's inner-core conditions. Earth Planet. Sci. Lett. **451**, 89–96 (2016)
179. M.C.M. Farquhar, H. Lipson, A.R. Weill, An X-ray study of iron-rich iron-silicon alloys. J. Iron Steel Inst. **152**, 457–472 (1945)
180. F. Lihl, H. Ebel, Röntgenographische Untersuchungen über den Aufbau eisenreicher Eisen-Silizium-Legierungen. Arch. Eisenhüttenwes. **32**(7), 489–491 (1961)
181. F. Richter, W. Pepperhoff, The lattice constant of ordered and disordered iron-silicon alloys. Arch. Eisenhuettenwes. **45**, 107–109 (1974)
182. E.A. Kümmerle et al., Thermal formation of vacancies in Fe_3Si. Phys. Rev. B **52**(10), R6947–R6950 (1995)
183. A. Gude, H. Mehrer, Diffusion in the DO_3-type intermetallic phase Fe_3Si. Philos. Mag. A **76**(1), 1–29 (1997)
184. G.E. Lakso, M.J. Marcinkowski, Plastic deformation behavior in the Fe_3Si superlattice. Trans. Metall. Soc. AIME **245**, 1111–1120 (1969)
185. G.E. Lakso, M.J. Marcinkowski, Plastic deformation in Fe-Si alloys. Metall. Trans. **5**(4), 839–845 (1974)
186. S.K. Ehlers, M.G. Mendiratta, Tensile behaviour of two DO_3-ordered alloys: Fe_3Si and Fe-20 at % Al-5 at % Si. J. Mater. Sci. **19**(7), 2203–2210 (1984)
187. T.D. Yensen, The magnetic properties of some iron alloys melted in Vacuo. Trans. Am. Inst. Electr. Eng. **34**(2), 2601–2670 (1915)
188. M.G. Corson, The constitution of the Fe-silicon alloys, particularly in connection with the properties of corrosion-resisting alloys of this composition. Trans. AIME **80**, 249–296 (1928)

189. L.K. Varga et al., Structural and magnetic properties of metastable $Fe_{1-x}Si_x$ (0.15< x <0.34) alloys prepared by a rapid-quenching technique. J. Phys. Condens. Matter. **14**(8), 1985 (2002)
190. W.F. Barrett, W. Brown, R.A. Hadfield, On the electrical conductivity and magnetic permeability of various alloys of iron. Sci. Trans. R. Dublin Soc. **7**(Series 2), 67–126 (1900)
191. J.L. Haughton, M.L. Becker, Alloys of iron research. IX. The constitution of the alloys of iron with silicon. J. Iron Steel Inst. **121**, 315–335 (1930)
192. M. Fallot, Ferromagnetism of iron alloys (in French). Ann. Phys. (Paris, Fr.) **11**(6), 305–387 (1936)
193. M.K. Kolel-Veetil et al., Carbon nanocapsule-mediated formation of ferromagnetic Fe_5Si_3 nanoparticles. J. Phys. Chem. C **113**(33), 14663–14671 (2009)
194. L.K. Varga et al., Magnetic properties of rapidly quenched $Fe_{100-x}Si_x$ (15<x<34) alloys. Mater. Sci. Eng. A **304-306**, 946–949 (2001)
195. J.E. Wittig, G. Frommeyer, Deformation and fracture behavior of rapidly solidified and annealed iron-silicon alloys. Metall. Mater. Trans. A **39**, 252–265 (2008)
196. D. Hawezy, The influence of silicon content on physical properties of non-oriented silicon steel. Mater. Sci. Technol. **33**(14), 1560–1569 (2017)
197. H.H. Uhlig, R.W. Revie, *Corrosion and Corrosion Control – An Introduction to Corrosion Science and Engineering*, 3rd edn. (Wiley, New York, 1985), p. 458
198. S. Yamaguchi, An electron diffraction study of the protective coating on metals and alloys (continued): the acid-proof surface of ferrosilicon (Duriron). Bull. Chem. Soc. Jpn. **16**(9), 332–335 (1941)
199. R. Wasmuht, Über Konstitution und Eigenschaften des säurebeständigen Silicium-Eisengusses (About constitution and properties of acid-proof silicon-iron castings). Angew. Chem. **45**(36), 569–573 (1932)
200. R.K. Roy et al., Development of rapidly solidified 6.5 wt% silicon steel for magnetic applications. Trans. Indian Inst. Met. **63**(4), 745–750 (2010)
201. Y. Takada et al., Commercial scale production of Fe-6.5 wt. % Si sheet and its magnetic properties. J. Appl. Phys. **64**(10), 5367–5369 (1988)
202. F. Fiorillo, Advances in Fe-Si properties and their interpretation. J. Magn. Magn. Mater. **157–158**, 428–431 (1996)
203. K. Fujita, M. Namikawa, Y. Takada, Magnetic properties and workability of 6.5% Si steel sheet manufactured by siliconizing process. J. Mater. Sci. Technol. **16**(2), 137–140 (2000)
204. T. Ros-Yañez, Y. Houbaert, V. Gómez Rodríguez, High-silicon steel produced by hot dipping and diffusion annealing. J. Appl. Phys. **91**(10), 7857–7859 (2002)
205. C. Bolfarini et al., Magnetic properties of spray-formed Fe–6.5%Si and Fe–6.5%Si–1.0%Al after rolling and heat treatment. J. Magn. Magn. Mater. **320**(20), e653–e656 (2008)
206. J.H. Yu et al., The effect of heat treatments and Si contents on B2 ordering reaction in high-silicon steels. Mater. Sci. Eng. A **307**(1), 29–34 (2001)
207. R. Li et al., Magnetic properties of high silicon iron sheet fabricated by direct powder rolling. J. Magn. Magn. Mater. **281**(2), 135–139 (2004)
208. K. Arai, N. Tsuya, Ribbon-form silicon-iron alloy containing around 6.5 percent silicon. IEEE Trans. Magn. **16**(1), 126–129 (1980)
209. K.N. Kim et al., The effect of boron content on the processing for Fe–6.5wt% Si electrical steel sheets. J. Magn. Magn. Mater. **277**(3), 331–336 (2004)
210. T. Doi, H. Ninomiya, High silicon steel sheet realizing excellent high frequency reactor performance, in *Twenty-Seventh Annual IEEE Applied Power Electronics Conference and Exposition (APEC)*, (2012), pp. 1740–1746
211. K. Senda, M. Namikawa, Y. Hayakawa, Electrical steels for advanced automobiles – core materials for motors, generators, and high-frequency reactors. JFE Techn. Rep. No. **4**, 1–7 (2004)
212. P.-W. Jang, B.-H. Lee, G.-B. Choi, Variation of magnetic properties of Fe-Si compressed cores with Si content. J. Korean Magn. Soc. **20**(1), 13–17 (2010)
213. P. Jang, G. Choi, Acoustic noise characteristics and magnetostriction of Fe-Si powder cores. IEEE Trans. Magn. **48**(4), 1549–1552 (2012)

214. B. You et al., Optimization of powder core inductors of buck-boost converters for Hybrid Electric Vehicles, in *2009 IEEE Vehicle Power and Propulsion Conference*, (2009)

215. B. You et al., Experimental comparison of Mega Flux and JNEX inductors in high power dc-dc converter of Hybrid Electric Vehicles, in *2011 IEEE Vehicle Power and Propulsion Conference*, (2011)

216. M.S. Rylko, J.G. Hayes, M.G. Egan, Experimental investigation of high-flux density magnetic materials for high-current inductors in hybrid-electric vehicle DC-DC converters, in *2010 IEEE Vehicle Power and Propulsion Conference*, (2010)

217. K. Lee et al., A study on the performance of 10kW Grid-Connected Photovoltaic power conditioning system with characteristics variation in inductor core materials, in *The 2010 International Power Electronics Conference – ECCE ASIA*, (2010)

218. H. Masumoto, T. Yamamoto, On a new alloy "Sendust" and its magnetic and electric properties. J. Jpn. Inst. Met. **1**(3), 127–135 (1937)

219. H.H. Helms Jr., E. Adams, Sendust sheet-processing techniques and magnetic properties. J. Appl. Phys. **35**(3), 871–872 (1964)

220. N. Tsuya et al., Magnetic recording head using ribbon-sendust. IEEE Trans. Magn. **17**(6), 3111–3113 (1981)

221. T. Kobayashi et al., A tilted sendust sputtered ferrite video head. IEEE Trans. Magn. **21**(5), 1536–1538 (1985)

222. S. Yoshida et al., Permeability and electromagnetic-interference characteristics of Fe–Si–Al alloy flakes–polymer composite. J. Appl. Phys. **85**(8), 4636–4638 (1999)

223. K. Nomura et al., Mössbauer study on Fe-Si-Al flakes-polymer composites for noise filter at high frequency bands. J. Jpn. Soc. Powder Powder Metall. **50**(4), 260–265 (2003)

224. P. Winkler, W. Günther, Enlarging the standard permeability set of powder e-cores by combination of different perm core-halves, in *PCIM Europe 2018; International Exhibition and Conference for Power Electronics, Intelligent Motion, Renewable Energy and Energy Management*, (2018)

225. K. Yoo, B.K. Lee, D. Kim, Investigation of vibration and acoustic noise emission of powder core inductors. IEEE Trans. Power Electron., 1–1 (2018)

226. M.G. Mendiratta, H.A. Lipsitt, $D0_3$-domain structures in Fe_3Al-X alloys. Mat. Res. Soc. Symp. Proc. **39**, 155–162 (1985)

227. B.K. Grant, *International directory of company histories*, vol 17 (St. James Press, Detroit, 1997), p. 736

228. P.D. Schenck, Acid resisting irons. Chem. Metall. Eng. **28**, 678 (1923)

229. O. Schneeweiss et al., Fe_3Si surface coating on SiFe steel. J. Magn. Magn. Mater. **215–216**, 115–117 (2000)

230. A. Ionescu et al., Structural, magnetic, electronic, and spin transport properties of epitaxial $Fe_3Si/GaAs(001)$. Phys. Rev. B **71**(9), 094401 (2005)

231. J. Herfort et al., Epitaxial growth of $Fe_3Si/GaAs(001)$ hybrid structures for spintronic application. J. Cryst. Growth **278**(1), 666–670 (2005)

232. R. Mantovan et al., Synthesis and characterization of Fe_3Si/SiO_2 structures for spintronics. Phys. Status Solidi A **205**(8), 1753–1757 (2008)

233. R. Nakane, M. Tanaka, S. Sugahara, Preparation and characterization of ferromagnetic $D0_3$-phase Fe_3Si thin films on silicon-on-insulator substrates for Si-based spin-electronic device applications. Appl. Phys. Lett. **89**(19), 192503 (2006)

234. N. Dahal, V. Chikan, Phase-Controlled Synthesis of Iron Silicide (Fe_3Si and $FeSi_2$) Nanoparticles in Solution. Chem. Mater. **22**(9), 2892–2897 (2010)

235. M. Rudy, I. Jung, G. Sauthoff, Ferritic Fe-Al-Ni alloys for high temperature applications, in *High Temperature Alloys – Their Exploitable Potential*, (Elsevier, London/New York, 1987)

236. C.T. Sims, A history of superalloy metallurgy for superalloy metallurgists, in *Superalloys 1984*, (TMS-AIME, Warrendale, 1984)

237. A. Kracke, Superalloys, the most successful alloy system of modern times – past, present and future, in *7th International Symposium on Superalloy 718 and its Derivatives*, (TMS, Warrendale, 2010)

238. V. Seetharaman, M. Sundararaman, R. Krishnan, Precipitation hardening in a PH 13-8 Mo stainless steel. Mater. Sci. Eng. **47**, 1–11 (1981)

239. W.M. Garrison, R. Strychor, A preliminary study of the influence of separate and combined aluminum and nickel additions on the properties of a secondary hardening steel. Metal. Trans. **19A**(12), 3103–3107 (1988)

240. R. Hamano, The effect of the precipitation of coherent and incoherent precipitates on the ductility and toughness of high-strength steel. Metall. Trans. **24A**(1), 127–139 (1993)

241. S. Jiang et al., Ultrastrong steel via minimal lattice misfit and high-density nanoprecipitation. Nature **544**, 460–465 (2017)

242. Y. Liao, I. Baker, Microstructure and room-temperature mechanical properties of $Fe_{30}Ni_{20}Mn_{35}Al_{15}$. Mater. Charact. **59**, 1546–1549 (2008)

243. I. Jung, G. Sauthoff, Creep behaviour of the intermetallic B2 phase (Ni,Fe)Al with strengthening soft precipitates. Z. Metallkd. **80**(7), 484–489 (1989)

244. M.J.S. Rawlings et al., Effect of titanium additions upon microstructure and properties of precipitation-strengthened Fe-Ni-Al-Cr ferritic alloys. Acta Mater. **128**, 103–112 (2017)

245. S.I. Baik et al., Increasing the creep resistance of Fe-Ni-Al-Cr superalloys via Ti additions by optimizing the $B2/L2_1$ ratio in composite nano-precipitates. Acta Mater. **157**, 142–154 (2018)

246. H.A. Calderon, M.E. Fine, J.R. Weertman, Coarsening and morphology of β' particles in Fe-Ni-Al-Mo ferritic alloys. Metall. Trans. A **19**(5), 1135–1146 (1988)

247. G. Song et al., Ferritic alloys with extreme creep resistance via coherent hierarchical precipitates. Sci. Rep. **5**, 16327 (2015)

248. S.I. Baik, M. Rawlings, D.C. Dunand, Effect of hafnium micro-addition on precipitate microstructure and creep properties of a Fe-Ni-Al-Cr-Ti ferritic superalloy. Acta Mater. **153**, 126–135 (2018)

249. S.M. Zhu, S.C. Tjong, J.K.L. Lai, Creep behavior of a β'(NiAl) precipitation strengthened ferritic Fe-Cr-Ni-Al alloy. Acta Mater. **46**(9), 2969–2976 (1998)

250. Y. Zhao et al., Creep behavior as dislocation climb over NiAl nanoprecipitates in ferritic alloy: the effects of interface stresses and temperature. Int. J. Plast. **69**, 89–101 (2015)

251. C. Stallybrass, A. Schneider, G. Sauthoff, The strengthening effect of (Ni,Fe)Al precipitates on the mechanical properties at high temperatures of ferritic Fe-Al-Ni-Cr alloys. Intermetallics **13**(12), 1263–1268 (2005)

252. Z.K. Teng et al., Room temperature ductility of NiAl-strengthened ferritic steels: effects of precipitate microstructure. Mater. Sci. Eng. A **541**, 22–27 (2012)

253. Z.Q. Sun et al., Duplex precipitates and their effects on the room-temperature fracture behaviour of a NiAl-strengthened ferritic alloy. Mater. Res. Lett. **3**(3), 128–134 (2015)

254. M.D. Bhandarkar et al., Structure and elevated temperature properties of carbon-free ferritic alloys strengthened by a Laves phase. Metall. Trans. A **6**(6), 1281–1289 (1975)

255. S. Muneki et al., Creep characteristics in carbon free new martensitic alloys. Mater. Sci. Eng. A **406**, 43–49 (2005)

256. F. Wever, W. Peter, Ausscheidungshärtung und Dauerstandfestigkeit von Eisen-Niob-Legierungen und nioblegierten Stählen. Arch. Eisenhuettenwes. **15**(8), 357–363 (1942)

257. J.S. Lee et al., Causes of breakdown of creep strength in 9Cr–1.8W–0.5Mo–VNb steel. Mater. Sci. Eng. A **428**(1–2), 270–275 (2006)

258. F. Abe, Effect of fine precipitation and subsequent coarsening of Fe_2W Laves phase on the creep deformation behavior of tempered martensitic 9Cr-W steels. Metall. Mater. Trans. A **36**(2), 321–332 (2005)

259. M.G. Gemmill et al., Study of 7% and 8% chromium creep-resistant steels. J. Iron Steel Inst. **184**(10), 122–144 (1956)

260. V. Foldyna et al., Evaluation of structural stability and creep resistance of 9–12% Cr steels. Steel Res. **67**(9), 375–381 (1996)

261. J. Hald, Metallurgy and creep properties of new 9–12%Cr steels. Steel Res. **67**(9), 369–374 (1996)

262. O. Prat et al., The role of Laves phase on microstructure evolution and creep strength of novel 9%Cr heat resistant steels. Intermetallics **32**, 362–372 (2013)
263. B. Kuhn et al., Development of high chromium ferritic steels strengthened by intermetallic phases. Mater. Sci. Eng. A **594**, 372–380 (2014)
264. N. Kanno et al., Mechanical properties of austenitic heat-resistant Fe–20Cr–30Ni–2Nb steel at ambient temperature. Mater. Sci. Eng. A **662**, 551–563 (2016)
265. M. Shibuya et al., Effect of nickel and cobalt addition on the precipitation-strength of 15Cr ferritic steels. Mater. Sci. Eng. A **528**, 5387–5393 (2011)
266. J. Froitzheim et al., Development of high strength ferritic steel for interconnect application in SOFCs. J. Power Sources **178**, 163–173 (2008)
267. L. Niewolak et al., Temperature dependence of phase composition in W and Si-alloyed high chromium ferritic steels for SOFC interconnect applications. J. Alloys Compd. **717**, 240–253 (2017)
268. B. Kuhn et al., Effect of Laves phase strengthening on the mechanical properties of high Cr ferritic steels for solid oxide fuel cell interconnect application. Mater. Sci. Eng. A **528**, 5888–5899 (2011)
269. B. Kuhn et al., Development status of higher performance ferritic (HiperFer) steels, in *8th International Conference on Advances in Materials Technology for Fossil Power Plants (EPRI 2016)*, (ASM International, Materials Park, 2016)
270. I. Tarigan et al., Novel concept for creep strengthening mechanism using grain boundary Fe_2Nb Laves phase in austenitic heat resistant steel. Mater. Res. Soc. Symp. Proc. **1295**, 317–322 (2011)
271. M. Takeyama, Novel concept of austenitic heat resistant steels strengthened by intermetallics. Mater. Sci. Forum **539–543**, 3012–3017 (2007)
272. Lyta et al., Microstructure development of oxide scale during steam oxidation of the Fe20Cr30Ni2Nb (at%) austenitic steel at 1073K. Mater. Trans. JIM **54**(12), 2276–2284 (2013)

Chapter 11
Stainless Steels

David San-Martin, Carola Celada-Casero, Javier Vivas, and Carlos Capdevila

Abbreviations

AFA	Alumina-forming austenitic
AHSS	Advanced high-strength steel
AISI	American Iron and Steel Institute
AOD	Argon oxygen decarburization
APT	Atom probe tomography
ASME	American Society of Mechanical Engineers
ASS	Austenitic stainless steel
A-USC	Advanced ultra-supercritical
AUST SS	Austenitic stainless steel (Fig. 11.30)
BCC	Body-centred cubic
BCT	Body-centred tetragonal
BF-TEM	Bright field transmission electron microscopy
BH	Bake hardening
CAPL	Continuous annealing and pickling line
CNAs	Castable nanostructure alloys
CP	Complex phase

D. San-Martin (✉) · C. Capdevila
Physical Metallurgy Department, National Centre for Metallurgical Research (CENIM-CSIC), Madrid, Spain
e-mail: dsm@cenim.csic.es; ccm@cenim.csic.es

C. Celada-Casero
Tata Steel, IJmuiden, Noord-Holland, The Netherlands
e-mail: carola.alonso-de-celada-casero@tatasteeleurope.com

J. Vivas
Joining Processes, IK4 Lortek, Gipuzkoa, Ordizia, Spain
e-mail: jvivas@lortek.es

© Springer Nature Switzerland AG 2021
R. Rana (ed.), *High-Performance Ferrous Alloys*,
https://doi.org/10.1007/978-3-030-53825-5_11

459

Cr_{eq}	Chromium equivalent number
CRSS	Critical resolved shear stress
CSEF	Creep strength enhanced ferritic steels
CTE	Coefficient of thermal expansion
DBTT	Ductile-brittle transition temperature
DP	Dual phase
DSS	Duplex stainless steel
EBSD	Electron back scattering diffraction
ECAP	Equal channel angular pressing
ECC	Equal channel contrast
EDS	Energy dispersive spectroscopy
ESTEP	European Steel Technology Platform
FCC	Face-centred cubic
FSS	Ferritic stainless steel
HAB	High-angle grain boundary
HCP	Hexagonal close-packed
HPT	High-pressure torsion
HRTEM	High-resolution transmission electron microscopy
HSLA	High-strength low-alloyed
HSS	High-strength steel
HV	Vickers hardness
IF	Interstitial free
IPF	Inverse pole figure
LAB	Low-angle grain boundary
L-IP®	Lightweight steels with induced plasticity
LMP	Larson-Miller parameter
MDF	Multidirectional forging
MSS	Martensitic stainless steel
NG	Nanograined
Ni_{eq}	Nickel equivalent number
NTW-γ	Nanotwinned austenitic grains
ODS	Oxide dispersion-strengthened
PH	Precipitation hardening
PHSS	Precipitation hardening stainless steel
PT	Partitioning temperature
Q&P	Quenching and partitioning
Q&T	Quenching and tempering
QT	Quenching temperature
RAFM	Reduced activation ferritic/martensitic
RAP	Rolling, annealing and pickling

ROI	Region of interest
SADP	Selective area diffraction pattern
SEM	Scanning electron microscopy
SIM	Strain-induced martensite
SOFC	Solid oxide fuel cell
SPD	Severe plastic deformation
SSINA	The Specialty Steel Industry of North America
SSS	Solid solution strengthening
STEM	Scanning transmission electron microscopy
TE	Total elongation
TEM	Transmission electron microscopy
TMCP	Thermomechanical controlled processing
TRIP	Transformation-induced plasticity
TWIP	Twinning-induced plasticity steel
UE	Uniform elongation
UFG	Ultrafine-grained
USC	Power plants ultra-supercritical power plants
UTS	Ultimate tensile strength
V-AOD	Vacuum argon oxygen decarburization
VCR	Vacuum converter refiner
VOD	Vacuum oxygen decarburization
Wt	Weight
XRD	X-ray diffraction
YS	Yield strength

Symbols

α	Ferrite
α_1, α_2	Constants (Eq. 11.11)
α_ε	Factor (Table 11.7)
χ	Chi-phase
δ	ferrite and cell diameter in (Eq. 11.11)
ε	HCP martensite
ε_b	Misfit size parameter (Eq. 11.8)
ε_c	Constrained lattice parameter misfit (Table 11.7)
ε_{micro}	Elastic microstrains (Eq. 11.12)
ε_T	Total deformation
γ	Austenite
γ_{apb}	Antiphase boundary free energy of the precipitate (Table 11.7)
γ_{SFE}	Stacking fault energy (Eqs. 11.19 and 11.20)
γ_{SFET}	Temperature-dependent stacking fault energy (Eq. 11.20)
γ^{twin}	Twinned austenite

$\Delta\gamma_{sf}$	Stacking fault energy mismatch between matrix and precipitates (Table 11.7)
ν	Poisson's ratio (Eqs. 11.12 and 11.13)
ρ	Dislocation density (Eqs. 11.11, 11.12 and 11.13)
σ_y	Yield strength (Eq. 11.6)
σ_{SS}	Contribution of the solid solution strengthening to the yield strength (Eq. 11.6)
σ_{gs}	Contribution of the grain size to the yield strength (Eqs. 11.6 and 11.18)
σ_ρ	Contribution of the forest dislocation to the yield strength (Eq. 11.6)
σ_{ppt}	Contribution of the precipitation hardening to the yield strength (Eq. 11.6)
σ_0	Friction stress (Eq. 11.6)
a	Lattice constant (Table 11.6)
a_{bcc}	Lattice constant of ferrite (Table 11.6)
a_{fcc}	Lattice constant of austenite (Table 11.6)
A_i	Strengthening factor of element "i" in solid solution (Eq. 11.8)
A_{lath}	Lath martensite section (Eq. 11.13)
b	Burgers vector (Eqs. 11.11, 11.12, 11.14, 11.15 and Table 11.6)
C_i	Concentration of alloying element "i" in solid solution (Eq. 11.8)
C	Constant (Eq. 11.17)
d_{lath}	Lath martensite distance (related to its section)
d	Mean grain size (Eq. 11.18)
E	Young's modulus (Eqs. 11.12 and 11.13)
$f_{\alpha'}$	Volume fraction of martensite
G	Shear modulus (Eqs. 11.11, 11.12, 11.13, 11.14, 11.15; also used to defined a precipitate in maraging steels: $Ni_{16}M_6Si_7$, M=Nb,Ti; Table 11.2, Table 11.9)
G'	Shear modulus of a precipitate (Eq. 11.17)
ΔG	Shear modulus mismatch between matrix and precipitates (Table 11.7)
K	K_{LN} = factor (Eqs. 11.8–11.9)
k_{HP}	Hall-Petch coefficient (Eq. 11.18)
K_{IC}	Fracture toughness
L	Distance between particles (Eqs. 11.15 and 11.16)
M	Taylor factor (Eqs. 11.9 and 11.11)
m	Factor (Table 11.7)
M_d	Temperature above which no martensite forms upon loading
Md_{30} or $Md_{30/50}$	Temperature at which 50 % of martensite is obtained in a tension test for a true deformation of 0.3 (Eqs. 11.3, 11.4 and 11.21).

M_f	Martensite finish temperature
M_s	Martensite start temperature (Eq. 11.5)
M_{SI}	The temperature below which the isothermal formation of martensite is observed
R_m	Tensile strength (Table 3.1)
T	Absolute temperature
T_γ	Austenitization temperature
$V_{\alpha'}$	amount of martensite (%, Eq. 11.22)
V_f^p	Precipitate volume fraction (Eqs. 11.14 and 11.16)
w_{lath}	Lath martensite width (related to its section, Eq. 11.13)
X	Particle mean diameter (Eqs. 11.14, 11.15 and 11.16)

11.1   Introduction

Steels remain, up to date, as one of the most successful and cost-effective class of material in improving the quality of life, with around 1.6 billion tons of finished steel products manufactured in 2017 [1]. They are recyclable, possess great durability and require low amounts of energy to be produced compared to other materials. There are currently more than 3500 different grades, each with different properties. One reason for their overwhelming dominance is the endless variety of microstructures and properties that can be generated during its thermomechanical processing. However, one of the main limitations of plain or low-alloyed carbon steels is that they easily rust in air and aqueous environments, corrode in different acids and oxidize when exposed to high temperatures in unprotected (open air) atmospheres.

During the beginning of the twentieth century, several investigations carried out in Europe and America unveiled the positive influence of adding Cr, in quantities generally above 12.0 wt.-%, on the corrosion and oxidation resistance of plain carbon steels. The first patent and seminal work are usually attributed to the English metallurgist *Harry Brearley* (Sheffield, UK). He is also recognized as the discoverer of the rustless steel, later denominated as stainless steel by Ernest Stuart (manager of a cutlery firm in Sheffield and with whom *Brearly* collaborated). At that time *Brearly* was producing experimental alloys with chromium contents between 6 and 15 wt.-% trying to improve the wear resistance [2]. To disclose the microstructure, carbon steels are standardly etched with a solution composed of nitric acid and alcohol (so-called Nital). However, he discovered that this solution had a weak effect on the microstructure of high-chromium steels, as they were very difficult to be etched chemically in this way, i.e. they were resistant to diluted acidic solutions. He also noticed that the state of heat treatment affected their response to etching. The first commercial cast of a stainless steel was made by *Brearley* in 1913 with a composition of Fe-0.24C-12.80Cr-0.44Mn-0.20Si (in wt.-%). Initially this steel was intended for gun barrels, but it was later proposed by *Brearley* to produce cutlery and other food-related utensils/tools in collaboration with *Ernest Stuart* (Sheffield was the centre of the cutlery industry in England at that time). *Brearley's* discovery of stainless steels was announced in The New York Times in January 1915. Since this date, the world production of stainless steels has increased with ascending rate. While in 1950, the production of stainless steels was less than 1 million tons, the total production of stainless steels in 2018 has been around 50.7 million tons, growing at a rate of 5.8 % a year with China being the world leading producer with more than 53 % [6].

The addition of chromium to steel promotes the formation of a chromium oxide, $(Fe,Cr)_2O_3$, passive surface film of a few nanometres, which protects it from staining, rusting, corroding or oxidizing in environments where normal steels are susceptible [3]. However, to prevent pitting and rusting in more hostile environments, other alloying elements like Mo are also required. In addition to their stainlessness, stainless steels do not discolour in a normal atmospheric environment and keep a shiny surface [4]. The use of stainless steels ranges from the low-end products such as water buckets or cutlery to the very high-end ones, such as the spacecraft propellers and casings, constructions or surgical implants or tools. The ubiquity of stainless steels in our daily life makes it impossible to enumerate all their applications [5]. Stainless steels also outperform ordinary steels on high-temperature mechanical properties. Stainless steels are much better in terms of fire resistance and retention of strength and stiffness at elevated temperatures compared with carbon steels. For these reasons, steel is frequently the "gold standard" against which emerging structural materials are compared.

11.2   Overview of Stainless Steels

11.2.1   Stainless Steels Categories

Taking as a basis the three main matrix phases that may coexist in the microstructure (ferrite, austenite, martensite), stainless steels have been generally classified in five groups: (1) ferritic stainless steels (FSS), (2) austenitic stainless steels (ASS), (3) martensitic stainless steels (MSS), (4) duplex stainless steels (DSS, mixture of ferrite and austenite) and (5) precipitation hardening stainless steels (PHSS, generally contain martensite and small volume fraction of metastable austenite) [4, 6]. These different families are briefly described as follows.

Ferritic Stainless Steels (4XX Series)
FSS are iron-chromium alloys that have the same structure as pure iron at room temperature, body-centred cubic (BCC) ferrite (α) phase. Therefore, the solubility of carbon is rather limited, and most of the carbon forms carbides such as M_7C_3 and $M_{23}C_6$ (M = Cr,Fe). The Cr content varies between 10.5 and 30 wt.-%. The

C content is low and generally below or around 0.1 wt.-%. Other elements like Si or Mn are generally around or below 1 wt.-%. High strength, corrosion resistance or stabilized grades depend on the presence of different elements like Mo, Ti, Nb, Al, Cu and N in the composition. Typical ferritic grades are AISI 430 (Fe-17Cr); 405 (Al added); 415 (Ni added); 434; 409, 430Ti, 439 and 441 (Ti and/or Nb stabilized); 434, 436 and 444 (Mo and Ti and/or Nb stabilized). Other special grades like 446, 447 and 448, in addition to Ti/Nb and Mo, contain elements like Cu (max. 0.6 wt.-%), Ni (max. 4.5 wt.-%) and Cr contents of 25–30 wt.-%. Their corrosion resistance and toughness are moderate; however, they are affordable due to the low content of alloying elements compared to other stainless steels. Ferritic stainless steels are readily welded in thin sections but suffer grain growth with consequential loss of properties when welded in thicker sections. They have multiple applications in the automotive/transportation, building and construction, food equipment, home appliances or offices and other industries.

Austenitic Stainless Steels (2XX and 3XX Series)
The content of this steel in alloying elements like Ni or Mn allows ASS to retain the high-temperature face-centred cubic (FCC) austenite (γ) phase at room temperature. ASS are the most commonly used stainless steels as they comprise 70 % of the total production. Depending on the steel grade, the Cr content varies between 16 and 25 wt.-% and Ni between 1 and 22 wt.-%. The C is kept below 0.08 wt.-% (lower carbon grades, <0.03 wt.-%, are identified with an "L": AISI 304L). They may also contain Mn (1.0–15.5 wt.-%), Si (0.5–3 wt.-%), Mo ($\leq$5 wt.-%), Ti/Nb ($\leq$0.8 wt.-%), Cu ($\leq$4 wt.-%), N ($\leq$0.5 wt.-%) and other minor elements. This group can be divided into several subgroups: low Ni, Cr-Mn grades (also referred to as "200-series" grades), Cr-Ni grades (sometimes referred as 18-8 SS, indicating the approximate content in Cr and Ni, respectively), Cr-Ni-Mo grades, high-performance austenitic grades and high-temperature heat-resistant austenitic grades, for use at temperatures exceeding 550 °C. The most well-known 18Cr-8Ni grades are AISI 304, 316, 321 and 347; these later ones alloyed with Ti (AISI 321) [7] and Nb (AISI 347) to stabilize against $M_{23}C_6$ formation (to minimize the sensitization that leads to intergranular corrosion [8]) and to improve creep behaviour via the formation of TiC/NbC nanocarbides. ASS have superior corrosion resistance and toughness compared to FSS and MSS; the corrosion performance may be varied to suit a wide range of service environments by optimizing the C or Mo content. They are the most weldable of the stainless steels; however, some grades can be prone to sensitization of the weld heat-affected zone and weld metal hot cracking. They have greater thermal expansion coefficient than FSS and MSS, which makes them less resistant to cyclic oxidation. They possess excellent formability which can be tailored by adjusting the stacking fault energy via composition optimization. Besides, they can be hardened by cold-rolling. They are frequently used in kitchen sinks, chemical industry, food processing and as structural components in energy-generating power plants.

Martensitic Stainless Steels (MSS, 4XX Series)

Fe-Cr-C MSS generally have a higher content in different alloying elements than FSS and an optimum balanced composition among C, Ni, Cr and Mo that allow promoting the phase transformation from the high-temperature parent FCC austenite phase into the body-centred tetragonal (BCT) martensite (α') phase during air cooling. The tetragonality of the martensite phase depends on the C content of the alloy, and for low-carbon grades the crystal structure of martensite is generally close to BCC. After cooling to room temperature, carbon supersaturated martensite is very brittle and should be given a tempering treatment to increase the ductility and toughness. By this heat treatment, the strength can increase one order of magnitude (from 200 to 2000 MPa). These steels are generally divided depending on the carbon content: (1) <0.06 wt.-% ("L" grades); (2) 0.06–0.30 wt.-% (AISI 410, 420, 431); and (3) >0.30 wt.-% (AISI 440A, B, C). Typical Cr contents are between 11.5 and 18 wt.-%, Mn between 0.5 and 2.5 wt.-%, Si between 0.5 and 1.0 wt.-% in addition to other elements like Ni (0.6–3.0 wt.-%; AISI 414, 422, 431), Nb (0.05–0.3; AISI 410Cb), Mo (0.5–1.25 wt.-%; AISI 414L, 416, 422; 440), W (0.75–1.25 wt.-%: AISI 422) or V (0.15–0.30 wt.-%) and Se (0.15 wt.-%). In special 12 wt.-%Cr tool steels (D2, D3, D6), carbon is increased (1.5–2.15 wt.-%), and carbide forming elements like Mo, W and V are added to the composition to promote the formation of very hard and stable carbides. The corrosion resistance of this type of steels is moderate but lower than that of austenitic and ferritic grades due to the lower or null amount of Ni and other alloying elements. However, these lower alloying elements makes them less costly than other stainless steels. By adding some Ni and reducing the carbon content, the rather poor weldability of these steels can be improved. For applications where high corrosion resistance and creep strength are required (power generation, chemical and petrochemical industries), 12Cr-0.2C (wt.-%) grades are alloyed with Mo for corrosion, Ni to avoid delta ferrite, W/Co for solid solution hardening and B to stabilize and reduce the coarsening rate of the $M_{23}C_6$ and prevent cavitation at high temperatures. Finally, Nb/V are added for precipitation hardening via the formation of MCN nanocarbonitrides (M = Nb, V), finely dispersed in the martensitic microstructure. This heat-resistant MSS can be used in a wide range of operating temperatures but above 650 °C usually undergo a quick degradation due to the recovery of the martensitic microstructure and coarsening of the carbides present in the microstructure. Figure 11.1 summarizes the evolution of the composition in the development of 12 wt.-%Cr MSS for boilers. Common applications of MSS include steam, gas and jet engine turbine blades that operate at high temperatures, boilers, pipes and valves for the power generation, chemical and petroleum industries, surgical instruments, razor blades, cutlery, gears and shafts.

Duplex Stainless Steels (DSS)

The DSS have a mixed ferritic-austenitic microstructure (usually 60/40 %, respectively) and are typically low-carbon steels (0.03–0.06 wt.-%), containing Cr in

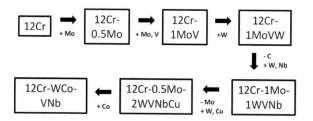

Fig. 11.1 Development of 12 wt.-% Cr MSS grades for boiler applications. (Information extracted from [9])

amounts between 21 and 26 wt.-% and Ni between 2 and 7 wt.-%. The additions of Mo ($\leq$4 wt.-%), N ($\leq$0.3 wt.-%) and Cu ($\leq$2 wt.-%) provide these steels with good corrosion resistance. The DSS have higher yield strength and greater stress corrosion cracking resistance to chloride than ASS. Service temperatures are limited to the range from $-40\,^{\circ}$C (below it poor impact toughness is seen) to $280\,^{\circ}$C (above it embrittling precipitates form). The most commonly used grades are 2205 (22Cr-5.5Ni) and 2507 (25Cr-5.5Ni); this latter one is also called as Super-Duplex, and it has higher corrosion resistance. Their common usage is in petrochemical industry, marine applications (desalination plants and any seawater bearing systems), and heat exchangers.

Precipitation Hardening Stainless Steels
These are commonly classified as austenitic, semi-austenitic and martensitic depending on the matrix of the steel. Their mechanical properties are obtained after a process of ageing (around 400–650 °C) in which nanoprecipitates (carbides and/or intermetallic phases) are allowed to form. Martensitic PHSS possess strengths higher than that of their martensitic counterparts of the 4XX series. In addition, these steels can exhibit good ductility and toughness depending on the heat treatment conditions, and their corrosion resistance is highly comparable to the austenitic grades. The need for special heat treatments to achieve the final mechanical properties generally makes them more expensive than other stainless steels and hampers their usage. They are not classified using any standard numbering system as other families of stainless steels presented before. The most widely known steels grades are PH 13–8, 15–5 PH, 17–4 PH, 17–7 PH, PH 15-7 Mo (the numbers refer to the usual amount of Cr and Ni in the composition, respectively), A286 (15Cr-25Ni), A350 (17Cr-5Ni) and AM355 (16Cr-5Ni). These steels have low-carbon contents (<0.1 wt.-%) and are alloyed with Mn ($\leq$2 wt.-%) and Si ($\leq$1 wt.-%) and different amounts of Mo, Al, Nb, Ti, Ta, V, Cu and N to, (i) improve the corrosion resistance or (ii) promote the formation of these strengthening nanocarbides (MX type; M = Ti, Nb, V, Ta; X = C,N), intermetallic phases (β-NiAl) or Cu-rich precipitates. Only using cutting-edge high-resolution transmission electron microscopy (HRTEM) and 3D atom probe tomography (APT) it has been recently unveiled that the sequence of precipitate formation of some commercial PHSS during isothermal treatments is very complex. For example, in 17-4 PH, it has been

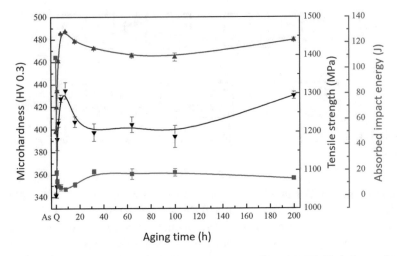

Fig. 11.2 Time evolution of the mechanical properties of alloy 17-4 PH SS during ageing at 450 °C after a solution treatment at 1040 °C for 1 h [10]

observed that Cu-rich precipitates (BCC) evolve to 9R and finally FCC; besides, Ni-(Mn,Nb)-Si-rich precipitates also form around some Cu-rich precipitates and Cr-rich (α') domains nucleated only after prolonged times. In addition, sometimes Nb/Cr nitrides have been observed [10–13]. Figure 11.2 shows the evolution of the mechanical properties during ageing at 450 °C of a 17-4 PH SS steel; the hardness and tensile strength initially increase sharply and then drop due to the coarsening of Cu-rich precipitates and finally increases again slightly after long ageing due to the formation of Cr-rich (α') domains. Figure 11.3 shows APT results obtained after ageing for 200 h at 450 °C [10]; the formation of Cu-rich precipitates and Ni-Mn-Si (NMS clusters) is shown. In 15-5 PH similar behaviour has been reported: formation of Cu-rich precipitates as BCC, 9R and FCC, clustering of Ni-Mn-Si and formation of some NbC [14–16]. In other alloys like the PH13-8, the strengthening is due to the formation of β-NiAl [17]). Common applications of these commercial steels include valves, shafts, gears, surgical tools, razor blades, aircraft frames and turbine blades.

11.2.2 Role of the Alloying Elements

Alloying plain carbon steels (Fe-C) with substitutional or interstitial chemical elements influence the microstructure and properties of steels in different ways and provide them with a great versatility. Figure 11.4 summarizes the chemical elements intentionally used in the composition of stainless steels [18]. The main ones, used in

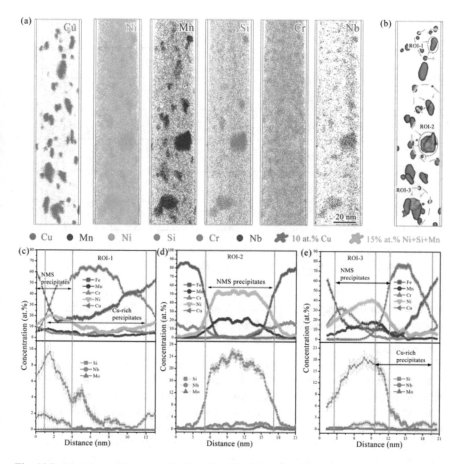

Fig. 11.3 (**a**) Atom probe maps of alloy 17-4 PH SS aged at 450 °C for 200 h after solution treatment at 1040 °C for 1 h; (**b**) Cu-rich precipitates and NMS (Ni-Mn-Si-rich precipitates) delineated by 10 % isoconcentration (yellow) and 15 % isoconcentration surfaces (green), respectively. (**c**) The 1D profiles across ROI-1 and (**e**) ROI-3 centre along the direction of the arrow. (**d**) The 1D profiles across ROI-2 are from centre and perpendicular to the map [10]. ROI stands for region of interest

all types of stainless steels, have been highlighted with bold font (Fe, Cr, Mn, Si, C). Nickel (Ni) is another very important alloying element of ASS and PHSS, although it is not present in MSS and some FSS. Other elements like P, S and O are generally found as impurities that should be kept to the minimum practical levels; however, elements like S could be added intentionally to improve the machinability. The rest of the elements (Ti, V, Nb, Ta, Mo, W, Cu, Al, N, B, Se, Co) are included in the composition to improve certain properties that vary among steel grades depending on the family and application [19]. More recently, some other alloying elements like rare earth elements (La/Ce) have been used in experimental alloys to link S and

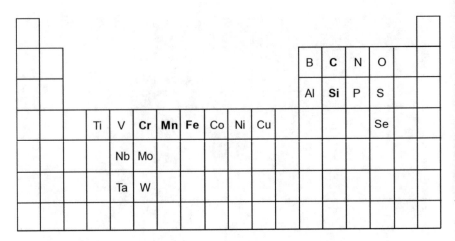

Fig. 11.4 Chemical elements used in the composition of stainless steels. In bold font are the elements commonly found in all types of stainless steels

improve grain boundary cohesion along with B addition [20, 21]. As follows, the role of some of these elements is described with more detail.

Chromium (Cr)

This is the most important alloying element since it makes steel "stainless". Its effectiveness stems from its high degree of reactivity, which leads to the formation of an adherent, insoluble and very fine layer (2/3 nm) of chromium oxide, $(Fe,Cr)_2O_3$ that shields the metal from uniform and localized attacks. The minimum content necessary to make the steel stainless varies a bit among available reports between 10.5 and 12.0 wt.-% [22]. The corrosion resistance of stainless steels increases with its content, and, under more aggressive environments, a much higher content is necessary to maintain the stability of the protective oxide layer. In addition, Cr also increases the resistance to oxidation at high temperatures and promotes a ferritic microstructure at room temperature. As it has been discussed before, this element may form stable carbides $M_{23}C_6$, formation of which at the grain boundaries may lead to localized intergranular corrosion (sensitization) and failure [8]. In addition, Cr also promotes the formation of intermetallic phases during ageing in service conditions (σ- and χ-phase) which could again reduce the corrosion resistance of the steel [23]. These phases may contribute to the mechanical embrittlement and a reduction in toughness.

Carbon (C)

It is an important alloying element in all ferrous metal-based materials. It is a very strong austenite stabilizer. In solid solution or precipitated as nanocarbides, carbon increases the strength of steel profoundly. In ASS, FSS, PHSS and DSS, it is kept to low levels (below 0.15–0.2 wt.-%). In MSS, the carbon content might be higher and intentionally added in amounts varying from 0.1 wt.-% to 1.2 wt.-% to promote the formation of martensite by heat treatment (austenitization plus quenching) and thus increase both the strength and hardness. The principal effect of carbon on corrosion resistance is determined by the way in which it exists in the alloy (solid solution or as precipitates). If it is combined with Cr as coarse carbides ($M_{23}C_6$, $M = Cr,Fe$), it may have a detrimental effect on corrosion resistance by removing some of the Cr from solid solution in the alloy and, thus, reducing the amount of Cr available to ensure the formation of the protecting oxide passive layer. Thus, if upon certain heat treatments or service conditions, the precipitation of Cr-carbides takes place at grain boundaries, this would lead to localized corrosion at those locations (intergranular corrosion), referred to as sensitization [24, 25]. Besides, the presence of these coarse chromium carbides of the type $M_{23}C_6$ at grain boundaries promotes the formation of cavities in service and the failure of the steel [26, 27]. In some heat-resistant stainless steels Nb/Ti/V/Ta ($= M$) are added with the aim of promoting the formation fine carbides (MX; $X = C,N$) that improve the creep strength. As the stoichiometry of microalloying carbides is TiC or NbC, to estimate the amount of Nb/Ti introduced in the alloy, the ratio between the atomic weight of C and Nb or Ti is used: 7.8 and 4, respectively. Thus, the amount of Nb and Ti is generally taken as equal or below $7.8 \times C$ or $4 \times C$, respectively, which assumes that all the carbon will be linked to Nb or Ti as carbides. The presence of nitrogen modifies slightly the Nb/Ti considered.

Nickel (Ni)

It is an essential element in the composition of ASS and PHSS. It is also part of the composition in some MSS and FSS. It promotes the stabilization of austenite at room temperature, prevents the formation of δ-ferrite at high temperatures and, thus, makes the steel nonmagnetic. It has no direct influence on the passive layer but exerts a beneficial effect, particularly in sulphuric acid environments. Besides, pitting corrosion develops slower in the presence of high Ni contents. It increases ductility and toughness (decreases the ductile-brittle transition temperature) and is a moderate strengthener even at cryogenic temperatures. In PHSS and some heat-resistant stainless steels, it is used to form intermetallic compounds that increase the strength when combined with elements like Ti or Al ($Ni_3(Al,Ti)$, NiAl) [28]. In novel heat-resistant FSS, creep strength is improved when this element is combined with Ti/Nb and Si to form the G-phase, $Ni_{16}M_6Si_7$, $M = Nb,Ti$ [29, 30]. In martensitic grades Ni addition combined with low-carbon contents improves the weldability.

Molybdenum (Mo)

It is used in stainless steels (specially ASS, FSS and DSS) in amounts up to 8 wt.-% but most commonly in the range of 2–4 wt.-%. This element has great effects in improving the resistance to pitting and crevice corrosion in chloride environments; the higher the content, the better the resistance to higher chloride levels. It also increases the resistance in environments with hydrochloric and dilute sulphuric acid.

Besides, Mo decreases the tendency of the passive (Cr oxide) films to break down. It increases the mechanical strength in solid solution, especially at high temperatures to improve creep strength in creep-resistant steels. However, it strongly promotes a ferritic microstructure which could be a problem for the toughness and ductility in some martensitic or austenitic grades. In addition, Mo also enhances the risk for the formation of deleterious secondary phases, some of which are also rich in Cr (Laves, σ- and χ-phase), which may cause embrittlement and failure in service [23]. It also promotes an unstable high-temperature oxide deleterious for the high-temperature oxidation resistance.

Nitrogen (N)

It strengthens and stabilizes the austenite phase at room temperature. It retards carbide sensitization and the formation of secondary phases. In combination with Mo (and Cr), it improves the resistance to chloride pitting and crevice corrosion. In low-carbon ASS used for welding operations, in which the strength is low, N is added in small amounts (0.1 wt.-%) to restore the mechanical properties and increase the yield strength. In Cr-Mn steels of the 200 series and in other experimental alloys in which Ni is substituted partially or completely by Mn, N is added in amounts up to 0.4 wt.-% to improve solid solution hardening and corrosion resistance [31, 32]. In these alloys, the higher the Cr/Mn content, the higher the solubility of this element in the austenite. However, amounts higher than 0.4 wt.-% cannot be introduced readily in solid solution using melting and processing under ambient pressure conditions; thus, more costly techniques are required. In this regard, powder metallurgical techniques and other more specialized and expensive techniques allow introducing higher amounts of N [33]. Besides, surface hardening techniques such as nitriding or nitrocarburizing could be also used to increase the amount beyond 0.4 wt.-% [34].

Copper (Cu)

It enhances the corrosion resistance in certain acids (sulphuric and phosphoric acids) and sea water environments. It promotes the retention of austenite in the microstructure at room temperature and prevents the formation of δ-ferrite at high temperature in ASS and PHSS. It can be added to decrease work-hardening and improve the machinability and the formability [35]. It also strengthens the microstructure by forming nanoprecipitates or atom clusters during the initial stages of ageing in FSS, ASS and PHSS [16, 36]. In this latter family, these clusters may also act as nucleation sites for Ni-rich intermetallic compounds in PHSS [37, 38]. Its content should be control and reduced if the steel is used for human implants.

Titanium (Ti)

It is a highly reactive element; it has a strong tendency to link with carbon, nitrogen, oxygen and sulphur. This element has been used to fix the C present in solid solution to prevent the formation of coarse chromium carbides ($M_{23}C_6$) in ASS and MSS; this way sensitization is minimized and weldability improved [8]. In some ASS, like AISI 321, Ti is used to provide high-temperature strength (boilers, refinery applications) through the formation of fine TiC precipitates in the microstructure. It has also been used in deoxidation practices during steel casting. As it was discussed above, in novel heat-resistant steels, creep is improved by combining Ti with Ni/Nb/Si to form the precipitation hardening G-phase, $Ni_{16}M_6Si_7$, M = Nb,Ti [29]. In PHSS this element is also used to promote the formation of intermetallic phases combined with Ni and Al (($Ni_3(Al,Ti)$)). Other uses or consequence of adding Ti are (i) it increases the hardenability and (ii) improves pitting corrosion by promoting the formation of Ti_2S instead of MnS (which are initiation sites for pitting).

Niobium (Nb)

It is used for similar reasons as Ti. Nb is a strong carbide and nitride-forming element used to fixed C in solid solution and prevent the formation of coarse chromium carbides ($M_{23}C_6$) in ASS and MSS [8]. It is extensively used to promote the formation of NbCN or Z-phase (NbCrN) to improve creep strength in some heat-resistant stainless steels (ASS and MSS), though contradictory reports can be found concerning the positive influence of this Z-phase on the creep strength [39, 40]. Long ageing during creep at high temperatures leads to the formation of the Nb-rich Laves phase (Fe_2Nb) from NbCN precipitates reported to be harmful for the creep strength. It prevents intergranular corrosion, particularly in the heat-affected zone after welding. In austenitic steel AISI 347, it plays the same role as Ti in alloy 321, improving the high-temperature strength (creep behaviour). Nb addition improves the thermal fatigue resistance in ferritic steels.

Vanadium (V)

Similarly to Nb and Ti, this element is also a strong carbide and nitride-forming element (VCN) and a ferrite stabilizer. Precipitated as carbonitrides it can contribute to the austenite grain refinement and also to high-temperature strengthening of MSS [41]. It has been proposed as an alternative element substituting Nb in heat-resistant steel to avoid the formation of the Laves. However, precipitated as Z-phase during the early stages of creep deformation has led to an improvement in creep strength at 700 °C [42]. Besides, its lower dissolution temperature compared to other Ti- or Nb-rich carbonitrides avoids the presence of undissolved coarse carbonitrides normally present after high-temperature solution heat treatments. It is also used as a deoxidizer to produce a gas-free ingot.

Tantalum (Ta)

Combined with Nb is used to improve the high-temperature creep behaviour in some austenitic (grades AISI 330 and 348) and ferritic (grade AISI 436) stainless steels. It is also used in PHSS to promote the formation of carbides (again combined with Nb) that strengthens the microstructure by precipitation. In some novel heat-resistant

MSS, the substitution of Nb by Ta has been proposed to promote the formation of TaCN instead of NbCN and avoid the formation Laves during service creep conditions [41]. Improved creep strength has been associated with the formation stable Z-phase (CrTaN) precipitates in Ta-alloyed MSS [43, 44]. Besides, the use of Nb has to be minimized in heat-resistant steels for nuclear applications.

Cobalt (Co)

It is employed in MSS to raise the M_s temperature and avoid the retention of austenite at room temperature. In non-stainless 9–10 wt.-% Cr steels and contrary to Mo and Cr, Co has been reported to reduce the corrosion resistance [45]. It is a solid solution strengthener and contributes to stabilize the martensitic microstructure at high temperatures, delaying its softening in heat-resistant MSS. In addition, it has recently been added to improve creep strength via the stimulation of a fine precipitation of intermetallic phases (Laves phase, χ-phase and μ-phase) in FSS [46]. However, it high prize hampers and limits its usage to some specific very demanding applications. Besides, steels for nuclear reactors should not be alloyed with Co because this element becomes highly radioactive when exposed to radiation.

Tungsten (W)

The use of tungsten with Cr strengthens the stability of the protective oxide layer against corrosion/oxidation by forming WO_3. It is also a strong solid solution strengthener even at high temperatures reason why it is used in MSS intended for high-temperature structural applications in energy-generating power plants. In these steels, long holding creep exposure may promote to the formation of the W-rich Laves phase, which coarsens very rapidly and may lead to the embrittlement of the alloy. On the contrary, in commercial FSS like Crofer® 22H and 22APU, the improvement in creep strength has been associated with the formation of intermetallic phases like the Laves phase of the type Fe_2W [47, 48]. Besides, the positive influence of W in solid solution on the corrosion resistance and mechanical behaviour is lost by the formation of this second phase.

Aluminium (Al)

It can provide superior oxidation resistance at high temperatures compared to Cr, and, as it will be described later in this chapter, it is used in certain heat-resistant FSS for this purpose [49]. In PHSS and heat-resistant stainless steels, Al is used to promote the formation of intermetallic compounds ($Ni_3(Al,Ti)$, NiAl) that increase the strength in the aged condition [50–52].

Silicon (Si)

It is a ferrite stabilizer added to steel to assist deoxidation. Small amounts of silicon, in combination with Mo, improve corrosion resistance in sulphuric acid in ASS. High-silicon content also improves the resistance to oxidation and prevents carburization and the formation of carbides at elevated temperatures. This element promotes the G-phase, along with Ni and Ti, which has been recently shown to improve the creep strength in some novel heat-resistant FSS [29, 30]. In addition (i) in small amounts, Si confers mild hardenability to steels; (ii) it may form iron silicides and Cr-rich intermetallic phases that could embrittle the microstructure. Its usage generally lies between 0.3 and 0.75 wt.-%.

Manganese (Mn)

It is found as an alloying element in all types of stainless steel. It is used as a deoxidizer (forming MnO) prior to the casting of the alloy. It is an austenite stabilizer used in grades of the 200 series in quantities from 4 to 16 wt.-% to replace partially the more expensive nickel element; however it is not as strong austenite stabilizer as Ni (approximately half effective) [53]. Other reasons for its usage as an alloying element are the following: (i) it improves the strength, toughness and workability; (ii) it forms MnS rather than iron sulphide inclusions, as these latter ones may cause hot cracking failure. However, the formation of too many MnS in the microstructure may be also detrimental to the mechanical properties; (iii) it increases the solubility of N in solid solution.

Sulphur (S)

The most important beneficial effect of S is to improve the machinability. However, it also leads to a reduction in the hot workability of stainless steels. In general, the use of S is minimized as much as possible because it has been for long know that it provokes grain boundary decohesion and failure in steels.

Phosphorus (P)

It is generally present as an impurity, and its content should be reduced to the lowest practical levels. It reduces the hot workability during high-temperature forming operations (forging, rolling) and promotes hot cracking during cooling after welding.

Selenium (Se)

It improves the machinability of steels; however, it is more expensive than S and its use should be considered only when other benefits are required. In this regard, Se improves ductility and toughness compared to S-added steels with the same content. The use of Se globalizes sulphide inclusions similarly as rare earth elements, which leads to an improvement in the impact toughness.

Boron (B)

It increases hardenability and grain boundary cohesion by preventing the segregation to grain boundaries of other elements like sulphur or phosphorous or by segregating on the cavity surface instead of S [20, 21]. In heat-resistant MSS, it is used to improve the thermal stability of $M_{23}C_6$ carbides and delay its coarsening rate by entering into the composition of this carbide. In heat-resistant ASS, when segregated to grain boundaries, it has been suggested to promote a fine precipitation of the Laves phase in Nb-alloyed grades (Fe_2Nb), enhancing the creep strength [54]. Similarly, in heat-resistant alumina-forming ASS, the addition of B (along with C) promoted a fine and higher coverage of the grain boundaries by Laves phase and B2-NiAl precipitates and suppressed their coarsening [55].

11.2.3 Phases and Phase Transformations

As it has been already mentioned above, three different phases may coexist as matrix phases in the microstructure of stainless steels: austenite, ferrite and martensite. Other secondary phases, generally in the form of micro- or nanoprecipitates (carbides, nitrides, oxides, sulphides and intermetallic phases) might be present in the microstructure depending on the composition, thermomechanical processing or service conditions. Some of these second phases are promoted intentionally with the aim of improving the mechanical properties, while others may appear in the microstructure and reduce the performance or lead to the failure of the steel under service conditions.

11.2.3.1 Matrix Phases: Ferrite, Austenite and Martensite

The main matrix phase present in the microstructure can be tailored by designing the steel composition. With respect to the role of the alloying elements in favouring the formation of any of these three main phases, they can be divided into two groups, those that stabilize or promote the formation of austenite (austenite stabilizer) and those that do the same with ferrite (ferrite stabilizers). It must be clarified that martensite is a metastable phase and a transformation product of austenite; thus, only if austenite is present in the microstructure of stainless steels, martensite could be also formed. In this regard, the stability of austenite in ASS, which is mainly dependent on the steel composition and also on other factors such as the grain size, will determine its inclination to transform to martensite when subjected to external mechanical (stresses/strains) or thermal (cooling) stimuli, respectively [56, 57]. Table 11.1 compiles the main austenite- and ferrite-forming elements used in stainless steels.

Traditionally, the effect of the main alloying elements (Cr, Ni, C, Mn, Si, Mo, N) on the microstructure of Fe-Cr-Ni stainless steels has been quantified by using the nickel and chromium equivalent numbers, Ni_{eq} and Cr_{eq}, respectively. These numbers are especially useful in ASS in which the ratio between Cr_{eq} and Ni_{eq}

Table 11.1 Main austenite and ferrite-stabilizing elements used in stainless steels

Austenite-stabilizing elements	Ferrite-stabilizing elements
Carbon	Chromium
Nitrogen	Molybdenum
Nickel	Silicon
Manganese	Titanium
Copper	Niobium
Cobalt	Aluminium
	Vanadium
	Tungsten
	Tantalum

has been employed to predict the solidification path followed by the melt during casting and anticipate whether martensite/ferrite, along with the austenite, will be present in the microstructure, at room temperature, or not. Depending on this ratio (Cr_{eq}/Ni_{eq}), the solidification modes can be divided into the following four types, namely, austenitic (A mode), austenitic-ferritic (AF mode), ferritic-austenitic (FA mode) and ferritic solidification (F mode) [58–61]. The Cr_{eq} and Ni_{eq} suggested by *Schaeffler and DeLong* are given in Eqs. (11.1) and (11.2), respectively:

$$Cr_{eq} = \%Cr + \%Mo + 1.5\%Si + 0.5\%Nb \tag{11.1}$$

$$Ni_{eq} = \%Ni + 30\%C + 30\%N + 0.5\%Mn \tag{11.2}$$

Afterwards, other researchers have incorporated the effects of more elements in such expressions [62].

Based on these equivalent numbers, one tool commonly used to predict the microstructure of high-alloyed steels at room temperature is the *Schaeffler* diagram (Fig. 11.5) [63–65]. In this diagram A, M and F stand for austenite, martensite and ferrite, respectively. Similar diagrams have been developed to undertake these predictions [66–69]. From this diagram it can be inferred that high Ni_{eq} numbers are required to ensure an austenitic microstructure, while high Cr_{eq} numbers promote the formation of ferrite in the microstructure as it would be expected. For Cr_{eq}

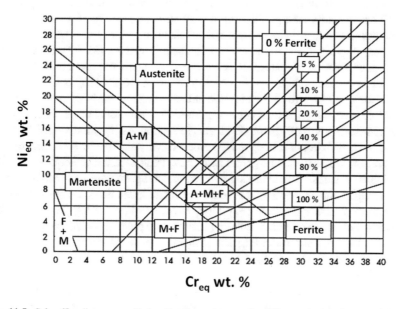

Fig. 11.5 *Schaeffler* diagram predicting the phase structure for different chemical compositions in stainless steels as a function of the Ni_{eq} and Cr_{eq}. A, M and F stand for austenite, martensite and ferrite, respectively [63, 65]

below around 18 wt.-%, the Ni_{eq} number can be used as a measure for assessing the stability of austenite in stainless steels against the martensitic transformation; the lower the Ni_{eq} is, the higher the tendency to promote the formation of martensite. This diagram should be only taken as an approximation to reality, as it does not take into account parameters like the cooling rate which may influence significantly the microstructure during casting. Nowadays there are more complete and precise computational tools or thermodynamic/kinetic software, like *ThermoCalc/DICTRA* [70], to predict temperature of phase evolution during casting or like *MAGMA* [71] that take other factors into account such as the geometry or size of the component to be cast.

11.2.3.2 Stability of Austenite

The stability of austenite and its tendency to form martensite in stainless steels have been the subject of multiple surveys because it has a prominent influence on the final properties of ASS. Besides, ASS comprise most of the world production of stainless steels (~75 %). In these metastable steels, the austenite-to-martensite transformation can be promoted in three different ways: (i) mechanically under the influence of external loads [72–74]; (ii) athermally during continuous cooling [72, 75] and (iii) isothermally (thermally activated) with time at a constant temperature [76–80]. The first two modes of transformation are better understood as they also take place in low alloys steels. For example, in transformation-induced plasticity (TRIP) or quenching and partitioning (Q&P) steels, which are of great interest for the automotive industry, the thermal and mechanical stability of retained austenite plays very important role in the final mechanical properties [81–83]. However, the isothermal formation of martensite is scarcely and mainly observed in high-alloyed steels or iron-based alloys Fe-Ni, Fe-Ni-Mn and Fe-Ni-Cr [76] but also in high carbon/nitrogen steels [84]. Besides, the amount transformed is generally very low (<5 vol.-%) [76, 80, 85] with some exceptions [77]. Additional problems associated with the investigation of this mode of transformation are the following: (i) it takes place at cryogenic temperatures and (ii) proceeds very slowly (several hours or even days), reason why external magnetic fields have been used to accelerate it [79, 86]. For all these reasons, general models able to predict this transformation as a function of the composition and temperature cannot be found in the literature besides the one proposed by *Ghosh and Olson* [87], though with the information provided in this paper, it is not possible to undertake any quantitative predictions. Models developed to predict the nucleation and kinetics of martensite formation under the first two modes have been investigated more extensively but, principally, for low-alloyed steels [88–92].

The easiest approach to evaluate the thermal or the mechanical stability of the austenite in stainless steels is through the employment of semiempirical equations that relate parameters like the M_s or the M_{d30} temperatures with the chemical composition and the grain size of the steel. These temperatures stand for the "martensite start temperature" during continuous cooling [92–94] and the "temperature at which

50 % of martensite is obtained in a tension test for a true deformation of 0.3"
in a fully austenitic steel [56, 95], respectively. The larger these values are, the
more unstable the austenite is. Another important parameter of interest is the M_f
temperature, which is the temperature at which the formation of martensite ends,
below the M_s during a continuous cooling treatment. The temperature below which
the isothermal formation of martensite is observed has been defined as M_{SI} or M_s^{iso}
[76, 77, 80]; nevertheless no empirical equations able to predict this or the M_f
temperature are available in the literature up to date. On the contrary, since the
1940s, empirical equations can be found in the literature for the M_s [92–94, 96] or
the M_{d30} [56, 97]. However, these equations seem to be valid generally for a narrow
range of compositions, and their application should be also made with caution.
Parameter M_{d30} was proposed by *Angel* [56] who came up with the following
equation for austenitic stainless steels (composition in wt.-%):

$$M_{d30} \left(^\circ C\right) = 413 - 462\,(C + N) - 9.2Si - 8.1Mn - 13.7Cr - 9.5Ni - 18.5Mo$$

$$(11.3)$$

Nohara et al. revised existing equations and propose a slightly different expression that included also the influence of the grain size through the inclusion of the
ASTM grain size number (*G*) [97]:

$$M_{d30} = 551 - 462\,(C + N) - 9.2Si - 8.1Mn - 13.7Cr - 29\,(Ni + Cu)$$
$$- 18.5Mo - 68Nb - 1.42\,(G - 8.0)$$

$$(11.4)$$

For the estimation of the M_s, *Ishida* revised the existing literature and proposed
the following general equation for iron-based alloys:

$$M_s \left(^\circ C\right) = 545 - 330C + 2Al + 7Co - 14Cr - 13Cu - 23Mn - 5Mo$$
$$- 4Nb - 13Ni - 7Si + 3Ti + 4V$$

$$(11.5)$$

This equation has been recently employed successfully to design martensitic
stainless steels [98, 99]. More complex approaches based on thermodynamic modelling or neural network predictions have been developed to predict this parameter
in steels [90, 92, 93].

Other important parameter to predict the mechanical stability and anticipate what
type of deformation processes will take place within the microstructure of ASS is
the stacking fault energy [73, 100, 101]. This parameter will be introduced more
extensively in Sect. 11.4.1 of this chapter.

11.2.3.3   Secondary Phases

The composition of stainless steels is designed with high amounts of different alloying elements because they are intended for very demanding applications, among which are those requiring (i) very high strength at low/high temperatures and/or (ii) corrosion/oxidation resistance. In PHSS and some MSS, FSS and ASS, the composition is tailored so as to allow the formation of different types of nano-precipitates which are induced via the application of thermal or thermomechanical treatments (carbonitrides, (TiNbV)(CN); intermetallics, η-Ni$_3$(Ti,Al), β-NiAl, G-phase, Laves phase, Z-phase, Cu-rich precipitates [16, 21, 36, 37, 50, 98, 102–105]). The presence of these precipitates may play the following roles:

(a) Represent pinning obstacles to the movement of dislocations, delaying crack formation and propagation, and, thus, increasing the yield/tensile and/or the creep strength.
(b) Prevent the sensitization of the microstructure that leads to stress corrosion cracking. Along with these desirable precipitates, others many form (M$_{23}$C$_6$, σ, χ, Laves) that generally lead to the fragilization of the microstructure [8, 24, 106], although phases like M$_{23}$C$_6$ or the Laves have been associated to creep strength improvements in heat-resistant steels [39, 47, 105]. When their presence is unavoidable, alloying strategies are used to minimize their detrimental influence on the properties, though this is not always possible.

Table 11.2 summarizes the different phases, crystal structure, lattice parameters, formation range of temperatures and chemical formula of the most important second phase precipitates that have been found in stainless steels. Some of them will be mentioned extensively throughout this chapter because their presence or absence affects the properties of these steels profoundly.

11.2.4   Production and Processing

Stainless steels are produced in an electric arc furnace where all the raw materials are melted together [111]: high carbon ferrochromium alloy (generally contains ~50 wt.-% Cr plus Si and C), stainless steel scrap, conventional steel scrap and sometimes Ni and Mo. The melting process requires up to 12 h; the arc typically reaches up to 3500 °C and the molten steel up to 1800 °C. The molten material is then transferred into an AOD (argon oxygen decarburization) vessel, where the carbon levels are reduced using a reduction mix that contains Si, Al and lime (the molten mixture generally contains around 1.5–2.0 wt.-% C and most stainless steels normally contain after casting low-carbon levels below 0.1 wt.-%). In this process, a mixture of argon and carbon is injected forcing the carbon to combine with oxygen and forming CO that is liberated from the melt [112]. Some alloying elements might be also added at this stage. An alternative decarburization method to obtain very low C and N levels is the vacuum oxygen decarburization (VOD). This method

Table 11.2 Secondary phases present in stainless steels [22, 107–110]

Phase	Symbol	Type	Formula	T (°C)	Structure	Lattice constants (nm)
Carbide		M_7C_3	$(Cr,Fe,Mo)_7C_3$	950–1050	Orthorhombic	$a = 0.452; b = 0.699; c = 1.211$
Carbide		$M_{23}C_6$	$(Cr,Fe,Mo)_{23}C_6$	600–950	Cubic fcc	$a = 1.057–1.068$
Carbide		M_6C	$(Cr,Fe,Mo,Cb)_6C$	700–950	Cubic diamond	$a = 1.093–1.128$
Nitride		M_2N	$(Cr,Fe)_2N$	650–950	Hexagonal	$a = 0.478; c = 0.446$
Nitride	Z	MN	$(Nb,Cr)N$	700–1000	Tetragonal	$a = 0.3037; c = 0.7391$
Nitride	π		$Fe_7Mo_{13}N_4$	550–600	Cubic	$a = 0.647$
Carbonitride		MC	Ti(CN)	700–1400	Cubic fcc	$a = 0.424–0.432$
Carbonitride		MC	Nb(CN)	700–1350	Cubic fcc	$a = 0.438–0.447$
Sigma	σ	AB	(Fe,Ni,Cr,Mo)	550–1050	Tetragonal	$a = 0.886–0.892; c = 0.454$
Chi	χ	$A_{48}B_{10}$	$Fe_{36}Cr_{12}Mo_{10}$	600–900	Cubic fcc	$a = 0.8807–0.8878$
Alpha prime	α′		CrFe(Cr61–83 %)	350–550	Cubic fcc	$a = 0.2877$
Beta	β	AB	NiAl	<600	Cubic	$a = 0.2887$
Laves	η	A_2B	$(Fe,Cr)_2(Mo,Nb,Ti,Si)$	550–900	Hexagonal	$a = 0.473–0.482; c = 0.726–0.785$
Intermetallic	η	A_3B	$Ni_3(Ti,Al)$	300–550	Hexagonal	$a = 0.5109–0.5115; c = 0.8299–0.8304$
Intermetallic	γ′	A_3B	$(Ni,Co,Fe,Cr)_3(Ti,Al)$	300–550	Cubic fcc	$a = 0.3565–0.3601$
Mu	μ		$(Cr,Fe)_7(Mo)_2$		Rhombohedral	$a = 0.4762; c = 2.5015$
R	R		(Fe,Mo,Cr,Ni)	550–650	Hexagonal	$a = 1.0903; c = 1.9347$
Tau	τ			551–650	Orthorhombic	$a = 0.05; b = 0.484; c = 0.286$
G	G	$A_{16}B_6D_7$	$Ni_{16}(Nb,Ti)_6Si_7$	500–850	Cubic fcc	$a = 1.12$

does not necessarily need the injection of Ar along with O and has the advantage of reducing the sulphur content down to very low levels (0.001 wt.-%). Other methods combine both systems or have developed a vacuum AOD or V-AOD (also called vacuum converter refiner, VCR), to take advantage of the benefits of the two systems [113]. Next, to obtain high-quality stainless steels, secondary metallurgy plays an important role. The aim is to make the final adjustments in the composition of the steel. In this process, the molten steel undergoes various metallurgical operations in the ladle, involving the removal of unwanted inclusions and the addition of minor alloying elements like Ti, Nb, Cu, etc., to fine tune the final composition.

Subsequently, to produce different semifinished forms, the ladle is transported to a continuous casting machine to obtain the slabs or casting area to obtain ingots/blooms/billets. During the continuous casting, the melt goes through a nozzle into the tundish and then through a submerge entry nozzle into a water cool copper mould; water spray cooling below the mould and between the rollers is used to solidify the melt and produce a solid strand of steel. The strand is continuously cooled until it reaches the straightening zone and the cutting station to make the slabs of the desired length.

After producing the slabs, blooms and billets, these products are subjected to different forming operations that involved, for example, hot rolling or hot forging to be flattened or reshaped. This process is carried out above the recrystallization temperature of the stainless steel, which allows obtaining a recrystallized and equiaxed microstructure after rolling. Slabs are heated above 1200 °C before putting into the rolling mill train. Mills are generally composed of several rolling stands. The finishing train is an exit roller, run-out table with a cooling system followed by a coiler. The cooling system is a set of lamellar water cooling banks with controlled water flow that ensures that the desired coiling temperature is obtained before reaching the coiler. After the hot rolling, the flatten slabs may also undergo a cold-rolling process to further reduce the thickness and create plates and sheets/strips. According to the standard practice, plates are generally referred to as thicker than 4.76 mm and sheets/strips thinner than this size [114]. The blooms/billets can be also heated in a furnace or by induction prior to hot rolling or drawing into smaller diameters to make, for example, bars and wires. In this case, the hot rolling process is very intense because all sides are rolled at the same time; so to prevent the temperature from increasing too much, rolling speed has to be controlled, and a cooling system has to be implemented. Bars could have different cross section geometries: circle, square, rectangle, hexagons, etc., or even complex shapes in the form of H, I, U or T [115].

After hot rolling, most stainless steels receive an annealing and pickling treatment. The annealing depends on the steel grade, and it is usually carried out in a continuous annealing line. With this heat treatment, internal stresses are relieved, and the structure is softened after the thermomechanical processing [109]. After the heat treatments, steels follow a descaling process. The scale that builds up on the surface of the steel during hot rolling and annealing is removed by different methods: (i) one is pickling; a mix of acids (nitric, hydrofluoric and, sometimes, sulphuric acids) is used to remove the furnace scale. This system is composed of

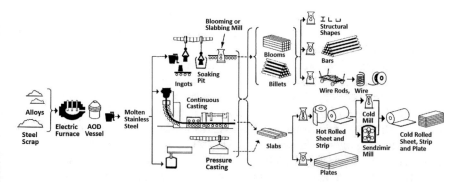

Fig. 11.6 The making of stainless steel. (Image taken from the webpage of "The Specialty Steel Industry of North America (SSINA)" voluntary trade association [116])

a series of baths followed by high-pressure water rising and drying; (ii) electro-cleaning uses an electric current and phosphoric acid to descale. The descaling helps promoting the passive surface film that naturally occurs in stainless steels [117]. Annealing and descaling may be repeated several times during the manufacturing process after different forming operations [114, 118, 119]. The most modern processing facilities include a continuous annealing and pickling lines (CAPL) or RAP lines (rolling, annealing and pickling processes are carried out one after the other in the same processing line [113, 115, 120]). After hot rolling, annealing and pickling, the cold-rolling mills employed to produce sheets/strips use a cluster configuration of small-diameter working rolls, backed by layers of larger supporting rolls. Typical configurations are the Sendzimir 20-high or Z-high mills [113]. These mills are equipped to measure the thickness, profile and flatness. Characteristic cold-rolling reductions are 50–90 % and depend on the work-hardening behaviour of each steel grade. As this process occurs below the recrystallization temperature of the steel, deformation accumulates inside the material, increasing its strength but reducing the ductility. Again, after cold-rolling, annealing and pickling are necessary to recrystallize the microstructure and descale the surface. However, descaling could be also mechanical in cases where the scale is resistant to acidic solutions used in pickling. Other finishing methods use a protective atmosphere (hydrogen/nitrogen) free of oxygen (so-called bright annealing furnace), no scale forms and the pickling step is not required. The hot rolling and cold-rolling lines can be part of the same line when the target plate thickness is above 2–3 mm. Finally, a skin or temper pass rolling may be applied; it involves a reduction of around 1 %, and it is used to remove surface defects, improve strip flatness and shape, and thus obtain a high-quality surface with optimum desired mechanical properties. To improve deep drawability and surface quality of FSS, continuous yielding is usually preferred for lower yielding ratio and elimination of Lüders bands, effect that is usually observed even in ultralow purified FSS [121, 122]. Skin pass rolling is very useful to FSS because it breaks the Cottrell atmospheres that link interstitials (C,N) and dislocations, preventing the yield point phenomenon and Lüders band formation during subsequent forming operations [123]. However, this process has also been associated with other problems, like longitudinal buckling that results in the formation of wrinkles [124].

To obtain the final size and shape, finishing operations are applied before sending the steels to the customers/end users. Some end users also implement these additional processing steps within their production plants depending on the final component. These may involve [115, 125–129] (i) levelling of the plates/sheets; (ii) edge trimming and slitting the coils to narrower dimensions; (iii) straight/circle shearing using guillotine knives; (iv) sawing with high speed blades; (v) blanking using punches or dies; (vi) nibbling; (vii) flame cutting or plasma jet cutting; (viii) bending, spinning or roll forming; (ix) press forming (deep drawing, stretch forming, ironing or hydroforming); (x) hole flanging; (xi) machining (milling, turning, drilling, threading); (xii) grinding application of a plastic film coat for surface protection; (xiii) edge preparation for welding or (xiv) welding.

11.3 Strengthening and Ductilization Mechanisms

Steels offer a wide range of mechanical properties because there are many ways of strengthening them. Although quantifying the different strengthening mechanism is not an easy task, the different contributions are generally described using the following mathematical expression for the yield strength (σ_y) [130]:

$$\sigma_y = \sigma_0 + \sigma_{ss} + \sigma_{gs} + \sigma_\rho + \sigma_{ppt} \tag{11.6}$$

where σ_0 is the friction stress for pure iron and σ_{ss}, σ_{ppt}, σ_{gs} and σ_ρ are the contributions to the strengthening due to the solid solution, the precipitation, the grain size and the dislocations, respectively. However, several authors have discussed that the previous expression might overestimate the dislocation forest and precipitation hardening terms [130–132]. Thus, alternative expressions have been suggested in which the overall strengthening obey a quadratic mixture law, resulting in this modified expression:

$$\sigma_y = \sigma_0 + \sigma_{ss} + \sigma_{gs} + \sqrt{\sigma_\rho^2 + \sigma_{ppt}^2} \tag{11.7}$$

In the following sections, the different strengthening mechanism will be discussed with more detail.

11.3.1 Friction Stress

The friction stress, σ_0, for pure BCC annealed iron, in the absence of defects, has been estimated or considered to have a very wide range of values in the literature, from ~219 MPa [133] to 50 MPa [132]. For FCC metals it has been taken as negligible and around 0–1 MPa [134].

11.3.2 Solid Solution Strengthening

Solid solution strengthening (SSS) has its origin in the presence of alloying elements other than Fe in the steel crystal structure. These foreign atoms, interstitial or substitutional solutes, possess different atomic radii compared to Fe which create distortion of the crystal lattice and, thus, stress fields. Thus, SSS occurs when the strain fields around solutes interfere with the motion of dislocations as they move through the lattice during the application of plastic deformation. Substitutional and interstitial atoms in solid solution produce different lattice distortions in the face-centred cubic (FCC) and body-centred cubic (BCC) crystal structures. Substitutional atoms produce isotropic strains in the BCC crystal structure of ferrite that only interact with the hydrostatic components of the strain fields of dislocations. In contrast, interstitial atoms (carbon or nitrogen) produce tetragonal distortions that interact with the shear, dominant component of dislocations strain field. This is why interstitial SSS is so potent in ferrite. On the contrary, an interstitial atom in austenite behaves like a substitutional solute in ferrite, which is why carbon is less effective as a strengthener of austenite [135].

Linear models have been considered by several authors to quantify the SSS and relate the yield strength with the composition of stainless steels and other alloys [136–140]. These models propose linear dependence equations as it was summarized by *Sieurin* et al., [140]. These authors, based on the classical theory of *Labusch-Nabarro* [141, 142] proposed Eq. (11.8), dependent on the misfit size parameter, ε_b, associated with each chemical element (i), to weight the influence of each element on the solid solution (Eq. (11.8)) and a factor K_{LN}:

$$\sigma_{ss} = K \sum_i \varepsilon_{bi}^n C_i^p + A_N N^{1/2} \tag{11.8}$$

With $p = 2/3$ or 1; $n = 4/3$ or 1, C_i the concentration of alloying element i in solid solution and A_N being the strengthening factor of nitrogen (N). K adopts different values depending on the value of exponents n and p. For $p = 4/3$ and $n = 2/3$, $K=K_{LN}$ which has a model value given by [140]:

$$K_{LN} = 1.1 \times 10^{-3} \frac{M\beta^{4/3} w^{1/3} G}{b^{1/3}} \tag{11.9}$$

This equation takes the influence of nitrogen separately. In Eq. (11.9), M is the Taylor factor (3.06), $\beta = 16$, and w is set to $5b$ and G the shear modulus (76 GPa). Introducing these values in Eq. (11.9) gives $K_{LN} = 17{,}100$ MPa. Other equations proposed by *Sieurin* et al. define a strengthening factor (A_i) to weight this influence of each chemical element (i) according to [140]:

$$\sigma_{ss} = K_1 + \sum_i A_i C^p + A_N N^{1/2} \qquad (11.10)$$

With a K_1 a constant. The strengthening factors combine both size misfit and modulus misfit effect and are usually determined from experiments. The above equations also contained three additional factors to consider the influence of the amount of delta ferrite (varying linearly), the thickness of the sample (plate/sheet) tested and the austenite grain size; following these last two terms is a Hall-Petch type of relationship. To obtain quantitative estimations using Eq. (11.8), misfit parameters for stainless steels are necessary and only available for ferrite but not for austenite, although they could be calculated using the Vegard's law [143]. Multiple regression analysis using data from 65 ASS was carried out by these authors to evaluate the hardening coefficients for different values of exponents n and k. The most simple and widely used expression in the literature takes $n = p = 1$; the values obtained in this case for the contribution to the yield strength (σ_{ss}), tensile strength (R_m) and elongation (ε_L) are given in Table 11.3.

Sieurin et al. reviewed the information available in the literature related to the strengthening factors, for different chemical elements in ASS, including the influence of the austenite grain size and the amount of delta ferrite. The strengthening factors (A_i) summarized in their paper to calculate σ_{ss} are provided in Table 11.4 along with some additional values; equations and strengthening factors to calculate the tensile strength (R_m) or the elongation can be also found in this paper. As it can be concluded after comparing the different values, the scatter found in the strengthening parameters is very large among the different researchers.

An equation similar to Eq. (11.10), with $p = 1$ and treating N similarly as other alloying elements, has been used extensively to describe the SSS in ferritic and martensitic steels [99, 130, 148, 149]. Table 11.5 provides K_i values for some alloying elements in ferrite that have been compiled by Lu et al. and used to design creep-resistant MSS [99]. The values shown in this table have been determined for ferritic binary Fe-M systems. The effect of Cr in solid solution (not given in this table) has been generally reported negligible (0) [150] or even negative (-31) [130];

Table 11.3 Strengthening factors (A_i) of some alloying elements in austenite for ASS using Eq. (11 10) with $p = 1$ to calculate yield strength (σ_{ss}) and tensile strength (R_m) [140]

Element	K_1	N	Cr	Ni	Mo	Mn	Si
σ_{ss}, A_i (MPa/at. fraction)	–	77	7	2.9	–	20	33
R_m, A_i (MPa/at. fraction)	172	123	19	−4.3	9.8	–	–

Table 11.4 Strengthening factors (A_i) of some alloying elements in austenite for ASS using Eq. (11.10) with $p = 1$ to calculate yield strength (σ_{ss}, MPa/wt.-%) according to different authors

Author	K_1	N	C	Cr	Ni	Mo	Mn	Si	Cu	V	W	Nb	Ti	Al
Irvine [144]	63.5	496	356.5	3.7	–	14.6	–	20.1	–	18.6	4.5	40.3	26.3	12.7
Pickering [145]	68	493	354	3.7	–	14	–	20	–	–	–	–	–	–
Leffler [146]	0	339	183	17.4	–9.6	18	–	–	41	–	–	–	–	–
Nordberg [147]	120	$210\sqrt{(N+0.02)}$	–	2	0	14	2	0	10	–	–	–	–	–
Ohkubo et al. [139]	–110	650	350	20	–4.2	7.3	–0.3	20	–5.6	–	–	–	–	–
Eliasson and Sandström [138]	81–111	877	598	–0.1	5.7	5.1	–1.4	23	–18	–	–	–	–	–
Skuin [140]	–	–	688	–	–	–	–	–	–	–	–	–	–	–
Lorenz et al. [140]	40	450	525	8.4	–	22	2	–	5	–	–	–	–	–
Kohl [140]	112	517	422	5.2	–	–	–	–	–	–	–	–	–	–

Table 11.5 Strengthening factors (A_i) of some alloying elements in ferrite [99]

Element	C, N	Ni	Ti	Mo	Al	Co	V	Mn	Si	W
K_i (MPa/at.-%)	1103.4	19.2	17.9	15.9	9.0	2.1	2.0	16.9	25.8	31.8

this is understandable given the small atomic misfit between Cr and Fe and their equal BCC crystal structure.

More complex computational models have been developed recently for different metals, high-alloyed steels, like high-entropy alloys that take these effects into account [151–155]. These could give predictions that are more accurate though at the expense of computational time.

11.3.3 Dislocation Strengthening

Dislocations can be introduced in the microstructure by promoting, for example, the martensitic transformation and/or by the application of plastic deformation. In this regard, Fig. 11.7 demonstrates how cold-working can significantly modify the strength-elongation balance of different commercial austenitic stainless steels of the 200 and 300 series. The results are compared with the mechanical properties of some selected commercial automotive steels [83]: TRIP (transformation-induced plasticity), DP (dual phase), CP (complex phase), IF (interstitial free), HSLA (high-strength low-alloyed), BH (bake hardening) steels and a commercial aluminium alloy Al6061-T4 (T4 = solution treated and naturally aged) [156]. Work-hardening of these ASS during plastic deformation not only occurs by storing deformation (dislocations) in the austenite but also by the formation of strain-induced martensite (SIM) from the austenite via the TRIP effect [91, 157, 158]. The austenite in these steels is metastable, and its tendency to transform to martensite depends on the composition and other parameters like the grain size. This will be further discussed in the examples exposed in Sect. 11.4 of this chapter.

The strengthening contribution from dislocations has been estimated in the past using Eq. (11.11) in its more general formulation, where δ has been taken as the cell diameter in FCC metals [159] or the lath width in martensitic steels [160], M is the Taylor factor (2.9–3.06), α_1 (0.2–0.3), α_2 (2–3) are constants, b is the Burgers vector, G is the shear modulus (76–80 GPa) and ρ is the dislocation density ([53, 134, 159, 161]):

$$\sigma_\rho = M\alpha_1 Gb\sqrt{\rho} + M\alpha_2 Gb\left(\frac{1}{\delta}\right) \tag{11.11}$$

However, the second term of this expression is generally not included in many formulations [91, 162]. Dislocation densities can be evaluated experimentally by means of neutron or X-ray diffraction, since the peak broadening of a particular line, corresponding to an interplanar d spacing, can be considered related only to the

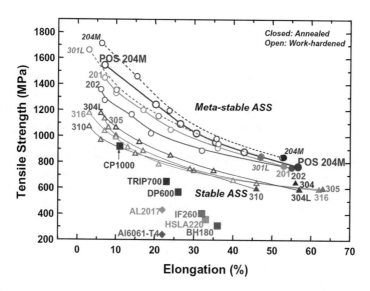

Fig. 11.7 Tensile strength versus elongation of various commercial ASS grades in annealed and work-hardened condition compared to automotive steels (TRIP700, CP1000, DP600, IF260, HSLA220, BH180) and aluminium alloys (Al6061-T4, Al2017) [156]

Table 11.6 Values for the lattice parameters and Burgers vector for FCC and BCC crystal lattice

Symbol	Meaning	Value
a	Lattice constant	$a_{bcc} = 0.286$ nm
		$a_{fcc} = 0.356$ nm
b	Magnitude of the Burgers vector	$b_{bcc} = a_{bcc}\sqrt{3}/2 = 0.249$ nm
		$b_{fcc} = a_{fcc}\sqrt{2}/2 = 0.254$ nm

elastic microstrains, ε_{micro}. Christien et al. derived an expression for the dislocation density [163]:

$$\rho = \frac{3E}{Gb^2 \left(1 + 2v^2\right)} \varepsilon_{micro}^2 \tag{11.12}$$

where E is the Young's modulus (200 ± 15 GPa) and v the Poisson's ratio of the steel (0.3). For martensitic microstructures, the following expression was derived recently as a function of the dimensions of the martensite laths (with the lath section $A_{lath} = w_{lath} \, d_{lath}$):

$$\rho = \frac{3E}{G \left(1 + 2v^2\right)} \frac{4\varepsilon^2 w_{lath}}{bd_{lath}^2} \tag{11.13}$$

Table 11.6 shows typical values for the lattice parameters and Burgers vector for FCC and BCC crystal lattices.

11.3.4 Precipitation Strengthening

This is a very important strengthening factor in PHSS and heat-resistant steels. The composition of these alloys is tailored to promote the formation of nanoprecipitates that strengthen and/or stabilize the microstructure at low and high temperatures during service conditions depending on the intended application [11, 48, 49, 164].

The precipitation strengthening effect relies on the interaction between precipitates and dislocations. Essentially, precipitation strengthening is achieved by promoting the formation of second phase nanoparticles via ageing treatment at intermediate temperatures (400–700 °C). These particles act as obstacles to dislocation movement, increasing the yield strength or critical resolved shear stress (CRSS) of the material; in other words, the shear stress at which permanent plastic deformation first occurs in an annealed single crystal. The degree of strengthening obtained is not only highly dependent upon the volume fraction and the size of particles but also on the structure of the particle (whether it is an ordered particle or not) and the nature of the dislocation-particle interaction [165]. There are two possible dislocation-particle interactions as a function of the particle nature:

(a) *Impenetrable particles*: The dislocation will avoid the particle by the well-known *Orowan* looping mechanism, by cross-slipping or climbing, and the particle will remain unchanged. In this case the actual strength of the particle is irrelevant since the bypassing operation depends only upon the interparticle spacing (Fig. 11.8a). Impenetrable particles are those whose interface has very different atomic configuration with the matrix, and, therefore, the misfit between their interatomic distances ($\varepsilon_{\text{misfit}}$) may be larger than 25 %. Since incoherent particles are characterized by a high interfacial energy (~0.5–1 J/m^2), their equilibrium shape will be roughly spherical. The contribution to the strengthening provided by the looping of dislocations assuming spherical precipitates can be approximated by employing the *Ashby-Orowan* Eq. [166]:

$$\sigma_{\text{ppt}} = \left(\frac{0.538 Gb \sqrt{V_f^p}}{X} \right) \ln \left(\frac{X}{2b} \right) \tag{11.14}$$

where V_f^p is the precipitate volume fraction and X is the particle mean diameter (mm). The other parameters have been defined before. In the literature other forms of this equation can be found:

$$\sigma_{\text{ppt}} = \left(\frac{2Gb\varnothing}{4\pi \, (L - X)} \right) \ln \left(\frac{L - X}{2b} \right) \tag{11.15}$$

with $\varnothing = (1 + 1/(1 - \nu))/2$ and L the distance between particles expressed by:

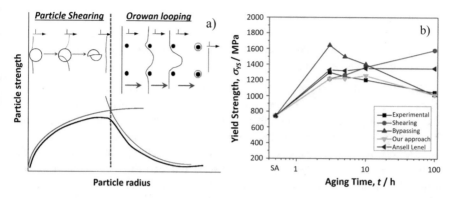

Fig. 11.8 (a) Schematic illustration of the strength variation versus particle size for the two possible interactions: (1) glide dislocations with penetrable ordered particles by particle shearing or cutting mechanism (this generates new interfaces and antiphase boundaries (APB) and (2) interaction with impenetrable particles by *Orowan* bowing/looping or bypassing [160]; (b) measured and calculated yield strength from the different models and the approach by *Schnitzer* et al. [169] for alloy PH 13-8

$$L = \left(\left(1.23 \cdot \left(2\pi / 3 V_f^p \right) \right)^{1/2} - X(2/3)^{1/2} \right) \qquad (11.16)$$

Ansell and Lenel [167] revised this theory and concluded that contrary to the mechanism proposed by *Orowan* in which yielding occurs when the dislocations move past the precipitates, leaving residual sloops surrounding them, they considered that yielding takes place only if the particles deform plastically or fracture. In their expression, the influence of the precipitate size was not considered. This expression has been employed by other authors [168, 169]:

$$\sigma_{ppt} = \left(\frac{2G'}{4C} \right) \left(\frac{\left(V_f^p \right)^{1/3}}{\left(0.82 - \left(V_f^p \right)^{1/3} \right)} \right) \qquad (11.17)$$

with C a constant (30) and G' the shear modulus of the precipitate ($Ni_3Ti = 55.16$ GPa [168]; $NiAl = 49.65$ MPa [169, 170]).

(b) *Penetrable particles:* If the strength of the particle is such that the maximum resistance force is attained, then particles will be sheared, and the dislocation will pass through the particle (Fig. 11.8a). When a dislocation shears a particle, it passes from the matrix into the particle due to the continuity of slip planes across the matrix/particle interface. This situation can only arise when the precipitate/matrix interface is fully coherent or semi-coherent, and, thus, there is no or little misfit ($\varepsilon_{misfit} < 25$ %) between the lattices. Whereas the strengthening after *Ashby-Orowan* is only dependent on the volume fraction and the diameter of the particle, the strength increment from dislocation shearing is influenced by the intrinsic material properties and crystal structure. There are five factors, listed in Table 11.7, that contribute to increase the yield strength, the first three being the most important.

To estimate the strength increment from precipitates, the operative mechanism must be first identified, although it may also change. Usually, at early stages of the precipitation sequence, particles nucleate coherently with the matrix, since it is energetically more favourable than the creation of an incoherent interface, enhancing a particle shearing mechanism. However, as the particles grow with the ageing time, they tend to become semi-coherent and incoherent as the energy of the strained coherent interface becomes greater than that for an incoherent interface. This loss of coherency occurs when the particle reaches its critical radius, which is of the order of 5–10 nm and results in a change in the mechanism from shearing to *Orowan* bypassing (Fig. 11.8a). *Schnitzer* et al. [169] applied these models to estimate the yield strength of PH 13–8Mo after ageing at 575 °C for different times and found that none of the models gave reliable predictions due to the consideration of an average particle size. Using a distribution of precipitate sizes and assuming that some are sheared and some bypassed significantly improved the model outcome (Fig. 11.8b and Table 11.7).

11.3.5   Strengthening Due to Grain Size Refinement

Grain size refinement is one of the predominant strengthening methods for improvement of strength and toughness in steels, as well as in other metals. The well-known *Hall-Petch* equation (Eq. (11.12)) describes the strengthening due to grain boundaries (σ_{gs}) and establishes that the strength is inversely proportional to the square root of the mean grain size (d) and proportional to the *Hall-Petch* coefficient (k_{HP}) [173, 174]:

$$\sigma_{gs} = k_{HP}d^{-1/2} \tag{11.18}$$

This microstructure property dependency has been validated for a wide range of grain sizes in metals in general and stainless steels in particular. However, the validity limit of this expression appears to be at about 20 nm of mean grain size, value beyond which the strengthening decreases. There exists extensive literature concerned with ASS [175–181]. In the case of martensitic steels (stainless or not), the microstructure is more complex, showing a hierarchical microstructure that consists of prior austenite grains, packets and blocks (~1–5 µm), all of which are high-angle grain boundaries (HABs) with an angle of misorientation generally considered as greater than 10–15 °. Blocks are subdivided into martensitic laths (<1 µm), defined as low-angle grain boundary (LAB) regions with similar angles of low misorientation (Fig. 11.9). Lath boundaries due to its low angle are generally

Table 11.7 Factors contributing to the increase in the yield strength for the shearing mechanism, where M, G and b have been defined before in the text; $\alpha_\varepsilon = 2.6$, $m = 0.85$, ΔG and $\Delta \gamma_{sf}$ are the modulus and stacking fault energy mismatches between the matrix (M) and the precipitate (P), respectively; ε_c is the constrained lattice parameter misfit; V_f^P is the volume fraction of the precipitates; and γ_{apb} is the antiphase boundary free energy of the precipitate [171, 172]

Strengthening mechanism	Meaning	Descriptive equation
Modulus hardening	Difference in the shear modulus G between M and P	$\sigma_{ms} = M 0.0055 (\Delta G)^{\frac{3}{2}} \left(\frac{2V_f^P}{G} \right)^{\frac{1}{2}} \left(\frac{X}{b} \right)^{\frac{3m}{2}-1}$
Coherency strengthening	Elastic interaction between the strain fields of the coherent misfitting precipitate and the dislocation	$\sigma_{cs} = M \alpha_\varepsilon (G \varepsilon_c)^{\frac{3}{2}} \left(\frac{X V_f^P}{0.5 G b} \right)^{\frac{1}{2}}$
Order strengthening	A mobile dislocation shears an ordered P and creates an APB on the slip plane	$\sigma_{os} = M 0.81 \frac{\gamma_{apb}}{2b} \left(\frac{3 \pi V_f^P}{8} \right)^{\frac{1}{2}}$
Chemical strengthening	New P-M interface is produced. There is a surface energy associated with this new interface	$\sigma_{cc} = \left(\frac{12 \gamma_s^3 V_f^P}{\pi G b X^2} \right)^{-\frac{1}{2}}$
Stacking-fault strengthening	Difference in the stacking fault energy between M and P impedes the motion of dislocations since the separation of partial dislocation varies depending on the phase	$\sigma_{sf} = (\Delta \gamma_{sf})^{\frac{3}{2}} \frac{1}{b^2} \left(\frac{3 \pi^2 V_f^P X}{16 G} \right)^{\frac{1}{2}}$

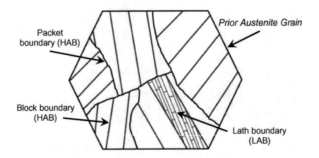

Fig. 11.9 Hierarchical microstructure of martensitic steels [160]

considered as transparent to the dislocation movement, and, thus, block size should be considered as the strengthening grain size. However, as discussed by *Kim* et al., the very high dislocation density of fresh or low-tempered martensite as well as the large number of lath boundaries could contribute significantly to strengthening [130].

The refinement of the microstructure is held as a unique mechanism for improving strength and toughness, and one of the key methods to achieve it is through thermomechanically controlled processing (TMCP), which involves controlling the hot deformation processes such as rolling, controlled cooling and direct quenching, in order to change its shape and refine the microstructure [182]. The deformation leads to a breaking down of the original coarse microstructure by repeated recrystallization of the steel. However, it is difficult to obtain nanograined/ultrafine-grained (NG/UFG) ASS by conventional TMCP [183]. Nowadays, these limitations can be overcome by using sever plastic deformation (SPD) processes such as equal channel angular pressing (ECAP) [184, 185], high-pressure torsion (HPT) [186, 187], multiple compression or multidirectional forging (MDF) [188–191], hydrostatic extrusion [192] or cold-rolling [178, 181, 193]. The refinement of the microstructure using different approaches and its impact on the mechanical properties will be introduced with more detail in the following section.

11.4 Latest Developments

This section reviews some recent investigations concerned with the development of novel stainless steels of different types and in the understanding of their transformation behaviour under different thermal and thermomechanical conditions: metastable ASS (Sect. 11.4.1), ultrafine or nanostructure stainless steels (mostly metastable ASS, Sect. 11.4.2), PH/maraging stainless steels (Sect. 11.4.3), quenching and partitioning (Q&P) stainless steels (mostly MSS, Sect. 11.4.4) and heat-resistant stainless steels (ASS, FSS and MSS; Sect. 11.4.5). It becomes difficult to include all steel grades that have been developed recently; the intention is to cover the main research trends regarding stainless steels with a main focus on their mechanical behaviour.

11.4.1 Metastable (TRIP) Stainless Steels

In this context, metastable stainless steels will be referred to as those having a fully (or almost fully) metastable austenitic FCC microstructure at room temperature. This metastable austenite, present at room temperature, can be transformed to martensite (α' or ε) under externally applied mechanical strains/stress [157, 158, 194] or/and upon cooling (athermal martensite) [72] or isothermal holding [79, 80] below room temperature. Several reports have also shown that the application of external magnetic fields can accelerate this transformation [77, 78]. Alloys that transform upon cooling to cryogenic temperatures show a faster transformation kinetics when subjected to mechanical strains/stresses as the mechanical energy provided adds to the thermodynamic chemical energy, speeding up the nucleation of martensite. Existing commercial ASS of the 300 and 200 series are metastable steels, and they would be included in this section [195]; however, the focus here will be placed on developments carried out outside the standard commercial grades (AISI 304, 316, 321, 347, 201, 204). There exists extensive research in the development of high- and medium-manganese metastable steels [196] and low-alloyed TRIP steels [182, 197]; as these are not stainless, they are outside the scope of this chapter and will not be considered.

The mechanical properties of metastable stainless steels depend strongly on the mechanical stability of the austenite. High strength is usually coupled with low ductility. Nevertheless, the transformation-induced plasticity (TRIP) associated with martensitic transformation during plastic deformation of this type of steels leads not only to an enhancement of the strength but also an increase in the uniform ductility, facture toughness, wear resistance and fatigue resistance [181, 198–200]. For this reason, understanding the parameters that influence the metastability is of key importance to control the mechanical behaviour of existing stainless steels and design novel TRIP steels. The occurrence of the martensitic transformation, the TRIP effect, and, thus, the stability of the austenite depend on the stacking fault energy (γ_{SFE}), the initial microstructure and deformation conditions. The γ_{SFE} depends mainly on the composition and temperature [201–203]. Depending on the γ_{SFE}, the shear bands generated during plastic deformation can be in the form of hexagonal close-packed ε martensite, mechanical twins and planar slip bands. In this regard, it has been reported that low values below 20 mJ m^{-2} yield the formation of martensite, ε (hcp) and/or α' (bcc), and if both phases are present, the sequence of transformation follows $\gamma \rightarrow \varepsilon \rightarrow \alpha'$ [32, 73, 75, 201, 204], with ε serving as nucleation sites for α'. For values between 15 and 20 mJ m^{-2}, TRIP (α') and twinning could coexist [205, 206], and the transformation sequence would now follow $\gamma \rightarrow \gamma^{twin} \rightarrow \alpha'$, though, in this case, there are contradictory reports regarding whether the twins would act as nucleation sites or not. *Tian* et al.

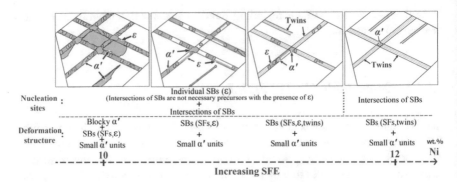

Fig. 11.10 Schematic diagram of the correlation between deformation structure and SFE in Fe-Cr-Ni alloys with Ni variation. *SB* shear band, *SF* stacking fault [73]

investigated Fe-Cr-Ni alloys with different Ni contents (10–12 wt.-%) and observed that the nucleation of α' takes place only at the intersections of shear bands in the presence of twins and cannot occur without the presence of ε [73]. Figure 11.10 provides a diagram in which these authors summarized the nucleation sites for α' as a function of the γ_{SFE} for Fe-Cr-Ni alloys based on their experimental observations (Fig. 11.11). In contrast, other authors have shown that, for example, in AISI 316 cold-rolled or cold-drawn samples, the nucleation of martensite occurred at the intersection of austenite grain boundaries with twins or at the intersection of twins, respectively [207]; whereas in AISI 304, *Shen* et al. verified that α' nucleated preferentially at ε martensite or at intersections of deformation twins [205]. In novel experimental Fe-Cr-Mn-Ni alloys in which high amounts of Mn (6–10 wt.-%) have been introduced substituting Ni, no sign of α' nucleating on twins is mentioned. For low cold-rolled reductions, nucleation occurs at grain boundaries and slip bands, and for large reductions, $\gamma \rightarrow \varepsilon \rightarrow \alpha'$ seem to be an additional mechanism, but no nucleation on twins has been addressed [32]. Finally, for γ_{SFE} values between 20 and 30–45 mJ m^{-2}, only mechanical twinning operates [91, 208], and, above 30–45 mJ m^{-2}, the deformation process is controlled mainly by dislocation glide, and no phase transformation takes place [196]. The threshold values provided above that set the end of the $\gamma \rightarrow \varepsilon$, $\gamma \rightarrow \gamma^{twin}$ or $\gamma \rightarrow \alpha'$ as the stacking fault energy increases varies slightly depending on different reports available in the literature; thus, those values should be taken as an approximation.

There exist several reports, some of them very recent, in which composition dependent expressions have been derived for γ_{SFE} in austenitic stainless steels. Besides a number of models (thermodynamic, ab initio) have been proposed to predict this parameter [91, 100, 101, 209–213]. *Meric de Bellefon* et al. recently reviewed several of these expressions and applied a multivariate linear regression with random intercepts to propose a new general expression that would overcome the limitations and problems of previous formulations [209]:

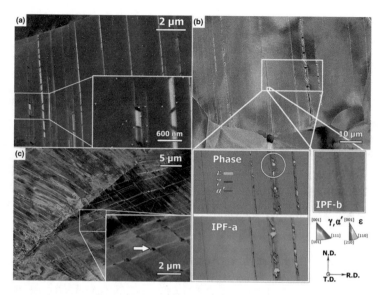

Fig. 11.11 Deformation structure at 10pct cold-rolling reduction for (**a**) alloy Fe-18Cr-10.5Ni, (**b**) alloy Fe-18Cr-11Ni (IPF-b: mechanical twinning) and (**c**) at 30 % cold-rolling reduction for alloy Fe-18Cr-11.5Ni [73]. Alloy chemistries are in wt.-%

$$\gamma_{\text{SFE}} \left(\text{mJ m}^{-2} \right) = 2.2 + 1.9\text{Ni} - 2.9\text{Si} + 0.77\text{Mo} + 0.5\text{Mn}$$
$$+ 40\text{C} - 0.016\text{Cr} - 3.6\text{N} \quad (11.19)$$

In this equation, the composition is in wt.-%. In addition, *Galindo-Nava and Rivera-Díaz-del-Castillo* proposed the following expression to estimate the temperature-dependent variation of γ_{SFE} (with T in K) [91]:

$$\gamma_{\text{SFET}} \left(\text{mJ m}^{-2} \right) = \gamma_{\text{SFE}} + 0.05 \ (T - 293) \quad (11.20)$$

Equation (11.19) shows that interstitial elements have a profound effect on γ_{SFE} and, therefore, on the mechanical stability of austenite, the deformation mechanisms, the sequence of nucleation of martensite and the mechanical properties [212, 214–217]. Especially the carbon content has been identified as the main source of error in the determination of semiempirical equations for the γ_{SFE} because small fluctuations in its content or inaccuracies in its determination may result in important variations in this parameter [209]. *Lee* et al. observed that increasing the amount of carbon (0.15–0.42 wt.-%) in a Fe-18Cr-10Mn-0.4 N (wt.-%) alloy changed the deformation products present in the microstructure following this sequence: $\alpha' \rightarrow \alpha' + \text{twins} \rightarrow \text{twins}$ [214]. A similar effect was observed by the authors for nitrogen in Fe-18Cr-10Mn-N (wt.-%) alloys [214, 218]. *Jeon and Chang* also observed the mechanical stabilizing effect of N when added to a AISI 301 steel and

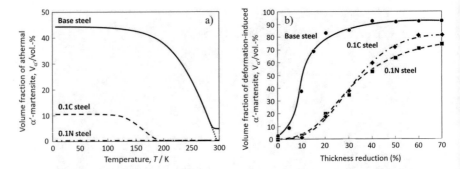

Fig. 11.12 (a) Changes in volume fraction of deformation-induced α'-martensite as a function of thickness reduction by cold-rolling in an AISI304 steel; (b) changes in volume fraction of athermal α'-martensite upon cryogenic cooling [215]

the inhibition of α' formation for high contents [74]. *Masumura* et al. determined the evolution of strain-induced martensite with the cold-rolled reduction in three steels: (i) AISI 304 alloyed with 0.1 wt.-% C, (ii) AISI 304 alloyed with 0.1 wt.-% N and (iii) non-alloyed AISI 304, concluding that the mechanical stability of an AISI 304 carbon-added steel is slightly higher than that of nitrogen-added steel but only for large reductions (Fig. 11.12a), while they observed that the thermal stabilization effect of carbon is much weaker than that of nitrogen. In the 0.1 wt.-% N steel, no athermal martensite was detected during down to 4 K, while for the 0.1 wt.-% C steels, the M_s was just below 200 K (base material around 300 K) (Fig. 11.12b) [215, 216]. These results highlight that the effect of interstitials on the mechanical and thermal stabilities might have opposite effects. Besides the stacking fault energy, other parameters used in the literature to estimate the mechanical stability of the austenite are the M_{d30} and $M_{d30/50}$, the temperature at which 50 % of martensite will form at 30 % true strain. The expression mostly used in the literature was proposed by *Nohara* et al. (Eq. (11.4)) [219], which modified the one initially proposed by Angel [56] to include the effect of the austenite grain size and the one proposed by *Sjoberg* (11.18)) [95]. In this latter one, contrary to the one proposed by *Nohara* et al., it was suggested that N has a more pronounced effect on the mechanical stability of austenite. Both equations seem to be in contradiction with the work of *Masumura* et al. discussed above [215], which found C to have a slightly greater influence.

$$M_{d30}\left(^{\circ}C\right) = 608 - 13\,(\%Cr) - 34\,(\%Ni) - 12\,(\%Mn) - 6.5\,(\%Mo)$$
$$- 7.8\,(\%Si) - 515\,(\%C) - 821\,(\%N) \tag{11.21}$$

Alloying ASS with C and/or N above 0.1 wt.-% has been also investigated from the point of view of its influence on precipitation of second phases and mechanical properties. A significant number of papers and reviews that have investigated the formation of carbides ($M_{23}C_6$, MC) and intermetallic phases (σ, χ, Laves) at high temperatures in Fe-Cr-Ni alloys showed that their formation generally deteriorate the mechanical and creep properties and, thus, the service life of components used in energy-generating power plants [22, 108, 220]. Since C promotes the formation of $M_{23}C_6$ which promotes intragranular corrosion and cavitation during creep, N has been suggested as a preferred alternative because it improves the corrosion resistance by delaying the nucleation of this phase as well as χ-phase and shifts the nucleation of the Laves phase to higher temperatures. However, older reports have also shown that the formation of Cr_2N during ageing at high temperatures may cause sensitization and fragilization [221]. *Kim* et al. investigated how the precipitation of $M_{23}C_6$ and Cr_2N during ageing at 800 °C determines the mechanical response of a Fe-15Cr-15Mn-4Ni (wt.-%), alloyed with different amounts of N and/or C [217]. Figure 11.13 depicts the size and density of precipitates as a function of the ageing time at 800 °C for three steels alloyed with (i) C2, 0.2 wt.-%C; (ii) N2, 0.2 wt.-%N; and (iii) CN20, 0.2 wt.-%C + 0.2 wt.-%N. Precipitates $M_{23}C_6$ and Cr_2N appear in the carbon alloyed and N-alloyed steels, respectively; however, the density of Cr_2N is much lower than that of $M_{23}C_6$. In CN20, the presence of N delays the formation of intragranular $M_{23}C_6$ precipitates; in this steel, the amount of Cr_2N is much lower than for N2 and has not been presented by the authors.

Figure 11.14 shows the evolution of yield strength (YS), ultimate tensile strength (UTS) and uniform elongation (UE) as a function of the ageing time at 800 °C for the same three steels. In as-received condition, the addition of N in steel N2, compared to the addition of carbon in C2, enhances the YS, while similar values are obtained for the UTS and UE. The synergistic effect of combining C and N (steel CN20) has a profound effect on raising the YS/UTS values, while the UE is not affected. The results show that the mechanical properties of N-alloyed steel are not affected by ageing; which means that the formation of chromium carbides have very little influence of the mechanical properties, likely due to the low density of these precipitates. While they are significantly modified in C-alloyed steels (the formation of carbides increases the UTS but deteriorates the UE). EBSD results carried out by *Kim* et al. [217] on deformed microstructures of the three steels show that mechanical twins, α' and ε martensites, are observed in all samples. Ageing affects more profoundly C2 steel microstructure in which the volume fraction of both martensites increases and mechanical twins decreases. This is due to the impoverishment of the matrix in carbon content (austenite stability decreases). Ageing does not affect the deformation microstructures of N2 steel. The amount of martensite and mechanical twins is lower in CN20 due to the higher amount of interstitials compared to the other two steels. As carbide precipitation kinetics is slower in this steel compared to C2, the effect of ageing on the deformed microstructures is less clear.

A major concern by the industry in the last decades has been the replacement of Ni in stainless steels to reduce the impact of nickel price fluctuations and produce more cost-effective alloys. In this regard, the research initiated in the 1930s led to the development of the 200 series ASS [53]. In these Fe-Cr alloys (Cr: 15–22 wt.-%), Ni (15–7.0 wt.-%) has been partially replaced with Mn (4–16 wt.-%), N (0.1–0.4 wt.-%) and, sometimes, Cu (1–4 wt.-%), like in 201LN, 203, 204 Cu

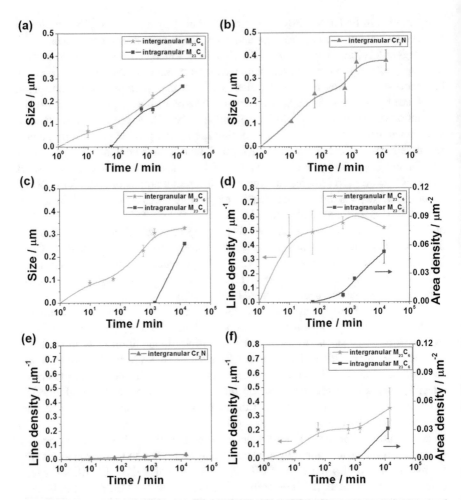

Fig. 11.13 Size of the observed precipitates ($M_{23}C_6$ and Cr_2N) during ageing at 800 °C in steel Fe-15Cr-15Mn-4Ni (wt.-%) alloyed with different amounts of C and/or N: (**a**) C2 (0.2 wt.-%), (**b**) N2 (0.2 wt.-%) and (**c**) CN20 (0.2 wt.-% of each element) [23], and density of the observed precipitates in (**d**) C2, (**e**) N2 and (**f**) CN20 [217]

[222]. In these steels, the nitrogen content is always below 0.4 wt.-%, which is the maximum amount that can be introduced by conventional steelmaking processes. Copper increases the stacking fault energy and, thus, the mechanical stability of austenite favouring twinning instead of α' martensite formation [35]. Besides, it has been observed that the addition of Cu to an AISI 304 steel hinders the formation of ε martensite, possibly due to the increase in the stacking fault energy [223]. In Fe-Cr-Mn-Ni-N steels, the combined addition of Cu and Ni seems to avoid

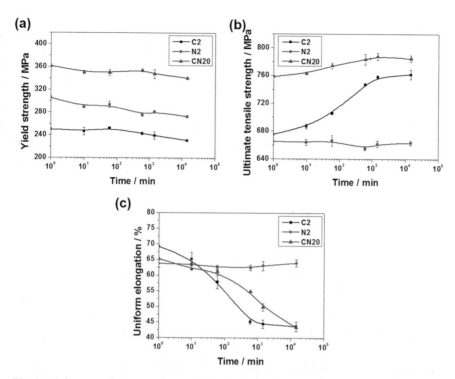

Fig. 11.14 Evolution of (**a**) yield strength, (**b**) ultimate tensile strength and (**c**) uniform elongation of steel Fe-15Cr-15Mn-4Ni alloyed with different amounts of C and/or N: C2 (0.2 wt.-%), N2 (0.2 wt.-%) and CN20 (0.2 wt.-% of each element) after ageing at 800 °C [23, 217]

the formation of δ-ferrite and decreases the ductile-brittle transition temperature (DBTT) by enhancing the stability of austenite and suppressing the formation of strain-induced martensite [224].

In addition to the existing commercial stainless steels of the 200 series, the increasing demands to develop the third generation of advanced high-strength steels, to meet the requirements of lightweight automobile body in the automotive industry [156, 225, 226], have placed its focus on the design of new TRIP/TWIP austenitic stainless steels due to their good combination of strength and ductility obtained by the formation of ε/α' martensite and/or twins as it has been discussed previously. By combining the formation of ε/α' martensite (phase transformation hardening, TRIP) and twins (deformation twinning, TWIP), the aim is to obtain a high-strength/ductility (TRIP) and high-formability (TWIP) steels [227, 228]. These steels could be also of interest in sea water systems, chemical and nuclear industries, power generation and ballistic applications as reviewed by Lo et al. [22] or biomedical applications [229]. Extensive recent work can be found in the literature concerning the replacement of Ni by Mn/N to produce novel Fe-Cr-Mn-Ni-N TRIP/TWIP ASS [31, 32, 202, 206, 230, 231].

Biermann et al. [231] investigated the influence of varying the Ni content from 3 to 9 wt.-% in a Fe-16Cr-6Mn-Ni (wt.-%) steel and concluded that the Ni

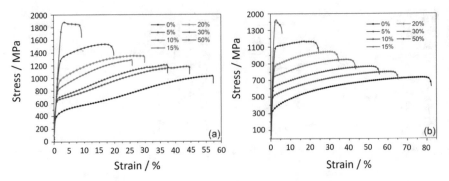

Fig. 11.15 Engineering stress-strain curves of (**a**) 14Cr-10Mn steel and (**b**) 16Cr-6Mn steel at different cold-rolling reductions [32]

content increases the stacking fault energy and the mechanical stability of austenite, resulting in a reduction in the amount of α' martensite formed in the microstructure. Increasing the temperature to 100 °C had a similar effect to increasing the amount of Ni on reducing the amount of α'. The influence of Ni content and temperature on the formation of ε martensite did not follow a clear trend. *Zhang* et al. [32] investigated the mechanical properties of two experimental alloys with different Cr, Mn and Ni contents (Fe-14Cr-10Mn-1Ni and Fe-16Cr-6Mn-4Ni, wt.-%) after different cold-rolling reductions (0, 5, 10, 15, 20, 30, 50 %) at room temperature (Fig. 11.15). The balance between strength and ductility could be greatly modified depending on the degree of deformation introduced by cold-rolling. When comparing both steels, the lower mechanical stability of alloy 14Cr-10Mn compared to 16Cr-6Mn produced significant amounts of ε/α' martensite for low cold reductions, which contributed to increasing the YS and UTS and to decreasing the UE.

Figure 11.16a shows the engineering stress-strain curves of tensile tests conducted between −196 and 400 °C till fracture in a Fe-19Cr-3Mn-4Ni (wt.-%) stainless steel [230]. The highest elongation was reached around the M_d temperature, 60 °C (73 %), which the authors defined as the temperature above which no martensite forms upon loading. The phase fraction of martensite after tensile deformation until fracture at different temperatures is shown in Fig. 11.16b. The steel does not form as-quenched martensite at temperatures as low as −196 °C. Therefore, the martensite contents in this figure represent the deformation-induced martensite (α') only. The highest martensite content of about 78 vol.-% was formed at −80 °C, temperature at which the UTS is the highest. By comparing Fig. 11.16a, b, it becomes evident that the UTS is closely related to the amount of α' transformed in the microstructure during deformation.

Martin et al. [202] investigated the sequence of transformation during deformation in a Fe-15.5Cr-6Mn-6Ni (wt.-%) alloy between −60 °C and 200 °C. At low temperatures only α' martensite was observed. As the temperature was increased, the transformation took place via ε martensite ($\gamma \rightarrow \varepsilon \rightarrow \alpha'$) in deformation bands. However, at 100 °C and 200 °C, the predominant mechanisms were deformation

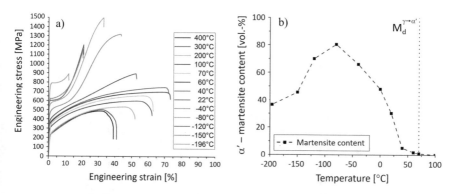

Fig. 11.16 (**a**) Engineering stress-strain curves of tensile specimens of Fe-19Cr-3Mn-4Ni (wt.-%) steel in the temperature range of −196 °C to 400 °C; (**b**) temperature dependence of deformation-induced martensite fraction after tensile tests until fracture [230]

twinning and dislocation glide. Figure 11.17 shows the temperature evolution of (a) the YS, UTS and UE and (b) amount of α' (athermal and strain-induced). The strength mainly increases as the amount of martensite increases, although for fully austenitic microstructure, the strength also decreases slightly as temperature increases. On the other hand, the elongation decreases abruptly as the martensite volume fraction increases and, for fully austenitic microstructures, as temperature increases. These authors proposed a scheme showing the temperature evolution of the transformation mechanism and stacking fault energy for the investigated steel (Fig. 11.18). The stacking fault energy at high temperatures was calculated assuming a temperature increase of 0.075 mJ K^{-1} m^2. The quantitative evolution of the UTS (dash line) and UE (solid line) is also shown as a function of the temperature. This scheme describes the same trends experimentally observed by many authors as discussed so far.

Other important parameter that influences the deformation mechanism and, thus, the martensitic transformation is the grain size, which can be modified through thermomechanical treatments [182]. The grain size plays an important role also on the nucleation of martensite and on the rate of the $\gamma \rightarrow \alpha'$ transformation which directly affects the mechanical properties of stainless steels as it has been discussed so far. *Tomimura* et al. [232], among other aspects, studied the influence of the grain size on parameters like the M_s temperature (Fig. 11.19a) or the athermal martensite α' content (Fig. 11.19b) in two Fe-Cr-Ni alloys, finding that the finer the grain size, the lower the Ms and the more stable the austenite were.

With the aim of unveiling the influence of grain size on the TRIP effect, *Kisko* et al. obtained different austenite grain sizes (0.5, 1.5, 4 and 18 μm) by reversion annealing of cold-rolled microstructures in a commercial Fe-Cr-Mn-Ni alloy 204Cu. They found that during tensile loading, at different strains (2, 10 and 20 %), martensite nucleation sites where influenced by the grain size [222]: (i) for the lowest ultrafine grain size (0.5 μm), martensite nucleated at grain boundaries and twins; (ii) while for coarser grain sizes, martensite nucleated at shear bands or ε

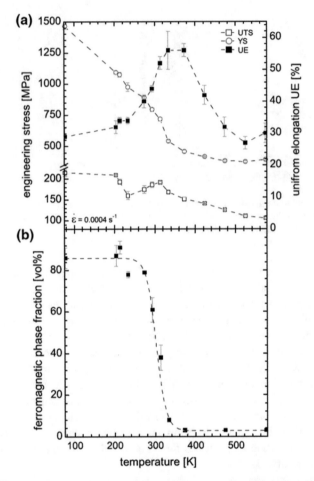

Fig. 11.17 Parameters obtained from the tensile tests performed at different testing temperatures for the TRIP/TWIP steel, Fe-15.5Cr-6Mn-6Ni (wt.-%) alloy: (**a**) YS, yield strength; UTS, ultimate tensile strength; UE, uniform elongation; (**b**) ferromagnetic phase fraction after testing until UE [202]

martensite. The rate of the transformation was also affected by the grain size; it decreased as the grain size dropped from 18 to 1.5 μm but then it was the fastest for the ultrafine grain (0.5 μm). In the same sense, *Xu* et al. investigated two austenitic microstructures of AISI 316LN with different grain sizes (2 and 12 μm) [233], the finest obtained by reversion annealing of cold-rolled microstructures. The refinement of the microstructure stabilized it against the γ → α′ transformation which was only observed for the coarser grain size (12 μm), while twinning was the operating deformation mechanism in the finer grain size (2 μm). *Challa* et al. [234] observed the same change in the deformation mechanism from γ → α′ to

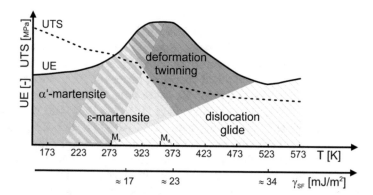

Fig. 11.18 Schematic overview of the activated deformation mechanisms as a function of testing temperature in correlation with the SFE and the mechanical properties of the TRIP/TWIP steel, Fe-15.5Cr-6Mn-6Ni (wt.-%) alloy [202]

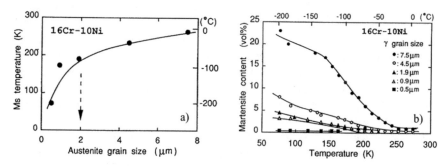

Fig. 11.19 (a) Effect of austenite grain size on the M_s temperature in a Fe-16Cr-Ni (wt.-%) alloy; (b) changes in martensite content during cooling from room temperature showing grain size dependence of austenite stability at low temperatures [232]

$\gamma \rightarrow \gamma_{TWIN}$ in AISI 301LN austenitic microstructures with different grain sizes from 22 μm to 225 nm.

The use of reversion annealing of deformed microstructures to obtained ultrafine grains and its influence on phase transformations and mechanical properties will be discussed with more detailed in the following section.

11.4.2 Nanostructured and Ultrafine-Grained Stainless Steels

A disadvantage of ASS is their relatively low YS, which limits their wide usage in structural applications, is the reason behind some commercial grades having to be hardened by cold-rolling [235]. As discussed in the previous section, other methodologies focus on adjusting the chemical composition by, for example, increasing the N content, which has beneficial effects on the YS and work-hardening behaviour [74, 217]. In addition, one of the strategies that have received extensive attention in recent years to obtain a good balance between strength and ductility [182] and fatigue resistance [236, 237] in ASS consists in the grain refinement of the grain size down to nano or ultrafine range. Several processing methodologies have been employed to refine the microstructure, which generally consists in applying deformation followed by a controlled annealing treatment. This deformation has been introduced in a number of ways: (i) cold-rolling at room temperature [57, 193, 233, 234, 236, 238–251]; (ii) cryo-rolling [181, 252–255]; (iii) warm-rolling (400 °C) [256]; (iv) HPT [186, 187, 256] or ECAP [184, 185, 237, 257]; (v) MDF [188–191]; (vi) hydrostatic extrusion [192]; and (vii) shot-peening [258, 259]. Severe plastic deformation SPD processes are still far from being applicable in large-scale production and shot-peening only influences a few microns below the steel surface; thus, the most widely used methodology consist in the reversion of microstructures containing high-volume fractions of deformed strain-induced α' martensite induced by prior one-step cold-, cryo- or warm-rolling. After the application of the deformation, the steel is annealed up to temperatures above the A_S temperature, at which the $\alpha' \rightarrow \gamma$ is stimulated. Most of these investigations have focused on commercial steel grades like 304, 316, 201 and 301 rather than on new experimental steel grades in which the steel chemistry has been modified. The application of consecutive cycles of cold-rolling + annealing has also given mixed nano/ultrafine-grained microstructures [260]. This strategy has also been used in duplex stainless steels with good strength-ductility balance and corrosion resistance [261]. Besides the composition of the ASS, other factors that determined the kinetics of the $\alpha' \rightarrow \gamma$ transformation and austenite grain size obtained after reversion annealing are (i) the amount of deformation (cold-rolled reduction), (ii) volume fraction of strain-induced martensite introduced by deformation, (iii) the heating rate, (iv) annealing temperature and (v) time [193, 232, 248, 262–264]. *Tomimura* et al. determined the amount of α' martensite induced by 90 % cold-rolling as a function of $Ni_{eq} = (Ni + 0.35Cr)$ in Cr-Ni stainless steels (Fig. 11.20a) and the evolution of the grain size with the annealing temperature (1123–1373 °C)

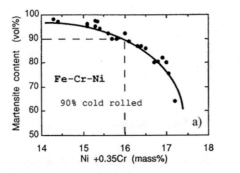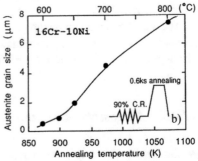

Fig. 11.20 (**a**) Relation between Ni equivalent (Ni + 0.35Cr) and the amount of martensite induced by 90 % cold-rolling in Cr-Ni stainless steel; (**b**) relation between temperature and grain size of austenite reversed from strain-induced martensite α' in a Fe-16Cr-10Ni (wt.-%) steel [232]

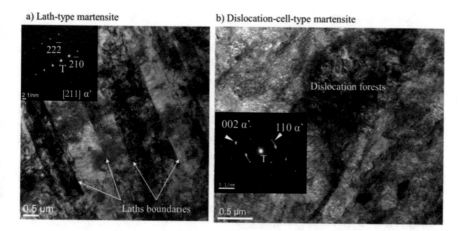

Fig. 11.21 Bright field (BF)-TEM micrographs and corresponding SADPs illustrating different morphologies of the martensite phase of the as-received state (92.5 % cold-reduced): (**a**) lath-type martensite, (**b**) dislocation cell-type martensite in a Fe-12Cr-9Ni-4Mo-2Cu (wt.-%) ASS [263]

in a Fe-16Cr-10Ni (wt.-%) steel (Fig. 11.20b) [232]. The fit to the experimental values in Fig. 11.20a can be described by the Eq. (11.22):

$$V_{\alpha'} \text{ (vol.-\%)} = 100 - 2.5 \times (\text{Ni} + 0.35 \times \text{Cr} - 14)^2 \qquad (11.22)$$

This study (Fig. 11.20a) predicts that 90 % of cold-rolling will transform more than 90 % of the microstructure to α' martensite for $\text{Ni}_{eq} \leq 16$. The retained austenite present in the microstructure after cold-rolling, prior to the annealing treatment, will result in coarser grains after the reversion treatment that would not contribute to enhancing the strength [232, 246]; for this reason, it is advisable to form as much α' martensite as possible during the TMCP step.

Depending on the alloy composition and thus the stability of the austenite, different studies have shown that different cold-rolling reductions are necessary to obtain a $f_{\alpha'} = 0.9$–1.0: AISI 304, 80 % [244]; AISI 301, 55 % [264]; and AISI 304 L, 50 % [177]. If the amount of deformation is high enough, the strain-induced martensitic microstructure develops from lath-like martensite to dislocation cell-type martensite in which the deformation breaks the lath microstructure [178]. Figure 11.21 shows examples of these two microstructures in a cold-rolled Fe-12Cr-9Ni-4Mo-2Cu (wt.-%) metastable ASS [263]. It is clear that in one case the lath microstructure remains visible, while for the other the lath-like microstructure has been broken by a mess of dislocations. The selective area diffraction pattern (SADP) also helps to identify both microstructures: (i) lath-like shows well-defined diffraction spots (Fig. 11.21a) and (ii) dislocation-type shows ring-like diffraction pattern (Fig. 11.21b). The dislocation cell-type microstructure contains a higher number density of nucleation sites for austenite to form during the annealing process, leading to a finer grain size and to higher strength [178]. A recent investigation has shown that if the two types of microstructures coexist, a bimodal austenite grain size distribution will be obtained, leading to a lower strength but to an improvement in ductility compared to those where only dislocation-type martensite is present [265]. Two additional concerns should be borne in mind: (i) avoiding grain coarsening and (ii) formation of carbides or intermetallic [263, 266]. If grains grow too much, the austenite will be less stable; as a result athermal α' martensite may be formed upon cooling to room temperature [247], deteriorating the toughness and ductility of the steel. The precipitation of carbides or intermetallic phases at grain boundaries could lead to intergranular corrosion of the steel. In summary, it is preferable that the annealing holding times are short; and the shorter the time, the higher the annealing temperature but keeping in mind that a fully $\alpha' \rightarrow \gamma$ transformation would be desirable [177].

In summary, the following conditions should be met to obtain a fully ultra-fine/nanometre austenitic microstructure:

1. A 90–100 % of the microstructure should be transformed to α' martensite during the TMCP treatment.
2. A dislocation-type microstructure is preferred compared to a lath-type to maximize the number of number of nucleation sites for the $\alpha' \rightarrow \gamma$ phase transformation.
3. The higher the heating rates, the finer the grain size for a given austenitization temperature above A_S temperature.
4. Low annealing temperatures and short holding times to avoid: (i) excessive grain growth that would lower the strength, decrease the austenite stability and promote the formation of athermal α' martensite upon cooling to room temperature and (ii) formation of intermetallics/carbides that could deteriorate the corrosion resistance.
5. Holding times should be long enough to complete the $\alpha' \rightarrow \gamma$ phase transformation.

Focusing on the mechanical properties of ultrafine microstructures, *Zhao and Jiang* [182] in a recent comprehensive review compiled information from several investigations concerning the austenite grain size and mechanical properties of some commercial and experimental ultrafine-grained and nanostructured ASS obtained by the application of deformation followed by annealing. This information, provided in Table 11.8, has been completed with other publications. In many of these publications, mechanical properties have been obtained for several processing conditions and ultrafine grain sizes; in these cases, only some significant results that show good combination of strength/ductility synergy have been provided.

In a similar sense, Fig. 11.22 provides mechanical property data compiled by *Hamada* et al. for different alloys [246]. The target strength-ductility range (the orange band) set by the European Commission for automotive steels is shown in this figure [270]. Besides, this figure also shows (full black squares) the data obtained for the commercial AISI 201 after various deformation followed by annealing treatments [271] along with information for other types of steels used nowadays in the automotive industry, as well as mechanical property data for Al and Mg

Table 11.8 Results reported in the literature for different ultrafine and nanostructured ASS

ASS	CR reduction (%)	Annealing T (°C)	Holding time (s)	Grain Size (μm)	YS (MPa)	Elongation (%)	Year	Refs.
16Cr-10Ni	90	600	600	0.5	700	–	1991	[232]
301LN	63	800	1	0.54	700	35	2007	[267]
301LN	62	800	1	0.7	711	37	2009	[264]
304L	90	700	18,000	0.33	1000	40	2010	[177]
301LN	77	800	1	0.6	880	44	2010	[178]
201L	95	850	30	0.06	1300	33	2011	[249]
201L	95	750	180	0.12	1490	17	2011	[249]
18Cr-12Mn-0.25N	80	900	100	0.24	1150	21	2014	[268]
201Nb	90	900	60	0.093	1000	35	2015	[247]
304	85	580	1800	0.08	1120	12	2015	[260]
201	60	800	10	1.5	800	50	2015	[246]
304L	65	850	60	0.62	855	44	2015	[248]
15Cr-9Mn-0.11Nb	60	800	10	0.6	813	51	2016	[269]
16Cr-10Ni	74	850	100	0.25	585	35	2016	[243]
15Cr-9Mn	60	800	10	1.2	542	63	2016	[269]
316LN	90	900	120	2	994	40	2017	[233]
Fe-14Cr-10Mn	30				1191	20	2018	[32]
Fe-16Cr-6Mn	30				1038	24	2018	[32]

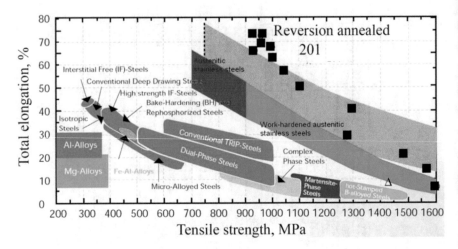

Fig. 11.22 Comparison of the mechanical properties between the reversion annealed type 201 austenitic stainless steel and other conventional steel grades. The European Steel Technology Platform (ESTEP) maps new generation of high-strength steels for lightweight construction [272]. Comparison of the mechanical properties (tensile strength and elongation) between the reversion annealed type 201 and some other current steel grades. (The data for work-hardened grades are from Refs. [246, 270])

alloys. These results unveil that the tensile strength-ductility balance achieved after the application of this thermomechanical treatments are excellent and exceed those of conventional annealed or work-hardened stainless steels (red and pink bands).

Other novel approach for strengthening ASS is to promote the formation of multiple nanoscale twins within some austenite grains (nanotwinned austenitic grains, NTW-γ) using dynamic plastic deformation [273, 274]. In this way, single (austenitic) phase composite steels containing twinned and non-twinned grains would be generated. No new phase boundaries are created. The methodologies described before in this section, which lead to the refinement of the microstructure down to the nanometre scale under certain processing conditions, enhance the strength but reduce the ductility and work-hardening capability, which is detrimental to the stability of the plastic flow required for forming operations. The idea behind using this nanotwinned grains as strengthening sources comes from the discovery that submicron-grained Cu alloys containing nanoscale twins exhibit a very high tensile strength and a considerable ductility and work-hardening capability [275, 276]. It is known that twin boundaries block dislocation motion and act as slip planes to accommodate dislocations [277]; thus, steels with such defects in the microstructure would result in strong, ductile materials with the capacity to work-harden.

As it has been discussed already in this section, twins in ASS can be promoted by plastic deformation when twinning conditions prevail over other deformation mechanisms. In these steels twinning is favoured for high stacking fault energies. The refinement of the microstructure also leads to a change in the deformation mechanism under applied strains, from $\gamma \to \alpha'$ to $\gamma \to \gamma_{TWIN}$, as it was discussed previously. In addition, *Li* et al. have demonstrated that twinning is also stimulated when high strain rates are employed (dynamic plastic deformation) in Cu alloys [278]. After deformation, the microstructure has to be annealed to recover the microstructure. In this way, coarse grains are generated, but the twinned regions remain unchanged because twin boundaries possess a much lower energy than conventional grain boundaries. Thus, a composite alloy containing fine twinned and coarser non-twinned grains would be generated. The nanotwinned regions can be regarded as grains containing nanoscale twins, which are much stronger than the surrounding recrystallized coarser grains. The quantity and distribution of these NTW-γ are controlled by the deformation parameters as well as the subsequent annealing conditions [274].

The seminal work carried out in Cu alloy has been applied successfully to ASS. Several reports can be found in the literature by the same group of researchers concerning the formation of NTW-γ in metastable commercial ASS: AISI 304 [279, 280] and AISI 316 L [274, 279, 281, 282]. *Yi* et al. [279, 280] have obtained a nanotwinned reinforced austenitic microstructure by applying dynamic plastic deformation following the procedure described in Ref. [278] at $-196\,^\circ\text{C}$ and $150\,^\circ\text{C}$ and using different strains. The deformation was applied by compression of cylindrical samples using strain rates of $10^2-10^3\ \text{s}^{-1}$ in different steps. After a deformation of $\varepsilon_T = 0.3$, the samples deformed at $-196\,^\circ\text{C}$ contained a volume fraction of α' and ε martensites of 0.8 and 0.07, respectively; the remaining microstructure was retained austenite. A subsequent annealing treatment at $650\,^\circ\text{C}$ for 1 h reduced the volume fraction of α' martensite to 0.17, whereas the microstructure deformed at $150\,^\circ\text{C}$ ($\varepsilon_T = 1.0$) contained only NTW-γ grains (58 %) within a very dislocated austenitic matrix (tangles, walls and cells). Twins are very fine, with average lateral dimensions of 10–20 nm (different values have been reported by the authors for the same microstructure). The annealing treatment at $650\,^\circ\text{C}$ (1 h) reduced the amount of nanotwins to 50 %, and the microstructure contained some recrystallized areas (10 %) with average sizes of 2.2 µm. Figure 11.23 shows SEM-ECC (SEM-equal channel contrast) (a) and TEM (b) images of the microstructure after annealing.

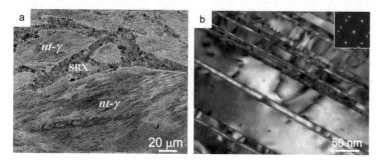

Fig. 11.23 Typical microstructures of the annealed nanotwinned austenite 304 SS at 650 °C for 1 h: (**a**) SEM-ECC image and (**b**) TEM image of the nanotwins with SAED pattern (inset) [280]

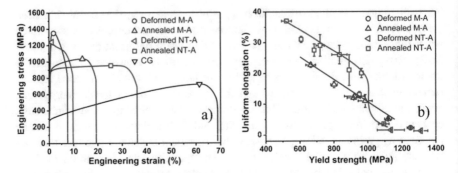

Fig. 11.24 (a) Tensile engineering stress-strain curves of the as-deformed and annealed nanotwinned austenite 304 SS in comparison with that of the as-deformed and annealed martensite-austenite 304 SS. The as-received sample is also included for comparison; (b) strength-ductility correlations obtained for the deformed and different annealing treatments (M-A, martensite austenite 304 SS; NT-A, nanotwinned austenite 304 SS) [280]

Figure 11.24a shows the mechanical behaviour of the samples after deformation at −196 °C (M-A) and 150 °C (NT-A) and subsequent annealing treatment. The deformed M-A sample has a YS = 1250 MPa, 100 MPa higher than the NT-A sample (1135 MPa). After annealing both samples have similar YS = 900 MPa, but the NT-A samples has a much larger uniform elongation (21 %) than the M-A sample (9 %). The work-hardening behaviour in the NT-A sample is attributed to the nanotwinned grains due to the limited work-hardening ability of ultrafine dislocation structures [283]. Figure 11.24b also provides different strength-ductility combinations obtained for M-A and NT-A samples in deformed and annealed states showing that better combination of mechanical properties can be obtained in NT-A samples due to the higher thermal stability of nanotwinned grains compared to α' martensite during the annealing treatments.

In ultrafine 304 ASS, the nucleation of α' martensite is inhibited or diminished due to the lack of nucleation sites such as ε martensite or twin boundaries, which reduces the uniform elongation. In contrast, in microstructures with NTW-γ the formation of α' martensite and stacking faults can be stimulated by the presence of twin boundaries as it has been observed [279] which would enhance the ductility. When 304 and 316 L alloys are compared, the much higher mechanical stability of 316 L (larger γ_{SFE}) impedes the formation of stacking faults and α' martensite even in microstructures containing NTW-γ, which reduces significantly the uniform elongation compared to 304 ASS [279].

11.4.3 Maraging/PH Stainless Steels

Commercial 18 wt.-%Ni maraging steels (200, 250, 300, 350) were developed in the 1960s for applications requiring ultrahigh strength and good fracture toughness but without paying much attention into the corrosion resistance, as they do not contain chromium in the composition [284]. In this type of alloys, a low-carbon soft martensitic matrix is strengthened during ageing by intermetallic nanoparticles. Along with the grain refinement, strengthening through coherent nanoprecipitation is the most effective way of increasing the strength of steels as it has been highlighted in a recent review [285], though this report was not focused on stainless steels.

The use of the terminology maraging and precipitation hardening in the literature can be found randomly even referring to the same experimental alloy in different publications. Sometimes the use of the name maraging has been avoided because it possesses a pejorative connotation as some of these alloys were conceived for military applications. It has been sometimes argued that maraging steels and PHSS differ mainly in the carbon content, which is usually an order of magnitude lower for maraging steels. In practice any steel that possesses a martensitic matrix strengthened by precipitates, including commercial PHSS, has been defined as maraging [285, 286]. Thus, due its low-carbon content, the strengthening of the martensitic matrix in the latter steels is achieved by promoting the precipitation of copper clusters/precipitates (bcc $\rightarrow$ 9R $\rightarrow$ fcc) [36] and/or Ni-rich and Mo-rich intermetallic phases: η-Ni_3(Ti,Al), B2-NiTi, Fe_7Mo_6 μ-phase, Laves (Fe_2Mo) and $Ti_6Si_7Ni_{16}$ G-phase [102] instead of carbides. Ni-rich precipitates along with Cu-rich phases have been referred to as the main sources for hardening in maraging steels during the first hours of ageing [162], while Mo-rich phases nucleate and contribute to the hardening after long ageing conditions due to the low diffusivity of Mo [287]. Several investigations have demonstrated the positive effect of Cu-alloying in contributing to the hardening during the initial stages of ageing of stainless steels or in accelerating the precipitation of Ni-rich intermetallic nanoprecipitates [285]. In this regard, this element has been used in commercial PH15–5 [36] and PH17–5 stainless steels along with other recently developed ones like NanoflexTM (1RK91) [28, 37], Fe-Cr-Ni-Ti-Al [288], Fe-Cr-Ni-Co-Cu [98] and Fe-Cr-Co-Ni-Mo-Ti alloys [289].

In recent years, efforts have been made to create new maraging stainless steels strengthened by multiple precipitates [52, 98, 103, 289, 290] as a way to improve the precipitation hardening contribution to the total strength and sustain also the hardening effect for long ageing times. Besides, atom probe tomography (APT) and high-resolution transmission electron microscopy (HRTEM) [52, 285, 291] have allowed studying precipitation sequences with a great detail and thus understanding the link between the chemical composition, the ageing conditions and the mechanical properties, among others. The development of corrosion resistance ultrahigh-strength (UHS) steels is of great importance in the automotive, aerospace, biomedical, nuclear, gear, bearing and other industries as they could be future key materials for lightweight engineering design strategies and associated savings in CO_2 emissions.

Researches from Sandvik Materials Technology have developed one of the most interesting maraging stainless steels in the last couple of decades. This alloy, termed NanoflexTM 1RK91, was conceived initially for surgical applications [292, 293]. Investigations around this steel have shown that (i) the pitting corrosion of this alloy is better than that of AISI 304 or 316; (ii) tensile strengths of up to 3 GPa

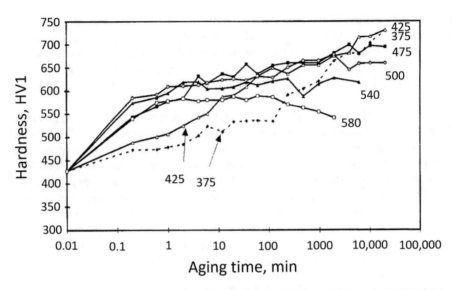

Fig. 11.25 Time evolution of the microhardness at different ageing temperatures in 1RK91 [292]

can be achieved in cold-drawn and aged wires; and (iii) continuous hardening up to 1000 h for temperatures below 500 °C and peak microhardness Vickers values above 700 HV not reported before in similar alloys have been achieved (Fig. 11.25) [28, 292, 294]. This unusually high hardness and strength was initially attributed to the presence of quasicrystalline R' phase precipitates [293, 294]. However, subsequent reports have unveiled that different precipitate species are involved and that the sequence of precipitation reactions is very complex. In fact, these involve initially the formation of copper precipitates/clusters, on top of which $Ni_3(Ti,Al)$ and $Ni_3(Ti,Al,Si)$ nucleate. With increasing ageing times above 40–100 h, Cr-rich α' phase, G-phase $Ti_6Si_7Ni_{16}$ and Mo-rich precipitates (quasicrystalline R'-phase, R-phase and Laves: Fe_2Mo) have been observed [37, 52, 293]. Another interesting feature of this alloy is that martensite is obtained (i) isothermally, at cryogenic temperatures, having the nose of the C-curve located around −40 °C [77, 295] or (ii) by the application of mechanical stresses/strains [296, 297].

Xu et al. have developed a general computational alloy design approach based on thermodynamic and physical metallurgical principles, which is coupled with a genetic optimization scheme to design new ultrahigh-strength martensitic stainless steels [103, 290]. The design approach has allowed obtaining martensitic stainless steels strengthened via promoting the formation of multiple precipitates: copper precipitates (Cu), carbides (NbTiC) and intermetallic phases (Ni_3Ti). Figure 11.26 shows the microstructural characterization undertaken using SEM-EDS (**a**) and STEM-EDS (**b–c**) analysis on a novel PHSS Fe-12Cr-4Ni-2Cu-2Co-0.5Ti-0.11Nb (wt.-%). Very few reported model alloys have been designed in which both carbides and intermetallic phases have been combined to enhance the hardening effect. The set of mechanical properties (tensile strength, total elongation and hardness) obtained for different ageing conditions is provided in Fig. 11.27 [98].

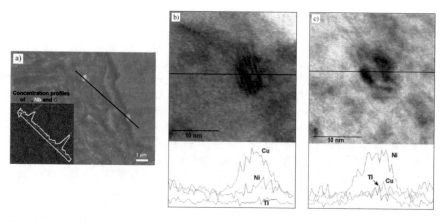

Fig. 11.26 (**a**) SEM micrographs of MC precipitates with EDS line scan results for Ni, Ti and C and (**b–c**) STEM micrograph of Cu- and Ni-rich precipitates with EDS line scan results of concentration profiles of Ni, Ti and Cu (horizontal line indicated) in a novel PHSS Fe-12Cr-4Ni-2Cu-2Co-Ti-Nb (wt.-%) after ageing at 500 °C for 24 h [98]

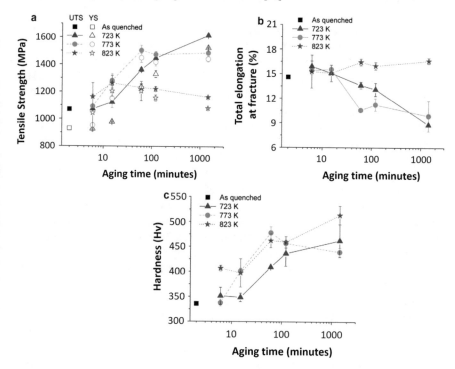

Fig. 11.27 Variations of (**a**) tensile strength (**b**) total elongation at fracture and (**c**) hardness with ageing time at ageing temperatures of 450 °C (723 K), 500 °C (773 K) and 550 °C (823 K), respectively, in a novel PHSS Fe-12Cr-4Ni-2Cu-2Co-Ti-Nb (wt.-%) [98]

Other authors have also contributed in this field. *Akhtar* et al. [298] proposed a set of four experimental maraging stainless steels Fe-12Cr-10Ni-2Mo-xTi-0.3V-0.3Al (wt.-%) with x varying between 1 and 2 wt.-%. The alloy with 1.5 wt.-% Ti to which 0.36 wt.-%Cu was added showed the best combination of mechanical properties (UTS = 1670 MPa; total elongation = 11 %; fracture toughness K_{IC} = 83.9 MPa m$^{1/2}$). The strengthening species were identified as Ti$_x$Ni plus some Al- and V-rich phases that were not clearly identified. The addition of Cu in this alloy was not discussed or did not have a significant effect.

Schnitzer et al. have produced several experimental Fe-Cr-Ni-Mo-Al-Ti (wt.-%) alloys [104, 288, 299]. They investigated the precipitation sequence using atom probe tomography (APT) in a Fe-12Cr-9Ni-2Mo-0.7Al-0.35Ti (wt.-%) alloy and found that during ageing at 525 °C, the hardening increases initially (0.25 h) abruptly due to the formation multicomponent (Ni,Al,Ti) clusters. After 3 h, these evolve towards two independent precipitates (spherical B2-NiAl, elongated η-Ni$_3$(Ti,Al)) that also contribute to the strengthening [104]. The addition of 1.9 wt.-% Cu to a Fe-13Cr-10Ni-1Al-1Mo-0.5Ti (wt.-%) alloy proved to accelerate precipitation reactions; after 0.25 h both intermetallic phases are already present [288]. Cu accelerates the precipitation in different ways: (i) for NiAl, Cu enters into the composition of the phase lowering the activation energy by reducing the lattice misfit; (ii) for Ni$_3$(Ti,Al), Cu precipitates serve as nucleation sites. Figure 11.28a–c shows APT analysis after 0.25 h showing the presence of all these three precipitates: Cu-rich precipitates in orange, NiAl in green and Ni$_3$(Ti,Al) in blue. Figure 11.28(d) provides the one-dimensional concentration profile through the cylinder marked in Fig. 11.28c (Cu and Ni$_3$(Ti,Al) precipitates).

These authors also found that the combined addition of 0.55Si and 0.83Ti (in wt.-%) to a model alloy Fe-13Cr-8Ni-1Al-0.5Mo (wt.-%) promotes the formation of spherical Ti$_6$Si$_7$Ni$_{16}$ precipitates (G-phase), instead of B2-NiAl, along with Ni$_3$(Ti,Al). This change in precipitation sequence reduces the peak hardness compared to a Ti-free PHSS alloy, PH13–8Mo, though the initial hardening after 15 min ageing at 525 °C is faster in the Ti/Si alloyed model steel [299]. After 3 h of ageing, a strong hardening response is due to the formation of NiAlTiSi clusters. The subsequent formation of the intermetallic phases continues to increase the hardening but to a lower extent. The coprecipitation of Ti$_6$Si$_7$Ni$_{16}$ and Ni$_3$(Ti,Al) had been observed before in Fe-Cr-Ni-Al-Ti-Si (wt.-%) alloys with 0.8–1.0 wt.-% Si and 0.9–1.0 wt.-% Ti [300].

An interesting, though very expensive, maraging stainless steel Fe-12Cr-13Co-5Ni-6Mo-0.4Ti-0.1Al (wt.-%) has been designed recently by *Li* et al. [289]. This alloy possesses a tensile strength of 1.9 GPa. Ageing at 500 °C stimulates a sequence of precipitation that includes, at the outset, the formation of (Ni,Ti,Al)-rich, Mo-rich and Cr-rich clusters that developed separately into η-Ni$_3$Ti, Mo-rich R′ and Cr-rich α′ phases. Besides, segregation of R′ and α′ phases to the interface of η-Ni$_3$(Ti,Al) is regarded as the reason for slow coarsening in this alloy, and the strength remains above 1.8 GPa for ageing times up to 100 h. Figure 11.29 shows the evolution of different mechanical properties with the ageing time at 500 °C in this steel.

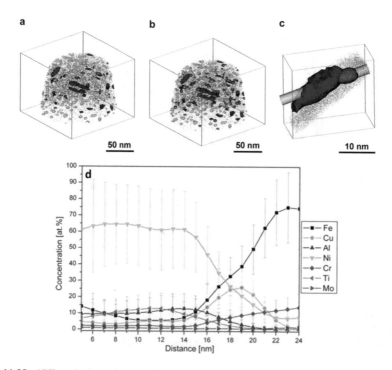

Fig. 11.28 APT analysis carried out in a Fe-13Cr-10Ni-2Cu-1Al-1Mo-0.5Ti (wt.-%) alloy: (**a**) isoconcentration surfaces aged for 0.25 h at 525 °C for regions containing more than 30 at.-% Ni + Al (spherical, green surfaces) and for regions containing more than 5 at.-% Ti (elongated, blue surfaces); (**b**) three types of isosurfaces are depicted: the green, spherical ones correspond to regions containing more than 30 at.-% Ni + Al; the blue, elongated surfaces correspond to precipitates containing more than 5 at.-% Ti; and the orange, spherical areas correspond to regions containing more than 10 at.-% Cu; (**c**) detail of the analyzed volume containing a Ti-enriched and a Cu-enriched area. The cylinder marks the volume evaluated for the one-dimensional concentration profile shown in (**d**) [288]

Table 11.9 summarizes the compositions of several recently developed maraging or precipitation hardening stainless steels reviewed in this section and the precipitating phases that have been observed in the microstructure during ageing.

11.4.4   Quenching and Partitioning (Q&P) Stainless Steels

The quenching and partitioning (Q&P) heat treatment was first proposed by *Speer* et al. [301] and designed to create microstructures composed by martensite and austenite. Since this seminal work in which the Q&P heat treatment concept was proposed, the main driving force to design and produce steel microstructures based on the application of this heat treatment has been the development of the third generation of advanced high-strength steels (AHSS) for the automotive industry [81, 83].

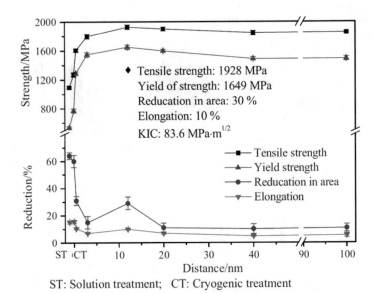

Fig. 11.29 Mechanical properties as a function of ageing time at 500 °C of a maraging stainless steel Fe-12Cr-13Co-5Ni-6Mo-0.4Ti-0.1Al (wt.-%). The UTS can reach 1928 MPa, and the fracture toughness is about 85 MPa m$^{1/2}$ when the ageing time is 12 h [289]

In previous decades, before the proposition of the Q&P concept, several high-strength low-alloyed steels with improved strength-ductility balance have been developed with application in the automotive industry. These steels have evolved with new alloying and processing strategies to tailor microstructures containing a mixture of various phases. Some of these automotive steels that belong to the first generation of low-alloyed advanced high-strength steels (AHSS) are dual-phase steels (DP), martensitic steels, complex phase steels (CP) or transformation-induced plasticity steels (TRIP) [302, 303]. The TRIP steels, which have been briefly introduced before, are based on the creation of a fine dispersion of retained austenite within a ferritic matrix (composed of ferrite and/or bainite and/or martensite). The presence of this retained phase improves toughness and ductility by making use of the so-called TRIP effect that has been already discussed and which involves the transformation of austenite to martensite. The tensile property combinations of the first generation of AHSS and some conventional HSS are shown in Fig. 11.30. These AHSS have already been implemented in existing auto-bodies [302, 304]. The second generation of AHSS, whose properties are also highlighted in this figure, includes highly alloyed steels such as twining-induced plasticity (TWIP), Al-added lightweight steels with induced plasticity (L-IP®), shear band-strengthened steels (SIP steels) or austenitic stainless steels (AUST SS). In this figure there is a clear property gap between both generations of AHSS. In the last 15 years, there has been increasing interest in the development of the "third generation" of AHSS with improved strength-ductility combinations compared to the first-generation AHSS

Table 11.9 Summary of recently developed maraging stainless steels (composition in wt.-%) and the precipitating phases observed in the microstructure during ageing (Fe to balance)

Cr	Ni	Ti	Al	Mo	Mn	Cu	C	Si	Nb	Co	Precipitates	Refs.
12.2	9.0	0.9	0.3	4.0	0.32	2.0	0.01	0.15	–	–	Cu- and (NiTiAlSi)-rich clusters; 9R; η-Ni$_3$(Ti,Al); γ'-Ni$_3$(Ti,Al,Si); R'; α'; Ti$_6$Si$_7$Ni$_{16}$ (G-phase)	[37, 38, 52]
12.0	4.1	0.5	0.03	0.5	0.61	2.2	0.09	0.57	0.11	2.09	Cu precipitates; MC; η-Ni$_3$Ti	[98]
12.1	9.0	0.4	0.7	2.0	–	–	–	0.05	–	–	(Ni,Al,Ti) clusters; B2-NiAl; η-Ni$_3$(Ti,Al)	[104]
12.8	10.3	0.5	1.1	1.3	0.06	1.9	0.01	0.03	–	–	Cu clusters; B2-NiAl; η-Ni$_3$(Ti,Al)	[288]
12.3	8.9	0.8	0.6	1.0	0.11	–	0.01	0.55	–	–	Ti$_6$Si$_7$Ni$_{16}$ (G-phase); η-Ni$_3$(Ti,Al)	[299]
12.3	4.5	0.4	0.09	5.6	–	–	0.001	–	–	13.1	Mo-rich, Cr-rich and (NiTiAl)-rich clusters; R'; α'; η-Ni$_3$(Ti,Al)	[289]

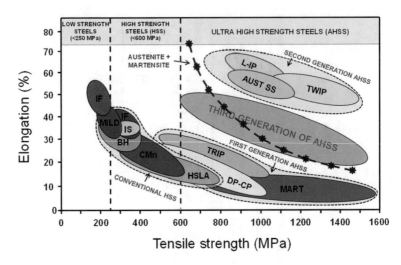

Fig. 11.30 Relationship between ultimate tensile strength and total elongation of conventional high-strength steels (HSS) and advanced high-strength steel (AHSS) families [305]

but at a lower cost than that required for the second-generation AHSS [82, 197]. It has been predicted that microstructures combining a mixture of stable austenite and martensite would generate properties similar to those targeted for this third generation of AHSS, as highlighted by a dashed violet line in Fig. 11.30. In conclusion, the presence of martensite as well as of austenite in the next generation of AHSS should be important constituents of the microstructure. Q&P steels possess a martensitic matrix in which many small retained austenite islands are embedded. Under straining this austenite would transform to martensite making use of the TRIP effect to enhance the ductility and toughness. These types of microstructures would help to fill in the gap.

The Q&P heat treatment is described in Fig. 11.31. After austenitizing, the high-temperature austenitic microstructure is partially transformed to martensite by quenching between M_s and M_f temperatures (start and end of the $\gamma \rightarrow \alpha'$ phase transformation). Subsequently, an isothermal annealing treatment is applied above M_s to promote carbon diffusion/partitioning from martensite to austenite and, thus, stimulate the carbon enrichment of this latter phase. This way, austenite is stabilized and retained after cooling to room temperature. The carbon enrichment leads to the austenite stabilization at room temperature. By optimizing the austenitization temperature, the quenching temperature (QT) and the partitioning temperature and time (PT,t), the volume fraction and stability of retained austenite present at room temperature can be tailored. Although this process might appear as simple, in addition to carbon partitioning, other processes can occur at the same time that may prevent or minimize the amount of carbon enrichment during this process and thus decrease the amount/stability of retained austenite at room temperature, for example, (i) carbon trapping by dislocations and interfaces in the martensite

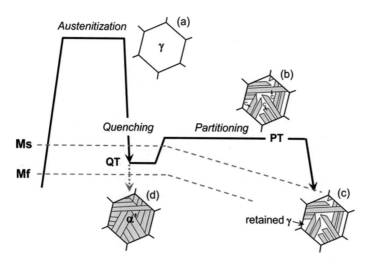

Fig. 11.31 Schematic heat treatment diagram of quenching and partitioning (Q&P) process [309]

phase, (ii) formation of carbides, (iii) decomposition of austenite and (iv) migration of the martensite/austenite interface. Previous research on bainitic and TRIP steels containing retained austenite has noticed that austenite stability is controlled mostly by the carbon content (levels around 1–2 wt.-%) but also by the size and morphology of the retained austenite phase [306]. In addition, alloying elements may also have an influence by preventing the precipitation of carbides (Si, P, Al) or stabilizing the austenite phase (Mn, Ni) [301, 307, 308]. Results obtained in low-alloyed, low-carbon Q&P steels have shown that this new type of steels are stronger but less ductile when compared to TRIP steels due to the generally lower amount of retained austenite [82]. Understanding and controlling the austenite stability by designing the proper composition and/or by optimizing the heat treatment are the key to tailoring the properties of new AHSS.

Compared to nickel-bearing ASS, low-/medium-carbon (<0.6 wt.-% C) chromium-rich (11.5–18 wt.-%) martensitic stainless steels are cheaper to produce and possess higher strength and hardness. These steels are generally supplied in the spheroidized annealed condition, with a microstructure composed by ferrite and carbides. According to several reports, these carbides are generally of the type M_3C, M_7C_3 or $M_{23}C_6$ [310, 311]. This ferritic microstructure is easily cold-worked or machined down to complex shapes. Typical processing routes of martensitic stainless steels involve cold-forming in this soft ferritic condition, followed by an austenitization step (950–1100 °C), air cooling to room temperature and tempering (250–350 °C) to decrease the brittleness of the martensite. During austenitization heat treatment above 950 °C, after forming, the initial ferritic matrix transforms to austenite. Carbides present in the initial microstructure dissolved (partially/completely) and carbon partitions to the austenitic phase formed. The state of carbide dissolution depends on the chemical composition of the steel.

Besides, the higher the carbon content of austenite at high temperatures, the greater are the hardness of martensite upon quenching but also the lower the value of M_s and M_f temperatures. As a result, the amount of retained austenite may increase if Mf temperature is below room temperature [312]. After cooling to room temperature, a final tempering treatment is applied to control the final mechanical properties of this type of steels [313]. A compromise is generally pursued to have high hardness with the minimum amount of retained austenite. In this condition, MSS show an excellent combination of corrosion resistance, high hardness and strength [24], good resistance to thermal and mechanical fatigue and excellent wear resistance. However compared to other low-alloyed martensitic steels, these grades show much poorer ductility and toughness under the same strength level.

In the last few years, some researchers have demonstrated that the Q&P heat treatment can be used in commercial ferritic transformable stainless steels [314], MSS [309, 315–317] and metastable ASS [318] to widen the application of this type of steels by improving the strength-ductility balance. Due to its high strength and corrosion resistance, MSS could be playing a role in the future as a structural component of automotive steels [319]. As it has been discussed above, the Q&P treatment involves quenching below the M_s temperature to transform only part of the austenite to martensite. During the partitioning step, at a temperature generally above the M_s, carbon is allowed to diffuse (partition) from martensite to austenite to enrich this latter phase, stabilize it and retain it upon further cooling to room temperature. However, besides the diffusion process, others may happen, such as precipitation of carbides, which would limit the carbon enrichment of the austenite. In this regard, during this partitioning step at low temperatures, Cr in stainless steels shows a very low diffusivity, and this would prevent the formation of carbides and nitrides, facilitating that interstitials are available for diffusion. In addition, Cr retards the bainitic transformation [320] which hinders the formation of this phase from the retained austenite during the partitioning steel. Finally, because the presence of Cr increases its solubility limit of N [321], this element could be also used as a stabilizing element for austenite in addition to carbon.

Mola et al. have been the first to investigate the Q&P process in stainless steels [314]. They used a commercial ferritic stainless steel grade AISI 430 which can be partially transformed to austenite during the austenitization heat treatment. Upon quenching below the M_s temperature, the microstructure would contain a mixture of ferrite, athermal α' martensite, and, depending on the quenching temperature, some retained austenite. Several Q&P heat treatment parameters were investigated in this work. The austenitization temperature was fixed to 1000 °C, holding for 5 s (a microstructure composed of ferrite and 38 vol.-% of austenite is obtained). It was found that the optimum QT temperature at which the amount of retained austenite in the microstructure was maximized (~9 vol.-%) was around 160 °C. Increasing the PT between 160 and 500 °C stabilized the retained austenite, while at 600 °C, the stabilizing effect decreases. Figure 11.32 shows the mechanical properties obtained from three different specimens of this steel: (i) water quenched from 1000 °C for 5 min (ferrite+martensite); (ii) annealed at 820 °C for 5 h (ferrite+M23C6); (iii)

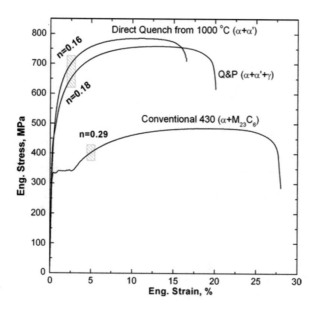

Fig. 11.32 Engineering stress-strain curves for quenched, Q&P-processed and conventionally processed AISI 430. The work-hardening exponents (n) estimated from the tensile tests, for the three different conditions investigated, are provided in the strain range, of the tensile curves, highlighted with shaded rectangular areas: (i) direct quench sample, $n = 0.16$ (for an eng. strain = 2–3 %); (ii) Q&P sample, $n = 0.18$ (for an eng. strain = 2–3 %); (iii) conventional sample, $n = 0.29$ (for an eng. strain = 4–5 %) [314]

Q&P treatment consisting of $T_\gamma = 1000\ ^\circ C$ (5 min), QT $= 150\ ^\circ C$ (30 min) and PT $= 400\ ^\circ C$ (30 min).

Due to the TRIP effect, the Q&P specimen shows higher ductility for a similar strength and a superior work-hardening exponent (n) at low strains compared to the sample directly quenched from 1000 °C. Besides, the authors observed that the conventionally processed specimen exhibits about 3 % of yield point elongation, not observed for the other two samples (direct quenching and Q&P). Since C is linked in the form of $(Cr,Fe)_{23}C_6$ in the sample annealed at 820 °C, the static strain ageing (yielding) in the conventionally processed sheet was attributed to N in solid solution. In the directly quenched and the Q&P samples, the partitioning of C and N to martensite and austenite, respectively, would leave the ferritic matrix free of interstitials and eliminate the strain ageing. These results would demonstrate that the Q&P processing can potentially also be used as a pre-treatment to eliminate the static strain ageing in transformable FSS.

Tsuchiyama et al. [309] investigated the Q&P processing in an AISI 410 MSS (12Cr-0.1C, wt.-%) and two additional similar steel grades in which the carbon content was increased to study the influence of carbon content (12Cr-0.2C and 12Cr-0.3C, wt.-%). The Q&P processing parameters consisted of austenitization temperature (T_γ) = 1000 °C (30 min), quenching temperature (QT) = 240 °C

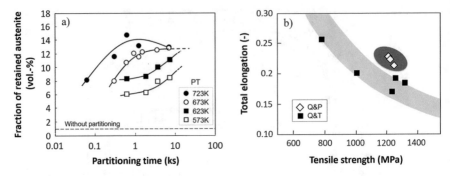

Fig. 11.33 (**a**) Change in volume fraction of austenite as a function of partitioning time for varying partitioning temperatures; (**b**) strength-ductility balance for Q&P and Q&T heat treatments in AISI 410 SS [309]

(1 min) and partitioning temperature (PT) = 300–450 °C (5, 10, 120 min). Figure 11.33a shows the influence of partitioning temperature and time on the volume fraction of retained austenite. Figure 11.33b shows that with Q&P heat treatments work-hardening enhances, and a better combination of mechanical properties (balance tensile-strength/total-elongation) can be achieved compared to conventional quenching and tempering (Q&T) heat treatments. In conditions (PT = 450 °C for 10 min) at which the amount of retained austenite was maximized (15 vol.-%), the longest elongation was obtained due to the TRIP effect. These authors discussed that the addition of Si was not necessary to prevent carbide precipitation because this would be retarded in high-chromium steel for two reasons: (i) chromium reduces the diffusivity of carbon by attracting carbon atoms and decreasing the thermodynamic activity of carbon [322] and (ii) the coarsening of cementite proceeds with distribution of chromium between cementite and ferrite [323]. However, they observed that during this partitioning heat treatment, small M_3C carbides precipitated in the martensite concluding that not all carbon could diffuse to the austenite during the partitioning step. In a parallel work [316], these authors investigated the role of Si (0.94, 0.21 and < 0.01 wt.-%) in the same type of stainless steels AISI 410 (12Cr-0.1C, wt.-%) subjected to Q&P and Q&T processing treatments. Figure 11.34a shows the variation in the amount of retained austenite with the partitioning time, at PT = 450 °C, for the three steels investigated (QT = 240–250 °C; 1 min). Figure 11.34b displays the strength-ductility balance for different heat treatment conditions and steels investigated. They concluded that (i) even for the non-alloyed Si steel, the Q&P heat treatment improves the ductility compared to Q&T heat treatments for the same strength value; (ii) 0.94 wt.-% Si retards the precipitation of cementite significantly, increases the amount of retained austenite and enhances its thermal stability after the Q&P heat treatment. The mechanical properties of the highest silicon steel are greatly improved almost doubling the strength (Q&P ~ 1300 MPa; Q&T ~ 750 MPa) for a similar value of the total elongation (~25–27 %).

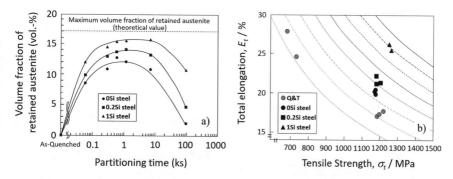

Fig. 11.34 (a) Change in volume fraction of retained austenite as a function of partitioning time at partitioning temperature of 723 K in Fe-12Cr–0.12C–Si (wt.-%) steels; (b) strength-ductility balance of Fe-12Cr–0.1C (wt.-%) steels with Q&P (PT = 450 °C – 600 s) and Q&T (450–120 s; 500–3600 s; 550 °C-3600 s) [316]

Mola et al. also produced a very complete and well-described report on the influence of different Q&P processing parameters in a higher carbon content MSS AISI 420 (13Cr-0.3C) and the influence of these heat treatments on the thermal and mechanical stability of austenite [317]. The conditions used in this study consisted of $T_\gamma = 1150$ °C (2–2.5 min), QT = 50–200 °C (3 min) and PT = 450 °C (3 min). At 1150 °C all carbides were dissolved. High-resolution dilatometry experiments showed that as the QT was increased, the amount of austenite transformed to α' martensite during quenching decreased (small dilatation). The subsequent partitioning step at 450 °C was able to stabilize the austenite as the M_s of the untransformed austenite decreased. For a QT > 140 °C, the M_s of the untransformed austenite was located below room temperature after the partitioning step. Direct quenching to room temperature retained 9.8 vol.-% of austenite. Figure 11.35a shows the influence of the quenching temperature on the volume fraction of retained austenite for two partitioning temperature conditions (300 °C-2 h; 450 °C-3 min) as estimated using different experimental techniques (dilatometry, XRD and magnetic saturation measurements). Figure 11.35b shows the evolution with the quenching temperature (QT) of the primary martensite (transformed after the first quenching and tempered during the partitioning step), secondary martensite (fresh and non-tempered martensite obtained upon quenching to room temperature after partitioning) and the retained austenite phase fractions in the Q&P specimens partition treated at 723 K (450 °C) from dilatometry calculations (dashed lines).

Figure 11.36a shows the stress-strain curves for Q&P specimens partition treated at 723 K (450 °C) as a function of the quenching temperature for the same AISI 420 MSS [317]. Figure 11.36b summarizes the mechanical properties of Q&P specimen partition treated at 723 K and 573 K (450 °C and 300 °C) for different quenching temperatures. Two groups of samples can be distinguished; those having fresh untempered martensite in the microstructure (QT ≥ 140 °C) and those who have not. The first group showed lower elongation and yield stress due to the presence of brittle fresh untempered martensite that lead to having a

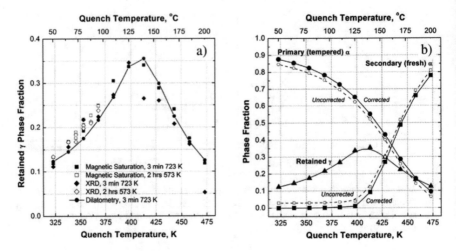

Fig. 11.35 Q&P heat treatments in an AISI 420 MSS: (**a**) retained austenite fractions as a function of quench temperature for Q&P-processed specimens partition treated at 723 K and 573 K (450 °C and 300 °C). The austenite fractions measured by XRD and magnetic saturation are included for comparison; (**b**) primary martensite, secondary martensite and retained austenite phase fractions in the Q&P specimens partition treated at 723 K (450 °C) from dilatometry calculations (dashed lines). The nonzero secondary martensite fractions at quench temperatures ranging from 323 K to 398 K (from 50 °C to 125 °C) originate from the nonlinear coefficient of thermal expansion (CTE) of ferrite and austenite. The corrected values after offsetting to zero of secondary martensite fractions in this range are represented by solid lines. The summation of primary martensite, secondary martensite and retained austenite fractions invariably gives 1 [317]

brittle fracture during the tensile tests. The best maximum elongation (15.7 %; UTS = 1570 MPa) was obtained for a quenching temperature of 80 °C at which the amount of retained austenite is around 20 vol.-%, far from the maximum amount obtained at QT = 140 °C. This result highlights that not only the amount of retained austenite but other factors, extensively reviewed by other authors, like carbon content, grain size and grain geometry, could affect the stability of austenite, the TRIP effect and its ability to enhance the ductility of steels. In addition, these authors found that in this steel, austenite stabilization also occurs due to the local carbon enrichment at austenite grain boundaries and austenite/martensite phase boundaries which deactivates nucleation sites for martensite formation. This effect is larger when higher amount of primary martensite formed (i.e. the lower the QT). However, these authors did not refer to this stabilization as being mechanical or thermal. But they explained that the lower the QT is, the higher the carbon content and the higher the mechanical stabilization in the range of QT investigated. Besides, the results of the authors showed that carbon enrichment of the austenite is much lower than that expected for full partitioning due to cementite precipitation in α' martensite. On the other hand, the thermal stability increased with the decrease in the amount of retained austenite; therefore, the variation in the thermal stability with QT does not necessarily correlate with the variation of the mechanical stability with

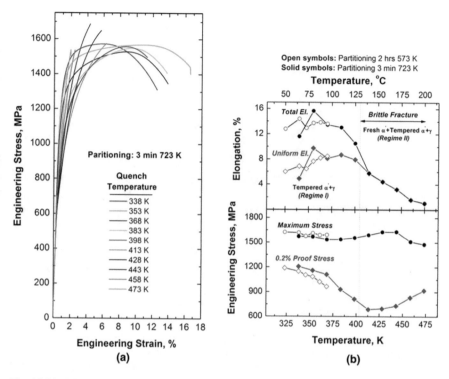

Fig. 11.36 Q&P heat treatments in an AISI 420 MSS: (**a**) room temperature engineering stress-strain curves for Q&P specimens partition treated at 723 K (450 °C). (**b**) Summarized tensile properties of Q&P specimen partition treated at 723 K and 573 K (450 °C and 300 °C) for different quenching temperatures [317]

QT, in the range of QT investigated. Similar conclusions to those observed in these previous works were also obtained by Huang et al. [315] in different stainless steels (13.1Cr-0.47C; 12.6Cr-0.46C-3.1Co; 12.8Cr-0.26C-0.17N, all in wt.-%).

Recently, the Q&P concept has also been used in a metastable ASS (14.9Cr-0.16C-0.12N-2.9Ni-3.0Mn-0.5Si, wt.-%) [318]. In this type of steels, the M_s temperature is usually around or below room temperature which makes it necessary to apply cryogenic treatments to promote the formation of α' martensite. A solution treatment at 1150 °C for 30 min followed by quenching to room temperature promoted the formation of an amount of 5.5 vol.-% of α' martensite. Introducing the steel in liquid nitrogen (QT = −196 °C) led to a microstructure composed of 58 vol.-% of α' martensite, the rest being retained austenite. Subsequently, the authors used the following partitioning conditions: PT = 450 °C (3 min). Around this range of temperatures and up to a holding time of 30 min, paraequilibrium M_3C carbides have been the only precipitated phase observed in the martensite [324] although some authors have suggested that the following precipitation sequence could be observed: $MC \rightarrow M_3C \rightarrow M_7C_3 + M_{23}C_6 + M_6C$ [325]; and, thus, different

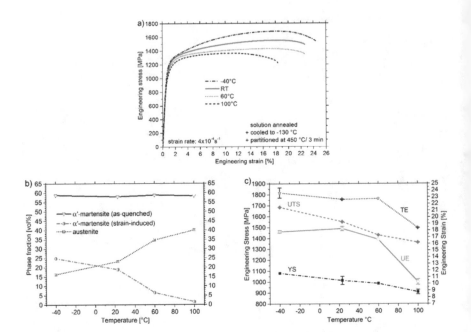

Fig. 11.37 (**a**) Engineering stress-strain curves in the Q&P condition (QT = −196 °C; PT = 450 °C, 3 min); (**b**) amount of austenite and α′ martensite after the tensile tests; (**c**) average mechanical properties in the Q&P condition for alloy Fe-14.9Cr-0.16C-0.12N-2.9Ni-3.0Mn-0.5Si (wt.-%) [318]

precipitation species could be found. This phase is less harmful to the corrosion resistance (sensitization) as its formation requires less Cr (removed from the solid solution). This would reduce the amount of interstitials available for partitioning from martensite to austenite and the stability of austenite, but the yield strength of the martensite would increase by nanoprecipitation [317]. The reduction in ductility due to nanoprecipitation could be counteracted, by the presence of high amounts of austenite (42 vol.-%) in the microstructure. XRD analysis of diffraction peak (311)γ before and after the partitioning step showed a shift in position to lower angles (larger lattice parameter) and an increase in asymmetry (nonuniform enrichment of the austenite). The authors correlated the observed increase in the lattice parameter of austenite with an increase of 0.1 wt.-% in the amount of interstitials in solid solution in the austenite. Thus, the average interstitial content of austenite was estimated to have increased from 0.28 wt.-% before partitioning to 0.38 wt.-% after partitioning. Figure 11.37a shows the tensile tests performed on the Q&P microstructure at different temperatures. The amount of strain-induced martensite (α′) transformed after the tensile tests increases progressively when the testing temperature is lowered from 100 to −40 °C (Fig. 11.37b). Both the strength (YS, UTS) and the elongation (UE, TE) improve as the temperature is decreased (Fig. 11.37c). As it has been discussed in previous sections, the stacking fault energy decreases with temperature, and the tendency to form strain-induced α' martensite increases, which tends to enhance the ductility and work-hardening rate as it is observed in these results. The following mechanical properties were obtained at room temperature: YS = 1050 MPa, UTS = 1550 MPa and TE = 22 %. After the test, the amounts of strain-induced martensite and retained austenite were around 19 vol.-% and 23 vol.-%, respectively.

11.4.5 High-Temperature Stainless Steels

The construction of new steam and nuclear (fission/fusion) power generation plants and chemical and petroleum plants relies on the development of high strength, corrosion, high-temperature oxidation and high thermal fatigue and creep-resistant steels that can stand for long times at constantly increasing operating temperatures and stresses. High-temperature stable microstructures, even under nuclear irradiation in some cases, are required for these applications. The driving force for the development of these steels is to improve the thermal efficiency and reduce CO_2 emissions and other environmentally dangerous gases that contribute to the greenhouse effect. In this regard, due to the increasing demands in electricity, which are expected to grow in the coming decade, it is becoming critical to replace current fossil fuel based electrical production (which in 2006 was, for example, 70 % of the total electricity generated in the US) with cleaner energy-generating sources that do not contribute to these emissions (nuclear, hydropower, wind, geothermal and solar, which constitute only the remaining 30 %) [326].

Numerous studies on heat-resistant steels actively conducted since the early 1970s have allowed great progress in 9–12 wt.-% Cr ferritic/martensitic steels, ASS, FSS and oxide dispersion-strengthened (ODS) steels. Because, formally, stainless steels are considered as those possessing a chromium content above 12 wt.-% (though some times as low as 10.5 wt.-%), the extensive research and development in producing novel 9 wt.-% Cr steels for steam and nuclear plants will be not mentioned in this chapter with detail except for some cases in which mainly the processing rather than the alloying could be translated to higher chromium steels to improved their creep properties. To have a more complete view on the development of this type of steels (9 wt.-% Cr), the readers are referred to some general books, encyclopaedias or recent reviews [305, 327, 328]. In this section, we will focus on existing trends in novel alloying strategies published in the recent scientific literature aimed at improving the creep behaviour of stainless steels.

11.4.5.1 Austenitic Stainless Steels

Zhou et al. recently summarized some of the most characteristic commercial heat-resistant ASS which comprise TP347H, Super304H, NF709, Sanicro25 and HR3C [329]. These alloys are typically 18Cr-8Ni, 20Cr-25Ni or 22Cr-25Ni combined with

0.8–1.6Mn and 0.1–0.6Si (wt.-%). The creep strength or heat-resistant ability of these alloys is based on alloying with elements like Ti, Nb, Cu, W, Mo and N. The first three elements (Ti, Nb, Cu) are added to improve the precipitation hardening through the formation of nanoprecipitates, a strategy that has already been discussed before for precipitation hardening steels: W for high-temperature solid solution strengthening, Mo for pitting resistance and solid solution strengthening and N to enhance the MX precipitation and increase strength and corrosion resistance. In the last decade, a similar alloying strategy with or without novel processing routes has been employed in experimental alloys to improve the creep strength and other high-temperature properties, like oxidation/corrosion resistance, compared to existing commercial alloys. From the point of view of improving the creep strength, two groups of alloying strategies can be found in the recent literature; where the strengthening is based on (1) the MX and Z-phase and (2) the MX and/or intermetallics such as the Laves (Fe_2Nb), NiAl-B2 and/or Ll2-ordered γ'-N_3(Al,Ti). In these two groups of alloys, the use of Cu was sometimes introduced to add the positive independent strengthening effect by Cu nanoprecipitates.

As an example of a seminal research corresponding to the first group, *Sawaragi* et al., by alloying with around 3 wt.-% Cu, 0.4 wt.-% Nb, 0.1 wt.-% N, conceived, 25 years ago, the novel Super304H ASS that improved the 10^5 h creep rupture strength at 600–700 °C by more than 20–30 % compared to the existing TP347H grade (which only contains 0.8 wt.-% Nb with no Cu) [330]. In this work, the main cause for the improvement of the creep strength was attributed to the intragranular coherent precipitation of copper nanoparticles, in addition to the contribution by other precipitates like Z-phase (NbCrN) and NbCN. *Park* et al. [331] suggested that adding 0.3 wt.-% V to the composition of this Super304H ASS could improve the creep strength. Lower reheating temperatures are necessary to dissolve coarse V-rich precipitates compared to Ti- and Nb-rich phases, and, thus, V can be more readily available for nanoprecipitation. This alloying element has been used extensively in 9–12 wt.-%Cr martensitic heat-resistant steel to promote the formation of MX nanoprecipitates which contribute to the enhancement of the creep strength [164]. *Park* et al. observed that the modified V-rich Z-phase (CrNbVN) and Cu nanoprecipitates formed during creep deformation, improving the creep strength. Other phases that were observed in the microstructure were NbCN and V-rich M_2X that transformed to Z-phase during creep.

Alternative alloying strategies to improve creep resistance include the use of B and rare earth elements (Ce, La) in addition to the microalloying elements. *Laha* et al. [21] who modified grade AISI 304 with different additions of Ti, Nb, Cu, B and Ce (9 alloys in total), found that creep resistance was improved by the addition of Nb, Ti and Cu. However, the major improvement was obtained by supressing of creep cavitation via alloying with B and Ce in addition to Nb/Ti/Cu. This was because (i) Ce links with S, preventing its segregation to the grain boundaries, which increases its cohesive strength and retards cavity nucleation; (ii) segregation of B instead of S on the cavity surfaces retards the cavity growth rate. The presence of Ti, Nb and Cu contributed to the precipitation hardening of the matrix by the formation of (Ti,Nb)(C,N) and Cu nanoprecipitates. A similar investigation was

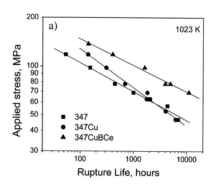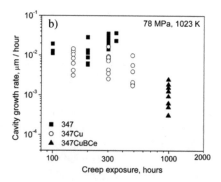

Fig. 11.38 (a) Variation of creep rupture life; (b) variation of cavity growth rate with creep exposure at 78 MPa and 750 °C (1023 K) for grade AISI 347, modified 347 with 3 wt.-% Cu (347Cu) and modified 347 with 3 wt.-% Cu, 0.06 wt.-% B and 0.015 wt.-%Ce (347CuBCe) [20]

carried out by these authors using grade AISI 347 alloyed with Cu and/or B and Ce, and similar conclusions were obtained [20]. Figure 11.38 shows the variation of the creep rupture life and of cavity growth rate with creep exposure at 78 MPa and 750 °C (1023 K).

Vu and Jung conceived a novel Fe-15Cr-15Ni-2.9Cu (wt.-%) heat-resistant steel alloyed with 0.40Nb-0.26N-4.0Mn-1.47Mo (in wt.-%) [39]. The proposed increase in N and reduction of C (0.04 wt.-%) aimed at promoting the more thermally stable NbN instead of NbC and minimizing the volume fraction of $M_{23}C_6$ carbides at grain boundaries. Besides, the addition of Mn increases the solubility of N and, along with Mo, reduces its activity, which slows down the formation kinetics of MX precipitates. Mn also reduces the price by allowing for a reduction of the Ni content. At the same time, these two elements (Mn, Mo) increase the activity of Nb which, along with the low activation of N, aims at stimulating a denser and finer precipitation of NbN. The relatively low amount of Nb compared to other commercial alloys also prevents the formation of coarse Nb(C,N) precipitates. The results show a better creep strength than commercial NF709 ASS at 750 °C using 78–200 MPa. The two main precipitates that played a key role were (Nb,Cr)N precipitates, at the early stage, that evolved to Z-phase during creep, at longer times. According to these authors, contrary to some investigations in other stainless steels [40], these Z-phase precipitates show a high coarsening resistance and contributed to improve the creep strength in this steel. Cu particles also contributed to the strengthening. Surprisingly, the presence of other phases like $M_{23}C_6$, χ-phase at grain boundaries or Cr_2N inside the grains did not have any detrimental influence. In a separate survey, these researchers also observed that by introducing a hot rolling step after a solution heat treatment at 1270 °C, the creep strength was improved by stimulating a more uniform distribution of nano-sized carbonitrides although its volume fraction was lower compared to a conventional heat treatment without the hot rolling step, due to the formation of $M_{23}C_6$. In addition, number density and distribution of the copper precipitates were not affected by the hot rolling; however, these particles are less thermally stable than the carbonitrides. The refinement of the microstructure and the coarser and higher volume fraction of $M_{23}C_6$ at austenite grain boundaries were likely the reason for a decrease in the creep ductility [332].

The applicability of Fe-Cr-Mn-N alloys (Ni-free or low-Ni), replacing the more expensive Fe-Cr-Ni alloys, for high-temperature applications has also been explored by some researchers. As it has been discussed before, Mn has been suggested to replace Ni because it is the most effective alternative as an austenite stabilizer to reduce costs [333, 334]. In some of these alloys, N is also added in the composition to improve (i) the corrosion resistance of low-Ni steels [335] and (ii) the creep strength by enhancing the formation of more stable MX nanoprecipitates. *Lee* et al. by adding Nb (0.43 wt.-%) and Cu (2.0 wt.-%) to a Fe-14Cr-10Mn-2.5Ni (wt.-%) alloy were able to suppress the nucleation of Cr_2N and promoted Z-phase and MX formation [336]. Besides, during creep, Cu alloying promoted the independent precipitation of strengthening nanometre-sized Cu particles that contributed significantly to improve the creep strength. Thus, alloying with these elements turned out to improve the creep resistance of the steels. The formation of $M_{23}C_6$ at the grain boundaries was also observed during creep deformation.

The second group of novel heat-resistant stainless steels under development for structural components in coal-fired power plants is the carbide/intermetallic strengthened alumina-forming austenitic (AFA) stainless steel [49, 337]. These alloys were initially conceived during the 1970s. *Yamamoto* et al. have recently unveiled that alloying this steel with Al using minimum amounts of 2.5 wt.-% promotes the formation of a protective Al_2O_3 layer at 600–900 °C provided that Ti and V are removed from the composition (if these two elements are also present, higher amounts of Al are necessary to promote the formation of the alumina layer). This layer confers superior oxidation and corrosion protection compared to the usual Cr_2O_3 layer formed in conventional stainless steels [337]. When Ti/V are present, only internal oxidation occurs; therefore, to overcome the removal of Ti/V, high amounts of Nb (>0.6 wt.-%) were considered, and this seemed to enhance the stability of the alumina layer. Concerning the character of Al; this is a ferrite-forming element that promotes the formation of δ-ferrite and σ-phase at high temperatures, which deteriorates the creep strength significantly. Thus, the addition of this element along with Cr and Ni has to be optimized to avoid the formation of these phases. The amount of Al and Cr in these alloys usually varies between 2.5–5.0 and 12–15 wt.-%, respectively. Due to the high amount of Al, high levels of Ni need to be employed (20–32 wt.-%), in spite of its high price, to avoid the formation of δ/σ-phase [49]. On the other hand, the additions of W, Mo, Cu, B, 0.3–3.5 wt.-% Nb, ≤3 wt.-% Ti, ≤0.3 wt.-% Zr, ≤0.2 wt.-% Si and ≤ 0.1 wt.-% C, which would also modify the stability of the austenite when present in solid solution, have been used to stimulate the formation of strengthening carbides and intermetallic phases. The strengthening species that contribute to the creep resistance of the AFA grades depends significantly on whether Nb and C are present or not. *Yamamoto* et al. developed a grade that was strengthened by nanocarbides (mostly NbC) and realized that maximizing the formation of nanoscale MC carbides was the key to optimizing creep resistance of these AFA alloys, with an additional minor contribution of NiAl-

B2 and Laves $Fe_2(Mo,Nb)$ [338]. Although NiAl-B2 is effective strengthener only below 425 °C in Fe-Cr-Ni-Al steels [339], recent reports in AFA steels suggest that they improve the creep resistance up to higher temperatures [340]. When C was eliminated, the Laves Fe_2Nb and/or NiAl-B2 phase were the key phases that improved the creep behaviour [341, 342]. In this case, the formation of Ll2-ordered γ'-N_3(A,Ti) occurred during service conditions, although their contribution to the creep strength was discussed to be negligible. These investigations have also reported that the presence of Si (0.2 wt.-%) refines the Laves phase and prevents its coarsening, enhancing the creep strength. In another report, *Yamamoto* et al. published that alloying with Ti and Zr increases the volume fraction and stabilizes Ll2, respectively, against B2, and this seems to improve the creep strength in Fe-14Cr-32Ni-3Nb-(3-4)Al (wt.-%) alloys [50]. The strategy of substituting Ni by Mn has also been explored by these authors [343]. However, in this type of alloys, Cu has been included in the composition as an additional austenite stabilizer because the inclusion of N, along with Mn, would lead to the formation of coarse AlN precipitates. Cu contributes to the strengthening through nanoprecipitation as it has been discussed extensively. The Mn/Cu-alloy steels could find application in chemical and petrochemical industries substituting alloy AISI 347, under service temperatures below 650 °C, but not in coal-fired steam power plants due to their lower creep resistance.

Zhao et al. investigated the influence of Cu (2.8 wt.-%) in a Fe-15Cr-20Ni-Mo-Al-Nb-V-B (wt.-%) alloy and found that this element also promotes the formation of Ll2 against B2 and the precipitation of Cu nanoparticles all together improved the creep behaviour [344]. Other investigations by *Hu* et al. have reported that the main strengthening phase in a Fe-14Cr-32Ni-3Nb-3Al-2Ti (wt.-%) alloy was the Ll2 precipitates. In this alloy the presence of B and C did not influence Ll2 precipitates but enhanced the grain boundary coverage by the Laves phase and B2-NiAl precipitates and suppressed their coarsening. By increasing the grain boundary cohesion and optimizing the grain boundary precipitate distribution, an improvement in the ductility was achieved [55]. Separately, *Jang* et al. reported that the addition of W to a Fe-14Cr-20Ni-2Mn-2.5Al-1Mo-0.8Nb-0.08C-B (wt.-%) steel improved the creep strength via the enhancement of the precipitation hardening associated with the Laves phase and suppressed the σ-phase formation [345]. Besides, as it would be expected, cold-working accelerates the precipitation kinetics of all these intermetallic phases [51].

Figure 11.39 shows the Larson-Miller parameter (LMP) of some AFA alloys developed by *Yamamoto* et al. together with those of commercially available chromia-forming heat-resistant stainless steels, plotted as a function of stress. The reason why the AFA alloy with a higher Ni content has a lower creep resistance is not clear yet [49]. However it should be borne in mind that a higher amount of Ni benefits the oxidation resistance [346].

The failure of heat-resistant ASS at high temperatures is generally due to nucleation and coarsening of carbides and/or intermetallic phases, mainly located at grain boundaries, after long-term exposure above 600 °C [329]. Failure due to $M_{23}C_6$, Z-phase, Laves or σ-phase has been reported, although in some occasions,

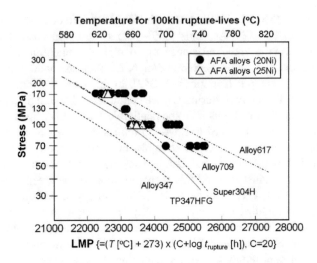

Fig. 11.39 LMP plot of AFA alloys together with commercially available heat-resistant steels and alloys [49]

as it has been mentioned above, some of these phases also have a positive influence on the creep behaviour:

(a) $M_{23}C_6$ present as fine precipitates can contribute to the strengthening of the steel, but as they coarse under service conditions at high temperatures, they serve as nucleation sites for void/crack formation after long high-temperature ageing. Besides, they may promote the sensitization of the steel leading to inter-granular corrosion. It has been recently reported that failure of Super304H ASS occurs due to the precipitation and coarsening of $M_{23}C_6$ at grain boundaries [26, 27] by creep cavity formation under service conditions. A reduction in creep ductility has also been observed to be associated with the presence of coarse $M_{23}C_6$ [39]. On the contrary, other works have shown that (1) the presence of phases like $M_{23}C_6$, χ-phase or Cr_2N did not have any detrimental influence during creep of a Fe-15Cr-15Ni-Mn-Nb-Mo-N (wt.-%) alloy at 750 °C (78–200 MPa) [332]; or (2) by applying a heat treatment around 700–800 °C to Cu alloyed AISI 304, the creep resistance was improved due to (i) the precipitation of $M_{23}C_6$ carbides at grain boundaries with an optimum interspacing and volume fraction; (ii) the serration of grain boundaries by the precipitation of this phase and (iii) increase in the number of $\Sigma 3$ twin boundaries [347].
(b) Z-phase (NbCrN) plays a similar role as $M_{23}C_6$ carbides; as fine nanoprecipitates it contributes to the strengthening [39, 331, 336], but when they coarsen, properties like fatigue suffer.
(c) σ-Phase precipitates present at grain boundaries deteriorate the creep strength, toughness and corrosion resistance. They may transform from austenite, δ-ferrite or $M_{23}C_6$ during creep. It has been reported that the failure of Super304H

ASS is due to the formation of this phase [106] by creep cavity formation under service conditions.

(d) Laves phase (Fe_2Nb, Fe_2Ti), similarly to the σ-phase, deteriorates the toughness, ductility and corrosion resistance (known as Laves phase embrittlement). However, some reports have suggested that nanoprecipitates of this phase, of which refinement is promoted by Si addition, may improve creep resistance [105]. *Chen* et al. [54] investigated two Fe-20Cr-30Ni-2Nb (wt.-%) alloys (with and without B) and found that the precipitation of the Laves phase (Fe_2Nb) at grain boundaries contributed to the strengthening and improved the creep resistance. In addition, the segregation of B at grain boundaries enhanced the precipitation of this phase and improved the creep behaviour even further.

Besides these advances in creep as well as corrosion/oxidation resistance, the use of austenitic steels in nuclear applications, at high temperature, even under moderate irradiation doses, is still a major problem. The influence of neutron irradiation in ASS has been studied for long time because this type of steels has broad applications as structural components in light water reactors and fuel cladding in fast spectrum nuclear reactors [348, 349]. However, a major limitation of these steels regards with having significant void swelling accompanied by precipitation at moderate neutron irradiation doses (>10 dpa) compared to ferritic or ferritic-martensitic steels at 300–700 °C [350]. Though this section is devoted to high-temperature resistance, the issue concerned with high-temperature resistance under irradiation conditions is of the major importance in the nuclear power-generating industry. To overcome this weakness and improve the swelling resistance or other properties like strength, stress corrosion cracking or oxidation resistance of ASS, a number of recent works have reported positive effects: (i) through the refinement of the microstructure down to the ultrafine range (<1 μm) [351] or (ii) by grain boundary engineering [352, 353]. *Sun* et al. [351] obtained ultrafine microstructures in a AISI 304 L grade (~100 nm) by using severe plastic deformation (ECAP) at 500 °C. They reported an improvement in the Fe ion irradiation resistance, at 500 °C, compared to the coarser (~35 μm) counterparts of the same alloy. The thermal stability of these microstructures was excellent up to 600 °C that led to having stable defect sinks (HABs) that contributed to having a reduction of void swelling. Figure 11.40 compares the results of this study concerning void swelling using Fe ions (blue and red data points) with neutron-irradiated AISI 304 [354]. At a similar radiation dose, 80 dpa, the swelling resistance improved by an order of magnitude. In addition, the average void swelling rate over 35–80 dpa in the coarse and ultrafine microstructures was ~0.18 %/dpa and ~0.03 %/dpa, which is a reduction by a factor of five. Besides, no $M_{23}C_6$ precipitation was observed in ultrafine grains compared to the coarse grains where their presence was extensive. These authors concluded that UFG austenitic stainless steels have promising applications in extreme radiation environment and the use of nanograins in combination with other defect sinks, such as high-density nanoscale oxide or nitride precipitates, could lead to even greater improvement in radiation resistance.

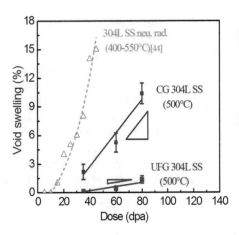

Fig. 11.40 Comparison of void swelling of 304L SS irradiated by Fe ions (this study) and fast neutrons spectrum [351]

11.4.5.2 Ferritic Stainless Steels

Ferritic heat-resistant steels have been widely used in high-temperature components in power plant applications. As it has been argued, new heat-resistant steels are required for advanced ultra-supercritical (A-USC) thermal power plants with operational temperatures ≥ 700 °C. The effort in developing ferritic stainless steels for structural components of these power plants is justified by their much smaller heat expansion coefficient which minimizes cracking during thermal-shocking, leading to a better thermal fatigue resistance [355]. Novel FSS should have high Cr content to offer high corrosion/oxidation resistance above 650 °C compared to existing creep strength enhanced ferritic steels (CSEF, 9–12 wt.-%), like P/T91–92, generally limited to $T < 620$ °C [356]. However, using high Cr contents would more likely develop Z-phase from MX precipitates, resulting in a drop in the creep strength [40, 357, 358]. It seems difficult to improve at the same time long-term creep strength and oxidation resistance using the same alloying approach as for 9–12 wt.-% ferritic-martensitic steels. Therefore, novel alloys are putting their focus on improving the mechanical properties via intermetallic rather than via MX precipitation. According to *Kuhn* et al. [48], this concept was raised for the first time in 2001 by the *National Institute of Materials Science, Japan* [359], and their patent considered Fe-Cr (>13 wt.-%) alloys in which the following intermetallic phases were present: Laves phase (Fe_2W, Fe_2Mo), μ phase, σ phase and/or Ni_3X (X = Al or Ti). Around the same time, the *Institute for Energy and Climate Research, Microstructure and Properties of Materials* (IEK-2) of *Forschungszentrum Jülich and ThyssenKrupp VDM* (now *Outokumpu VDM*) developed high-chromium (22 wt.-% Cr), ferritic stainless steels for the application in high-temperature (700–800 °C) automotive solid oxide fuel cell (SOFC) [360] that led to the development of Crofer® 22 APU and Crofer® 22 H (SOFC is an energy conversion device that produces electricity by electrochemically combining a fuel and an oxidant across an ionic conducting oxide electrolyte [361]). The addition of W, Nb and Si to Crofer 22 H compared to Crofer 22 APU leads to a much higher creep strength at 700 °C due to the intra- and intergranular formation of the Laves phase. At 800 °C after long exposure, Laves phase was only present at grain boundaries, and this led a decrease in creep strength of Crofer 22 H alloy; however, the strength of this alloy was still higher than for Crofer 22 APU due to the solid solution strengthening of W [47]. Other phases that were observed in these alloys are $(Nb,Ti)(C,N)$ and La_2O_3.

In the last decade, several FSS strengthened by intermetallics with high creep/oxidation/corrosion resistance have been proposed. These alloys add to the base Fe-Cr composition different amounts of Nb, Ti, Si, W or rare earth elements (La) to stimulate the formation of intermetallic nanoprecipitates with high-temperature stability that could substitute MX as strengthening precipitates and enhance the creep behaviour. The precipitates used to strengthen these novel FSS are the G-phase ($M_6Si_7Ni_{16}$, M = Nb,Ti) and the Laves phase ($(Fe,Cr,Si)_2(Nb,W,Ti)$); however, in alloys where microalloying elements (Ti, Nb) are employed, the formation of MC (M = Nb,Ti) carbides results, which is sometimes inevitable even for ultralow-carbon grades (<0.02 wt.-%C) [30]. This would prevent the formation of coarse $M_{23}C_6$ grain boundary carbides but not the long-term formation of the Z-phase during ageing.

Taking commercial alloys Crofer 22 APU and Crofer 22H [47, 360, 362, 363] as basis, *Kuhn* et al. have developed several Fe-(18–22)Cr FSS alloyed with different amounts of Cr (18–23), Nb (0.48–0.57), W (1.75–2.5) and Ti ($\leq$0.065) in wt.-% [48, 364–366]. Among the experimental alloys investigated, the one with the higher amount of W (2.5) and Nb (0.57) gave the best creep properties (Fig. 11.41) due to the formation of nanometre-sized highly stable Laves phase ($(Fe,Si,Cr)_2(Nb,W)$ that show very slow coarsening kinetics even after 30,000 h at 600 °C. The results were compared with the results obtained for ferritic-martensitic P92 alloy [367]. For the FSS with the higher chromium content (<22 wt.-%), coarse Laves and sigma phase areas were observed at grain boundaries after creep testing. The role of these coarse particles in the rupture of these alloys is yet to be investigated according to the authors. These alloys have also showed excellent steam oxidation resistance compared to ferritic-martensitic steels. All these results show the potential of FSS strengthened by intermetallic phase to be applied as structural materials in steam power plants at operating temperatures above 620 °C.

Yang et al. have recently developed novel Fe-20Cr-3Ni-3Si-1Mn (wt.-%) FSS alloyed with different amounts of microalloying elements Ti (1.5–2.0 wt.-%) and Nb (2 wt.-%) with the main target of using, for the first time, the G-phase [368] as a strengthening phase in FSS [29, 30]. In the early investigation, two phases were detected in the steels alloyed with 2Ti and 2Nb, after ageing at 660 °C, Laves phase (Fe_2Nb, Fe_2Ti) and G-phase [29]. In the Ti-alloyed steel, the strengthening effect measured at room temperature was much more pronounced compared to the Nb-alloyed one and mainly due to the more pronounced formation of the G-phase. In a more recent report [30], the same steel alloyed with Ti (now referred to as having 1.5 wt.-%), was subjected to a wider range of isothermal heat treatments at low (<550–700 °C) and high (800–1100 °C) temperatures. The following precipitates

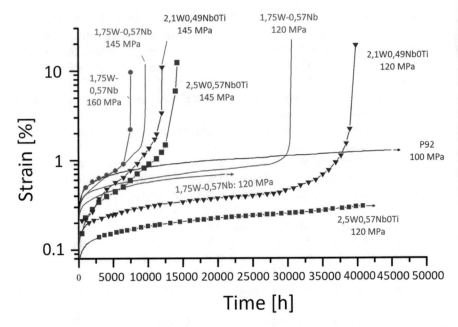

Fig. 11.41 Examples of creep curves for alloys 1.75W-0.57Nb, 2.1W-0.49NbOTi (wt.-%) and 2.5W-0.57Nb-0Ti (wt.-%) at 600 °C in air (stress levels indicated). Arrows show experiments in progress. All other plots reached rupture. The symbols mark the interruptions during discontinuous creep testing in comparison to P92 (1040 °C–1070 °C, 2 h/730 °C–800 °C, 2 h) at 100 MPa [48]

were now reported: (1) at high temperatures, ultrafine TiC and coarse Laves (Fe$_2$Ti) phases were formed mainly intragranularly and at grain boundaries, respectively, while at low ageing temperatures, ultrafine TiC formed inter- and intragranularly and nanometre-sized G-phase appeared intragranularly. The stability of the G-phase was up to around 750 °C although thermodynamic calculations showed that the synergistic effect of adding increasing amounts of Ni, Si and Ti could raise the volume fraction of this phase, improving its high-temperature thermal stability and minimizing the formation of the Laves phase. These results suggest that these steels could be used for high-temperature applications. Preliminary room temperature experiments showed that ageing at 560 °C for 3–8 h increased the compression YS at room temperature from 700 to 1700 MPa.

In contrast to these approaches, *Lu* et al. have designed a computational model focused on maximizing the solid solution strengthening in FSS instead of the precipitation hardening [369], though they have also developed a model to include precipitation hardening in creep-resistant MSS [99]. These authors discuss that solid solution strengthening is a time-independent factor during creep testing provided that no precipitation takes place in the microstructure during high-temperature exposure. Precipitation would remove atoms from the solid solution and lead to the failure of the steel depending on the phases formed and amount of elements removed from the solid solution. A generic alloy design approach coupled with thermodynamics software is used to design both the optimum composition and the solution temperature that can perform well at service temperatures around 650 °C. Alloying elements Mn, Al, Si and Ni are found to give the highest strength contribution; however, elements like W, which are employed in novel compositions as a potent solid solution strengthener at high temperatures were not considered in this investigation. This approach has been applied to existing commercial FSS compositions to validate the model predictions and approach. The authors concluded that with the current model novel alloys could be designed with higher solid solution strengthening compared to existing commercial alloys. Among the six compositions designed in this research, the authors argued that the optimum one would be Fe-12.3Cr-5.5Ni-10.7Mn-3.8Mo-3.3Si-5.2Al-1.9Ti-1.8 V (wt.-%) alloy.

Recently, *Shibuya* et al. have developed a different kind of FSS Fe-15Cr-6W-1Mo-0.2V-0.06Nb-0.05C (wt.-%) alloy with different amounts of Ni (up to 2.15 wt.-%) and Co (2–3 wt.-%) which show a ferritic matrix with some islands of martensite at the grain boundaries depending on the amount of Ni/Co [46, 370, 371]. For the alloy with the highest amount of Ni (2 wt.-%) and Co (3 wt.-%), the maximum amount of martensite was obtained (~30 %), in good agreement with the thermodynamic calculations. The results showed that the addition of Co improves the creep strength. These alloys increased the rupture lives significantly compared to alloy T92 at 650–750 °C. After creep rupture at 650 °C, the microstructure contained Laves phase (Fe_2M), χ-phase ($Fe_{36}Cr_{12}M_{10}$) and μ-phase under 1 μm in size precipitated intragranularly in ferrite, while $M_{23}C_6$ carbides were precipitated in the martensite or at the grain boundaries. At this temperature the creep rupture times were short, and no precipitation within the martensite was observed. In a more recent report, the authors have investigated with more detail the influence of Ni (0.9–2.15 wt.-%) for a fixed Co value of 2 wt.-% on the creep behaviour at 650–750 °C [371]. They have found that the creep rupture lifetimes were 100 times that of T92 ferritic-martensitic steel and the creep strength after 10,000 h at 650 °C was double of that of this steel. In this case, the results showed that contrary to a previous work on these steels, creep rupture samples above 700 °C show precipitation of intermetallic phases both within ferrite and martensite phases. Besides, it has been mentioned that V-rich Z-phase (Cr(V,Nb)N) was observed to be precipitated at the grain boundaries. According to ThermoCalc calculations, MX ((V,Nb)(C,N)) was also expected but not observed. The authors discussed that the improvement in creep strength could be attributed to the precipitation of intermetallics and carbides at grain boundaries. Figure 11.42 shows the creep rupture strengths of 15 wt.-% Cr steels with various nickel contents at 650, 700 and 750 °C. The authors proposed these intermetallic strengthened Fe-15 wt.-% Cr steels as promising creep-resistant material for the next-generation fossil power plants.

11.4.5.3 Martensitic Stainless Steels

Nowadays, the energy conservation and the environmental protection are very important topics and are the driving forces for the development and improvement of

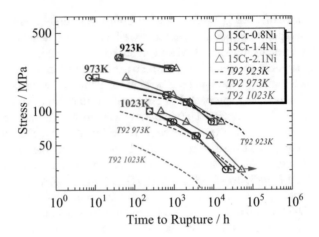

Fig. 11.42 Creep rupture strength of 15 wt.-% Cr steels with various Ni contents compared with a conventional heat-resistant steel (ASME T92, dashed line) at 650 °C (923 K), 700 °C (973 K) and 750 °C (1023 K) [371]

12 wt.-%Cr (12Cr) heat-resistant steels. For the future power plants, the reduction in fuel consumption and the CO_2 emission will be achieved by increasing the efficiency. This implies an increase in the operation temperature and the steam pressure. For the 12Cr heat-resistant steels employed for high-temperature applications in components of power plants good corrosion resistance, mechanical properties and fabricability are mandatory. Among mechanical properties, creep strength is the most highlighted property to validate these steels for high-pressure and high-temperature applications. The required improvement in the creep strength has led different research to be divided in this chapter into two lines as presented below.

Improvement in Creep Strength by Chemical Composition Modification
The creep strength of the 12Cr heat-resistant steels can be improved considering the different strengthening sources available at high temperature that allow stabilizing the microstructure for a long time. The most important sources are the solid solution strengthening and precipitation strengthening.

The solid solution strengthening of W and Mo has been widely demonstrated for these steels. However, it is important to control adequately the amount of these elements because at high contents they can promote the formation of delta ferrite which decreases the creep strength. Another problem produced by the high content of these elements is the formation of Laves phase [372, 373]. This phase coarsens very fast during operation and reduces the life of the component and its ductility. Re is another element often used due to its very similar behaviour to Mo and W [374]. Although, when a high content of these delta ferrite promoting elements is employed to increase the solid solution strengthening, other elements can be added to inhibit the formation of this undesired phase. Ni, Cu and Co are austenite formers and are able to inhibit the formation of delta ferrite by decreasing the chromium equivalent.

Concerning the precipitation strengthening, it is important to consider, first of all, the two most important precipitates present in these steels after tempering, the $M_{23}C_6$ carbides and the MX carbonitrides. The $M_{23}C_6$ carbides are located on prior austenite grain boundaries, block boundaries and lath boundaries. They contain a high amount of Cr and other metals that can help to its stabilization such as W. The MX carbonitrides are located within the lath and are enriched in V, Ti and Nb. The main advantage of these carbonitrides compared to $M_{23}C_6$ carbides is their higher thermal stability. By contrast they are found in a very low volume fraction, around ten times lower than that for the $M_{23}C_6$ carbides.

Currently, two main methodologies exist to improve the creep strength in steels strengthened by MX precipitates. The first one consists in increasing the addition of MX precipitate formers. This strategy aims at increasing the volume fraction of these precipitates and, consequently, stabilizing the microstructure by pinning dislocations during creep [375, 376]. The increase of MX promoting elements implies the increment in V, Nb and N. In the case of fusion application, Nb is replaced by Ta [44, 377]. However, the C content is not increased because the increment of C affects directly the formation of $M_{23}C_6$ carbides and increases their volume fraction, while it does not affect the MX carbonitride volume fractions significantly [378]. Following this line of research, *Tan* et al. developed a new alloy grade called CNAs (castable nanostructure alloys) for advanced fusion reactors [379]. These alloys contain just 9 wt.-% Cr, so they cannot be considered stainless steel, but the results shown in this work tried to demonstrate the potential of this route to be explored in 12 wt.-% Cr alloys. In these alloys, the increase of MX volume fraction is achieved by adding N, Ta, V and Ti. Some of these alloys showed significantly refined MX nanoprecipitates with a high number density, two orders of magnitude higher than that obtained for the conventional reduced activation ferritic/martensitic (RAFM) steels. These optimized compositions lead to improvements in YS, creep resistance and Charpy impact toughness compared to these conventional RAFM steels.

The second methodology explores the stabilization of the $M_{23}C_6$ carbides to prevent or delay their coarsening during creep. This stabilization is promoted by adding B and N. *Abe* et al. concluded in their work [380] that the addition of B and N without the formation of BN allows stabilizing the $M_{23}C_6$ carbides in the vicinity of prior austenite grain boundaries, which results in a retardation of the onset of acceleration creep. It is important to mention that the B and N addition has to be controlled to avoid the formation of BN, which decreases the B available for the stabilization of the $M_{23}C_6$ carbides. The stabilization of these carbides by B occurs because B atoms occupy vacancies in the vicinity of growing carbide interfaces near prior austenite grain boundaries, which makes it difficult to accommodate local volume change around growing carbides and hence causes the reduction in the coarsening rate. These carbides, when coarsen, have been associated with the formation of cavities at grain boundaries during creep, leading to the failure of the steel. This same approach of B alloying has been proven to be also successful in preventing or delaying cavitation in ASS [21].

The investigations mentioned so far have focused on modifying the stability or increasing the volume fraction of precipitates obtained after tempering, prior to the operation. The problem with the approach that focuses on increasing the volume fraction of MX nanoprecipitates is that these precipitates are not in equilibrium with the matrix and tend to dissolve during operation to form a most stable phase, the Z-phase. The Z-phase particles are slow to nucleate, but once formed they coarsen very quickly and consume the beneficial MX precipitates. The coarse Z-phase particles formed during service do not contribute to the strength of the material and, thus, result in poor long-term creep resistance. Trying to overcome this problem, *Danielsen and Hald* [381] have proposed to use the Z-phase precipitates as strengthening dispersion instead of MX nanoprecipitates for 12 wt.-% Cr martensitic steels. To have a positive effect on the creep strength, the Z-phase precipitates have to be fine and homogeneously distributed with a high number density. Besides, its coarsening rate has to be reduced by controlling the alloying elements. Many works have been focused on studying the effect of C and other elements such as Cu to stabilize the Z-phase. The most promising results concerning the improvement of the creep strength, have been obtained when the C content has been reduced dramatically, down to 0.005 C wt.-%. According to *Rashidi* et al. [44], low C addition promotes mainly the formation of MX nitrides instead of carbides, which transform to Z-phase quickly. As a result of this fast transformation during the early stages of creep, Z-phase precipitates are very fine and remain as such after long ageing times.

Improvement in Creep Strength by Alternative Processing Routes

An alternative approach to improve the creep strength that has been studied widely by many authors consists in introducing a TMCP step during the manufacturing of the steel [382–385]. Among these authors, *Klueh* et al. developed an optimized TMCP that involves different processing steps [164, 386]. The first is the austenitizing, during which the steel is heated up to 1200–1300 °C. This temperature should be high enough to dissolve most of the MX nanoprecipitates but not too high to avoid the delta ferrite formation (eluding the formation of δ-ferrite is mandatory as this phase decreases the creep strength). Then, the steel is cooled down to an ausforming temperature. In this step deformation is applied to the austenite to obtain, after cooling to room temperature, a martensitic microstructure with a higher dislocation density compared to that obtained in the absence of an ausforming step. Increasing the dislocation density in the martensite would increase the number density and reduce the size of the MX nanoprecipitates after tempering. As it has been discussed above, the presence of these precipitates delays the degradation of the microstructures under service conditions. The results reported showed that high number density of MX precipitates is obtained by the TMCP compared to that obtained after conventional heat treatment that do not include ausforming steps. These distribution and number density of precipitates are very similar to that obtained in ODS steels. However, ODS alloys are produced by much more complicated and expensive powder-metallurgy/mechanical-alloying procedures. The main disadvantage of this processing route is the low creep ductility of the microstructures obtained, which is considerably lower than that obtained after conventional heat treatment [387]. The risk of fast burst fracture for the power plants components has to be avoided, so future work in the optimization of this processing route is needed in order to avoid the drop in creep ductility.

Another alternative processing route that has been postulated as a promising heat treatment to increase the operation temperature of 12 wt.-% Cr heat-resistant steels is via the application of Q&P type of heat treatments. The research carried out so far has only been applied to 9 wt.-% Cr steels as far as the authors know: *Plesiutsching et al.* [388] investigated two heat-resistant steels typically used in power plants. These authors performed Q&P and Q&T (quenching and tempering) processing treatments using the same austenitization in four different conditions, i.e. $T_\gamma = 1100$ & 1150 °C for 30 and 80 min. Different austenite grain sizes were obtained (50, 100, 300 and 700 μm) in this way. Following were the QT and PT parameters of the Q&P treatment. Steel 1: 0.07 wt.-%C, QT = 360 °C and PT = 700 °C-3 h. Steel 2: 0.16 wt.-%C, QT = 320 °C and PT = 700 °C-3 h. The Q&T samples after quenching were fully martensitic and were tempered at 700 °C for 3 h for the both steels. The microstructure of the Q&P samples transformed partially from austenite to α' martensite during quenching to QT, and, during partitioning at 700 °C, austenite decomposed in ferrite+carbides while martensite was tempered. Thus, although the authors described this heat treatment as Q&P, the target of the proposed heat treatments did not focus on the partitioning of carbon from martensite to the austenite but on the decomposition of the austenite. Creep rupture tests were performed at 650 °C under a load of 150 MPa. The results showed that the applied Q&P treatment enhances the creep rupture time for both steels for all prior austenite grain sizes investigated although it was much greater for the lower carbon steel. However, the creep rupture strain was only enhanced for the higher carbon content grade and for the larger grain sizes. The authors concluded that the Q&P heat treatment offers the possibility to improve the creep strength of an entire class of high-temperature materials. However, it should be emphasized again that the heat treatment presented cannot be catalogued as quenched and partitioning because the objective of the partitioning step had nothing to do with the promotion of carbon diffusion from martensite to the austenite with the aim of stabilizing austenite at room temperature. However, the proposed heat treatments and microstructures obtained are reported to improve the creep behaviour of the steels compared to conventional Q&T which could be of interest for researchers in the field.

Altogether, the martensitic steels used in today's power plants do exhibit more or less serious disadvantages, and for this reason it seems hardly possible to achieve further improvement in creep strength and oxidation resistance, necessary for a highly efficient 650 °C steam power plant, on this technological base. A paradigm shift in alloy development – away from improving creep resistance with steam oxidation resistance as a secondary goal – towards combined improvement of oxidation and creep resistance, therefore, seems mandatory.

11.5   Industrial Applications

In this section, some industrial applications concerned with stainless steels will be introduced. Because of the extension of the applications list, the focus will be placed on the most popular applications and those recently reported in research journals.

11.5.1   Transportation

Stainless steels are widely used in industries like aerospace, automotive, shipbuilding and railroad transportations.

In the automotive industry stainless steels, AISI 409, 439, 441 and 304 have been used in parts of the exhaust system components (manifold, silencer, diesel particle filter, catalytic converter). Other applications are brakes disks, shafts and electric motors (420), clamps (441 or 430) and small-diameter shafts (416). In buses, modified type 409 is used as a structural component, and types 430 and 304 are used for exposed functional parts. AISI 409 has been also employed to manufacture the core of fuels pumps taking advantage of their good behaviour to oxidation and wet corrosion, good resistance to thermal fatigue and high-temperature strength [389]. It has been suggested that ultrafine or nanocrystalline ASS developed by severe cold-working and annealing would allow improving corrosion resistance and crashworthiness, together with a potential weight reduction. The AISI 301 steel highlights among the austenitic steels for their highest work-hardening rate and could be a suitable candidate that could be implemented in the automotive industry [226].

For tank trucks, type 304 has been the most frequently used stainless steel, and type 316 and higher-alloyed grades have been used to carry corrosive chemicals. Stainless steels are also used to fabricate chemical tankers in the shipping industry to carry aggressive chemicals and hazardous or toxic materials. In many cases, duplex stainless steels (grade 2205) have replaced the use of austenitic stainless steels (AISI 304, 316, 317) due to its better pitting and crevice corrosion resistance in chloride containing media and better mechanical properties. Weight reduction of the carrier up to 10 % in total can be also obtained replacing ASS by DSS [53, 390].

Some stainless steels are used for the body of railcars like grades AISI 410, 301 LN, 304 and 201 [391]. Its resistance to corrosion and low-cost maintenance (does not have to be painted to protect it from the weather) are benefits, and its fire resistance is also a significant safety advantage.

PHSS such as 13-8 PH, 15-5 PH and 17-4 PH, ASS like 304 and MSS like AISI 403, 410, 440C and 610 are used in aircraft tubing, landing gear components, engine structural parts, bearings, cooler sections of the engine or critical fasteners throughout aircrafts [53, 305].

11.5.2   Construction

Stainless steels have certain advantages for construction applications [392, 393]; however, there are still limited examples in construction because of the higher price compared to carbon steels [394]. Nevertheless, their combination of corrosion resistance, strength, ductility and formability allow making thinner and durable structures. Nowadays, stainless steels are used in a very wide range of structural and architectural elements; from small but intricate glazing castings to loadbearing girders and arches in bridges and loading-sustaining structural components in general. Most widely used stainless steels are ASS AISI 304 and 316, but DSS are expanding due to their similar or superior corrosions resistance, higher strength and lower nickel content, enabling a reduction in section sizes leading to lighter and more cost-effective structures. However, their higher strength limits their formability compared to ASS. Stainless steels are especially found in bridges where they are usually chosen for its durability combined with high ductility and strength. For example, DSS AISI 2205 has been used in the Millennium bridge footbridge (York, UK), the Cala Galdana highway bridge on the island of Menorca, the Marina Bay footbridge in Singapore and the Stonecutters bridge towers in Hong Kong. AISI 2304 has been used for the Padre Arrupe Bridge linking the Guggenheim Museum to the University of Deusto in Bilbao and for the Celtic Gateway footbridge in Holyhead, UK. AISI 2101 has been used for the Siena Bridge in Ruffolo and the Likholefossen Bridge in Norway [392, 393]. In airports, railway and bus stations and shopping centres, stainless steels have been used for roofs, railings, stairways, escalators and barriers. Seatings in public areas, playground equipment and other urban furniture are also increasingly being made from stainless steels. For example, one of the most used stainless steel in the past has been the 304 stainless steel. It is employed in urban environments for roofs and facades along with grade AISI 439. In the case of marine environments, the steel used was the 316. However, in the present the 316 is used in both environments [395].

Stainless steels are very much used as reinforcements in concrete structures. However, corrosion protection by galvanizing provides just short-term protection [396]. For this reason, novel research lines are developing coatings for stainless steel reinforcements with epoxy resin [397] or using the electrodeposition of nickel followed by galvanizing [398].

One of the last applications for the stainless steels is in sustainable buildings. The desire for energy-conserving buildings is key in the schedule for the new construction. Traditionally the improvement in the thermal insulation was carried out by thicker external walls. This is a problem in modern commercial building, where the target is to maximize the rental space. Stainless steel cladding systems used with insulation systems, such as low-cost nanotechnology, can provide very high levels of thermal insulation [399].

11.5.3 Biomedical

Metallic materials often find applications as biomaterials in human body because as compared to ceramics and polymers, they display superior tensile strength, fracture toughness and fatigue behaviour. The most widely used metallic biomaterials are stainless steels, Co-Cr alloys and Ti-alloys. Although Ti and Co-Cr alloys possess certain properties which are superior compared to stainless steels (biocompatibility, corrosion resistance, specific strength, wear resistance), these are used in large quantities because they are inexpensive by a factor of one-tenth to one-fifth [400].

Stainless steels are used in numerous medical devices such as kidney dialysis machines, hip joints, bone plates, screws and stents catheters [400]. The stainless steel that is mostly used up to date is AISI 316L; for example, this ASS is still the favoured one in the fabrication of stents [401, 402], and it can be used simultaneously as a structural support in the vessel and as an antenna for wireless telemetry capable of monitoring any cardiovascular disease [403]. However, there has been an increasing interest in producing biodegradable implants polymeric or metallic (Mg- or Fe-based) that would make unnecessary their need to be corrosion-resistant [402, 404]. AISI 316 is also employed for orthopaedic implants due to its good mechanical properties, acceptable corrosion resistance and low cost compared to other materials. Ultrafine-grained AISI 316 would also provide superior mechanical properties, corrosion resistance without affecting the biocompatibility of coarse counterparts [405, 406]. However, this austenitic steel contains a high content of Ni, and this element produces several incompatibility problems like allergic reaction with skin [407]. Furthermore, other problems have been associated with the implant of 316L, for example, pitting corrosion and stress corrosion [408]. Nonmagnetic steels are necessary for magnetic resonance imaging, since magnetic material influences the tomographic images of tissues.

Then, other austenitic steels as nitrogen-strengthened stainless steels and nickel-free austenitic stainless steels are widely employed in the last years for surgical implants [409, 410]. For example, novel Ni-free ASS for coronary stents seem to be a potential replacement for AISI 316 [411]. Many researchers have studied the effect of nitrogen on blood compatibility for these high-nitrogen nickel-free stainless steels. These work indicated that different nitrogen content had an influence on the surface energy value, which in turn influenced the haemocompatibility [411]. Other works have been focused on the study of cytocompatibility demonstrating that increasing nitrogen content allows increasing the cytocompatibility [412]. One of these low Ni and low N stainless steels, NAS 106N, is used as the implanted stainless steels electrode for functional electrical stimulation. NAS 106N has excellent fatigue strength, and a smaller amount of Ni is released in saline compared to AISI 316L [413].

Finally, a novel approach that makes use of new manufacturing techniques consists in improving the surface finishing to enhance some mechanical properties and creates the best environment for cell adhesion using additive manufacturing techniques [414–416].

11.5.4 Power Plants

Stainless steels are extensively used in the components of the power plants for high-temperature applications. The stainless steels mostly used nowadays for power plants can be divided in two main groups, ferritic steels and austenitic steels. In the ferritic steels are included the high-chromium high-alloyed steels, such as 12 wt.-% Cr martensitic steels and 12–18 wt.-% Cr ferritic steels of the AISI 400 series. In the case of austenitic steels are included 18Cr-8Ni steels and 25Cr-20Ni steels of the AISI 300 series, 21Cr-32Ni steels such as Alloy 800H and Cr–Mn steels of the AISI 200 series [355].

Alloys 12Cr-MoV (Lapelloy), 12Cr-MoVNbW (AL419 and AISI 422) and 12Cr-MoVNb (H46, FV448) have been used for steam turbine blades. However, they present some limitation of creep strength above 550 °C, which limits their application in the ultra-supercritical (USC) power plants. In Japan, the 12Cr-MoVNbWNB (TF2) was developed to increase this operation temperature.

For cases and valve bodies with steam temperatures of 600 °C and 630 °C, the 12Cr-MoVNbW (TOS 302) and 12Cr-MoVNbWCo (TOS 303) cast steels have been proposed, respectively [417].

On the other hand, high-chromium martensitic steels such as 12Cr-MoNiV (X12CrNiMo), 12Cr-MoV (X22CrMoV) and 11Cr-MoVNbN (X19CrMoVNb) have been employed in Europe for bolts in steam turbines operating at temperatures in the range over 450–565 °C. Other steels considered for bolting are 12Cr-MoNiV (M152) and 12Cr-MoVNb-WCu (FV535); though in the USA the use of the steels 12Cr-MoVNb (H46) and 12Cr- MoVW (AISI 422) has been favoured. In Japan the steel 12Cr-MoVNbW (TF3) is used [418].

Stainless steels also will play an important role in the future fusion reactors. The 12Cr-W-Mn and 12Cr-V-Mn investigated in the future materials programmes in the USA [419] appear to have good resistance to radiation damage, stable microstructures and favourable combinations of strength and toughness.

11.5.5 Household Appliances and Miscellaneous Applications

Stainless steels are in these years a highly desirable and competitive material for indoor and outdoor use. Nowadays, they are improving the appearance and functionality and are widely used in our kitchens due to their hygienic properties, which protect the food. The reduction in costs has extended the use of these steels in a variety of products such as cutlery and tableware and, more recently, by ranges, ovens, work surfaces and mixing appliances and has made these products more accessible for low-end customers.

To cite some examples, the austenitic stainless steels AISI 304 and AISI 304 L are employed in belt buckles, bottle openers, candle holders, hip flasks, key ring fobs, money clips, torch bodies and casings for pens and pencils. Concerning, ferritic stainless steels the AISI 430 is used for kitchen cabinets and worktops, sinks and drainers. A good overview can be found in reference [305].

Based on the good corrosion resistance, aesthetic appearance, ease of cleaning and for their good properties to accept any pleasing finish, stainless steels have a big impact in leisure products. Some of the typical applications are for golf clubs (17–4 PH), swimming pool filter grills, pipelines, water treatment and filter units (AISI 316 and AISI 316L) and computer discs (AISI 304 and AISI 304L).

11.6 Summary and Outlook

Since the discovery of the excellent corrosion resistance of chromium alloyed steels at the end of the nineteenth century and the patent and commercial manufacturing of the first stainless steels in the beginning of the twentieth century, these alloys have made a significant contribution to the development of mankind already for more than 100 years. In this regard, stainless steels have been sometimes defined as one of the miracle materials discovered during the last century. This statement, which may seem exaggerated, could be supported by their omnipresence of in our daily life. They are used in a number of important industries and/or environments: healthcare (including implants, surgical instruments, syringes), clean and renewable energy (nuclear, biomass, wind, solar, hydroelectric), petrochemical, food/water processing and storage (including desalination, water treatment), transport (automotive, aerospace, railcar), construction (bridges, buildings) and household appliances and utensils. All these applications are justified based on their interesting combination of mechanical properties: corrosion/oxidation resistance and mechanical (wear, strength, ductility, toughness), which can be easily tailored based on composition and processing and which make them very durable (can last tens of years without being replaced or repaired). Besides, they are hygienic (easy to keep clean and disinfected, so it does not contaminate the products they touch) and 100 % recyclable without loss of quality.

The annual production growth rate of stainless steels is currently 5.8 %, mainly driven by the Chinese and other Asian economies. This production is mainly focused on commercial grades, especially austenitic (~75 %), like AISI 304 or 316. Nevertheless, there exist a great number of investigations published yearly on new stainless steels, intended to meet the increasing and more severe industrial demands. As it has been reviewed in this chapter, current or recent ventures concerning stainless steels focus both on (i) finding alternative processing routes in existing commercial stainless steels and on (ii) optimizing the chemistry of the alloy using standard or new thermomechanical processing routes. Some try to improve the mechanical properties of existing grades, trying not to degrade other properties, by keeping the same chemistry (Q&P heat treatments and reversion heat treatments of deformed microstructures); other, on improving the mechanical properties along with the corrosion resistance (heat-resistant steels, alumina-forming steels) or finding cheaper chemistries but with the same properties as the original alloys (Cr-

Mn-N stainless steels to minimize/replace the use of Ni); or then those focused on improving the mechanical properties via the modification of the chemistry and processing route.

In summary, thanks to their singular combination of properties, stainless steels have transformed our daily life. For some applications they are unrivalled. Stainless steels arrived late to our society compared to other materials but are without doubt essential nowadays, contributing critically to having a much higher quality of life.

Acknowledgements The authors would like to acknowledge the financial support from the following research projects: DIMMAT project of Madrid region under programme S2013/MIT-2775, Ferro-Ness project of the Spanish Ministerio de Economia y Competitividad (MINECO) through the form of a Coordinate Project (MAT2016-80875-C3-1-R). PressPerfect project, RFSR-CT-2012-00021 funded by the Research Fund for Coal & Steel (RFCS Programme, Commission of the European Communities). Besides, Mr. Javier Vivas acknowledges financial support in the form of a FPI (PhD.) Grant BES-2014-069863.

The authors greatly acknowledge Prof. Carlos Garcia-de-Andres, former head of our group, for his important contributions in the field of stainless steels and Mr. Javier Vara Miñambres and Mr. Jesus Chao Hermida, former technicians, for their continuous support in the development of this research direction in our group in the last 30 years.

References

1. *World Steel in Figures 2018*. 2018 Accessed October 2018; Available from: https://www.worldsteel.org/media-centre/press-releases/2018/world-steel-in-figures-2018.html
2. H.M. Cobb, The life of Harry Brearly (1871–1948), in *The History of Stainless Steel*, (ASM International, Materials Park, 2010), pp. 33–54
3. S. Lozano-Perez et al., Atom-probe tomography characterization of the oxidation of stainless steel. Scripta Materialia **62**(11), 855–858 (2010)
4. Joseph Ki Lai Leuk, C.H.S., Kin Ho Lo, *Stainless Steels: An Introduction and Their Recent Developments* (Bentham Science Publishers, Hong Kong, 2012)
5. *Outokumpu Stainless Corrosion Handbook*, (Outokumpu Stainless Steel Oy, Finland, 2004)
6. *The Stainless Steel Family*. [Octubre 2018]; Available from: http://www.worldstainless.org/Files/issf/non-image-files/PDF/TheStainlessSteelFamily.pdf
7. M.S. Ghazani, B. Eghbali, Characterization of the hot deformation microstructure of AISI 321 austenitic stainless steel. Mater. Sci. Eng. A **730**, 380–390 (2018)
8. A.S. Lima et al., Sensitization evaluation of the austenitic stainless steel AISI 304L, 316L, 321 and 347. J. Mater. Sci. **40**(1), 139–144 (2005)
9. F. Masuyama, New developments in steels for power generation boilers, in *Advanced Heat Resistant Steels for Power Generation*, ed. by R. V. A. J. Nutting, (The Institute of Materials, London, 1998), pp. 33–48
10. Z. Wang et al., Nano-precipitates evolution and their effects on mechanical properties of 17-4 precipitation-hardening stainless steel. Acta Materialia **156**, 158–171 (2018)
11. G. Yeli et al., Sequential nucleation of phases in a 17-4PH steel: microstructural characterisation and mechanical properties. Acta Materialia **125**, 38–49 (2017)
12. C. Servant, E.H. Gherbi, G. Cizeron, TEM investigation of the tempering behaviour of the maraging PH 17.4 Mo stainless steel. J. Mater. Sci. **22**(7), 2297–2304 (1987)
13. U.K. Viswanathan, P.K.K. Nayar, R. Krishnan, Kinetics of precipitation in 17-4 PH stainless steel. Mater. Sci. Technol. **5**(4), 346–349 (1989)

14. T. Zhou et al., Quantitative electron microscopy and physically based modelling of Cu precipitation in precipitation-hardening martensitic stainless steel 15-5 PH. Mater. Des. **143**, 141–149 (2018)
15. L. Couturier et al., Evolution of the microstructure of a 15-5PH martensitic stainless steel during precipitation hardening heat treatment. Mater. Des. **107**, 416–425 (2016)
16. H.R. Habibi, The effect of ageing upon the microstructure and mechanical properties of type 15-5 PH stainless steel. Mater. Sci. Eng. A **338**(1), 142–159 (2002)
17. V. Seetharaman, M. Sundararaman, R. Krishnan, Precipitation hardening in a PH 13-8 Mo stainless steel. Mater. Sci. Eng. **47**(1), 1–11 (1981)
18. *Design Guidelines for the selection and use of Stainless Steels*, ed. A.I.a.S.I.S.S.I.o.N. America. Nickel Development Institute
19. H.K.D.H. Bhadeshia, R.W.K.H. *The* Effects of alloying elements in iron-carbon alloys, in *Steels: Microstructure and Properties*, ed. by R.W.K.H. H.K.D.H. Bhadeshia, (Butterworth-Heinemann, 2006), pp. 71–93
20. K. Laha, J. Kyono, N. Shinya, An advanced creep cavitation resistance Cu-containing 18Cr–12Ni–Nb austenitic stainless steel. Scripta Materialia **56**(10), 915–918 (2007)
21. K. Laha, J. Kyono, N. Shinya, Suppression of creep cavitation in precipitation-hardened austenitic stainless steel to enhance creep rupture strength. Trans. Indian Inst. Metals **63**(2), 437–441 (2010)
22. K.H. Lo, C.H. Shek, J.K.L. Lai, Recent developments in stainless steels. Mater. Sci. Eng. R Rep. **65**(4), 39–104 (2009)
23. C.-C. Hsieh, W. Wu, Overview of intermetallic sigma (Σ) phase precipitation in stainless steels. ISRN Metall. **2012**, 16 (2012)
24. T.M. Devine, The mechanism of sensitization of austenitic stainless steel. Corros. Sci. **30**(2), 135–151 (1990)
25. A.J. Sedriks, *Corrosion of Stainless Steel*, 2nd edn. (Wiley, New York, 1996)
26. X. Huang et al., Microstructure and property evolutions of a novel Super304H steel during high temperature creeping. Mater. High Temp. **35**(5), 438–450 (2018)
27. Q. Zhou et al., Precipitation kinetics of M23C6 carbides in the Super304H austenitic heat-resistant steel. J. Mater. Eng. Perform. **26**(12), 6130–6139 (2017)
28. C. Celada-Casero et al., Continuous hardening during isothermal aging at 723 K (450 °C) of a precipitation hardening stainless steel. Metall. Mater. Trans. A **47**(11), 5280–5287 (2016)
29. M. Yang et al., New insights into the precipitation strengthening of ferritic steels: nanoscale G-phase particle. Mater. Lett. **209**, 134–137 (2017)
30. M. Yang et al., Microstructural evolution and precipitation strengthening in a new 20Cr ferritic trial steel. Mater. Sci. Eng. A, 2018
31. P. Behjati et al., Design of a new Ni-free austenitic stainless steel with unique ultrahigh strength-high ductility synergy. Mater. Des. **63**, 500–507 (2014)
32. Y. Zhang et al., Martensite transformation behavior and mechanical properties of cold-rolled metastable Cr-Mn-Ni-N austenitic stainless steels. Mater. Sci. Eng. A **724**, 411–420 (2018)
33. J.W. Simmons, Overview: high-nitrogen alloying of stainless steels. Mater. Sci. Eng. A **207**(2), 159–169 (1996)
34. M.A.J. Somers, T.L. Christiansen, 14 – Low temperature surface hardening of stainless steel, in *Thermochemical Surface Engineering of Steels*, ed. by E. J. Mittemeijer, M. A. J. Somers, (Woodhead Publishing, Oxford, 2015), pp. 557–579
35. B.M. Gonzalez et al., The influence of copper addition on the formability of AISI 304 stainless steel. Mater. Sci. Eng. A **343**(1), 51–56 (2003)
36. H.R. Habibi, Atomic structure of the Cu precipitates in two stages hardening in maraging steel. Mater. Lett. **59**(14), 1824–1827 (2005)
37. M. Hättestrand et al., Precipitation hardening in a 12%Cr–9%Ni–4%Mo–2%Cu stainless steel. Acta Materialia **52**(4), 1023–1037 (2004)
38. M. Thuvander, M. Andersson, K. Stiller, Precipitation process of martensitic PH stainless steel Nanoflex. Mater. Sci. Technol. **28**(6), 695–701 (2012)

39. V.T. Ha, W.S. Jung, Creep behavior and microstructure evolution at 750 °C in a new precipitation-strengthened heat-resistant austenitic stainless steel. Mater. Sci. Eng. A **558**, 103–111 (2012)

40. H. Yu, W. Xu, S. van der Zwaag, On the relationship between the chromium concentration, the Z-phase formation and the creep strength of ferritic-martensitic steels. Steel Res. Int. **89**(10), 1800177 (2018)

41. X. Xiao et al., Microstructure stability of V and Ta microalloyed 12%Cr reduced activation ferrite/martensite steel during long-term aging at 650 °C. J. Mater. Sci. Technol. **31**(3), 311–319 (2015)

42. D.-B. Park et al., Effect of vanadium addition on the creep resistance of 18Cr9Ni3CuNbN austenitic stainless heat resistant steel. J. Alloys Compd. **574**, 532–538 (2013)

43. F. Liu et al., A new 12% chromium steel strengthened by Z-phase precipitates. Scripta Materialia **113**, 93–96 (2016)

44. M. Rashidi et al., Microstructure and mechanical properties of two Z-phase strengthened 12%Cr martensitic steels: the effects of Cu and C. Mater. Sci. Eng. A **694**, 57–65 (2017)

45. S. Peissl et al., Influence of chromium, molybdenum and cobalt on the corrosion behaviour of high carbon steels in dependence of heat treatment. Mater. Corros. **57**(10), 759–765 (2006)

46. M. Shibuya et al., Effect of nickel and cobalt addition on the precipitation-strength of 15Cr ferritic steels. Mater. Sci. Eng. A **528**(16), 5387–5393 (2011)

47. B. Kuhn et al., Effect of Laves phase strengthening on the mechanical properties of high Cr ferritic steels for solid oxide fuel cell interconnect application. Mater. Sci. Eng. A **528**(18), 5888–5899 (2011)

48. B. Kuhn et al., Development of high chromium ferritic steels strengthened by intermetallic phases. Mater. Sci. Eng. A **594**, 372–380 (2014)

49. Y. Yamamoto et al., Overview of strategies for high-temperature creep and oxidation resistance of alumina-forming austenitic stainless steels. Metall. Mater. Trans. A **42**(4), 922–931 (2011)

50. Y. Yamamoto, G. Muralidharan, M.P. Brady, Development of L12-ordered Ni3(Al,Ti)-strengthened alumina-forming austenitic stainless steel alloys. Scripta Materialia **69**(11), 816–819 (2013)

51. G. Trotter et al., Accelerated precipitation in the AFA stainless steel Fe–20Cr–30Ni–2Nb–5Al via cold working. Intermetallics **53**, 120–128 (2014)

52. M. Thuvander, M. Andersson, K. Stiller, Atom probe tomography investigation of lath boundary segregation and precipitation in a maraging stainless steel. Ultramicroscopy **132**, 265–270 (2013)

53. *ASM Handbook, Volume 1: Properties and selection: irons, Steels, and High-Performance Alloys*, (ASM International, 1993)

54. S.W. Chen et al., Precipitation behavior of Fe2Nb Laves phase on grain boundaries in austenitic heat resistant steels. Mater. Sci. Eng. A **616**, 183–188 (2014)

55. B. Hu et al., Effect of boron and carbon addition on microstructure and mechanical properties of the aged gamma-prime strengthened alumina-forming austenitic alloys. Intermetallics **90**, 36–49 (2017)

56. T. Angel, Formation of martensite in austenitic stainless steels. J. Iron Steel Inst. Jpn. **177**, 165–174 (1954)

57. A. Järvenpää et al., Stability of grain-refined reversed structures in a 301LN austenitic stainless steel under cyclic loading. Mater. Sci. Eng. A **703**, 280–292 (2017)

58. A. Di Schino et al., Solidification mode and residual ferrite in low-Ni austenitic stainless steels. J. Mater. Sci. **35**(2), 375–380 (2000)

59. J.Q. Guo et al., Nucleation process control of undercooled stainless steel by external nucleation seed. Acta Materialia **47**(14), 3767–3778 (1999)

60. J.C. Ma et al., Microstructural evolution in AISI 304 stainless steel during directional solidification and subsequent solid-state transformation. Mater. Sci. Eng. A **444**(1), 64–68 (2007)

61. C. Celada et al., Chemical banding revealed by chemical etching in a cold-rolled metastable stainless steel. Mater. Character. **84**, 142–152 (2013)
62. E. Folkhard, *Welding Metallurgy of Stainless Steels* (Springer, New York, 1988)
63. *Welding of Stainless steels and Other Joining Methods*. Vol. N° 9002, (American Iron and Steel Institute/Nickel Institute, Toronto, Washington, DC, 1979)
64. A.L. Schaeffler, Selection of austenitic electrodes for welding dissimilar metals. Weld. J. **26**(10), 1–20 (1947)
65. A.L. Schaeffler, Constitution diagram for stainless steel weld metal. Metal Prog. **56**(11), 680–680B (1949)
66. S. Brandi, C. Schön, A thermodynamic study of a constitutional diagram for duplex stainless steels. J. Phase Equilib. Diffus. **38**(3), 268–275 (2017)
67. W.T. DeLong, E.R. Szumachowski, Measurement and calculation of ferrite in stainless-steel weld metal. Weld. J. **35**(11), 526 (1956)
68. C.J. Long, The ferrite content of austenitic stainless steel weld metal. Weld. J. **52**(7), 281 (1973)
69. D.J. Kotecki, WRC-1992 constitution diagram for stainless steel weld metals: a modification of the WRC-1988 diagram. Weld. J. **71**(5), 171s–178s (1992)
70. Thermo-Calc Software. [cited October 2018]; Available from: https://thermocalc.com
71. MAGMA Software. [cited October 2018]; Available from: https://www.magmasoft.com
72. Y. Tian, A. Borgenstam, P. Hedström, Comparing the deformation-induced martensitic transformation with the athermal martensitic transformation in Fe-Cr-Ni alloys. J. Alloys Compd. **766**, 131–139 (2018)
73. Y. Tian et al., Deformation microstructure and deformation-induced martensite in austenitic Fe-Cr-Ni alloys depending on stacking fault energy. Metall. Mater. Trans. A **48**(1), 1–7 (2017)
74. J. Jeon, Y. Chang, Effect of nitrogen on deformation-induced martensitic transformation in an austenitic 301 stainless steels. Metals **7**(11), 503 (2017)
75. Y. Tian et al., Martensite formation during incremental cooling of Fe-Cr-Ni alloys: An in-situ bulk X-ray study of the grain-averaged and single-grain behavior. Scripta Materialia **136**, 124–127 (2017)
76. A. Borgenstam, M. Hillert, Activation energy for isothermal martensite in ferrous alloys. Acta Materialia **45**(2), 651–662 (1997)
77. D. San Martin et al., Real-time martensitic transformation kinetics in maraging steel under high magnetic fields. Mater. Sci. Eng. A **527**(20), 5241–5245 (2010)
78. D. San Martín et al., Isothermal martensitic transformation in a 12Cr–9Ni–4Mo–2Cu stainless steel in applied magnetic fields. J. Magn. Magn. Mater. **320**(10), 1722–1728 (2008)
79. D. San Martin et al., Real-time synchrotron X-ray diffraction study on the isothermal martensite transformation of maraging steel in high magnetic fields. J. Appl. Crystallogr. **45**(4), 748–757 (2012)
80. M. Villa, M.A.J. Somers, Thermally activated martensite formation in ferrous alloys. Scripta Materialia **142**, 46–49 (2018)
81. J.G. Speer et al., Analysis of microstructure evolution in quenching and partitioning automotive sheet steel. Metall. Mater. Trans. A **42**(12), 3591 (2011)
82. D.K. Matlock, J.G. Speer, Third generation of AHSS: microstructure design concepts, in *Microstructure and Texture in Steels*, (Springer, London, 2009)
83. D.K. Matlock, E. De Moor, P.J. Gibbs, Recent developments in advanced high strength sheet steels for automotive applications: an overview. JESTECH **15**(1), 1–12 (2012)
84. M. Villa, M.F. Hansen, M.A.J. Somers, Martensite formation in Fe-C alloys at cryogenic temperatures. Scripta Materialia **141**, 129–132 (2017)
85. T. Fukuda, T. Kakeshita, K. Kindo, Effect of high magnetic field and uniaxial stress at cryogenic temperatures on phase stability of some austenitic stainless steels. Mater. Sci. Eng. A **438–440**, 212–217 (2006)
86. J.-y. Choi, T. Fukuda, T. Kakeshita, Effect of magnetic field on isothermal martensitic transformation in a sensitized SUS304 austenitic stainless steel. J. Alloys Compd. **577**, S605–S608 (2013)

87. G. Ghosh, G.B. Olson, Kinetics of F.c.c. → b.c.c. heterogeneous martensitic nucleation—II. Thermal activation. Acta Metallurgica et Materialia **42**(10), 3371–3379 (1994)

88. H.K. Yeddu, Phase-field modeling of austenite grain size effect on martensitic transformation in stainless steels. Comput. Mater. Sci. **154**, 75–83 (2018)

89. H.K. Yeddu et al., Multi-length scale modeling of martensitic transformations in stainless steels. Acta Materialia **60**(19), 6508–6517 (2012)

90. E.I. Galindo-Nava, On the prediction of martensite formation in metals. Scripta Materialia **138**, 6–11 (2017)

91. E.I. Galindo-Nava, P.E.J. Rivera-Díaz-del-Castillo, Understanding martensite and twin formation in austenitic steels: a model describing TRIP and TWIP effects. Acta Materialia **128**, 120–134 (2017)

92. S.M.C. van Bohemen, L. Morsdorf, Predicting the Ms temperature of steels with a thermodynamic based model including the effect of the prior austenite grain size. Acta Materialia **125**, 401–415 (2017)

93. C. Capdevila, D.S. Martin, F.G. Caballero, C.G. de Andres, Martensite start temperature of steel: effect of alloying elements, in *Encyclopedia of iron, steel, and their alloys*, ed. by G. E. T. R. Colás, (CRC Press, New York, 2016)

94. K. Ishida, Calculation of the effect of alloying elements on the Ms temperature in steels. J. Alloys Compd. **220**(1), 126–131 (1995)

95. J. Sjoberg, Influence of analysis on the properties of stainless spring steel. Wire **23**, 155–158 (1973)

96. K.W. Andrews, Empirical formulae for the calculation of some transformation temperatures. J. Iron Steel Inst. Jpn. **203**, 721–727 (1965)

97. K. Nohara, Y. Ono, N. Ohashi, Composition and grain size dependencies of strain-induced martensitic transformation in metastable austenitic stainless steels. J. Iron Steel Inst. Jpn. **63**, 772–782 (1977)

98. W. Xu et al., A new ultrahigh-strength stainless steel strengthened by various coexisting nanoprecipitates. Acta Materialia **58**(11), 4067–4075 (2010)

99. Q. Lu, W. Xu, S. van der Zwaag, The design of a compositionally robust martensitic creep-resistant steel with an optimized combination of precipitation hardening and solid-solution strengthening for high-temperature use. Acta Materialia **77**, 310–323 (2014)

100. J. Lu et al., Stacking fault energies in austenitic stainless steels. Acta Materialia **111**, 39–46 (2016)

101. A. Das, Revisiting stacking fault energy of steels. Metall. Mater. Trans. A **47**(2), 748–768 (2016)

102. W. Sha, A. Cerezo, G.D.W. Smith, Phase chemistry and precipitation reactions in maraging steels: Part IV. Discussion and conclusions. Metall. Mater. Trans. A **24**(6), 1251–1256 (1993)

103. W. Xu, P.E.J.R.D.D. Castillo, S.V.D. Zwaag, A combined optimization of alloy composition and aging temperature in designing new UHS precipitation hardenable stainless steels. Comput. Mater. Sci. **45**(2), 467–473 (2009)

104. H. Leitner, M. Schober, R. Schnitzer, Splitting phenomenon in the precipitation evolution in an Fe–Ni–Al–Ti–Cr stainless steel. Acta Materialia **58**(4), 1261–1269 (2010)

105. Y. Yamamoto et al., Alumina-forming austenitic stainless steels strengthened by Laves phase and MC carbide precipitates. Metall. Mater. Trans. A **38**(11), 2737–2746 (2007)

106. Y.-S. Ji et al., Long-term evolution of σ phase in 304H austenitic stainless steel: experimental and computational investigation. Mater. Character. **128**, 23–29 (2017)

107. D. San Martin et al., A new etching route for revealing the austenite grain boundaries in an 11.4% Cr precipitation hardening semi-austenitic stainless steel. Mater. Character. **58**(5), 455–460 (2007)

108. T. Sourmail, Precipitation in creep resistant austenitic stainless steels. Mater. Sci. Technol. **17**(1), 1–14 (2001)

109. R.L.P. Angelo Fernando Padilha, P.R. Rios, Stainless steel: heat treatment, in *Encyclopedia of iron, steel, and their alloys*, ed. by R. C. George, E. Totten, (Taylor and Francis, New York, 2016), pp. 3255–3282

110. L. Karlsson, Intermetallic phase precipitation in duplex stainless steels and weld metals: metallurgy, influence on properties and welding aspects. Weld. Res. Counc. Bull. **43**(5), 1–23 (1999)
111. E.B.P.A.R.C. Nunnington, Stainless steel slag fundamentals: from furnace to Tundish. Ironmak. Steelmak. **29**(2), 133–139 (2002)
112. H.-J. Odenthal et al., Simulation of fluid flow and oscillation of the argon oxygen decarburization (AOD) process. Metall. Mater. Trans. B **41**(2), 396–413 (2010)
113. S. Ikeda, Technical progress of stainless steel and its future trend, in *Nippon steel technical report. 2010, Nippon Steel: Otemachi, Chiyoda-ku*, (Tokyo), pp. 1–7
114. *Standard Specification for General Requirements for Flat-Rolled Stainless and Heat-Resisting Steel Plate, Sheet, and Strip, ASTM A480/A480M-18*, (ASTM International, West Conshohocken, 2018), pp. 1–26
115. *Handbook of Stainless Steel*, (Outokumpu Stainless AB, Riihitontuntie, Finland, 2013)
116. (SSINA), S.S.I.o.N.A. Stainless Steel Production. 2018 [cited 2018 October 2018]; Available from: http://www.ssina.com/overview/stainless-production.html
117. *Standard Practice for Cleaning, Descaling, and Passivation of Stainless Steel Parts, Equipment, and Systems, ASTM A380–06*, (ASTM International: West Conshohocken, Pennsylvania, EEUU, 2006), pp. 1–13
118. America, S.S.I.o.N. Stainless Steel Production. 2018 [cited October 2018; Available from: http://www.ssina.com/overview/stainless-production.html
119. H.J. Cross, L.S. Levy, S. Sadhra, T. Sorahan, C. McRoy, Manufacture, processing and use of stainless steel: a review of the health effects, in *European Confederation of Iron and Steel Industries (EUROFER)*, (1999)
120. P. Erkkilä, Trends and challenges in the stainless steel industry. Ironmak. Steelmak. **31**(4), 277–284 (2004)
121. A. Hishinuma, S. Takaki, K. Abiko, Recent progress and future R & D for high-chromium iron-base and chromium-base alloys. Physica Status Solidi (a) **189**(1), 69–78 (2002)
122. Z.Y. Liu et al., The correlation between yielding behavior and precipitation in ultra purified ferritic stainless steels. Mater. Sci. Eng. A **527**(16), 3800–3806 (2010)
123. V. Talyan, R.H. Wagoner, J.K. Lee, Formability of stainless steel. Metall. Mater. Trans. A **29**(8), 2161–2172 (1998)
124. K. Komori, Analysis of longitudinal buckling in temper rolling. ISIJ Int. **49**(3), 408–415 (2009)
125. *ASM Handbook Volume 14A: Metalworking: Bulk Forming*, (ASM International, 2005), pp. 1–888
126. *ASM Handbook Volume 14B: Metalworking: Sheet Forming*, (ASM International, 2006)
127. Kotecki, J.C.L.A.D.J., *Welding Metallurgy and Weldability of Stainless Steels*. 2005
128. *ASM Handbook Volume 6: Welding, Brazing, and Soldering* (ASM International, 1993)
129. *ASM Handbook Volume 16: Machining*, (ASM International, 1989), pp. 1–944
130. B. Kim et al., The influence of silicon in tempered martensite: Understanding the microstructure–properties relationship in 0.5–0.6wt.% C steels. Acta Materialia **68**, 169–178 (2014)
131. S. Queyreau, G. Monnet, B. Devincre, Orowan strengthening and forest hardening superposition examined by dislocation dynamics simulations. Acta Materialia **58**(17), 5586–5595 (2010)
132. P.E.J. Rivera-Díaz-del-Castillo, K. Hayashi, E.I. Galindo-Nava, Computational design of nanostructured steels employing irreversible thermodynamics. Mater. Sci. Technol. **29**(10), 1206–1211 (2013)
133. H.K.D.H. Bhadeshia, Modelling of microstructure and properties, in *Steels: Microstructure and Properties*, (Butterworth-Heinemann, 2006), pp. 307–334
134. M. Huang et al., Modelling strength and ductility of ultrafine grained BCC and FCC alloys using irreversible thermodynamics. Mater. Sci. Technol. **25**(7), 833–839 (2009)
135. H.K.D.H. Bhadeshia, The strengthening of iron and its alloys, in *Steels: Microstructure and Properties*, ed. by R. W. K. H. H. K. D. H. Bhadeshia, (Butterworth-Heinemann, 2006), pp. 17–38

136. H. Suzuki, *Dislocations and Mechanical Properties of Crystals* (Wiley, New York, 1957)
137. N.F. Mott, F.R.N.N., Dislocation theory and transient creep, in *Report of a conference on strength of solids, The Physical Society*, (University of Bristol, 1948), pp. 1–19
138. J. Eliasson, R. Sandström, Proof strength values for austenitic stainless steels at elevated temperatures. Steel Res. **71**(6–7), 249–254 (2000)
139. N. Ohkubo et al., Effect of alloying elements on the mechanical properties of the stable austenitic stainless steel. ISIJ Int. **34**(9), 764–772 (1994)
140. H. Sieurin, J. Zander, R. Sandström, Modelling solid solution hardening in stainless steels. Mater. Sci. Eng. A **415**(1), 66–71 (2006)
141. R. Labusch, Statistische theorien der mischkristallhärtung. Acta Metallurgica **20**(7), 917–927 (1972)
142. F.R.N. Nabarro, The theory of solution hardening. Philos. Mag. A **35**(3), 613–622 (1977)
143. L. Vegard, Die Konstitution der Mischkristalle und die Raumfüllung der Atome. Zeitschrift für Physik **5**(1), 17–26 (1921)
144. K.J. Irvine, F.B. Pickering, The strength of austenitic stainless steels. J. Iron Steel Inst. Jpn. **119**, 1017–1028 (1969)
145. F.B. Pickering, *The Metallurgical Evolution of Stainless Steel* (Metals Park, 1979)
146. B. Leffler, *Nordic Symposium on Mechanical Properties of Stainless Steels (SIMR)* (Sigtuna, 1990)
147. H. Nordberg, La Metallurgia Italiana. **85**, 147–154 (1994)
148. Y. Nakada, A.S. Keh, Solid solution strengthening in Fe-N single crystals. Acta Metallurgica **16**(7), 903–914 (1968)
149. R.D.K. Misra et al., Ultrahigh strength hot rolled microalloyed steels: microstructural aspects of development. Mater. Sci. Technol. **17**(9), 1119–1129 (2001)
150. M.J. Kelley, N.S. Stoloff, Effect of chromium on low-temperature deformation of high-purity iron. Metall. Trans. A **7**(2), 331–333 (1976)
151. I. Toda-Caraballo, P.E.J. Rivera-Díaz-del-Castillo, Modelling solid solution hardening in high entropy alloys. Acta Materialia **85**, 14–23 (2015)
152. I. Toda-Caraballo, A general formulation for solid solution hardening effect in multicomponent alloys. Scripta Materialia **127**, 113–117 (2017)
153. C. Varvenne et al., Solute strengthening in random alloys. Acta Materialia **124**, 660–683 (2017)
154. C. Varvenne, A. Luque, W.A. Curtin, Theory of strengthening in fcc high entropy alloys. Acta Materialia **118**, 164–176 (2016)
155. M. Walbrühl et al., Modelling of solid solution strengthening in multicomponent alloys. Mater. Sci. Eng. A **700**, 301–311 (2017)
156. Y.H. Kim, K.Y. Kim, Y.D. Lee, Nitrogen-alloyed, metastable austenitic stainless steel for automotive structural applications. Mater. Manuf. Process. **19**(1), 51–59 (2004)
157. W.S. Park et al., Strain-rate effects on the mechanical behavior of the AISI 300 series of austenitic stainless steel under cryogenic environments. Mater. Des. **31**(8), 3630–3640 (2010)
158. M. Moallemi et al., Deformation-induced martensitic transformation in a 201 austenitic steel: the synergy of stacking fault energy and chemical driving force. Mater. Sci. Eng. A **653**, 147–152 (2016)
159. E. Nes, Modelling of work hardening and stress saturation in FCC metals. Prog. Mater. Sci. **41**(3), 129–193 (1997)
160. R.A. Barrett, P.E. O'Donoghue, S.B. Leen, A dislocation-based model for high temperature cyclic viscoplasticity of 9–12Cr steels. Comput. Mater. Sci. **92**, 286–297 (2014)
161. G.E. Dieter, *Mechanical Metallurgy*, Crystal Research and Technology, 3rd edn. (Mc Graw-Hill Book Co., New York, 1986)
162. E.I. Galindo-Nava, W.M. Rainforth, P.E.J. Rivera-Díaz-del-Castillo, Predicting microstructure and strength of maraging steels: elemental optimisation. Acta Materialia **117**, 270–285 (2016)
163. F. Christien, M.T.F. Telling, K.S. Knight, Neutron diffraction in situ monitoring of the dislocation density during martensitic transformation in a stainless steel. Scripta Materialia **68**(7), 506–509 (2013)

164. R.L. Klueh, N. Hashimoto, P.J. Maziasz, Development of new nano-particle-strengthened martensitic steels. Scripta Materialia **53**(3), 275–280 (2005)
165. J.W. Martin, *Precipitation Hardening: Theory and Applications*, 2nd edn. (Butterworth-Heinemann, Oxford, 1998)
166. T. Gladman, Precipitation hardening in metals. Mater. Sci. Technol. **15**(1), 30–36 (1999)
167. G.S. Ansell, F.V. Lenel, Criteria for yielding of dispersion-strengthened alloys. Acta Metallurgica **8**(9), 612–616 (1960)
168. E.V. Pereloma et al., Ageing behaviour of an Fe–20Ni–1.8Mn–1.6Ti–0.59Al (wt%) maraging alloy: clustering, precipitation and hardening. Acta Materialia **52**(19), 5589–5602 (2004)
169. R. Schnitzer, S. Zinner, H. Leitner, Modeling of the yield strength of a stainless maraging steel. Scripta Materialia **62**(5), 286–289 (2010)
170. R. Taillard, A. Pineau, Room temperature tensile properties of Fe-19wt.%Cr alloys precipitation hardened by the intermetallic compound NiAl. Mater. Sci. Eng. **56**(3), 219–231 (1982)
171. A.J. Ardell, Precipitation hardening. Metall. Trans. A **16**(12), 2131–2165 (1985)
172. J.S. Zhang, *High Temperature Deformation and Fracture of Materials*, 1st edn. (Woodhead Publishing, 2010)
173. E.O. Hall, The deformation and ageing of mild steel: III discussion of results. Proc. Phys. Soc. Sect. B **64**(9), 747 (1951)
174. N.J. Petch, The cleavage strength of polycrystals. J. Iron Steel Inst. Jpn. **174**(1), 25–28 (1953)
175. B.P. Kashyap, K. Tangri, Hall-Petch relationship and substructural evolution in boron containing type 316L stainless steel. Acta Materialia **45**(6), 2383–2395 (1997)
176. A. Di Schino, M. Barteri, J.M. Kenny, Grain size dependence of mechanical, corrosion and tribological properties of high nitrogen stainless steels. J. Mater. Sci. **38**(15), 3257–3262 (2003)
177. F. Forouzan et al., Production of nano/submicron grained AISI 304L stainless steel through the martensite reversion process. Mater. Sci. Eng. A **527**(27), 7334–7339 (2010)
178. R.D.K. Misra et al., On the significance of nature of strain-induced martensite on phase-reversion-induced nanograined/ultrafine-grained austenitic stainless steel. Metall. Mater. Trans. A **41**(1), 3 (2009)
179. I. Shakhova et al., Effect of large strain cold rolling and subsequent annealing on microstructure and mechanical properties of an austenitic stainless steel. Mater. Sci. Eng. A **545**, 176–186 (2012)
180. C.X. Huang et al., Mechanical behaviors of ultrafine-grained 301 austenitic stainless steel produced by equal-channel angular pressing. Metall. Mater. Trans. A **42**(7), 2061–2071 (2011)
181. J. Huang et al., Enhanced mechanical properties of type AISI301LN austenitic stainless steel through advanced thermo mechanical process. Mater. Sci. Eng. A **532**, 190–195 (2012)
182. J. Zhao, Z. Jiang, Thermomechanical processing of advanced high strength steels. Prog. Mater. Sci. **94**, 174–242 (2018)
183. Y. Ma, J.-E. Jin, Y.-K. Lee, A repetitive thermomechanical process to produce nano-crystalline in a metastable austenitic steel. Scripta Materialia **52**(12), 1311–1315 (2005)
184. S.V. Dobatkin et al., Structural changes in metastable austenitic steel during equal channel angular pressing and subsequent cyclic deformation. Mater. Sci. Eng. A **723**, 141–147 (2018)
185. F.Y. Dong et al., Optimizing strength and ductility of austenitic stainless steels through equal-channel angular pressing and adding nitrogen element. Mater. Sci. Eng. A **587**, 185–191 (2013)
186. M. El-Tahawy et al., Exceptionally high strength and good ductility in an ultrafine-grained 316L steel processed by severe plastic deformation and subsequent annealing. Mater. Lett. **214**, 240–242 (2018)
187. H. Wang et al., Annealing behavior of nano-crystalline austenitic SUS316L produced by HPT. Mater. Sci. Eng. A **556**, 906–910 (2012)
188. A. Belyakov et al., Substructures and internal stresses developed under warm severe deformation of austenitic stainless steel. Scripta Materialia **42**(4), 319–325 (2000)

189. Y. Nakao, H. Miura, Nano-grain evolution in austenitic stainless steel during multi-directional forging. Mater. Sci. Eng. A **528**(3), 1310–1317 (2011)
190. M. Tikhonova et al., Grain boundary assembles developed in an austenitic stainless steel during large strain warm working. Mater. Character. **70**, 14–20 (2012)
191. B. Wang, Z. Liu, J. Li, Microstructure evolution in AISI201 austenitic stainless steel during the first compression cycle of multi-axial compression. Mater. Sci. Eng. A **568**, 20–24 (2013)
192. W. Pachla et al., Nanostructurization of 316L type austenitic stainless steels by hydrostatic extrusion. Mater. Sci. Eng. A **615**, 116–127 (2014)
193. G.S. Sun et al., The significant role of heating rate on reverse transformation and coordinated straining behavior in a cold-rolled austenitic stainless steel. Mater. Sci. Eng. A **732**, 350–358 (2018)
194. G.B. Olson, Effects of stress and deformation on martensite formation, in *Encyclopedia of Materials: Science and Technology*, ed. by R. W. C. K. H. Jürgen Buschow, M. C. Flemings, B. Ilschner, E. J. Kramer, S. Mahajan, P. Veyssière, (Pergamon, 2001), pp. 2381–2384
195. M.F. McGuire, *Stainless Steels for Design Engineers* (ASM International, 2008)
196. B.C. De Cooman, Y. Estrin, S.K. Kim, Twinning-induced plasticity (TWIP) steels. Acta Materialia **142**, 283–362 (2018)
197. A. Grajcar, R. Kuziak, W. Zalecki, Third generation of AHSS with increased fraction of retained austenite for the automotive industry. Arch. Civil Mech. Eng. **12**(3), 334–341 (2012)
198. G. Fargas, J.J. Roa, A. Mateo, Influence of pre-existing martensite on the wear resistance of metastable austenitic stainless steels. Wear **364–365**, 40–47 (2016)
199. J.Y. Yun et al., Effect of strain-induced ε- and α'-martensitic transformations on the sliding wear resistance in austenitic Fe–Cr–C–Mn alloys. Wear **368–369**, 124–131 (2016)
200. P. Hedström et al., Stepwise transformation behavior of the strain-induced martensitic transformation in a metastable stainless steel. Scripta Materialia **56**(3), 213–216 (2007)
201. J. Talonen, H. Hänninen, Formation of shear bands and strain-induced martensite during plastic deformation of metastable austenitic stainless steels. Acta Materialia **55**(18), 6108–6118 (2007)
202. S. Martin et al., Deformation mechanisms in austenitic TRIP/TWIP Steel as a function of temperature. Metall. Mater. Trans. A **47**(1), 49–58 (2016)
203. G. Meric de Bellefon, J.C. van Duysen, Tailoring plasticity of austenitic stainless steels for nuclear applications: review of mechanisms controlling plasticity of austenitic steels below 400 °C. J. Nucl. Mater. **475**, 168–191 (2016)
204. M. Hatano et al., Presence of ε-martensite as an intermediate phase during the strain-induced transformation of SUS304 stainless steel. Philos. Mag. Lett. **96**(6), 220–227 (2016)
205. Y.F. Shen et al., Twinning and martensite in a 304 austenitic stainless steel. Mater. Sci. Eng. A **552**, 514–522 (2012)
206. S. Martin, C. Ullrich, D. Rafaja, Deformation of austenitic CrMnNi TRIP/TWIP steels: nature and role of the ε-martensite. Mater. Today Proc. **2**, S643–S646 (2015)
207. N. Nakada et al., Deformation-induced martensitic transformation behavior in cold-rolled and cold-drawn type 316 stainless steels. Acta Materialia **58**(3), 895–903 (2010)
208. S. Curtze, V.T. Kuokkala, Dependence of tensile deformation behavior of TWIP steels on stacking fault energy, temperature and strain rate. Acta Materialia **58**(15), 5129–5141 (2010)
209. G. Meric de Bellefon, J.C. van Duysen, K. Sridharan, Composition-dependence of stacking fault energy in austenitic stainless steels through linear regression with random intercepts. J. Nucl. Mater. **492**, 227–230 (2017)
210. S. Curtze et al., Thermodynamic modeling of the stacking fault energy of austenitic steels. Acta Materialia **59**(3), 1068–1076 (2011)
211. L. Vitos, J.O. Nilsson, B. Johansson, Alloying effects on the stacking fault energy in austenitic stainless steels from first-principles theory. Acta Materialia **54**(14), 3821–3826 (2006)
212. P.J. Brofman, G.S. Ansell, On the effect of carbon on the stacking fault energy of austenitic stainless steels. Metall. Trans. A **9**(6), 879–880 (1978)
213. R.E. Schramm, R.P. Reed, Stacking fault energies of seven commercial austenitic stainless steels. Metall. Trans. A **6**(7), 1345 (1975)

214. T.-H. Lee et al., Correlation between stacking fault energy and deformation microstructure in high-interstitial-alloyed austenitic steels. Acta Materialia **58**(8), 3173–3186 (2010)
215. T. Masumura et al., The difference in thermal and mechanical stabilities of austenite between carbon- and nitrogen-added metastable austenitic stainless steels. Acta Materialia **84**, 330–338 (2015)
216. T. Masumura et al., Difference between carbon and nitrogen in thermal stability of metastable 18%Cr-8%Ni austenite. Scripta Materialia **154**, 8–11 (2018)
217. K.-S. Kim, J.-H. Kang, S.-J. Kim, Effects of carbon and nitrogen on precipitation and tensile behavior in 15Cr-15Mn-4Ni austenitic stainless steels. Mater. Sci. Eng. A **712**, 114–121 (2018)
218. T.-H. Lee, C.-S. Oh, S.-J. Kim, Effects of nitrogen on deformation-induced martensitic transformation in metastable austenitic Fe–18Cr–10Mn–N steels. Scripta Materialia **58**(2), 110–113 (2008)
219. K. Nohara, Y. Ono, N. Ohashi, Composition and grain size dependencies of strain-induced martensitic transformation in metastable austenitic stainless steels. Tetsu-to-Hagane **63**(5), 772–782 (1977)
220. M. Vach et al., Evolution of secondary phases in austenitic stainless steels during long-term exposures at 600, 650 and 800 °C. Mater. Character. **59**(12), 1792–1798 (2008)
221. J.W. Simmons et al., Effect of nitride (Cr_2N) precipitation on the mechanical, corrosion, and wear properties of austenitic stainless steel. ISIJ Int. **36**(7), 846–854 (1996)
222. A. Kisko et al., The influence of grain size on the strain-induced martensite formation in tensile straining of an austenitic 15Cr–9Mn–Ni–Cu stainless steel. Mater. Sci. Eng. A **578**, 408–416 (2013)
223. M. Cristobal et al., Rapid fabrication and characterization of AISI 304 stainless steels modified with Cu additions by additive alloy melting (ADAM). J. Mater. Res. Technol. (2018)
224. B. Hwang, Influence of Cu and Ni on ductile-brittle transition behavior of metastable austenitic Fe-18Cr-10Mn-N alloys. Korean J. Mater. Res. **23**(7), 7 (2013)
225. J. Charles, The new 200-series: an alternative answer to Ni surcharge Risks or opportunities? Rev. Met. Paris. **104**, 10 (2007)
226. M. Eskandari et al., Potential application of nanocrystalline 301 austenitic stainless steel in lightweight vehicle structures. Mater. Des. **30**(9), 3869–3872 (2009)
227. O. Bouaziz et al., High manganese austenitic twinning induced plasticity steels: a review of the microstructure properties relationships. Curr. Opin. Solid State Mater. Sci. **15**(4), 141–168 (2011)
228. L. Chen, Y. Zhao, X. Qin, Some aspects of high manganese twinning-induced plasticity (TWIP) steel, a review. Acta Metallurgica Sinica (English Letters) **26**(1), 1–15 (2013)
229. M. Talha et al., Effect of cold working on biocompatibility of Ni-free high nitrogen austenitic stainless steels using Dalton's Lymphoma cell line. Mater. Sci. Eng. C **35**, 77–84 (2014)
230. M. Hauser et al., Anomalous stabilization of austenitic stainless steels at cryogenic temperatures. Mater. Sci. Eng. A **675**, 415–420 (2016)
231. H. Biermann, J. Solarek, A. Weidner, SEM investigation of high-alloyed austenitic stainless cast steels with varying austenite stability at room temperature and 100 °C. Steel Res. Int. **83**(6), 512–520 (2012)
232. K. Tomimura et al., Optimal chemical composition in Fe-Cr-Ni alloys for ultra grain refining by reversion from deformation induced martensite. ISIJ Int. **31**(7), 721–727 (1991)
233. D.M. Xu et al., Deformation behavior of high yield strength – high ductility ultrafine-grained 316LN austenitic stainless steel. Mater. Sci. Eng. A **688**, 407–415 (2017)
234. V.S.A. Challa et al., Significance of interplay between austenite stability and deformation mechanisms in governing three-stage work hardening behavior of phase-reversion induced nanograined/ultrafine-grained (NG/UFG) stainless steels with high strength-high ductility combination. Scripta Materialia **86**, 60–63 (2014)
235. L.P. Karjalainen et al., Some strengthening methods for austenitic stainless steels. Steel Res. Int. **79**(6), 404–412 (2008)

236. J. Liu et al., High-cycle fatigue behavior of 18Cr-8Ni austenitic stainless steels with grains ranging from nano/ultrafine-size to coarse. Mater. Sci. Eng. A **733**, 128–136 (2018)
237. H. Ueno et al., Enhanced fatigue properties of nanostructured austenitic SUS 316L stainless steel. Acta Materialia **59**(18), 7060–7069 (2011)
238. D.M. Xu et al., The significant impact of cold deformation on structure-property relationship in phase reversion-induced stainless steels. Mater. Character. **145**, 157–171 (2018)
239. D.M. Xu et al., The effect of annealing on the microstructural evolution and mechanical properties in phase reversed 316LN austenitic stainless steel. Mater. Sci. Eng. A **720**, 36–48 (2018)
240. C. Lei et al., Deformation mechanism and ductile fracture behavior in high strength high ductility nano/ultrafine grained Fe-17Cr-6Ni austenitic steel. Mater. Sci. Eng. A **709**, 72–81 (2018)
241. A. Järvenpää et al., Austenite stability in reversion-treated structures of a 301LN steel under tensile loading. Mater. Character. **127**, 12–26 (2017)
242. N. Gong et al., Effect of martensitic transformation on nano/ultrafine-grained structure in 304 austenitic stainless steel. J. Iron Steel Res. Int. **24**(12), 1231–1237 (2017)
243. V.S.A. Challa et al., Influence of grain structure on the deformation mechanism in martensitic shear reversion-induced Fe-16Cr-10Ni model austenitic alloy with low interstitial content: Coarse-grained versus nano-grained/ultrafine-grained structure. Mater. Sci. Eng. A **661**, 51–60 (2016)
244. M. Naghizadeh, H. Mirzadeh, Microstructural evolutions during annealing of plastically deformed AISI 304 austenitic stainless steel: martensite reversion, grain refinement, recrystallization, and grain growth. Metall. Mater. Trans. A **47**(8), 4210–4216 (2016)
245. R.D.K. Misra et al., Relationship of grain size and deformation mechanism to the fracture behavior in high strength–high ductility nanostructured austenitic stainless steel. Mater. Sci. Eng. A **626**, 41–50 (2015)
246. A.S. Hamada et al., Enhancement of mechanical properties of a TRIP-aided austenitic stainless steel by controlled reversion annealing. Mater. Sci. Eng. A **628**, 154–159 (2015)
247. H.S. Baghbadorani et al., Influence of Nb-microalloying on the formation of nano/ultrafine-grained microstructure and mechanical properties during martensite reversion process in a 201-type austenitic stainless steel. Metall. Mater. Trans. A **46**(8), 3406–3413 (2015)
248. M. Shirdel, H. Mirzadeh, M.H. Parsa, Nano/ultrafine grained austenitic stainless steel through the formation and reversion of deformation-induced martensite: mechanisms, microstructures, mechanical properties, and TRIP effect. Mater. Character. **103**, 150–161 (2015)
249. A. Rezaee et al., The influence of reversion annealing behavior on the formation of nanograined structure in AISI 201L austenitic stainless steel through martensite treatment. Mater. Des. **32**(8), 4437–4442 (2011)
250. B. Ravi Kumar, S. Sharma, B. Mahato, Formation of ultrafine grained microstructure in the austenitic stainless steel and its impact on tensile properties. Mater. Sci. Eng. A **528**(6), 2209–2216 (2011)
251. A. Poulon et al., Fine grained austenitic stainless steels: the role of strain induced α' martensite and the reversion mechanism limitations. ISIJ Int. **49**(2), 293–301 (2009)
252. Y. Xiong et al., Cryorolling impacts on microstructure and mechanical properties of AISI 316 LN austenitic stainless steel. Mater. Sci. Eng. A **709**, 270–276 (2018)
253. C. Zheng et al., Microstructure and mechanical behavior of an AISI 304 austenitic stainless steel prepared by cold- or cryogenic-rolling and annealing. Mater. Sci. Eng. A **724**, 260–268 (2018)
254. R. Singh et al., Mechanical behavior of 304 Austenitic stainless steel processed by cryogenic rolling. Mater. Today Proc. **5**(9, Part 1), 16880–16886 (2018)
255. P. Mallick et al., Microstructure-tensile property correlation in 304 stainless steel after cold deformation and austenite reversion. Mater. Sci. Eng. A **707**, 488–500 (2017)
256. Y.F. Shen et al., Suppression of twinning and phase transformation in an ultrafine grained 2GPa strong metastable austenitic steel: Experiment and simulation. Acta Materialia **97**, 305–315 (2015)

257. C. Sun et al., Thermal stability of ultrafine grained Fe–Cr–Ni alloy. Mater. Sci. Eng. A **542**, 64–70 (2012)
258. S. Pour-Ali et al., Correlation between the surface coverage of severe shot peening and surface microstructural evolutions in AISI 321: A TEM, FE-SEM and GI-XRD study. Surf. Coat. Technol. **334**, 461–470 (2018)
259. Y. He et al., Study of the austenitic stainless steel with gradient structured surface fabricated via shot peening. Mater. Lett. **215**, 187–190 (2018)
260. G.S. Sun et al., Ultrahigh strength nano/ultrafine-grained 304 stainless steel through three-stage cold rolling and annealing treatment. Mater. Character. **110**, 228–235 (2015)
261. L. Jinlong et al., Effect of ultrafine grain on tensile behaviour and corrosion resistance of the duplex stainless steel. Mater. Sci. Eng. C **62**, 558–563 (2016)
262. C. Celada Casero, D. San Martín, Austenite formation in a cold-rolled semi-austenitic stainless steel. Metall. Mater. Trans. A **45**(4), 1767–1777 (2014)
263. C. Celada-Casero et al., Mechanisms of ultrafine-grained austenite formation under different isochronal conditions in a cold-rolled metastable stainless steel. Mater. Character. **118**, 129–141 (2016)
264. M.C. Somani et al., Enhanced mechanical properties through reversion in metastable austenitic stainless steels. Metall. Mater. Trans. A **40**(3), 729–744 (2009)
265. S. Sabooni et al., The role of martensitic transformation on bimodal grain structure in ultrafine grained AISI 304L stainless steel. Mater. Sci. Eng. A **636**, 221–230 (2015)
266. D.L. Johannsen, A. Kyrolainen, P.J. Ferreira, Influence of annealing treatment on the formation of nano/submicron grain size AISI 301 Austenitic stainless steels. Metall. Mater. Trans. A **37**(8), 2325–2338 (2006)
267. S. Rajasekhara et al., Hall–Petch behavior in ultra-fine-grained AISI 301LN stainless steel. Metall. Mater. Trans. A **38**(6), 1202–1210 (2007)
268. P. Behjati et al., Effect of annealing temperature on nano/ultrafine grain of Ni-free austenitic stainless steel. Mater. Sci. Eng. A **592**, 77–82 (2014)
269. A. Kisko et al., Effects of reversion and recrystallization on microstructure and mechanical properties of Nb-alloyed low-Ni high-Mn austenitic stainless steels. Mater. Sci. Eng. A **657**, 359–370 (2016)
270. E.S.T. Platform, *From a Strategic Research Agenda to Implementation* (European Communities, 2006)
271. D. Peckner, *I.M.B., Handbook of Stainless Steels* (McGraw-Hill Book Company, New York, 1977)
272. J.W. Christian, S. Mahajan, Deformation twinning. Prog. Mater. Sci. **39**(1), 1–157 (1995)
273. K. Lu, L. Lu, S. Suresh, Strengthening materials by engineering coherent internal boundaries at the nanoscale. Science **324**(5925), 349–352 (2009)
274. K. Lu et al., Strengthening austenitic steels by using nanotwinned austenitic grains. Scripta Materialia **66**(11), 878–883 (2012)
275. L. Lu et al., Ultrahigh strength and high electrical conductivity in copper. Science **304**(5669), 422–426 (2004)
276. Y.F. Shen et al., Tensile properties of copper with nano-scale twins. Scripta Materialia **52**(10), 989–994 (2005)
277. Z.H. Jin et al., Interactions between non-screw lattice dislocations and coherent twin boundaries in face-centered cubic metals. Acta Materialia **56**(5), 1126–1135 (2008)
278. Y.S. Li, N.R. Tao, K. Lu, Microstructural evolution and nanostructure formation in copper during dynamic plastic deformation at cryogenic temperatures. Acta Materialia **56**(2), 230–241 (2008)
279. H.Y. Yi et al., Work hardening behavior of nanotwinned austenitic grains in a metastable austenitic stainless steel. Scripta Materialia **114**, 133–136 (2016)
280. H.Y. Yi et al., Comparison of strength–ductility combinations between nanotwinned austenite and martensite–austenite stainless steels. Mater. Sci. Eng. A **647**, 152–156 (2015)
281. F.K. Yan, N.R. Tao, K. Lu, Tensile ductility of nanotwinned austenitic grains in an austenitic steel. Scripta Materialia **84–85**, 31–34 (2014)

282. F.K. Yan et al., Strength and ductility of 316L austenitic stainless steel strengthened by nanoscale twin bundles. Acta Materialia **60**(3), 1059–1071 (2012)
283. T.R. Lee, C.P. Chang, P.W. Kao, The tensile behavior and deformation microstructure of cryorolled and annealed pure nickel. Mater. Sci. Eng. A **408**(1), 131–135 (2005)
284. Maraging steels, in *ASM Handbook, Volume 1: Properties and Selection: Irons, Steels, and High-Performance Alloys*, (ASM International, 1993), pp. 1869–1887
285. Z.B. Jiao et al., Co-precipitation of nanoscale particles in steels with ultra-high strength for a new era. Materials Today **20**(3), 142–154 (2017)
286. W. Sha, Maraging steels: microstructure during thermal processing, in *Encyclopedia of Iron, Steel, and Their Alloys*, ed. by G. E. T. Rafael Colás, (Taylor and Francis, New York, 2016), pp. 2128–2139
287. R. Tewari et al., Precipitation in 18 wt% Ni maraging steel of grade 350. Acta Materialia **48**(5), 1187–1200 (2000)
288. R. Schnitzer et al., Effect of Cu on the evolution of precipitation in an Fe–Cr–Ni–Al–Ti maraging steel. Acta Materialia **58**(10), 3733–3741 (2010)
289. Y. Li et al., A new 1.9GPa maraging stainless steel strengthened by multiple precipitating species. Mater. Des. **82**, 56–63 (2015)
290. W. Xu, P.E.J. Rivera-Díaz-del-Castillo, S. van der Zwaag, Designing nanoprecipitation strengthened UHS stainless steels combining genetic algorithms and thermodynamics. Comput. Mater. Sci. **44**(2), 678–689 (2008)
291. K. Stiller, H.O. Andrén, M. Andersson, Precipitation in maraging and martensitic chromium steels – what can we learn using 3-DAP and EFTEM. Mater. Sci. Technol. **24**(6), 633–640 (2008)
292. P. Liu, Relationships between microstructure and properties of stainless steels – a few working examples. Mater. Character. **44**(4), 413–424 (2000)
293. P. Liu, A.H. Stigenberg, J.O. Nilsson, Isothermally formed quasicrystalline precipitates used for strengthening in a new maraging stainless steel. Scripta Metallurgica et Materialia **31**(3), 249–254 (1994)
294. J.O. Nilsson, A. Hultin Stigenberg, P. Liu, Isothermal formation of quasicrystalline precipitates and their effect on strength in a 12Cr-9Ni-4Mo maraging stainless steel. Metall. Mater. Trans. A **25**(10), 2225–2233 (1994)
295. M. Holmquist, J.O. Nilsson, A.H. Stigenberg, Isothermal formation of martensite in a 12Cr-9Ni-4Mo maraging stainless steel. Scripta Metallurgica et Materialia **33**(9), 1367–1373 (1995)
296. J. Post, C. de Vries, J. Huetink, Validation tool for 2D multi-stage metal-forming processes on meta-stable stainless steels. **209**, 5558–5572 (2009)
297. K. Datta et al., A low-temperature study to examine the role of ε-martensite during strain-induced transformations in metastable austenitic stainless steels. Acta Materialia **57**(11), 3321–3326 (2009)
298. F. Akhtar et al., A new kind of age hardenable martensitic stainless steel with high strength and toughness. Ironmak. Steelmak. **34**(4), 285–289 (2007)
299. M. Schober, R. Schnitzer, H. Leitner, Precipitation evolution in a Ti-free and Ti-containing stainless maraging steel. Ultramicroscopy **109**(5), 553–562 (2009)
300. A. Gemperle et al., Aging behaviour of cobalt free chromium containing maraging steels. Mater. Sci. Technol. **8**(6), 546–554 (1992)
301. J. Speer et al., Carbon partitioning into austenite after martensite transformation. Acta Materialia **51**(9), 2611–2622 (2003)
302. C. Lesch, N. Kwiaton, F.B. Klose, Advanced High Strength Steels (AHSS) for automotive applications – tailored properties by smart microstructural adjustments. Steel Res. Int. **88**(10), 1700210 (2017)
303. *Advanced High-Strength Steels Application Guidelines (AHSS Guidelines) Version 6.0*, ed. M. K. Stuart Keeler, Peter J. Mooney. (WorldAutoSteel, 2017)
304. *Ultra Light Steel Autobody-Advanced Vehicle Concepts (ULSAB-AVC). Overview Report.* 2002: Autosteel.org

305. C.C. Carlos García de Andrés, D.S. Martín, Structural steels, in *Encyclopedia of Iron, Steel, and Their Alloys*, ed. by G. E. T. R. Colás, (CRC Press, New York, 2016), pp. 3388–3409
306. E. Jimenez-Melero et al., Characterization of individual retained austenite grains and their stability in low-alloyed TRIP steels. Acta Materialia **55**(20), 6713–6723 (2007)
307. E. De Moor et al., Austenite stabilization through manganese enrichment. Scripta Materialia **64**(2), 185–188 (2011)
308. E. Jimenez-Melero et al., The effect of aluminium and phosphorus on the stability of individual austenite grains in TRIP steels. Acta Materialia **57**(2), 533–543 (2009)
309. T. Tsuchiyama et al., Quenching and partitioning treatment of a low-carbon martensitic stainless steel. Mater. Sci. Eng. A **532**, 585–592 (2012)
310. C. García de Andrés, G. Caruana, L.F. Alvarez, Control of M23C6 carbides in 0.45C–13Cr martensitic stainless steel by means of three representative heat treatment parameters. Mater. Sci. Eng. A **241**(1), 211–215 (1998)
311. A. Bjärbo, M. Hättestrand, Complex carbide growth, dissolution, and coarsening in a modified 12 pct chromium steel – an experimental and theoretical study. Metall. Mater. Trans. A **32**(1), 19–27 (2001)
312. C. Garcia de Andres et al., Effects of carbide-forming elements on the response to thermal treatment of the X45Cr13 martensitic stainless steel. J. Mater. Sci. **33**(16), 4095–4100 (1998)
313. K.J. Irvine, *The Metallurgical Evolution of Stainless Steels: A Discriminative Selection of Outstanding Articles and Papers from the Scientific Literature* (American Society for Metals, Ohio, EEUU, 1979), pp. 43–62
314. J. Mola, B.C. De Cooman, Quenching and partitioning processing of transformable ferritic stainless steels. Scripta Materialia **65**(9), 834–837 (2011)
315. Q. Huang et al., Influence of martensite fraction on tensile properties of quenched and partitioned (Q&P) martensitic stainless steels. Steel Res. Int. **87**(8), 1082–1094 (2016)
316. J. Tobata et al., Role of silicon in quenching and partitioning treatment of low-carbon martensitic stainless steel. ISIJ Int. **52**(7), 1377–1382 (2012)
317. J. Mola, B.C. De Cooman, Quenching and partitioning (Q&P) processing of martensitic stainless steels. Metall. Mater. Trans. A **44**(2), 946–967 (2013)
318. M. Wendler et al., Quenching and partitioning (Q&P) processing of fully austenitic stainless steels. Acta Materialia **133**, 346–355 (2017)
319. Steels, K.S., *Save the Weight*. 2018; Available from: http://www.kvastainless.com/automotive.html
320. C. Bruno, J.G.S. De Cooman, *Fundamentals of Steel Product Physical Metallurgy (AIST)* (Association for Iron and Steel Technology, Warrendale, EEUU, 2011), p. 642
321. A.P.A.R. Botte, Nitrogen addition in steelmaking using nitriding ferroalloys, in *High Nitrogen Steels*, (The Institute of Metals, Lille, 1988)
322. S.-J. Lee, D.K. Matlock, C.J. Van Tyne, Carbon diffusivity in multi-component austenite. Scripta Materialia **64**(9), 805–808 (2011)
323. S. Björklund, L.F. Donaghey, M. Hillert, The effect of alloying elements on the rate of ostwald ripening of cementite in steel. Acta Metallurgica **20**(7), 867–874 (1972)
324. L. Yuan et al., Nanoscale austenite reversion through partitioning, segregation and kinetic freezing: example of a ductile 2GPa Fe–Cr–C steel. Acta Materialia **60**(6), 2790–2804 (2012)
325. F. Yan et al., An investigation of secondary carbides in the spray-formed high alloyed Vanadis 4 steel during tempering. Mater. Character. **59**(7), 883–889 (2008)
326. R. Szilard, J. Busby, The case for extended nuclear reactor operation. J. Metal. **61**, 24–27 (2009)
327. R.L.K.A.D.R. Harries, *High-Chromium Ferritic and Martensitic Steels for Nuclear Applications* (American Society for Testing and Materials (ASTM, West Conshohocken, 2001)
328. L. Tan et al., Recent status and improvement of reduced-activation ferritic-martensitic steels for high-temperature service. J. Nucl. Mater. **479**, 515–523 (2016)
329. Y. Zhou et al., Precipitation and hot deformation behavior of austenitic heat-resistant steels: a review. J. Mater. Sci. Technol. **33**(12), 1448–1456 (2017)

330. Y. Sawaragi, H. Senba, S. Yamamoto, Properties of a new 18-8 austenitic steel tube (SUPER 304H) for fossil fired boilers after service exposure with high elevated temperature strength. Sumitomo Search **56**, 34–43 (1994)

331. D.-B. Park et al., High-temperature creep behavior and microstructural evolution of an 18Cr9Ni3CuNbVN austenitic stainless steel. Mater. Character. **93**, 52–61 (2014)

332. V.T. Ha, W.S. Jung, Effects of heat treatment processes on microstructure and creep properties of a high nitrogen 15Cr–15Ni austenitic heat resistant stainless steel. Mater. Sci. Eng. A **528**(24), 7115–7123 (2011)

333. J. Bakajová et al., Influence of annealing conditions on microstructure and phase occurrence in high-alloy CrMnN steels. Mater. Character. **61**(10), 969–974 (2010)

334. G. Franke, C. Altstetter, Low-cycle fatigue behavior of Mn/N stainless steels. Metall. Trans. A **7**(11), 1719–1727 (1976)

335. H.-b. Li et al., High nitrogen austenitic stainless steels manufactured by nitrogen gas alloying and adding nitrided ferroalloys. J. Iron Steel Res. Int. **14**(3), 64–69 (2007)

336. K.-H. Lee et al., Effect of Nb and Cu on the high temperature creep properties of a high Mn–N austenitic stainless steel. Mater. Character. **83**, 49–57 (2013)

337. Y. Yamamoto et al., Creep-resistant, Al_2O_3-forming austenitic stainless steels. Science **316**(5823), 433–436 (2007)

338. H. Bei et al., Aging effects on the mechanical properties of alumina-forming austenitic stainless steels. Mater. Sci. Eng. A **527**(7), 2079–2086 (2010)

339. D.V.V. Satyanarayana, G. Malakondaiah, D.S. Sarma, Characterization of the age-hardening behavior of a precipitation-hardenable austenitic steel. Mater. Character. **47**(1), 61–65 (2001)

340. Y. Yamamoto et al., Effect of alloying additions on phase equilibria and creep resistance of alumina-forming austenitic stainless steels. Metall. Mater. Trans. A **40**(8), 1868–1880 (2009)

341. Y. Yamamoto et al., Alloying effects on creep and oxidation resistance of austenitic stainless steel alloys employing intermetallic precipitates. Intermetallics **16**(3), 453–462 (2008)

342. I. Baker et al., Preliminary creep testing of the alumina-forming austenitic stainless steel Fe-20Cr-30Ni-2Nb-5Al. Mater. Sci. Eng. A **718**, 492–498 (2018)

343. Y. Yamamoto et al., Evaluation of Mn substitution for Ni in alumina-forming austenitic stainless steels. Mater. Sci. Eng. A **524**(1), 176–185 (2009)

344. B. Zhao et al., Formation of L12-ordered precipitation in an alumina-forming austenitic stainless steel via Cu addition and its contribution to creep/rupture resistance. Scripta Materialia **109**, 64–67 (2015)

345. M.-H. Jang et al., Improved creep strength of alumina-forming austenitic heat-resistant steels through W addition. Mater. Sci. Eng. A **696**, 70–79 (2017)

346. M.P. Brady et al., Composition, microstructure, and water vapor effects on internal/external oxidation of alumina-forming austenitic stainless steels. Oxid. Metal. **72**(5), 311 (2009)

347. G. Singh et al., Enhancing the high temperature plasticity of a Cu-containing austenitic stainless steel through grain boundary strengthening. Mater. Sci. Eng. A **602**, 77–88 (2014)

348. L.K. Mansur et al., Materials needs for fusion, Generation IV fission reactors and spallation neutron sources – similarities and differences. J. Nucl. Mater. **329–333**, 166–172 (2004)

349. K.L. Murty, I. Charit, Structural materials for Gen-IV nuclear reactors: challenges and opportunities. J. Nucl. Mater. **383**(1), 189–195 (2008)

350. P.J. Maziasz, Overview of microstructural evolution in neutron-irradiated austenitic stainless steels. J. Nucl. Mater. **205**, 118–145 (1993)

351. C. Sun et al., Superior radiation-resistant nanoengineered austenitic 304L stainless steel for applications in extreme radiation environments. Sci. Rep. **5**, 7801 (2015)

352. L. Tan et al., Microstructure optimization of austenitic alloy 800H (Fe–21Cr–32Ni). Mater. Sci. Eng. A **528**(6), 2755–2761 (2011)

353. L. Tan, T.R. Allen, J.T. Busby, Grain boundary engineering for structure materials of nuclear reactors. J. Nucl. Mater. **441**(1), 661–666 (2013)

354. F.G.A.D.L. Porter, *Dimensional Stability and Mechanical Behaviour of Irradiated Metals and Alloys* (British Nuclear Energy Society, Brighton, 1983)

355. F. Masuyama, History of power plants and progress in heat resistant steels. ISIJ Int. **41**(6), 612–625 (2001)

356. R. Viswanathan, W. Bakker, Materials for ultrasupercritical coal power plants—Boiler materials: Part 1. J. Mater. Eng. Perform. **10**(1), 81–95 (2001)

357. H.K. Danielsen, J. Hald, Behaviour of Z phase in 9–12%Cr steels. Energy Mater. **1**(1), 49–57 (2006)

358. H.K. Danielsen, J. Hald, Influence of Z-phase on long-term creep stability of martensitic 9 to 12% Cr steels. VGB PowerTech **5**, 68–73 (2009)

359. K. K., High-chromium containing ferrite based heat resistant steel N.R.I.f. Metals. 2001: Japan

360. P. Lamp et al., Development of an auxiliary power unit with solid oxide fuel cells for automotive applications. Fuel Cells **3**(3), 146–152 (2003)

361. N.Q. Minh, Solid oxide fuel cell technology – features and applications. Solid State Ionics **174**(1), 271–277 (2004)

362. Y.-T. Chiu, C.-K. Lin, J.-C. Wu, High-temperature tensile and creep properties of a ferritic stainless steel for interconnect in solid oxide fuel cell. J. Power Sourc. **196**(4), 2005–2012 (2011)

363. Y.-T. Chiu, C.-K. Lin, Effects of Nb and W additions on high-temperature creep properties of ferritic stainless steels for solid oxide fuel cell interconnect. J. Power Sourc. **198**, 149–157 (2012)

364. J. Lopez Barrilao, B. Kuhn, E. Wessel, Identification, size classification and evolution of Laves phase precipitates in high chromium, fully ferritic steels. Micron **101**, 221–231 (2017)

365. J. Lopez Barrilao et al., Microstructure of intermetallic particle strengthened high-chromium fully ferritic steels. Mater. Sci. Technol. **33**(9), 1056–1064 (2017)

366. J. Lopez Barrilao, B. Kuhn, E. Wessel, Microstructure evolution and dislocation behaviour in high chromium, fully ferritic steels strengthened by intermetallic Laves phases. Micron **108**, 11–18 (2018)

367. P.J. Ennis et al., Microstructural stability and creep rupture strength of the martensitic steel P92 for advanced power plant. Acta Materialia **45**(12), 4901–4907 (1997)

368. H.J. Beattie Jun, F.L. Versnyder, A new complex phase in a high-temperature alloy. Nature **178**, 208 (1956)

369. Q. Lu, W. Xu, S.V.D. Zwaag, Designing new corrosion resistant ferritic heat resistant steel based on optimal solid solution strengthening and minimisation of undesirable microstructural components. Comput. Mater. Sci. **84**, 198–205 (2014)

370. M. Shibuya et al., Effect of precipitation behavior on creep strength of 15%Cr ferritic steels at high temperature between 923 and 1023K. Mater. Sci. Eng. A **592**, 1–5 (2014)

371. M. Shibuya et al., Improving the high-temperature creep strength of 15Cr ferritic creep-resistant steels at temperatures of 923–1023K. Mater. Sci. Eng. A **652**, 1–6 (2016)

372. F. Abe, Creep rates and strengthening mechanisms in tungsten-strengthened 9Cr steels. Mater. Sci. Eng. A **319–321**, 770–773 (2001)

373. A. Fedoseeva et al., Effect of tungsten on creep behavior of 9%Cr–3%Co martensitic steels. Metals **7**(12), 573 (2017)

374. A. Fedoseeva et al., On effect of rhenium on mechanical properties of a high-Cr creep-resistant steel. Mater. Lett. **236**, 81–84 (2019)

375. J. Vivas et al., Importance of austenitization temperature and ausforming on creep strength in 9Cr ferritic/martensitic steel. Scripta Materialia **153**, 14–18 (2018)

376. M. Tamura et al., Effect of MX type particles on creep strength of ferritic steel. J. Nucl. Mater. **321**(2–3), 288–293 (2003)

377. M. Rashidi et al., Mechanistic insights into the transformation processes in Z-phase strengthened 12% Cr steels. Mater. Des. **158**, 237–247 (2018)

378. F. Abe, M. Taneike, K. Sawada, Alloy design of creep resistant 9Cr steel using a dispersion of nano-sized carbonitrides. Int. J. Press. Vessels Piping **84**(1–2), 3–12 (2007)

379. L. Tan, L.L. Snead, Y. Katoh, Development of new generation reduced activation ferritic-martensitic steels for advanced fusion reactors. J. Nucl. Mater. **478**, 42–49 (2016)

380. F. Abe, Effect of boron on microstructure and creep strength of advanced ferritic power plant steels. Proc. Eng. **10**, 94–99 (2011)

381. H.K. Danielsen, J. Hald, Influence of Z-phase on long-term creep stability of martensitic 9–12%Cr steels. VGB PowerTech **5**, 68–73 (2009)

382. J. Vivas et al., Effect of ausforming temperature on creep strength of G91 investigated by means of Small Punch Creep Tests. Mater. Sci. Eng. A **728**, 259–265 (2018)

383. S. Hollner et al., High-temperature mechanical properties improvement on modified 9Cr–1Mo martensitic steel through thermomechanical treatments. J. Nucl. Mater. **405**(2), 101–108 (2010)

384. L. Tan, Y. Yang, J.T. Busby, Effects of alloying elements and thermomechanical treatment on 9Cr Reduced Activation Ferritic–Martensitic (RAFM) steels. J. Nucl. Mater. **442**(1–3) Supplement 1, S13–S17 (2013)

385. L. Tan et al., Microstructure control for high strength 9Cr ferritic–martensitic steels. J. Nucl. Mater. **422**(1–3), 45–50 (2012)

386. R.L. Klueh, N. Hashimoto, P.J. Maziasz, New nano-particle-strengthened ferritic/martensitic steels by conventional thermo-mechanical treatment. J. Nucl. Mater. (Part A), **367**, 48–370, 53 (2007)

387. J. Vivas et al., Microstructural degradation and creep fracture behavior of conventionally and thermomechanically treated 9% chromium heat resistant steel. Metal Mater Int (2018)

388. E. Plesiutschnig et al., Ferritic phase transformation to improve creep properties of martensitic high Cr steels. Scripta Materialia **122**, 98–101 (2016)

389. R.S. Vidyarthy, D.K. Dwivedi, A comparative study on creep behavior of AISI 409 ferritic stainless steel in as-received and as-welded condition (A-TIG and M-TIG). Mater. Today Proc. **5**(9, Part 1), 17097–17106 (2018)

390. J. Charles, B.V. Duplex stainless steels for chemical tankers, in *5th World Conference on Duplex Stainless Steels*, (KCI Publishing, Maastricht, 1997)

391. *Railcars in Stainless Steel, A Sustainable Solution for Sustainable Public Transport.* 2017 [cited 2018 October]; Available from: http://www.worldstainless.org/Files/ISSF/non-image-files/PDF/ISSF_Railcars_in_Stainless_Steel.pdf

392. N.R. Baddoo, Stainless steel in construction: a review of research, applications, challenges and opportunities. J. Constr. Steel Res. **64**(11), 1199–1206 (2008)

393. G. Gedge, Structural uses of stainless steel – buildings and civil engineering. J. Constr. Steel Res. **64**(11), 1194–1198 (2008)

394. B. Rossi, Discussion on the use of stainless steel in constructions in view of sustainability. Thin-Walled Struct. **83**, 182–189 (2014)

395. I. Odnevall Wallinder et al., Release rates of chromium and nickel from 304 and 316 stainless steel during urban atmospheric exposure – a combined field and laboratory study. Corros. Sci. **44**(10), 2303–2319 (2002)

396. J.A. González, C. Andrade, Effect of carbonation, chlorides and relative ambient humidity on the corrosion of galvanized rebars embedded in concrete. Br. Corros. J. **17**(1), 21–28 (1982)

397. K. Miyauchi et al., A study of adhesion on stainless steel in an epoxy/dicyandiamide coating system: influence of glass transition temperature on wet adhesion. Prog. Organ. Coat. **99**, 302–307 (2016)

398. K. Sarkar et al., Investigation of microstructure and corrosion behaviour of prior nickel deposited galvanised steels. Surf. Coat. Technol. **348**, 64–72 (2018)

399. S. Brunner et al., Vacuum insulation panels for building applications – continuous challenges and developments. Energy Build. **85**, 592–596 (2014)

400. M. Niinomi, M. Nakai, J. Hieda, Development of new metallic alloys for biomedical applications. Acta Biomaterialia **8**(11), 3888–3903 (2012)

401. G. Mani et al., Coronary stents: a materials perspective. Biomaterials **28**(9), 1689–1710 (2007)

402. S.H. Im, Y. Jung, S.H. Kim, Current status and future direction of biodegradable metallic and polymeric vascular scaffolds for next-generation stents. Acta Biomaterialia **60**, 3–22 (2017)

403. E.Y. Chow et al., Evaluation of cardiovascular stents as antennas for implantable wireless applications. IEEE Trans. Microw. Theory Tech. **57**(10), 2523–2532 (2009)

404. H. Hermawan, D. Dubé, D. Mantovani, Developments in metallic biodegradable stents. Acta Biomaterialia **6**(5), 1693–1697 (2010)

405. S.V. Muley et al., An assessment of ultra fine grained 316L stainless steel for implant applications. Acta Biomaterialia **30**, 408–419 (2016)

406. R.D.K. Misra et al., Understanding the impact of grain structure in austenitic stainless steel from a nanograined regime to a coarse-grained regime on osteoblast functions using a novel metal deformation–annealing sequence. Acta Biomaterialia **9**(4), 6245–6258 (2013)

407. D.A. Basketter et al., Nickel, cobalt and chromium in consumer products: a role in allergic contact dermatitis? Contact Dermatitis **28**(1), 15–25 (1993)

408. M. Sivakumar, U. Kamachi Mudali, S. Rajeswari, Investigation of failures in stainless steel orthopaedic implant devices: fatigue failure due to improper fixation of a compression bone plate. J. Mater. Sci. Lett. **13**(2), 142–145 (1994)

409. V. Vats, T. Baskaran, S.B. Arya, Tribo-corrosion study of nickel-free, high nitrogen and high manganese austenitic stainless steel. Tribol. Int. **119**, 659–666 (2018)

410. M. Li et al., Study of biocompatibility of medical grade high nitrogen nickel-free austenitic stainless steel in vitro. Mater. Sci. Eng. C **43**, 641–648 (2014)

411. Y. Ren et al., In vitro study on a new high nitrogen nickel-free austenitic stainless steel for coronary stents. J. Mater. Sci. Technol. **27**(4), 325–331 (2011)

412. T. Ma et al., Cytocompatibility of high nitrogen nickel-free stainless steel for orthopedic implants. J. Mater. Sci. Technol. **28**(7), 647–653 (2012)

413. M. Niinomi, Recent metallic materials for biomedical applications. Metall. Mater. Trans. A **33**(3), 477 (2002)

414. S. Bose, S.F. Robertson, A. Bandyopadhyay, Surface modification of biomaterials and biomedical devices using additive manufacturing. Acta Biomaterialia **66**, 6–22 (2018)

415. A. Riemer et al., On the fatigue crack growth behavior in 316L stainless steel manufactured by selective laser melting. Eng. Fract. Mech. **120**, 15–25 (2014)

416. V.K. Balla et al., Laser surface modification of 316L stainless steel with bioactive hydroxyapatite. Mater. Sci. Eng. C **33**(8), 4594–4598 (2013)

417. R.L. Klueh, *High-chromium ferritic and martensitic steels for nuclear applications* (American Society for Testing and Materials, 2001)

418. F. Abe, *Creep resistant steel* (Woodhead Publishing, 2008)

419. R.L. Klueh, Developing steels for service in fusion reactors. JOM **44**(4), 20–24 (1992)

Chapter 12
Electrical Steels

Carlos Capdevila

Abbreviations and Symbols

Ac1	Temperature at which the austenite is fully transformed during continuous cooling
Ac3	Temperature at which the austenite starts to decompose during continuous cooling
AlN	Aluminium nitride
B	Magnetic induction
ECC	Electron channelling contrast
FRT	Finishing rolling temperature
GO	Grain-oriented electrical steels
H	Magnetic intensity
HiB	High-permeability steel
HT	Reheating temperature
ICR	intermediate cold rolling
ND	Normal direction
NGO	Non-grain-oriented electrical steels
P_h	hysteresis losses
P_{tot}	Total loss
PVD	Physical vapour deposition
RD	Rolling direction
RT	Roughing temperature
TD	Transverse direction
TEP	Thermoelectric power

C. Capdevila (✉)
Physical Metallurgy Department, National Centre for Metallurgical Research (CENIM-CSIC), Madrid, Spain
e-mail: ccm@cenim.csic.es

© Springer Nature Switzerland AG 2021
R. Rana (ed.), *High-Performance Ferrous Alloys*,
https://doi.org/10.1007/978-3-030-53825-5_12

567

w_h	Energy per unit volume spent on describing the hysteresis cycle
μ_o	Vacuum permeability
ρ	Electrical resistivity
χ	Magnetic susceptibility

12.1 Introduction

Electric steels are a special type of flat products characterized by a low carbon content, less than 0.005 wt.%, and silicon contents between 0.3 and 4 wt.%. Within the production of steels, they do not represent more than 1% worldwide [27] but, despite this, it is one of the most important materials for electromagnetic components, such as generators, electric motors or transformers, where heat or so-called core loss is produced by eddy currents and hysteresis.

12.2 Development Chronology of Electrical Steels

Electrical steels belong to the group of so-called soft magnetic materials (soft magnetic materials are considered those whose coercive field is less than 1 kA/m). Amorphous or nanocrystalline materials, base alloys FeNi and FeCo, Ferrites, sintered magnetic materials and composite materials are also placed in this same group of materials. All these materials are used to conduct the magnetic flux and are characterized by their ability to magnetize and demagnetize, and their low coercive field.

The application of electrical steels comprises machines and devices for the generation and distribution of electricity, where the most important parameters are power loss, permeability, orientation, insulation or stress and temperature sensitivity of properties. It has been estimated than 5% of all electricity generated is dissipated by heat or so-called core loss produced by eddy currents and hysteresis, and that the annual energy loss produced in this way is 20 times the energy initially required to produce electrical steels [159]. The importance of electrical steels is closely related to the development of industrial equipment such as pumps, compressors, air conditioning equipment, lighting, automation, transportation, electromagnetic components for audio and video, etc. [27]. Table 12.1 lists the typical applications of electrical steels [97].

The first transformers in the late nineteenth century were mainly manufactured with wrought-iron cores. In the beginning of the twentieth century, the works of Barret and Brown [161] highlighted the benefits of the addition of silicon to iron. But it was not until the publication of the work of Hadfield and co-workers on the production of 3 wt.% silicon steel in 1903 with a significant effect on loss reduction, that the technology of electrical steels really began [35, 36]. The addition of small amounts of silicon to the iron demonstrated to reduce the magnetic losses significantly. Silicon allows, among others, to reduce the currents induced by

Table 12.1 Typical applications of electrical steels [97]

Application	Non grain-oriented steel			Grain-Oriented steel	
	Silicon free	Low silicon	High silicon	Conventional	High permeability
Small motors					
Lamp ballasts					
Medium AC motors					
Welding transformers					
Audio transformers					
Small power transformers					
Large rotating machines					
Medium generators					
Distribution transformers					
Power transformers					

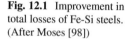

Fig. 12.1 Improvement in total losses of Fe-Si steels. (After Moses [98])

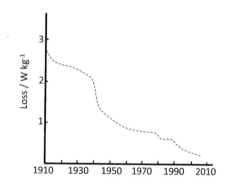

increasing the resistivity, increase the magnetic permeability and reduce the harmful effects of carbon, since it makes difficult the formation of carbides that impede the movement of the magnetic domains. Figure 12.1 illustrates the evolution of core losses at magnetic field of 1 T, 50 Hz, with time.

Beyond purely economic indicators, there are also environmental reasons for the development and improvement of this type of material. Thus, the United Nations Framework Convention on Climate Change, meeting in 1997 in Kyoto, establishes the reduction in carbon dioxide emissions as an urgent process for the resolution of environmental problems and the promotion of sustainable development. In this framework, special attention must be paid to the improvement of magnetic properties, since this leads to a decrease in energy consumption and, consequently, to a reduction in energy consumption, and therefore, the carbon dioxide emissions.

Iron and iron-based alloys are the main soft magnetic materials used today. Figure 12.2 shows the evolution of electrical steel in the twentieth century. The improvement of the magnetic properties of these materials has been carried out, mainly, through the elimination of the impurities in the material and the control of the crystallographic orientation.

In the early times, the lamination was carried out through 'manual' laminators, where the operators introduced the preheated steel sheets between the rolls of the laminator initially driven by steam and then electrically. The lamination

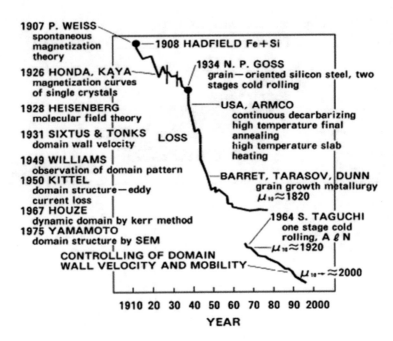

Fig. 12.2 Historical development of core loss reduction in electrical steel. (After Xia et al. [162])

temperatures were between 800 and 1000 °C. The scale that emerged on the surface was removed using of acid baths.

Development of cold rolling process in 1930 further reduced the thickness and better magnetic properties could be achieved in electrical steels. The production of long thin sheets allowed the possibility of eliminating one of the most harmful impurities for electrical steels: carbon. It was necessary to obtain thin sheets so that, when subjected to a decarburizing atmosphere, the carbon could diffuse to the surface. In this way carbon is removed from the material to amounts less than 0.002 wt.%, where the magnetic ageing due to carbon no longer occurs. Cold rolling allowed new developments of electrical steels since 1980 further on.

It was not until 1934 when Goss developed the idea from the work of Beck [3] and Ruder [133] that iron has high-magnetic permeability in certain crystallographic directions and planes. In this type of steel, the grains are oriented such that the <100> direction is parallel to the direction of rolling and the (110) planes are parallel to the surface of sheet. American company Armco and later Nippon Steel developed a process that allowed to control the texture of the cold-rolled material [128]. The process involved two-stage cold reduction and intermediate annealing, followed by decarburization, batch annealing and flattening. Modern day production of electrical steels still relies on this texture which is commonly known as Goss texture and the steel is popularly called grain oriented (GO) steel. These steels have superior magnetic properties in the direction of rolling and, thus, are applicable to transformers where the magnetic flux always has the same direction.

Nippon Steel produced a new type of high-permeability steel known as HiB [150]. The manufacturing process was different from the Goss's steel as only one cold rolling reduction was applied, and using aluminium nitride (AlN) as the primary inhibitor along with manganese sulphide (MnS) as the secondary. The angle of orientation with the rolling direction was decreased from 7° to 3° but the grain size increased from 0.3 mm to 1 cm.

Introduction of stress coatings facilitated further the reduction in power loss and magnetostriction. When the coating of the sheets obtained was required to achieve electrical insulation between stacked sheets, it was resorted to spraying an aqueous suspension of kaolin on the surface thereof. The ever increasing demand to increase efficiency led to new domain refining techniques, and Nippon Steel developed laser scribing [60] where losses were found to be 0.85 W/Kg at 1.7 T for a thickness of 0.23 mm. These losses were 5–8% less than the unscratched HiB steel [97]. A recoat on the damaged surface was necessary as the coating vaporized. Other methods have also been employed to achieve the same where grooves were made on steel on which the ceramic coating was deposited [59].

12.3   Brief Introduction to Magnetic Properties in Steels

Iron, together with nickel and cobalt, is one of the few ferromagnetic metals at room temperature. This property, the ferromagnetism, is intimately linked to the electronic structure of the metal. Smallman and Bhisop [147] presented a nice review which describes the reasons for the theory of ferromagnetism even today not being completely understood. Nevertheless, from the electron theory of metals, it is possible to build up a broad picture of ferromagnetic materials which explain not only their ferromagnetic properties but also the associated high resistivity and electronic specific heat of those metals compared to copper. In recent years considerable experimental work has been done on the electronic behaviour of the transition elements, and this suggests that the electronic structure of iron is somewhat different to that of cobalt and nickel.

Ferromagnetism, like paramagnetism, has its origin in the electron spin. In ferromagnetic materials, however, permanent magnetism is obtained, and this indicates that there is a tendency for electron spins to remain aligned in one direction even when the field has been removed. Therefore, microscopically, the ferromagnetic materials constitute of small magnets or saturated magnetic domains within which the magnetic moments of the atoms are aligned in the same direction. On a macroscopic scale this magnetization is not observable due to the random orientation of the magnetic domains of the material. These magnetic domains are known as Weiss domains and are separated from one another by the Bloch walls [15]. In general, each of the crystals that form a polycrystalline material can have its own domain structure, as shown in Fig. 12.3.

When a steel is subjected to an external magnetic field, part of the energy is invested in magnetizing and another is lost by heat generation. The loss of energy

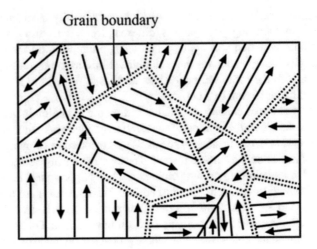

Fig. 12.3 Domain structure in a polycrystal

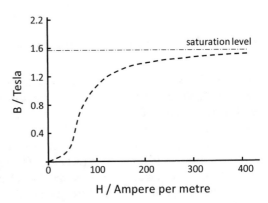

Fig. 12.4 Magnetization curve of commercial annealed iron [15]

related to this heat dissipation is known as total loss (P_{tot}) and is traditionally separated into two terms: hysteresis losses (P_h) and dynamic losses (P_d) [15, 127, 131, 132, 134]. Hysteresis losses derive from the inherent mechanism of magnetization of the material, while dynamic losses are caused by currents induced in the material by applying a variable magnetic field.

The word hysteresis has its origin in the Greek 'hysteros', which means 'delay', in reference to the phenomenon according to which the state of a material depends on its previous history and is made evident by the delay of the effect on the cause that produces it. Within the field of magnetism, hysteresis is manifested in the relationship between magnetic induction $\vec{B}$ and magnetic intensity $\vec{H}$. Consider a sample of ferromagnetic material. If the magnetic intensity, initially zero, is increased, then the relation $\vec{B} - \vec{H}$ will describe a curve similar to that in Fig. 12.4, which is the curve of the first magnetization of the material.

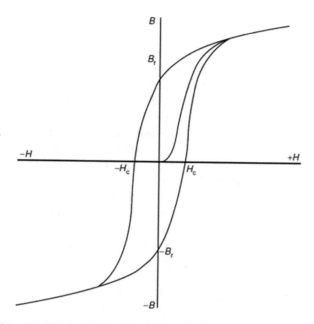

Fig. 12.5 Schematic of the hysteresis cycle of a ferromagnetic material

The slope of this curve is called magnetic permeability of the material, μ, and indicates the ease of magnetization of a material. The relationship between the magnetization of the material $\vec{M}$, the magnetic induction $\vec{B}$ and the magnetic intensity $\vec{H}$ is given by

$$\vec{B} = \mu_o \left(\vec{H} + \vec{M} \right) = \mu \vec{H} \tag{12.1}$$

where μ_o is the vacuum permeability and $\vec{H} = \chi \vec{M}$ where χ is the magnetic susceptibility of the environment. If the field $\vec{H}$ is increased to produce the saturation of the material (saturation magnetization $\vec{M}$ s) and then $\vec{H}$ is decreased, the relation $\vec{B} - \vec{H}$ does not return down the same curve as in Fig. 12.4, but now moves on the new curve of Fig. 12.5.

The magnetization $\vec{M}$, (and consequently $\vec{B}$) once established, does not disappear with the elimination of $\vec{H}$; in fact, an inverted magnetic intensity is required to reduce the magnetization to zero. If $\vec{H}$ continues to increase in the opposite direction then the magnetization will be established in the opposite direction. Finally, when $\vec{H}$ increases again, the relationship $\vec{B} - \vec{H}$ follows the lower curve until the cycle is completely closed. This curve is called the hysteresis cycle of the material. In Fig. 12.5, B_r stands for the so-called remnant magnetization and the magnitude of H_c is the so-called coercive field. The area enclosed by the hysteresis curve (w_h) represents the energy per unit volume spent on describing the hysteresis cycle.

The discontinuous process of magnetization causes the hysteresis losses: the small magnetic domains, separated by Bloch walls, have to grow by movement of the domain walls and later rotate towards the direction of the applied field. In this process, the interaction among the walls of the magnetic domains, the particles and defects present in the steel occurs. Precisely it is the existence of these defects and precipitates that causes that the magnetization is not a continuous process but takes place in jumps each time the walls find an obstacle and overcome it. In this way, hysteresis losses depend on metallurgical factors such as the presence of impurities, grain boundaries, dislocations, precipitates, etc. Also, the texture of the material, that is, the orientation of the grains in the material, influences the losses by hysteresis because, as will be seen later, the magnetic anisotropy of the material leads to easier magnetization in certain crystallographic directions.

Another important source of losses is caused by induced currents in the material when exposed to variable magnetic fields. These currents, named as Foucault currents, take the form of cycles or turns in the plane perpendicular to the magnetic flux, in such a way that they create a magnetic field opposite to the inductor field. The losses due to this phenomenon are proportional to the square of the frequency of the applied field and the thickness of the material. Therefore, besides reducing the frequency of work form the common 50 or 60 Hz, there are three other techniques for decreasing this effect [4]:

- Decrease the steel sheet thickness, which reduces the extension of the induced currents. This is a commonly used practice, and thus most of the magnetic loops are formed by stacked plates.

  Electrically isolate some plates from others, in order that the induced currents do not propagate between them. This insulation is achieved in semi-finished electrical steels by a final oxidation in the last annealing at high temperature, or in finished products, by varnishing each of the sheets. In general, this last technique is much more effective. There are already a great variety of coatings depending on the needs to be covered. When choosing a coating, factors such as corrosion resistance, weldability, heat resistance, compressive strength, roughness, etc. [75] should be taken into account.
- Use elements that produce an increase in resistivity in the material, such as silicon, phosphorus, aluminium and manganese, since the dissipated power is proportional to the inverse of the resistivity.

Earlier in this section above, the total losses were described as the sum of the losses by hysteresis and the losses by induced currents. The hysteresis losses are proportional to the frequency of the applied field that is, to the frequency of the current, whereas the losses by eddy currents are proportional to the square of the frequency. Therefore, the total losses can be expressed as a function of the current frequency, f, according to the relation:

$$P_{tot}(f) = P_h + P_d = w_h f + K f^2 \qquad (12.2)$$

where K is a constant. It is to be expected, therefore, that $P_{tot}(f)$ varies linearly with frequency and takes the value of w_h at zero frequency. However, the curve of $P_{tot}(f)$ versus f does not keep linearity at low frequencies. This discrepancy is a consequence of differences between the eddy current losses. This difference is named anomalous losses and is usually expressed by an anomaly factor (η) defined as the ratio of the obtained to the expected losses.

12.4 Factors Affecting the Magnetic Properties of Electrical Steels

As explained in the previous section, a controlled texture characterized by the coincidence of the direction of easy magnetization in the plane of the sheet is one of the factors that influence the magnetic properties of electrical steels. However, this is not the only factor that affects the magnetic behaviour of the final product. Other factors such as grain size and chemical composition exert a significant influence on the magnetic properties of electrical steels.

12.4.1 Influence of Texture and Grain Size

The optimum properties, in addition to optimizing the composition of the material as well as the distribution of precipitates, are achieved by large grain size and controlled texture. The desirable texture for an easy magnetization consists on the crystallographic direction [100] in the rolling plane, and more specifically for the non-grain-oriented (NGO) steels, the plane (001) parallel to the rolling plane. Regarding the final grain size, numerous studies that correlate coarse grain with good magnetic properties [14, 19, 23, 37, 40, 42, 46, 49, 55, 65, 80, 85, 87, 94, 113, 114, 118, 122, 125, 135, 136, 142, 167]. This is because an increase in the ferrite grain size leads to an increase in the size of the magnetic domains. A smaller number of magnetic domains lead to a lower number of domain walls to be moved during the magnetization processes. This in turn causes a decrease in magnetic losses [99, 101]. On the other hand, if the magnetic domains are too large, the domain walls have to move more quickly during the magnetization processes. This results in an increase in losses when the ferritic grains are greater than 120–130 μm [142]. Besides, the losses by induced currents also increase with coarse grains, since the induced currents find less difficulties (grain limits) to propagate [100]. Therefore, as the total losses result from the sum of P_h and P_d, these are minimized for an ideal grain size [90].

12.4.2 Influence of Alloying Elements

The importance of the role played by the alloying elements in the soft magnetic materials is an evident fact since a pure substance rarely meets the requirements necessary for its application in a magnetic device [103, 109, 121]. The reason is that the pure material does not simply possess all the necessary properties, whether magnetic or non-magnetic, for its application in electric motors, transformers, generators, etc. [12, 120]. One of the most significant examples is iron itself: it has attractive magnetic properties (saturation magnetization, relative permeability, coercive field [12]); however, its electrical resistivity is not high enough to keep the losses due to induced currents low, and its resistance to corrosion is very poor. That is why it is alloyed with elements such as silicon, aluminium, manganese and phosphorus to increase the resistivity and improve the magnetic properties and permeability. On the other hand, impurities such as carbon, nitrogen, sulphur and oxygen form various inclusions, such as carbides and/or nitrides, sulphides or oxides, that impede the movement of the magnetic domains during magnetization and are considered to be detrimental to the magnetic properties [143].

Among the alloying elements that are added to the iron to improve its performance in magnetic devices, silicon is the mostly used one and its use is so much that the electrical steels are, on numerous occasions, called simply Fe-Si alloys. The addition of silicon not only improves the magnetic properties but also other properties that make it especially attractive for application in magnetic devices. The main advantages are the following [12, 90]:

- Decrease of the coercive field. For instance, the addition of 2–2.5 wt.% of silicon decreases the coercive field by a half with respect to that of pure iron and, consequently, the losses by hysteresis are also reduced.
- Increase of the electrical resistivity (ρ). This effect is even more important than the effect on the magnetic properties. For a silicon content of 6 wt. %, the resistivity at room temperature linearly increases to ~$8 \times 10\text{–}7 \, \Omega$ m from $1 \times 10\text{–}7 \, \Omega$ m for a pure iron. The losses by induced currents are proportionally reduced by following the relation $P_d \sim 1/\rho$.

Silicon is an element of the so-called alphagens, and therefore it favours the formation of ferrite, α, in detriment to austenite, γ. Moreover, as indicated by the Fe-Si phase diagram of Fig. 12.6, silicon content greater than 2.5 wt.% results in the suppression of the $\alpha \to \gamma$ phase transformation. This allows achieving higher temperatures during annealing without the formation of austenite. The presence of austenite in the microstructure during the annealing would lead to a refinement of the microstructure and, thus, to a deterioration of the magnetic properties [71–73].

- Improvement of mechanical resistance. The elastic limit reaches values of $5.1 \times 108 \, \text{N m}^{-2}$ for a content of 4.5% of silicon in mass compared to $1.8 \times 108 \, \text{N m}^{-2}$ for pure iron.
- Decrease in density by 0.813% per wt.% of silicon added, which implies a decrease in the weight of the devices.

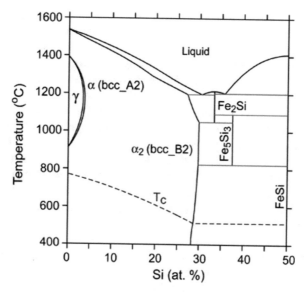

Fig. 12.6 Fe-Si phase diagram. (After ThermoCalc®)

However, the addition of silicon also carries with it some disadvantages:

- Curie temperature decrease. It is the temperature from which the material stops being ferromagnetic, as shown in Fig. 12.6.
- Increase in brittleness. This fact limits the manufacturing of electrical steels by rolling to silicon contents below 6.5 wt.%. This embrittlement is due to ordering processes (ordered to disordered phase transition). This is particularly unfortunate since Fe-Si alloys with a silicon content of 6.5 wt.% presents almost a negligible magnetostriction[1] in the [100] direction.

In spite of these drawbacks, and due to the potentially good characteristics of a material with a high-silicon content, numerous studies seek to optimize processing with high-silicon contents. Some researchers have employed surface diffusion techniques, either by chemical vapour deposition [154], or by immersion in Si-rich baths and annealing at high temperatures (1000–1250 °C) to allow the diffusion of Si into the material [129–132, 156]. Studies of rolling schemes have also been carried out to obtain sheets with a high-silicon content [64, 129, 130, 132].

Aluminium, like silicon, is incorporated into electrical steels to increase the resistivity and, thus, decrease the induced currents. This element is also added to combine with the nitrogen to promote the formation of aluminium nitride (AlN). In this way, the nitrogen is 'trapped' in the precipitates and no free nitrogen remains

---

[1]Magnetostriction is a property of magnetic materials by which a change of shape occurs in the presence of a magnetic field. This causes a vibration at the frequency of the fluctuations of the field and, as a consequence, vibrations in electrical machines such as motors and transformers.

in solid solution. The presence of these precipitates hinders the movement of the domain walls in the processes of magnetization and/or demagnetization. According to this, the aluminium content must be greater than the stoichiometric value required for the precipitation of all nitrogen. In this way the nitrogen of the solid solution is eliminated and aluminium remains which allows to increase the resistivity of the material. This phenomenon is known as magnetic ageing [88].

However, the formation of these aluminium nitrides also has some disadvantages. During the hot rolling processing, the deformation-induced precipitation would lead to a pinning of grain size in the hot band. This, consequently, could lead to a grain refinement before cold rolling and thus to a deterioration of the magnetic properties [11, 53, 142]. The aluminium content must be, therefore, high enough to produce the formation of coarse precipitates that do not impede grain growth [11, 86]. Nakayama and Honju [104] established limits for the content of aluminium in an electric steel with 0.3 wt.% of silicon so that if the content of AlN is less than 0.0024 wt.% the magnetic losses are not affected. If the AlN content is greater than 0.0024 wt.%, and the aluminium content is less than 0.1 wt.%, the losses are greater due to grain refinement by the pinning effect of AlN small precipitates. On the other hand, if the amount of AlN is greater than 0.0024 wt.% and aluminium content greater than 0.1 wt.%, coarse AlN precipitates (average diameter greater than 1 μm) are formed and no deterioration of the magnetic properties is observed. Hou et al. [49, 53] investigated the effect of aluminium content on non-grain-oriented electrical steels and concluded that magnetic losses are reduced by increasing the aluminium content in the range of 0.022–0.32 wt.% due to the thickening of the precipitates. Besides, the formation of these precipitates can have an effect on the texture, promoting the formation of [111]-type textures and is therefore frequently used in deep-drawing steels [57, 93, 160]. On the contrary, this texture is highly detrimental from the point of view of the magnetic properties and hence coarse precipitate or specific thermomechanical treatments are promoted to prevent its formation [84].

Along with aluminium and silicon, phosphorus is another element that is usually added to electric steels to increase the value of resistivity and thereby reduce the losses due to induced currents [84]. Phosphorus is, of the whole chemistry of steels, the one with biggest influence on increasing resistivity. However, phosphorus contents greater than 0.14 wt.% can induce cracks in the plates during rolling. There are some work that also indicate a detrimental effect of phosphorus due to the increase of the losses through the decrease in the recrystallized grain size [121]. Park et al. [121] point to another possible harmful effect of phosphorus by favouring the [111]-texture. This fact was also observed by Hou [54], although in subsequent work by the same author the opposite effect was obtained [51].

The incorporation of manganese to the electric steels improves the magnetic properties through the increase of the resistivity and the improvement of the texture [83]. However, it can affect the grain size due to the formation of inclusions such as manganese sulphides (MnS) that pin the grain growth during annealing. Therefore, manganese contents of about >0.3 wt.% are required to avoid as much as possible the formation of fine particles of MnS that can pin the grain coarsening [74, 105]. It has also been observed that if the contents of Mn are too high, a deterioration of the magnetic properties could be produced through the refinement of the grain size by precipitation of inclusions such as $MnSiN_2$ [166].

In addition to the elements mentioned above, there are numerous studies about the influence of other elements on the properties of electrical steels. Among them, perhaps, antimony and tin are the two that have received the most attention. Their interest lies in the fact that the segregation of these elements at the grain boundaries induces the nucleation of grains with certain orientations which, in the end, has an impact on the obtaining of a favourable texture from the magnetic properties point of view [30].

There are also studies on the influence of titanium, vanadium and niobium [50, 106, 107]. The main role of those elements consists of carbon removal from solid solution by means of the corresponding carbide/nitride precipitation during hot rolling stage of steel manufacturing route. Therefore, the removal of carbon from solid solution would avoid problems of magnetic ageing during the service of the material. However, the presence of these precipitates delays the recrystallization and grain growth of the cold-rolled sheets, and therefore, smaller grain size is obtained. In addition, it favours the development of textures with the direction [111] parallel to the rolling direction during continuous annealing after cold rolling. All these lead to the degradation of the magnetic properties of the steel.

12.4.3   Influence of Impurity Elements

The presence of impurities such as carbon, nitrogen, sulphur or oxygen induces the formation of small particles (of the order of tens or hundreds of nanometres) detrimental from the point of view of the magnetic properties. Their effects can be observed at two different times. These particles can form, on the one hand, during thermomechanical processing and inhibit the grain growth obtaining a finer microstructure and, as a result, higher magnetic losses. In addition, precipitation during the restoration process – recrystallization of the material can give rise to textures of [111]-type which are very detrimental to the magnetic properties of steel [57]. On the other hand, the heating of the magnetic device itself during service, induces the precipitation of particles by magnetic ageing. This hinders the movement of the magnetic domains during the processes of magnetization with the consequent increase of the magnetic losses [92].

In general, a steel with a low carbon content (less than 0.005 wt.%) allows to reduce the addition of alloying elements keeping an acceptable level of magnetic properties (Table 12.2). In addition, low carbon contents simplify the thermome-chanical processing as they avoid the final decarburization annealing stage. At the same time, the effect of magnetic ageing by precipitation of carbides is reduced.

Table 12.2 Magnetic properties of several grades of semi-finished electrical steels [83]

Steel	Losses (W/kg)
Ultra-low carbon Fe-Si (Al + Si ~ 1.3 wt.%)	5.2
M-45 low carbon (Al + Si ~ 1.85 wt.%)	6.2
M-43 low carbon (Al + Si ~ 2.35 wt.%)	5.1

12.4.4 Influence of Thermomechanical Processing on Electrical Steels

From the analysis carried out so far, it can be deduced that different thermomechanical processes give rise to different types of electrical steels [6, 21, 44, 86, 102, 111, 144, 155]. Traditionally, during steel processing, a final annealing of decarburization was carried out to decrease the carbon content below 0.005 wt.% and obtain a coarse-grained microstructure. In this way it was possible to improve the magnetic properties of semi-finished Fe-Si alloys [48].

Currently, the development of steel production technology has allowed, through the installation of so-called Ruhrstahl-Heraeus vacuum degasifiers that the carbon can be removed from the molten steel so that the process of decarburization is no longer necessary [48, 52, 148]. In addition, the development of continuous annealing equipment during the last decades has led to increase the annealing temperatures, to reduce the annealing time to a few minutes and to improve the surface characteristics of the cold-rolled steels. Therefore, it is possible to manufacture a NGO electrical steels nowadays from hot rolled with or without subsequent annealing, cold rolling, continuous annealing and a continuous coating line (surface oxidation, organic, inorganic or mixed). As a consequence, there are so many parameters during processing to optimize in order to achieve the best ever NGO electrical steel. This is the reason there are numerous studies driven to optimize each of these parts of the thermomechanical processing, i.e. to obtain the best possible properties in the final product [27, 48, 85, 151].

12.4.5 Influence of the Microstructure Before the Cold Rolling

The process of hot rolling and annealing determines the microstructure of the steel before cold rolling. Such is the importance of this process that some authors affirm that, simply with an adequate hot rolling and annealing scheme, it is possible to obtain suitable microstructures for magnetic devices from steels whose composition was originally intended for deep drawing [86]. In order to achieve good magnetic properties, a large grain size is sought before cold rolling, whereby the density of recrystallized nuclei at the grain is limited during the subsequent annealing decreases. It is convenient to minimize the nucleation in these zones since it is precisely there where the nucleation of grains predominates with the direction [111] parallel to the direction of rolling, very unfavourable from the point of view of the magnetic properties. On the other hand, the hot or cold rolling of a coarse grain microstructure causes a non-uniform distribution of the deformation through the microstructure, which leads to the formation of deformation bands where the nucleation of Goss and Cube grains takes place.

In order to obtain the largest possible grain size after hot rolling, the highest possible annealing temperatures that promote recrystallization and grain growth are sought [10, 21, 86, 167]. Moreover, a high temperature causes the coarsening of particles such as manganese sulphides and aluminium nitrides reducing or eliminating their effect as refiners of the microstructure by grain pinning [48]. In fact, the temperature drop after the hot rolling can be in such a way that the microstructure does not recrystallize and, in this case, a subsequent annealing can be chosen to control the recrystallization process in a better way. However, the temperature of this annealing should not be so high that it promotes the formation of austenite in place of ferrite, since this would entail the refinement of the microstructure [20, 21]. Some studies also point out that a large grain together with a texture with the [111] direction parallel to the transverse to the rolling direction, and particularly, with abundant (111) [112] component, favour the formation of Cube-oriented grain nuclei during annealing after cold rolling [17, 116, 120].

Figure 12.7 shows some of the rolling schemes used. In general, it is preferred that most of the rolling stages take place in the $\alpha + \gamma$ biphase field for low-silicon steels while, for higher silicon contents, a 'mixed' rolling is preferred.

Regarding the final rolling temperature, there does not seem to be a clear consensus. While some authors advocate a decrease in this temperature to achieve a larger ferritic grain size during annealing [142, 167], there are studies that reveal the opposite effect provided that the finishing rolling temperature is in the ferritic field [10].

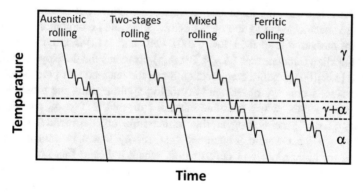

Fig. 12.7 Hot rolling schemes for electrical steels

12.4.6 Influence of Cold Rolling and Final Annealing

The amount of cold rolling is subject to the need to obtain a steel sheet with a certain thickness which, in the case of electrical steels, is between 0.35 mm and 1 mm. Thus, when cutting hot-rolled sheets of 2–3 mm thickness, the reduction of this parameter during cold rolling is between 50% and 85% [27]. In principle, even greater reductions could be beneficial, not only because a smaller thickness would reduce the dynamic losses, as was seen in Sect. 2.3, but also for the obtaining of more favourable textures [12, 27]. However, high reductions can lead to a smaller recrystallized grain size and thus to the increase of hysteresis losses [54].

In general, the deformation texture must be such that it contains a strong component [111] parallel to the normal lamination, ND, and, particularly, the component (111) [112], to favour the formation of the texture with [001] parallel to the direction of rolling during recrystallization [17]. This is because the nucleation of this type of grains takes place in the bands of deformation of the grains with [111] parallel to the normal to the rolling direction and especially of the component (111) [112] [58, 120, 155].

12.4.7 Influence of Temper Rolling

Salinas-Beltran and co-workers [141] reported recently the effect that tensile deformation, similar to that experienced in temper rolling, and annealing conditions have on the microstructure and magnetic properties of NGO electrical steel. The authors found that the final recrystallized grain size is inversely proportional to the deformation level imposed before the annealing, i.e. the higher the deformation, the smaller is the resulting grain size. Besides, the authors demonstrated that if the annealing is performed at temperatures below cementite precipitation temperature, the texture measured is of $\langle 011 \rangle$//RD, $\langle 001 \rangle$//ND and $\langle 111 \rangle$//ND type. In contrast, higher annealing temperatures ($A_{e1} < T < A_{e3}$) promotes the development of [332] [$1\bar{1}3$]and [110][001] texture components. Since the magnetic properties depend on the grain size, secondary phases and crystallographic texture, the most beneficial textures are obtained when the materials are annealed at $T < A_{e1}$; however, the secondary phases precipitating at this temperature are detrimental to magnetic properties. All these studies concluded that energy losses in annealed samples are higher for higher levels of deformation, which indicates that grain size is the main microstructural parameter that affects the magnetic behaviour of annealed samples. The lowest energy losses are obtained when the material is strained to 8% engineering strain.

12.5   Grain-Oriented (GO) Electrical Steels

The GO electrical steels are obtained from a complex thermomechanical processing (Fig. 12.8) and give excellent results in terms of permeability and electrical consumption in certain conditions (unidirectional magnetic field, such as in an electrical transformer). The hot-rolled strip is side trimming and pickled to remove surface oxides. Then, it is cold rolled to about 0.6 mm thickness from the initial hot band thickness of 2–2.5 mm. The material is given a decarburizing anneal down to less than 0.003 wt.% in carbon followed by coating with a thin layer of magnesium oxide that will react with the steel surface during the next recrystallization anneal at 1200 °C to form a thin magnesium silicate layer called forsterite layer. Finally, the material is given a flattening anneal, when the excess of magnesium oxide is removed.

In his seminal paper, Moses [97] described how the GO electrical steels were developed for the first time. As a summary, it might be said that Goss [32] was able to develop a grain texture in silicon-iron sheet, by means of cold rolling and annealing, that enhanced magnetic properties when magnetized along its rolling

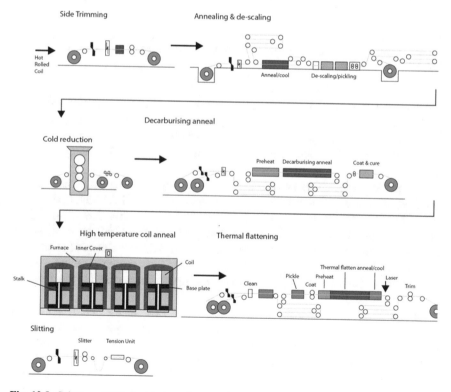

Fig. 12.8 Scheme of GO electrical steel processing according to Cogent Power. (https://cogent-power.com/)

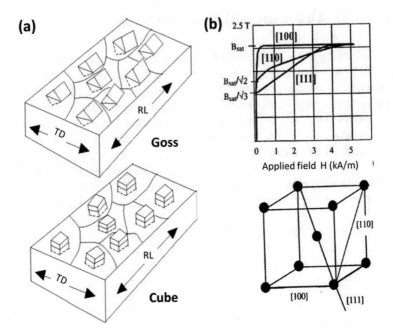

Fig. 12.9 (a) Crystallographic orientation according to Goss and Cube textures in a polycrystalline steel sheet; (b) magnetic anisotropy of iron. Magnetization curves for a monocrystal. (After Ros-Yañez et al. [132])

direction. Goss managed to produce a texture with a high proportion of grains having [001] directions close to the rolling direction and (110) planes close to the sheet plane. Armco, in the USA, commercialized the process and the first strip was produced in 1939. It was 0.32 mm thick, with a loss around 1.5 W/kg, at 1.5 T, 50 Hz.

The particularity of these materials lies in that they are formed by coarse grains oriented in such a way that their crystallographic planes (110) remain parallel to the rolling plane with the direction [100] parallel to this same direction. This is known as Goss texture (Fig. 12.9a). In iron, this direction [100] is the direction of easy magnetization, which explains why a grain-oriented electrical steel plate magnetized in the direction of rolling is characterized by a strong permeability and a weak coercive field. Hence the interest of obtaining the address [100] in the direction of the exciter field.

Figure 12.9b shows the magnetic anisotropy of iron. In the lower part of the figure, the unit cell of the iron α or ferrite is shown, indicating the crystallographic directions [100], [110] and [111]; in the upper part of that same figure, the relation between the applied magnetic field ΔH and the magnetic induction ΔB is shown when applying ΔH in each of those three directions. It can be observed that the simplest way to saturate the material is by applying the magnetic field in the direction [100], that is, in that direction the magnetic field necessary to saturate the material is weaker than in the other two directions. Figure 12.9a shows two of the orientations, Goss and Cube, which, because they have the direction [100] parallel to the rolling direction, are desirable from the point of view of magnetic properties.

In the early stages of development of electrical steels, the mechanism of formation of the Goss texture was not understood until the work of May and Turnbull [91]. They reported that it is necessary the presence of dispersed precipitates for the formation of the key-microstructure of GO electrical steels. They suggested a function whereby the precipitates inhibit normal grain growth and thus maintain the driving force for secondary recrystallization. In the early GO electrical steel grades, those particles were the small MnS particles formed during the hot rolling stage that are precipitated as the steel cools and, meanwhile at the same time, some crystals with the Goss texture are formed along with many other orientations. After the cold rolling, nuclei with the Goss texture recrystallize during the decarburization anneal and grow during the high-temperature anneal when the MnS retards the growth of other grains (secondary recrystallization process). All grains formed in such way do not have the ideal Goss orientation, but most are within 6° of the ideal [100] (110); this is the best that can be achieved with MnS as a grain growth inhibitor [91].

Since then, many have carried out research to find out the role of different impurities on the development of microstructure in electrical steels. Hayakawa [39] reported an historical overview on this matter which is summarized in Fig. 12.10. Hayakawa [39] detailed the work of Saito [139, 140] and his extensive investigation of the effects of solute such as Pb, Sb, Nb, Ag, Te, Se and S to boost the magnetic properties by promoting a higher accumulation of orientations towards the Goss orientation in the secondary recrystallization texture.

Grenoble and Fiedler [34] showed that sulphur and nitrogen, when present together as solutes, enable secondary recrystallization to occur. Grenoble [33] also reported secondary recrystallization with a combination of solute S, N and B. Use of a combination of inhibitors and solute elements contributed to the stability of secondary recrystallization and improved magnetic properties.

This research allowed the development of a revolutionary new electrical steel by Nippon Steel Corporation in 1965: high-permeability grain-oriented silicon iron (HiB). The production route of this new steel grade was simplified by the addition of around 0.025 wt.% aluminium that led to the precipitation of aluminium nitride to act as grain growth inhibitor. The final product not only had a better orientation than conventional steel but also a coarser grain size which, at flux densities of 1.7 T and higher, had permeability three times higher than that of the best conventional steel. Besides, the stress sensitivity of loss and magnetostriction were lower because of the improved orientation and the presence of a high tensile stress introduced by a so-called stress coating. The stress coating technology was an important milestone in the manufacturing of electrical steels, and it will be described separately in this chapter. The stress coating helps to reduce eddy current loss which is high in coarse-grained electrical steels, and moreover, this coating leads to a further decrease of losses by a reduction in hysteresis loss.

Nowadays, considerable amounts of the GO electrical steels produced worldwide are high-permeability steels with stress coatings consisting mainly of colloidal

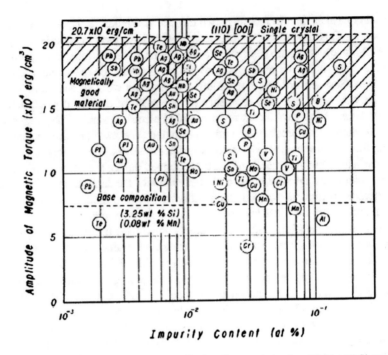

Fig. 12.10 Effect of impurity content on amplitudes of magnetic torque of 3.25 wt% Si-steel strip annealed at 1100 °C. (Reproduced with permission from Saito [140] © 1963 The Japan Institute of Metals and Materials)

silica with low thermal expansion and amorphous structure. The application of this coating at the last stage of the steel production process has an indirect beneficial effect refining the main domain widths in high-permeability steels [137]. However, this effect is greatest in materials whose grains are mainly within 2° of the ideal Goss texture [149]. The conventional forsterite layer itself induces the domain wall pinning [95, 145] which produces a loss increase, and the stress coating must cancel this deterioration before it starts improving the properties. The major improvement in this issue found inspiration from the early work with GO electrical steels where the losses can be reduced (by over 40% [61, 68]) cutting fine grooves into the surface. This phenomenon is even more pronounced in high-permeability, large-grained material [157] because of, as shown by domain studies, the beneficial effect of internal stresses generated during the scratching [16]. Therefore, the application of a continuous laser beam direct to the forsterite layer was found to decrease the overall losses [67, 126].

The main efforts, at present aimed at mass production of even better silicon steels, seem to be devoted to improving the secondary recrystallization technique and to combining it with very highly oriented thin gauge material with better domain refinement [28, 29, 82, 117]. This needs to be combined with increasing the steel resistivity, without the addition of more substitutional elements and also improving the steel/coating interface to optimize the built-in tension in the steel.

12.5.1 Case Study: Innovative Processing for Improved Electrical Steel Properties

Texture control is the core technology of production of GO electrical steels. Xia et al. reported recently a nice review of developments of GO electrical steels and the mechanism of secondary recrystallization on grain-oriented electrical steels to improve its magnetic properties [162]. However, the work carried out by Houbaert and co-workers [129–132, 156] established a novel strategy to decrease the total losses in the GO electrical steel taking advantage of the improvement in magnetic properties and permeability by the addition of silicon and aluminium. This section tries to summarize this strategy.

As it was mentioned above, the addition of Si and/or Al increases the electrical resistivity, which reduces the eddy current losses. Besides, the higher the value of Si and/or Al, the lower the values of the magnetostriction and of the magnetocrystalline anisotropy. Therefore, there has been a lot of interest, recently, in trying to produce steels with an increased concentration of Si and/or Al to values above 3 wt %. However, it is difficult to achieve due to the cracking of the steel during cold rolling because of brittle Fe-Si ordered phases DO3 or B2.

Okada et al. [110] reported that it is possible to enrich conventional steel (Si < 3.2 wt %) up to 6.5 wt % Si using silicon from an applied coating by means of chemical vapour deposition (CVD) followed by diffusion annealing. Other alternatives were studied by He et al. [43] who used physical vapour deposition (PVD).

Verbeken et al. [156] described a novel processing first reported by Ros-Yañez et al. [129] based on an increase of Si and/or Al obtained by surface deposition by a short hot dipping in a molten hypo- or hypereutectic Al-Si bath carried out in an equipment originally designed for hot-dip galvanizing, i.e. a Rhesca® hot-dip simulator [132], followed by a diffusion annealing. The authors described the growth of an ordered phase Fe_3Si (DO3) in contact with the Fe-Si substrate, and concluded that diffusion reaction is the controlling mechanism for the intermetallic compounds. Therefore, the most deformed structures reacted faster because of the faster diffusion through high-diffusivity paths such as grain boundaries and dislocations. Figure 12.11 shows the composition profiles through the coating and the electrical steel substrate.

Using these basics, ThyssenKrupp patented a procedure which is applied to fully processed electrical steels as well as hot rolled, especially with low thickness, or cold rolled steel qualities used as substrates [26]. Cold rolling of electrical steel qualities thinner than 0.35 mm, after dipping and before the annealing treatment to get the desired final thickness, appears to be an interesting variant in the production

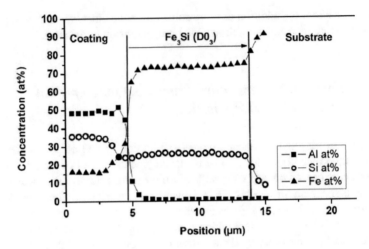

Fig. 12.11 Composition profile across the coating thickness on a Fe substrate with 3.7 wt % Si, hot dipped for 100 s and fast cooled with a N_2 flow. (After Verbeken et al. [156])

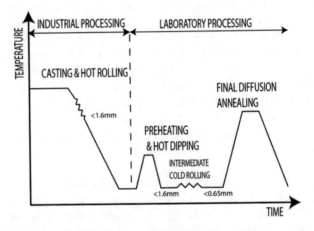

Fig. 12.12 Overview of the complete experimental scheme, including industrial processing, followed by the processing with intermediate cold rolling (ICR). (After Verbeken et al. [156])

route defined as intermediate cold rolling (ICR). Figure 12.12 gives a schematic overview of the ICR processing route.

12.6 Non-grain-Oriented (NGO) Electrical Steels

The NGO electrical steels are used for electrical machines that involve some type of rotation, such as motors or generators, and in which the magnetic flux does not follow a certain direction but normally rotates. That is why the type of texture desired in these cases is different from that required in GO electrical steels. When the rotation of the magnetic flux takes place in the plane of the steel sheet, it is necessary to obtain the direction [100] parallel to the plane of the sheet but this time with a greater anisotropy than in the case of grain-oriented electrical steels. Therefore, those textures with the plane (001) parallel to the direction of rolling, such as the texture known as Cube, [100] (001), are desired in this type of materials.

The NGO electrical steels are obtained by simpler thermomechanical processing than grain-oriented steels. Fully processed strip must be decarburized and continuously annealed to minimize residual stress and to obtain good shape and microstructure. Decarburization is carried out by annealing in moist hydrogen, and the goal of annealing is to replace the fine recrystallized grain structure by a large grain structure, which helps to give high permeability and low hysteresis loss [97]. These steels have undergone a gradual but unspectacular development since Hadfield's seminal work [1, 2, 27, 35, 36, 48]. By progressive elimination of impurities, losses at 1.5 T, 50 Hz, have been cut from around 7 W/kg in the early times to less than 2 W/kg in the best grades today.

The NGO electrical steels can be classified into two categories:

- Semi-finished: This type of NGO steels undergoes a continuous annealing after the cold rolling, followed by a weak deformation in the cold rolling stand, and finally an annealing of decarburization and grain growth [79, 89]. Semi-processed strip is sometimes provided in a partly decarburized state, because then the material has the required degree of hardness, shape and surface conditions to ensure good dimensional stability [90]. The term 'semi-finished' derives from the fact that, in general, the material is supplied to the customer after the first annealing, and this gives the final cold rolling and decarburizing anneal at between 800 and 840 °C (depending on the composition) needed to optimize the magnetic properties. It is very important to produce uniform stress relaxation needed to prevent rolling distortion. Also, the decarburizing atmosphere must be in contact with both surfaces of all strips to ensure adequate removal of carbon.
- Finished products, that is, hot rolled to a thickness of 2–3 mm and then cold rolled to thicknesses of 0.35 mm, 0.50 mm and 0.65 mm. The processing ends with a final annealing at high temperature.

In accordance with this, the scheme of the complete thermomechanical treatment on the study material is represented in Fig. 12.13, and it is possible to divide it into two well-differentiated parts: hot rolling and cold rolling.

The NGO electrical steels are almost invariably covered with a thin insulating layer, towards the end of the production process. Important coating properties are, among other properties, electrical insulation, corrosion resistance, temperature stability and hardness. The coating may just be a thin oxide film, formed by the introduction of air at the end of the annealing cycle, or it may be a complex organic/inorganic layer especially chosen for good insulation, punchability, weldability, etc. More details on coating of electrical steels will be described in subsequent section further in this chapter.

Fig. 12.13 Processing
scheme of NGO electrical
steels

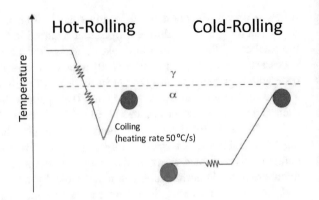

The niche of NGO electrical steels is fitted between the low-grade cold-rolled steels or expensive high-silicon steels. These steels have been developing following a twofold strategy: on one hand, towards low cost for use in cheap devices and, on the other hand, towards high-magnetic efficiency for use in applications where increased material cost was offset by higher efficiency. Now, low-silicon alloys, without much silicon (or aluminium) to increase their resistivity and hence reduce eddy current losses, are produced with improved steelmaking processes, so that a large decrease in hysteresis loss more than compensates for the lower resistivity.

Basically, the goal of modern production of NGO electrical steels is seeking for higher permeability and lower losses, in what is essentially a cheap product with good mechanical strength [13, 27, 151]. As reported by Moses [97], the loss in NGO electrical steels is mainly dominated by hysteresis loss, which is around 60–70% of the total loss in the best grades. This has been reduced by minimizing impurities [9, 69, 70, 84, 96, 104–106, 168] and can be further reduced by optimizing the grain size [20, 27, 41, 56, 63, 119, 146]. Coarse grains (equiaxed grains of about 1 mm in average diameter) experience a significant drop of the losses, but, in the counterpart, anomalous eddy current loss predominates and the total losses increase [47]. To reduce the eddy current component of loss, high-silicon or aluminium content is needed [38, 104, 105], although manganese also shows potential for development [76–78]. However, in general, the higher the resistivity becomes, the lower will be the permeability because additions reduce the saturation magnetization. Although silicon can reduce the coercive force (and resistivity as stated earlier), present trends are mainly to aim for lower additives but cleaner steels, to keep induction high and losses low [84]. Therefore, the desired direction for future development of NGO electrical steels is the implementation of low-loss and high-permeability electrical steel for motor materials. The development trend for higher permeability at lower cost is obviously a requirement which can be achieved by pursuing low alloy material with low impurities, to reduce the high hysteresis loss.

12.6.1   Case Study: Control of Texture and Grain Size in Two NGO Electrical Steel Qualities

This section is devoted to illustrate the microstructural optimization in two NGO electrical steel qualities (Table 12.3) with the same processing history. In this sense, the processing route is schematically illustrated in Fig. 12.14. As mentioned above, there are two stages in the processing route, i.e. hot rolling and reheating, and a second stage formed by cold rolling and annealing. The key temperatures are summarized in Table 12.4.

In order to promote the recrystallization of the hot-rolled microstructure after hot rolling, rapid reheating cycles were performed at temperatures indicated in Table 12.4. The benefits of obtaining a big grain size before cold rolling in this NGO electrical steel qualities is twofold [21]. Firstly, in order to obtain a lower grain boundary surface, and so to avoid a high nucleation density of undesirable textures

Table 12.3 Chemistry of the NGO steel qualities studied (in 10^{-3} wt%) [24]

	C	N	S	Ti	Mn	Al	Cr	Si	V	P
ES1	<4	<22	<8	<40	300	100–200	30–60	300–400	<10	<50
ES2	<4	<39	<8	<40	300	100–200	30–60	1200–1500	<20	<190

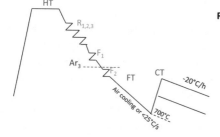

Rolling conditions :

HT : 1180°C
R$_{1,2,3}$: Roughing (to 1050°C)
F1 : 950°C
F2 : 880°C
FT : 860°C
CT : from 700 to 950°C

Fig. 12.14 Processing route of semi-finished NGO electrical steels with chemistries in Table 12.3. The HT stands for reheating temperature; R1,2,3 stands for roughing pass temperatures; F1,2 stands for the entrance temperature (F1) and exit temperature (F2) of finishing rolling, i.e. FRT which stands for finishing rolling temperature is the average of F1 and F2; FT stands for the temperature at which the fast cooling starts; and CT stands for the coiling temperature

Table 12.4 Silicon levels and finish rolling temperatures [24]

	Si, wt.- %	A_{r1}[a], °C	Finish rolling temperature (FRT), °C	
			Entry (F1)	Exit (F2)
ES1B	0.3	949	899	850
ES1C	0.3	839	777	
ES2A	1.3	1010	832	752
ES2B	1.3	904	814	

[a]A_{r1} stands for the highest temperature avoiding the austenite transformation

Fig. 12.15 Near surface optical image from as-received ES1C [24]

during recrystallization after cold rolling. Secondly, to promote the appearance of shear bands, since in coarse grained specimens, the imposed strain during cold rolling is accommodated by breaking into several blocks separated each other by shear bands [58], where the nucleation of cube grains takes place [120]. Finally, the annealing after cold rolling was carried out.

The samples with higher finish rolling temperature (FRT), i.e. ES1B and ES2B steels, present a recrystallized surface (Fig. 12.15) and deformed interior but highly recovered with subgrain structure, finer in ES2B steel, as it is shown in Fig. 12.16. The subgrain structure is illustrated in electropolished samples by means of electron channelling contrast (ECC) imaging in scanning electron microscopy (SEM). This technique is based on the fact that the yield of the backscattered electrons by the material depends on the crystal structure and orientation of the crystallites in the sample. This effect allows to clearly differentiate between recrystallized (i.e. strain-free) and deformed grains.

On the other hand, those samples with lower FRT, i.e. ES1C and ES2A, also show a recrystallized surface, in a lower level than ES2B and ES1B steels and with smaller grain size. Subgrain structure is less developed and in any case with smaller subgrains.

Figure 12.17 shows the mid-thickness textures after hot rolling. It has been generally agreed that deformation at high ferrite – range temperatures produces a texture consisting of an α-fibre with [110] stretching from (001)[110] to (111)[110], and a γ-fibre. From the present data it can be inferred that the α-component is stronger at higher FRT while the γ-fibre remains roughly constant or more uniform when the FRT decreases.

In order to promote a coarser grain structure before cold rolling, a reheating treatment is performed. The maximum temperature selected for avoiding the austenite transformation, since the austenite to ferrite transformation results in grain

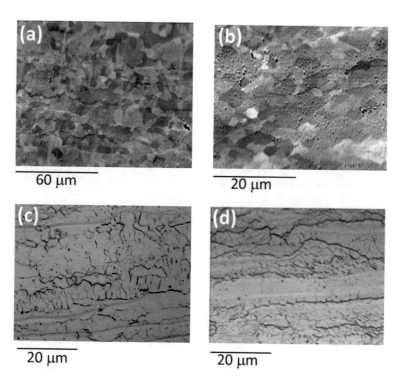

Fig. 12.16 As hot-rolled structures. (**a**) ES1B ECC image, (**b**) ES2B ECC image, (**c**) ES1B optical image, (**d**) ES2B optical image [24]

refinement [21], was 900 °C for ES1 and 950 °C for ES2 according with A_{r1} values listed in Table 12.4. Subsequently, ES1 samples after hot rolling were reheated for 1.5 h at 800, 850 and 900 °C and ES2 samples at 800, 850, 900 and 950 °C in order to simulate coil cooling from these temperatures. The recrystallization and grain growth processes were studied by interrupting reheating cycles at 800 and 900 °C by quenching after 10, 30, 100, 300 and 1000 s in all the studied steels. From the grain size measure after 1.5 hour (Fig. 12.18) it is clear that bigger grain size is obtained with the highest FRT in both Si steel grades. Maximum grain size is obtained after 1.5 hour at 850 °C for ES1B and ES2B steels. Grain size change with temperature is less noticeable in ES1C and ES2A steel, with a maximum at 850 °C as well for ES1C steel and at 900 °C for ES2A steel.

Metallographic analysis revealed that recrystallization produce coarser and elongated grains in material with higher FRT values (ES1B and ES2B). This effect is more pronounced after reheating at low annealing temperatures (Fig. 12.19a, b). However, material with lower FRT values (ES1C and ES2A) recrystallize in finer and more equiaxed grains, as it is clearly shown in Fig. 12.19c. This phenomenon does not depend on reheating temperature. Likewise, abnormal grain growth is also observed in all samples near the surface as shown in Fig. 12.19d.

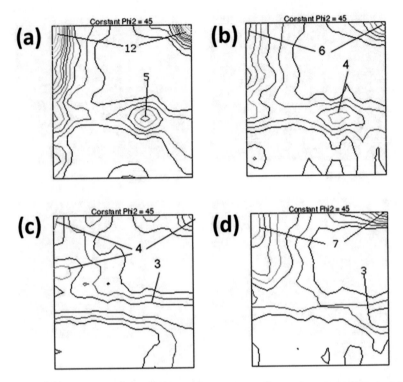

Fig. 12.17 Mid-thickness textures after hot rolling: (**a**) ES1-B, (**b**) ES1-C, (**c**) ES2-A and (**d**) ES2-B [24]

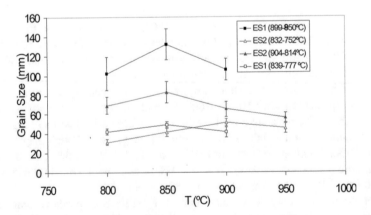

Fig. 12.18 Grain size after reheating treatments at different temperatures (T). The temperatures between brackets in the legends stands for entry (F1) and exit (F2) temperatures in the finishing rolling strand [24]

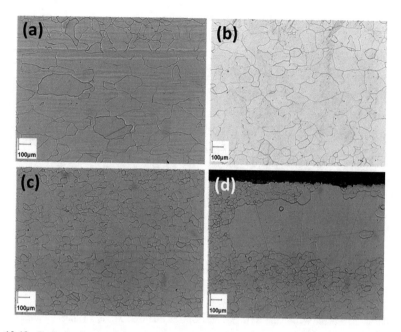

Fig. 12.19 Optical micrographs after 1.5 h cycles. (**a**) ES1B, 850 °C; (**b**) ES1B, 900 °C; (**c**) and (**d**) ES1C 850 °C [24]. Images in (**c**) and (**d**) are from different locations of the sample

Detailed observation shows precipitates inside recrystallized grains. Precipitates tend to form alignments or 'cells'. This 'cell' or 'network' structure is especially evident for the lower annealing temperatures in the steels with higher FRT. Energy-dispersive X-ray spectroscopy (XEDS) analysis allows to conclude that these precipitates are AlN (Fig. 12.20).

The recrystallization processes in these steels might be better understood with the assistance of EBSD analysis. Metallographic and electron back scatter diffraction (EBSD) results show that recrystallization takes place with the development of a subgrain structure and big anisotropic grains in steels with higher FRT values (Figs. 12.21a, b and 12.22a). To the contrary, usual nucleation and growth mechanisms were observed in steels with lower FRT values (Figs. 12.21c and 12.22b). Therefore, it is clear from the results obtained that two different recrystallization mechanisms takes place depending on the FRT value.

Furthermore, mid-thickness textures of recrystallized samples (Fig. 12.23) shows weaker textures than the hot-rolled material: electrical steels with the lowest FRT (ES1C and ES2A) are almost not textured, whereas main texture components remain when material with high-FRT is annealed.

The results reported here for NGO electrical steels with higher FRT are consistent with those reported by Kestens et al. [62] who showed that recrystallization proceeds by a continuous recrystallization plus subgrain coalescence mechanism, which prompts to a bigger anisotropic grain. The temperature at which the last

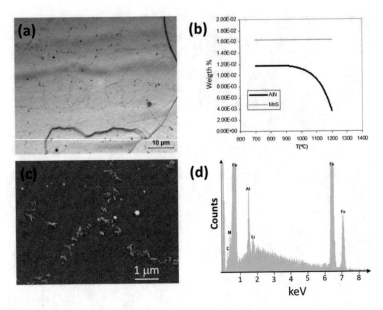

Fig. 12.20 AlN precipitation in ES1B steel. (**a**) Optical micrograph; (**b**) MTDATA data; (**c**) SEM; (**d**) XEDS spectrum [24]

hot rolling reduction occurs is higher enough to ensure that most of the AlN precipitation occurs during the subsequent recovery – recrystallization processes. Thus, subgrain boundaries are pinned, and the subsequent migration of high angle boundaries are stopped by the particles (continuous recrystallization) [45, 81, 138]. Further rearrangement of boundaries is controlled by dissolution and growth of AlN particles. Since the FRT is high, the stored energy due to deformation is low enough to avoid the nucleation of recrystallized grain on the deformed grain boundaries. The recrystallization process involved is then well described by subgrain coalescence which prompts the big anisotropic grains in the microstructure.

On the other hand, steels with lower FRT values show the usual nucleation plus growth mechanism. Meanwhile most of the AlN precipitation events occur after the last deformation step in material with higher FRT, some precipitation will occur between deformation steps in material with lower FRT. Therefore, the amount of AlN able to precipitate after the last deformation step, and hence during the subsequent recovery – recrystallization processes, is reduced. This can explain that no subgrain boundary pinning by AlN particles occurs in steels with lower FRT. This fact, together with higher stored energy because of lower deformation temperature, promotes the nucleation and growth mechanism for recrystallization in detriment of continuous recrystallization in steels with lower FRT.

The higher stored energy because of the lower FRT together with the AlN precipitation during the hot rolling schedule ensures a faster recrystallization and avoids the precipitation process into the subgrain structure during recovery. The

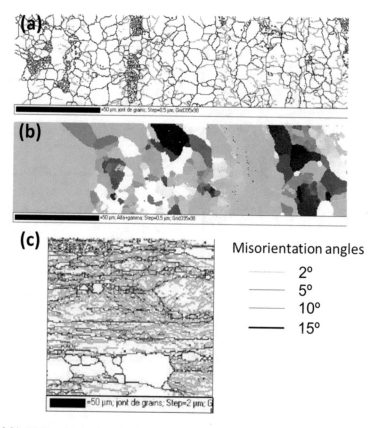

Fig. 12.21 EBSD analysis. Recrystallization during reheating: (**a**) grain boundaries in ES1B; (**b**) α-fibre (green) and γ-fibre (red) in ES1B steel; and (**c**) grain boundaries in ES1C steel [24]

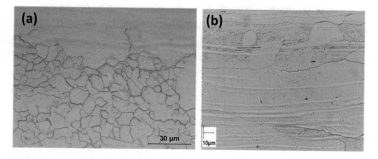

Fig. 12.22 Recrystallization during reheating. Optical micrographs (**a**) ES1B and (**b**) ES1C steels. Etch: Nital 5% [24]

resultant grains are smaller and equiaxed. These assumptions are consistent with the optical micrographs shown in Fig. 12.24.

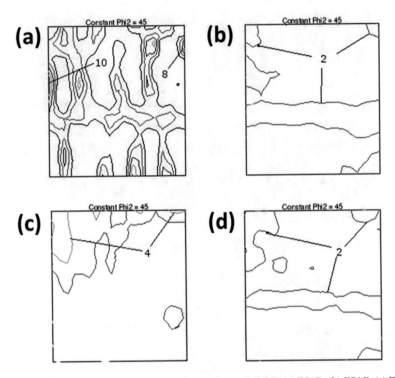

Fig. 12.23 Mid-thickness textures after 1.5 h annealing at 800 °C: (**a**) ES1B, (**b**) ES1C, (**c**) ES2B and (**d**) ES2A steels [24]

The decrease in grain size detected during recrystallization in ES1B and ES2B steels (Fig. 12.18) could be explained in the following. Because the higher reheating temperature, the thermal activation is larger, and so more grains can grow after continuous recrystallization. Variation of hardness with thermoelectric power (TEP) in Fig. 12.25 [5, 7, 8, 22, 25] shows that at 900 °C, recrystallization takes place before almost any precipitation, and this fact together with the observation of well equiaxed smaller grains at this temperature seems to point to a discontinuous recrystallization process, i.e. nucleation and growth, which takes place simultaneously with continuous recrystallization mechanism.

Taking the foregoing results into consideration, the reheating cycles with coarser grain size were chosen for the subsequent cold rolling and annealing. ES1 and ES2 with fine recrystallized grains, as well as not reheated ES1 samples were also cold rolled in order to compare the final material. Cold rolling schedules are summarized in Table 12.5.

Cold rolling of recrystallized samples leads to a deformed grain with shear bands, both in coarse and fine grain microstructures, whereas no shears bands were reported in hot rolled material without reheating (Fig. 12.26).

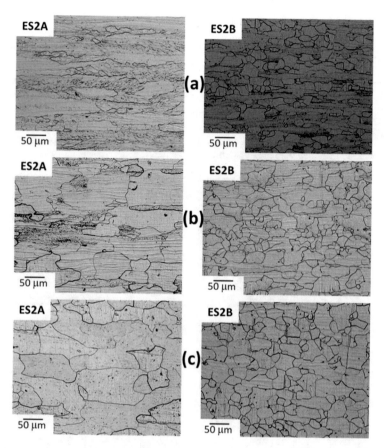

Fig. 12.24 Optical micrographs after reheating at 850 °C for (**a**) 30, (**b**) 100 and (**c**) 1000 s in ES2A and ES2B steels Etch: Nital 5% [24]

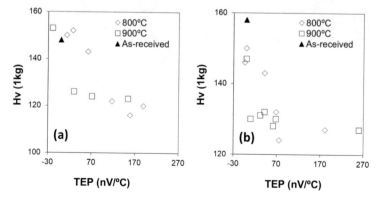

Fig. 12.25 TEP vs. hardness plots for different reheating temperatures of (**a**) ES1B and (**b**) ES2B steels [24]

Table 12.5 Cold rolling schedules [24]

Sample	FRT (°C)	Reheating temperature (°C)	Cold rolling reduction (%)
ES1B80	900	No reheating	80
ES1BT80	900	850	80
ES1C80	850	No reheating	80
ES1CT80	850	850	80
ES2AT60	850	950	60
ES2BT60	900	850	60

Specimens are identified with their cold rolling reductions

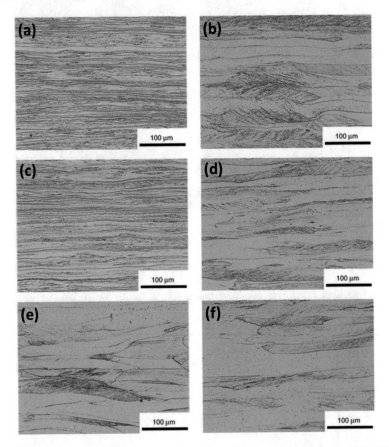

Fig. 12.26 Optical micrographs of cold-rolled electrical steels: (**a**) ES1B80, (**b**) ES1BT80, (**c**) ES1C80, (**d**) ES1CT80, (**e**) ES2AT60 and (**f**) ES2BT60. Etch Nital 5% [24]

Cold-rolled material was reheated at 10 °C/s, and after 60 s of holding at 650 °C it was cooled at 10 °C/s to room temperature. Severe shear band nucleation was reported in electrical steels during reheating. Nucleation of new grains in shear bands was found mainly in ES1BT80 and ES2BT60 samples, because of the

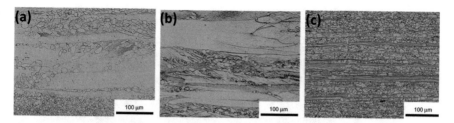

Fig. 12.27 Nucleation of recrystallized grains after annealing of cold rolled material for 60 s at 650 °C in (**a**) ES1BT80, (**b**) ES2BT60 and (**c**) ES1B80 steels. Etch: Nital 5% [24]

lower grain boundary area (Fig. 12.27a, b). This was also observed in the cold rolled material of fine grain size, i.e. in ES1CT80 and ES2AT60 samples. A more homogeneous nucleation through the whole microstructure occurred in the non-annealed material, ES1B80 and ES1C80 (Fig. 12.27c).

Recrystallized grains were equiaxed and homogeneous through thickness as shown in Fig. 12.28.

Despite the different grain size prior to cold rolling, minor differences between recrystallized grain size after cold rolling and annealing were observed in ES1 material and even a bigger grain size was achieved for finer microstructure before cold rolling. Nevertheless, massive nucleation in shear bands was detected, which may explain the evolution of grain size. This fact may lead to a more desirable texture as compared with the unrecrystallized or finer microstructure previous to cold rolling, since shear bands are reported to be nucleation sites of desirable textures in electrical steels (Fig. 12.29). The texture analysis might conclude that (001)<1–10> desirable component is more intense after annealing at 700 °C.

12.7 Coatings for Electrical Steels

Electrical steel coatings are pigmented coatings that insulate silicon steel sheets of motors and generators [18]. The ability of motors or generators to be magnetized and demagnetized is crucial, and besides the use of electrical steels as core material, the heat-curing electrical steel coating is the next key factor in the efficiency of motors and generators. The main function of the coating is to insulate the electrical steel sheets in order to prevent the flow of electricity and to reduce the eddy current. Electrical steel coatings also serve to lengthen the shelf life of the required punching tools.

As mentioned in sections above, the first electrical steel was developed by Goss [32] in the 1930s of the last century, and it was commercialized by Armco in the early 1940s. The basics of the processing route consist on hot rolling and cold rolling to reduce the initial hot-band down to 6 mm thick strip, followed by a decarburization treatment to reduce the amount of carbon down 0.003 wt.%, which

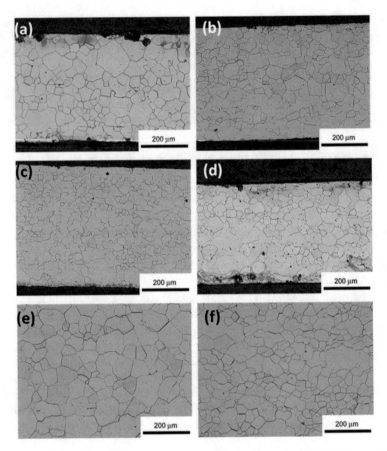

Fig. 12.28 Optical microstructure after annealing at 900 °C (ES1) and 950 °C (ES2) revealing the grain structure as described in Table 12.5: (**a**) ES1B80, (**b**) ES1BT80, (**c**) ES1C80, (**d**) ES1CT80, (**e**) ES2AT60 and (**f**) ES2BT60 steel. Etch: Nital 5% [[24]

is followed by coating with a thin MgO layer. Then, the Fe-Si alloy is annealed at 1200 °C to promote secondary recrystallization of large grains with [001](110) texture, which is predominant because the presence of MnS particles suppresses the growth of grains with other orientations. During this annealing, the addition of magnesium oxide (MgO), which is added as a separator to prevent the risk of sheets sticking together, will react with the steel surface to form a thin magnesium silicate layer called the glass film (Mg_2SiO_4) or forsterite layer. The forsterite layer facilitates sulphur removal which is absorbed in the coating. Finally, the material is given a flattening anneal, when excess MgO is removed and a thin phosphate coating is applied which reacts with the magnesium silicate to form a strong, highly insulating coating.

In the early 1970s, the role of phosphate coating was finally clarified by the work of Pfutzner and Zehetbauer [123, 124]. This work illustrated the way of taking

Fig. 12.29 Nucleation of
(001)<1–10> grains (in red)
on shear bands in ES1 during
annealing at 650 °C [24]

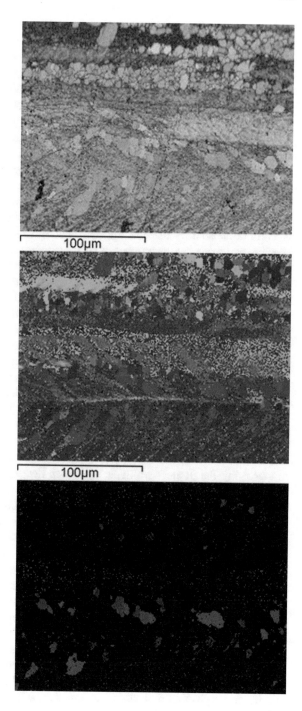

advantage of the stress sensitivity of the power loss and magnetostriction to reduce the total losses by the coating. The annealing and biaxial stress of coating stage in the processing route of electrical steels generate a complex state of tension. The mechanism on which coating applies tension on steel is twofold. Firstly, isotropic tension is applied by the coating on steel due to the strain set up by the difference in thermal contraction of the steel and the coating. Secondly, the uniaxial tension applied to steel in the rolling direction is held by the coating and a tension of 1.2 MPa was measured. In addition to these, there are other two causes of stress in the steel. During high temperature coil annealing, the coil due to large mass causes widening of the base, but this is restricted due to the tight wound and this causes transverse compression. Furthermore, the interlaminar compression perpendicular to the plane acting on waviness in the strip causes longitudinal compression.

Further work of Washko and Choby [158] demonstrated that better oriented steel is more influenced by the coating that induces a reduction in 180° domain wall spacing, and a drop in the number of supplementary domain structures. An applied stress of 6.9 MPa on conventional GO steels showed loss improvement of less than 3% whereas the same stress applied on HiB steels showed an improvement of 10–14% [164, 165]. These findings led to the development of the so-called stress coatings. Goel [31] reported an extensive literature review on the development of stress coatings considering the actual coating technology used. In summary, coating helps the steel in reducing both the losses and the magnetostriction. The eddy current losses are reduced by the insulation resistance supplied by the coating between stacked steel sheets. Besides, the coating also reduces the hysteresis and anomalous loss by improving the surface roughness that would increase the number of mobile domains with the concomitant reduction in energy consumption on moving the domains. The existence of a beneficial tensile stress in the substrate because of the difference in thermal contraction between steel and the coating eliminates the surface closure domains whereas losses are reduced as tensile stress helps in narrowing the domain wall spacing which decreases the anomalous loss. The refinement in hysteresis loss is due to reduction in the surface closure domains. In this sense, the Nishiike et al. [108] investigated the influence of surface properties on the magnetic properties of grain-oriented silicon steel was investigated through measurements of iron losses and magnetic domain patterns in an oxide film-coated and mirror surface-finished specimen, with tensile stress applied in the rolling direction. They reported that the hysteresis loss falls rapidly when the surface roughness is reduced, presumably because of the enhanced mobility of the domain wall. When the tensile stress is increased to about 10 MPa, the eddy current loss is reduced due to the refinement of magnetic domains, but above 10 MPa or so it gradually declines, presumably due to the rotation of the magnetization direction.

Nippon steel corporation [153] currently applies a coating with 4–16 weight % colloidal silica, 3–24 weight % aluminium phosphate and 0.2–4.5 weight % of compound from the group of chromic anhydride. The coating is cured at 350 °C to achieve a deposition of 4 gm/m^2 on one side with a power loss reduction of 5% and magnetostriction reduction of 15 $\mu\epsilon$ at compressive stress of 30 kg/cm^2. The reason for reduction in power loss and magnetostriction by the coating was due to the low thermal expansion coefficient of silica in colloidal silica and also due to the super microgranular nature of it which removes the unevenness in the surface.

On the other hand, Armco Steel Corporation [66] manufacture a coating of silica and potassium oxide in the ratio 2:1 to 2.5:1 on glass coating with or without the phosphate coating and annealed at 550–900 °C in the atmosphere of dry nitrogen and hydrogen in the ratio 9:1 to develop a coating of thickness 0.04–0.2 mm. The reduction in power loss recorded is by 4–5% and magnetostriction reduction of 50 $\mu\epsilon$ at no applied stress. The improvement was due to the low coefficient of thermal expansion that imparts tensile stress and high resistivity of the coating.

JFE Steel Corporation [152] produced a chromium-free coating with a solution of phosphates from the group of Al, Mg, Ca, Ba, Sr, Zn and Mn with colloidal silica of 0.5 to 10 mol and one element from Mg, Sr, Zn, Ba and Ca with 0.02 to 2.5 mol in relation to PO_4:1 mol in phosphates with a thickness of 3 μm on one side and achieved 8 MPa of tension in the coating. The tension developed in the coating was due to the difference in thermal expansion coefficient and the amount of tension developed depended upon the thickness of the coating with thinner coating applying less tension than thick coating.

ThyssenKrupp manufactures an inorganic coating on the glass film layer which forms during annealing. A coating thickness of 2–5 μm provides good electrical resistance and high-stacking factor. The coating, which is annealing resistant up to 840 °C, enables wound cores and punched laminations to be stress relief annealed. The coating is chemically resistant to any fluid it may be exposed to during the production process [115].

AK Steel manufactures an inorganic coating equivalent to ASTM A976 C-5, commercially named as CARLITE® (https://www.aksteel.de/), consisting of a glass-like film formed during high-temperature hydrogen anneal as the result of the reaction of an applied coating of MgO and silicates in the surface of the steel, and subsequently phosphate, and with ceramic fillers added to enhance the interlaminar resistance. CARLITE® coating provides other important advantages such as potential for reduced transformer building factor from added resistance to elastic strain damage, potential for reduction of magnetostriction related transformer noise, high stacking factor, and low coefficient of friction.

Cogent Power Ltd. manufactures the coating during a high-temperature coil anneal (HTCA) at 1100 °C in hydrogen for 5 days, where the steel is coated in a magnesium oxide slurry which reacts with fayalite (Fe_2SiO_4) on the steel surface to form forsterite (Mg_2SiO_4). After the HTCA, the steel sheets pass to the thermal flattening line where the steel strip is held under tension while the final phosphate coating is applied (see https://cogent-power.com/downloads).

There are also a number of alternative methods, not industrially established yet for the case of electrical steels that can be used to apply a coating on the surface of steel sheet such as sol gel [112], CVD [163], PVD [43], plasma spraying, wet coating, printing, electroless and electrochemical plating.

12.8 Future of Electrical Steels

It is clear that the immediate future of electric steels is to optimize the existing product ranges, both NGO and GO electrical steels. Manufacturing cost savings will continue to be a high priority and may offer scope for the development of new production methods, such as spray forming. Research will continue on how to increase the resistivity of the material without incorporating large amounts of alloying elements, due to the need to reduce eddy current losses while maintaining a high saturation. The use of manganese or aluminium should be considered in any new production route and even the use of more expensive elements, such as chromium and nickel, could be justified in some cases in which the demand for reduced losses continues.

The customer demand for reduced losses will boost the research to produce high-performance electrical steels through better secondary recrystallization methods, grain orientation control, increasing the electrical resistivity, gauge reduction and understanding the magnetic domain structure [16, 40, 41], and of course, the development of better and more effective stress coatings. All of these will definitively play a dominant role in minimizing losses and magnetostriction.

Obviously, the incorporation of the chain of production of electrical steels to a productive sector within a circular economy will pay more attention to the environment and the recycling capacity of both steels and manufacturing routes, which can lead to changes in the manufacturing route to be more sustainable from the ecological point of view. It is also a need to produce steel qualities that reduce the noise emission of the magnetic cores, hence the need for a lower magnetostriction, which can only be achieved through a more careful domain control.

A better understanding of normal magnetization conditions in products that use non-oriented steels will help manufacturers to develop more appropriate products through better texture control, e.g. providing more isotropic materials for small motors and transformers or textured steel for lamp ballasts. The relationship between core loss and permeability, in the stators of rotating machines, should be better understood.

Acknowledgements The author acknowledges financial support to Spanish Ministerio de Economia y Competitividad (MINECO) through in the form of a Coordinate Project (MAT2016-80875-C3-1-R). Authors also acknowledge financial support to Comunidad de Madrid through DIMMAT-CM_S2013/MIT-2775 project.

References

1. B.K. Bae, J.S. Woo, J.K. Kim, Effect of heating rate on properties of non-oriented electrical steel containing 0.4% Si. J. Magn. Magn. Mater. **254**, 373–375 (2003)
2. P. Baudouin, A. Belhadj, Y. Houbaert, Effect of the rapid heating on the magnetic properties of non-oriented electrical steels. J. Magn. Magn. Mater. **238**(2–3), 221–225 (2002)

3. K. Beck, Das magnetische Verhalten von Eisenkristallen bei gewöhnlicher Temperatur. Naturforschende Gesellschaft in Zürich 63(1–2), 116–186 (1918)
4. G. Béranger, G. Henry, G. Sanz, *The Book of Steel* (Lavoisier, Paris, 1996)
5. F.G. Caballero, C. Capdevila, L.F. Alvarez, C.G. de Andres, Thermoelectric power studies on a martensitic stainless steel. Scr. Mater. 50(7), 1061–1066 (2004)
6. P.R. Calvillo, T. Ros-Yanez, D. Ruiz, R. Colás, Y. Houbaert, Plane strain compression of high silicon steel. Mater. Sci. Technol. 22(9), 1105–1111 (2006)
7. C. Capdevila, T. De Cock, F.G. Caballero, C. Garcia-Mateo, C.G. de Andres, Evaluation of the austenitic grain growth by thermoelectric power measurements. Mater. Sci. Forum 467–470, 863–867 (2004)
8. C. Capdevila, J.P. Ferrer, F.G. Caballero, C.G. De Andres, Influence of processing parameters on the recrystallized microstructure of extra-low-carbon steels. Metall. Mater. Trans. A-Phys. Metall. Mater. Sci. 37A(7), 2059–2068 (2006)
9. L. Chang, Y.S. Hwang, Effect of vanadium content and annealing temperature on recrystallisation, grain growth, and magnetic properties in 0.3%Si electrical steels. Mater. Sci. Technol. 14(7), 608–618 (1998)
10. L. Chang, Y.S. Hwang, Effect of finish rolling temperature on static recrystallisation in hot bands of electrical steel containing 1-3% silicon. Mater. Sci. Technol. 18(2), 151–159 (2002).
11. L. Chang, Y.S. Hwang, C.K. Hou, S.S.O.C. Iron, *Effect of Second Phase Particles on Recrystallization and Subsequent Grain Growth in Cold-Rolled Electrical Steels, vol 33. 37th Mechanical Working and Steel Processing Conference Proceedings* (Iron & Steel Soc Aime, Warrendale, 1996)
12. C.W. Chen, *Magnetism and Metallurgy of Soft Magnetic Materials* (Dover Publications, 2013)
13. N. Chen, S. Zaefferer, L. Lahn, K. Günther, D. Raabe, Effects of topology on abnormal grain growth in silicon steel. Acta Mater. 51(6), 1755–1765 (2003)
14. L. Cheng, N. Zhang, P. Yang, W.M. Mao, Retaining {100} texture from initial columnar grains in electrical steels. Scr. Mater. 67(11), 899–902 (2012).
15. S. Chikazumi, S. Chikazumi, C.D. Graham, *Physics of Ferromagnetism* (Clarendon Press, Oxford, 1997)
16. S. Chikazumi, K. Suzuki, On the maze domain of silicon-iron crystal (I). J. Phys. Soc. Jpn. 10(7), 523–534 (1955).
17. M.A. Cunha, S.C. Paolinelli, Non-oriented silicon steel recrystallization texture study, in *Textures of Materials, Pts 1 and 2*, ed. by D. N. Lee, vol. 408-4, (Materials Science Forum. Trans Tech Publications Ltd, Zurich-Uetikon, 2002), pp. 779–784
18. M.J. Davies, A.J. Moses, D. Snell, Measurement of coatings and surface oxide layers on grain oriented electrical steel. Ironmak. Steelmak. 25(2), 124–130 (1998)
19. B. De Boer, J. Wieting, Formation of a near {001}?110? Recrystallization texture in electrical steels. Scr. Mater. 37(6), 753–760 (1997)
20. M.F. de Campos, F.J.G. Landgraf, I.G.S. Falleiros, G.C. Fronzaglia, H. Kahn, Texture evolution during the processing of electrical steels with 0.5% Si and 1.25% Si. ISIJ Int. 44(10), 1733–1737 (2004)
21. M.F. de Campos, F.J.G. Lanegraf, R. Takanohashi, F.C. Chagas, I.G.S. Falleiros, G.C. Fronzaglia, H. Kahn, Effect of the hot band grain size and intermediate annealing on the deformation and recrystallization textures in low silicon electrical steels. ISIJ Int. 44(3), 591–597 (2004)
22. T. De Cock, J.P. Ferrer, C. Capdevila, F.G. Caballero, V. Lopez, C. de Andres, Austenite retention in low Al/Si multiphase steels. Scr. Mater. 55(5), 441–443 (2006)
23. A.L. Etter, T. Baudin, R. Penelle, Influence of the Goss grain environment during secondary recrystallisation of conventional grain oriented Fe-3%Si steels. Scr. Mater. 47(11), 725–730 (2002)
24. J.P. Ferrer, T. De Cock, C. Capdevila, F.G. Caballero, Thermomechanical processing optimization of non grain-oriented electrical steels (2012)

25. J.P. Ferrer, T. De Cock, C. Capdevila, F.G. Caballero, C.G. de Andres, Comparison of the annealing behaviour between cold and warm rolled ELC steels by thermoelectric power measurements. Acta Mater. **55**(6), 2075–2083 (2007)

26. O. Fischer, Y. Houbaert, J. Schneider, T. Ros-Yanez, Verfahren zum Herstellen von magnetischem Band oder Tafeln. Germany Patent DE102005004037B3, 2005

27. O. Fischer, J. Schneider, Influence of deformation process on the improvement of non-oriented electrical steel. J. Magn. Magn. Mater. **254–255**, 302–306 (2003)

28. M. Furtkamp, G. Gottstein, D.A. Molodov, V.N. Semenov, L.S. Shvindlerman, Grain boundary migration in Fe-3.5% Si bicrystals with [001] tilt boundaries. Acta Mater. **46**(12), 4103–4110 (1998)

29. M. Furtkamp, P. Lejček, S. Tsurekawa, Grain boundary migration in Fe-3wt.%Si alloys. Interface Sci. **6**(1), 59–66 (1998).

30. M. Godec, M. Jenko, R. Mast, H.J. Grabke, Texture measurements on electrical steels alloyed with tin. Vacuum **61**(2–4), 151–155 (2001).

31. V. Goel, *Novel Coating Technologies for Electrical Steels* (Cardiff University, Wales, 2016)

32. N.P. Goss, Electrical sheet and method and apparatus for its manufacture and test. USA Patent US1965559, 1934

33. H.E. Grenoble, The role of solutes in the secondary recrystallization of silicon iron. IEEE Trans. Magn. **13**(5), 1427–1432 (1977).

34. H.E. Grenoble, H.C. Fiedler, Secondary recrystallization of silicon-iron by solute-induced boundary restraint. J. Appl. Phys. **40**(3), 1575–1576 (1969).

35. R. Hadfield, E. Newbery, The corrosion and electrical properties of steels. Proc. R. Soc. Lond. Ser. A-Cont. Pap Math. Phys. Character **93**(647), 56–67 (1917).

36. R.A. Hadfield, Magnetic composition and method of making same. USA Patent US745829A, 1903

37. J.Harase, R. Shimizu, K. Kuroki, T. Nakayama, T. Wada, T.Watanabe, Effect of primary recrystallization texture, orientation distribution and grain boundary characteristics on the secondary recrystallization behaviour of grain-oriented silicon steel having aln and mns as inhibitors. In: Proceedings of Fourth JIM International Symposium on Grain Boundary Structure and Related Phenomena, Sendai, Japan, Minakami, Japan. Japan Institute of Metals, pp. 563–570, 1986

38. D. Hawezy, The influence of silicon content on physical properties of non-oriented silicon steel. Mater. Sci. Technol. **33**(14), 1560–1569 (2017).

39. Y. Hayakawa, Mechanism of secondary recrystallization of Goss grains in grain-oriented electrical steel. Sci. Technol. Adv. Mater. **18**(1), 480–497 (2017).

40. Y. Hayakawa, M. Kurosawa, Orientation relationship between primary and secondary recrystallized texture in electrical steel. Acta Mater. **50**(18), 4527–4534 (2002)

41. Y. Hayakawa, M. Muraki, J.A. Szpunar, The changes of grain boundary character distribution during the secondary recrystallization of electrical steel. Acta Mater. **46**(3), 1063–1073 (1998)

42. Y. Hayakawa, J.A. Szpunar, A new model of Goss texture development during secondary recrystallization of electrical steel. Acta Mater. **45**(11), 4713–4720 (1997)

43. X.D. He, X. Li, Y. Sun, Microstructure and magnetic properties of high silicon electrical steel produced by electron beam physical vapor deposition. J. Magn. Magn. Mater. **320**(3–4), 217–221 (2008).

44. N.H. Heo, Cold rolling texture, nucleation and direction distribution of {110} grains in 3% Si–Fe alloy strips. Mater. Lett. **59**(22), 2827–2831 (2005).

45. H. Homma, B. Hutchinson, Orientation dependence of secondary recrystallisation in silicon-iron. Acta Mater. **51**(13), 3795–3805 (2003)

46. H. Homma, K. Murakami, T. Tamaki, N. Shibata, T. Yamamoto, Y. Ikuhara, Effects of grain boundary characters for secondary recrystallisation in grain oriented silicon steel. Mater. Sci. Forum **558–559**, 633–640 (2007)

47. K. Honma, T. Nozawa, H. Kobayashi, Y. Shimoyama, I. Tachino, K. Miyoshi, Development of non-oriented and grain-oriented silicon steel (invited). IEEE Trans. Magn. **21**(5), 1903–1908 (1985).
48. C.K. Hou, Effect of hot band annealing temperature on the magnetic properties of low-carbon electrical steels. ISIJ Int. **36**(5), 563–571 (1996)
49. C.K. Hou, C.T. Hu, S. Lee, Effect of residual aluminium on the microstructure and magnetic properties of low carbon electrical steels. Mater. Sci. Eng. A **125**(2), 241–247 (1990).
50. C.K. Hou, C.T. Hu, S. Lee, The effect of titanium and niobium on degradation of magnetic properties of lamination steels. J. Magn. Magn. Mater. **87**(1–2), 44–50 (1990).
51. C.K. Hou, C.T. Hu, S. Lee, The effect of phosphorus on the core loss of lamination steels. J. Magn. Magn. Mater. **109**(1), 7–12 (1992).
52. C.K. Hou, C.T. Hu, S.B. Lee, Effect of residual aluminum on the microstructure and magnetic-properties of low-carbon electrical steels. Mater. Sci. Eng. A-Struct. Mater. Prop. Microstruct. Processing **125**(2), 241–247 (1990c).
53. C.K. Hou, S. Lee, The effect of aluminum on the magnetic properties of lamination steels. IEEE Trans. Magn. **27**(5), 4305–4309 (1991).
54. C.K. Hou, P.C. Wang, Effects of composition and process variables on core loss and hardness of low carbon electrical steels. J. Magn. Magn. Mater. **92**(1), 109–115 (1990).
55. Y. Houbaert, T. Ros-Yáñez, A. Monsalve, J. Barros Lorenzo, Texture evolution in experimental grades of high-silicon electrical steel. Phys. B Condens. Matter **384**(1–2), 310–312 (2006)
56. Y. Hu, V. Randle, T. Irons, Macrotexture and microtexture evolution in cold rotted non-oriented electrical steel sheets during annealing. Mater. Sci. Technol. **22**(11), 1333–1337 (2006)
57. W.B. Hutchinson, Development and control of annealing textures in low-carbon steels. Int. Met. Rev. **29**(1), 25–40 (1984).
58. H. Inagaki, Fundamental aspect of texture formation in low-carbon steel. ISIJ Int. **34**(4), 313–321 (1994).
59. Y. Inokuti, Ultra-low iron loss unidirectional silicon steel sheet. Japan Patent EP0910101 (A4), 1998
60. T. Iuchi, S. Yamaguchi, T. Ichiyama, M. Nakamura, T. Ishimoto, K. Kuroki, Laser processing for reducing core loss of grain oriented silicon steel. J. Appl. Phys. **53**(3), 2410–2412 (1982).
61. K. Iwayama, S. Taguchi, K. Kuroki, T. Wada, Relation between orientation and core loss in grain-oriented 3% silicon steel with high permeability. J. Magn. Magn. Mater. **26**(1–3), 37–39 (1982).
62. L. Kestens, J. Jonas, P. Van Houtte, E. Aernoudt, Orientation selective recrystallization of nonoriented electrical steels. Metall. Mater. Trans. A **27**(8), 2347–2358 (1996)
63. L. Kestens, J.J. Jonas, P. VanHoutte, E. Aernoudt, Orientation selection during static recrystallization of cross-rolled non-oriented electrical steel. Textures Microtextures **26–27**, 321–335 (1998)
64. K.N. Kim, L.M. Pan, J.P. Lin, Y.L. Wang, Z. Lin, G.L. Chen, The effect of boron content on the processing for Fe-6.5 wt% Si electrical steel sheets. J. Magn. Magn. Mater. **277**(3), 331–336 (2004).
65. K.-J. Ko, J.-T. Park, J.-K. Kim, N.-M. Hwang, Morphological evidence that Goss abnormally growing grains grow by triple junction wetting during secondary recrystallization of Fe-3% Si steel. Scr. Mater. **59**(7), 764–767 (2008)
66. D.M. Kohler, Potassium silicate coated silicon steel. USA Patent US3522113A, 1968
67. R.F. Krause, G.C. Rauch, W.H. Kasner, R.A. Miller, Effect of laser scribing on the magnetic properties and domain structure of high-permeability 3% Si-Fe. J. Appl. Phys. **55**(6), 2121–2123 (1984).
68. K. Kuroki, K. Fukawa, T. Wada, Artificial domain control of the grain oriented silicon steel. Nippon Kinzoku Gakkaishi **45**(4), 379–383 (1981).
69. K. Kuroki, T. Sato, T. Wada, Inhibitors for grain oriented silicon steel. Nippon Kinzoku Gakkaishi **43**(3), 175–181 (1979)

70. F. Landgraf, Nonoriented electrical steels. JOM J. Miner. Met. Mater. Soc. **64**(7), 764–771 (2012).

71. F.J.G. Landgraf, M. Emura, J.C. Teixeira, M.F. de Campos, Effect of grain size, deformation, aging and anisotropy on hysteresis loss of electrical steels. J. Magn. Magn. Mater. **215**, 97–99 (2000)

72. F.J.G. Landgraf, M. Emura, J.C. Teixeira, M.F. de Campos, C.S. Muranaka, Anisotropy of the magnetic losses components in semi-processed electrical steels. J. Magn. Magn. Mater. **197**, 380–381 (1999)

73. F.J.G. Landgraf, R. Takanohashi, F.C. Chagas, M.F. De Campos, I.G.S. Falleiros, The origin of grain size inhomogeneity in semi-processed electrical steels. J. Magn. Magn. Mater. **215– 216**, 92–93 (2000).

74. K.C. Liao, The effect of manganese and sulfur contents on the magnetic properties of cold rolled lamination steels. Metall. Trans. A **17**(8), 1259–1266 (1986).

75. M. Lindenmo, A. Coombs, D. Snell, Advantages, properties and types of coatings on non-oriented electrical steels. J. Magn. Magn. Mater. **215**, 79–82 (2000).

76. M.F. Littmann, Iron and silicon – iron alloys. IEEE Trans. Magn. **MAG7**(1), 48-+ (1971).

77. M.F. Littmann, Development of improved cube-on-edge texture from strand cast 3pct silicon-iron. Metall. Trans. A. **6**(5), 1041–1048 (1975).

78. M.F. Littmann, Grain-oriented silicon steel sheets. J. Magn. Magn. Mater. **26**(1–3), 1–10 (1982).

79. H. Liu, Z. Liu, C. Li, G. Cao, G. Wang, Solidification structure and crystallographic texture of strip casting 3 wt.% Si non-oriented silicon steel. Materials Charact. **62**(5), 463–468 (2011).

80. J.L. Liu, Y.H. Sha, F. Zhang, J.C. Li, Y.C. Yao, L. Zuo, Development of {2 1 0}<0 0 1> recrystallization texture in Fe-6.5 wt.% Si thin sheets. Scr. Mater. **65**(4), 292–295 (2011).

81. H.T. Liu, H.L. Li, H. Wang, Y. Liu, F. Gao, L.Z. An, S.Q. Zhao, Z.Y. Liu, G.D. Wang, Effects of initial microstructure and texture on microstructure, texture evolution and magnetic properties of non-oriented electrical steel. J. Magn. Magn. Mater. **406**, 149–158 (2016).

82. T. Lizzi, K.S.V.L. Narasimhan, E.J. Dulis, Development of goss texture in fe-6. 5% si-2% ni. IEEE Trans. Magn. **MAG-22**(5) (1986)

83. G. Lyudkovsky, P.K. Rastogi, M. Bala, Nonoriented electrical steels. J. Metals **38**(1), 18–25 (1986).

84. G. Lyudkovsky, P.K. Rastogi, M. Bala, Nonoriented electrical steels. JOM **38**(1), 18–26 (1986).

85. G. Lyudkovsky, P.D. Southwick, The effect of thermomechanical history upon the microstructure and magnetic properties of nonoriented silicon steels. Metall. Trans. A **17**(8), 1267–1275 (1986).

86. S. Mager, J. Wieting, Influence of the hot rolling conditions on texture formation in Fe-Si sheets. J. Magn. Magn. Mater. **133**(1–3), 170–173 (1994)

87. W.M. Mao, P. Yang, Control of {100} textures and improvement of magnetic. Cailiao Rechuli Xuebao **37**(4), 1–4 (2016)

88. K.M. Marra, E.A. Alvarenga, V.T.L. Buono, Magnetic aging a nisotropy of a semi-processed non-oriented electrical steel. Mater. Sci. Eng. A **390**(1–2), 423–426 (2005).

89. K.M. Marra, E.A. Alvarenga, V.T.L. Buono, Magnetic aging anisotropy of a semi-processed non-oriented electrical steel. Mater. Sci. Eng. A-Struct. Mater. Prop. Microstruct. Processing **390**(1–2), 423–426 (2005)

90. K. Matsumura, B. Fukuda, Recent developments of non-oriented electrical steel sheets. IEEE Trans. Magn. **20**(5), 1533–1538 (1984).

91. J.E. May, D. Turnbull, Effect of impurities on the temperature dependence of the (110) [001] texture in silicon-Iron. J. Appl. Phys. **30**(4), S210–S212 (1959).

92. G.M. Michal, J.A. Slane, The kinetics of carbide precipitation in silicon-aluminum steels. Metall. Trans. A **17**(8), 1287–1294 (1986).

93. J.T. Michalak, R.D. Schoone, Recrystallization and texture development in a low-carbon aluminum-killed steel. Trans. Metall. Soc. Aime **242**(6), 1149-& (1968)

94. A. Morawiec, On abnormal growth of Goss grains in grain-oriented silicon steel. Scripta Materialia **64**(5), 466–469 (2011).
95. W.G. Morris, P. Rao, J.W. Shilling, D.R. Fecich, Effect of Forsterite coatings on the domain structure of grain-oriented 3-percent Si-Fe. IEEE Trans. Magn. **14**(1), 14–17 (1978).
96. D. Moseley, Y. Hu, V. Randle, T. Irons, Role of silicon content and final annealing temperature on microtexture and microstructure development in non-oriented silicon steel. Mater. Sci. Eng. A-Struct. Mater. Prop. Microstruct. Processing **392**(1–2), 282–291 (2005)
97. A.J. Moses, Electrical steels: past, present and future developments. IEE Proc. A (Phys. Sci. Meas. Instrum. Manage. Educ.) **137**(5), 233–245 (1990)
98. A.J. Moses, Energy efficient electrical steels: magnetic performance prediction and optimization. Scr. Mater. **67**(6), 560–565 (2012).
99. A.J. Moses, Relevance of microstructure and texture to the accuracy and interpretation of 1 and 2 directional characterisation and testing of grain-oriented electrical steels. Int. J. Appl. Electromagn. Mech. **55**, S3–S13 (2017).
100. A.J. Moses, S.N. Konadu, Some effects of grain boundaries on the field distribution on the surface of grain oriented electrical steels. Int. J. Appl. Electromagn. Mech. **13**(1–4), 339–342 (2001)
101. A.J. Moses, W.A. Pluta, Anisotropy influence on hysteresis and additional loss in silicon steel sheets. Steel Res. Int. **76**(6), 450–454 (2005).
102. S. Nakashima, K. Takashima, J. Harase, Effect of cold-rolling reduction on secondary recrystallization in grain-oriented electrical steel produced by single-stage rolling process. ISIJ Int. **31**(9), 1013–1019 (1991)
103. S. Nakashima, K. Takashima, J. Harase, Effect of silicon content on secondary recrystallization in grain-oriented electrical steel produced by single-stage cold-rolling process. ISIJ Int. **31**(9), 1007–1012 (1991)
104. T. Nakayama, N. Honjou, Effect of aluminum and nitrogen on the magnetic properties of non-oriented semi-processed electrical steel sheet. J. Magn. Magn. Mater. **213**(1), 87–94 (2000)
105. T. Nakayama, N. Honjou, T. Minaga, H. Yashiki, Effects of manganese and sulfur contents and slab reheating temperatures on the magnetic properties of non-oriented semi-processed electrical steel sheet. J. Magn. Magn. Mater. **234**(1), 55–61 (2001)
106. T. Nakayama, M. Takahashi, Effects of vanadium on magnetic properties of semi-processed non-oriented electrical steel sheets. J. Mater. Sci. **30**(23), 5979–5984 (1995)
107. T. Nakayama, T. Tanaka, Effects of titanium on magnetic properties of semi-processed non-oriented electrical steel sheets. J. Mater. Sci. **32**(4), 1055–1059 (1997).
108. U. Nishiike, T. Kan, A. Honda, Influence of surface properties on the stress magnetization properties of grain-oriented silicon steel. IEEE Transl. J. Magn. Jpn. **8**(11), 777–782 (1993).
109. Y. Oda, Y. Tanaka, A. Chino, K. Yamada, The effects of sulfur on magnetic properties of non-oriented electrical steel sheets. J. Magn. Magn. Mater. **254**, 361–363 (2003)
110. K. Okada, T. Yamaji, K. Kasai, Basic investigation of CVD method for manufacturing 6.5% Si steel sheet. ISIJ Int. **36**(6), 706–713 (1996).
111. C. Oldani, S.P. Silvetti, Microstructure and texture evolution during the annealing of a lamination steel. Scr. Mater. **43**(2), 129–134 (2000)
112. T. Olding, M. Sayer, D. Barrow, Ceramic sol-gel composite coatings for electrical insulation. Thin Solid Films **398**, 581–586 (2001).
113. Y. Onuki, R. Hongo, K. Okayasu, H. Fukutomi, Texture development in Fe–3.0 mass% Si during high-temperature deformation: Examination of the preferential dynamic grain growth mechanism. Acta Mater. **61**(4), 1294–1302 (2013).
114. Y. Ozaki, M. Muraki, T. Obara, Microscale texture and recrystallized orientations of coarse grained 3%Si-steel rolled at high temperature. ISIJ Int. **38**(6), 531–538 (1998)
115. U. Paar, S. Sepeur, S. Goedicke, K. Steinhoff, Method for coating metal surfaces. Germany Patent, 2010
116. S.C. Paolinelli, M.A. Cunha, A.B. Cota, Effect of hot band grain size on the texture evolution of 2% Si non-oriented steel during final annealing. IEEE Trans. Magn. **51**(6), 4 (2015).

117. H.-K. Park, J.-H. Kang, C.-S. Park, C.-H. Han, N.-M. Hwang, Pancake-shaped growth of abnormally-growing Goss grains in Fe-3%Si steel approached by solid-state wetting. Mater. Sci. Eng. A **528**(7–8), 3228–3231 (2011).

118. H.-K. Park, S.-D. Kim, S.-C. Park, J.-T. Park, N.-M. Hwang, Sub-boundaries in abnormally growing Goss grains in Fe-3% Si steel. Scr. Mater. **62**(6), 376–378 (2010)

119. J.-T. Park, J.A. Szpunar, Effect of initial grain size on texture evolution and magnetic properties in nonoriented electrical steels. J. Magn. Magn. Mater. **321**(13), 1928–1932 (2009).

120. J.T. Park, J.A. Szpunar, Evolution of recrystallization texture in nonoriented electrical steels. Acta Mater. **51**(11), 3037–3051 (2003)

121. J.T. Park, J.S. Woo, S.K. Chang, Effect of phosphorus on the magnetic properties of non-oriented electrical steel containing 0.8 wt% silicon. J. Magn. Magn. Mater. **182**(3), 381–388 (1998)

122. J.Y. Park, K.S. Han, J.S. Woo, S.K. Chang, N. Rajmohan, J.A. Szpunar, Influence of primary annealing condition on texture development in grain oriented electrical steels. Acta Mater. **50**(7), 1825–1834 (2002)

123. H. Pfutzner, C. Bengtsson, M. Zehetbauer, Effects of torsion on domains of coated si-fe sheets. IEEE Trans. Magn. **20**(5), 1554–1556 (1984).

124. H. Pfutzner, M. Zehetbauer, On the mechanism of domain refinement due to scratching. Jpn. J. Appl. Phys. Part 2 Lett. **21**(9), L580–L582 (1982).

125. R. PremKumar, I. Samajdar, N.N. Viswanathan, V. Singal, V. Seshadri, Relative effect(s) of texture and grain size on magnetic properties in a low silicon non-grain oriented electrical steel. J. Magn. Magn. Mater. **264**(1), 75–85 (2003)

126. P. Rauscher, B. Betz, J. Hauptmann, A. Wetzig, E. Beyer, C. Grünzweig, The influence of laser scribing on magnetic domain formation in grain oriented electrical steel visualized by directional neutron dark-field imaging. Sci. Rep. **6**, 38307 (2016). https://www.nature.com/articles/srep38307#supplementary-information

127. J.R. Reitz, F.J. Milford, R.W. Christy, *Foundations of Electromagnetic Theory* (Addison-Wesley, New York, 1993)

128. A. Rollett, F. Humphreys, G.S. Rohrer, M. Hatherly, *Recrystallization and Related Annealing Phenomena*, 2nd edn. (Elsevier Ltd., Amsterdam, 2004).

129. T. Ros-Yanez, Y. Houbaert, O. Fischer, J. Schneider, Production of high silicon steel for electrical applications by thermomechanical processing. J. Mater. Process. Technol. **143**, 916–921 (2003)

130. T. Ros-Yanez, D. Ruiz, J. Barros, Y. Houbaert, R. Colás, Study of deformation and aging behaviour of iron-silicon alloys. Mater. Sci. Eng. A **447**(1–2), 27–34 (2007)

131. T. Ros-Yañez, Y. Houbaert, M. De Wulf, Evolution of magnetic properties and microstructure of high-silicon steel during hot dipping and diffusion annealing, in *IEEE International Magnetics Conference, INTERMAG Europe 2002*, ed. by J. Fidler, B. Hillebrands, C. Ross, et al., (Institute of Electrical and Electronics Engineers Inc., Piscataway, 2002).

132. T. Ros-Yañez, Y. Houbaert, V. Gómez Rodríguez, High-silicon steel produced by hot dipping and diffusion annealing. J. Appl. Phys. **91**(10 I), 7857–7859 (2002).

133. W.E. Ruder, Magnetisation and crystal orientation. Trans. Am. Soc. Steel Treat. **8**, 23–29 (1920)

134. D. Ruiz, T. Ros Yáñez, Y. Houbaert, R.E. Vandenberghe, Order in high silicon electrical steel characterized by Mössbauer spectroscopy, in *IEEE International Magnetics Conference-2002 IEEE INTERMAG, Amsterdam*, (Institute of Electrical and Electronics Engineers Inc., Piscataway, 2002)

135. G. Rusakov, M. Lobanov, A. Redikultsev, I. Kagan, Model of 110⟨001⟩ texture formation in shear bands during cold rolling of Fe-3 Pct Si alloy. Metall. Mater. Trans. A **40**(5), 1023–1025 (2009)

136. G.M. Rusakov, A.A. Redikultsev, M.L. Lobanov, Formation mechanism for the orientation relationship between 110⟨001⟩ and 111⟨112⟩ grains during twinning in Fe-3 Pct Si alloy. Metall. Mater. Trans. A **39**(10), 2278–2280 (2008)

137. T. Sadayori, Y. Iida, B. Fukuda, K. Iwamoto, K. Sato, Y. Shimizu, Developments of grain-oriented silicon steel sheets with low iron loss. Kawasaki Steel Tech. Rep. **22**, 84–91 (1990)
138. G. Sahoo, C.D. Singh, M. Deepa, S.K. Dhua, A. Saxena, Recrystallization behaviour and texture of non-oriented electrical steels. Mater. Sci. Eng. A **734**, 229–243 (2018).
139. T. Saito, Effect of minor elements on normal grain growth rate in singly oriented Si-steel. J. Jpn. Inst. Metals **27**(4), 186–191 (1963).
140. T. Saito, Effect of minor elements on secondary recrystallization in singly oriented Si-steel. J. Jpn. Inst. Metals **27**(4), 191–195 (1963).
141. J. Salinas-Beltrán, A. Salinas-Rodríguez, E. Gutiérrez-Castañeda, R. Deaquino Lara, Effects of processing conditions on the final microstructure and magnetic properties in non-oriented electrical steels. J. Magn. Magn. Mater. **406**, 159–165 (2016).
142. A. Saxena, S.K. Chaudhuri, Correlating the aluminum content with ferrite grain size and core loss in non-oriented electrical steel. ISIJ Int. **44**(7), 1273–1275 (2004).
143. A. Saxena, A. Sengupta, S.K. Chaudhuri, Effect of absorbed nitrogen on the microstructure and core loss property of non-oriented electrical steel. ISIJ Int. **45**(2), 299–301 (2005).
144. A. Schoppa, J. Schneider, C.D. Wuppermann, Influence of the manufacturing process on the magnetic properties of non-oriented electrical steels. J. Magn. Magn. Mater. **215**, 74–78 (2000)
145. J.W. Shilling, W.G. Morris, M.L. Osborn, P. Rao, Orientation dependence of domain wall spacing and losses in 3-percent Si-Fe single crystals. IEEE Trans. Magn. **14**(3), 104–111 (1978).
146. Y. Sidor, F. Kovac, T. Kvackaj, Grain growth phenomena and heat transport in non-oriented electrical steels. Acta Mater. **55**(5), 1711–1722 (2007)
147. R.E. Smallman, R.J. Bishop, Chapter 6 – The physical properties of materials, in *Modern Physical Metallurgy and Materials Engineering*, ed. by R. E. Smallman, R. J. Bishop, 6th edn., (Butterworth-Heinemann, Oxford, 1999), pp. 168–196.
148. N. Sumida, T. Fujii, Y. Oguchi, H. Morishita, K. Yoshimura, F. Sudo, Production of ultra-low carbon steel by combined process of bottom-blown converter and rh degasser. Kawasaki Steel Tech. Rep. **8**, 69–76 (1983)
149. S. Taguchi, K. Kuroki, T. Wada, K. Iwayama, Effect of degree of grain orientation alignment on core loss for grain-oriented electrical steel. Nippon Kinzoku Gakkai-si **46**(6), 609–615 (1982)
150. S. Taguchi, A. Sakakura, The effects of AlN on secondary recrystallization textures in cold rolled and annealed (001)[100] single crystals of 3% silicon iron. Acta Metall. **14**(3), 405–423 (1966).
151. M. Takashima, M. Komatsubara, N. Morito, {001}(210) texture development by two-stage cold rolling method in non-oriented electrical steel. ISIJ Int. **37**(12), 1263–1268 (1997)
152. M. Takashima, M. Muraki, M. Watanabe, T. Shigekuni, Treatment solution for insulation coating for grain oriented electrical steel sheet and method for producing grain oriented electrical steel sheet having insulation coating. Japan Patent US8535455B2, 2007
153. O. Tanaka, T. Yamamoto, T. Takata, Method for forming an insulating film on an oriented silicon steel sheet. Japan Patent US3856568A, 1971
154. Y. Tanaka, Y. Takada, M. Abe, S. Masuda, Magnetic properties of 6.5% Si-Fe sheet and its applications. J. Magn. Magn. Mater. **83**(1–3), 375–376 (1990).
155. K. Ushioda, W.B. Hutchinson, Role of shear bands in annealing texture formation in 3%Si-Fe (111)[112?] single crystals. ISIJ Int. **29**(10), 862–867 (1989)
156. K. Verbeken, I. Infante-Danzo, J. Barros-Lorenzo, J. Schneider, Y. Houbaert, Innovative processing for improved electrical steel properties. Rev. Metal (Madrid) **46**(5), 458–468 (2010).
157. T. Wada, K. Kuroki, K. Iwayama, Effect of cold rolling on the left brace 110 right brace lt an br 001 rt an br secondary recrystallization in 3% silicon steel. Tetsu To Hagane **70**(15), 2065–2072 (1984).

158. S.D. Washko, E.G. Choby, Evidence for the effectiveness of stress coatings in improving the magnetic-properties of high permeability 3-percent si-fe. IEEE Trans. Magn. **15**(6), 1586–1591 (1979).
159. F.E. Werner, Electrical steels: 1970–1990, in *Energy Efficient Electrical Steels*, (The Metallurgical Society, Warrendale, 1981)
160. F.G. Wilson, T. Gladman, Aluminium nitride in steel. Int. Mater. Rev. **33**(1), 221–286 (1988).
161. E.P. Wohlfarth, *Ferro-Magnetic Materials, volume 2. Handbook of Magnetic Materials* (North Holland, New York, 1980)
162. Z. Xia, Y. Kang, Q. Wang, Developments in the production of grain-oriented electrical steel. J. Magn. Magn. Mater. **320**(23), 3229–3233 (2008).
163. H. Yamaguchi, M. Muraki, M. Komatsubara, Application of CVD method on grain-oriented electrical steel. Surf. Coat. Technol. **200**(10), 3351–3354 (2006).
164. T. Yamamoto, T. Nozawa, Effects of tensile stress on total loss of single crystals of 3 percent silicon-iron. J. Appl. Phys. **41**(7), 2981-& (1970).
165. T. Yamamoto, S. Taguchi, A. Sakakura, T. Nozawa, Magnetic properties of grain-oriented silicon steel with high permeability orientcore hi-b. IEEE Trans. Magn. **MAG8**(3), 677–681 (1972).
166. H. Yashiki, T. Kaneko, Effects of Mn and S on the grain growth and texture in cold rolled 0.5% Si steel. ISIJ Int. **30**(4), 325–330 (1990).
167. H. Yashiki, A. Okamoto, Effect of hot-band grain size on magnetic properties of non-oriented electrical steels. IEEE Trans. Magn. **MAG-23**(5), 3086–3088 (1987)
168. A. Zaveryukha, C. Davis, An investigation into the cause of inhomogeneous distributions of aluminium nitrides in silicon steels. Mat. Sci. Eng. A-Struct. Mater. Prop. Microstruct. Processing **345**(1–2), 23–27 (2003)

Index

© Springer Nature Switzerland AG 2021
R. Rana (ed.), *High-Performance Ferrous Alloys*,
https://doi.org/10.1007/978-3-030-53825-5

Printed in the United States
by Baker & Taylor Publisher Services